Horst Reichel Sophie Tison (Eds.)

STACS 2000

17th Annual Symposium
on Theoretical Aspects of Computer Science
Lille, France, February 17-19, 2000
Proceedings

Series Editors

Gerhard Goos, Karlsruhe University, Germany
Juris Hartmanis, Cornell University, NY, USA
Jan van Leeuwen, Utrecht University, The Netherlands

Volume Editors

Horst Reichel
Universität Dresden, Faculty of Computer Science
Dept. of Theoretical Computer Science
01062 Dresden, Germany
E-mail: reichel@tcs.inf.tu-dresden.de

Sophie Tison
Université de Lille, LIFL - UPRESA 8022, CNRS, Bât. M3 - UFR IEEA
59655 Villeneuve d'ASCQ Cedex, France
E-mail: Sophie.Tison@lifl.fr

Cataloging-in-Publication Data applied for

Die Deutsche Bibliothek - CIP-Einheitsaufnahme

STACS <17, 2000, Lille>:
Proceedings / STACS 2000 / 17th Annual Symposium on Theoretical Aspects of Computer Science, Lille, France, February 17 - 19, 2000. Horst Reichel ; Sophie Tison. - Berlin ; Heidelberg ; New York ; Barcelona ; Hong Kong ; London ; Milan ; Paris ; Singapore ; Tokyo : Springer, 2000
(Lecture notes in computer science ; Vol. 1770)
ISBN 3-540-67141-2

CR Subject Classification (1991): F, D, G.1-2, I.3.5, E.1, E.3-4

ISSN 0302-9743
ISBN 3-540-67141-2 Springer-Verlag Berlin Heidelberg New York

Typesetting: Camera-ready by author, data conversion by Boller Mediendesign
Printed on acid-free paper SPIN 10719766 57/3144 5 4 3 2 1 0

Preface

STACS, the Symposium on Theoretical Aspects of Computer Science, is held annually, alternating between France and Germany. STACS is organized jointly by the Special Interest Group for Theoretical Computer Science of the Gesellschaft für Informatik (GI) in Germany and the Maison de l'Informatique et des Mathématiques Discrètes (MIMD) in France. STACS 2000 was the 17th in the series. It was held in Lille from February 17th to 19th, 2000. Previous STACS symposia took place in Paris (1984), Saarbrücken (1985), Orsay (1986), Passau (1987), Bordeaux (1988), Paderborn (1989), Rouen (1990), Hamburg (1991), Cachan (1992), Würzburg (1993), Caen (1994), München (1995), Grenoble (1996), Lübeck (1997), Paris (1998), and Trier (1999). All STACS proceedings have been published in the Lecture Notes in Computer Science of Springer-Verlag.

STACS has become one of the most important annual meetings in Europe for the theoretical computer science community. It covers a wide range of topics in the area of foundations of computer science. This time, 146 submissions from 30 countries were received, all in electronic form. Jochen Bern designed the electronic submission procedure, which performed marvelously. Many thanks to Jochen.

The submitted papers address fundamental problems from many areas of computer science: algorithms and data structure, automata and formal languages, complexity, verification, logic, cryptography. Many tackled new areas, including mobile computing and quantum computing. During the program committee meeting in Lille, 51 papers were selected for presentation. Most of the papers were evaluated by five members of the program committee, partly with the assistance of subreferees for a total of 700 reports. We thank the program committee for its demanding work in the evaluation process. We also thank all the reviewers whose names are listed on the next pages.

We are specially grateful to the invited speakers Pascal Koiran, Thomas Henzinger, and Amin Schokrollahi for accepting our invitation and presenting us their insights on their research area.

We would like to express our sincere gratitude to Anne-Cécile Caron, Rémi Gilleron, and Marc Tommasi who invested their time and energy to organize this conference. Thanks also to all members of the *Laboratoire d'Informatique Fondamentale de Lille*, especially to our secretaries, Annie Dancoisne and Michèle Driessens.

The conference was made possible by the financial support of the following institutions: European Community, Ministère des Affaires Etrangères, Ministère de l'Education Nationale de la Recherche et de la Technologie, Ministère de la Défense, Région Nord/Pas-de-Calais, Ville de Lille, Université de Lille 1, and other organizations.

December 1999

Horst Reichel
Sophie Tison

Program Committee

H. Alt (Berlin)
P. Crescenzi (Firenza)
A. Czumaj (Paderborn)
V. Diekert (Stuttgart)
M. Habib (Montpellier)
D. Krob (Paris 6)
M. Mitzenmacher (Harvard)
M. Ogihara (Rochester)
H. Reichel (Dresden)
Y. Robert (Lyon)
Ph. Schnoebelen (Cachan)
S. Tison (Lille)
H. Vollmer (Wurzburg)
I. Walukiewicz (Warsaw)
G. Woeginger (Graz)

Local Arrangements Committee

A.C. Caron
R. Gilleron
S. Tison
M. Tommasi

Referees

Farid Ablayev
Susanne Albers
Eric Allender
Jorge Almeida
Andris Ambainis
Klaus Ambos-Spies
Jerome Amilhastre
André Arnold
Alfons Avermiddig
Yossi Azar
Luitpold Babel
Olivier Bailleux
Soeren Balko
Euripides Bampis
Ulrike Baumann
Sven Baumer
Cristina Bazgan
Marie-Pierre Beal
Olivier Beaumont
Luca Becchetti
Bernd Becker
Wolfgang Bein
S. Bellantoni
Petra Berenbrink
Vincent Berry
Francois Blanchard
Stephen Bloch
Hans L. Bodlaender
Dominique Bolignano
Beate Bollig
Bernd Borchert
Francis Bossut
Vincent Bouchitté
Vincent Boudet
Yacine Boufkhad
Luc Bougé
Pierre Boulet
Peter Braß
Björn Brodèn
H. Kleine Buening
Harry Buhrman
Gerhard Buntrock
Costas Busch
Béatrice Bérard
Roland Büschkes
Jan Camenisch
Salvatore Caporaso
Anne-Cécile Caron
Olivier Carton
Giuseppe Castagna
Didier Caucal
Nicolò Cesa-Bianchi
Bogdan Chlebus
Marek Chrobak
Valentina Ciriani
Andrea Clementi
Mireille Clerbout
Peter Clote
Bruno Codenotti
Olivier Cogis
Jean-Francois Collard
Hubert Comon
Anne Condon
Derek Corneil
B. Courcelle
Maxime Crochemore
Janos Csirik
Carsten Damm
T. Decker
Jean-Paul Delahaye
Marianne Delorme
Stéphane Demri
Frédéric Desprez
Philippe Devienne
Henning Dierks
Martin Dietzfelbinger
Krzysztof Diks
F. Dupont de Dinechin
Yevgeniy Dodis
Frank Drewes
Manfred Droste

Stefan Droste
R. Ducournau
Bruno Durand
Christoph Durr
Ronald Dutton
Todd Ebert
Stephan Eidenbenz
Robert Elsaesser
Lars Engebretsen
Leah Epstein
Thomas Erlebach
Javier Esparza
J. Mark Ettinger
Philippe Ezequel
Sandor Fekete
Stefan Felsner
Henning Fernau
Paolo Ferragina
Afonso Ferreira
Guillaume Fertin
Amos Fiat
Jean-Christophe Filliâtre
Paul Fischer
Roger Fischlin
Caroline Fontaine
Cédric Fournet
Hervé Fournier
Daniel Fridlender
Thom Fruehwirth
Sibylle Fröschle
Ulrich Furbach
Didier Galmiche
Giorgio Gambosi
Harald Ganzinger
Max Garzon
Leszek Gasieniec
Paul Gastin
Stéphane Gaubert
Ricard Gavalda
Cyril Gavoille
Karin Genther
Dora Giammarresi
Christian Glasser
Andy Gordon
Jean Goubault-Larrecq
Erich Grädel
Roberto Grossi
Grzegorz Grudzinski
Jozef Gruska
Alessio Guglielmi
Mesut Günes
Stefan Haar
Bjarni V. Halldorsson
Mikael Hammar
Dagmar Handke
Michael Hanus
Jin-Kao Hao
S. Harari
Jan Heering
F. Meyer auf der Heide
Laura Heinrich-Litan
Keijo Heljanko
Sybille Hellebrand
Lisa Hellerstein
Lane A. Hemaspaandra
Harald Hempel
Montserrat Hermo
Ulrich Hertrampf
Andréas Herzig
Clemens Heuberger
Colin de la Higuera
Daniel Hirschkoff
Frank Hoffmann
Thomas Hofmeister
H. Hollatz
Markus Holzer
Han Hoogeveen
Everett Howe
Peter Hoyer
Juraj Hromkovic
Gullaume Huard
M. Huchard
Thomas Hune
Heinrich Hussmann
Jean-Michel Hélary
Gabriel Istrate
Kazuo Iwama
Purush Iyer
Andreas Jakoby
Petr Jancar
David Janin
Thomas Jansen
Matthias Jantzen
Alain Jean-Marie
H.-D. Hecker
Tao Jiang
Colette Johnen
David Juedes
Astrid Kaffanke
Reinhard Kahle
Juhani Karhumäki
Lila Kari
Marek Karpinski
Claire Kenyon
Astrid Kiehn
Gunnar W. Klau
Ton Kloks
Christian Knauer
Ulrich Kohlenbach
Pascal Koiran
Jean-Claude Konig
Jan Korst
Sven Kosub
Dexter Kozen
Marcus Kracht
Jan Krajicek
Evangelos Kranakis
Dieter Kratsch
Klaus Kriegel
Gregory Kucherov
Manfred Kunde
Dietrich Kuske
Mirosław Kutyłowski
Marta Kwiatkowska
Barbara König
Anna Labella
Grégory Lafitte
Jean-Marc Lanlignel
Sophie Laplante
Lawrence Larmore
Slawomir Lasota
Michel Latteux
Christian Lavault
Thierry Lecroq
Vincent Lefèvre
Hanno Lefmann
Stefano Leonardi
Pierre Lescanne
Paolo Liberatore
Wolfgang Lindner
Andrzej Lingas
Claudia Linnhoff-Popien
Maciej Liskiewicz
C. Loeding
Markus Lohrey
Satyanarayana V. Lokam
Stefano Lonardi
Krzysztof Lorys
Denis Lugiez
Martin Löbbing

Haiko Müller
Ioan Macarie
Rupak Majumdar
Bernard Malfon
Yannis Manoussakis
A. Marchetti-Spaccamela
Jerzy Marcinkowski
V. W. Marek
Maurice Margenstern
Jean-Yves Marion
Bruno Martin
Philippe Mathieu
Elvira Mayordomo
Ernst W. Mayr
Richard Mayr
Jacques Mazoyer
Pierre McKenzie
Jean-François Mehaut
Lutz Meißner
Guy Melançon
Patrice Ossona de Mendez
Patrizia Mentrasti
Wolfgang Merkle
Yves Metivier
Christian Michaux
Eugenio Moggi
B. Monien
Cristopher Moore
F. Morain
Nicole Morawe
Till Mossakowski
Rustam Mubarakzjanov
Martin Mundhenk
Anca Muscholl
Raymond Namyst
Moni Naor
Codrin Nichitiu
Rolf Niedermeier
Joachim Niehren
Mogens Nielsen
Damian Niwinski
Richard Nock
Michael Noecker
John Noga
Andre Osterloh
Anna Östlin
Linda Pagli
Luigi Palopoli
Alessandro Panconesi
Tomi Pasanen
Vangelis Th. Paschos
Andrzej Pelc
David Peleg
François Pellegrini
Jean-Guy Penaud
Stephane Perennes
Holger Petersen
Antoine Petit
Serge G. Petiton
Michel Petitot
C. Picouleau
Giovanni Pighizzini
Jean-Eric Pin
Ramon Pino-Perez
Wojciech Plandowski
Corinne Plourde
David Pointcheval
Natacha Portier
Frédéric Prost
Kirk Pruhs
Christophe Raffalli
Rajeev Ranan
Christophe Rapine
Fabrice Rastello
Kenneth W. Regan
John Reif
Klaus Reinhardt
Rüdiger Reischuk
Eric Remila
Marco Riedel
Soren Riis
John Michael Robson
John D. Rogers
Yves Roos
Kristoffer Rose
Peter Rossmanith
Günter Rote
Joerg Rothe
Catherine Roucairol
James S. Royer
Danuta Rutkowska
Jan Rutten
Isabelle Ryl
Wojciech Rytter
Harald Räcke
Géraud Sénizergues
Gunter Saake
Amit Sahai
Nasser Saheb
Lakhdar Sais
Jacques Sakarovitch
Grażyna Salbierz
Kai Salomaa
Miklos Santha
Martin Sauerhoff
Francesco Scarcello
Marcus Schaefer
Christian Scheideler
Isaac D. Scherson
Alexander Schill
Christian Schindelhauer
Manfred Schmidt-Schauss
Heinz Schmitz
Arnold Schoenhage
Christoph Scholl
Klaus Schroeder
Rainer Schuler
Klaus U. Schulz
Nicole Schweikardt
Thomas Schwentick
Sebastian Seibert
Helmut Seidl
Joel Seiferas
Rocco Servedio
Peter Sewell
Jiri Sgall
Detlef Sieling
Riccardo Silvestri
Inre Simon
David Skillicorn
Martin Skutella
Christian Sohler
O. Spaniol
Mike Sperber
Renzo Sprugnoli
Christian Stangier
Bernhard Steffen
Angelika Steger
R. Stenzel
Frank Stephan
Jacques Stern
Lorna Stewart
Leen Stougie
Howard Straubing
Martin Strauss
Grégoire Sutre
Jean-Marc Talbot
Jean-Pierre Talpin
Sovanna Tan
Alain Tapp

Alain Terlutte
Pascal Tesson
Denis Therien
Michael Thielscher
Thomas Thierauf
E. Thierry
Wolfgang Thomas
Dirk Thissen
Karsten Tinnefeld
Ioan Todinca
Jacobo Torán
Thierry Boy de la Tour
Ralf Treinen
Luca Trevisan
Denis Trystram
Tomasz Tyksiński
Brigitte Vallee
Moshe Y. Vardi
Stefano Varricchio
Helmut Veith
L. Viennot
Heiko Vogler
Stephan Waack
Alan S. Wagner
Klaus W. Wagner
Johannes Waldmann
Michael Wallbaum
Osamu Watanabe
John Watrous
Gerd Wechsung
Ingo Wegener
Pascal Weil
Jean-Christophe Weill
Carola Wenk
Matthias Westermann
Klaus Wich
Peter Widmayer
Thomas Wilke
Pierre Wolper
J. Wu
Jun Yang
Sergio Yovine
Christof Zalka
Martin Ziegler
Wieslaw Zielonka
Marius Zimand
Justin Zobel

Sponsoring Institutions

- The European Community;
- The European Association for Theoretical Computer Science;
- Le Ministère de l'Education Nationale, de l'Enseignement Supérieur et de la Recherche;
- Le Ministère des Affaires Étrangères;
- Le Centre National de la Recherche Scientifique (CNRS);
- L'Ecole Nouvelle d'Ingénieurs en Communication (ENIC);
- La Région Nord/Pas-de-Calais;
- Le Département du Nord;
- Les villes de Lille et Villeneuve d'Ascq;
- L'Université des Sciences et Technologies de Lille;
- Le Laboratoire d'Informatique Fondamentale de Lille;
- Le Ministère de la Défense.

Table of Contents

Codes and Graphs

M. Amin Shokrollahi

Bell Laboratories, Room 2C-353, 700 Mountain Ave, Murray Hill, NJ, 07974, USA
amin@research.bell-labs.com

Abstract. In this paper, I will give a brief introduction to the theory of low-density parity-check codes, and their decoding. I will emphasize the case of correcting erasures as it is still the best understood and most accessible case. At the end of the paper, I will also describe more recent developments.

1 Introduction

In this paper, I want to give a brief introduction to the theory of low-density parity-check codes, or LDPC codes, for short. These codes were first introduced in the early 1960's by Gallager in his PhD-thesis [6]. They are built using sparse bipartite graphs in a manner that we will describe below. As it turns out, an analysis of these codes requires tools and methods from graph theory most of which were not common knowledge in the early 1960's. This fact may explain to some extent why LDPC codes were almost completely forgotten after their invention. As it turned out, major impacts on the theoretical analysis of these codes came not from coding theory, but from Theoretical Computer Science, and this is why this paper appears in the proceedings of a conference on Theoretical Aspects of Computer Science.

I will deliberately be very brief on the history of LDPC-codes since I would like to concentrate more on very recent developments. But no paper on this topic would be complete without mentioning the names of Zyablov and Pinsker [20,21] and Margulis [12] from the Russian school who had realized the potential of LDPC codes in the 1970's, and the name of Tanner [19], who re-invented and extended LDPC codes. In fact, re-invention seems to be a recurring theme: with the advent of the powerful class of Turbo codes [4], many researchers started to study other types of codes which have fast encoders and decoders and can perform at rates very close to theoretical upper bounds derived by Shannon [15]. For instance, MacKay re-invented some versions of LDPC codes and derived many interesting and useful properties. His paper [11] is a must for anybody who wants to work in this field as it gives a fresh and detailed look at various aspects of LDPC codes. At the same time when coding theorists were struck by the performance of Turbo codes and were starting to remember LDPC codes, these codes were again re-invented, this time in the Theoretical Computer Science community. Graphs are very useful objects in this field (and one may sometimes get the impression that everything in this field is either based on, or motivated

H. Reichel and S. Tison (Eds.): STACS 2000, LNCS 1770, pp. 1–12, 2000.

by, or derived from graphs!). On the other hand, linear codes had been shown to be very useful in the construction of Probabilistically Checkable Proofs [2]. It seems like a good idea to combine these concepts. Sipser and Spielman did exactly this [17]. One of the (many) very interesting results of this work is that it shows how the performance of the codes constructed is directly related to the expansion properties of the underlying graph. The Russian School seems to have known this too, though Sipser and Spielman were completely unaware of this.

A very useful and extremely advantageous property of LDPC codes is that it is very easy to design various efficient encoders and decoders for them. This was already done by Gallager, though largely without a rigorous analysis. His algorithms were re-invented, improved, and analyzed later. For instance, Sipser and Spielman give a simple encoder and decoder which run in time $O(n^2)$ and time $O(n)$ respectively, where n is the block-length of the code (see below for a definition of this parameter). The expansion property described above actually translates into the error performance of the *algorithm*, rather than that of the abstract code. Spielman [18] applies another idea to decrease the encoding time to linear as well, and uses the best known explicit expanders to construct codes that not only have very efficient encoders and decoders, but are also "asymptotically good." I will not describe further what this technical term means, and leave it by the remark that construction of asymptotically good codes is very difficult in itself even if one does not assume that they are efficiently encodable and decodable.

Motivated by the practical problem of sending packets through high-speed computer networks, a group at UC Berkeley and the International Computer Science Institute in Berkeley consisting of Luby, Mitzenmacher, Spielman, Stemann, and myself designed a very simple algorithm for correcting erasures for LDPC codes [10]. Similar codes had already been constructed by Alon and Luby [1], but they performed poorly in practice. The work [10] had several major impacts on subsequent work on LDPC codes. First, it contained for the first time a rigorous analysis of a probabilistic decoding algorithm for LDPC codes. This analysis was later greatly simplified by Luby, Mitzenmacher, and myself [7], and this later method developed into the core of the analysis of LDPC codes under other, more complicated error models. Second, the paper [10] proves that highly irregular bipartite graphs perform much better than regular graphs (which were the method of choice up to then) if the particular simple decoder of that paper is used. The paper goes even further: such codes not only perform better, but if the graphs are sampled randomly from the set of graphs with a particular degree distribution, then the codes can be used to transmit at rates arbitrarily close to the capacity of the erasure channel. In other words, the decoder can recover from a portion of the encoding which is arbitrarily close to the lower bound dictated by the length of the (uncoded) message. In short, we say that these sequences of degree distributions are *capacity-achieving.*

The model of erasures is very realistic for applications such as data transfer on high-speed networks, but codes are typically used in situations where one does not know the positions of the errors. Here the problem is much harder. Based on the approach in [10] and equipped with the new analysis in [7], Luby, Mitzenmacher, Spielman, and myself rigorously analyzed some of Gallager's original "flipping" decoders and invented methods to design appropriate degree distributions so that the corresponding graphs could recover from as many errors as possible [8]. To obtain a better performance, the decoder had to be changed. The most powerful efficient decoder known to date is the so-called belief-propagation. These were tested on good erasure codes, and the results were reported in [9]. The hope was that since codes that can decode many erasures are capable of decoding many errors if the (exponential time) maximum-likelihood decoder is used, and since belief propagation is a very good decoder, then good erasure codes should perform very good under belief-propagation as well. As it turns out, they perform good in experiments, but they do not beat Turbo codes. Moreover, we did not have a method to analyze the asymptotic performance of these codes, and had to rely on heuristic experiments to judge their quality.

Such an analysis was derived by Richardson and Urbanke [14] by generalizing the analysis in [8] (which was itself based on that of [7]). Based on this analysis and using the methods in the pervious papers by Luby et al. enriched with some new weapons, Richardson, Urbanke, and myself [13] were able to construct codes that perform at rates much closer to the capacity than Turbo codes.

In the rest of the paper I will try to describe some of the details left out in the above discussion. I will define most of the objects that we have to work with, though sometimes rigor is traded against clarity.

2 Channels and Codes

For most of what we will be describing in this paper, the following definition of a communication channel will be sufficient: A channel is a finite labeled directed bipartite graph between a set A called the *code-alphabet* and a set B called the *output-alphabet* such that the labels are nonnegative real numbers and satisfy the following property: for any element $a \in A$, the sum of the labels of edges emanating from a is 1. Semantically, the graph describes a communication channel in which elements from A are transmitted and those of B are received. The label of an edge from a to b is the (conditional) probability of obtaining b given that a was transmitted. Examples of Channels are given in Figure 1.

Our aim is to reliably transmit information through an unreliable channel, i.e., a channel which can cause errors with nonzero probability. We want, thereby, to reduce the error in the communication. A first idea to do this is to transmit blocks of symbols rather than individual symbols from A. If the symbols are chosen uniformly at random from A (an assumption commonly made), then this scheme does not provide more protection than the original one. The main idea behind reducing the error is that of adding *redundancy.* The computation of redundant symbol from a message is called *encoding.* This operation produces a *codeword* from a *message word* and we assume that it is an injective map from the set of message words to the set of codewords. A *code* is then the set of codewords obtained this way. The counter-operation to encoding is something we

Fig. 1. (a) The binary erasure channel, and (b) the binary symmetric channel

call *de-encoding.* This operation computes from a codeword the original message word (usually by forgetting the redundant symbol). A more important operation is that of *decoding.* It assigns to a word x over the alphabet A a codeword c. In applications, x is the corrupted version of a codeword c', and the decoder is successful if $c = c'$.

It should be intuitively clear that adding redundancy does indeed reduce the error. For instance, by repeating each transmitted symbol r times and using majority rule for decoding, one can easily reduce the decoding error below any constant ϵ by increasing r. This technique, called *repetition coding,* is used by many teachers who repeat their material several times to reach all their students.

If k symbols of the alphabet are encoded to n symbols, then n is called the *block-length* and the fraction k/n is called the *rate* of the code. The rate equals the fraction of real information symbols in a codeword. The repetition code described above has block-length rk where k is the length of the message. Its rate is thus $1/r$, which decreases to zero as r increases to infinity. Since sending information over a channel is often times expensive, it is desirable to have sequences of codes of constant rate for which the decoding error probability decreases to zero as the block-length goes to infinity.

In a fundamental paper which marks the birth of modern coding and information theory, Shannon [15] completely answered questions of this type. He showed that if the "maximum-likelihood decoding algorithm" is used (which decodes a word over the alphabet to a codeword of minimal Hamming distance), then for a given channel there is a critical rate, called the *capacity of the channel,* such that the decoding error probability for any code of rate larger than that approaches 1. Furthermore, he showed using a random coding argument, that for any rate below the capacity there are codes of that rate for which the decoding error probability decreases to zero exponentially fast in the block-length, as the latter goes to infinity. Computing the capacity is not an easy task in general. Figure 1 shows the capacities of the binary symmetric channel [15] and the binary erasure channel [5].

Shannon's paper answered many old questions and generated many new ones. Because of the nature of random coding used in his proofs, the first question was about how to explicitly construct the codes promised by that theorem.

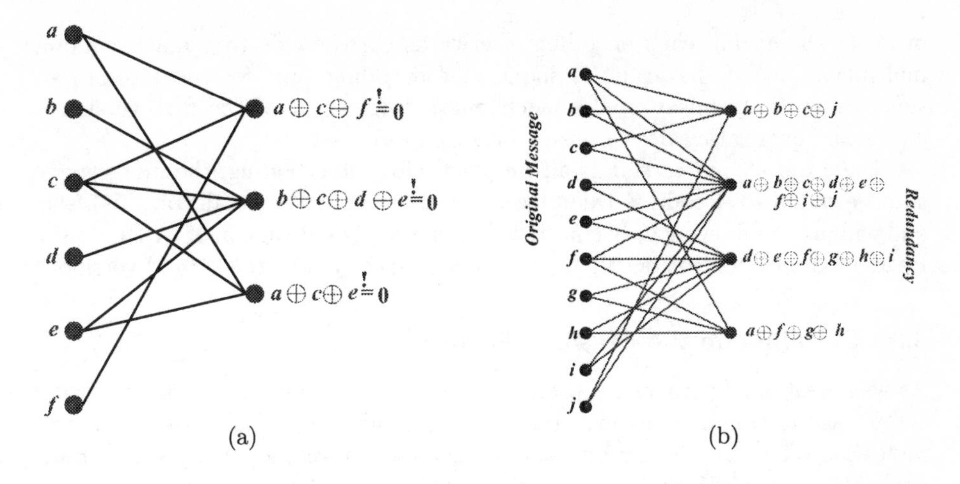

Fig. 2. The two versions of LDPC codes: (a) Original version, and (b) dual version

The second more serious question was that of efficient decoding of such codes, as maximum-likelihood decoding is a very hard task in general. (It was shown many years later that a corresponding decision problem is NP-hard [3].)

Low-density parity check (LDPC) codes, described in the next section, are very well suited to (at least partially) answering both of these questions.

3 LDPC Codes

3.1 Code Construction

In the following we will assume that the code-alphabet A is the binary field GF(2). Let G be a bipartite graph between n nodes on the right called *message nodes* and r nodes on the right called *constraint (or check) nodes.* The graph gives rise to a code in (at least) two different ways, see Figure 2: in the first version (which is Gallager's original version), the coordinates of a codeword are indexed by the message nodes $1, \ldots, n$ of G. A vector $(x_1, \ldots, x_n)$ is a valid codeword if and only if for each constraint node the sum (over GF(2)) of the values of its adjacent message nodes is zero. Since each constraint node imposes one linear condition on the x_i, the rate of the code is at least $(n-r)/n$.

In the second version, the message nodes are indexed by the original message. The constraint nodes contain the redundant information: the value of each such node is equal to the sum (over GF(2)) of the values of its adjacent message nodes. The block-length of this code is $n+r$, and its rate is $n/(n+r)$.

These two versions look quite similar, but differ fundamentally from a computational point of view. The encoding time of the second version is proportional to the number of edges in the graph G, while it is not clear how to encode the first version without solving systems of linear equations. (This needs to be done once for the graph; each encoding afterwards corresponds to a matrix/vector multiplication.) If the graph is sparse, the encoding time for the second version is essentially linear in the block-length, while that of the first version is essentially quadratic (after a pre-processing step).

While the second version is advantageous for the encoding, the first version is more suited to decoding. I don't want to go into further details on this issue, and will in the following only consider Gallager's original version of LDPC codes. Readers are invited to consult [18,10] to learn more about the second version.

3.2 Decoding on the Erasure Channel

As was mentioned earlier, the principal motivation behind our work on erasure codes was the design of forward error-correction schemes in high-speed computer networks. When data is sent over such a network, it is divided into packets. Each packet has an identifier which uniquely describes the entity it comes from and its location within that entity. Packets are then routed through the network from a sender to a recipient. Often, certain packets do not arrive their destination; in certain protocols like the TCP/IP the recipient requests in this case a retransmission of the packets that have not arrived, upon which the sender initiates the retransmission. These steps are iterated several times until the receiver has obtained the complete data. This protocol is excellent in certain cases, but is very poor in scenarios in which feedback channels do not exist (satellite links), or when one sender has to serve a large number of recipients (multicast). The channel corresponding to this scenario is very well modeled by an erasure channel, and corresponding codes can be used to remedy the mentioned shortcomings.

A *linear code* over a field $\mathbb{F}$ of block-length n and dimension k is a k-dimensional subspace of the standard vector space $\mathbb{F}^n$. The minimal Hamming weight of a nonzero element in a linear code is called the *minimum distance* of the code, usually denoted by d. It is not hard to see [5] that a linear code of minimum distance d is capable of correcting any pattern of $d-1$ or less erasures, essentially by solving a system of linear equations of size $O(d)$ over $\mathbb{F}$. Further, Elias showed that random linear codes achieve capacity of the erasure channel with high probability. The running time of $O(d^3)$ of the decoder is, however, very slow for our applications in which d has to be very large (in the 100,000's).

The decoder that we use for the LDPC codes is extremely simple; we will describe it in the case of the binary erasure channel in the following. The decoder maintains a register for each of the message and constraint nodes. All of these registers are initially set to zero. In the first round of the decoding, the value of each received message node is added to the values of all of its adjacent constraint nodes, and then the message nodes and all the edges emanating from it are deleted. Once this *direct recovery step* is complete, the second *substitution recovery phase* kicks in. Here, one looks for a constraint node of degree one. Note that since the value of a constraint node in an intact codeword should be zero, a constraint node of degree one contains the value of its unique adjacent message node. This value is copied into the corresponding message nodes, that value is added to those of all its adjacent constraint nodes, and the message node together with all edges emanating from it are deleted from the graph. If there are no nodes left, or if there are no constraint nodes of degree one left, then the decoder stops. Note that the decoding time is proportional to the number of edges in the graph. If the graph is sparse, i.e., if the number edges is linear in the number of nodes, then the decoder is linear time (at least on a RAM with unit cost measure).

The hope is that there is always enough supply of degree one constraint nodes so that the decoder finishes successfully. Whether or not this is the case depends on the original fraction of erasures and on the graph. Surprisingly, however, the only important parameter of the underlying graph is the distribution of nodes of various degrees. This analysis is the topic of the next section.

3.3 The Analysis

To describe the conditions for successful decoding concisely, we need one further piece of notation. We call an edge in the graph G *of left (right) degree* i if it is connected to a message (constraint) node of degree i. Let λ_i and ρ_i denote the fraction of edges of left degree i and right degree i, respectively. Further, we define the generating functions $\lambda(x) = \sum_i \lambda_i x^{i-1}$ and $\rho(x) = \sum_i \rho_i x^{i-1}$. The rather peculiar look of the exponent of x in these polynomials is an artifact of the particular *message passage decoding* that we are using. This is best explained by the analysis itself, which I will now describe in an informal way.

Let e be an edge between the message node m and the constraint node c. What is the probability that this edge is deleted at the ℓth round of the algorithm? This is the probability that the check node c is of degree one at the ℓth round, and, equivalently, it is the probability that the message node m is corrected at that round. To compute this probability, we unroll the graph in the neighborhood of the node m and consider the subgraph obtained by the neighborhood of depth ℓ of m. This is the subgraph of all the nodes in the graph except those that are connected to m via the edge e, for which there is a path of length at most 2ℓ connecting them to m. In the following we will assume that this graph is a tree. Suppose that the graph is sampled uniformly at random from the set of graphs which have an edge distribution according to the polynomials $\lambda(x)$ and $\rho(x)$. Let p_ℓ denote the probability that m is *not* corrected at round ℓ. Further, let δ denote the original fraction of erasures. Then, obviously $p_0 = \delta$. Further, because we have assumed that the neighborhood of m is a tree, at each level ℓ of the tree the message nodes are still erased with independent probability $p_{i\ell}$. (We assume that only the message nodes contribute to levels in the tree, so that the message nodes forming the leaves are at level 0 and the root m is at level ℓ.) ¿From this, we can establish a recursion for p_ℓ. A message node at level $\ell+1$ is not corrected if and only if it has not been received directly, and all the constraint nodes it is connected to have degree larger than 1. A constraint node has degree one if and only if all its descending message nodes at level ℓ have already been corrected. This happens with independent probability $1-p_\ell$, and since the message node has j edges emanating from it with probability ρ_j, and $j-1$ of them are descending message nodes in the tree, the probability that such a check node is of degree one is $\rho(1-p_\ell)$. Hence, the probability that a message node at level $\ell+1$ is connected only to descending constraint nodes of degree larger than 1 is $\lambda(1-\rho(1-p_\ell))$. That node is thus not corrected with probability $\delta\lambda(1-\rho(1-p_\ell))$, where the factor δ explains the probability that the node has not been received directly. Hence, this gives $p_{\ell+1} = \delta\lambda(1-\rho(1-p_\ell))$. Altogether, we obtain the condition

$$\delta\lambda(1-\rho(1-p_\ell)) < p_\ell \tag{1}$$

for successful decoding. More precisely, this says that if neighborhoods of depth ℓ of message nodes are trees, and if $\delta\lambda(1-\rho(1-x)) < (1-\epsilon)x$ for $x \in (0,\delta)$, then after ℓ rounds of the algorithm the probability that a message node has not been corrected is at most $(1-\epsilon)^\ell\delta$. For large random graphs the probability that the neighborhood of a message node is not a tree is small, and the argument shows that the decoding algorithm reduces the probability of undecoded message node below any constant. To show that the process finishes successfully, one needs expansion [10].

The above informal discussion can be made completely rigorous using proper martingale arguments [8,14]. Summarizing, the condition for successful decoding after a δ-fraction of erasures is

$$\delta\lambda(1-\rho(1-x)) < x \quad \text{for} \quad x \in (0,\delta). \tag{2}$$

3.4 Capacity Achieving Sequences

The condition (2) is very handy if one wants to analyse the performance of random graphs with a *given* degree distribution. For instance, it turns out that the performance of regular graphs deteriorates as the degree of the message nodes increases [10]. In fact, the best performance is obtained if all message nodes have degree three. On the other hand, this condition does not give a clue on how to design good degree distributions λ and ρ. Our aim is to construct sequences that asymptotically achieve the capacity of the erasure channel. In other words, we want δ in (2) to be arbitrarily close to $1-R$, where R is the rate of the code. To make this definition more rigorous, we call a sequence $(\lambda_\ell, \rho_\ell)_{\ell\geq 0}$ *capacity-achieving of rate* R if (a) the corresponding graphs give rise to codes of rate at least R, and (b) for all $\epsilon > 0$ there exists an ℓ_0 such that for all $\ell \geq \ell_0$ we have

$$(1-R)(1-\epsilon)\lambda(1-\rho(1-x)) < x \quad \text{for} \quad x \in (0,(1-R)(1-\epsilon)).$$

It is surprising that such sequences do really exist. The first such sequence was discovered in [10]. To describe it, we first need to mention that, given λ and ρ, the average left and right degree of the graph is $1/\sum_i \lambda_i/i$ and $1/\sum_i \rho_i/i$, respectively. These quantities can be conveniently expressed as $1/\int_0^1 \lambda(x)\mathrm{d}x$ and $1/\int_0^1 \rho(x)\mathrm{d}x$. As a result, the rate of the code is at least $1-\int_0^1 \rho(x)\mathrm{d}x/\int_0^1 \lambda(x)\mathrm{d}x$. It is a nice exercise to deduce from the equation (2) alone that δ is always less than or equal to $1-R$, i.e., less than or equal to $\int_0^1 \rho(x)\mathrm{d}x/\int_0^1 \lambda(x)\mathrm{d}x$.

The first examples of capacity-achieving sequences of any given rate R were discovered in [10], and I will describe them in the following: fix a parameter D and let $\lambda_D(x) := \frac{1}{H(D)}\sum_{i=1}^{D} x^i/i$, where $H(D)$ is the harmonic sum $\sum_{i=1}^{D} 1/i$. Since $\int_0^1 \lambda_D(x)\mathrm{d}x = \frac{1}{H(D)}(1-1/(D+1))$. Let $\rho_D(x) := \mathrm{e}^{\mu(x-1)}$, where μ is the unique solution to the equation

$$\frac{1}{\mu}(1-\mathrm{e}^{-\mu}) = \frac{1-R}{H(D)}\left(1-\frac{1}{D+1}\right).$$

Then the sequence $(\lambda_D(x), \rho_D(x))_{D\geq 1}$ gives rise to codes of rate at least R. Further, we have

$$\begin{aligned}\delta\lambda_D(1-\rho_D(1-x)) &= \delta\lambda_D(1-\mathrm{e}^{-\mu x}) \\ &\leq \frac{-\delta}{H(D)}\ln(\mathrm{e}^{-\mu x}) \\ &= \frac{\delta\mu x}{H(D)}.\end{aligned}$$

Hence, successful decoding is possible if the fraction of erasures is no more than $H(D)/\mu$. Note that this quantity equals $(1-R)(1-1/(D+1))/(1-\mathrm{e}^{-\mu})$, and that this quantity is larger than $(1-R)(1-1/D)$. Hence, we have that

$$(1-R)(1-1/D)\lambda_D(1-\rho_D(1-x)) < x \quad \text{for} \quad x \in (0,(1-R)(1-1/D)).$$

This shows that the sequence is indeed capacity achieving. We have named these sequences the *Heavy-Tail/Poisson* sequences, or, more commercially oriented, *Tornado codes.*

In the meantime, I have obtained yet another capacity achieving sequence whose left side is closely related to the power series expansion of $(1-x)^{1/D}$, and which is right-regular, i.e., all nodes on the right have the same degree [16]. More precisely, the new sequence is defined as follows. For integers $a \geq 2$ and $n \geq 2$ let

$$\rho_a(x) := x^{a-1}, \qquad \lambda_{a,n}(x) := \frac{\sum_{k=1}^{n-1}\binom{\alpha}{k}(-1)^{k+1}x^k}{1-n\binom{\alpha}{n}(-1)^{n+1}},$$

where $\alpha := 1/(a-1)$. For the correct choice of the parameter n and other properties of these sequences we refer the reader to [16].

I would like to close this section with a few comments on the trade-off between proximity to the channel capacity and the running time of the decoder. For the Heavy-Tail/Poisson sequence the average degree of a message node was less than $H(D)$, and it could tolerate up to $(1-R)(1-1/D)$ fraction of erasures. Hence, to get close to within $1-\epsilon$ of the capacity $1-R$, we needed codes of average degree $O(\log(1/\epsilon))$. This is shown to be essentially optimal in [16]. In other words, to get within $1-\epsilon$ of the channel capacity, we need graphs of average degree $\Omega(\log(1/\epsilon))$. The same relation also holds for the right-regular sequences. Hence, these codes are essentially optimal for our simple decoders.

4 Codes on Other Channels

In this section we will *briefly* describe some of the most recent developments on the field of LDPC-codes. Already in 1998, Luby, Mitzenmacher, Spielman, and myself [8] started to adapt the analysis of [7] to the situation of simple decoders of Gallager for transmission over a binary symmetric channel. To my knowledge, this was the first rigorous analysis of a probabilistic decoder for LDPC codes. One common feature between these decoders and our simple erasure decoder described above is the following: at each round of the iteration, one has to keep track of only one real variable; in the case of the erasure decoder this variable describes the probability of a message node being still erased. In the case of Gallager's decoders it equals the probability of a message node being in error.

The analysis of more powerful decoders like the belief-propagation decoder is more complicated, as, at each round, one has to keep track of a density function describing the distribution of various values at a message node. (For a description of the belief-propagation algorithm, we refer the reader to [11].) Nevertheless, Richardson and Urbanke managed to generalize the analysis of [14] to this case as well. One of the main results of that paper is the derivation of a recursion for the (common) density functions of the message nodes at each iteration of the algorithm. The analysis was further simplified in [13], and will be described below. First, we assume that the input alphabet is the set $\{\pm 1\}$. At each round, the algorithm passes messages from message nodes to check nodes, and then from check nodes to message nodes. We assume that at the message nodes the messages are represented as log-likelihood ratios

$$\log \frac{p(y|x=1)}{p(y|x=-1)},$$

where y represents all the observations conveyed to the message node at that time. Now let f_ℓ denote the probability density function at the message nodes at the ℓth round of the algorithm. f_0 is then the density function of the error which the message bits are originally exposed to. It is also denoted by P_0. These density functions are defined on the set $\mathbb{R}\cup\{\pm\infty\}$. It turns out that they satisfy a *symmetry condition* [13] $f(-x) = f(x)\mathrm{e}^{-x}$. As a result, the value of any of these density functions is determined from the set of its values on the set $\mathbb{R}_{\geq 0} \cup \{\infty\}$. The restriction of a function f to this set is denoted by $f^{\geq 0}$. (The technical difficulty of defining a function at ∞ could be solved by using distributions instead of functions, but we will not further discuss it here.)

For a function f defined on $\mathbb{R}_{\geq 0} \cup \{\infty\}$ we define a *hyperbolic change of measure* γ via

$$\gamma(f)(x) := f(\ln\coth x/2)\mathrm{csch}(x).$$

If f is a function satisfying the symmetry condition, then $\gamma(f^{\geq 0})$ defines a function on $\mathbb{R}_{\geq 0}\cup\{\infty\}$ which can be uniquely extended to a function F on $\mathbb{R}\cup\{\pm\infty\}$. The transformation mapping f to F is denoted by $\mathit{\Gamma}$. It is a bijective mapping from the set of density functions on $\mathbb{R}\cup\{\pm\infty\}$ satisfying the symmetry condition into itself. Let f_ℓ denote the density of the common density function of the messages passed from message nodes to check nodes at round ℓ of the algorithm. f_0 then denotes the density of the original error, and is also denoted by P_0. Suppose that the graph has a degree distribution given by $\lambda(x)$ and $\rho(x)$. Then we have the following:

$$f_\ell = P_0 \otimes \lambda(\Gamma^{-1}(\rho(\Gamma(f_{\ell-1})))), \quad \ell \geq 1. \tag{3}$$

Here, $\otimes$ denotes the convolution, and for a function f, $\lambda(f)$ denotes the function $\sum_i \lambda_i f^{\otimes(i-1)}$. In the case of the erasure channel, the corresponding density functions are two-point mass functions, with a mass p_ℓ at zero and a mass $(1 - p_\ell)$ at infinity. In this case, the iteration translates to [13]

$$p_\ell = \delta\lambda(1 - \rho(1 - p_{\ell-1})),$$

where δ is the original fraction of erasures. This is exactly the same as in (1).

5 Conclusion and Open Problems

There are still a large number of important open questions about LDPC codes. Among them I would like to single out two topics which I call the "asymptotic theory" and "short codes". As the name suggests, the asymptotic theory deals with the asymptotic performance of various decoding algorithms for LDPC codes. As discussed above, a lot of progress has been made in the asymptotic analysis of belief propagation and other algorithms. One of the main important open questions is, for each given algorithm, the design of the degree structure of the underlying graphs such that the corresponding codes perform *optimally* with respect to the given decoding algorithm. This is particularly important for the case of belief-propagation. Here we go even further, and ask about *capacity-achieving* sequences of degree distributions. In other words, given a channel with capacity C, we want for any given ϵ explicit degree distributions $\lambda_\epsilon(x)$ and $\rho_\epsilon(x)$ such that, asymptotically, LDPC codes obtained from sampling graphs with these distributions perform at rates that are within ϵ of C when decoded using belief-propagation. The only case for which we know such sequences is that of the erasure channel [10,16]. We conjecture that such sequences exist for other channels like the AWGN channel, or the BSC, as well.

The topic of "short codes" deals with the construction and analysis of good "short" codes. Here we only mention the question of a rigorous analysis of codes of finite length (rather than the asymptotic analysis discussed in the last paragraph). The problem that arises here is that for most of the message nodes, the neighborhood is a tree only for very small depths. In other words, the decoder works on graphs that have cycles, and the analysis described above is not adequate.

References

1. N. Alon and M. Luby. A linear time erasure-resilient code with nearly optimal recovery. *IEEE Trans. Inform. Theory*, 42:1732–1736, 1996.

2. S. Arora and S. Safra. Probabilistic checking of proofs: a new characterization of NP. *J. ACM*, 45:70–122, 1998.
3. E.R. Berlekamp, R.J. McEliece, and H.C.A. van Tilborg. On the inherent intractability of certain coding problems. *IEEE Trans. Inform. Theory*, 24:384–386, 1978.
4. C. Berroux, A. Glavieux, and P. Thitimajshima. Near Shannon limit error-correcting coding and decoding. In *Proceedings of ICC'93*, pages 1064–1070, 1993.
5. P. Elias. Coding for two noisy channels. In *Information Theory, Third London Symposium*, pages 61–76, 1955.
6. R. G. Gallager. *Low Density Parity-Check Codes.* MIT Press, Cambridge, MA, 1963.
7. M. Luby, M. Mitzenmacher, and M.A. Shokrollahi. Analysis of random processes via and-or tree evaluation. In *Proceedings of the 9th Annual ACM-SIAM Symposium on Discrete Algorithms*, pages 364–373, 1998.
8. M. Luby, M. Mitzenmacher, M.A. Shokrollahi, and D. Spielman. Analysis of low density codes and improved designs using irregular graphs. In *Proceedings of the 30th Annual ACM Symposium on Theory of Computing*, pages 249–258, 1998.
9. M. Luby, M. Mitzenmacher, M.A. Shokrollahi, and D. Spielman. Improved low-density parity-check codes using irregular graphs and belief propagation. In *Proceedings 1998 IEEE International Symposium on Information Theory*, page 117, 1998.
10. M. Luby, M. Mitzenmacher, M.A. Shokrollahi, D. Spielman, and V. Stemann. Practical loss-resilient codes. In *Proceedings of the 29th annual ACM Symposium on Theory of Computing*, pages 150–159, 1997.
11. D.J.C. MacKay. Good error-correcting codes based on very sparse matrices. *IEEE Trans. Inform. Theory*, 45:399–431, 1999.
12. G. A. Margulis. Explicit constructions of graphs without short cycles and low density codes. *Combinatorica*, 2:71–78, 1982.
13. T. Richardson, M.A. Shokrollahi, and R. Urbanke. Design of provably good low-density parity check codes. *IEEE Trans. Inform. Theory (submitted)*, 1999.
14. T. Richardson and R. Urbanke. The capacity of low-density parity check codes under message-passing decoding. *IEEE Trans. Inform. Theory (submitted)*, 1998.
15. C. E. Shannon. A mathematical theory of communication. *Bell System Tech. J.*, 27:379–423, 623–656, 1948.
16. M.A. Shokrollahi. New sequences of linear time erasure codes approaching the channel capacity. To appear in the Proceedings of AAECC'13, 1999.
17. M. Sipser and D. Spielman. Expander codes. *IEEE Trans. Inform. Theory*, 42:1710–1722, 1996.
18. D. Spielman. Linear-time encodable and decodable error-correcting codes. *IEEE Trans. Inform. Theory*, 42:1723–1731, 1996.
19. M. R. Tanner. A recursive approach to low complexity codes. *IEEE Trans. Inform. Theory*, 27:533–547, 1981.
20. V. V. Zyablov. An estimate of the complexity of constructing binary linear cascade codes. *Probl. Inform. Transm.*, 7:3–10, 1971.
21. V. V. Zyablov and M. S. Pinsker. Estimation of error-correction complexity of Gallager low-density codes. *Probl. Inform. Transm.*, 11:18–28, 1976.

A Classification of Symbolic Transition Systems*

Thomas A. Henzinger and Rupak Majumdar

Department of Electrical Engineering and Computer Sciences
University of California at Berkeley, CA 94720-1770, USA
{tah,rupak}@eecs.berkeley.edu

Abstract. We define five increasingly comprehensive classes of infinite-state systems, called STS1–5, whose state spaces have finitary structure. For four of these classes, we provide examples from hybrid systems.

STS1 These are the systems with finite *bisimilarity* quotients. They can be analyzed symbolically by (1) iterating the predecessor and boolean operations starting from a finite set of observable state sets, and (2) terminating when no new state sets are generated. This enables model checking of the μ-calculus.

STS2 These are the systems with finite *similarity* quotients. They can be analyzed symbolically by iterating the predecessor and positive boolean operations. This enables model checking of the existential and universal fragments of the μ-calculus.

STS3 These are the systems with finite *trace-equivalence* quotients. They can be analyzed symbolically by iterating the predecessor operation and a restricted form of positive boolean operations (intersection is restricted to intersection with observables). This enables model checking of linear temporal logic.

STS4 These are the systems with finite *distance-equivalence* quotients (two states are equivalent if for every distance d, the same observables can be reached in d transitions). The systems in this class can be analyzed symbolically by iterating the predecessor operation and terminating when no new state sets are generated. This enables model checking of the existential conjunction-free and universal disjunction-free fragments of the μ-calculus.

STS5 These are the systems with finite *bounded-reachability* quotients (two states are equivalent if for every distance d, the same observables can be reached in d or fewer transitions). The systems in this class can be analyzed symbolically by iterating the predecessor operation and terminating when no new states are encountered. This enables model checking of reachability properties.

* This research was supported in part by the DARPA (NASA) grant NAG2-1214, the DARPA (Wright-Patterson AFB) grant F33615-C-98-3614, the MARCO grant 98-DT-660, the ARO MURI grant DAAH-04-96-1-0341, and the NSF CAREER award CCR-9501708.

H. Reichel and S. Tison (Eds.): STACS 2000, LNCS 1770, pp. 13–34, 2000.

0 Introduction

To explore the state space of an infinite-state transition system, it is often convenient to compute on a data type called "region," whose members represent (possibly infinite) sets of states. Regions might be implemented, for example, as constraints on the integers or reals. We say that a transition system is "symbolic" if it comes equipped with an algebra of regions which permits the effective computation of certain operations on regions. For model checking, we are particularly interested in boolean operations on regions as well as the predecessor operation, which, given a target region, computes the region of all states with successors in the target region. While a region algebra supports individual operations on regions, the iteration of these operations may generate an infinite number of distinct regions. In this paper, we study restricted classes of symbolic transition systems for which certain forms of iteration, if terminated after a finite number of operations, still yield sufficient information for checking interesting, unbounded temporal properties of the system.

0.1 Symbolic Transition Systems

Definition: Symbolic transition system A *symbolic transition system* $\mathcal{S} = (Q, \delta, R, \ulcorner\cdot\urcorner, P)$ consists of a (possibly infinite) set Q of *states*, a (possibly nondeterministic) *transition* function $\delta : Q \to \mathbf{2}^Q$ which maps each state to a set of successor states, a (possibly infinite) set R of *regions*, an *extension* function $\ulcorner\cdot\urcorner : R \to \mathbf{2}^Q$ which maps each region to a set of contained states, and a finite set $P \subseteq R$ of *observables*, such that the following six conditions are satisfied:

1. The set P of observables covers the state space Q; that is, $\bigcup\{\ulcorner p\urcorner \mid p \in P\} = Q$.
2. For each region $\sigma \in R$, there is a region $Pre(\sigma) \in R$ such that
$$\ulcorner Pre(\sigma)\urcorner = \{s \in Q \mid (\exists t \in \delta(s) : t \in \sigma)\};$$
furthermore, the function $Pre : R \to R$ is computable.
3. For each pair $\sigma, \tau \in R$ of regions, there is a region $And(\sigma, \tau) \in R$ such that $\ulcorner And(\sigma,\tau)\urcorner = \ulcorner\sigma\urcorner \cap \ulcorner\tau\urcorner$; furthermore, the function $And : R \times R \to R$ is computable.
4. For each pair $\sigma, \tau \in R$ of regions, there is a region $Diff(\sigma, \tau) \in R$ such that $\ulcorner Diff(\sigma,\tau)\urcorner = \ulcorner\sigma\urcorner \backslash \ulcorner\tau\urcorner$; furthermore, the function $Diff : R \times R \to R$ is computable.
5. All emptiness questions about regions can be decided; that is, there is a computable function $Empty : R \to \mathbb{B}$ such that $Empty(\sigma)$ iff $\ulcorner\sigma\urcorner = \emptyset$.
6. All membership questions about regions can be decided; that is, there is a computable function $Member : Q \times R \to \mathbb{B}$ such that $Member(s, \sigma)$ iff $s \in \ulcorner\sigma\urcorner$.

The tuple $\mathcal{R}_{\mathcal{S}} = (P, Pre, And, Diff, Empty)$ is called the *region algebra* of $\mathcal{S}$. □

Remark: Duality We take an existential view of symbolic transition systems. The dual, universal view requires (1) $\bigcap\{\ulcorner p \urcorner \mid p \in P\} = \emptyset$, (2–4) closure of R under computable functions $\overline{Pre}$, $\overline{And}$, and $\overline{Diff}$ such that

$$\ulcorner \overline{Pre}(\sigma) \urcorner = \{s \in Q \mid (\forall t \in \delta(s) \colon t \in \sigma)\},$$

$\ulcorner \overline{And}(\sigma, \tau) \urcorner = \ulcorner \sigma \urcorner \cup \ulcorner \tau \urcorner$, and $\ulcorner \overline{Diff}(\sigma, \tau) \urcorner = Q \backslash \ulcorner Diff(\tau, \sigma) \urcorner$, and (5) a computable function $\overline{Empty}$ for deciding all universality questions about regions (that is, $\overline{Empty}(\sigma)$ iff $\ulcorner \sigma \urcorner = Q$). All results of this paper have an alternative, dual formulation. □

0.2 Example: Polyhedral Hybrid Automata

A *polyhedral hybrid automaton H of dimension m*, for a positive integer m, consists of the following components [AHH96]:

Continuous variables A set $X = \{x_1, \ldots, x_m\}$ of real-valued variables. We write $\dot{X}$ for the set $\{\dot{x}_1, \ldots, \dot{x}_m\}$ of dotted variables (which represent first derivatives during continuous change), and we write X' for the set $\{x'_1, \ldots, x'_m\}$ of primed variables (which represent values at the conclusion of discrete change). A *linear constraint* over X is an expression of the form $k_0 \sim k_1 x_1 + \cdots + k_m x_m$, where $\sim \in \{<, \leq, =, \geq, >\}$ and $k_0, \ldots, k_m$ are integer constants. A *linear predicate* over X is a boolean combination of linear constraints over X. Let L^m be the set of linear predicates over X.

Discrete locations A finite directed multigraph (V, E). The vertices in V are called *locations*; the edges in E are called *jumps*.

Invariant and flow conditions Two vertex-labeling functions *inv* and *flow*. For each location $v \in V$, the invariant condition $inv(v)$ is a conjunction of linear constraints over X, and the flow condition $flow(v)$ is a conjunction of linear constraints over $\dot{X}$. While the automaton control resides in location v, the variables may evolve according to $flow(v)$ as long as $inv(v)$ remains true.

Update conditions An edge-labeling functions *update*. For each jump $e \in E$, the update condition $update(e)$ is a conjunction of linear constraints over $X \cup X'$. The predicate $update(e)$ relates the possible values of the variables at the beginning of the jump (represented by X) and at the conclusion of the jump (represented by X').

The polyhedral hybrid automaton H is a *rectangular automaton* [HKPV98] if

- —all linear constraints that occur in invariant conditions of H have the form $x \sim k$, for $x \in X$ and $k \in \mathbb{Z}$;
- —all linear constraints that occur in flow conditions of H have the form $\dot{x} \sim k$, for $x \in X$ and $k \in \mathbb{Z}$;
- —all linear constraints that occur in jump conditions of H have the form $x \sim k$ or $x' = x$ or $x' \sim k$, for $x \in X$ and $k \in \mathbb{Z}$;
- —if e is a jump from location v to location v', and $update(e)$ contains the conjunct $x' = x$, then both $flow(v)$ and $flow(v')$ contain the same constraints on $\dot{x}$.

The rectangular automaton H is a *singular automaton* if each flow condition of H has the form $\dot{x}_1 = k_1 \wedge \ldots \wedge \dot{x}_m = k_m$. The singular automaton H is a *timed automaton* [AD94] if each flow condition of H has the form $\dot{x}_1 = 1 \wedge \ldots \wedge \dot{x}_m = 1$.

The polyhedral hybrid automaton H defines the symbolic transition system $\mathcal{S}_H = (Q_H, \delta_H, R_H, \ulcorner\cdot\urcorner_H, P_H)$ with the following components:

- $Q_H = V \times \mathbb{R}^m$; that is, every state $(v, \mathbf{x})$ consists of a location v (the discrete component of the state) and values $\mathbf{x}$ for the variables in X (the continuous component).
- $(v', \mathbf{x}') \in \delta_H(v, \mathbf{x})$ if either (1) there is a jump $e \in E$ from v to v' such that the closed predicate $update(e)[X, X' := \mathbf{x}, \mathbf{x}']$ is true, or (2) $v' = v$ and there is a real $\Delta \geq 0$ and a differentiable function $f \colon [0, \Delta] \to \mathbb{R}^m$ with first derivative $\dot{f}$ such that $f(0) = \mathbf{x}$ and $f(\Delta) = \mathbf{x}'$, and for all reals $\varepsilon \in (0, \Delta)$, the closed predicates $inv(v)[X := f(\varepsilon)]$ and $flow(v)[\dot{X} := \dot{f}(\varepsilon)]$ are true. In case (2), the function f is called a *flow function.*
- $R_H = V \times L^m$; that is, every region (v, ϕ) consists of a location v (the discrete component of the region) and a linear predicate ϕ over X (the continuous component).
- $\ulcorner(v, \phi)\urcorner_H = \{(v, \mathbf{x}) \mid \mathbf{x} \in \mathbb{R}^m \text{ and } \phi[X := \mathbf{x}] \text{ is true}\}$; that is, the extension function maps the continuous component ϕ of a region to the values for the variables in X which satisfy the predicate ϕ. Consequently, the extension of every region consists of a location and a polyhedral subset of $\mathbb{R}^m$.
- $P_H = V \times \{true\}$; that is, only the discrete component of a state is observable.

It requires some work to see that $\mathcal{S}_H$ is indeed a symbolic transition system. First, notice that the linear predicates over X are closed under all boolean operations, and that satisfiability is decidable for the linear predicates. Second, the *Pre* operator is computable on R_H, because all flow functions can be replaced by straight lines [AHH96].

0.3 Background Definitions

The symbolic transition systems are a special case of transition systems. A *transition system* $\mathcal{S} = (Q, \delta, \cdot, \ulcorner\cdot\urcorner, P)$ has the same components as a symbolic transition system, except that no regions are specified and the extension function is defined only for the observables (that is, $\ulcorner\cdot\urcorner \colon P \to \mathbf{2}^Q$).

State equivalences A *state equivalence* $\cong$ is a family of relations which contains for each transition system $\mathcal{S}$ an equivalence relation $\cong^{\mathcal{S}}$ on the states of $\mathcal{S}$. The $\cong$ *equivalence problem* for a class C of transition systems asks, given two states s and t of a transition system $\mathcal{S}$ from the class C, whether $s \cong^{\mathcal{S}} t$. The state equivalence $\cong_a$ is *as coarse as* the state equivalence $\cong_b$ if $s \cong_a^{\mathcal{S}} t$ implies $s \cong_b^{\mathcal{S}} t$ for all transition systems $\mathcal{S}$. The equivalence $\cong_a$ is *coarser than* $\cong_b$ if $\cong_a$ is as coarse as $\cong_b$, but $\cong_b$ is not as coarse as $\cong_a$. Given a transition system $\mathcal{S} = (Q, \delta, \cdot, \ulcorner\cdot\urcorner, P)$ and a state equivalence $\cong$, the *quotient system* is the transition system $\mathcal{S}/_{\cong} = (Q/_{\cong}, \delta/_{\cong}, \cdot, \ulcorner\cdot\urcorner/_{\cong}, P)$ with the following components:

—the states in $\mathcal{S}/_{\cong}$ are the equivalence classes of $\cong_{\mathcal{S}}$;
—$\tau \in \delta/_{\cong}(\sigma)$ if there is a state $s \in \sigma$ and a state $t \in \tau$ such that $t \in \delta(s)$;
—$\sigma \in \ulcorner p \urcorner/_{\cong}$ if there is a state $s \in \sigma$ such that $s \in \ulcorner p \urcorner$.

The quotient construction is of particular interest to us when it transforms an infinite-state system $\mathcal{S}$ into a finite-state system $\mathcal{S}/_{\cong}$.

State logics A *state logic* L is a logic whose formulas are interpreted over the states of transition systems; that is, for every L-formula φ and every transition system $\mathcal{S}$, there is a set $[\![\varphi]\!]_{\mathcal{S}}$ of states of $\mathcal{S}$ which satisfy φ. The L *model-checking problem* for a class C of transition systems asks, given an L-formula φ and a state s of a transition system $\mathcal{S}$ from the class C, whether $s \in [\![\varphi]\!]_{\mathcal{S}}$. Two formulas φ and ψ of state logics are *equivalent* if $[\![\varphi]\!]_{\mathcal{S}} = [\![\psi]\!]_{\mathcal{S}}$ for all transition systems $\mathcal{S}$. The state logic L_a is *as expressive as* the state logic L_b if for every L_b-formula φ, there is an L_a-formula ψ which is equivalent to φ. The logic L_a is *more expressive than* L_b if L_a is as expressive as L_b, but L_b is not as expressive as L_a. Every state logic L *induces* a state equivalence, denoted $\cong_L$: for all states s and t of a transition system $\mathcal{S}$, define $s \cong_L^{\mathcal{S}} t$ if for all L-formulas φ, we have $s \in [\![\varphi]\!]_{\mathcal{S}}$ iff $t \in [\![\varphi]\!]_{\mathcal{S}}$. The state logic L *admits abstraction* if for every L-formula φ and every transition system $\mathcal{S}$, we have $[\![\varphi]\!]_{\mathcal{S}} = \bigcup\{\sigma \mid \sigma \in [\![\varphi]\!]_{\mathcal{S}/_{\cong_L}}\}$; that is, a state s of $\mathcal{S}$ satisfies an L-formula φ iff the $\cong_L$ equivalence class of s satisfies φ in the quotient system. Consequently, if L admits abstraction, then every L model-checking question on a transition system $\mathcal{S}$ can be reduced to an L model-checking question on the induced quotient system $\mathcal{S}/_{\cong_L}$. Below, we shall repeatedly prove the L model-checking problem for a class C to be decidable by observing that for every transition system $\mathcal{S}$ from C, the quotient system $\mathcal{S}/_{\cong_L}$ has finitely many states and can be constructed effectively.

Symbolic semi-algorithms A *symbolic semi-algorithm* takes as input the region algebra $\mathcal{R}_{\mathcal{S}} = (P, \mathit{Pre}, \mathit{And}, \mathit{Diff}, \mathit{Empty})$ of a symbolic transition system $\mathcal{S} = (Q, \delta, R, \ulcorner\cdot\urcorner, P)$, and generates regions in R using the operations P, *Pre*, *And*, *Diff*, and *Empty*. Depending on the input $\mathcal{S}$, a symbolic semi-algorithm on $\mathcal{S}$ may or may not terminate.

0.4 Preview

In sections 1–5 of this paper, we shall define five increasingly comprehensive classes of symbolic transition systems. In each case $i \in \{1, \ldots, 5\}$, we will proceed in four steps:

1 Definition: Finite characterization We give a state equivalence $\cong_i$ and define the class $\mathsf{STS}(i)$ to contain precisely the symbolic transition systems $\mathcal{S}$ for which the equivalence relation $\cong_i^{\mathcal{S}}$ has finite index (i.e., there are finitely many $\cong_i^{\mathcal{S}}$ equivalence classes). Each state equivalence $\cong_i$ is coarser than its predecessor $\cong_{i-1}$, which implies that $\mathsf{STS}(i-1) \subsetneq \mathsf{STS}(i)$ for $i \in \{2, \ldots, 5\}$.

2 Algorithmics: Symbolic state-space exploration We give a symbolic semi-algorithm that terminates precisely on the symbolic transition systems in the class $\mathsf{STS}(i)$. This provides an operational characterization of the class $\mathsf{STS}(i)$ which is equivalent to the denotational definition of $\mathsf{STS}(i)$. Termination of the semi-algorithm is proved by observing that if given the region algebra of a symbolic transition system $\mathcal{S}$ as input, then the extensions of all regions generated by the semi-algorithm are $\cong_i^{\mathcal{S}}$ blocks (i.e., unions of $\cong_i^{\mathcal{S}}$ equivalence classes). If $\mathcal{S}$ is in the class $\mathsf{STS}(i)$, then there are only finitely many $\cong_i^{\mathcal{S}}$ blocks, and the semi-algorithm terminates upon having constructed a representation of the quotient system $\mathcal{S}/_{\cong_i}$. The semi-algorithm can therefore be used to decide all $\cong_i$ equivalence questions for the class $\mathsf{STS}(i)$.

3 Verification: Decidable properties We give a state logic L_i which admits abstraction and induces the state equivalence $\cong_i$. Since $\cong_i$ quotients can be constructed effectively, it follows that the L_i model-checking problem for the class $\mathsf{STS}(i)$ is decidable. However, model-checking algorithms which rely on the explicit construction of quotient systems are usually impractical. Hence, we also give a symbolic semi-algorithm that terminates on the symbolic transition systems in the class $\mathsf{STS}(i)$ and directly decides all L_i model-checking questions for this class.

4 Example: Hybrid systems The interesting members of the class $\mathsf{STS}(i)$ are those with infinitely many states. In four out of the five cases, following [Hen96], we provide certain kinds of polyhedral hybrid automata as examples.

1 Class-1 Symbolic Transition Systems

Class-1 systems are characterized by finite bisimilarity quotients. The region algebra of a class-1 system has a finite subalgebra that contains the observables and is closed under *Pre*, *And*, and *Diff* operations. This enables the model checking of all μ-calculus properties. Infinite-state examples of class-1 systems are provided by the singular hybrid automata.

1.1 Finite Characterization: Bisimilarity

Definition: Bisimilarity Let $\mathcal{S} = (Q, \delta, \cdot, \ulcorner\cdot\urcorner, P)$ be a transition system. A binary relation $\preceq$ on the state space Q is a *simulation* on $\mathcal{S}$ if $s \preceq t$ implies the following two conditions:

1. For each observable $p \in P$, we have $s \in \ulcorner p \urcorner$ iff $t \in \ulcorner p \urcorner$.
2. For each state $s' \in \delta(s)$, there is a state $t' \in \delta(t)$ such that $s' \preceq t'$.

Two states $s, t \in Q$ are *bisimilar*, denoted $s \cong_1^{\mathcal{S}} t$, if there is a symmetric simulation $\preceq$ on $\mathcal{S}$ such that $s \preceq t$. The state equivalence $\cong_1$ is called *bisimilarity.* □

Definition: Class $\mathsf{STS1}$ A symbolic transition system $\mathcal{S}$ belongs to the class $\mathsf{STS1}$ if the bisimilarity relation $\cong_1^{\mathcal{S}}$ has finite index. □

Symbolic semi-algorithm Closure1
Input: a region algebra $\mathcal{R} = (P, Pre, And, Diff, Empty)$.

$T_0 := P$;
for $i = 0, 1, 2, \ldots$ **do**
$T_{i+1} := T_i$
$\cup \{Pre(\sigma) \mid \sigma \in T_i\}$
$\cup \{And(\sigma, \tau) \mid \sigma, \tau \in T_i\}$
$\cup \{Diff(\sigma, \tau) \mid \sigma, \tau \in T_i\}$
until $\ulcorner T_{i+1} \urcorner \subseteq \ulcorner T_i \urcorner$.

The termination test $\ulcorner T_{i+1} \urcorner \subseteq \ulcorner T_i \urcorner$, which is shorthand for $\{\ulcorner \sigma \urcorner \mid \sigma \in T_{i+1}\} \subseteq \{\ulcorner \sigma \urcorner \mid \sigma \in T_i\}$, is decided as follows: for each region $\sigma \in T_{i+1}$ check that there is a region $\tau \in T_i$ such that both $Empty(Diff(\sigma, \tau))$ and $Empty(Diff(\tau, \sigma))$.

Fig. 1. Partition refinement

1.2 Symbolic State-Space Exploration: Partition Refinement

The bisimilarity relation of a finite-state system can be computed by partition refinement [KS90]. The symbolic semi-algorithm Closure1 of Figure 1 applies this method to infinite-state systems [BFH90, Hen95]. Suppose that the input given to Closure1 is the region algebra of a symbolic transition system $\mathcal{S} = (Q, \delta, R, \ulcorner \cdot \urcorner, P)$. Then each T_i, for $i \geq 0$, is a finite set of regions; that is, $T_i \subseteq R$. By induction it is easy to check that for all $i \geq 0$, the extension of every region in T_i is a $\cong_1^{\mathcal{S}}$ block. Thus, if $\cong_1^{\mathcal{S}}$ has finite index, then Closure1 terminates. Conversely, suppose that Closure1 terminates with $\ulcorner T_{i+1} \urcorner \subseteq \ulcorner T_i \urcorner$. From the definition of bisimilarity it follows that if for each region $\sigma \in T_i$, we have $s \in \ulcorner \sigma \urcorner$ iff $t \in \ulcorner \sigma \urcorner$, then $s \cong_1^{\mathcal{S}} t$. This implies that $\cong_1^{\mathcal{S}}$ has finite index.

Theorem 1A *For all symbolic transition systems $\mathcal{S}$, the symbolic semi-algorithm* Closure1 *terminates on the region algebra $\mathcal{R}_{\mathcal{S}}$ iff $\mathcal{S}$ belongs to the class* STS1.

Corollary 1A *The $\cong_1$ (bisimilarity) equivalence problem is decidable for the class* STS1 *of symbolic transition systems.*

1.3 Decidable Properties: Branching Time

Definition: μ-calculus The formulas of the *μ-calculus* are generated by the grammar

$$\varphi ::= p \mid \overline{p} \mid x \mid \varphi \vee \varphi \mid \varphi \wedge \varphi \mid \exists\bigcirc \varphi \mid \forall\bigcirc \varphi \mid (\mu x \colon \varphi) \mid (\nu x \colon \varphi),$$

for constants p from some set Π, and variables x from some set X. Let $\mathcal{S} = (Q, \delta, \cdot, \ulcorner \cdot \urcorner, P)$ be a transition system whose observables include all constants; that is, $\Pi \subseteq P$. Let $\mathcal{E} : X \to \mathbf{2}^Q$ be a mapping from the variables to sets of states. We write $\mathcal{E}[x \mapsto \rho]$ for the mapping that agrees with $\mathcal{E}$ on all variables, except that $x \in X$ is mapped to $\rho \subseteq Q$. Given $\mathcal{S}$ and $\mathcal{E}$, every formula φ of the μ-calculus defines a set $[\![\varphi]\!]_{\mathcal{S},\mathcal{E}} \subseteq Q$ of states:

$$\begin{aligned}
[\![p]\!]_{\mathcal{S},\mathcal{E}} &= \ulcorner p \urcorner; \\
[\![\overline{p}]\!]_{\mathcal{S},\mathcal{E}} &= Q \backslash \ulcorner p \urcorner; \\
[\![x]\!]_{\mathcal{S},\mathcal{E}} &= \mathcal{E}(x); \\
[\![\varphi_1 \{ {}^{\vee}_{\wedge} \} \varphi_2]\!]_{\mathcal{S},\mathcal{E}} &= [\![\varphi_1]\!]_{\mathcal{S},\mathcal{E}} \{ {}^{\cup}_{\cap} \} [\![\varphi_2]\!]_{\mathcal{S},\mathcal{E}}; \\
[\![\{ {}^{\exists}_{\forall} \} \bigcirc \varphi]\!]_{\mathcal{S},\mathcal{E}} &= \{ s \in Q \mid (\{ {}^{\exists}_{\forall} \} t \in \delta(s) \colon t \in [\![\varphi]\!]_{\mathcal{S},\mathcal{E}}) \}; \\
[\![\{ {}^{\mu}_{\nu} \} x \colon \varphi]\!]_{\mathcal{S},\mathcal{E}} &= \{ {}^{\cap}_{\cup} \} \{ \rho \subseteq Q \mid \rho = [\![\varphi]\!]_{\mathcal{S},\mathcal{E}[x \mapsto \rho]} \}.
\end{aligned}$$

If we restrict ourselves to the closed formulas of the μ-calculus, then we obtain a state logic, denoted L_1^μ: the state $s \in Q$ *satisfies* the L_1^μ-formula φ if $s \in [\![\varphi]\!]_{\mathcal{S},\mathcal{E}}$ for any variable mapping $\mathcal{E}$; that is, $[\![\varphi]\!]_{\mathcal{S}} = [\![\varphi]\!]_{\mathcal{S},\mathcal{E}}$ for any $\mathcal{E}$. □

Remark: Duality For every L_1^μ-formula φ, the *dual* L_1^μ-formula $\overline{\varphi}$ is obtained by replacing the constructors p, $\overline{p}$, $\vee$, $\wedge$, $\exists\bigcirc$, $\forall\bigcirc$, μ, and ν by $\overline{p}$, p, $\wedge$, $\vee$, $\forall\bigcirc$, $\exists\bigcirc$, ν, and μ, respectively. Then, $[\![\overline{\varphi}]\!]_{\mathcal{S}} = Q \backslash [\![\varphi]\!]_{\mathcal{S}}$. It follows that the answer of the model-checking question for a state $s \in Q$ and an L_1^μ-formula φ is complementary to the answer of the model-checking question for s and the dual formula $\overline{\varphi}$. □

The following facts about the μ-calculus are relevant in our context [AH98]. First, L_1^μ admits abstraction, and the state equivalence induced by L_1^μ is $\cong_1$ (bisimilarity). Second, L_1^μ is very expressive; in particular, L_1^μ is more expressive than the temporal logics CTL* and CTL, which also induce bisimilarity. Third, the definition of L_1^μ naturally suggests a model-checking method for finite-state systems, where each fixpoint can be computed by successive approximation. The symbolic semi-algorithm ModelCheck of Figure 2 applies this method to infinite-state systems.

Suppose that the input given to ModelCheck is the region algebra of a symbolic transition system $\mathcal{S} = (Q, \delta, R, \ulcorner \cdot \urcorner, P)$, a μ-calculus formula φ, and any mapping $E : X \to \mathbf{2}^R$ from the variables to sets of regions. Then for each recursive call of ModelCheck, each T_i, for $i \geq 0$, is a finite set of regions from R, and each recursive call returns a finite set of regions from R. It is easy to check that all of these regions are also generated by the semi-algorithm Closure1 on input $\mathcal{R}_{\mathcal{S}}$. Thus, if Closure1 terminates, then so does ModelCheck. Furthermore, if it terminates, then ModelCheck returns a set $[\varphi]_E \subseteq R$ of regions such that $\bigcup \{ \ulcorner \sigma \urcorner \mid \sigma \in [\varphi]_E \} = [\![\varphi]\!]_{\mathcal{S},\mathcal{E}}$, where $\mathcal{E}(x) = \bigcup \{ \ulcorner \sigma \urcorner \mid \sigma \in E(x) \}$ for all $x \in X$. In particular, if φ is closed, then a state $s \in Q$ satisfies φ iff $Member(s, \sigma)$ for some region $\sigma \in [\varphi]_E$.

Theorem 1B. *For all symbolic transition systems $\mathcal{S}$ in* STS1 *and every L_1^μ-formula φ, the symbolic semi-algorithm* ModelCheck *terminates on the region algebra $\mathcal{R}_{\mathcal{S}}$ and the input formula φ.*

Corollary 1B *The L_1^μ model-checking problem is decidable for the class* STS1 *of symbolic transition systems.*

1.4 Example: Singular Hybrid Automata

The fundamental theorem of timed automata [AD94] shows that for every timed automaton, the (time-abstract) bisimilarity relation has finite index. The proof

Symbolic semi-algorithm ModelCheck
Input: a region algebra $\mathcal{R} = (P, Pre, And, Diff, Empty)$, a formula $\varphi \in L_1^\mu$, and a mapping E with domain X.

Output: $[\varphi]_E :=$
if $\varphi = p$ **then return** $\{p\}$;
if $\varphi = \overline{p}$ **then return** $\{Diff(q,p) \mid q \in P\}$;
if $\varphi = (\varphi_1 \vee \varphi_2)$ **then return** $[\varphi_1]_E \cup [\varphi_2]_E$;
if $\varphi = (\varphi_1 \wedge \varphi_2)$ **then**
return $\{And(\sigma, \tau) \mid \sigma \in [\varphi_1]_E$ and $\tau \in [\varphi_2]_E\}$;
if $\varphi = \exists\bigcirc\, \varphi'$ **then return** $\{Pre(\sigma) \mid \sigma \in [\varphi']_E\}$;
if $\varphi = \forall\bigcirc\, \varphi'$ **then return** $P\backslash\backslash\{Pre(\sigma) \mid \sigma \in (P\backslash\backslash[\varphi']_E)\}$;
if $\varphi = (\mu x\colon \varphi')$ **then**
$T_0 := \emptyset$;
for $i = 0, 1, 2, \ldots$ **do**
$T_{i+1} := [\varphi']_{E[x \mapsto T_i]}$
until $\bigcup\{\ulcorner\sigma\urcorner \mid \sigma \in T_{i+1}\} \subseteq \bigcup\{\ulcorner\sigma\urcorner \mid \sigma \in T_i\}$;
return T_i;
if $\varphi = (\nu x\colon \varphi')$ **then**
$T_0 := P$;
for $i = 0, 1, 2, \ldots$ **do**
$T_{i+1} := [\varphi']_{E[x \mapsto T_i]}$
until $\bigcup\{\ulcorner\sigma\urcorner \mid \sigma \in T_{i+1}\} \supseteq \bigcup\{\ulcorner\sigma\urcorner \mid \sigma \in T_i\}$;
return T_i.

The *pairwise-difference* operation $T\backslash\backslash T'$ between two finite sets T and T' of regions is computed inductively as follows:

$$T\backslash\backslash\emptyset = T;$$
$$T\backslash\backslash(\{\tau\} \cup T') = \{Diff(\sigma,\tau) \mid \sigma \in T\}\backslash\backslash T'.$$

The termination test $\bigcup\{\ulcorner\sigma\urcorner \mid \sigma \in T\} \subseteq \bigcup\{\ulcorner\sigma\urcorner \mid \sigma \in T'\}$ is decided by checking that $Empty(\sigma)$ for each region $\sigma \in (T\backslash\backslash T')$.

Fig. 2. Model checking

can be extended to the singular automata [ACH+95]. It follows that the symbolic semi-algorithm ModelCheck, which has been implemented for polyhedral hybrid automata in the tool HyTech [HHWT95], decides all L_1^μ model-checking questions for singular automata. The singular automata form a maximal class of hybrid automata in STS1. This is because there is a 2D (two-dimensional) rectangular automaton whose bisimilarity relation is state equality [Hen95].

Theorem 1C *The singular automata belong to the class* STS1. *There is a 2D rectangular automaton that does not belong to* STS1.

2 Class-2 Symbolic Transition Systems

Class-2 systems are characterized by finite similarity quotients. The region algebra of a class-2 system has a finite subalgebra that contains the observables and is closed under *Pre* and *And* operations. This enables the model checking of all existential and universal μ-calculus properties. Infinite-state examples of class-2 systems are provided by the 2D rectangular hybrid automata.

2.1 Finite Characterization: Similarity

Definition: Similarity Let $\mathcal{S}$ be a transition system. Two states s and t of $\mathcal{S}$ are *similar*, denoted $s \cong_2^{\mathcal{S}} t$, if there is a simulation $\preceq$ on $\mathcal{S}$ such that both $s \preceq t$ and $t \preceq s$. The state equivalence $\cong_2$ is called *similarity*. □

Definition: Class STS2 A symbolic transition system $\mathcal{S}$ belongs to the class STS2 if the similarity relation $\cong_2^{\mathcal{S}}$ has finite index. □

Since similarity is coarser than bisimilarity [vG90], the class STS2 of symbolic transition systems is a proper extension of STS1.

2.2 Symbolic State-Space Exploration: Intersection Refinement

The symbolic semi-algorithm Closure2 of Figure 3 is an abstract version of the method presented in [HHK95] for computing the similarity relation of an infinite-state system. Suppose that the input given to Closure2 is the region algebra of a symbolic transition system $\mathcal{S} = (Q, \delta, R, \ulcorner\cdot\urcorner, P)$. Given two states $s, t \in Q$, we say that t *simulates* s if $s \preceq t$ for some simulation $\preceq$ on $\mathcal{S}$. For $i \geq 0$ and $s \in Q$, define

$$Sim_i(s) \;=\; \bigcap\{\ulcorner\sigma\urcorner \mid \sigma \in T_i \text{ and } s \in \ulcorner\sigma\urcorner\},$$

where the set T_i of regions is computed by Closure2. By induction it is easy to check that for all $i \geq 0$, if t simulates s, then $t \in Sim_i(s)$. Thus, the extension of every region in T_i is a $\cong_2^{\mathcal{S}}$ block, and if $\cong_2^{\mathcal{S}}$ has finite index, then Closure2 terminates. Conversely, suppose that Closure2 terminates with $\ulcorner T_{i+1}\urcorner \subseteq \ulcorner T_i\urcorner$. From the definition of simulations it follows that if $t \in Sim_i(s)$, then t simulates s. This implies that $\cong_2^{\mathcal{S}}$ has finite index.

Theorem 2A *For all symbolic transition systems* $\mathcal{S}$, *the symbolic semi-algorithm* Closure2 *terminates on the region algebra* $\mathcal{R}_{\mathcal{S}}$ *iff* $\mathcal{S}$ *belongs to the class* STS2.

Corollary 2A *The* $\cong_2$ *(similarity) equivalence problem is decidable for the class* STS2 *of symbolic transition systems.*

2.3 Decidable Properties: Negation-Free Branching Time

Definition: Negation-free μ-calculus The *negation-free μ-calculus* consists of the μ-calculus formulas that are generated by the grammar

$$\varphi \;::=\; p \mid x \mid \varphi \vee \varphi \mid \varphi \wedge \varphi \mid \exists\bigcirc\varphi \mid (\mu x\colon \varphi) \mid (\nu x\colon \varphi),$$

Symbolic semi-algorithm Closure2
Input: a region algebra $\mathcal{R} = (P, Pre, And, Diff, Empty)$.

$$
\begin{aligned}
&T_0 := P; \\
&\textbf{for } i = 0, 1, 2, \ldots \textbf{ do} \\
&\quad T_{i+1} := T_i \\
&\qquad\qquad \cup \{Pre(\sigma) \mid \sigma \in T_i\} \\
&\qquad\qquad \cup \{And(\sigma, \tau) \mid \sigma, \tau \in T_i\} \\
&\quad \textbf{until } \ulcorner T_{i+1} \urcorner \subseteq \ulcorner T_i \urcorner.
\end{aligned}
$$

The termination test $\ulcorner T_{i+1} \urcorner \subseteq \ulcorner T_i \urcorner$ is decided as in Figure 1.

Fig. 3. Intersection refinement

for constants $p \in \Pi$ and variables $x \in X$. The state logic L_2^μ consists of the closed formulas of the negation-free μ-calculus. The state logic $\overline{L_2^\mu}$ consists of the duals of all L_2^μ-formulas. □

The following facts about the negation-free μ-calculus and its dual are relevant in our context [AH98]. First, both L_2^μ and $\overline{L_2^\mu}$ admit abstraction, and the state equivalence induced by both L_2^μ and $\overline{L_2^\mu}$ is $\cong_2$ (similarity). It follows that the logic L_1^μ with negation is more expressive than either L_2^μ or $\overline{L_2^\mu}$. Second, the negation-free logic L_2^μ is more expressive than the existential fragments of CTL* and CTL, which also induce similarity, and the dual logic $\overline{L_2^\mu}$ is more expressive than the universal fragments of CTL* and CTL, which again induce similarity.

If we apply the symbolic semi-algorithm ModelCheck of Figure 2 to the region algebra of a symbolic transition system $\mathcal{S}$ and an input formula from L_2^μ, then the cases $\varphi = \overline{p}$ and $\varphi = \forall\bigcirc\varphi'$ are never executed. It follows that all regions which are generated by ModelCheck are also generated by the semi-algorithm Closure2 on input $\mathcal{R}_\mathcal{S}$. Thus, if Closure2 terminates, then so does ModelCheck.

Theorem 2B *For all symbolic transition systems $\mathcal{S}$ in* STS2 *and every L_2^μ-formula φ, the symbolic semi-algorithm* ModelCheck *terminates on the region algebra $\mathcal{R}_\mathcal{S}$ and the input formula φ.*

Corollary 2B *The L_2^μ and $\overline{L_2^\mu}$ model-checking problems are decidable for the class* STS2 *of symbolic transition systems.*

2.4 Example: 2D Rectangular Hybrid Automata

For every 2D rectangular automaton, the (time-abstract) similarity relation has finite index [HHK95]. It follows that the symbolic semi-algorithm ModelCheck, as implemented in HyTech, decides all L_2^μ and $\overline{L_2^\mu}$ model-checking questions for 2D rectangular automata. The 2D rectangular automata form a maximal class of hybrid automata in STS2. This is because there is a 3D rectangular automaton whose similarity relation is state equality [HK96].

Theorem 2C *The 2D rectangular automata belong to the class* STS2. *There is a 3D rectangular automaton that does not belong to* STS2.

3 Class-3 Symbolic Transition Systems

Class-3 systems are characterized by finite trace-equivalence quotients. The region algebra of a class-3 system has a finite subalgebra that contains the observables and is closed under *Pre* operations and those *And* operations for which one of the two arguments is an observable. This enables the model checking of all linear temporal properties. Infinite-state examples of class-3 systems are provided by the rectangular hybrid automata.

3.1 Finite Characterization: Traces

Definition: Trace equivalence Let $\mathcal{S} = (Q, \delta, \cdot, \ulcorner\cdot\urcorner, P)$ be a transition system. Given a state $s_0 \in Q$, a *source-s_0 trace* π of $\mathcal{S}$ is a finite sequence $p_0 p_1 \ldots p_n$ of observables $p_i \in P$ such that

1. $s_0 \in \ulcorner p_0 \urcorner$;
2. for all $0 \leq i < n$, there is a state $s_{i+1} \in (\delta(s_i) \cap \ulcorner p_{i+1} \urcorner)$.

The number n of observables (minus 1) is called the *length* of the trace π, the final state s_n is the *sink* of π, and the final observable p_n is the *target* of π. Two states $s, t \in Q$ are *trace equivalent*, denoted $s \cong_3^{\mathcal{S}} t$, if every source-$s$ trace of $\mathcal{S}$ is a source-t trace of $\mathcal{S}$, and vice versa. The state equivalence $\cong_3$ is called *trace equivalence.* □

Definition: Class STS3 A symbolic transition system $\mathcal{S}$ belongs to the class STS3 if the trace-equivalence relation $\cong_3^{\mathcal{S}}$ has finite index. □

Since trace equivalence is coarser than similarity [vG90], the class STS3 of symbolic transition systems is a proper extension of STS2.

3.2 Symbolic State-Space Exploration: Observation Refinement

Trace equivalence can be characterized operationally by the symbolic semi-algorithm Closure3 of Figure 4. We shall show that, when the input is the region algebra of a symbolic transition system $\mathcal{S} = (Q, \delta, R, \ulcorner\cdot\urcorner, P)$, then Closure3 terminates iff the trace-equivalence relation $\cong_3^{\mathcal{S}}$ has finite index. Furthermore, upon termination, $s \cong_3^{\mathcal{S}} t$ iff for each region $\sigma \in T_i$, we have $s \in \ulcorner\sigma\urcorner$ iff $t \in \ulcorner\sigma\urcorner$.

Theorem 3A *For all symbolic transition systems $\mathcal{S}$, the symbolic semi-algorithm* Closure3 *terminates on the region algebra $\mathcal{R}_{\mathcal{S}}$ iff $\mathcal{S}$ belongs to the class* STS3.

Proof [HM99] We proceed in two steps. First, we show that Closure3 terminates on the region algebra $\mathcal{R}_{\mathcal{S}}$ iff the equivalence relation $\cong_{L_3^\mu}^{\mathcal{S}}$ induced by the linear-time μ-calculus (defined below) has finite index. Second, we show that $\cong_{L_3^\mu}$

Symbolic semi-algorithm Closure3
Input: a region algebra $\mathcal{R} = (P, Pre, And, Diff, Empty)$.

$T_0 := P$;
for $i = 0, 1, 2, \ldots$ **do**
$T_{i+1} := T_i$
$\cup \{Pre(\sigma) \mid \sigma \in T_i\}$
$\cup \{And(\sigma, p) \mid \sigma \in T_i \text{ and } p \in P\}$
until $\ulcorner T_{i+1} \urcorner \subseteq \ulcorner T_i \urcorner$.

The termination test $\ulcorner T_{i+1} \urcorner \subseteq \ulcorner T_i \urcorner$ is decided as in Figure 1.

Fig. 4. Observation refinement

coincides with trace equivalence. The proof of the first part proceeds as usual. It can be seen by induction that for all $i \geq 0$, the extension of every region in T_i, as computed by Closure3, is a $\cong^{\mathcal{S}}_{L_3^\mu}$ block. Thus, if $\cong^{\mathcal{S}}_{L_3^\mu}$ has finite index, then Closure3 terminates. Conversely, suppose that Closure3 terminates with $\ulcorner T_{i+1} \urcorner \subseteq \ulcorner T_i \urcorner$. It can be shown that if two states are not $\cong^{\mathcal{S}}_{L_3^\mu}$-equivalent, then there is a region in T_i which contains one state but not the other. It follows that if for each region $\sigma \in T_i$, we have $s \in \ulcorner \sigma \urcorner$ iff $t \in \ulcorner \sigma \urcorner$, then $s \cong^{\mathcal{S}}_{L_3^\mu} t$. This implies that $\cong^{\mathcal{S}}_{L_3^\mu}$ has finite index.

For the second part, we show that L_3^μ is as expressive as the logic $\exists$BÜCHI, whose formulas are the existentially interpreted Büchi automata, and that $\exists$BÜCHI is as expressive as L_3^μ. This result is implicit in a proof by [EJS93]. By induction on the structure of an L_3^μ-formula φ, we can construct a Büchi automaton B_φ such that for all transition systems $\mathcal{S}$, a state s of $\mathcal{S}$ satisfies φ iff for some infinite source-s trace of $\mathcal{S}$ is accepted by B_φ. Conversely, given a Büchi automaton B, we can construct an L_3^μ-formula which is equivalent to $\exists B$ [Dam94]. Since the state equivalence induced by $\exists$BÜCHI is trace equivalence, it follows that $\cong_{L_3^\mu}$ is also trace equivalence. □

Corollary 3A *The* $\cong_3$ *(trace) equivalence problem is decidable for the class* STS3 *of symbolic transition systems.*

3.3 Decidable Properties: Linear Time

Definition: Linear-time μ-calculus The *linear-time μ-calculus* (also called "L_1" in [EJS93]) consists of the μ-calculus formulas that are generated by the grammar

$$\varphi \ ::= \ p \mid x \mid \varphi \vee \varphi \mid p \wedge \varphi \mid \exists\bigcirc\varphi \mid (\mu x \colon \varphi) \mid (\nu x \colon \varphi),$$

for constants $p \in \Pi$ and variables $x \in X$. The state logic L_3^μ consists of the closed formulas of the linear-time μ-calculus. The state logic $\overline{L_3^\mu}$ consists of the duals of all L_3^μ-formulas. □

The following facts about the linear-time μ-calculus and its dual are relevant in our context (cf. the second part of the proof of Theorem 3A). First, both L_3^μ and $\overline{L_3^\mu}$ admit abstraction, and the state equivalence induced by both L_3^μ and $\overline{L_3^\mu}$ is $\cong_3$ (trace equivalence). It follows that the logic L_2^μ with unrestricted conjunction is more expressive than L_3^μ, and $\overline{L_2^\mu}$ is more expressive than $\overline{L_3^\mu}$. Second, the logic L_3^μ with restricted conjunction is more expressive than the existential interpretation of the linear temporal logic LTL, which also induces trace equivalence. For example, the existential LTL formula $\exists(p\mathcal{U}q)$ ("on some trace, p until q") is equivalent to the L_3^μ-formula $(\mu x\colon q \vee (p \wedge \exists\bigcirc x))$ (notice that one argument of the conjunction is a constant). The dual logic $\overline{L_3^\mu}$ is more expressive than the usual, universal interpretation of LTL, which again induces trace equivalence. For example, the (universal) LTL formula $p\mathcal{W}q$ ("on all traces, either p forever, or p until q") is equivalent to the $\overline{L_3^\mu}$-formula $(\nu x\colon p \wedge \forall\bigcirc(q \vee x))$ (notice that one argument of the disjunction is a constant).

If we apply the symbolic semi-algorithm ModelCheck of Figure 2 to the region algebra of a symbolic transition system $\mathcal{S}$ and an input formula from L_3^μ, then all regions which are generated by ModelCheck are also generated by the semi-algorithm Closure3 on input $\mathcal{R}_\mathcal{S}$. Thus, if Closure3 terminates, then so does ModelCheck.

Theorem 3B *For all symbolic transition systems $\mathcal{S}$ in* STS3 *and every L_3^μ-formula φ, the symbolic semi-algorithm* ModelCheck *terminates on the region algebra $\mathcal{R}_\mathcal{S}$ and the input formula φ.*

Corollary 3B *The L_3^μ and $\overline{L_3^\mu}$ model-checking problems are decidable for the class* STS3 *of symbolic transition systems.*

Remark: LTL model checking These results suggest, in particular, a symbolic procedure for model checking LTL properties over STS3 systems [HM99]. Suppose that $\mathcal{S}$ is a symbolic transition system in the class STS3, and φ is an LTL formula. First, convert $\neg\varphi$ to a Büchi automaton $B_{\neg\varphi}$ using a tableau construction, and then to an equivalent L_3^μ-formula ψ (introduce one variable per state of $B_{\neg\varphi}$). Second, run the symbolic semi-algorithm ModelCheck on inputs $\mathcal{R}_\mathcal{S}$ and ψ. It will terminate with a representation of the complement of the set of states that satisfy φ in $\mathcal{S}$. □

3.4 Example: Rectangular Hybrid Automata

For every rectangular automaton, the (time-abstract) trace-equivalence relation has finite index [HKPV98]. It follows that the symbolic semi-algorithm ModelCheck, as implemented in HyTech, decides all L_3^μ and $\overline{L_3^\mu}$ model-checking questions for rectangular automata. The rectangular automata form a maximal class of hybrid automata in STS3. This is because for simple generalizations of rectangular automata, the reachability problem is undecidable [HKPV98].

Theorem 3C *The rectangular automata belong to the class* STS3.

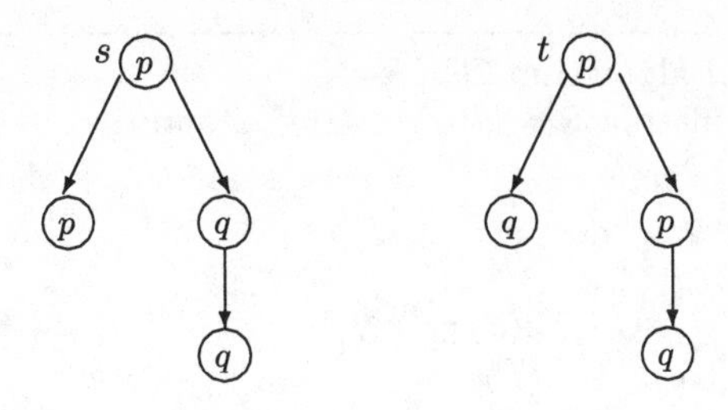

Fig. 5. Distance equivalence is coarser than trace equivalence

4 Class-4 Symbolic Transition Systems

We define two states of a transition system to be "distance equivalent" if for every distance d, the same observables can be reached in d transitions. Class-4 systems are characterized by finite distance-equivalence quotients. The region algebra of a class-4 system has a finite subalgebra that contains the observables and is closed under *Pre* operations. This enables the model checking of all existential conjunction-free and universal disjunction-free μ-calculus properties, such as the property that an observable can be reached in an even number of transitions.

4.1 Finite Characterization: Equi-distant Targets

Definition: Distance equivalence Let $\mathcal{S}$ be a transition system. Two states s and t of $\mathcal{S}$ are *distance equivalent*, denoted $s \cong_4^{\mathcal{S}} t$, if for every source-$s$ trace of $\mathcal{S}$ with length n and target p, there is a source-t trace of $\mathcal{S}$ with length n and target p, and vice versa. The state equivalence $\cong_4$ is called *distance equivalence*. □

Definition: Class STS4 A symbolic transition system $\mathcal{S}$ belongs to the class STS4 if the distance-equivalence relation $\cong_4^{\mathcal{S}}$ has finite index. □

Figure 5 shows that distance equivalence is coarser than trace equivalence (s and t are distance equivalent but not trace equivalent). It follows that the class STS4 of symbolic transition systems is a proper extension of STS3.

4.2 Symbolic State-Space Exploration: Predecessor Iteration

The symbolic semi-algorithm Closure4 of Figure 6 computes the subalgebra of a region algebra $\mathcal{R}_{\mathcal{S}}$ that contains the observables and is closed under the *Pre* operation. Suppose that the input given to Closure4 is the region algebra of a symbolic transition system $\mathcal{S} = (Q, \delta, R, \ulcorner\cdot\urcorner, P)$. For $i \geq 0$ and $s, t \in Q$, define $s \sim_i^{\mathcal{S}} t$ if for every source-s trace of $\mathcal{S}$ with length $n \leq i$ and target p, there is a source-t trace of $\mathcal{S}$ with length n and target p, and vice versa. By induction it is easy to check that for all $i \geq 0$, the extension of every region in T_i, as computed by Closure4, is a $\sim_i^{\mathcal{S}}$ block. Since $\sim_i^{\mathcal{S}}$ is as coarse as $\sim_{i+1}^{\mathcal{S}}$ for all $i \geq 0$, and $\cong_2^{\mathcal{S}}$ is

Symbolic semi-algorithm Closure4
Input: a region algebra $\mathcal{R} = (P, Pre, \cdot, Diff, Empty)$.

$T_0 := P$;
for $i = 0, 1, 2, \ldots$ **do**
$T_{i+1} := T_i$
$\cup \{Pre(\sigma) \mid \sigma \in T_i\}$
until $\ulcorner T_{i+1} \urcorner \subseteq \ulcorner T_i \urcorner$.

The termination test $\ulcorner T_{i+1} \urcorner \subseteq \ulcorner T_i \urcorner$ is decided as in Figure 1.

Fig. 6. Predecessor iteration

equal to $\bigcap\{\sim_i^S \mid i \geq 0\}$, if $\cong_2^S$ has finite index, then $\cong_2^S$ is equal to $\sim_i^S$ for some $i \geq 0$. Then, Closure2 will terminate in i iterations. Conversely, suppose that Closure4 terminates with $\ulcorner T_{i+1} \urcorner \subseteq \ulcorner T_i \urcorner$. In this case, if for all regions $\sigma \in T_i$, we have $s \in \ulcorner \sigma \urcorner$ iff $t \in \ulcorner \sigma \urcorner$, then $s \cong_4^S t$. This is because if s can reach an observable p in n transitions, but t cannot, then there is a region in T_i, namely, $Pre^n(p)$, such that $s \in \ulcorner Pre^n(p) \urcorner$ and $t \notin \ulcorner Pre^n(p) \urcorner$. It follows that $\cong_4^S$ has finite index.

Theorem 4A *For all symbolic transition systems $\mathcal{S}$, the symbolic semi-algorithm* Closure4 *terminates on the region algebra $\mathcal{R}_\mathcal{S}$ iff $\mathcal{S}$ belongs to the class* STS4.

Corollary 4A *The $\cong_4$ (distance) equivalence problem is decidable for the class* STS4 *of symbolic transition systems.*

4.3 Decidable Properties: Conjunction-Free Linear Time

Definition: Conjunction-free μ-calculus The *conjunction-free μ-calculus* consists of the μ-calculus formulas that are generated by the grammar

$$\varphi \ ::= \ p \mid x \mid \varphi \vee \varphi \mid \exists\bigcirc \varphi \mid (\mu x \colon \varphi)$$

for constants $p \in \Pi$ and variables $x \in X$. The state logic L_4^μ consists of the closed formulas of the conjunction-free μ-calculus. The state logic $\overline{L_4^\mu}$ consists of the duals of all L_4^μ-formulas. □

Definition: Conjunction-free temporal logic The formulas of the *conjunction-free temporal logic $L_4^\diamond$* are generated by the grammar

$$\varphi \ ::= \ p \mid \varphi \vee \varphi \mid \exists\bigcirc \varphi \mid \exists\diamond_{\leq d} \varphi \mid \exists\diamond \varphi,$$

for constants $p \in \Pi$ and nonnegative integers d. Let $\mathcal{S} = (Q, \delta, \cdot, \ulcorner\cdot\urcorner, P)$ be a transition system whose observables include all constants; that is, $\Pi \subseteq P$. The $L_4^\diamond$-formula φ defines the set $[\![\varphi]\!]_\mathcal{S} \subseteq Q$ of satisfying states:

$$[\![p]\!]_\mathcal{S} = \ulcorner p \urcorner;$$
$$[\![\varphi_1 \vee \varphi_2]\!]_\mathcal{S} = [\![\varphi_1]\!]_\mathcal{S} \cup [\![\varphi_2]\!]_\mathcal{S};$$

$$
\begin{aligned}
[\![\exists\bigcirc\varphi]\!]_S &= \{s \in Q \mid (\exists t \in \delta(s){:}\ t \in [\![\varphi]\!]_S)\};\\
[\![\exists\Diamond_{\leq d}\varphi]\!]_S &= \{s \in Q \mid \text{there is a source-}s\text{ trace of }\mathcal{S}\text{ with}\\
&\qquad\qquad \text{length at most } d \text{ and sink in } [\![\varphi]\!]_S\};\\
[\![\exists\Diamond\varphi]\!]_S &= \{s \in Q \mid \text{there is a source-}s\text{ trace of }\mathcal{S}\text{ with sink in }[\![\varphi]\!]_S\}.
\end{aligned}
$$

(The constructor $\exists\Diamond_{\leq d}$ is definable from $\exists\bigcirc$ and $\vee$; however, it will be essential in the $\exists\bigcirc$-free fragment of $L_4^\Diamond$ we will consider below.) □

Remark: Duality For every $L_4^\Diamond$-formula φ, the *dual* formula $\overline{\varphi}$ is obtained by replacing the constructors p, $\vee$, $\exists\bigcirc$, $\exists\Diamond_{\leq d}$, and $\exists\Diamond$ by $\overline{p}$, $\wedge$, $\forall\bigcirc$, $\forall\Box_{\leq d}$, and $\forall\Box$, respectively. The semantics of the dual constructors is defined as usual, such that $[\![\overline{\varphi}]\!]_S = Q\backslash[\![\varphi]\!]_S$. The state logic $\overline{L_4^\Diamond}$ consists of the duals of all $L_4^\Diamond$-formulas. It follows that the answer of the model-checking question for a state $s \in Q$ and an $\overline{L_4^\Diamond}$-formula $\overline{\varphi}$ is complementary to the answer of the model-checking question for s and the $L_4^\Diamond$-formula φ. □

The following facts about the conjunction-free μ-calculus, conjunction-free temporal logic, and their duals are relevant in our context. First, both L_4^μ and $\overline{L_4^\mu}$ admit abstraction, and the state equivalence induced by both L_4^μ and $\overline{L_4^\mu}$ is $\cong_4$ (distance equivalence). It follows that the logic L_3^μ with restricted conjunction is more expressive than L_4^μ, and $\overline{L_3^\mu}$ is more expressive than $\overline{L_4^\mu}$. Second, the conjunction-free μ-calculus L_4^μ is more expressive than the conjunction-free temporal logic $L_4^\Diamond$, and $\overline{L_4^\mu}$ is more expressive than $\overline{L_4^\Diamond}$, both of which also induce distance equivalence. For example, the property that an observable can be reached in an even number of transitions can be expressed in L_4^μ but not in $L_4^\Diamond$.

If we apply the symbolic semi-algorithm ModelCheck of Figure 2 to the region algebra of a symbolic transition system $\mathcal{S}$ and an input formula from L_4^μ, then all regions which are generated by ModelCheck are also generated by the semi-algorithm Closure4 on input $\mathcal{R}_S$. Thus, if Closure4 terminates, then so does ModelCheck.

Theorem 4B *For all symbolic transition systems $\mathcal{S}$ in* STS4 *and every L_4^μ-formula φ, the symbolic semi-algorithm* ModelCheck *terminates on the region algebra $\mathcal{R}_S$ and the input formula φ.*

Corollary 4B *The L_4^μ and $\overline{L_4^\mu}$ model-checking problems are decidable for the class* STS4 *of symbolic transition systems.*

5 Class-5 Symbolic Transition Systems

We define two states of a transition system to be "bounded-reach equivalent" if for every distance d, the same observables can be reached in d or fewer transitions. Class-5 systems are characterized by finite bounded-reach-equivalence quotients. Equivalently, for every observable p there is a finite bound n_p such that all states that can reach p can do so in at most n_p transitions. This enables the model checking of all reachability and (by duality) invariance properties. The

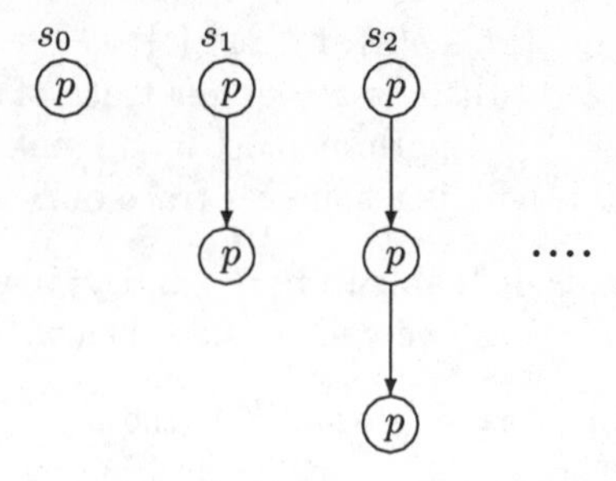

Fig. 7. Bounded-reach equivalence is coarser than distance equivalence

transition systems in class 5 have also been called "well-structured" [AČJT96]. Infinite-state examples of class-5 systems are provided by networks of rectangular hybrid automata.

5.1 Finite Characterization: Bounded-Distance Targets

Definition: Bounded-reach equivalence Let $\mathcal{S}$ be a transition system. Two states s and t of $\mathcal{S}$ are *bounded-reach equivalent*, denoted $s \cong_5^{\mathcal{S}} t$, if for every source-$s$ trace of $\mathcal{S}$ with length n and target p, there is a source-t trace of $\mathcal{S}$ with length at most n and target p, and vice versa. The state equivalence $\cong_5$ is called *bounded-reach equivalence.* □

Definition: Class STS5 A symbolic transition system $\mathcal{S}$ belongs to the class STS5 if the bounded-reach-equivalence relation $\cong_{\mathcal{S}}^5$ has finite index. □

Figure 7 shows that bounded-reach equivalence is coarser than distance equivalence (all states s_i, for $i \geq 0$, are bounded-reach equivalent, but no two of them are distance equivalent). It follows that the class STS5 of symbolic transition systems is a proper extension of STS4.

5.2 Symbolic State-Space Exploration: Predecessor Aggregation

The symbolic semi-algorithm Reach of Figure 8 starts from the observables and repeatedly applies the *Pre* operation, but its termination criterion is more easily met than the termination criterion of the semi-algorithm Closure4; that is, Reach may terminate on more inputs than Closure4. Indeed, we shall show that, when the input is the region algebra of a symbolic transition system $\mathcal{S} = (Q, \delta, R, \ulcorner\cdot\urcorner, P)$, then Reach terminates iff $\mathcal{S}$ belongs to the class STS5. Furthermore, upon termination, $s \cong_5^{\mathcal{S}} t$ iff for each observation $p \in P$ and each region $\sigma \in T_i^p$, we have $s \in \ulcorner\sigma\urcorner$ iff $t \in \ulcorner\sigma\urcorner$.

An alternative characterization of the class STS5 can be given using well-quasi-orders on states [AČJT96, FS98]. A *quasi-order* on a set A is a reflexive and transitive binary relation on A. A *well-quasi-order* on A is a quasi-order $\preceq$ on A

Symbolic semi-algorithm Reach
Input: a region algebra $\mathcal{R} = (P, Pre, And, Diff, Empty)$.

for each $p \in P$ **do**
$T_0 := \{p\}$;
for $i = 0, 1, 2, \dots$ **do**
$T_{i+1} := T_i \cup \{Pre(\sigma) \mid \sigma \in T_i\}$
until $\bigcup\{\ulcorner\sigma\urcorner \mid \sigma \in T_{i+1}\} \subseteq \bigcup\{\ulcorner\sigma\urcorner \mid \sigma \in T_i\}$
end.

The termination test $\bigcup\{\ulcorner\sigma\urcorner \mid \sigma \in T_{i+1}\} \subseteq \bigcup\{\ulcorner\sigma\urcorner \mid \sigma \in T_i\}$ is decided as in Figure 2.

Fig. 8. Predecessor aggregation

such that for every infinite sequence $a_0, a_1, a_2, \dots$ of elements $a_i \in A$ there exist indices i and j with $i < j$ and $a_i \preceq a_j$. A set $B \subseteq A$ is *upward-closed* if for all $b \in B$ and $a \in A$, if $b \preceq a$, then $a \in B$. It can be shown that if $\preceq$ is a well-quasi-order on A, then every infinite increasing sequence $B_0 \subseteq B_1 \subseteq B_2 \subseteq \cdots$ of upward-closed sets $B_i \subseteq A$ eventually stabilizes; that is, there exists an index $i \geq 0$ such that $B_j = B_i$ for all $j \geq i$.

Theorem 5A. *For all symbolic transition systems $\mathcal{S}$, the following three conditions are equivalent:*

1. *$\mathcal{S}$ belongs to the class* STS5.
2. *The symbolic semi-algorithm* Reach *terminates on the region algebra $\mathcal{R}_\mathcal{S}$.*
3. *There is a well-quasi-order $\preceq$ on the states of $\mathcal{S}$ such that for all observations p and all nonnegative integers d, the set $[\![\exists\Diamond_{\leq d}\, p]\!]_\mathcal{S}$ is upward-closed.*

Proof (2 $\Rightarrow$ 1) Define $s \sim^\mathcal{S}_{\leq n} t$ if for all observations p, for every source-s trace with length n and target p, there is a source-t trace with length at most n and target p, and vice versa. Note that $\sim^\mathcal{S}_{\leq n}$ has finite index for all $n \geq 0$. Suppose that the semi-algorithm Reach terminates in at most i iterations for each observation p. Then for all $n \geq i$, the equivalence relation $\sim^\mathcal{S}_{\leq n}$ is equal to $\sim^\mathcal{S}_{\leq i}$. Since $\cong^\mathcal{S}_5$ is equal to $\bigcap\{\sim^\mathcal{S}_{\leq n} \mid n \geq 0\}$, it has finite index.

(1 $\Rightarrow$ 3) Define the quasi-order $s \preceq^\mathcal{S}_5 t$ if for all observables p and all $n \geq 0$, for every source-s trace with length n and target p, there is a source-t trace with length at most n and target p. Then each set $[\![\exists\Diamond_{\leq d}\, p]\!]_\mathcal{S}$, for an observable p and a nonnegative integer d, is upward-closed with respect to $\preceq^\mathcal{S}_5$. Furthermore, if $\cong^\mathcal{S}_5$ has finite index, then $\preceq^\mathcal{S}_5$ is a well-quasi-order. This is because $s \cong^\mathcal{S}_5 t$ implies $s \preceq^\mathcal{S}_5 t$: if there were an infinite sequence $s_0, s_1, s_2, \dots$ of states such that for all $i \geq 0$ and $j < i$, we have $s_j \not\preceq^\mathcal{S}_5 s_i$, then no two of these states would be $\cong^\mathcal{S}_5$ equivalent.

(3 ⇒ 2) This part of the proof follows immediately from the stabilization property of well-quasi-orders [AČJT96]. □

5.3 Decidable Properties: Bounded Reachability

Definition: Bounded-reachability logic The *bounded-reachability logic* $L_5^\diamond$ consists of the $L_4^\diamond$-formulas that are generated by the grammar

$$\varphi ::= p \mid \varphi \vee \varphi \mid \exists\diamond_{\leq d}\varphi \mid \exists\diamond\varphi,$$

for constants $p \in \Pi$ and nonnegative integers d. The state logic $\overline{L_5^\diamond}$ consists of the duals of all $L_5^\diamond$-formulas. □

The following facts about bounded-reachability logic and its dual are relevant in our context. Both $L_5^\diamond$ and $\overline{L_5^\diamond}$ admit abstraction, and the state equivalence induced by both $L_5^\diamond$ and $\overline{L_5^\diamond}$ is $\cong_5$ (bounded-reach equivalence). It follows that the conjunction-free temporal logic $L_4^\diamond$ is more expressive than $L_5^\diamond$, and $\overline{L_4^\diamond}$ is more expressive than $\overline{L_5^\diamond}$. For example, the property that an observable can be reached in exactly d transitions can be expressed in $L_4^\diamond$ but not in $L_5^\diamond$. Since $L_5^\diamond$ admits abstraction, and for STS5 systems the induced quotient can be constructed using the symbolic semi-algorithm Reach, we have the following theorem.

Theorem 5B *The $L_5^\diamond$ and $\overline{L_5^\diamond}$ model-checking problems are decidable for the class* STS5 *of symbolic transition systems.*

A direct symbolic model-checking semi-algorithm for $L_5^\diamond$ and, indeed, $L_4^\diamond$ is easily derived from the semi-algorithm Reach. Then, if Reach terminates, so does model checking for all $L_4^\diamond$-formulas, including unbounded $\exists\diamond$ properties. The extension to $L_4^\diamond$ is possible, because $\exists\bigcirc$ properties pose no threat to termination.

5.4 Example: Networks of Rectangular Hybrid Automata

A *network of timed automata* [AJ98] consists of a finite state controller and an arbitrarily large set of identical 1D timed automata. The continuous evolution of the system increases the values of all variables. The discrete transitions of the system are specified by a set of synchronization rules. We generalize the definition to rectangular automata. Formally, a *network of rectangular automata* is a triple (C, H, R), where C is a finite set of controller locations, H is a 1D rectangular automaton, and R is a finite set of rules of the form $r = (\langle c, c'\rangle, e_1, \ldots, e_n)$, where $c, c' \in C$ and $e_1, \ldots, e_n$ are jumps of H. The rule r is enabled if the controller state is c and there are n rectangular automata $H_1, \ldots, H_n$ whose states are such that the jumps $e_1, \ldots, e_n$, respectively, can be performed. The rule r is executed by simultaneously changing the controller state to c' and the state of each H_i, for $1 \leq i \leq n$, according to the jump e_i. The following result is proved in [AJ98] for networks of timed automata. The proof can be extended to rectangular automata using the observation that every rectangular automaton is simulated by an appropriate timed automaton [HKPV98].

Theorem 5C *The networks of rectangular automata belong to the class* STS5. *There is a network of timed automata that does not belong to* STS4.

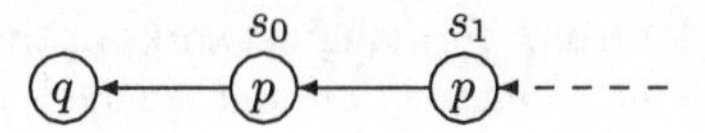

Fig. 9. Reach equivalence is coarser than bounded-reach equivalence

6 General Symbolic Transition Systems

For studying reachability questions on symbolic transition systems, it is natural to consider the following fragment of bounded-reachability logic.

Definition: Reachability logic The *reachability logic* $L_6^\diamond$ consists of the $L_5^\diamond$-formulas that are generated by the grammar

$$\varphi ::= p \mid \varphi \vee \varphi \mid \exists\diamond\varphi,$$

for constants $p \in \Pi$. □

The reachability logic $L_6^\diamond$ is less expressive than the bounded-reachability logic $L_5^\diamond$, because it induces the following state equivalence, $\cong_6$, which is coarser than bounded-reach equivalence (see Figure 9: all states s_i, for $i \geq 0$, are reach equivalent, but no two of them are bounded-reach-equivalent).

Definition: Reach equivalence Let $\mathcal{S}$ be a transition system. Two states s and t of $\mathcal{S}$ are *reach equivalent*, denoted $s \cong_6^{\mathcal{S}} t$, if for every source-$s$ trace of $\mathcal{S}$ with target p, there is a source-t trace of $\mathcal{S}$ with target p, and vice versa. The state equivalence $\cong_6$ is called *reach equivalence.* □

For every symbolic transition system $\mathcal{R}$ with k observables, the reach-equivalence relation $\cong_6^{\mathcal{R}}$ has at most 2^k equivalence classes and, therefore, finite index. Since the reachability problem is undecidable for many kinds of symbolic transition systems (including Turing machines and polyhedral hybrid automata [ACH+95]), it follows that there cannot be a general algorithm for computing the reach-equivalence quotient of symbolic transition systems.

References

[ACH+95] R. Alur, C. Courcoubetis, N. Halbwachs, T.A. Henzinger, P.-H. Ho, X. Nicollin, A. Olivero, J. Sifakis, and S. Yovine. The algorithmic analysis of hybrid systems. *Theoretical Computer Science*, 138:3–34, 1995.

[AČJT96] P. A. Abdulla, K. Čerāns, B. Jonsson, and Y.-K. Tsay. General decidability theorems for infinite-state systems. In *Proceedings of the 11th Annual Symposium on Logic in Computer Science*, pages 313–321. IEEE Computer Society Press, 1996.

[AD94] R. Alur and D.L. Dill. A theory of timed automata. *Theoretical Computer Science*, 126:183–235, 1994.

[AH98] R. Alur and T.A. Henzinger. *Computer-aided Verification: An Introduction to Model Building and Model Checking for Concurrent Systems.* Draft, 1998.

[AHH96] R. Alur, T.A. Henzinger, and P.-H. Ho. Automatic symbolic verification of embedded systems. *IEEE Transactions on Software Engineering*, 22:181–201, 1996.

[AJ98] P. Abdulla and B. Jonsson. Verifying networks of timed automata. In *TACAS 98: Tools and Algorithms for Construction and Analysis of Systems*, Lecture Notes in Computer Science 1384, pages 298–312. Springer-Verlag, 1998.

[BFH90] A. Bouajjani, J.-C. Fernandez, and N. Halbwachs. Minimal model generation. In *CAV 90: Computer-aided Verification*, Lecture Notes in Computer Science 531, pages 197–203. Springer-Verlag, 1990.

[Dam94] M. Dam. CTL* and ECTL* as fragments of the modal μ-calculus. *Theoretical Computer Science*, 126:77–96, 1994.

[EJS93] E.A. Emerson, C.S. Jutla, and A.P. Sistla. On model checking for fragments of μ-calculus. In *CAV 93: Computer-aided Verification*, Lecture Notes in Computer Science 697, pages 385–396. Springer-Verlag, 1993.

[FS98] A. Finkel and Ph. Schnoebelen. *Well-structured Transition Systems Everywhere.* Technical Report LSV-98-4, Laboratoire Spécification et Vérification, ENS Cachan, 1998.

[Hen95] T.A. Henzinger. Hybrid automata with finite bisimulations. In *ICALP 95: Automata, Languages, and Programming*, Lecture Notes in Computer Science 944, pages 324–335. Springer-Verlag, 1995.

[Hen96] T.A. Henzinger. The theory of hybrid automata. In *Proceedings of the 11th Annual Symposium on Logic in Computer Science*, pages 278–292. IEEE Computer Society Press, 1996.

[HHK95] M.R. Henzinger, T.A. Henzinger, and P.W. Kopke. Computing simulations on finite and infinite graphs. In *Proceedings of the 36rd Annual Symposium on Foundations of Computer Science*, pages 453–462. IEEE Computer Society Press, 1995.

[HHWT95] T.A. Henzinger, P.-H. Ho, and H. Wong-Toi. HyTech: the next generation. In *Proceedings of the 16th Annual Real-time Systems Symposium*, pages 56–65. IEEE Computer Society Press, 1995.

[HK96] T.A. Henzinger and P.W. Kopke. State equivalences for rectangular hybrid automata. In *CONCUR 96: Concurrency Theory*, Lecture Notes in Computer Science 1119, pages 530–545. Springer-Verlag, 1996.

[HKPV98] T.A. Henzinger, P.W. Kopke, A. Puri, and P. Varaiya. What's decidable about hybrid automata? *Journal of Computer and System Sciences*, 57:94–124, 1998.

[HM99] T.A. Henzinger and R. Majumdar. Symbolic model checking for rectangular hybrid systems. Submitted for publication, 1999.

[KS90] P.C. Kanellakis and S.A. Smolka. CCS expressions, finite-state processes, and three problems of equivalence. *Information and Computation*, 86:43–68, 1990.

[vG90] R.J. van Glabbeek. *Comparative Concurrency Semantics and Refinement of Actions.* PhD thesis, Vrije Universiteit te Amsterdam, The Netherlands, 1990.

Circuits versus Trees in Algebraic Complexity

Pascal Koiran

LIP, Ecole Normale Supérieure de Lyon
46 allée d'Italie, 69364 Lyon Cedex 07, France
http://www.ens-lyon.fr/~koiran

Abstract. This survey is devoted to some aspects of the "P = NP ?" problem over the real numbers and more general algebraic structures. We argue that given a structure M, it is important to find out whether NP_M problems can be solved by polynomial depth computation trees, and if so whether these trees can be efficiently simulated by circuits. Point location, a problem of computational geometry, comes into play in the study of these questions for several structures of interest.

1 Introduction

In algebraic complexity one measures the complexity of an algorithm by the number of basic operations performed during a computation. The basic operations are usually arithmetic operations and comparisons, but sometimes transcendental functions are also allowed [21,22,23,26]. Even when the set of basic operations has been fixed, the complexity of a problem depends on the particular model of computation considered. The two main categories of interest for this paper are circuits and trees. In section 2 and 3 we present a general framework for studying these questions, in the spirit of Poizat's theory of computation over arbitrary structures [16,35]. The focus is therefore on superpolynomial lower bounds for decision problems, and in particular for NP_M-complete problems. This line of research was initiated by Blum, Shub and Smale [5]; the main emphasis of their paper was on the case where M is a ring (see [4] for a recent account).

We will ignore completely the large body of work on (sub)polynomial lower bounds, which has been an active area of research for decades ([7] is a comprehensive text on this topic). A consequence of our higher ambition is that we actually have very few lower bounds to present. The main result of this type, presented in Theorem 2, was obtained by Meer [29] for the reals with addition and equality. The transfer theorems of section 4 show that there may be good reasons for this relative scarcity of definitive results. The bright side of this state of affairs is that there are plenty of difficult open problems to capture the attention of present and future researchers.

As explained in section 5, point location, a problem of computational geometry, plays an important role in the proofs of two transfer theorems. The branching complexity of point location in arrangements of real or complex hypersurfaces is also discussed in that section. The upper bound of Theorem 8 on point location in arrangements of complex hypersurfaces seems to be new; in the real case, this

H. Reichel and S. Tison (Eds.): STACS 2000, LNCS 1770, pp. 35–52, 2000.

is a recent result of Grigoriev [20]. Finally, we show in section 6 that a solution to the "computation tree alternative" would lead to a much better understanding of the "P = NP ?" question over the real and complex numbers.

2 Computation Models

2.1 Arbitrary Structures

We first recall some elementary definitions from logic, which should be familiar to most of our readers. By "structure", we mean a set M equipped with a finite set of functions $f_i : M^{n_i} \to M$ and relations $r_i \subseteq M^{m_i}$. A function of arity 0 is called a *constant*. We always assume that our structure contains the equality relation.

Terms are built from the basic functions of M by composition. More precisely, we have the following inductive definition:

(i) Variables and elements of M are terms of depth 0;
(ii) A term of depth $d \geq 1$ is of the form $f_i(t_1, \ldots, t_{n_i})$, where f_i is a function of M and $t_1, \ldots, t_{n_i}$ are terms of maximal depth $d - 1$.

A term in which n distinct variables $x_1, \ldots, x_n$ occur computes a function from M^n to M.

An *atomic formula* is of the form $r_i(t_1, \ldots, t_{m_i})$ where $t_1, \ldots, t_{m_i}$ are terms. A quantifier-free formula is a boolean combination of atomic formulas. To be completely precise, one can give an inductive definition:

(i) Atomic formulas are formulas of depth 0.
(ii) If F is a formula of depth $d-1$, $\neg F$ is a formula of depth d.
(iii) If $\max(\text{depth}(F), \text{depth}(G)) = d-1$, $F \vee G$ and $F \wedge G$ are formulas of depth d.

A formula in which n distinct variables $x_1, \ldots, x_n$ occur defines a subset of M^n. The elements of M occurring in terms or formulas (or in trees, circuits, etc...) are called *parameters*.

Example 1. If M is a field, terms of $(M, +, \times, =)$ represent polynomials; the sets defined by quantifier-free formulas are called *constructible*. If M is an ordered field, the definable sets of $(M, +, \times, \leq)$ are by definition the semi-algebraic sets.

An existential formula is of the form $\exists x_1 \, \exists x_2 \cdots \exists x_n \, F$, where F is a quantifier-free formula. M is said to admit quantifier elimination if every existential formula is equivalent to a quantifier-free formula (of course, as long as complexity issues are not taken into account, it is sufficient to consider the case $n = 1$). As explained in section 4, this notion plays an important role in the study of the "$\text{P}_M = \text{NP}_M$?" problem.

2.2 Trees

The simplest and most powerful computation model that we shall consider is the *branching tree*. A branching tree with n input variables $x_1, \ldots, x_n$ recognizes a subset of M^n in the following way. Each internal node g is labeled by some atomic formula $F_g(x_1, \ldots, x_n)$ of M, and has two children. If the input satisfies F_g we go left, otherwise we go right. Leaves are labeled *accept* or *reject*. Alternatively, leaves could be labeled by terms of M. The tree would then compute a function from M^n to M. As with other tree models, complexity is measured by depth, i.e., the branching complexity of a subset of M^n is the smallest depth of a branching tree that recognizes it.

The branching tree model is obviously not very realistic since any term, no matter how complex, can be evaluated for free. Consequently, this model is perhaps more suitable for proving lower bounds than upper bounds (some upper bounds are nevertheless presented in section 5.2). Lower bounds for approximating roots of complex polynomials in the structure $(\mathbb{R}, +, -, \times, /, \leq)$ can be found in [39,41]. In this context, the terms *topological complexity* and *topological decision tree* have been used instead of *branching complexity* and *branching tree.*

The *computation tree* is a more realistic model in which the complexity of terms is taken into account. In addition to branching nodes, a computation tree has unary computation nodes. A computation node g computes a term of the form $f_i(t_1, \ldots, t_{n_i})$ where f_i is a function of M and $t_1, \ldots, t_{n_i}$ are variables, parameters from M or terms computed by computation nodes located between the current node g and the root of the tree. The branching nodes of a computation tree are labeled by atomic formulas of the form $r_i(t_1, \ldots, t_{n_i})$ where r_i is a relation of M and again $t_1, \ldots, t_{n_i}$ are variables, parameters from M or terms computed by computation nodes located between the current node and the root of the tree.

Note that a computation tree is nothing but a special way of representing a formula, and therefore the subsets of M^n accepted by computation trees (or by branching trees, or by circuits as defined in section 2.3) are simply the definable sets. One may again argue that the computation tree model is mostly suitable for lower bounds as it is still too powerful. Note for instance than in the standard structure $M = \{0, 1\}$ any subset of M^n can be recognized in depth n. As far as we are concerned, the "fully realistic" model is the circuit model of section 2.3. Computation trees are nevertheless suitable for upper bounds if preprocessing is allowed. That is, if we have to recognize very quickly a subset X of M^n for some fixed value of n, we might first spend a lot of time (and space) to construct a computation tree recognizing X. The depth of this tree is a reasonable measure of the number of elementary operations needed to decide whether an element of M^n belongs to X. Preprocessing is a quite common technique in computational geometry, see for instance [17].

2.3 Circuits

The input gates of a circuit are labeled by variables or parameters from M. There are several types of computation gates in a circuit over M:

1. For each function f_i of M, gates of type f_i apply this function to their n_i inputs.
2. For each relation r_i of M, gates of type r_i apply the characteristic function of r_i to their inputs. For this definition to make sense, we assume that M contains two distinguished elements called 0 and 1.
3. Finally, selection gates compute a function $s(x, y, z)$ of their three inputs such that $s(0, y, z) = y$ and $s(1, y, z) = z$. The behaviour of s on an input (x, y, z) with $x \notin \{0, 1\}$ is not important. We shall assume that $s(x, y, z)$ is equal in this case to some fixed term $t(x, y, z)$ of M.

A circuit is *parameter-free* if it uses only the parameters 0 and 1. Strangely enough, selection gates already appear in [42], in a context where they are not really needed. Indeed, in any field the term $t(x, y, z) = xz + (1-x)y$ is a selection function.

The number of computation gates in a circuit is its *size*. A circuit C with n input variables and m output gates computes a function from M^n to M^m. We are mostly interested in the case $m = 1$. For such a circuit, the set of accepted inputs is by definition the set of inputs $(x_1, \ldots, x_n) \in M^n$ such that $C(x_1, \ldots, x_n) = 1$. As mentioned before, circuits with n input variables accept exactly the definable subsets of M^n.

Computation trees are at least as powerful as circuits since a circuit of size s can be simulated by a computation tree of depth $O(s)$. As we shall see in section 4, the converse is not so clear.

3 Complexity Classes

Our main complexity classes, such as P and NP, will be defined in terms of circuits. Formally, a problem is simply a subset of $M^\infty = \bigcup_{n \geq 1} M^n$. We first define the class $\mathbb{P}_M$ of non-uniform polynomial-time problems. A problem $X \subseteq M^\infty$ is in $\mathbb{P}_M$ if there exists parameters $\alpha_1, \ldots, \alpha_p \in M$, a polynomial $p(n)$ and a family of parameter-free circuits $(C_n)_{n \geq 1}$ where C_n has $n + p$ inputs, is of size at most $p(n)$ and satisfies the following condition:

$$\forall x \in M^n \; x \in X \Leftrightarrow C_n(\alpha_1 \ldots \alpha_p, x_1, \ldots, x_n) = 1 \tag{1}$$

$\mathbb{NP}_M$ is the non-deterministic version of $\mathbb{P}_M$. That is, a problem X is in $\mathbb{NP}_M$ if there exists a polynomial $q(n)$ and a problem $Y \in \mathbb{P}_M$ such that for any $n \geq 1$ and any $x \in M^n$,

$$x \in X \Leftrightarrow \exists y \in M^{q(n)} \; \langle x, y \rangle \in Y. \tag{2}$$

Several equivalent choices for the pairing function $\langle ., . \rangle : M^\infty \times M^\infty \to M^\infty$ are possible, for instance:

$$\langle (x_1, \ldots, x_n), (y_1, \ldots, y_m) \rangle = (0, x_1, 0, x_2, \ldots, 0, x_n, 1, y_1, 1, y_2, \ldots, 1, y_m).$$

One could also choose q so that the map $n \mapsto n + q(n)$ is injective, and simply concatenate x and y.

A problem X is in the class P_M of polynomial-time problems if $X \in \mathbb{P}_M$ and the corresponding circuit family (C_n) in (1) is uniform in the following sense: there exists a (classical) Turing machine which on input n constructs C_n in time polynomial in n. This definition makes sense since a parameter-free circuit is a purely boolean object. To be completely precise one should specify how circuits are encoded in binary words [35]; there is no significant difference with the classical case [1]. This complexity class can also be defined with Turing machines over M instead of circuits [16,35]. In the case where M is a ring, these Turing machines are equivalent to Blum-Shub-Smale machines [5].

The class NP_M of non-deterministic polynomial time problems is obtained from P_M in the same way as $\mathbb{NP}_M$ is obtained from $\mathbb{P}_M$: just replace the condition $Y \in \mathbb{P}_M$ in (2) by $Y \in \mathrm{P}_M$.

Example 2. For the standard structure $M = \{0, 1\}$ the usual complexity classes are recovered: $\mathrm{P}_M = \mathrm{P}$, $\mathrm{NP}_M = \mathrm{NP}$, $\mathbb{P}_M = \mathrm{P/poly}$, $\mathbb{NP}_M = \mathrm{NP/poly}$.

For $M = (\mathbb{R}, +, -, \times, \leq)$ or $M = (\mathbb{R}, +, -, \leq)$, $\mathrm{P}_M = \mathbb{P}_M$ and $\mathrm{NP}_M = \mathbb{NP}_M$ since circuit families can be encoded in parameters.

One can also define the class $\mathbb{P}^0_M$ of parameter-free non-uniform polynomial time problems (set $p = 0$ in (1)), and from this class we obtain the classes $\mathbb{NP}^0_M$, P^0_M and NP^0_M, which are the parameter-free versions of $\mathbb{NP}_M$, P_M and NP_M. These parameter-free classes appear in section 6.

The problems "$\mathbb{P}_M = \mathbb{NP}_M$?" and "$\mathrm{P}_M = \mathrm{NP}_M$?" are open for most structures of interest (but for any structure M, the latter equality implies the former). The main open problem in this general theory is whether there exists a structure M satisfying $\mathrm{P}_M = \mathrm{NP}_M$.

There is more than P and NP in the world of structural complexity, and lots of classical complexity classes can be redefined in our general framework. For instance, the reader can easily imagine how the k-th levels Σ^k_M and Π^k_M of the uniform polynomial hierarchy are defined [4,35] (replace the existential quantifiers in (2) by k alternating quantifier blocks). Of course, there is also a non-uniform polynomial hierarchy.

One of the main differences with the classical theory comes from the role of space: as shown by Michaux [33], a Turing machine over $(\mathbb{R}, +, -, \times, \leq)$ can perform any computation in constant work space. The classical definition of PSPACE is therefore of little use in our general framework. We will instead work with the class PAR_M of parallel polynomial time problems. It is again easier to first define as in [8] a non-uniform version of this complexity class: a problem X is $\mathbb{PAR}_M$ if there exist parameters $\alpha_1, \ldots, \alpha_p \in M$, a polynomial $p(n)$ and a family of parameter-free circuits $(C_n(x_1, \ldots, x_n, y_1, \ldots, y_p))$ such that $C_n(x_1, \ldots, x_n, \alpha_1, \ldots, \alpha_p)$ solves X for inputs in M^n, and is of depth at most $p(n)$. We say that X is PAR_M if there exists a Turing machine which, on input n, construct C_n in work space polynomially bounded in n.

Example 3. For $M = \{0, 1\}$, $\mathrm{PAR}_M = \mathrm{PSPACE}$; for $M = (\mathbb{R}, +, -, \times, \leq)$, our definition is equivalent to those of [4] and [11].

4 Is P Equal to NP ?

If one wishes to investigate the "$\mathrm{P}_M = \mathrm{NP}_M$?" in some structure M, the first thing to do is to find out whether M admits quantifier elimination. Indeed, if M does not admit quantifier elimination NP_M problems cannot be solved by circuits (or equivalently by quantifier-free formulas) of *any* size. One should therefore consult one's favourite model theory book (e.g. [24,34]) for a list of structures that admit or do not admit quantifier elimination. For instance, the following structures admit quantifier elimination:

(i) vector spaces over $\mathbb{Q}$, e.g., $\mathbb{R}_{vs} = (\mathbb{R}, +, -, =)$;
(ii) ordered vector spaces over $\mathbb{Q}$, e.g., $\mathbb{R}_{ovs} = (\mathbb{R}, +, -, \leq)$;
(iii) algebraically closed fields of any characteristic, e.g., $\mathbb{C} = (\mathbb{C}, +, -, \times, =)$;
(iv) real closed fields, e.g., $\mathbb{R} = (\mathbb{R}, +, -, \times, \leq)$;
(v) differentially closed fields of characteristic zero (of which we can give no natural example).

The reals with exponentiation $(\mathbb{R}, +, -, \times, \exp, \leq)$ and the integers $(\mathbb{N}, +, -, \times, \leq)$ do not admit quantifier elimination. However, the "$\mathrm{NP}_M = \mathrm{coNP}_M$?" question makes sense in the former structure since it is model-complete (i.e., every existential formula is equivalent to a universal formula).

For structures (i) through (iv), elimination of quantifiers can be performed in a relatively efficient manner in the sense that $\mathrm{NP}_M \subseteq \mathrm{PAR}_M$. Proofs of this fact for $M = \mathbb{R}_{vs}$ and $M = \mathbb{R}_{ovs}$ can be found in [25] (these two structures satisfy the stronger property that $\mathrm{NP}_M = \mathrm{BNP}_M$, i.e., existential quantification over M is polynomially equivalent to existential quantification over $\{0, 1\}$). For algebraically closed fields one may consult [10,13] and for real closed fields [2,18,37]. The complexity theory of differentially closed fields is not so well understood. One may cite a triply exponential quantifier elimination algorithm [19] and a study of the "P = NP ?" problem for these structures in [36].

Quantifier elimination is even more closely connected to the "$\mathrm{P}_M = \mathrm{NP}_M$?" question than explained at the beginning of this section. Roughly speaking, $\mathrm{P}_M = \mathrm{NP}_M$ means that existential quantifiers can be eliminated in polynomial time, if we represent the eliminating formula by a circuit:

Theorem 1. $\mathrm{P}_M = \mathrm{NP}_M$ *if and only if there exist parameters* $\alpha_1, \ldots, \alpha_k$ *of* M *and a polynomial time algorithm (in the classical sense) which, given a parameter-free existential formula* $\exists y_1 \cdots \exists y_p\ F(x_1, \cdots, x_n, y_1, \ldots, y_p)$*, outputs a circuit* C *over* M *such that* $C(x_1, \ldots, x_n, \alpha_1, \ldots, \alpha_k)$ *is equivalent to the existential formula.*

The "$\mathrm{P}_M = \mathrm{NP}_M$?" question can therefore be viewed as a purely classical question, that is, a question about algorithms over the structure $\{0, 1\}$. This theorem remains true if we replace the classical elimination algorithm by an algorithm over M, which is the way it is stated in [35].

We have precious few examples of structures that admit quantifier elimination but where P_M is provably different from NP_M. The main example is due to Klaus Meer [29]:

Theorem 2. $P_{\mathbb{R}_{vs}} \neq NP_{\mathbb{R}_{vs}}$.

Another example is the structure of arborescent dictionaries constructed in [16] (see also [35]). Meer's proof of Theorem 2 is based on a multidimensional version of the knapsack problem. The proof given below is based on Shub and Smale's [38] Twenty Questions (TQ) problem. An input $(x_1, \ldots, x_n)$ is in TQ if x_1 is an integer between 0 and $2^n - 1$ (x_1 is therefore the only truly "numerical" input; the only role of $x_2, \ldots, x_n$ is to specify the value of n).

Twenty Questions is in $NP_{\mathbb{R}_{vs}}$ since $(x_1, \ldots, x_n) \in TQ$ if and only if

$$\exists u_1, \ldots, u_n \in \{0, 1\} \; x_1 = \sum_{i=1}^{n} 2^{i-1} u_i.$$

However, TQ is not only outside $P_{\mathbb{R}_{vs}}$ but its branching complexity is exactly equal to 2^n. Indeed, this problem can obviously be solved in depth 2^n (just perform the 2^n tests "$x_1 = 0$?", ..., "$x_n = 2^n - 1$?" sequentially). For the converse, let T_n be a branching tree of depth d which solves Twenty Questions for inputs of the form $x_2 = 0, \ldots, x_n = 0$. This tree has a single real-valued input x_1 and recognizes a finite subset of $\mathbb{R}$, of cardinality 2^n. We claim that $d \geq 2^n$. Each internal node of T_n performs a test of the form "$l(x_1) = 0$?" where l is an affine function; we may assume without loss of generality that l is not identically 0. The *generic path* of T_n is obtained by answering *no* to all these tests. Observe that inputs which follow the generic path are rejected since T_n recognizes a finite subset of $\mathbb{R}$. Accepted inputs must therefore satisfy one of the test performed along the canonical path, which proves the claim, and Theorem 2.

One might be led by the simplicity of this proof to the belief that other results of this type should not be too difficult to obtain. For instance, one could try to:

(i) separate higher levels of the polynomial hierarchy over $\mathbb{R}_{vs}$;
(ii) obtain separation results for richer structures than $\mathbb{R}_{vs}$. For instance, is $P_{\mathbb{R}_{ovs}}$ different from $NP_{\mathbb{R}_{ovs}}$?

It turns out that both questions are quite difficult, however. In the first direction, it is actually possible to separate the first few levels of the polynomial hierarchy [15]: a variation on the proof of Theorem 2 shows that Twenty Questions is not in $coNP_{\mathbb{R}_{vs}}$, which separates $NP_{\mathbb{R}_{vs}}$ from $coNP_{\mathbb{R}_{vs}}$; and a similar problem can be used to separate $\Sigma^2_{\mathbb{R}_{vs}} \cap \Pi^2_{\mathbb{R}_{vs}}$ from $\Sigma^1_{\mathbb{R}_{vs}} \cup \Pi^1_{\mathbb{R}_{vs}}$. Separating higher levels is essentially "impossible", as shown by a transfer theorem of [15]:

Theorem 3. *For all* $k \geq 0$*:*

$$PH = \Sigma^k \Rightarrow PH_{\mathbb{R}_{vs}} = \Sigma^{k+2}_{\mathbb{R}_{vs}}.$$

We showed in the same paper that parallel polynomial time cannot be separated from the polynomial hierarchy:

Theorem 4. $P = PSPACE \Rightarrow PAR_{\mathbb{R}_{vs}} = \Sigma^2_{\mathbb{R}_{vs}} \cap \Pi^2_{\mathbb{R}_{vs}}$

and that P cannot be separated from NP $\cap$ coNP:

Theorem 5. $\mathrm{P} = \mathrm{NP} \Rightarrow \mathrm{P}_{\mathbb{R}_{vs}} = \Sigma^1_{\mathbb{R}_{vs}} \cap \Pi^1_{\mathbb{R}_{vs}}$.

Bourgade has extended theorems 3 and 4 to infinite Abelian groups of prime exponent [6].

Twenty Questions cannot be used to separate P from NP in $\mathbb{R}_{ovs}$ since this problem is in $\mathrm{P}_{\mathbb{R}_{ovs}}$ by binary search. In fact, obtaining a proof of this separation is all but hopeless by another transfer theorem from [15]:

Theorem 6. $\mathrm{P} = \mathrm{NP} \Rightarrow \mathrm{P}_{\mathbb{R}_{ovs}} = \mathrm{NP}_{\mathbb{R}_{ovs}}$.

In a previous paper [14] we obtained a similar result for the "P = PAR ?" problem:

Theorem 7. $\mathrm{P} = \mathrm{PSPACE} \Rightarrow \mathrm{P}_{\mathbb{R}_{ovs}} = \mathrm{PAR}_{\mathbb{R}_{ovs}}$.

Theorems 3 through 7 show that a number of separations between real complexity classes are (at least) as hard to prove as outstanding conjectures from discrete complexity theory. Yet we know that these separations must hold true since an equality of two real complexity classes would imply the equality of the corresponding discrete complexity classes. This follows from the "boolean parts" results of [25,12]. For instance, $\Sigma^k_{\mathbb{R}_{vs}} = \Pi^k_{\mathbb{R}_{vs}}$ would imply $\Sigma^k = \Pi^k$, and $\mathrm{P}_{\mathbb{R}_{ovs}} = \mathrm{NP}_{\mathbb{R}_{ovs}}$ would imply P/poly = NP/poly. Non-uniformity comes into play in the latter statement only because arbitrary real parameters are allowed: P/poly can be replaced by P and NP/poly by NP if we work with parameter-free machines, or in the structure $\mathbb{Q}_{ovs}$ instead of $\mathbb{R}_{ovs}$.

We will see in the next section that Theorems 6 and 7 hinge on the following question:

Question 1. Are computation trees more powerful than circuits ?

If this deliberately fuzzy question is interpreted in its broadest sense, it is possible to give a positive answer for many structures of interest. For instance, as pointed out before, in the standard structure $M = \{0, 1\}$ any boolean function of n variables can be computed by a tree of depth n, but a simple counting argument shows that most boolean functions have exponential circuit complexity [1]. Bounds on the number of consistent sign conditions (à la Thom-Milnor, see end of section 5.1) show that most boolean functions also have exponential circuit complexity over $\mathbb{R}$ and $\mathbb{C}$, even if arbitrary real or complex parameters are allowed [28]. It is nevertheless possible to construct a special-purpose structure in which polynomial size computation trees are equivalent to polynomial size circuits (hint: try to encode a computation tree in a circuit's parameter; this can be done efficiently with Poizat's arborescent dictionaries).

This question becomes more interesting if we compare the power of trees and circuits on a restricted class of problems, for instance on NP_M problems. In this case the situation is dramatically different since, as pointed out by Poizat (personal communication), we do not have a single example of a structure M where NP_M problems can be solved by polynomial depth computation trees, but

$P_M \neq NP_M$. For instance, computation trees over $\mathbb{R}_{ovs}$ are likely to be more powerful than circuits on $NP_{\mathbb{R}_{ovs}}$ problems (in fact, as explained in the next section any such problem can be solved polynomial depth computation trees) but by Theorem 6 proving this is essentially "impossible."

Shub and Smale's invention of Twenty Questions was motivated by an attempt to separate $P_{\mathbb{C}}$ from $NP_{\mathbb{C}}$. The plausible number-theoretic conjecture that "$k!$ is ultimately hard to compute" was put forward in [38], and was shown to imply that Twenty Questions is not $P_{\mathbb{C}}$. Since Twenty Questions is $NP_{\mathbb{C}}$, a proof of this conjecture would indeed separate $P_{\mathbb{C}}$ from $NP_{\mathbb{C}}$. We point out however that Shub and Smale's proof of this implication is based on a canonical path argument similar to the argument of Theorem 2 (but more involved). As a result, the conjecture in fact implies that Twenty Questions cannot be solved by polynomial depth computation trees. Although of great interest, its proof would therefore shed no light on Question 1.

5 Complexity and Point Location

5.1 Point Location by Computation Trees

There are two main steps in the proofs of Theorems 6 and 7:

(i) Show that any $NP_{\mathbb{R}_{ovs}}$ (or $PAR_{\mathbb{R}_{ovs}}$) problem can be solved by polynomial depth computation trees.

(ii) Show that under an appropriate complexity-theoretic assumption (P = NP or P = PSPACE), these trees can be "transformed" into polynomial size circuits.

Two proofs of (i) are known. The first one is essentially due to Meyer auf der Heide [31,32]. In fact this author showed that $PAR_{\mathbb{R}_{ovs}}$ problems can be solved by $\mathbb{R}_{ovs}$-branching trees of polynomial depth. In order to obtain polynomial depth computation trees, we just have to check that the coefficients of affine functions labeling the nodes of these branching trees are integers of polynomial size [15]. Meyer auf der Heide's proof exploits the geometrical structure of a $PAR_{\mathbb{R}_{ovs}}$ problem: its restriction to inputs in $\mathbb{R}^n$ is a union of faces of an arrangement of $2^{n^{O(1)}}$ hyperplanes. We recall that any finite set of m hyperplanes of equations $h_1(x) = 0, \ldots, h_m(x) = 0$ partitions $\mathbb{R}^n$ in a finite number of faces, where each face is the set of points satisfying a system of equations of the form

$$\text{sign}(h_1(x)) = \epsilon_1, \ldots, \text{sign}(h_m(x)) = \epsilon_m \tag{3}$$

for some fixed sign vector $(\epsilon_1, \ldots, \epsilon_m) \in \{-1, 0, 1\}^m$. In order to decide whether an input point should be accepted we just need to locate it in the arrangement, that is, to determine to which face it belongs. Meyer auf der Heide [32] asked whether any union of m hyperplanes of $\mathbb{R}^n$ can be recognized by $\mathbb{R}_{ovs}$-branching trees of depth $(n \log m)^{O(1)}$. His construction does not quite yield that result because it uses certain bounds on the size of the hyperplanes' equations. In [14] we used a construction of Meiser [30] to give a second proof of (i); this construction also yields a positive answer to Meyer auf der Heide's question. In fact Meiser's construction almost answers that question, except that he has a non-degeneracy assumption on the arrangement, and that he allows multiplications in his computation model (that is, he works with $\mathbb{R}$-branching tree instead of $\mathbb{R}_{ovs}$-branching trees). These problems can be fixed as explained in [14]. To complete the second proof of (i) we also need to analyze the size of coefficients in the resulting $\mathbb{R}_{ovs}$-branching trees; this is done in the same paper.

The transformation of polynomial depth trees into polynomial size circuits under the assumption P = PSPACE is based on an exhaustive search procedure among all polynomial depth trees. This procedure can be made to run in polynomial space, and thus in polynomial time under the assumption P = PSPACE (see [14] or section 6). In order to use the weaker assumption P = NP, the mere knowledge that $\text{PAR}_{\mathbb{R}_{ovs}}$ can be solved by polynomial depth trees is not sufficient: we also need to know how these trees are constructed. We chose to work with Meyer auf der Heide's construction because it yields a stronger result than Meiser's. Namely, we obtained the unconditional result that $\text{NP}_{\mathbb{R}_{ovs}}$ problems can be solved by $\text{P}_{\mathbb{R}_{ovs}}$ algorithms with the help of a boolean NP oracle [15]; with Meiser's construction one would obtain an oracle in a higher level of the polynomial hierarchy.

Nothing like Theorem 6 or Theorem 7 is known if we replace $\mathbb{R}_{ovs}$ by $\mathbb{R}$ or $\mathbb{C}$. It is not even known whether $\text{NP}_{\mathbb{R}}$ or $\text{NP}_{\mathbb{C}}$ problems can be solved by polynomial depth computation trees (as shown in section 6, this question is very much related to the "P = NP ?" problem over the real and complex numbers). In fact, it is natural to conjecture that Twenty Questions *cannot* be solved by polynomial depth computation trees (as pointed out in section 4, this would follow from the conjecture that $k!$ is ultimately hard to compute). The bold conjecture that all $\text{PAR}_{\mathbb{R}}$ problems *can* be solved by polynomial depth computation tree was put forward in [14]. Like in the $\mathbb{R}_{ovs}$ case, there is an intimate relationship between $\text{PAR}_{\mathbb{R}}$ problems and point location: the restriction to $\mathbb{R}^n$ of such a problem is a union of faces of an arrangement defined by $m = 2^{n^{O(1)}}$ polynomials $f_1, \ldots, f_m$ of degree $2^{n^{O(1)}}$. Faces are defined as in the $\mathbb{R}_{ovs}$ case: just replace $h_1, \ldots, h_m$ in (3) by $f_1, \ldots, f_m$. A similar property is also true of $\text{PAR}_{\mathbb{C}}$ problems. Here we need to redefine the sign function so that it only takes the values 0 (if its input is equal to zero) or 1 (if it is nonzero). As explained in [14], point location in an arrangement defined by m real polynomials of fixed degree in n variables can be performed by a computation tree of depth polynomial in $n \log m$ by reduction to the linear case.

At least we know that $\text{PAR}_{\mathbb{R}}$ and $\text{PAR}_{\mathbb{C}}$ problems have polynomial branching complexity. In the real case, this follows from the well-known result that m polynomials of degree d in n variables define an arrangement with $(md)^{O(n)}$ faces (see [2] for a sharper bound) and from a recent result of Grigoriev [20] who, answering a question of [14], showed that point location in an arrangement with N faces has branching complexity $O(\log N)$. The $(md)^{O(n)}$ bound on the number of faces still holds in the complex case, and point location in an arrangement with N faces has branching complexity $O(n \log N)$ as explained in section 5.2. This shows that $\mathrm{PAR}_{\mathbb{C}}$ problems indeed have polynomial branching complexity.

5.2 Branching Complexity of Point Location

Recall that a tree solves the point location problem for a given arrangement if two input points arriving at the same leaf of the tree always belong to the same cell. In this section we prove the $O(n \log N)$ upper bound on the complexity of point location in the complex case, and give some hints for the proof of Grigoriev's $O(\log N)$ bound. In fact, his proof applies not only to $\mathbb{R}$ but to any ordered field. Likewise, we state and prove our $O(n \log N)$ bound not just for the field of complex numbers, but for an arbitrary field.

We need the following fact for the proof of Theorem 8 (see [9] for a constructive version).

Proposition 1. *Let K be an infinite field and V a variety of K^n defined by polynomials $f_1, \ldots, f_s \in K[X_1, \ldots, X_n]$ of degree at most d. This variety can be defined by $n+1$ polynomials of $K[X_1, \ldots, X_n]$ of degree at most d.*

Proof. We shall see that V can be defined by $n+1$ "generic" linear combinations of the input equations. More precisely, for a matrix $\alpha = (\alpha_{ij})_{1 \leq i \leq n+1, 1 \leq j \leq s}$ of elements of K, let us denote by V_α the variety of K^n defined by the polynomials $g_i = \sum_{j=1}^{s} \alpha_{ij} f_j$ $(1 \leq i \leq n+1)$. Obviously, $V \subseteq V_\alpha$ for any α. It turns out that the converse inclusion holds for "most" α's.

Let p be the characteristic of K and F_p the prime field of characteristic p (i.e., $F_0 = \mathbb{Q}$ and $F_p = \mathbb{Z}/p\mathbb{Z}$ for $p \geq 2$). We shall assume for now that K is algebraically closed, of infinite transcendence degree over F_p.

The coefficients of the f_i's lie in a subfield $k \subseteq K$ of finite transcendence degree. We claim that $V = V_\alpha$ if the entries of α are algebraically independent over k. Assume indeed that $\sum_{j=1}^{s} \alpha_{ij} f_j(x) = 0$ for $i = 1, \ldots, n+1$. Since the tuple $x\hat{}\alpha$ is of transcendence degree at least $s(n+1) - n$ over $k(x)$, any transcendence base of $k(x, \alpha)$ over $k(x)$ which is made up of entries of α must contain at least one row of α. If i is such a row, the equality $\sum_{j=1}^{s} f_j(x) \alpha_{ij} = 0$ implies that $f_j(x) = 0$ for all j since $f_j(x) \in k(x)$ and $\alpha_{i1}, \ldots, \alpha_{is}$ are algebraically independent over that field by choice of i. We therefore conclude that $x \in V$, and that $V_\alpha \subseteq V$ since x was an arbitrary point of V_α.

To complete the proof of the proposition, we just need to remove the assumption on K. In the general case, K can be embedded in an algebraically closed field $\hat{K}$ of infinite transcendence degree. The f_i's define a variety $\hat{V}$ of $\hat{K}^n$, and likewise for any matrix α with entries in $\hat{K}$, the g_i's define a variety $\hat{V}_\alpha$ of $\hat{K}^n$ which contains $\hat{V}$. Let G be the (constructible) set of all $\alpha \in \hat{K}^{s(n+1)}$ such that $\hat{V} = \hat{V}_\alpha$. We have seen that $\alpha \in G$ if its entries are algebraically independent over k. This implies that G is dense in $\hat{K}^{s(n+1)}$. It follows that for any infinite subset E of $\hat{K}$, and in particular for $E = K$, there exists a matrix $\alpha \in G$ with entries from E (this can be proved by induction). For such a matrix, $V = V_\alpha$ since $\hat{V} = \hat{V}_\alpha$.

Theorem 8. *Let $\mathcal{F} = \{f_1, \ldots, f_s\}$ be a family of polynomials of $K[X_1, \ldots, X_n]$, where K is an arbitrary field. The point location problem for $\mathcal{F}$ has branching complexity $O(n \log N)$, where N is the number of cells in the arrangement.*

Proof. We will assume that K is infinite since the result is trivial for finite fields (in that case, any function on K^n can be computed in depth $O(n)$).

For any subset I of $[n] = \{1, \ldots, n\}$, we denote by C_I the sign condition

$$(f_i = 0)_{i \in I} \wedge (f_i \neq 0)_{i \notin I}.$$

We say that C_I is feasible if the set of points satisfying this sign condition is nonempty (in which case it is a cell of the arrangement). We shall also denote by $\leq$ a (fixed) total order on the subsets of $[n]$ which is compatible with inclusion (i.e., $I \subseteq J$ implies $I \leq J$). Our construction is based on these two observations:

1. For any $I \subseteq [n]$, $V_I = \bigcup_{J \geq I} C_J$ is a variety of K^n: it is the union for $J \geq I$ of the varieties $(f_i = 0)_{i \in J}$.
2. Any variety of K^n can be defined by $n + 1$ equations, as shown in Proposition 1.

Note that $I \leq J$ if and only if $V_J \subseteq V_I$. Let us say that $I \subseteq [n]$ is feasible if C_I is a feasible sign condition. On input $x \in K^n$, our point location algorithm finds by binary search the largest feasible I such that $x \in V_I$. Since x satisfies C_I, this is the desired sign condition.

There are N satisfiable sign conditions, so that we only need $\log N$ steps of binary search. At each step the algorithm performs a test of the form: "$x \in V_J$?". By the second observation, this test can be performed in depth $n+1$. The overall depth of the corresponding tree is therefore $O(n \log N)$.

Note that $\log N$ is an obvious lower bound on the branching complexity of point location (a binary tree with at least N leaves must have depth at least $\log N$), so that the $O(n \log N)$ of Theorem 8 is not too far from the optimum.

If K is the field of real numbers (or more generally is an ordered field), the $\log N$ lower bound is in fact tight. Indeed, we no longer need $n + 1$ polynomials to define a variety as in Proposition 1: by the usual "sum of squares" trick, a single polynomial suffices (see [20] for a more direct proof). The construction of Theorem 8 therefore yields a branching tree of depth $O(\log N)$. As shown in [20], it possible within the same depth not only to find out whether each polynomial f_i is zero or non-zero at the input point, but to find out whether it is positive or negative:

Theorem 9. *Let $\mathcal{F} = \{f_1, \ldots, f_s\}$ be a family of polynomials of $R[X_1, \ldots, X_n]$ where R is an ordered field. The point location problem for $\mathcal{F}$ has branching complexity $O(\log N)$, where N is the number of cells of the arrangement.*

The proof of Theorem 9 relies on a two-stage construction. In the first stage we determine for each f_i whether it vanishes at the input point x. This can be done in depth $O(\log N)$ as explained before Theorem 9. In the second stage, we determine the sign of $f_i(x)$ for each polynomial f_i which does not vanish at x. The proof that this can also be done in depth $O(\log N)$ relies on a nice combinatorial lemma:

Lemma 1. *Let $u_1, \ldots, u_m$ be pairwise distinct vectors of $(\mathbb{Z}/2\mathbb{Z})^k$. If $m \geq 6$ there exists a vector $v \in (\mathbb{Z}/2\mathbb{Z})^k$ such that*

$$m/3 \leq |\{1 \leq i \leq m;\ \langle v, u_i \rangle = 0\}| \leq 2m/3.$$

For the proof of this lemma and its application to Theorem 9, we refer the reader to [20].

6 The Computation Tree Alternative

In this section we show that Question 1 plays a crucial role in the study of the "P = NP?" problem over the real and complex numbers. The proofs are based on techniques from [14].

The $\mathrm{NP}_{\mathbb{C}}$-complete $\mathrm{HN}_{\mathbb{C}}$ appears in the statement of our computation tree alternative for the complex numbers. This is the problem of deciding whether a system of polynomial equations $f_1(x_1, \ldots, x_n) = 0, \ldots, f_s(x_1, \ldots, x_n) = 0$ in n complex variables has a solution.

Theorem 10. *If* $\mathrm{HN}_{\mathbb{C}}$ *can be solved by a family of parameter-free computation trees of polynomial depth we have the following transfer theorem:* P = PSPACE *implies* $\mathrm{P}_{\mathbb{C}} = \mathrm{NP}_{\mathbb{C}}$. *Otherwise,* $\mathrm{P}_{\mathbb{C}} \neq \mathrm{NP}_{\mathbb{C}}$.

Proof. If $\mathrm{HN}_{\mathbb{C}}$ cannot be solved by a family of parameter-free computation trees of polynomial depth then this problem cannot be solved by a family of parameter-free circuits of polynomial size. By elimination of parameters [3,4,27], this implies that $\mathrm{P}_{\mathbb{C}} \neq \mathrm{NP}_{\mathbb{C}}$.

It is well known that the field of complex numbers satisfies the hypothesis of Theorem 12 below [13]. If P = PSPACE and $\mathrm{HN}_{\mathbb{C}}$ can be solved by a family of parameter-free computation trees of polynomial depth, $\mathrm{HN}_{\mathbb{C}}$ is therefore $\mathrm{P}_{\mathbb{C}}$ (even $\mathrm{P}^0_{\mathbb{C}}$) by Theorem 12 and thus $\mathrm{P}_{\mathbb{C}} = \mathrm{NP}_{\mathbb{C}}$.

Theorem 10 can be interpreted as follows. If we can prove (unconditionally) that $\mathrm{HN}_{\mathbb{C}}$ cannot be solved by a family of parameter-free polynomial depth computation trees, we have obtained an unconditional separation of $\mathrm{P}_{\mathbb{C}}$ from $\mathrm{NP}_{\mathbb{C}}$. If $\mathrm{HN}_{\mathbb{C}}$ *can* be solved by a family of parameter-free polynomial depth computation trees, $\mathrm{P}_{\mathbb{C}}$ is still very likely to be different from $\mathrm{NP}_{\mathbb{C}}$ (otherwise, NP would be included in BPP [40,28]). However, obtaining a proof of this separation would be a hopeless problem, at least in the current state of discrete complexity theory.

For the reals we can replace $\mathrm{HN}_{\mathbb{C}}$ by the $\mathrm{NP}_{\mathbb{R}}$-complete problem $4\mathrm{FEAS}_{\mathbb{R}}$: this is the problem of deciding whether a polynomial of degree at most 4 in n real variables has a root. We can only deal with parameter-free algorithms since elimination of parameters is not known to hold in the real case (see [8] for some results in this direction and more references).

Theorem 11. *If* $4\mathrm{FEAS}_{\mathbb{R}}$ *can be solved by a family of parameter-free computation trees of polynomial depth we have the following transfer theorem:* P $=$ PSPACE *implies* $\mathrm{P}^0_{\mathbb{R}} = \mathrm{NP}^0_{\mathbb{R}}$ *(which implies* $\mathrm{P}_{\mathbb{R}} = \mathrm{NP}_{\mathbb{R}}$*). Otherwise,* $\mathrm{P}^0_{\mathbb{R}} \neq \mathrm{NP}^0_{\mathbb{R}}$.

Proof. It is similar to the proof of Theorem 10 (the decision algorithm of [13] can be replaced by the algorithm of [2] or [37]).

In the proof of Theorem 12 it is convenient to work with branching trees of a restricted form instead of computation trees. That is, we will use the observation that a computation tree of depth d can be simulated by a branching tree of depth at most d whose nodes are labeled by atomic formulas $r_i(t_1, \ldots, t_{m_i})$ satisfying the following condition: $t_1, \ldots, t_{m_i}$ can be computed by straight-line programs of size at most d. We recall that a straight-line program over M is a circuit in which all gates are labeled by functions of M (relation and selection gates are not used).

Theorem 12. *Let* M *be a structure whose parameter-free* Σ^2 *formulas can be decided in polynomial space. Let* X *be a* NP^0_M *problem which can be solved by a family of parameter-free computation trees of polynomial depth. If* P $=$ PSPACE *this problem is* P^0_M.

For the proof of this theorem we need to associate to X a boolean problem $\tilde{X}$. An instance of $\tilde{X}$ is described by three integers n, L, d (written in unary), and by a conjunction $F(x_1, \ldots, x_n)$ of (parameter free) atomic formulas. For the terms in F we assume a straight-line representation. This conjunction defines a subset S_F of $\mathbb{R}^n$. An instance is positive if there exists a branching tree T satisfying the following properties:

(i) T is of depth at most d and solves X for inputs in S_F.
(ii) The nodes of T are labeled by atomic formulas of the form $r_i(t_1, \ldots, t_{m_i})$ where the terms $t_1, \ldots, t_{m_i}$ are computed by straight-line programs of length at most L.

We need an algorithm to solve $\tilde{X}$, and for positive instances of this problem we also need to compute the label l_r of the root of a corresponding tree T (this tree might not be unique, but any solution will do). Thus l_r is just a boolean value if T is reduced to a leaf, and an atomic formula otherwise.

Lemma 2. *If* $X \in \mathrm{NP}^0_M$ *then* $\tilde{X} \in$ PSPACE*. Moreover, for a positive instance* l_r *can be constructed in polynomial space.*

Proof. We first determine whether T can be of depth 0, i.e., reduced to a leaf. In that case, T recognizes either $\mathbb{R}^n$ or $\emptyset$ depending on the label of that leaf. Label 0 is suitable if the formula

$$\exists x \in \mathbb{R}^n \; F(x) \wedge (x \in X)$$

is false. By hypothesis on M, this Σ^1 formula can be decided in polynomial space (note that we can introduce additional existentially quantified variables in order to move from the straight-line representation of F to the standard representation). Label 1 is suitable if the Σ^2 formula

$$\exists x \in \mathbb{R}^n \; F(x) \wedge (x \notin X)$$

is false. If there is a solution in depth 0, we accept the instance of $\tilde{X}$ and output the corresponding label. Otherwise, for $d > 0$ we look for solutions of depth between 1 and d (for $d = 0$ we exit and reject the instance). To do this we enumerate (e.g. in lexicographic order) all atomic formulas $A(x_1, \ldots, x_n)$ where the terms in A are given by straight-line programs of length at most L. For each such formula we do the following.

1. Decide by a recursive call whether $(n, L, d-1, F(x) \wedge A(x))$ is a positive instance of $\tilde{X}$.
2. Decide by a recursive call whether $(n, L, d-1, F(x) \wedge \neg A(x))$ is a positive instance of $\tilde{X}$.
3. In case of a positive answer to both questions, exit the loop, accept (n, L, d, F) and output $l_r = A$.

The instance is rejected if it is not accepted in the course of this enumeration procedure.

In addition to the space needed to solve the depth 0 case, we just need to maintain a stack to keep track of recursive calls. Hence this algorithm runs in polynomial space, showing that $\tilde{X} \in$ PSPACE. For positive instances, the algorithm also outputs l_r as needed.

Proof (of Theorem 12). For inputs of size n, X can be solved by a computation tree of depth bounded by an^b, where a and b are constants. The idea is to use Lemma 2 to move down that tree. The hypothesis P = PSPACE implies that $\tilde{X} \in$ P. Moreover, for positive instances l_r can be constructed in polynomial time (one should argue that each bit of l_r is in PSPACE, and therefore in P). Thus we set $L = d = an^b$ and $F \equiv$ True. By hypothesis (n, L, d, F) is a positive instance of $\tilde{X}$ and therefore l_r can be computed in polynomial time. If l_r is a boolean value we stop and output that value. Otherwise A is an atomic formula, and we can determine in polynomial time whether the input $x \in \mathbb{R}^n$ to X satisfies A (because straight-line programs can be evaluated efficiently). If so, we set $F' = F \wedge A(x)$. Otherwise, we set $F' = F \wedge \neg A(x)$. In any case, we set $d' = d - 1$, and feed (n, L, d', F') to the algorithm deciding $\tilde{X}$. This process continues until a leaf is reached. This requires at most an^b steps.

Bruno Poizat pointed out that it would be natural to work with families of computation trees using the same parameters for all input sizes, since we use this convention for our polynomial-time algorithms. The reader may try to formulate and prove the counterparts of Theorems 10 and 11 in that setting.

Acknowledgements

Thanks to Dima Grigoriev for the discussions on the branching complexity of point location. The results of section 5.2 were obtained while we were visiting the Mathematical Sciences Research Institute in Berkeley. Useful comments were made by Hervé Fournier, Bruno Poizat, and Natacha Portier.

References

1. J.L. Balcázar, J. Díaz, and J. Gabarró. *Structural Complexity I.* EATCS Monographs on Theoretical Computer Science. Springer-Verlag, 1988.
2. S. Basu, R. Pollack, and M.-F. Roy. On the combinatorial and algebraic complexity of quantifier elimination. *Journal of the ACM*, 43(6):1002–1045, 1996.
3. L. Blum, F. Cucker, M. Shub, and S. Smale. Algebraic settings for the problem "P≠NP?". In J. Renegar, M. Shub, and S. Smale, editors, *The Mathematics of Numerical Analysis*, volume 32 of *Lectures in Applied Mathematics*, pages 125–144. American Mathematical Society, 1996.
4. L. Blum, F. Cucker, M. Shub, and S. Smale. *Complexity and Real Computation.* Springer-Verlag, 1998.
5. L. Blum, M. Shub, and S. Smale. On a theory of computation and complexity over the real numbers: NP-completeness, recursive functions and universal machines. *Bulletin of the American Mathematical Society*, 21(1):1–46, July 1989.
6. M. Bourgade. Séparations et transferts dans la hiérarchie polynomiale des groupes abéliens infinis. preprint, 1999.
7. P. Bürgisser, M. Clausen, and M. A. Shokrollahi. *Algebraic Complexity Theory.* Springer, 1997.
8. O. Chapuis and P. Koiran. Saturation and stability in the theory of computation over the reals. *Annals of Pure and Applied Logic*, 99:1–49, 1999.
9. A. Chistov and D. Grigoriev. Subexponential-time solving systems of algebraic equations I, II. Preprints LOMI E-9-83, E-10-83, Leningrad, 1983.
10. A. Chistov and D. Grigoriev. Complexity of quantifier elimination in the theory of algebraically closed fields. In *Proc. MFCS'84*, volume 176 of *Lectures Notes in Computer Science*, pages 17–31. Springer-Verlag, 1984.
11. F. Cucker and D. Grigoriev. On the power of real Turing machines with binary inputs. *SIAM Journal on Computing*, 26(1):243–254, 1997.
12. F. Cucker and P. Koiran. Computing over the reals with addition and order: Higher complexity classes. *Journal of Complexity*, 11:358–376, 1995.
13. N. Fichtas, A. Galligo, and J. Morgenstern. Precise sequential and parallel complexity bounds for quantifier elimination over algebraically closed fields. *Journal of Pure and Applied Algebra*, 67:1–14, 1990.
14. H. Fournier and P. Koiran. Are lower bounds easier over the reals ? In *Proc. 30th ACM Symposium on Theory of Computing*, pages 507–513, 1998.
15. H. Fournier and P. Koiran. Lower bounds are not easier over the reals: Inside PH. LIP Research Report 99-21, Ecole Normale Supérieure de Lyon, 1999.
16. J. B. Goode. Accessible telephone directories. *Journal of Symbolic Logic*, 59(1):92–105, 1994.
17. J. E. Goodman and J. O'Rourke, editors. *Handbook of Discrete and Computational Geometry.* CRC Press, 1997.

18. D. Grigoriev. Complexity of deciding Tarksi algebra. *Journal of Symbolic Computation*, 5:65–208, 1988.
19. D. Grigoriev. Complexity of quantifier elimination in the theory of ordinary differential equations. In *Proc. Eurocal'87*, volume 378 of *Lectures Notes in Computer Science*, pages 11–25. Springer-Verlag, 1989.
20. D. Grigoriev. Topological complexity of the range searching. To appear in *Journal of Complexity*.
21. D. Grigoriev, M. Karpinski, and R. Smolensky. Randomization and the computational power of algebraic and analytic decision trees. *Computational Complexity*, 6(4):376–388, 1996/97.
22. D. Grigoriev, M. Singer, and A. Yao. On computing algebraic functions using logarithms and exponentials. *SIAM Journal on Computing*, 24(2):242–246, 1995.
23. D. Grigoriev and N. Vorobjov. Complexity lower bounds for computation trees with elementary transcendental function gates. *Theoretical Computer Science*, 2:185–214, 1996.
24. W. Hodges. *Model Theory*. Encyclopedia of Mathematics and its Applications **42**. Cambridge University Press, 1993.
25. P. Koiran. Computing over the reals with addition and order. *Theoretical Computer Science*, 133(1):35–48, 1994.
26. P. Koiran. VC dimension in circuit complexity. In *Proc. 11th IEEE Conference on Computational Complexity*, pages 81–85, 1996.
27. P. Koiran. Elimination of constants from machines over algebraically closed fields. *Journal of Complexity*, 13(1):65–82, 1997. Erratum on *http://www.ens-lyon.fr/~koiran*.
28. P. Koiran. A weak version of the Blum, Shub & Smale model. *Journal of Computer and System Sciences*, 54:177–189, 1997.
29. K. Meer. A note on a $P \neq NP$ result for a restricted class of real machines. *Journal of Complexity*, 8:451–453, 1992.
30. S. Meiser. Point location in arrangements of hyperplanes. *Information and Computation*, 106(2):286–303, 1993.
31. F. Meyer auf der Heide. A polynomial linear search algorithm for the n-dimensional knapsack problem. *Journal of the ACM*, 31(3):668–676, 1984.
32. F. Meyer auf der Heide. Fast algorithms for n-dimensional restrictions of hard problems. *Journal of the ACM*, 35(3):740–747, 1988.
33. C Michaux. Une remarque à propos des machines sur $\mathbb{R}$ introduites par Blum, Shub et Smale. *C. R. Acad. Sci. Paris*, 309, Série I:435–437, 1989.
34. B. Poizat. *Cours de Théorie des Modèles*. Nur Al-Mantiq Wal-Ma'rifah **1**. 1985.
35. B. Poizat. *Les Petits Cailloux*. Nur Al-Mantiq Wal-Ma'rifah **3**. Aléas, Lyon, 1995.
36. N. Portier. Stabilité polynomiale des corps différentiels. *Journal of Symbolic Logic*, 64(2):803–816, 1999.
37. J. Renegar. On the computational complexity and geometry of the first-order theory of the reals. parts I, II, III. *Journal of Symbolic Computation*, 13(3):255–352, March 1992.
38. M. Shub and S. Smale. On the intractability of Hilbert's Nullstellensatz and an algebraic version of "P=NP". *Duke Mathematical Journal*, 81(1):47–54, 1996.
39. S. Smale. On the topology of algorithms. I. *Journal of Complexity*, 3:81–89, 1987.
40. S. Smale. On the P=NP problem over the complex numbers. Lecture given at the MSRI workshop on Complexity of Continuous and Algebraic Mathematics, November 1998. Lecture on video at www.msri.org.

41. V. A. Vassiliev. *Complements of Discriminants of Smooth Maps: Topology and Applications.* Translations of Mathematical Monographs **98**. American Mathematical Society, 1994.
42. J. von zur Gathen. Parallel arithmetic computations: a survey. In *Mathematical Foundations of Computer Science*, volume 233 of *Lecture Notes in Computer Science*, pages 93–112. Springer-Verlag, 1986.

On the Many Faces of Block Codes

Kaustubh Deshmukh[1], Priti Shankar[2], Amitava Dasgupta[2*], and B. Sundar Rajan[3]

[1] Department of Computer Science and Engineering, Indian Institute of Technology Bombay, Mumbai 400076, India
[2] Department of Computer Science and Automation, Indian Institute of Science, Bangalore 560012, India
[3] Department of Electrical Communication Engineering, Indian Institute of Science, Bangalore 560012, India

Abstract. Block codes are first viewed as finite state automata represented as trellises. A technique termed subtrellis overlaying is introduced with the object of reducing decoder complexity. Necessary and sufficient conditions for subtrellis overlaying are next derived from the representation of the block code as a group, partitioned into a subgroup and its cosets. Finally a view of the code as a graph permits a combination of two shortest path algorithms to facilitate efficient decoding on an overlayed trellis.

1 Introduction

The areas of system theory, coding theory and automata theory have much in common, but historically have developed largely independently. A recent book[9] elaborates some of the connections. In block coding, an *information sequence* of symbols over a finite alphabet is divided into *message blocks* of fixed length; each message block consists of k information symbols. If q is the size of the finite alphabet, there are a total of q^k distinct messages. Each message is encoded into a distinct *codeword* of n ($n > k$) symbols. There are thus q^k codewords each of length n and this set forms a *block code* of length n. A block code is typically used to correct errors that occur in transmission over a communication channel. A subclass of block codes, the *linear block codes* has been used extensively for error correction. Traditionally such codes have been described *algebraically*, their algebraic properties playing a key role in *hard decision* decoding algorithms. In hard decision algorithms, the signals received at the output of the channel are *quantized* into one of the q possible transmitted values, and decoding is performed on a block of symbols of length n representing the received codeword, possibly corrupted by some errors. By contrast, *soft decision* decoding algorithms do not require quantization before decoding and are known to provide significant coding gains when compared with hard decision decoding algorithms. That block codes have efficient *combinatorial* descriptions in the form of *trellises* was discovered in

* Presently at Nokia Research Center, Helsinki, Finland

H. Reichel and S. Tison (Eds.): STACS 2000, LNCS 1770, pp. 53–64, 2000.

1974 [1]. Two other early papers in this subject were [19] and [11]. A landmark paper by Forney [3] in 1988 began an active period of research on the trellis structure of block codes. It was realized that the well known Viterbi Algorithm [16] (which is actually a dynamic programming shortest path algorithm) could be applied to soft decision decoding of block codes. Most studies on the trellis structure of block codes confined their attention to linear codes for which it was shown that unique minimal trellises exist [13]. Trellises have been studied from the viewpoint of linear dynamical systems and also within an algebraic framework [18] [4] [7] [8]. An excellent treatment of the trellis structure of codes is available in [15].

This paper introduces a technique called subtrellis overlaying. This essentially splits a single well structured finite automaton representing the code into several smaller automata, which are then *overlayed*, so that they share states. The motivation for this is a reduction in the size of the trellis, in order to improve the efficiency of decoding. We view the block code as a group partitioned into a subgroup and its cosets, and derive necessary and sufficient conditions for overlaying. The conditions turn out to be simple constraints on the coset leaders. We finally present a two-stage decoding algorithm where the first stage is a Viterbi algorithm performed on the overlayed trellis. The second stage is an adaption of the A^* algorithm well known in the area of artificial intelligence. It is shown that sometimes decoding can be accomplished by executing only the first phase on the overlayed trellis(which is much smaller than the conventional trellis). Thus overlaying may offer significant practical benefits. Section 2 presents some background on block codes and trellises; section 3 derives the conditions for overlaying. Section 4 describes the new decoding algorithm; finally section 5 concludes the paper.

2 Background

We give a very brief background on subclasses of block codes called linear codes. Readers are referred to the classic text [10].

Let F_q be the field with q elements. It is customary to define linear codes algebraically as follows:

Definition 1. *A* linear block code C *of length* n *over a field* F_q *is a* k*-dimensional subspace of an* n*-dimensional vector space over the field* F_q *(such a code is called an* (n, k) *code).*

The most common algebraic representation of a linear block code is the generator matrix G. A $k \times n$ matrix G where the rows of G are linearly independent and which generate the subspace corresponding to C is called a *generator matrix* for C. Figure 1 shows a generator matrix for a $(4, 2)$ linear code over F_2.

A general block code also has a *combinatorial* description in the form of a *trellis.* We borrow from Kschischang et al [7] the definition of a trellis for a block code.

$$\mathbf{G} = \begin{bmatrix} 0\,1\,1\,0 \\ 1\,0\,0\,1 \end{bmatrix}$$

Fig. 1. Generator matrix for a (4, 2) linear binary code

Definition 2. *A trellis for a block code C of length n, is an edge labeled directed graph with a distinguished* root *vertex s, having in-degree 0 and a distinguished* goal *vertex f having out-degree 0, with the following properties:*

1. *All vertices can be reached from the root.*
2. *The goal can be reached from all vertices.*
3. *The number of edges traversed in passing from the root to the goal along any path is n.*
4. *The set of n-tuples obtained by "reading off" the edge labels encountered in traversing all paths from the root to the goal is C.*

The length of a path (in edges) from the root to any vertex is unique and is sometimes called the *time index* of the vertex. One measure of the size of a trellis is the total number of vertices in the trellis. It is well known that minimal trellises for linear block codes are unique [13] and constructable from a generator matrix for the code [7]. Such trellises are known to be *biproper*. Biproperness is the terminology used by coding theorists to specify that the finite state automaton whose transition graph is the trellis, is deterministic, and so is the automaton obtained by reversing all the edges in the trellis. In contrast, minimal trellises for non-linear codes are, in general, neither unique, nor deterministic [7]. Figure 2 shows a trellis for the linear code in Figure 1.

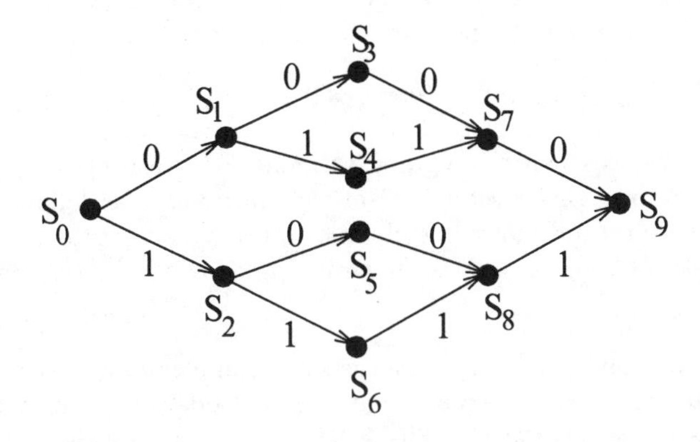

Fig. 2. A trellis for the linear block code of figure 1 with $S_0 = s$ and $S_9 = f$

Willems [18] has given conditions under which an arbitrary block code (which he refers to as a dynamical system) has a unique minimal realization.

Biproper trellises minimize a wide variety of structural complexity measures. McEliece [12] has defined a measure of Viterbi decoding complexity in terms of the number of edges and vertices of a trellis, and has shown that the biproper trellis is the "best" trellis using this measure, as well as other measures based on the maximum number of states at any time index, and the total number of states.

3 Overlaying of Subtrellises

We now restrict our attention to linear block codes. As we have mentioned earlier, every linear code has a unique minimal biproper trellis, so this is our starting point. Our object is to describe an operation which we term *subtrellis overlaying*, which yields a smaller trellis. Reduction in the size of a trellis is a step in the direction of reducing decoder complexity.

Let C be a linear (n, k) code with minimal trellis T_C. A *subtrellis* of T_C is a connected subgraph of T_C containing nodes at every time index $i, 0 \leq i \leq n$ and all edges between them. Partition the states of T_C into $n+1$ groups, one for each time index. Let S_i be the set of states corresponding to time index i, and $|S_i|$ denote the cardinality of the set S_i. Define $S_{max} = \max_i(|S_i|)$. The *state-complexity profile* of the code is defined as the sequence $(|S_0|, |S_1|, \cdots |S_n|)$. Minimization of S_{max} is often desirable and S_{max} is referred to as the *maximum state-complexity*. Our object here, is to partition the code C into disjoint subcodes, and "overlay" the subtrellises corresponding to these subcodes to get a reduced "shared" trellis. An example will illustrate the procedure.

Example 1. Let C be the linear $(4, 2)$ code defined by the generator matrix in Figure 1. C consists of the set of codewords $\{0000, 0110, 1001, 1111\}$ and is described by the minimal trellis in Figure 2. The state-complexity profile of the code is $(1, 2, 4, 2, 1)$. Now partition C into subcodes C_1 and C_2 as follows:

$$C = C_1 \cup C_2; \quad C_1 = \{0000, 0110\}; \quad C_2 = \{1001, 1111\};$$

with minimal trellises shown in figures 3(a) and 3(b) respectively.

The next step is the "overlaying" of the subtrellises as follows. There are as many states at time index 0 and time index n as partitions of C. States $(s_2, s_2'), (s_3, s_3'), (s_1, s_1'), (s_4, s_4')$ are superimposed to obtain the trellis in Figure 4.

Note that overlaying may increase the state-complexity at some time indices (other than 0 and n), and decrease it at others. Codewords are represented by (s_0^i, s_f^i) paths in the overlayed trellis, where s_0^i and s_f^i are the start and final states of subtrellis i. Thus paths from s_0 to s_5 and from s_0' to s_5' represent codewords in the overlayed trellis of figure 3. Overlaying forces subtrellises for

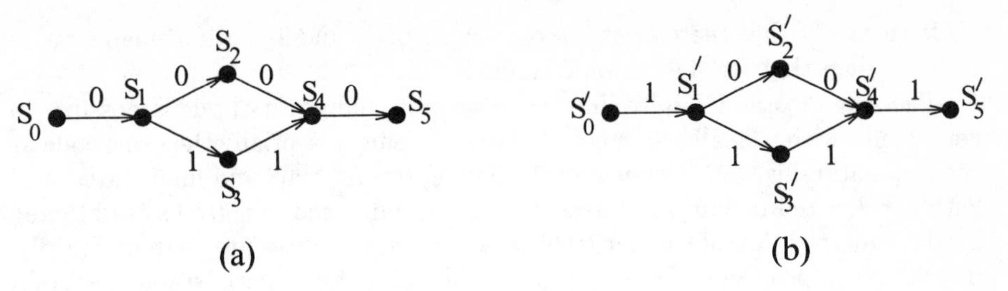

Fig. 3. Minimal trellises for (a) $C_1 = \{0000, 0110\}$ and (b) $C_2 = \{1001, 1111\}$

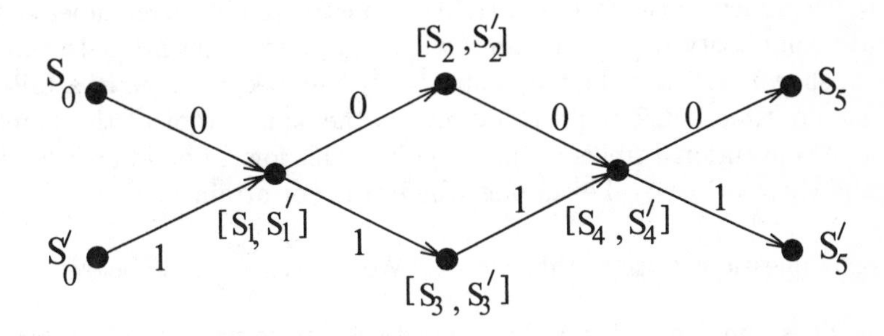

Fig. 4. Trellis obtained by overlaying trellis in figures 3(a) and 3(b)

subcodes to "share" states. Note that the shared trellis is also two way proper, with $S_{max} = 2$ and state-complexity profile $(2, 1, 2, 1, 2)$.

Not all partitions of the code permit overlaying to obtain biproper trellises with a reduced value of S_{max}. For instance, consider the following partition of the code.

$$C = C_1 \cup C_2; \quad C_1 = \{0000, 1001\}; \quad C_2 = \{0110, 1111\};$$

with minimal trellis T_1 and T_2 given in figures 5(a) and 5(b) respectively.

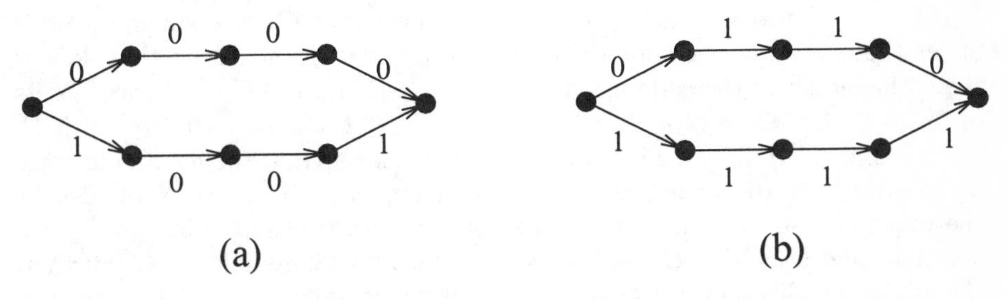

Fig. 5. Minimal subtrellis for (a) $C_1 = \{0000, 1001\}$ and (b) $C_2 = \{0110, 1111\}$

It turns out that there exists no overlaying of T_1 and T_2 with a smaller value of S_{max} than that for the minimal trellis for C.

The small example above illustrates several points. Firstly, it is possible to get a trellis with a smaller number of states to define essentially the same code as the original trellis, with the new trellis having several start and final states, and with a restricted definition of acceptance. Secondly, the new trellis is obtained by the superposition of smaller trellises so that some states are shared. Thirdly, not all decompositions of the original trellis allow for superposition to obtain a smaller trellis. The new trellises obtained by this procedure belong to a class termed *tail-biting trellises* described in a recent paper [2]. This class has assumed importance in view of the fact that trellises constructed in this manner can have low state complexity when compared with equivalent conventional trellises. It has been shown [17] that the maximum of the number of states in a tail-biting trellis at any time index could be as low as the square root of the number of states in a conventional trellis at its midpoint. This lower bound however, is not tight, and there are several examples where it is not attained.

Several questions arise in this context. We list two of these below.

1. How does one decide for a given coordinate ordering, whether there exists an overlaying that achieves a given lower bound on the maximum state complexity at any time index, and in particular, the square root lower bound?
2. Given that there exists an overlaying that achieves a given lower bound how does one find it? That is, how does one decide which states to overlay at each time index?

While, to the best of our knowledge, there are no published algorithms to solve these problems efficiently, in the general case, there are several examples of constructions of minimal tailbiting trellises for specific examples from generator matrices in specific forms in [2].

In the next few paragraphs, we define an object called an *overlayed trellis* and examine the conditions under which it can be constructed so that it achieves certain bounds.

Let C be a linear code over a finite alphabet. (Actually a group code would suffice, but all our examples are drawn from the class of linear codes.) Let $C_0, C_1, \ldots C_l$ be a partition of the code C, such that C_0 is a subgroup of C under the operation of componentwise addition over the structure that defines the alphabet set of the code(usually a field or a ring), and $C_1, \ldots C_l$ are cosets of C_0 in C. Let $C_i = C_0 + h_i$ where $h_i, 1 \leq h_i \leq l$ are coset leaders, with C_i having minimal trellis T_i. The subcode C_0 is chosen so that the maximum state complexity is N (occurring at some time index, say, m), where N divides M the maximum state complexity of the conventional trellis at that time index. The subcodes $C_0, C_1, \ldots C_l$ are all disjoint subcodes whose union is C. Further, the minimal trellises for $C_0, C_1, \ldots C_l$ are all structurally identical and two way proper. (That they are structurally identical can be verified by relabeling a path labeled $g_1g_2 \dots g_n$ in C_0 with $g_1 + h_{i_1}, g_2 + h_{i_2} \dots g_n + h_{i_n}$ in the trellis corresponding to $C_0 + h_i$ where $h_i = h_{i_1}h_{i_2} \dots h_{i_n}$.) We therefore refer to $T_1, T_2, \dots T_l$ as *copies* of T_0.

Definition 3. *An* overlayed *proper trellis is said to exist for C with respect to the partition $C_0, C_1, \dots C_l$ where $C_i, 0 \leq i \leq l$ are subcodes as defined above, corresponding to minimal trellises $T_0, T_1, \dots T_l$ respectively,with $S_{max}(T_0) = N$, iff it is possible to construct a proper trellis T_v satisfying the following properties:*

1. *The trellis T_v has $l+1$ start states labeled $[s_0, \emptyset, \emptyset, \dots \emptyset], [\emptyset, s_1, \emptyset \dots \emptyset] \dots [\emptyset, \emptyset, \dots \emptyset, s_l]$ where s_i is the start state for subtrellis $T_i, 0 \leq i \leq l$.*
2. *The trellis T_v has $l+1$ final states labeled $[f_0, \emptyset, \emptyset, \dots \emptyset], [\emptyset, f_1, \emptyset, \dots \emptyset], \dots [\emptyset, \emptyset, \dots \emptyset, f_l]$, where f_i is the final state for subtrellis T_i, $0 \leq i \leq l$.*
3. *Each state of T_v has a label of the form $[p_0, p_1, \dots p_l]$ where p_i is either $\emptyset$ or a state of $T_i, 0 \leq i \leq l$. Each state of T_i appears in exactly one state of T_v.*
4. *There is a transition on symbol a from state labeled $[p_0, p_1, \dots p_l]$ to $[q_0, q_1, \dots q_l]$ in T_v if and only if there is a transition from p_i to q_i in T_i, provided neither p_i nor q_i is $\emptyset$, for at least one value of i in the set $\{0, 1, 2, \dots l\}$.*
5. *The maximum width of the trellis T_v at an arbitrary time index $i, 1 \leq i \leq n-1$ is at most N.*
6. *The set of paths from $[\emptyset, \emptyset, \dots s_j, \dots \emptyset]$ to $[\emptyset, \emptyset, \dots, f_j, \dots \emptyset]$ is exactly $C_j, 0 \leq j \leq l$.*

Let the *state projection* of state $[p_0, p_1, \dots, p_i, \dots, p_l]$ into subcode index i be p_i if $p_i \neq \emptyset$ and empty if $p_i = \emptyset$. The *subcode projection* of T_v into subcode index i is defined by the symbol $|T_v|_i$ and consists of the subtrellis of T_v obtained by retaining all the non $\emptyset$ states in the state projection of the set of states into subcode index i and the edges between them. An overlayed trellis satisfies the property of *projection consistency* which stipulates that $|T_v|_i = T_i$. Thus every subtrellis T_j is embedded in T_v and can be obtained from it by a projection into the appropriate subcode index. We note here that the conventional trellis is equivalent to an overlayed trellis with $M/N = 1$.

To obtain the necessary and sufficient conditions for an overlayed trellis to exist, critical use is made of the fact that C_0 is a group and $C_i, 1 \leq i \leq l$ are its cosets. For simplicity of notation, we denote by G the subcode C_0 and by T, the subtrellis T_0. Assume T has state complexity profile $(m_o, m_1, \dots m_n)$, where $m_r = m_t = N$, and $m_i < N$ for all $i < r$ and $i > t$. Thus r is the first time index at which the trellis attains maximum state complexity and t is the last. Note that it is not necessary that this complexity be retained between r and t, i.e., the state complexity may drop between r and t. Since each state of T_v is an M/N-tuple, whose state projections are states in individual subtrellises, it makes sense to talk about a state in T_i corresponding to a state in T_v.

We now give a series of lemmas leading up to the main theorem which gives the necessary and sufficient conditions for an overlayed trellis to exist for a given decomposition of C into a subgroup and its cosets. The proofs of these are available in [14]

Lemma 1. *Any state v of T_v at time index in the range 0 to $t-1$, cannot have more outgoing edges than the corresponding state in T. Similarly, any state v at time index in the range $r+1$ to n in T_v cannot have more incoming edges than the corresponding state in T.*

We say that subtrellises T_a and T_b *share* a state v of T_v at level i if v has non $\emptyset$ state projections in both T_a and T_b at time index i.

Lemma 2. *If the trellises T_a and T_b share a state, say v at level $i \leq t$ then they share states at all levels j such that $i \leq j \leq t$. Similarly, if they share a state v at level $i \geq r$, then they share states at all levels j such that $r \leq j \leq i$.*

Lemma 3. *If trellises T_a and T_b share a state at time index i, then they share all states at time index i.*

Lemma 4. *If T_a and T_b share states at levels $i-1$ and i, then their coset leaders have the same symbol at level i.*

We use the following terminology. If h is a codeword say $h_1 h_2 \ldots h_n$, then for $i < t$, $h_{i+1} \ldots h_t$ is called the *tail* of h at i; for $i > r$ $h_r \ldots h_i$ is called the *head* of h at level i.

Lemma 5. *If T_a and T_b have common states at level $i < t$, then there exist coset leaders h_a and h_b of the cosets corresponding to T_a and T_b such that h_a and h_b have the same* tails *at level i. Similarly, if $i > r$ there exist h_a and h_b such that they have the same heads at level i.*

Now each of the M/N copies of T has m_i states at level i. Since the width of the overlayed trellis cannot exceed N for $1 \leq i \leq n-1$, at least $(M/N^2) \times m_i$ copies of trellis T must be overlayed at time index i. Thus there are at most N/m_i (i.e. $(M/N)/((M/N^2 \times m_i)))$ groups of trellises that are overlayed on one another at time index i. From Lemma 5 we know that if S is a set of trellises that are overlayed on one another at level $i, i < t$, then the coset leaders corresponding to these trellises have the same tails at level i. Similarly, if $i > r$ the coset leaders have the same heads at level i. This leads us to the main theorem.

Theorem 1. *Let G be a subgroup of the group code C under componentwise addition over the appropriate structure, with $S_{max}(T_C) = M$, $S_{max}(T) = N$ and let G have M/N cosets with coset leaders $h_0, h_1, \ldots h_{M/N-1}$. Let t, r be the time indices defined earlier. Then C has an overlayed proper trellis T_v with respect to the cosets of G if and only if:*
For all i in the range $1 \leq i \leq n-1$ there exist at most N/m_i collections of coset leaders such that
(i) If $1 \leq i < t$, then the coset leaders within a collection have the same tails at level i.
(ii)If $r < i < n$, the coset leaders within a collection have the same heads at level i.

Corollary 1. *If $M = N^2$ and the conditions of the theorem are satisfied, we obtain a trellis which satisfies the square root lower bound.*

Theorem 1 and corollary 1 answer both the questions about overlayed trellises posed earlier. However, the problem of the existence of an efficient algorithm for the decomposition of the code into a subgroup and its cosets remains open. In the next section we describe the decoding algorithm on an overlayed trellis.

4 Decoding

Decoding refers to the process of forming an estimate of the transmitted codeword $\mathbf{x}$ from a possibly garbled received version $\mathbf{y}$. The received vector $\mathbf{y}$ consists of a sequence of n real numbers, where n is the length of the code. The soft decision decoding algorithm can be viewed as a shortest path algorithm on the trellis for the code. Based on the received vector, a cost $l(u, v)$ can be associated with an edge from node u to node v. The well known Viterbi decoding algorithm [16] is essentially a dynamic programming algorithm, used to compute a shortest path from the source to the goal node.

4.1 The Viterbi Decoding Algorithm

For purposes of this discussion, we assume that the cost is a non negative number. Since the trellis is a regular layered graph, the algorithm proceeds level by level, computing a *survivor* at each node; this is a shortest path to the node from the source. For each branch b, leaving a node at level i, the algorithm updates the survivor at that node by adding the cost of the branch to the value of the survivor. For each node at level $i+1$, it compares the values of the path cost for each branch entering the node and chooses the one with minimum value. There will thus be only one survivor at the goal vertex, and this corresponds to the decoded codeword. For an overlayed trellis we are interested only in paths that go from s_i to f_i, $0 \leq i \leq l$.

4.2 The A^* Algorithm

The A^* algorithm is well known in the literature on artificial intelligence [6] and is a modification of the Dijkstra shortest path algorithm . That the A^* algorithm can be used for decoding was demonstrated in [5]. The A^* algorithm uses, in addition to the path length from the source to the node u, an estimate $h(u, f)$ of the shortest path length from the node to the goal node in guiding the search. Let $L_T(u, f)$ be the shortest path length from u to f in T. Let $h(u, f)$ be any lower bound such that $h(u, f) \leq L_T(u, f)$, and such that $h(u, f)$ satisfies the following inequality, i.e, for u a predecessor of v, $l(u, v) + h(v, f) \geq h(u, f)$. If both the above conditions are satisfied, then the algorithm A^*, on termination, is guaranteed to output a shortest path from s to f. The algorithm is given below.

Algorithm A^*
Input : A trellis $T = (V, E, l)$ where V is the set of vertices, E is the set of edges and $l(u, v) \geq 0$ for edge (u, v) in E, a source vertex s and a destination vertex f.
Output : The shortest path from s to f.
/* $p(u)$ is the cost of the current shortest path from s to u and $P(u)$ is a current shortest path from s to u */

```
begin
    S ← ∅,  S̄ ← {s},  p(s) ← 0,  P(s) ← ()
    repeat
        Let u be the vertex in S̄ with minimum value of p(u) + h(u, f).
        S ← S ∪ {u};   S̄ ← S̄ \ {u};
        if u = f then return P(f);
        for each (u, v) ∈ E do
            if v ∉ S then
                begin
                    p(v) ← min(p(u) + l(u, v), previous(p(v)));
                    if p(v) ≠ previous(p(v)) then append (u, v) to P(u)
to give P(v);
                    (S̄) ← (S̄) ∪ {v};
                end
    forever
end
```

4.3 Decoding on an Overlayed Trellis

Decoding on an overlayed trellis needs at most two phases. In the first phase, a conventional Viterbi algorithm is run on the overlayed trellis T_v . The aim of this phase is to obtain estimates $h()$ for each node, which will subsequently be used in the A^* algorithm that is run on subtrellises in the second phase. The winner in the first phase is either an $s_j - f_j$ path, in which case the second phase is not required, or an $s_i - f_j$ path, $i \neq j$, in which case the second phase is necessary. During the second phase, decoding is performed on one subtrellis at a time, the *current* subtrellis, say T_j (corresponding to subcode C_j) being presently the most promising one, in its potential to deliver the shortest path. If at any point, the computed estimate of the shortest path in the current subtrellis exceeds the minimum estimate among the rest of the subtrellises, currently held by, say, subtrellis T_k, then the decoder switches from T_j to T_k, making T_k the current subtrellis. Decoding is complete when a final node is reached in the current subtrellis. The two phases are described below. (All assertions in italics have simple proofs given in [14]).
Phase 1. Execute a Viterbi decoding algorithm on the shared trellis, and obtain survivors at each node. *Each survivor at a node u has a cost which is a lower bound on the cost of the least cost path from s_j to u in an $s_j - f_j$ path passing through u, $1 \leq j \leq N$. If there exists a value of k for which an $s_k - f_k$ path is*

an overall winner then this is the shortest path in the original trellis T_C. If this happens decoding is complete. If no such $s_k - f_k$ path exists go to Phase 2.

Phase 2

1. Consider only subtrellises T_j such that the winning path at T_j is an $s_i - f_j$ path with $i \neq j$ (i.e at some intermediate node a prefix of the $s_j - f_j$ path was "knocked out" by a shorter path originating at s_i), and such that there is no $s_k - f_k$ path with smaller cost. Let us call such trellises *residual trellises.* Initialize a sequence P_j for each residual trellis T_j to the empty sequence. P_j, in fact stores the current candidate for the shortest path in trellis T_j. Let the estimate $h(s_j, f_j)$ associated with the empty path be the cost of the survivor at f_j obtained in the first phase.
2. Create a heap of r elements where r is the number of residual trellises, with current estimate $h()$ with minimum value as the top element. Let j be the index of the subtrellis with the minimum value of the estimate. Remove the minimum element corresponding to T_j from the heap and run the A^* algorithm on trellis T_j (called the *current trellis*). For a node u, take $h(u, f_j)$ to be $h(s_i, f_j) - cost(survivor(u))$ where $cost(survivor(u))$ is the cost of the survivor obtained in the first phase. $h()$ *satisfies the two properties required of the estimator in the* A^* *algorithm.*
3. At each step, compare $p(u) + h(u, f_j)$ in the current subtrellis with the top value in the heap. If at any step the former exceeds the latter (associated with subtrellis, say, T_k), then make T_k the current subtrellis. Insert the current value of $p(u) + h(u, f_j)$ in the heap (after deleting the minimum element) and run the A^* algorithm on T_k either from start node s_k (if T_k was not visited earlier) or from the node which it last expanded in T_k. Stop when the goal vertex is reached in the current subtrellis.

In the best case (if the algorithm needs to execute Phase 2 at all) the search will be restricted to a single residual subtrellis; the worst case will involve searching through all residual subtrellises.

5 Conclusions

This paper offers a new perspective from which block codes may be fruitfully viewed. A technique called subtrellis overlaying is proposed , which reduces the size of the trellis representing the block code. Necessary and sufficient conditions for overlaying are derived from the representation of the code as a group. Finally a decoding algorithm is proposed which requires at most two passes on the overlayed trellis. For transmission channels with high signal to noise ratio, it is likely that decoding will be efficient. This is borne out by simulations on a code called the hexacode[2] on an additive white Gaussian noise(AWGN) channel, where it was seen that the decoding on the overlayed trellis was faster than that on the conventional trellis for signal to noise ratios of 2.5 dB or more[14]. Future work will concentrate on investigating the existence of an efficient algorithm for finding a good decomposition of a code into a subgroup and its cosets, and on obtaining overlayed trellises for long codes.

References

1. L.R.Bahl, J.Cocke, F.Jelinek, and J. Raviv, Optimal decoding of linear codes for minimizing symbol error rate, *IEEE Trans. Inform. Theory* **20**(2), March 1974, pp 284-287.
2. A.R.Calderbank, G.David Forney,Jr., and Alexander Vardy, Minimal Tail-Biting Trellises: The Golay Code and More, *IEEE Trans. Inform. Theory* **45**(5) July 1999,pp 1435-1455.
3. G.D. Forney, Jr.,Coset codes II: Binary lattices and related codes, *IEEE Trans. Inform. Theory* **36**(5), Sept. 1988,pp 1152-1187.
4. G.D. Forney, Jr. and M.D. Trott, The dynamics of group codes:State spaces, trellis diagrams and canonical encoders, *IEEE Trans. Inform. Theory* **39**(5) Sept 1993,pp 1491-1513.
5. Y.S.Han, C.R.P.Hartmann, and C.C.Chen, Efficient Priority-First Search Maximum-Likelihood Soft-Decision Decoding of Linear Block Codes, *IEEE Trans. Inform.Theory* **39**(5),Sept. 1993,pp 714-729.
6. P.E. Hart, N.J. Nilsson, and B. Raphael, A formal basis for the heuristic determination of minimum cost paths, *IEEE Trans. Solid-State Circuits* **SSC-4**, 1968, pp 100-107.
7. F.R.Kschischang and V.Sorokine, On the trellis structure of block codes, *IEEE Trans. Inform Theory* **41**(6), Nov 1995,pp 1924-1937.
8. F.R.Kschischang, The trellis structure of maximal fixed cost codes, *IEEE Trans. Inform Theory* **42**(6), Nov. 1996, pp 1828-1837.
9. D.Lind and M.Marcus, *An Introduction to Symbolic Dynamics and Coding*, Cambridge University Press, 1995.
10. F.J. MacWilliams and N.J.A. Sloane, *The Theory of Error Correcting Codes*, North-Holland, Amsterdam, 1981.
11. J.L.Massey, Foundations and methods of channel encoding, in *Proc. Int. Conf. on Information Theory and Systems* **65**(Berlin, Germany) Sept 1978.
12. R.J. McEliece, On the BCJR trellis for linear block codes, *IEEE Trans. Inform Theory* **42**, 1996, pp 1072-1092
13. D.J. Muder, Minimal trellises for block codes, *IEEE Trans. Inform Theory* **34**(5), Sept 1988,pp 1049-1053.
14. Amitava Dasgupta, Priti Shankar, Kaustubh Deshmukh, and B.S.Rajan, *On Viewing Block Codes as Finite Automata*, Technical Report IISc-CSA-1999-7, Department of Computer Science and Automation, Indian Institute of Science, Bangalore 560012, India, November, 1999.
15. A.Vardy, Trellis structure of codes, in *Handbook of Coding Theory*, R.A.Brualdi, W.C. Huffman, V.S. Pless, Eds., Vol.2, Chap. 24, Elsevier, 1998.
16. A.J. Viterbi, Error bounds for convolutional codes and an asymptotically optimum decoding algorithm, *IEEE Trans. Inform Theory* **13**, April 1967, pp 260-269.
17. N.Wiberg, H.-A. Loeliger and R.Kotter, Codes and iterative decoding on general graphs, *Euro. Trans. Telecommun.,***6** pp 513-526, Sept 1995.
18. J.C. Willems, Models for Dynamics, in *Dynamics Reported*, **2**, U. Kirchgraber and H.O. Walther, Eds. New York: Wiley, 1989, pp 171-269.
19. J.K. Wolf, Efficient maximum-likelihood decoding of linear block codes using a trellis, *IEEE Trans. Inform Theory* **24**(1), January 1978, pp 76-80.

A New Algorithm for MAX-2-SAT*

Edward A. Hirsch

Steklov Institute of Mathematics at St.Petersburg
27 Fontanka, St.Petersburg 191011, Russia
hirsch@pdmi.ras.ru
http://logic.pdmi.ras.ru/~hirsch

Abstract. Recently there was a significant progress in proving (exponential-time) worst-case upper bounds for the propositional satisfiability problem (SAT) and related problems. In particular, for MAX-2-SAT Niedermeier and Rossmanith recently presented an algorithm with worst-case upper bound $O(K \cdot 2^{K/2.88...})$, and the bound $O(K \cdot 2^{K/3.44...})$ is implicit from the paper by Bansal and Raman (K is the number of clauses). In this paper we improve this bound to $p(K)2^{K_2/4}$, where K_2 is the number of 2-clauses, and p is a polynomial. In addition, our algorithm and the proof are much simpler than the previous ones. The key ideas are to use the symmetric flow algorithm of Yannakakis and to count only 2-clauses (and not 1-clauses).

1 Introduction

SAT (the problem of satisfiability of a propositional formula in conjunctive normal form (*CNF*)) can be easily solved in time of the order 2^N, where N is the number of variables in the input formula. In the early 1980s this trivial bound was improved for formulas in 3-CNF by Monien and Speckenmeyer [18] (see also [19]) and independently by Dantsin [4] (see also [5,7]). After that, many upper bounds for SAT and its NP-complete subproblems were obtained ([12,13,15,16,22,25] are the most recent). Most authors consider bounds w.r.t. three main parameters: the length L of the input formula (i.e. the number of literal occurrences), the number K of its clauses and the number N of the variables occurring in it. In this paper we consider bounds w.r.t. the parameters K and L. The best such bounds for SAT are $p(L)2^{K/3.23...}$ [12,13] and $p(L)2^{L/9.7...}$ [13] (p is a polynomial).

The maximum satisfiability problem (*MAX-SAT*) is an important generalization of SAT. In this problem we are given a formula in CNF, and the answer is the maximal number of simultaneously satisfiable clauses. This problem is NP-complete[1] even if each clause contains at most two literals (*MAX-2-SAT*; see, e.g., [21]). This problem was widely studied in the context of approximation algorithms (see, e.g., [1,10,11,14,26]). As to the worst-case time bounds for the exact solution of MAX-SAT, Niedermeier and Rossmanith [20] recently proved two

* Supported by INTAS (project No. 96-0760) and RFBR (project No. 99-01-00113).

[1] A more precise NP-formulation is, of course, "given a formula in CNF and an integer k, decide whether there is an assignment that satisfies at least k clauses".

H. Reichel and S. Tison (Eds.): STACS 2000, LNCS 1770, pp. 65–73, 2000.

worst-case upper bounds: $O(L \cdot 2^{K/2.15\ldots})$ for MAX-SAT, and $O(K \cdot 2^{K/2.88\ldots})$ for MAX-2-SAT. For the latter bound, they presented an algorithm for MAX-SAT running in $O(L \cdot 2^{L/5.76\ldots})$ time; the desired bound follows since $L \leq 2K$. Bansal and Raman [2] have recently improved the MAX-SAT bounds to $O(L \cdot 2^{K/2.35\ldots})$ and $O(L \cdot 2^{L/6.89\ldots})$, which leads to the $O(K \cdot 2^{K/3.44\ldots})$ bound for MAX-2-SAT. Niedermeier and Rossmanith posed a question whether the bound for MAX-2-SAT can be improved by a direct algorithm (and not by an algorithm for general MAX-SAT for a bound w.r.t. L). In this paper, we answer this question by giving an algorithm which solves MAX-2-SAT in the $p(K)2^{K/4}$ time (p is a polynomial). In addition, our algorithm and the proof are much simpler then those in [2,20].

Most of the algorithms/bounds mentioned above use the Davis-Putnam procedure [8,9]. In short, this procedure allows to reduce the problem for a formula F to the problem for two formulas $F[v]$ and $F[\overline{v}]$ (where v is a propositional variable). This is called "splitting". Before the algorithm splits each of the obtained two formulas, it can transform them into simpler formulas F_1 and F_2 (using some *transformation rules*). The algorithm does not split a formula if it is trivial to solve the problem for it; these formulas are the leaves of the *splitting tree* which corresponds to the execution of such algorithm. For most known algorithms, the leaves are trivial formulas (i.e. the formulas containing no non-trivial clauses).

In the algorithm presented in this paper, the leaves are satisfiable formulas and formulas for which the (polynomial time) "symmetric flow" algorithm of Yannakakis [26] finds an optimal solution (this algorithm either finds an optimal solution or simplifies the input formula). Transformation rules include the pure literal rule, a slightly generalized resolution rule (using these two rules one can solve MAX-SAT in a polynomial time in the case that each variable occurs at most twice; it was already observed in, e.g., [23]), the frequent 1-clause rule [20], and the elimination of 1-clauses. Although in MAX-SAT 1-clauses cannot be eliminated by the usual unit propagation technique, in the case of MAX-2-SAT they can be eliminated by the symmetric flow algorithm of Yannakakis [26]. Thus, before each splitting we can transform a formula into one which consists of 2-clauses, and each variable occurs at least three times. Therefore, each splitting eliminates at least three 2-clauses in each branch. This observation would already improve the bound of [20] to $p(K)2^{K_2/3}$, where K_2 is the number of 2-clauses in the input formula, p is a polynomial. However, by careful choice of a variable for splitting, we get a better bound $p(K)2^{K_2/4}$, which implies the bound $p(L)2^{L/8}$ (since $L \geq 2K_2$).

In Sect. 2 we give basic definitions and formulate in our framework the known results we use. In Sect. 3 we present the algorithm and the proof of its worst-case upper bound.

2 Background

Let V be a set of Boolean variables. The negation of a variable v is denoted by $\overline{v}$. Given a set U, we denote $\overline{U} = \{\overline{u} \mid u \in U\}$. *Literals* (usually denoted by $l, l', l_1, l_2, \ldots$) are the members of the set $W = V \cup \overline{V}$. *Positive literals* are the members of the set V. *Negative literals* are their negations. If w denotes a negative literal $\overline{v}$, then $\overline{w}$ denotes the variable v.

Algorithms for finding the exact solution of MAX-SAT are usually designed for the unweighted MAX-SAT problem. However, the formulas are usually represented by multisets (i.e., formulas in CNF with integer positive weights). In this paper we consider the weighted MAX-SAT problem with positive integer weights. A *(weighted) clause* is a pair (ω, S) where ω is a strictly positive integer number, and S is a nonempty finite set of literals which does not contain simultaneously any variable together with its negation. We call ω the *weight* of a clause (ω, S).

An *assignment* is a finite subset of W which does not contain any variable together with its negation. Informally speaking, if an assignment A contains a literal l, it means that l has the value $True$ in A. In addition to usual clauses, we allow a special *true clause* $(\omega, \mathbb{T})$ which is satisfied by every assignment. (We also call it a $\mathbb{T}$-*clause*.)

The length of a clause (ω, S) is the cardinality of S. A k-*clause* is a clause of the length exactly k. In this paper a *formula in (weighted) CNF* (or simply *formula*) is a finite set of (weighted) clauses (ω, S), at most one for each S. The *length of a formula* is the sum of the lengths of all its clauses. The total weight of all clauses of a formula F is denoted by $\mathfrak{K}(F)$. The total weight of all 2-clauses of a formula F is denoted by $\mathfrak{K}_2(F)$.

The pairs $(0, S)$ are *not* clauses, however, for simplicity we write $(0, S) \in F$ for all S and all F. Therefore, the operators $+$ and $-$ are defined:

$$F + G = \{(\omega_1 + \omega_2, S) \mid (\omega_1, S) \in F \text{ and } (\omega_2, S) \in G, \text{ and } \omega_1 + \omega_2 > 0\},$$
$$F - G = \{(\omega_1 - \omega_2, S) \mid (\omega_1, S) \in F \text{ and } (\omega_2, S) \in G, \text{ and } \omega_1 - \omega_2 > 0\}.$$

Example 1. If

$$F = \{\ (2, \mathbb{T}),\ (3, \{x, y\}),\ (4, \{\overline{x}, \overline{y}\})\ \}$$

and

$$G = \{\ (2, \{x, y\}),\ (4, \{\overline{x}, \overline{y}\})\ \},$$

then

$$F - G = \{\ (2, \mathbb{T}),\ (1, \{x, y\})\ \}.$$

□

For a literal l and a formula F, the formula $F[l]$ is obtained by setting the value of l to $True$. More precisely, we define

$$\begin{aligned} F[l] = (&\{(\omega, S) \mid (\omega, S) \in F \text{ and } l, \overline{l} \notin S\} + \\ &\{(\omega, S \setminus \{\overline{l}\}) \mid (\omega, S) \in F \text{ and } S \neq \{\overline{l}\}, \text{ and } \overline{l} \in S\} + \\ &\{(\omega, \mathbb{T}) \mid \omega \text{ is the sum of the weights } \omega' \text{ of all clauses } (\omega', S) \text{ of } F \\ &\qquad \text{such that } l \in S\}. \end{aligned}$$

(Note that no $(\omega, \emptyset)$ or $(0, S)$ is included in $F[l]$, $F + G$ or $F - G$.) For an assignment $A = \{l_1, \ldots, l_s\}$ and a formula F, we define $F[A] = F[l_1][l_2] \ldots [l_s]$ (evidently, $F[l][l'] = F[l'][l]$ for every literals l, l' such that $l \neq \overline{l'}$). For short, we write $F[l_1, \ldots, l_s]$ instead of $F[\{l_1, \ldots, l_s\}]$.

Example 2. If

$$F = \{ \, (1, \mathbb{T}), \; (1, \{x, y\}), \; (5, \{\overline{y}\}), \; (2, \{\overline{x}, \overline{y}\}), \; (10, \{\overline{z}\}), \; (2, \{\overline{x}, z\}) \, \},$$

then

$$F[x, \overline{z}] = \{ \, (12, \mathbb{T}), \; (7, \{\overline{y}\}) \, \}.$$

□

The optimal value $\mathrm{OptVal}(F) = \max_A \{ \omega \,|\, (\omega, \mathbb{T}) \in F[A] \}$. An assignment A is *optimal* if $F[A]$ contains only one clause $(\omega, \mathbb{T})$ (or does not contain any clauses, in this case $\omega = 0$) and $\mathrm{OptVal}(F) = \omega \;\; (= \mathrm{OptVal}(F[A])\,)$.

A formula is in *2-CNF* if it contains only 2-clauses, 1-clauses and a $\mathbb{T}$-clause. A formula is in *2E-CNF* if it contains only 2-clauses and a $\mathbb{T}$-clause.

If we say that a (positive or negative) *literal v occurs* in a clause or in a formula, we mean that this clause (more formally, its second component) or this formula (more formally, one of its clauses) contains the literal v. However, if we say that a *variable v occurs* in a clause or in a formula, we mean that this clause or this formula contains the literal v, or it contains the literal $\overline{v}$. A variable v *occurs positively*, if the literal v occurs, and *occurs negatively*, if the literal $\overline{v}$ occurs. A literal l is an (i,j)-literal if l occurs exactly i times in the formula and the literal $\overline{l}$ occurs exactly j times in the formula. A literal is *pure* in a formula F if it occurs in F, and its negation does not occur in F. The following lemma is well-known and straightforward.

Lemma 1. *If l is a pure literal in F, then* $\mathrm{OptVal}(F) = \mathrm{OptVal}(F[l])$.

In this paper, the *resolvent* $\mathfrak{R}(C, D)$ of clauses $C = (\omega_1, \{l_1, l_2\})$ and $D = (\omega_2, \{\overline{l_1}, l_3\})$ is the formula

$$\{ \, (\max(\omega_1, \omega_2), \, \mathbb{T}), \;\; (\min(\omega_1, \omega_2), \, \{l_2, l_3\}) \, \}$$

if $l_2 \neq \overline{l_3}$, and the formula $\{(\omega_1 + \omega_2, \mathbb{T})\}$ otherwise. This definition is not traditional, but it is very useful in MAX-SAT context.

The following lemma is a straightforward generalization of the resolution correctness (see, e.g., [24]) for the case when there are weights, but the literal on which we are resolving does not occur in other clauses of the formula.

Lemma 2. *If F contains clauses $C = (\omega_1, \{v, l_1\})$ and $D = (\omega_2, \{\overline{v}, l_2\})$ such that the variable v does not occur in other clauses of F, then*

$$\mathrm{OptVal}(F) = \mathrm{OptVal}(\, (F - \{C, D\}) + \mathfrak{R}(C, D) \,).$$

The following simple observation is also well-known (see, e.g., [17,6]).

Lemma 3. *Let F be a formula in weighted CNF, and v be a variable. Then*

$$\text{OptVal}(F) = \max(\,\text{OptVal}(F[v]),\ \text{OptVal}(F[\overline{v}])\,).$$

We also note that a polynomial time algorithm for 2-SAT is known. In our context, a formula F is satisfiable if $\text{OptVal}(F)$ is equal to the sum of the weights of all clauses occurring in F.

Lemma 4 (see, e.g. [3]). *There is a polynomial time algorithm for 2-SAT.*

Yannakakis presented in [26] an algorithm which transforms a formula in 2-CNF into a formula in $2E$-CNF which has the same optimal value. This algorithm consists of two stages. The first stage is a removal of a maximum symmetric flow from a graph corresponding to the formula; this stage can be considered as a combination of three transformation rules (it is not important for us now which combination):

1. Replacing[2] of a "cycle"

$$\{\ (\omega, \{l_1, \overline{l_2}\}),\ (\omega, \{l_2, \overline{l_3}\}),\ \ldots,\ (\omega, \{l_k, \overline{l_1}\})\ \}$$

   by another cycle

$$\{\ (\omega, \{\overline{l_1}, l_2\}),\ (\omega, \{\overline{l_2}, l_3\}),\ \ldots,\ (\omega, \{\overline{l_k}, l_1\})\ \};$$

2. Replacing of a set

$$\{\ (\omega, \{\overline{l_1}\}),\ (\omega, \{l_1, \overline{l_2}\}),\ (\omega, \{l_2, \overline{l_3}\}),\ \ldots,\ (\omega, \{l_{k-1}, \overline{l_k}\})\ \}$$

   by the set

$$\{\ (\omega, \{\overline{l_1}, l_2\}),\ (\omega, \{\overline{l_2}, l_3\}),\ \ldots,\ (\omega, \{\overline{l_{k-1}}, l_k\}),\ (\omega, \{\overline{l_k}\})\ \};$$

3. Replacing two contradictory clauses $(\omega, \{l\})$ and $(\omega, \{\overline{l}\})$ by a true clause of the weight ω.

The second stage is replacing of the obtained formula F' by the formula $F'[A]$ for some assignment A (it is not important for us now which assignment). Evidently, this algorithm does not increase the total weight of all 2-clauses. Therefore, we can formulate the result of Yannakakis in the following form.

Lemma 5 ([26]). *There is a polynomial time algorithm which given an input formula F in weighted 2-CNF, outputs a formula G in weighted $2E$-CNF, such that $\mathfrak{K}_2(G) \leq \mathfrak{K}_2(F)$, and* $\text{OptVal}(F) = \text{OptVal}(G)$.

The following fact was observed by Niedermeier and Rossmanith.

Lemma 6 ([20]). *If the weight of a 1-clause $(\omega, \{l\})$ of a formula F is not less than the total weight of all clauses of F containing the literal $\overline{l}$, then* $\text{OptVal}(F) = \text{OptVal}(F[l])$.

[2] This replacing is made by subtracting the weights: e.g., if a formula contains a clause $(\omega', \{l_1, \overline{l_2}\})$ with $\omega' \geq \omega$, then it is split into two clauses $(\omega' - \omega, \{l_1, \overline{l_2}\})$ and $(\omega, \{l_1, \overline{l_2}\})$, and the latter clause is replaced as formulated.

3 Results

In this section we present Algorithm 1 which solves MAX-2-SAT in the time $p(K)2^{K_2/4}$, where p is a polynomial, K is the total weight of all clauses in the input formula, and K_2 is the total weight of 2-clauses in it (in the case of unweighted MAX-2-SAT, K_2 is the number of 2-clauses).

Algorithm 1.

Input: *A formula* F *in weighted* 2-*CNF.*
Output: $\mathrm{OptVal}(F)$.

Method.

(1) Apply the symmetric flow algorithm from [26] (see Lemma 5) to F.
(2) If there is a pure literal l in F, assume $F := F[l]$.
(3) If there is a variable that occurs in F exactly once positively in a clause C and exactly once negatively in a clause D, then $F := (F - \{C, D\}) + \mathfrak{R}(C, D)$.
(4) If F has been changed at steps (2)–(3), then go to step (1).
(5) If F is satisfiable[3], return the sum of the weights of all its clauses.
(6) If there is a variable v such that the total weight of the clauses of F in which this variable occurs, is at least 4, then execute Algorithm 1 for the formulas $F[v]$ and $F[\overline{v}]$, and return the maximum of its answers.
(7) Find[4] in F a clause $(\omega, \{l_1, l_2\})$ such that l_1 and l_2 are (2,1)-literals, and the two other clauses C and D containing the literals $l_2, \overline{l_2}$ do not contain the literals $l_1, \overline{l_1}$. Execute Algorithm 1 for the formulas $(F[l_1] - \{C, D\}) + \mathfrak{R}(C, D)$, and $F[\overline{l_1}, l_2]$, and return the maximum of its answers.

□

Theorem 1. *Given a formula* F *in* 2-*CNF, Algorithm 1 always correctly finds* $\mathrm{OptVal}(F)$ *in time* $p(\mathfrak{K}(F)) \cdot 2^{\mathfrak{K}_2(F)/4}$, *where* p *is a polynomial.*

Proof. Correctness. If Algorithm 1 outputs an answer, then its correctness follows from the lemmata of Sect. 2 (step (1): Lemma 5; step (2): Lemma 1; step (3): Lemma 2; step (6): Lemma 3; step (7): Lemmata 3, 2 and 6, where Lemma 6 is applied to the clause $\{l_2\}$ in the formula $F[\overline{l_1}]$, note that at step (7) the formula F consists of clauses of weight 1).

Since any change at steps (2) and (3) decreases the total weight of 2-clauses in F, and the step (1) does not increase it, the cycle (1)–(4) is repeated a polynomial number of times. Now it remains to show that at step (7) Algorithm 1 always can find a clause satisfying its conditions.

Note that at step (7) the formula F is not satisfiable, consists only of 2-clauses (and, maybe, a $\mathbb{T}$-clause), does not contain pure literals, and each variable occurs

[3] We can check it in a polynomial time [3], see Lemma 4.
[4] Theorem 1 proves that it is possible to find a clause satisfying the conditions of this step.

in it exactly three times, i.e. F contains only (2,1)-literals and (1,2)-literals. Since F is not satisfiable, there exists at least one clause in it that contains two (1,2)-literals (otherwise each clause contains a (2,1)-literal, i.e. the assignment consisting of all (2,1)-literals is satisfying; cf. "Extended Sign Principle" of [16]). Thus, (1,2)-literals occur in at most $N-1$ clauses of F, where N is the number of variables occurring in F. There are $3N/2$ 2-clauses in F. Hence, F contains more than $N/2$ 2-clauses consisting only of (2,1)-literals. There are at least $N+1$ literals in these clauses, thus, there is at least one (2,1)-literal occurring in *two* such clauses. This literal, and (at least) one of the two literals occurring with it in these clauses, satisfy the condition of the step (7).

Running time. Each of the steps of Algorithm 1 (not including recursive calls) takes only a polynomial time (Lemmata 5 and 4). The steps (1)–(5) do not increase the total weight of 2-clauses in F. By the above argument, each of these steps is executed a polynomial number of times during one execution of Algorithm 1 (again not including recursive calls). It suffices to show that for each formula F' which is an argument of a recursive call, $\mathfrak{K}_2(F') \leq \mathfrak{K}_2(F) - 4$.

Note that at the moment before a splitting (which precedes a recursive call), the formula F consists only of 2-clauses (and, maybe, a $\mathbb{T}$-clause). Then the statement follows from the conditions of the steps (6) and (7). □

Corollary 1. *Given a formula F in unweighted*[5] *2-CNF of length L, Algorithm 1 always correctly finds* OptVal(F) *in time $p(L)2^{L/8}$, where p is a polynomial.*

Remark 1. Of course, in Corollary 1 only the number of literal occurrences in 2-clauses is essential in the exponent.

Remark 2. Algorithm 1 can be easily redesigned so that it finds the optimal assignment (or one of them, if there are several assignments satisfying the same number of clauses) instead of OptVal(F).

4 Conclusion

In this paper we improved the existing upper bound for MAX-2-SAT with integer weights to $p(K)2^{K_2/4}$, where K_2 is the total weight of 2-clauses of the input formula (or the number of 2-clauses for unweighted MAX-2-SAT), K is the total weight of all clauses, and p is a polynomial. This also implies the $p(L)2^{L/8}$ bound for unweighted MAX-2-SAT, where L is the number of literal occurrences.

One of the key ideas of our algorithm is to count only 2-clauses (since MAX-1-SAT instances are trivial). It would be interesting to apply this idea to SAT, for example, by counting only 3-clauses in 3-SAT (since 2-SAT instances are easy). Also, it remains a challenge to find a "less-than-2^N" algorithm for MAX-SAT or even MAX-2-SAT, where N is the number of variables.

[5] I.e., all weights equal 1.

We did not investigate the practical performance of our algorithm and even did not give "practical" implementation. The "abstract" polynomial factor in the bound could be replaced by a concrete (low degree) polynomial if such implementation is given.

5 Acknowledgement

I am grateful to two anonymous referees who pointed out an annoying typo in the last step of Algorithm 1.

References

1. T. Asano, D. P. Williamson, *Improved Approximation Algorithms for MAX SAT*, To appear in Proc. of the 11th Annual ACM-SIAM Symposium on Discrete Algorithms, 2000.
2. N. Bansal, V. Raman, *Upper Bounds for MaxSat: Further Improved,* To appear in Proc. of ISAAC'99.
3. S. A. Cook, *The complexity of theorem-proving procedure,* Proc. 3rd Annual ACM Symposium on the Theory of Computing, 1971, pp. 151–159.
4. E. Dantsin, *Tautology proof systems based on the splitting method*, Leningrad Division of Steklov Institute of Mathematics (LOMI), PhD Dissertation, Leningrad, 1982 (in Russian).
5. E. Dantsin and L. O. Fuentes and V. Kreinovich, *Less than* 2^n *satisfiability algorithm extended to the case when almost all clauses are short*, Computer Science Department, University of Texas at El Paso, UTEP-CS-91-5, 1991.
6. E. Dantsin, M. Gavrilovich, E. A. Hirsch, B. Konev, *Approximation algorithms for Max Sat: a better performance ratio at the cost of a longer running time*, PDMI preprint 14/1998, available from `ftp://ftp.pdmi.ras.ru/pub/publicat/preprint/1998/14-98.ps`
7. E. Ya. Dantsin, V. Ya. Kreinovich, *Exponential upper bounds for the satisfiability problem*, Proc. of the IX USSR conf. on math. logic, Leningrad, 1988 (in Russian).
8. M. Davis, G. Logemann, D. Loveland, *A machine program for theorem-proving*, Comm. ACM, vol. 5, 1962, pp. 394–397.
9. M. Davis, H. Putnam, *A computing procedure for quantification theory*, J. ACM, vol. 7, 1960, pp. 201–215.
10. U. Feige, M. X. Goemans, *Approximating the value of two prover proof systems, with applications to MAX 2SAT and MAX DICUT*, Proc. of the Third Israel Symposium on Theory of Computing and Systems, 1995, pp. 182–189.
11. J. Håstad, *Some optimal inapproximability results*, Proc. of the 29th Annual ACM Symposium on Theory of Computing, 1997, pp. 1–10.
12. E. A. Hirsch, *Two new upper bounds for SAT*, Proc. of the 9th Annual ACM-SIAM Symposium on Discrete Algorithms, 1998, pp. 521–530.
13. E. A. Hirsch, *New Worst-Case Upper Bounds for SAT*, To appear in the special issue SAT-2000 and in the Journal of Automated Reasoning, Kluwer Academic Publishers, 2000.
14. H. Karloff, U. Zwick, *A 7/8-approximation algorithm for MAX 3SAT?*, In Proc. of the 38th Annual IEEE Symposium on Foundations of Computer Science, pp. 406–415, 1997.

15. O. Kullmann, *New methods for 3-SAT decision and worst-case analysis*, Theoretical Computer Science 223 (1-2), 1999, pp.1-71.
16. O. Kullmann, H. Luckhardt, *Deciding propositional tautologies: Algorithms and their complexity*, Preprint, 1997, 82 pages; The electronic version can be obtained at `http://www.cs.utoronto.ca/~kullmann`. A journal version, *Algorithms for SAT/TAUT decision based on various measures*, is to appear in Information and Computation.
17. M. Mahajan and V. Raman, *Parametrizing above guaranteed values: MaxSat and MaxCut*, Technical Report TR97-033, Electronic Colloquium on Computational Complexity, 1997. To appear in Journal of Algorithms.
18. B. Monien, E. Speckenmeyer, *3-satisfiability is testable in $O(1.62^r)$ steps*, Bericht Nr. 3/1979, Reihe Theoretische Informatik, Universität-Gesamthochschule-Paderborn.
19. B. Monien, E. Speckenmeyer, *Solving satisfiability in less then 2^n steps*, Discrete Applied Mathematics, vol. 10, 1985, pp. 287–295.
20. R. Niedermeier and P. Rossmanith. *New upper bounds for MaxSat*, Technical Report KAM-DIMATIA Series 98-401, Charles University, Praha, Faculty of Mathematics and Physics, July 1998. Extended abstract appeared in Proc. of the 26th International Colloquium on Automata, Languages, and Programming, LNCS 1644, pp. 575–584, 1999. A journal version is to appear in Journal of Algorithms.
21. Ch. H. Papadimitriou, *Computational Complexity*, Addison–Wesley, 1994, 532 p.
22. R. Paturi, P. Pudlak, M. E. Saks, F. Zane, *An Improved Exponential-time Algorithm for k-SAT*, Proc. of the 39th Annual Symposium on Foundations of Computer Science, 1998, pp. 628–637.
23. V. Raman, B. Ravikumar and S. Srinivasa Rao, *A simplified NP-complete MAXSAT problem*, Information Processing Letters 65(1), 1998, pp. 1–6.
24. J. A. Robinson, *Generalized resolution principle*, Machine Intelligence, vol. 3, 1968, pp. 77–94.
25. U. Schöning, *A probabilistic algorithm for k-SAT and constraint satisfaction problems,* Proc. of the 40th Annual Symposium on Foundations of Computer Science, 1999. To appear.
26. M. Yannakakis, *On the Approximation of Maximum Satisfiability*, Journal of Algorithms 17, 1994, pp. 475–502.

Bias Invariance of Small Upper Spans
(Extended Abstract)*

Jack H. Lutz[1] and Martin J. Strauss[2]

[1] Department of Computer Science, Iowa State Univeristy, Ames, Iowa, 50011, USA,
lutz@cs.iastate.edu,
http://www.cs.iastate.edu/~lutz/

[2] AT&T Labs—Research, 180 Park Ave., Florham Park, NJ, 07932 USA
mstrauss@research.att.com,
http://www.research.att.com/~mstrauss/

Abstract. The resource-bounded measures of certain classes of languages are shown to be invariant under certain changes in the underlying probability measure. Specifically, for any real number $\delta > 0$, any polynomial-time computable sequence $\boldsymbol{\beta} = (\beta_0, \beta_1, \dots)$ of biases $\beta_i \in [\delta, 1-\delta]$, and any class $\mathcal{C}$ of languages that is closed *upwards or downwards* under positive, polynomial-time truth-table reductions with linear bounds on number and length of queries, it is shown that the following two conditions are equivalent.
(1) $\mathcal{C}$ has p-measure 0 relative to the probability measure given by $\boldsymbol{\beta}$.
(2) $\mathcal{C}$ has p-measure 0 relative to the uniform probability measure.
The analogous equivalences are established for measure in E and measure in E_2. (Breutzmann and Lutz [5] established this invariance for classes $\mathcal{C}$ that are closed downwards under slightly more powerful reductions, but nothing was known about invariance for classes that are closed upwards.) The proof introduces two new techniques, namely, the *contraction* of a martingale for one probability measure to a martingale for an induced probability measure, and a new, improved *positive bias reduction* of one bias sequence to another. Consequences for the BPP versus E problem and small span theorems are derived.

1 Introduction

Until recently, all research on the measure-theoretic structure of complexity classes has been restricted to the uniform probability measure. This is the probability measure μ that intuitively corresponds to a random experiment in which a language $A \subseteq \{0,1\}^*$ is chosen probabilistically, using an independent toss of a fair coin to decide whether each string is in A. When effectivized by the methods of resource-bounded measure [15], μ induces measure-theoretic structure on

* This research was supported in part by National Science Foundation Grants CCR-9157382 (with matching funds from Rockwell, Microware Systems Corporation, and Amoco Foundation) and CCR-9610461. This work was done while the second author was at Iowa State University.

H. Reichel and S. Tison (Eds.): STACS 2000, LNCS 1770, pp. 74–86, 2000.

$\mathrm{E} = \mathrm{DTIME}(2^{\mathrm{linear}})$, $\mathrm{E}_2 = \mathrm{DTIME}(2^{\mathrm{polynomial}})$, and other complexity classes. Investigations of this structure by a number of researchers have yielded many new insights over the past seven years. The recent surveys [3, 16, 6] describe much of this work.

There are several reasons for extending our investigation of resource-bounded measure to a wider variety of probability measures. First, such variety is essential in cryptography, computational learning, algorithmic information theory, average-case complexity, and other potential application areas. Second, applications of the probabilistic method [2] often require use of non-uniform probability measures, and this is likely to hold for the resource-bounded probabilistic method [18, 16] as well. Third, resource-bounded measure based on non-uniform probability measures provides new methods for proving results about resource-bounded measure based on the uniform probability measure [5].

Motivated by such considerations, Breutzmann and Lutz [5] initiated the study of resource-bounded measure based on an arbitrary (Borel) probability measure ν on the Cantor space $\mathbf{C}$ (the set of all languages). (Precise definitions of these and other terms appear in the expanded version of this paper.) Kautz [13] and Lutz [17] have furthered this study in different directions, and the present paper is another contribution.

The principal focus of the paper [5] is the circumstances under which the ν-measure of a complexity class $\mathcal{C}$ is invariant when the probability measure ν is replaced by some other probability measure ν'. For an *arbitrary* class $\mathcal{C}$ of languages, such invariance can only occur if ν and ν' are fairly close to one another: Extending results of Kakutani [12], Vovk [24], and Breutzmann and Lutz [5], Kautz [13] has shown that the "square-summable equivalence" of ν and ν' is sufficient to ensure $\nu_{\mathrm{p}}(\mathcal{C}) = 0 \Longleftrightarrow \nu'_{\mathrm{p}}(\mathcal{C}) = 0$, but very little more can be said when $\mathcal{C}$ is arbitrary.

Fortunately, complexity classes have more structure than arbitrary classes. Most complexity classes of interest, including P, NP, coNP, R, BPP, AM, P/Poly, PH, etc., are closed downwards under positive, polynomial-time truth-table reductions ($\leq^{\mathrm{P}}_{\mathrm{pos-tt}}$-reductions), and their intersections with E are closed downward under $\leq^{\mathrm{P}}_{\mathrm{pos-tt}}$-reductions with linear bounds on the length of queries ($\leq^{\mathrm{P,lin}}_{\mathrm{pos-tt}}$-reductions). Breutzmann and Lutz [5] proved that every class $\mathcal{C}$ with these closure properties enjoys a substantial amount of invariance in its measure. Specifically, if $\mathcal{C}$ is any such class and $\boldsymbol{\beta}$ and $\boldsymbol{\beta}'$ are strongly positive, P-sequences of biases, then the equivalences

$$\begin{aligned} \mu^{\boldsymbol{\beta}}_{\mathrm{p}}(\mathcal{C}) = 0 &\Longleftrightarrow \mu^{\boldsymbol{\beta}'}_{\mathrm{p}}(\mathcal{C}) = 0, \\ \mu^{\boldsymbol{\beta}}(\mathcal{C}|\mathrm{E}) = 0 &\Longleftrightarrow \mu^{\boldsymbol{\beta}'}(\mathcal{C}|\mathrm{E}) = 0, \\ \mu^{\boldsymbol{\beta}}(\mathcal{C}|\mathrm{E}_2) = 0 &\Longleftrightarrow \mu^{\boldsymbol{\beta}'}(\mathcal{C}|\mathrm{E}_2) = 0 \end{aligned} \tag{1}$$

hold, where $\mu^{\boldsymbol{\beta}}$ and $\mu^{\boldsymbol{\beta}'}$ are the probability measures corresponding to the bias sequences $\boldsymbol{\beta}$ and $\boldsymbol{\beta}'$, respectively. (Intuitively, if $\boldsymbol{\beta} = (\beta_0, \beta_1, \dots)$ is a sequence of biases $\beta_i \in [0, 1]$, then the measure $\mu^{\boldsymbol{\beta}}$ corresponds to a random experiment in which a language $A \subseteq \{0, 1\}^*$ is chosen by tossing for each string s_i, independently of all other strings, a special coin whose probability of heads is β_i. If the toss comes up heads, then $s_i \in A$; otherwise $s_i \notin A$.)

Our primary concern in the present paper is to extend this bias invariance to classes that are closed *upwards* under some type $\leq_r^{\mathrm{P}}$ of polynomial reductions. We have two reasons for interest in this question. First and foremost, many recent investigations in complexity theory focus on the resource-bounded measure of the *upper* $\leq_r^{\mathrm{P}}$*-span*

$$\mathrm{P}_r^{-1}(A) = \{B | A \leq_r^{\mathrm{P}} B\} \tag{2}$$

of a language A. Such investigations include work on small span theorems [9, 14, 4, 11, 7] and work on the BPP versus E question [1, 7, 8]. In general, the upper $\leq_r^{\mathrm{P}}$-span of a language is closed upwards, but not downwards, under $\leq_r^{\mathrm{P}}$-reductions.

Our second reason for interest in upward closure conditions is that the above-mentioned results of Breutzmann and Lutz [5] do *not* fully establish the invariance of measures of complexity classes under the indicated changes of bias sequences. For example, if $\boldsymbol{\beta}$ is an arbitrary strongly positive P-sequence of biases, the results of [5] show that

$$\mu^{\boldsymbol{\beta}}(\mathcal{C}|\mathrm{E}) = 0 \Longleftrightarrow \mu(\mathcal{C}|\mathrm{E}) = 0, \tag{3}$$

but they do *not* show that

$$\mu^{\boldsymbol{\beta}}(\mathcal{C}|\mathrm{E}) = 1 \Longleftrightarrow \mu(\mathcal{C}|\mathrm{E}) = 1 . \tag{4}$$

In general, the condition $\nu(\mathcal{C}|\mathrm{E}) = 1$ is equivalent to $\nu(\mathcal{C}^c|\mathrm{E}) = 0$, where $\mathcal{C}^c$ is the complement of $\mathcal{C}$. Since $\mathcal{C}$ is closed downwards under $\leq_r^{\mathrm{P}}$-reductions if and only if $\mathcal{C}^c$ is closed upwards under $\leq_r^{\mathrm{P}}$-reductions, we are again led to consider upward closure conditions.

Our main theorem, the Bias Invariance Theorem, states that, if $\mathcal{C}$ is any class of languages that is closed *upwards or downwards* under positive, polynomial-time, truth-table reductions with linear bounds on number and length of queries ($\leq_{\mathrm{pos-lin-tt}}^{\mathrm{P,lin}}$-reductions), and if $\boldsymbol{\beta}$ and $\boldsymbol{\beta}'$ are strongly positive P-sequences of biases, then the equivalences (1) above hold. The proof introduces two new techniques, namely, the *contraction* of a martingale for one probability measure to a martingale for an induced probability measure (dual to the martingale *dilation* technique introduced in [5]) and a new, improved *positive bias reduction* of one bias sequence to another.

We also note three easy consequences of our Bias Invariance Theorem. First, in combination with work of Allender and Strauss [1] and Buhrman, van Melkebeek, Regan, Sivakumar, and Strauss [8], it implies that, if there is *any* strongly positive P-sequence of biases $\boldsymbol{\beta}$ such that the complete $\leq_{\mathrm{T}}^{\mathrm{P}}$-degree for E_2 does not have $\mu^{\boldsymbol{\beta}}$-measure 1 in E_2, then $\mathrm{E} \not\subseteq \mathrm{BPP}$. Second, in combination with the work of Regan, Sivakumar, and Cai [19], it implies that, for any reasonable complexity class $\mathcal{C}$, if there exists a strongly positive P-sequence of biases $\boldsymbol{\beta}$ such that $\mathcal{C}$ has $\mu^{\boldsymbol{\beta}}$-measure 1 in E, then $\mathrm{E} \subseteq \mathcal{C}$ (and similarly for E_2). Third, if $\leq_r^{\mathrm{P}}$

is any polynomial reducibility such that $A \leq^{\mathrm{P,lin}}_{\mathrm{pos-lin-tt}} B$ implies $A \leq^{\mathrm{P}}_{r} B$, and if $\boldsymbol{\beta}$ is a strongly positive P-sequence of biases, then the small span theorem for $\leq^{\mathrm{P}}_{r}$-reductions holds with respect to $\mu^{\boldsymbol{\beta}}$ if and only if it holds with respect to μ. Tantalizingly, this hypothesis places $\leq^{\mathrm{P}}_{r}$ "just beyond" the small span theorem of Buhrman and van Melkebeek [7], which is the strongest small span theorem proven to date for exponential time.

Due to space limitations in these proceedings, all proofs are omitted from this version of our paper. An expanded version, available at http://www.cs.iastate.-edu/~lutz/papers.html, includes proofs of our results.

2 Preliminaries

We write $\{0,1\}^*$ for the set of all (finite, binary) *strings*, and we write $|x|$ for the length of a string x. The empty string, λ, is the unique string of length 0. The *standard enumeration* of $\{0,1\}^*$ is the sequence $s_0 = \lambda, s_1 = 0, s_2 = 1, s_3 = 00, \ldots$, ordered first by length and then lexicographically. For $x, y \in \{0,1\}^*$, we write $x < y$ if x precedes y in this standard enumeration. For $n \in \mathbb{N}$, $\{0,1\}^n$ denotes the set of all strings of length n, and $\{0,1\}^{\leq n}$ denotes the set of all strings of length at most n.

If x is a string or an (infinite, binary) *sequence*, and if $0 \leq i \leq j < |x|$, then $x[i..j]$ is the string consisting of the ith through j^{th} bits of x. In particular, $x[0..i-1]$ is the *i-bit prefix* of x. We write $x[i]$ for $x[i..i]$, the ith bit of x. (Note that the leftmost bit of x is $x[0]$, the 0th bit of x.)

If w is a string and x is a string or sequence, then we write $w \sqsubseteq x$ if w is a prefix of x, i.e., if there is a string or sequence y such that $x = wy$.

The *Boolean value* of a condition ϕ is $[\![\phi]\!] = \textbf{if } \phi \textbf{ then } 1 \textbf{ else } 0$.

We work in the *Cantor space* $\mathbf{C}$, consisting of all languages $A \subseteq \{0,1\}^*$. We identify each language A with its *characteristic sequence*, which is the infinite binary sequence χ_A defined by

$$\chi_A[n] = [\![s_n \in A]\!] \tag{5}$$

for each $n \in \mathbb{N}$. Relying on this identification, we also consider $\mathbf{C}$ to be the set of all infinite binary sequences. The *complement* of a set X of languages is $X^c = \mathbf{C} - X$.

For each string $w \in \{0,1\}^*$, the *cylinder generated by* w is the set

$$\mathbf{C}_w = \{A \in \mathbf{C} \mid w \sqsubseteq \chi_A\} \ . \tag{6}$$

2.1 Resource-Bounded ν-Measure

Next, we briefly present the basic elements of resource-bounded measure based on an arbitrary probability measure ν on $\mathbf{C}$. The remaining material in this section is excerpted, with permission, with permission, from [5].

Definition 1. *A probability measure on* $\mathbf{C}$ *is a function* $\nu : \{0,1\}^* \longrightarrow [0,1]$ *such that* $\nu(\lambda) = 1$*, and for all* $w \in \{0,1\}^*$,

$$\nu(w) = \nu(w0) + \nu(w1) \ . \tag{7}$$

Definition 2. *A probability measure* ν *on* $\mathbf{C}$ *is positive if, for all* $w \in \{0,1\}^*$, $\nu(w) > 0$.

Definition 3. *If* ν *is a positive probability measure and* $u, v \in \{0,1\}^*$*, then the conditional* ν*-measure of* u *given* v *is*

$$\nu(u|v) = \begin{cases} 1 & \text{if } u \sqsubseteq v \\ \frac{\nu(u)}{\nu(v)} & \text{if } v \sqsubseteq u \\ 0 & \text{otherwise} \ . \end{cases} \tag{8}$$

Note that $\nu(u|v)$ is the conditional probability that $A \in \mathbf{C}_u$, given that $A \in \mathbf{C}_v$, when $A \in \mathbf{C}$ is chosen according to the probability measure ν.

Definition 4. *A probability measure* ν *on* $\mathbf{C}$ *is strongly positive if (*ν *is positive and) there is a constant* $\delta > 0$ *such that, for all* $w \in \{0,1\}^*$ *and* $b \in \{0,1\}$, $\nu(wb|w) \geq \delta$.

Definition 5. *A sequence of biases* $\boldsymbol{\beta} = (\beta_0, \beta_1, \beta_2, \ldots)$ *is strongly positive if there is a constant* $\delta > 0$ *such that, for all* $i \in \mathbb{N}$, $\beta_i \in [\delta, 1-\delta]$.

Definition 6. *The* $\boldsymbol{\beta}$*-coin-toss probability measure (also called the* $\boldsymbol{\beta}$*-product probability measure) is the probability measure* $\mu^{\boldsymbol{\beta}}$ *defined by*

$$\mu^{\boldsymbol{\beta}}(w) = \prod_{i=0}^{|w|-1} ((1-\beta_i) \cdot (1 - w[i]) + \beta_i \cdot w[i]) \tag{9}$$

for all $w \in \{0,1\}^*$.

We next review the well-known notion of a martingale over a probability measure ν. Computable martingales were used by Schnorr [20, 21, 22, 23] in his investigations of randomness, and have more recently been used by Lutz [15] in the development of resource-bounded measure.

Definition 7. *Let* ν *be a probability measure on* $\mathbf{C}$*. Then a* ν*-martingale is a function* $d : \{0,1\}^* \longrightarrow [0, \infty)$ *such that, for all* $w \in \{0,1\}^*$,

$$d(w)\nu(w) = d(w0)\nu(w0) + d(w1)\nu(w1).$$

To satisfy space constraints, we omit discussion of the success and computability of a ν-martingale, which are similar to the correponding notions for μ-martingales.

Definition 8. *Let ν be a positive probability measure on* $\mathbf{C}$, *let* $A \subseteq \{0,1\}^*$, *and let* $i \in \mathbb{N}$. *Then the* i^{th} *conditional ν-probability along A is*

$$\nu_A(i+1|i) = \nu(\chi_A[0..i] \mid \chi_A[0..i-1]) \ . \tag{10}$$

Definition 9. *Two positive probability measures ν and ν' on* $\mathbf{C}$ *are summably equivalent, and we write* $\nu \approx \nu'$, *if for every* $A \subseteq \{0,1\}^*$,

$$\sum_{i=0}^{\infty} |\nu_A(i+1|i) - \nu'_A(i+1|i)| < \infty \ . \tag{11}$$

Definition 10. *1. A P-sequence of biases is a sequence* $\boldsymbol{\beta} = (\beta_0, \beta_1, \beta_2, \dots)$ *of biases* $\beta_i \in [0,1]$ *for which there is a function*

$$\hat{\beta} : \mathbb{N} \times \mathbb{N} \longrightarrow \mathbb{Q} \cap [0,1] \tag{12}$$

with the following two properties.
(i) For all $i, r \in \mathbb{N}$, $|\hat{\beta}(i,r) - \beta_i| \leq 2^{-r}$.
(ii) There is an algorithm that, for all $i, r \in \mathbb{N}$, *computes* $\hat{\beta}(i,r)$ *in time polynomial in* $|s_i| + r$ *(i.e., in time polynomial in* $\log(i+1) + r$*).*

2. A P-exact sequence of biases is a sequence $\boldsymbol{\beta} = (\beta_0, \beta_1, \beta_2, \dots)$ *of (rational) biases* $\beta_i \in \mathbb{Q} \cap [0,1]$ *such that the function* $i \longmapsto \beta_i$ *is computable in time polynomial in* $|s_i|$.

Definition 11. *If* $\boldsymbol{\alpha}$ *and* $\boldsymbol{\beta}$ *are sequences of biases, then* $\boldsymbol{\alpha}$ *and* $\boldsymbol{\beta}$ *are summably equivalent, and we write* $\boldsymbol{\alpha} \approx \boldsymbol{\beta}$, *if* $\sum_{i=0}^{\infty} |\alpha_i - \beta_i| < \infty$.

It is clear that $\boldsymbol{\alpha} \approx \boldsymbol{\beta}$ if and only if $\mu^{\boldsymbol{\alpha}} \approx \mu^{\boldsymbol{\beta}}$.

Lemma 1 (Breutzmann and Lutz [5]). *For every P-sequence of biases* $\boldsymbol{\beta}$, *there is a P-exact sequence of biases* $\boldsymbol{\beta}'$ *such that* $\boldsymbol{\beta} \approx \boldsymbol{\beta}'$.

2.2 Truth-Table Reductions

A *truth-table reduction* (briefly, a $\leq_{\text{tt}}$-reduction) is an ordered pair (f, g) of total recursive functions such that for each $x \in \{0,1\}^*$, there exists $n(x) \in \mathbb{Z}^+$ such that the following two conditions hold.

(i) $f(x)$ is (the standard encoding of) an $n(x)$-tuple $(f_1(x), \dots, f_{n(x)}(x))$ of strings $f_i(x) \in \{0,1\}^*$, which are called the *queries* of the reduction (f, g) on input x. We use the notation $Q_{(f,g)}(x) = \{f_1(x), \dots, f_{n(x)}(x)\}$ for the set of such queries.

(ii) $g(x)$ is (the standard encoding of) an $n(x)$-input, 1-output Boolean circuit, called the *truth table* of the reduction (f, g) on input x. We identify $g(x)$ with the Boolean function computed by this circuit, i.e.,

$$g(x) : \{0,1\}^{n(x)} \longrightarrow \{0,1\} \ . \tag{13}$$

A truth-table reduction (f, g) *induces* the function

$$F_{(f,g)} : \mathbf{C} \longrightarrow \mathbf{C} \tag{14}$$

$$F_{(f,g)}(A) = \left\{x \in \{0,1\}^* \mid g(x)\left([\![f_1(x) \in A]\!] \cdots [\![f_{n(x)}(x) \in A]\!]\right) = 1\right\} .$$

Similarly, the *inverse image* of the cylinder $\mathbf{C}_z$ under the reduction (f, g) is

$$F^{-1}_{(f,g)}(\mathbf{C}_z) = \left\{A \in \mathbf{C} \mid z \sqsubseteq F_{(f,g)}(A)\right\} . \tag{15}$$

The following well-known fact is easily verified.

Lemma 2. *If ν is a probability measure on $\mathbf{C}$ and (f, g) is a $\leq_{\mathrm{tt}}$-reduction, then the function*

$$\nu^{(f,g)} : \{0,1\}^* \longrightarrow [0,1] \tag{16}$$

$$\nu^{(f,g)}(z) = \nu(F^{-1}_{(f,g)}(\mathbf{C}_z))$$

is also a probability measure on $\mathbf{C}$.

The probability measure $\nu^{(f,g)}$ of Lemma 2 is called the *probability measure induced by ν and (f, g).*

In this paper, we use the following special type of $\leq_{\mathrm{tt}}$-reduction.

Definition 12. *A $\leq_{\mathrm{tt}}$-reduction (f, g) is orderly if, for all $x, y, u, v \in \{0,1\}^*$, if $x < y$, $u \in Q_{(f,g)}(x)$, and $v \in Q_{(f,g)}(y)$, then $u < v$. That is, if x precedes y (in the standard ordering of $\{0,1\}^*$), then every query of (f, g) on input x precedes every query of (f, g) on input y.*

3 Martingale Contraction

Given a positive coin-toss probability measure ν, an orderly truth-table reduction (f, g), and a $\nu^{(f,g)}$-martingale d (where $\nu^{(f,g)}$ is the probability measure induced by ν and (f, g)), Breutzmann and Lutz [5] showed how to construct a ν-martingale $(f, g)\widehat{\ }d$, called the (f, g)-*dilation* of d, such that $(f, g)\widehat{\ }d$ succeeds on A whenever d succeeds on $F_{(f,g)}(A)$. In this section we present a dual of this construction. Given ν and (f, g) as above and a ν-martingale d, we show how to construct a $\nu^{(f,g)}$-supermartingale $(f, g)\smile d$, called the (f, g)-*contraction* of d, such that $(f, g)\smile d$ succeeds on A whenever d succeeds strongly on every element of $F^{-1}_{(f,g)}(\{A\})$.

The notion of an (f, g)-step, introduced in [5], will also be useful here.

Definition 13. *Let (f, g) be an orderly $\leq_{\mathrm{tt}}$-reduction.*

1. *An (f, g)-step is a positive integer l such that $F_{(f,g)}(0^{l-1}) \neq F_{(f,g)}(0^l)$.*
2. *For $k \in \mathbb{N}$, we let step(k) be the least (f, g)-step l such that $l \geq k$.*

3. *For $v, w \in \{0,1\}^*$, we write $v \succ w$ to indicate that $w \sqsubseteq v$ and $|v| = step(|w|+1)$. (That is, $v \succ w$ means that v is a proper extension of w to the next step.)*

Our construction makes use of a special-purpose inverse of $F_{(f,g)}$ that depends on both (f, g) and d.

Definition 14. *Let (f, g) be an orderly $\leq_{\text{tt}}$-reduction, let ν be a positive probability measure on $\mathbf{C}$, and let d be a ν-martingale. Then the partial function*

$$F^{-1}_{(f,g),d} : \{0,1\}^* \longrightarrow \{0,1\}^* \tag{17}$$

is defined recursively as follows.

(i) $F^{-1}_{(f,g),d}(\lambda) = \lambda$.
(ii) For $w \in \{0,1\}^*$ and $b \in \{0,1\}$, $F^{-1}_{(f,g),d}(wb)$ is the lexicographically first string $v \succ F^{-1}_{(f,g),d}(w)$ such that $F_{(f,g)}(v) = wb$ and, for all $v' \succ F^{-1}_{(f,g),d}(w)$ such that $F_{(f,g)}(v') = wb$, we have $d(v) \leq d(v')$. (That is, v minimizes $d(v)$ on the set of all $v \succ F^{-1}_{(f,g),d}(w)$ satisfying $F_{(f,g)}(v) = wb$.)

Note that the function $F^{-1}_{(f,g),d}$ is strictly monotone (i.e., $w \sqsubsetneq w'$ implies that $F^{-1}_{(f,g),d}(w) \sqsubsetneq F^{-1}_{(f,g),d}(w')$, provided that these values exist), whence it extends naturally to a partial function

$$F^{-1}_{(f,g),d} : \mathbf{C} \longrightarrow \mathbf{C} \,. \tag{18}$$

It is easily verified that $F^{-1}_{(f,g),d}$ inverts $F_{(f,g)}$ in the sense that, for all $x \in \{0,1\}^* \cup \mathbf{C}$, $F^{-1}_{(f,g),d}$ finds a preimage of $F_{(f,g)}(x)$, i.e.,

$$F_{(f,g)}(F^{-1}_{(f,g),d}(F_{(f,g)}(x))) = F_{(f,g)}(x) \,. \tag{19}$$

We now define the (f, g)-contraction of a ν-martingale d.

Definition 15. *Let (f, g) be an orderly $\leq_{\text{tt}}$-reduction, let ν be a positive probability measure on $\mathbf{C}$, and let d be a ν-martingale. Then the* (f, g)-contraction *of d is the function*

$$(f,g)_{\smile}d : \{0,1\}^* \longrightarrow \{0,1\}^* \tag{20}$$

defined as follows.

(i) $(f,g)_{\smile}d(\lambda) = d(\lambda)$.
(ii) For $w \in \{0,1\}^*$ and $b \in \{0,1\}$,

$$(f,g)_{\smile}d(wb) = \begin{cases} d(F^{-1}_{(f,g),d}(wb)) & \text{if } d(F^{-1}_{(f,g),d}(wb)) \text{ is defined} \\ 2 \cdot (f,g)_{\smile}d(w) & \text{otherwise.} \end{cases} \tag{21}$$

Theorem 1 (Martingale Contraction Theorem). *Assume that ν is a positive probability measure on $\mathbf{C}$, (f, g) is an orderly $\leq_{\text{tt}}$-reduction, and d is a ν-martingale. Then $(f,g)_{\smile}d$ is a $\nu^{(f,g)}$-supermartingale. Moreover, for every language $A \subseteq \{0,1\}^*$, if $F^{-1}_{(f,g)}(\{A\}) \subseteq S^{\infty}_{\text{str}}[d]$, then $A \in S^{\infty}[(f,g)_{\smile}d]$.*

4 Bias Invariance

In this section we present our main results.

Definition 16. *Let (f,g) be a $\leq_{\mathrm{tt}}$-reduction.*

1. *(f,g) is positive (briefly, a $\leq_{\mathrm{pos-tt}}$-reduction) if, for all $A, B \subseteq \{0,1\}^*$, $A \subseteq B$ implies $F_{(f,g)}(A) \subseteq F_{(f,g)}(B)$.*
2. *(f,g) is polynomial-time computable (briefly, a $\leq_{\mathrm{tt}}^{\mathrm{P}}$-reduction) if the functions f and g are computable in polynomial time.*
3. *(f,g) is polynomial-time computable with linear-bounded queries (briefly, a $\leq_{\mathrm{tt}}^{\mathrm{P,lin}}$-reduction) if (f,g) is a $\leq_{\mathrm{tt}}^{\mathrm{P}}$-reduction and there is a constant $c \in \mathbb{N}$ such that, for all $x \in \{0,1\}^*$, $Q_{(f,g)}(x) \subseteq \{0,1\}^{\leq c(1+|x|)}$.*
4. *(f,g) is polynomial-time computable with a linear number of queries (briefly, a $\leq_{\mathrm{lin-tt}}^{\mathrm{P}}$-reduction) if (f,g) is a $\leq_{\mathrm{tt}}^{\mathrm{P}}$-reduction and there is a constant $c \in \mathbb{N}$ such that, for all $x \in \{0,1\}^*$, $|Q_{(f,g)}(x)| \leq c(1+|x|)$.*

Of course, a $\leq_{\mathrm{pos-tt}}^{\mathrm{P,lin}}$-reduction is a $\leq_{\mathrm{tt}}$-reduction with properties 1–3, and a $\leq_{\mathrm{pos-lin-tt}}^{\mathrm{P,lin}}$-reduction is a $\leq_{\mathrm{tt}}$-reduction with properties 1–4.

We now present the Positive Bias Reduction Theorem. This strengthens the identically-named result of Breutzmann and Lutz [5] by giving a $\leq_{\mathrm{pos-lin-tt}}^{\mathrm{P,lin}}$-reduction in place of a $\leq_{\mathrm{pos-tt}}^{\mathrm{P,lin}}$-reduction. This technical improvement, which is essential for our purposes here, requires a substantially different construction. Details are omitted.

Theorem 2 (Positive Bias Reduction Theorem). *Let $\boldsymbol{\beta}$ and $\boldsymbol{\beta}'$ be strongly positive, P-exact sequences of biases. Then there exists an orderly $\leq_{\mathrm{pos-lin-tt}}^{\mathrm{P,lin}}$-reduction (f,g), and the probability measure induced by $\mu^{\boldsymbol{\beta}}$ and (f,g) is a coin-toss probability measure $\mu^{\boldsymbol{\beta}''}$, where $\boldsymbol{\beta}'' \approx \boldsymbol{\beta}'$.*

The following result is our main theorem.

Theorem 3 (Bias Invariance Theorem). *Assume that $\boldsymbol{\beta}$ and $\boldsymbol{\beta}'$ are strongly positive P-sequences of biases, and let $\mathcal{C}$ be a class of languages that is closed upwards or downwards under $\leq_{\mathrm{pos-lin-tt}}^{\mathrm{P,lin}}$-reductions. Then*

$$\mu_{\mathrm{p}}^{\boldsymbol{\beta}}(\mathcal{C}) = 0 \Longleftrightarrow \mu_{\mathrm{p}}^{\boldsymbol{\beta}'}(\mathcal{C}) = 0 \ . \tag{22}$$

The "downwards" part of Theorem 3 is a technical improvement of the Bias Equivalence Theorem of [5] from $\leq_{\mathrm{pos-tt}}^{\mathrm{P,lin}}$-reductions to $\leq_{\mathrm{pos-lin-tt}}^{\mathrm{P,lin}}$-reductions. The proof of this improvement is simply the proof in [5] with Theorem 2 used in place of its predecessor in [5].

The "upwards" part of Theorem 3 is entirely new. The proof of this result is similar to the proof of the Bias Equivalence Theorem in [5], but now in addition to using our improved Positive Bias Reduction Theorem, we use the Martingale Contraction Theorem of section 3 in place of the Martingale Dilation Theorem of [5]. We also note that the linear bound on number of queries in Theorem 2 is essential for the "upwards" direction.

If $\leq_r^{\mathrm{P}}$ is a polynomial reducibility, then a class $\mathcal{C}$ is closed upwards under $\leq_r^{\mathrm{P}}$-reductions if and only if $\mathcal{C}^c$ is closed downwards under $\leq_r^{\mathrm{P}}$-reductions. We thus have the following immediate consequence of Theorem 3.

Corollary 1. *Assume that $\boldsymbol{\beta}$ and $\boldsymbol{\beta}'$ are strongly positive P-sequences of biases, and let $\mathcal{C}$ be a class of languages that is closed upwards or downwards under $\leq_{\mathrm{pos-lin-tt}}^{\mathrm{P,lin}}$-reductions. Then*

$$\mu_{\mathrm{p}}^{\boldsymbol{\beta}}(\mathcal{C}) = 1 \iff \mu_{\mathrm{p}}^{\boldsymbol{\beta}'}(\mathcal{C}) = 1 \ . \tag{23}$$

We now mention some consequences of Theorem 3, beginning with a discussion of the measure of the complete $\leq_{\mathrm{T}}^{\mathrm{P}}$-degree for exponential time, and its consequences for the BPP versus E problem.

For each class $\mathcal{D}$ of languages, we use the notations

$$\mathcal{H}_{\mathrm{T}}(\mathcal{D}) = \{A | A \text{ is } \leq_{\mathrm{T}}^{\mathrm{P}}\text{-hard for } \mathcal{D}\}, \tag{24}$$

$$\mathcal{C}_{\mathrm{T}}(\mathcal{D}) = \{A | A \text{ is } \leq_{\mathrm{T}}^{\mathrm{P}}\text{-complete for } \mathcal{D}\}, \tag{25}$$

and similarly for other reducibilities. The following easy observation shows that every consequence of $\mu(\mathcal{C}_{\mathrm{T}}(\mathrm{E}_2)|\mathrm{E}_2) \neq 1$ is also a consequence of $\mu(\mathcal{C}_{\mathrm{T}}(\mathrm{E})|\mathrm{E}) \neq 1$.

Lemma 3. $\mu(\mathcal{C}_{\mathrm{T}}(\mathrm{E})|\mathrm{E}) \neq 1 \Longrightarrow \mu(\mathcal{C}_{\mathrm{T}}(\mathrm{E}_2)|\mathrm{E}_2) \neq 1$.

Proof. Juedes and Lutz [10] have shown that, if X is a set of languages that is closed downwards under $\leq_{\mathrm{m}}^{\mathrm{P}}$-reductions, then $\mu(X|\mathrm{E}_2) = 0 \Longrightarrow \mu(X|\mathrm{E}) = 0$. Applying this result with $X = \mathcal{H}_{\mathrm{T}}(\mathrm{E})^c = \mathcal{H}_{\mathrm{T}}(\mathrm{E}_2)^c$ yields the lemma.

Allender and Strauss [1] have proven that $\mu_{\mathrm{p}}(\mathcal{H}_{\mathrm{T}}(\mathrm{BPP})) = 1$. Buhrman, van Melkebeek, Regan, Sivakumar, and Strauss [8] have noted that this implies that $\mu(\mathcal{C}_{\mathrm{T}}(\mathrm{E}_2)|\mathrm{E}_2) \neq 1 \Longrightarrow \mathrm{E} \not\subseteq \mathrm{BPP}$. Combining this argument with Corollary 1 yields the following extension.

Corollary 2. *If there exists a strongly positive P-sequence of biases $\boldsymbol{\beta}$ such that $\mu^{\boldsymbol{\beta}}(\mathcal{C}_{\mathrm{T}}(\mathrm{E}_2)|\mathrm{E}_2) \neq 1$, then $\mathrm{E} \not\subseteq \mathrm{BPP}$.*

Regan, Sivakumar, and Cai [19] have proven a "most is all" lemma, stating that if $\mathcal{C}$ is any class of languages that is either closed under finite unions and intersections or closed under symmetric difference, then $\mu(\mathcal{C}|\mathrm{E}) = 1 \Longrightarrow \mathrm{E} \subseteq \mathcal{C}$. Combining this with Corollary 1 gives the following extended "most is all" result.

Corollary 3. *Let $\mathcal{C}$ be a class of languages that is closed upwards or downwards under $\leq_{\mathrm{pos-lin-tt}}^{\mathrm{P,lin}}$-reductions, and is also closed under either finite unions and intersections or symmetric difference. If there is any strongly positive, P-sequence of biases $\boldsymbol{\beta}$ such that $\mu^{\boldsymbol{\beta}}(\mathcal{C}|\mathrm{E}) = 1$, then $\mathrm{E} \subseteq \mathcal{C}$.*

Of course, the analagous result holds for E_2.

We conclude with a brief discussion of small span theorems. Given a polynomial reducibility $\leq_r^{\mathrm{P}}$, the *lower* $\leq_r^{\mathrm{P}}$*-span* of a language A is

$$\mathrm{P}_r(A) = \{B | B \leq_r^{\mathrm{P}} A\} , \tag{26}$$

and the *upper* $\leq_r^{\mathrm{P}}$*-span* of A is

$$\mathrm{P}_r^{-1}(A) = \{B | A \leq_r^{\mathrm{P}} B\} . \tag{27}$$

We will use the following compact notation.

Definition 17. *Let* $\leq_r^{\mathrm{P}}$ *be a polynomial reducibility type, and let* ν *be a probability measure on* $\mathbf{C}$. *Then the small span theorem for* $\leq_r^{\mathrm{P}}$*-reductions in the class* E *over the probability measure* ν *is the assertion*

$$\mathrm{SST}_\nu(\leq_r^{\mathrm{P}}, \mathrm{E}) \tag{28}$$

stating that, for every $A \in \mathrm{E}$, $\nu(\mathrm{P}_r(A)|\mathrm{E}) = 0$ *or* $\nu_{\mathrm{p}}(\mathrm{P}_r^{-1}(A)) = \nu(\mathrm{P}_r^{-1}(A)|\mathrm{E}) = 0$. *When the probability measure is* μ, *we omit it from the notation, writing* $\mathrm{SST}(\leq_r^{\mathrm{P}}, \mathrm{E})$ *for* $\mathrm{SST}_\mu(\leq_r^{\mathrm{P}}, \mathrm{E})$. *Similar assertions for other classes, for example,* $\mathrm{SST}_\nu(\leq_r^{\mathrm{P}}, \mathrm{E}_2)$, *are defined in the now-obvious manner.*

Juedes and Lutz [9] proved the first small span theorems, $\mathrm{SST}(\leq_{\mathrm{m}}^{\mathrm{P}}, \mathrm{E})$ and $\mathrm{SST}(\leq_{\mathrm{m}}^{\mathrm{P}}, \mathrm{E}_2)$, and noted that extending either to $\leq_{\mathrm{T}}^{\mathrm{P}}$ would establish $\mathrm{E} \not\subseteq \mathrm{BPP}$. Lindner [14] established $\mathrm{SST}(\leq_{1-\mathrm{tt}}^{\mathrm{P}}, \mathrm{E})$ and $\mathrm{SST}(\leq_{1-\mathrm{tt}}^{\mathrm{P}}, \mathrm{E}_2)$, and Ambos-Spies, Neis, and Terwijn [4] proved $\mathrm{SST}(\leq_{k-\mathrm{tt}}^{\mathrm{P}}, \mathrm{E})$ and $\mathrm{SST}(\leq_{k-\mathrm{tt}}^{\mathrm{P}}, \mathrm{E}_2)$ for all fixed $k \in \mathbb{N}$. Very recently, Buhrman and van Melkebeek [7] have taken a major step forward by proving $\mathrm{SST}(\leq_{g(n)-\mathrm{tt}}^{\mathrm{P}}, \mathrm{E}_2)$ for every function $g(n)$ satisfying $g(n) = n^{o(1)}$. We note that the Bias Invariance Theorem implies that small span theorems lying "just beyond" this latter result are somewhat robust with respect to changes of biases.

Theorem 4. *If* $\leq_r^{\mathrm{P}}$ *is a polynomial reducibility such that* $A \leq_{\mathrm{pos-lin-tt}}^{\mathrm{P,lin}} B$ *implies* $A \leq_r^{\mathrm{P}} B$, *then for every strongly positive P-sequence of biases* $\boldsymbol{\beta}$,

$$\mathrm{SST}_{\mu^{\boldsymbol{\beta}}}(\leq_r^{\mathrm{P}}, \mathrm{E}) \Longleftrightarrow \mathrm{SST}(\leq_r^{\mathrm{P}}, \mathrm{E}) , \tag{29}$$

and similarly for E_2.

Acknowledgment. The first author thanks Steve Kautz for a very useful discussion.

References

[1] E. Allender and M. Strauss. Measure on small complexity classes with applications for BPP. In *Proceedings of the 35th Symposium on Foundations of Computer Science*, pages 807–818, Piscataway, NJ, 1994. IEEE Computer Society Press.
[2] N. Alon and J. H. Spencer. *The Probabilistic Method*. Wiley, 1992.

[3] K. Ambos-Spies and E. Mayordomo. Resource-bounded measure and randomness. In A. Sorbi, editor, *Complexity, Logic and Recursion Theory*, Lecture Notes in Pure and Applied Mathematics, pages 1–47. Marcel Dekker, New York, N.Y., 1997.
[4] K. Ambos-Spies, H.-C. Neis, and S. A. Terwijn. Genericity and measure for exponential time. *Theoretical Computer Science*, 168:3–19, 1996.
[5] J. M. Breutzmann and J. H. Lutz. Equivalence of measures of complexity classes. *SIAM Journal on Computing*, 29:302–326, 2000.
[6] H. Buhrman and L. Torenvliet. Complete sets and structure in subrecursive classes. In *Proceedings of Logic Colloquium '96*, pages 45–78. Springer-Verlag, 1998.
[7] H. Buhrman and D. van Melkebeek. Hard sets are hard to find. In *Proceedings of the 13th IEEE Conference on Computational Complexity*, pages 170–181, New York, 1998. IEEE.
[8] H. Buhrman, D. van Melkebeek, K. Regan, D. Sivakumar, and M. Strauss. A generalization of resource-bounded measure, with an application. In *Proceedings of the 15th Annual Symposium on Theoretical Aspects of Computer Science*, pages 161–171, Berlin, 1998. Springer-Verlag.
[9] D. W. Juedes and J. H. Lutz. The complexity and distribution of hard problems. *SIAM Journal on Computing*, 24(2):279–295, 1995.
[10] D. W. Juedes and J. H. Lutz. Weak completeness in E and E_2. *Theoretical Computer Science*, 143:149–158, 1995.
[11] D. W. Juedes and J. H. Lutz. Completeness and weak completeness under polynomial-size circuits. *Information and Computation*, 125:13–31, 1996.
[12] S. Kakutani. On the equivalence of infinite product measures. *Annals of Mathematics*, 49:214–224, 1948.
[13] S. M. Kautz. Resource-bounded randomness and compressibility with repsect to nonuniform measures. In *Proceedings of the International Workshop on Randomization and Approximation Techniques in Computer Science*, pages 197–211. Springer-Verlag, 1997.
[14] W. Lindner. On the polynomial time bounded measure of one-truth-table degrees and p-selectivity, 1993. Diplomarbeit, Technische Universität Berlin.
[15] J. H. Lutz. Almost everywhere high nonuniform complexity. *Journal of Computer and System Sciences*, 44:220–258, 1992.
[16] J. H. Lutz. The quantitative structure of exponential time. In L.A. Hemaspaandra and A.L. Selman, editors, *Complexity Theory Retrospective II*, pages 225–254. Springer-Verlag, 1997.
[17] J. H. Lutz. Resource-bounded measure. In *Proceedings of the 13th IEEE Conference on Computational Complexity*, pages 236–248, New York, 1998. IEEE.
[18] J. H. Lutz and E. Mayordomo. Measure, stochasticity, and the density of hard languages. *SIAM Journal on Computing*, 23:762–779, 1994.
[19] K. W. Regan, D. Sivakumar, and J. Cai. Pseudorandom generators, measure theory, and natural proofs. In *36th IEEE Symposium on Foundations of Computer Science*, pages 26–35. IEEE Computer Society Press, 1995.
[20] C. P. Schnorr. Klassifikation der Zufallsgesetze nach Komplexität und Ordnung. *Z. Wahrscheinlichkeitstheorie verw. Geb.*, 16:1–21, 1970.
[21] C. P. Schnorr. A unified approach to the definition of random sequences. *Mathematical Systems Theory*, 5:246–258, 1971.
[22] C. P. Schnorr. Zufälligkeit und Wahrscheinlichkeit. *Lecture Notes in Mathematics*, 218, 1971.

[23] C. P. Schnorr. Process complexity and effective random tests. *Journal of Computer and System Sciences*, 7:376–388, 1973.
[24] V. G. Vovk. On a randomness criterion. *Soviet Mathematics Doklady*, 35:656–660, 1987.

The Complexity of Planarity Testing

Eric Allender*[1] and Meena Mahajan**[2]

[1] Dept. of Computer Science, Rutgers University, Piscataway, NJ, USA.
allender@cs.rutgers.edu
[2] The Institute of Mathematical Sciences, Chennai 600 113, INDIA.
meena@imsc.ernet.in

Abstract. We clarify the computational complexity of planarity testing, by showing that planarity testing is hard for L, and lies in SL. This nearly settles the question, since it is widely conjectured that L = SL [25]. The upper bound of SL matches the lower bound of L in the context of (nonuniform) circuit complexity, since L/poly is equal to SL/poly.
Similarly, we show that a planar embedding, when one exists, can be found in $\mathsf{FL}^{\mathsf{SL}}$.
Previously, these problems were known to reside in the complexity class AC^1, via a $O(\log n)$ time CRCW PRAM algorithm [22], although planarity checking for degree-three graphs had been shown to be in SL [23, 20].

1 Introduction

The problem of determining if a graph is planar has been studied from several perspectives of algorithmic research. From most perspectives, optimal algorithms are already known. Linear-time sequential algorithms were presented by Hopcroft and Tarjan [10] and (via another approach) by combining the results of [16, 4, 8]. In the context of parallel computation, a logarithmic-time CRCW-PRAM algorithm was presented by Ramachandran and Reif [22] that performs almost linear work.

From the perspective of computational complexity theory, however, the situation has been far from clear. The best upper bound on the complexity of planarity that has been published so far is the bound of AC^1 that follows from the logarithmic-time CRCW-PRAM algorithm of Ramachandran and Reif [22]. In a recent survey of problems in the complexity class SL [2], the planarity problem for graphs of *bounded degree* is listed as belonging to SL, but this is based on the claim in [23] that checking planarity for bounded degree graphs is in the "Symmetric Complementation Hierarchy", and on the fact that SL is closed under complement [20] (and thus this hierarchy collapses to SL). However, the algorithm presented in [23] actually works only for graphs of degree 3, and no straightforward generalization to graphs of larger degree is known. (This is

* Supported in part by NSF grant CCR-9734918.
** Part of this work was done when this author was supported by the NSF grant CCR-9734918 on a visit to Rutgers University during summer 1999.

H. Reichel and S. Tison (Eds.): STACS 2000, LNCS 1770, pp. 87–98, 2000.

implicitly acknowledged in [22, pp. 518–519].) Interestingly, Mario Szegedy has pointed out to us (personal communication) that an algebraic structure proposed by Tutte [28], when combined with more recent results about span programs and counting classes [14], gives a $\oplus\mathsf{L}$ algorithm for planarity testing. It is listed as an open question by Ja'Ja' and Simon [13] if planarity is in NL, although the subsequent discovery that NL is closed under complementation [11, 27] allows one to verify that one of the algorithms of [12, 13] can in fact be implemented in NL. It remains an open question if their algorithm can be implemented in SL, but in this paper we observe that the algorithm of Ramachandran and Reif can be implemented in SL.

We also show that the planarity problem is hard for L under projection reducibility.

Recall that

$$\mathsf{L} \subseteq \mathsf{SL} \subseteq \mathsf{NL} \subseteq \mathsf{AC}^1$$

$$\mathsf{SL} \subseteq \oplus\mathsf{L}.$$

(See [14].) L (respectively SL, NL) denotes deterministic (respectively symmetric, nondeterministic) logarithmic space, AC^1 denotes problems solvable by polynomial size AND-OR circuits of logarithmic depth, where the gates are allowed to have any number of inputs. The class $\oplus\mathsf{L}$ consists of problems solvable by nondeterministic logspace machines with an odd number of accepting paths. Although it is not known if NL is contained in $\oplus\mathsf{L}$, it is known that NL is contained in $\oplus\mathsf{L}$/poly [9].

This essentially solves the question of planarity from the complexity-theoretic point of view. To see this, it is sufficient to recall that it is widely conjectured that SL = L. This conjecture is based on the following considerations:

- The standard complete problem for SL is the graph accessibility problem for undirected graphs (*UGAP*). Upper bounds on the space complexity of *UGAP* have been dropping, from $\log^2 n$ [26], through $\log^{1.5} n$ [19], to $\log^{4/3} n$ [3]. It is suspected that this trend will continue to eventually reach $\log n$.
- *UGAP* can be solved in randomized logspace [1]. Recent developments in derandomization techniques have led many researchers to conjecture that randomized logspace is equal to L [25].

In the context of nonuniform complexity theory (for example, as explored in [15, 5]), the corresponding nonuniform complexity classes L/poly and SL/poly are equal.[1] Hence in this setting, the computational complexity of planarity is resolved; it is complete for L/poly under projections.

One consequence of our result is that counting the number of perfect matchings in a planar graph is reducible to the determinant, when the graph is presented as an adjacency matrix. More precisely, it follows from this paper and from [17] that there is a (nonuniform) projection that takes as input the adjacency matrix of a graph G, and produces as output a matrix M with the

[1] That is, a *universal traversal sequence* [1] can be used as an "advice string" to enable a logspace-bounded machine to solve *UGAP*.

property that if G is planar then the absolute value of $det(M)$ is the number of perfect matchings in G. (*Sketch*: Given the proper advice strings, a GapL algorithm can take as input the matrix M, compute its planar embedding (since this is in L/poly), then compute its "normal form embedding" along a unique computation path (since $\mathsf{NL} \subseteq \mathsf{UL/poly}$ [24]), and then use the algorithm in [17] to compute a number whose absolute value is the number of perfect matchings in M. Since the determinant is complete for GapL under projections, the result follows.)

The paper is organized as follows. In Section 2 we present our hardness result for planarity. In Section 3 we sketch the main parts of our analysis, showing that the algorithm of [22] can be implemented in SL.

2 Hardness of Planarity Testing

The following problem is known to be complete for L:

Definition 1. *Undirected Forest Accessibility (UFA): Given an undirected forest G and vertices u, v, decide if u and v are in the same tree.*

The hardness of this problem for L was shown in [6], where only an NC^1 reduction is claimed. However, it is easy to see that this problem is actually hard under uniform projections as well. (To see this, consider any L machine M; assume that configurations are time-stamped, that there is a unique accepting configuration c_h whose successor is itself (with the next time-stamp), and that M decides within n^k time for some constant k. Construct the polynomially sized computation graph G where time-stamped configurations are the vertices, and edges denoting machine moves are labeled by input bits or their negations. M accepts its input if and only if $(G, (c_0, 0), (c_h, n^k))$ is an instance of UFA, where c_0 is the initial configuration.)

Let G' be the complete graph on 5 vertices, minus any one edge (p, q).

The graph H is obtained by identifying vertices u and v of G (from the UFA instance) with vertices p and q of G'. Clearly, H is planar if and only if (G, u, v) is not in UFA.

We have thus proved the following theorem:

Theorem 1. *Planarity testing is hard for* L *under projections.*

It is worth noting that planarity testing remains hard for L even for graphs of degree 3. This does not follow from the construction given above, but can be shown by modifying a construction in [7]. Details will appear in the final paper.

3 The SL Algorithm for Planarity Testing and Embedding

We describe here how the algorithm of Ramachandran and Reif [22] can be implemented in SL. The algorithm of Ramachandran and Reif is complex, and it involves a number of fairly involved technical definitions. Due to space limitations, it is not possible to present all of the necessary technical definitions here. Therefore, we have written this section so that it can be read as a companion to [22]. We use the same notation as is used in [22], and we will show that each step of their algorithm can be computed in L^{SL}, or by NC^1 circuits with oracle gates for the reachability problem in undirected graphs. Since SL is closed under NC^1 reducibility and logspace-Turing reducibility [20], it follows that the entire algorithm can be implemented in SL.

Our approach will be as follows. First, we present some general-purpose algorithms for operating on graphs and trees. Next, we show how an open ear decomposition can be computed in SL; the parallel algorithm to perform this step is also fairly complex, and space limitations prevent us from presenting all of the details for this step. Therefore, we present this section so that it can be used as a companion to the presentation of the open ear decomposition algorithm of Ramachandran as given in [21]. Finally, we go through the other steps of the algorithm of [22].

3.1 Elementary Graph Computations in SL

Our method of exposition in this subsection is to give a statement of the subproblem to be solved, and then in parentheses give an indication of how this subproblem can be restated in a way that makes it clear that it can be solved using an oracle for undirected reachability, or by making use of primitive operations that have already been discussed.

Given a graph G, the following conditions can be checked in SL:

1. Are u and v in the same 2-component? (Algorithm: for each vertex x, check if the removal of x separates u and v. This can be tested using *UGAP*.)
2. Let each 2-component be labeled by the smallest two vertices in the 2-component. Is (u, v) the "name" of a 2-component? (First check that u and v are in the same 2-component, and then check that no $x < \max(u, v)$ with $x \notin \{u, v\}$ is in the same 2-component.)
3. Is u a cut-vertex? (Are there vertices v, w connected in G but not in $G-\{u\}$?)
4. Is there is a path (not necessarily simple) of odd length between vertices s and t? (Make two copies of each vertex. Replace edge (u, v) by edges $(u0, v1)$ and $(u1, v0)$. Check if $s0, t1$ are connected in this new graph.)
5. Is G bipartite (i.e. 2-colorable)? [23, 20, 2].
6. If G is connected, 2-colorable, and vertex 1 is colored 1, is vertex i colored 2? (Test if there is a path of odd length from 1 to i.)
7. Is edge e in the lexicographically first spanning tree T of G (under the standard ordering of edges)? [20]

Given a graph G and a spanning tree T, the following conditions can be checked in SL:

1. For $e \in T$ with $e = (x, y)$, does $x \longrightarrow y$ occur at position i of the lexicographically first Euler tour rooted at r, ET_r? Does $x \longrightarrow y$ precede $y \longrightarrow x$? (In

logspace, one can compute the lexicographically-first Euler tour by starting at r and following the edge $r \longrightarrow x$, where x is the smallest neighbor of r in T. At any stage in the tour, if the most recent edge traversed was $u \longrightarrow v$, the next edge in the Euler tour is $v \longrightarrow z$ where z is the smallest neighbor of v greater than u in T if such a neighbor exists, and z is the smallest neighbor of v otherwise.)

2. Is $u = parent(v)$ when T is rooted at r? (Equivalently, is $u \longrightarrow v$ the first edge of ET_r to touch v? This can be checked in L.)
3. If T is rooted at r, is u a descendant of v? (Equivalently, does the first occurrence of u in ET_r lie between the first and last occurrences of v?)
4. Is z the least common ancestor (lca) of vertices x and y in T? (Check that x and y are both descendants of z, and check that this property is not shared by any descendant of z.)
5. Is i the preorder number of vertex u? (Count the number of vertices with first occurrence before that of u in ET_r.)
6. Is vertex u on the fundamental cycle C_e created by non-tree edge e with T? (Let $e = (p, q)$. Vertex u is on C_e iff the graph $T - \{u\}$ has no path from p to q.)
7. Is edge f on C_e? (This holds iff $f = e$ or $f \in T$ and both endpoints of f are on C_e.)
8. Are vertices u, v on the same bridge with respect to C_e? (See [22] for a definition of "bridge". Vertices u and v are on the same bridge iff there is a path from u to v in G, with no internal vertices of the path belonging to C_e.)
9. Are edges f, g on the same bridge with respect to C_e? (This holds if $f, g \notin C_e$, neither f nor g is a trivial bridge (i.e. a chord of C_e), and the endpoints of f, g which are not on C_e are on the same bridge with respect to C_e.)
10. Is vertex u a point of attachment of the bridge of C_e that contains edge f? (Let $f = (f_1, f_2)$. If both f_1 and f_2 are on C_e, then these are the only points of attachment of the trivial bridge $\{f\}$. Otherwise, if f_i is not on C_e, then u is a point of attachment iff $u \in C_e$ and u, f_i are on the same bridge with respect to C_e.)
11. Given vertices u, v, on C_e, and given a sequence $\langle w_1, w_2, \ldots \rangle$, is there a path from u to v along C_e avoiding vertices $\langle w_1, w_2, \ldots \rangle$? (This is simply the question of connectivity in $C_e - \{w_1, w_2, \ldots\}$.)
12. Relative to C_e, do the bridges containing edges f and g interlace? (See [22] for a definition of "interlacing". Either there is a triple u, v, w where all three vertices are points of attachments of both bridges, or there is a 4-tuple u, v, w, x where (1) u, w are attachment points of the bridge containing f, (2) v, x are attachment points of the bridge containing g, and (3) u, v, w, x occur in cyclic order on C_e. To check cyclic order, use the previous test.)

3.2 Finding an Open Ear Decomposition

We follow the exposition from [21]. The algorithm in Figure 1 finds an open ear decomposition of a graph: it labels each edge e by the number of the first ear containing e.

input: biconnected graph G; vertices v, r; edge e
1: Find a spanning tree T, and number the vertices in preorder from 0 to $n-1$ with respect to root r.
2: Label the non-tree edges:
 2.1 For each vertex v other than r, find $low(v)$. Mark v if $low(v) < parent(v)$.
 2.2 Construct multigraph H: $V_H = V \setminus \{r\}$;
 For each $e \notin T$ with distinct base vertices x, y, put edge (x, y) in E_H;
 For each $e \notin T$ with a single base vertex x, put edge (x, x) in E_H;
 2.3 Find the connected components of H. Label a component C by $preorder(parent(a))$ for some (arbitrary) $a \in C$.
 2.4 Within each component C, find a spanning tree T_C, root it at a marked vertex if one exists, and preorder the vertices $0, 1, \ldots, k$.
 2.5 For $e = (parent(y), y) \in T_C$, label e with the pair $(label(C), y)$.
 For $e \notin T_C$, label e with the pair $(label(C), k+1)$.
 2.6 For $e \notin T$, label e with the label of the corresponding edge in H.
 2.7 Sort labels in non-decreasing order and relabel as $1, 2, \ldots$.
3: Label a tree edge $(parent(v), v)$ by the smallest label of any non-tree edge incident on a descendant of v (including v).
4: Relabel the non-tree edge labeled 1 by the label 0.

Fig. 1. Open Ear Decomposition Algorithm

In this procedure, most of the computations involve computing spanning trees, finding connected components, preordering a tree, and sorting labels, all of which can be performed in SL. The only new steps here are the computation of base vertices and low vertices. These are also easily seen to be computable in SL using the operations from Subsection 3.1; note that

1. z is a base vertex of non-tree edge $e = (x, y)$ if $parent(z) = lca(x, y)$ and either x or y is a descendant of z.
2. $low(v)$ is the smallest w such that $w \in C_e$ for a non-tree edge $e = (e_1, e_2)$ with e_1 or e_2 a descendant of v.

3.3 An Overview of the Algorithm

The planarity testing algorithm of Ramachandran and Reif [22] is outlined in Figure 2. If G^* is not 2-colorable (step 2.7), or if step 2.8 yields an embedding that is not planar, then the input graph is not planar. Otherwise, this procedure gives a planar combinatorial embedding of the input graph. For the complete algorithm and definitions of the terms used above, see [22].

The emphasis in [22] is to find a fast parallel algorithm that performs almost optimal work. However, for our purpose, any procedure that can be implemented in SL will do. Step 1 can be accomplished by determining, for each (u, v), if u and v are in the same biconnected component. Step 2.1 was addressed in subsection 3.2. Step 2.4 has been discussed in subsection 3.1. The remaining steps are discussed in the following subsections.

1: Decompose the input graph into its biconnected components.
2: For each biconnected component G, do
  2.1 Find an open ear decomposition $D = (P_0, P_1, \ldots, P_{r-1})$ with $P_0 = (s,t)$.
  2.2 Direct the ears to get directed acyclic graph G_{st}.
  2.3 Construct the local replacement graph G_l and the associated spanning tree T'_{st} and paths $D' = (P'_0, P'_1, \ldots, P'_{r-1})$.
  2.4 Compute the bridges of each fundamental cycle C'_i.
  2.5 Compute a bunch collection for each P'_i, and a hook for each bunch.
  2.6 For each P'_i, construct its bunch graph and the corresponding interlacing parity graph.
  2.7 Construct the constraint graph G^* and 2-color it, if possible.
  2.8 From the 2-coloring, obtain a combinatorial embedding of G_l and hence G. Test if this embedding is planar.
3: Piece together the embeddings of the biconnected components.

Fig. 2. Planarity Testing Algorithm

3.4 Constructing the Directed *st*-Numbering Graph G_{st}

Given an open ear decomposition $D = [P_0, \ldots P_{r-1}]$ of a biconnected graph G, where P_0 consists of the edge (s,t), the graph G_{st} is the result of orienting each edge of G, so that

- The edge (s,t) is oriented $s \longrightarrow t$.
- Let the two endpoints of an ear P_i be the vertices u and v.
  - If $ear(v) < ear(u)$, then all edges on P_i are oriented to form a path from v to u.
  - If $ear(u) < ear(v)$, then all edges on P_i are oriented to form a path from u to v.
  - If $ear(u) = ear(v)$, then all edges on P_i are oriented to form a path from u to v if u comes before v in the orientation on $P_{ear(v)}$, and are oriented to form a path from v to u otherwise.

G_{st} is acyclic, and every vertex lies on a path from s to t [18].

We show that G_{st} can be computed from G and D in logarithmic space.

Orienting the edges in ear P_i is easy if $ear(u) \neq ear(v)$. The routine shown in Figure 3 shows how to orient the edges if $ear(u) = ear(v) = i'$. It is clear that this can be implemented in L.

3.5 Constructing the Local Replacement Graph G_l

In G_l, each vertex v is replaced by a rooted tree T_v with $d(v) - 1$ vertices, one for each ear containing v. The construction exploits the fact that in the directed graph G_{st}, deleting the last edge of each path P_i for $i > 0$ gives a spanning tree T_{st}. The construction introduces new vertices, and maps each P_i to a path P'_i

Input (D, i)
Find the endpoints u and v of P_i. Let $ear(u) = ear(v) = i'$.
Note that P_i is oriented from u to v iff u comes before v in $P_{i'}$.

1. Let u' and v' be the endpoints of $P_{i'}$. Compute the bit B such that
   (u comes before v in P_i) iff [(u' comes before v' in i') $\oplus$ B].

(This can be done in logspace since $P_{i'}$ is a path. The routine can start at u' and see if it encounters u or v first.)

2. If $ear(v') \neq ear(u')$
   Then we can compute the orientation directly.
   Else
     Let $i'' = ear(v') = ear(u')$
     Let the endpoints of $P_{i''}$ be u'' and v''.
     Find and remember the bit B' such that
       (u' comes before v' in $P_{i''}$) iff [(u'' comes before v'' in $P_{i''}$) $\oplus B'$]
     (At this point, we can forget about u' and v'.)
     $u' := u''$; $v' := v''$; $i' := i''$; $B := B \oplus B'$
     GO TO statement 2.

end.procedure

Fig. 3. Orienting an ear.

which is essentially the same as P_i, but has an extra edge involving a new vertex at each end.

The construction of G_l proceeds in 3 phases. In the first / second phase, the first / last edge of each ear is rerouted to a possibly new endpoint via one of the new vertices. In the last phase, some of the new edges are further rerouted to account for parallel ears.

The entire construction uses only the elementary operations described in subsection 3.1, and so can be implemented in $\mathsf{FL}^{\mathsf{SL}}$. The implementation immediately yields the new directed graph G'_{st}, and a listing of the new left and right endpoints $L(P'_i)$ and $R(P'_i)$ of each path.

3.6 Bunch Collections and Hooks

In the spanning tree T'_{st} of the graph G'_{st}, each path P'_i has a unique non-tree edge, which forms the fundamental cycle C'_i with respect to T'_{st}. In [22], each bridge of C'_i is classified as spanning, anchor or non-anchor depending on how the attachment points of C'_i are placed with respect to P'_i. Since bridges can be computed in $\mathsf{FL}^{\mathsf{SL}}$ (see subsection 3.1), this classification is also in SL.

In the nomenclature of [22], bunches are approximations to bridges: bunches contain only the attachment edges of bridges. A bridge is represented by at least one and possibly more than one bunch, subject to certain conditions. The conditions are: (1) A non-anchor bunch must be the entire bridge. (2) A spanning bunch must contain all attachment points of the corresponding bridge on internal vertices of P_i' and at least one edge attaching on $L(P_i')$. (3) Edges within a bunch must be connected in G_l without using vertices from C_i' or from the other bunches. (4) The bunch collection for each P_i' must contain all attachments of bridges on its internal vertices and some attachment edges incident on $L(P_i')$. Bunch collections are computed using operations described in Subsection 3.1.

A representative edge for each anchor bunch B is the hook $H(B)$, which also is used to determine a planar embedding if G turns out to be planar. $H(B)$ is usually an attachment on $C_i' - P_i'$ of the bridge of C_i' that contains B. The exception is when $L(P_i')$ is the lca of the non-tree edge of P_i', in which case $H(B)$ may be the incoming tree edge to $L(P_i')$. Again, the entire procedure for computing hooks uses operations shown to be in SL in subsection 3.1, so $H(B)$ can be computed in $\mathsf{FL}^{\mathsf{SL}}$.

3.7 Bunch Graphs and Interlacing Parity Graphs

Once the bunch collections are formed, the bunch graphs are constructed as follows: extend each path P_i' to a path Q_i by introducing a new edge between $L(P_i')$ and a new vertex $U(P_i')$. Collapse each bunch B of P_i' to a single node v_B (which now has edges to some vertices of P_i'); thus B becomes a star S_B with center v_B. Further, if B is an anchor bunch, include edge $(U(P_i'), v_B)$, and if B is a spanning bunch, include edge $(R(P_i'), v_B)$. This gives the so-called bunch graph $J_i(Q_i)$, which can clearly be constructed in $\mathsf{FL}^{\mathsf{SL}}$.

For each $J_i(Q_i)$, an interlacing parity graph $G_{i,I}$ is constructed as follows: There is a vertex v_B for each star S_B, and a vertex for each triple (u, v, B) where u, v are attachment vertices of S_B on Q_i, and u is an extreme (leftmost / rightmost) attachment. Edges connect (1) a bunch vertex v_B to all its chords (u, v, B), (2) bunch vertices v_S, v_T which share an internal (non-extreme) attachment vertex on Q_i, and (3) each chord to its left and right chords, when they exist. The left and right chords are defined as follows: For chord (u, v, B), consider the set of chords $\{(u', v', B') \mid B' \neq B, u' < u < v' < v\}$; intuitively, these are chords of other bunches that interlace with B. The left chord of (u, v, B) is the chord from this set with minimum u'; ties are broken in favor of largest v'. Right chords are analogously defined.

All the information needed to construct $G_{i,I}$ can be extracted from $J_i(Q_i)$ by a logspace computation.

3.8 The Constraint Graph G^*

The constraint graph contains two parts. One is the union over all i of the interlacing parity graphs $G_{i,I}$, and thus can be constructed in $\mathsf{FL}^{\mathsf{SL}}$. The other part accounts for the fact that more than one bunch may belong to the same bridge, and hence all such bunches must be placed consistently (on the same side) with respect to a path or fundamental cycle. This part has paths of length 1 or 2, called links, between anchor bunches and related bunches. Determining for each anchor bunch the length of the link, and its other endpoint, requires information about $G_{i,I}$ and computations described in subsection 3.1, and so the constraint graph G^* can also be constructed in $\mathsf{FL}^{\mathsf{SL}}$.

If G^* is not 2-colorable, then G is not planar. If G^* is 2-colorable, then the 2-coloring yields a combinatorial embedding of G_l. Testing whether G^* is 2-colorable (i.e. bipartite), and obtaining a 2-coloring if one exists, is known to be in $\mathsf{FL}^{\mathsf{SL}}$; see for instance [2].

3.9 The Combinatorial Embedding of G_l and of G

Given an undirected graph, a combinatorial embedding ϕ is a cyclic ordering of the edges around each vertex. Replace each edge (u,v) by directed arcs $\langle u,v\rangle$ and $\langle v,u\rangle$ to give the arc set A. Then ϕ is a permutation on A satisfying $\phi(\langle u,v\rangle) = \langle u,w\rangle$ for some w; i.e. ϕ cyclically permutes the arcs leaving each vertex. Let R be the permutation mapping each arc to its inverse. The combinatorial embedding ϕ is planar iff the number of cycles f in $\phi^* \triangleq \phi \circ R$ satisfies Euler's formula $n + f = m + 1 + c$. (n, m, c are the number of vertices, undirected edges, connected components respectively.)

The 2-coloring of G^* partitions the non-P'_i edges with respect to P'_i in the obvious way (those that are to be embedded inside, and those that go outside). To further fix the cyclic ordering within each set, the algorithm of [22] computes, for each vertex v, a set of "tufts", which are the connected components of a graph that is easy to compute using the operations provided in subsection 3.1. Each tuft is labeled with a pair of vertices (again, these labels are easy to compute), and then the tufts are ordered by sorting these labels. (Sorting can be accomplished in logspace.) The cyclic ordering for tufts is either increasing or decreasing by labels, determined by the 2-coloring. This cyclic ordering then yields an ordering ϕ for all the arcs in G_l via a simple calculation.

To check planarity of ϕ, note that c can be computed in SL [20], n and m are known, so the only thing left to compute is f. This can be computed in L as follows: Count the number of arcs a for which $a = c(a)$, where $c(a)$ is the lexicographically smallest arc on the cycle of ϕ^* containing a.

Since G_l is obtained from G by local replacements only, an embedding ϕ' of G can be easily extracted from the embedding ϕ of G_l: just collapse vertices of T_v back into v.

3.10 Merging Embeddings of 2-Components

It is well-known that a graph is planar iff its biconnected components are planar; see for instance [29]. To constructively obtain a planar combinatorial embedding of G from planar combinatorial embeddings of its 2-components, note that the ordering of edges around each vertex which is not a cut-vertex is fixed within the corresponding 2-component. At cut-vertices, adopt the following strategy: Let w be a cut-vertex present in 2-components $(u_1,v_1),(u_2,v_2),\ldots(u_d,v_d)$. The edges of w in each of these components are ordered according $\phi_1,\ldots,\phi_d$. Let x_i

be the smallest neighbor of w in the 2-component (u_i, v_i). The orderings can be pasted together in $\mathsf{FL}^{\mathsf{SL}}$ as follows:

$$\begin{aligned} \phi(w,z) &= \phi_j(w,z) \qquad \text{if } z \text{ is in the 2-component } (u_j,v_j) \text{ and } z \neq x_j \\ \phi(w,x_j) &= \phi_{j+1}^{-1}(w,x_{j+1}), \\ \phi(w,x_d) &= \phi_1^{-1}(w,x_1). \end{aligned}$$

4 Open Problems

- Is planarity testing hard for SL? Is it in L? Until these classes are proved to coincide, there still remains some room for improvement in the bounds we present in this paper.
- Can any of the techniques used here be extended to construct embeddings of small genus graphs? For instance, what is the parallel complexity of checking if a graph has genus 1, and if so, constructing a toroidal embedding?

Acknowledgments

We thank Pierre McKenzie for directing us to a simplification in our hardness proof.

References

[1] R. Aleliunas, R. M. Karp, R. J. Lipton, L. Lovász, and C. Rackoff. Random walks, universal traversal sequences, and the complexity of maze problems. In *Proceedings of the 20th Annual Symposium on Foundations of Computer Science*, pages 218–223. IEEE, 1979.

[2] C. Àlvarez and R. Greenlaw. A compendium of problems complete for symmetric logarithmic space. Technical Report ECCC-TR96-039, Electronic Colloquium on Computational Complexity, 1996.

[3] R. Armoni, A. Ta-Shma, A. Wigderson, and S. Zhou. $SL \subseteq L^{\frac{4}{3}}$. In *Proceedings of the 29th Annual Symposium on Theory of Computing*, pages 230–239. ACM, 1997.

[4] K. Booth and G. Lueker. Testing for the consecutive ones property, interval graphs, and graph planarity using pq-tree algorithms. *Journal of Computer and System Sciences*, 13:335–379, 1976.

[5] A. Chandra, L. Stockmeyer, and U. Vishkin. Constant depth reducibility. *SIAM Journal on Computing*, 13(2):423–439, 1984.

[6] S. A. Cook and P. McKenzie. Problems complete for L. *Journal of Algorithms*, 8:385–394, 1987.

[7] K. Etessami. Counting quantifiers, successor relations, and logarithmic space. *Journal of Computer and System Sciences*, 54(3):400–411, Jun 1997.

[8] S. Even and R. Tarjan. Computing an *st*-numbering. *Theoretical Computer Science*, 2:339–344, 1976.

[9] A. Gál and A. Wigderson. Boolean vs. arithmetic complexity classes: randomized reductions. *Random Structures and Algorithms*, 9:99–111, 1996.
[10] J. Hopcroft and R. Tarjan. Efficient planarity testing. *Journal of the ACM*, 21:549–568, 1974.
[11] N. Immerman. Nondeterministic space is closed under complementation. *SIAM Journal on Computing*, 17(5):935–938, Oct 1988.
[12] J. Ja'Ja' and J. Simon. Parallel algorithms in graph theory: Planarity testing. *SIAM Journal on Computing*, 11:314–328, 1982.
[13] J. Ja'Ja' and J. Simon. Space efficient algorithms for some graph-theoretic problems. *Acta Informatica*, 17:411–423, 1982.
[14] M. Karchmer and A. Wigderson. On span programs. In *Proceedings of the 8th Conference on Structure in Complexity Theory*, pages 102–111. IEEE Computer Society Press, 1993.
[15] R. M. Karp and R. J. Lipton. Turing machines that take advice. *L' Ensignement Mathématique*, 28:191–210, 1982.
[16] A. Lempel, S. Even, and I. Cederbaum. An algorithm for planarity testing in graphs. In *Theory of Graphs: International Symposium*, pages 215–232, New York, 1967. Gordon and Breach.
[17] M. Mahajan, P. R. Subramanya, and V. Vinay. A combinatorial algorithm for Pfaffians. In *Proceedings of the Fifth Annual International Computing and Combinatorics Conference COCOON, LNCS Volume 1627*, pages 134–143. Springer-Verlag, 1999. DIMACS Technical Report 99-39.
[18] Y. Maon, B. Schieber, and U. Vishkin. Parallel ear decomposition search (EDS) and st-numbering in graphs. *Theoretical Computer Science*, 47:277–296, 1986.
[19] N. Nisan, E. Szemeredi, and A. Wigderson. Undirected connectivity in $O(\log^{1.5} n)$ space. In *Proceedings of the 33rd Annual Smposium on Foundations of Computer Science*, pages 24–29. IEEE Computer Society Press, 1992.
[20] N. Nisan and A. Ta-Shma. Symmetric Logspace is closed under complement. *Chicago Journal of Theoretical Computer Science*, 1995.
[21] V. Ramachandran. Parallel open ear decomposition with applications to graph biconnectivity and triconnectivity. In J. Reif, editor, *Synthesis of Parallel Algorithms*. Morgan Kaumann, 1993.
[22] V. Ramachandran and J. Reif. Planarity testing in parallel. *Journal of Computer and System Sciences*, 49:517–561, 1994.
[23] J. Reif. Symmetric complementation. *Journal of the ACM*, 31(2):401–421, 1984.
[24] K. Reinhardt and E. Allender. Making nondeterminism unambiguous. In *38 th IEEE Symposium on Foundations of Computer Science (FOCS)*, pages 244–253, 1997. to appear in SIAM J. Comput.
[25] M. Saks. Randomization and derandomization in space-bounded computation. In *Proceedings of the 11th Annual Conference on Computational Complexity*, pages 128–149. IEEE Computer Society, 1996.
[26] W. J. Savitch. Relationships between nondeterministic and deterministic tape complexities. *Journal of Computer and System Sciences*, 4(2):177–192, April 1970.
[27] R. Szelepcsényi. The method of forced enumeration for nondeterministic automata. *Acta Informatica*, 26(3):279–284, 1988.
[28] W. T. Tutte. Toward a theory of crossing numbers. *Journal of Combinatorial Theory*, 8:45–53, 1970.
[29] H. Whitney. Non-separable and planar graphs. *Transactions of the American Mathematical Society*, 34:339–362, 1932.

About Cube-Free Morphisms
(Extended Abstract)

Gwénaël Richomme and Francis Wlazinski

LaRIA, Université de Picardie Jules Verne
5 rue du Moulin Neuf, 80000 Amiens, France
{richomme,wlazinsk}@laria.u-picardie.fr

Abstract. We address the characterization of finite test-sets for cube-freeness of morphisms between free monoids, that is, the finite sets T such that a morphism f is cube-free if and only if $f(T)$ is cube-free. We first prove that such a finite test-set does not exist for morphisms defined on an alphabet containing at least three letters. Then we prove that for binary morphisms, a set T of cube-free words is a test-set if and only if it contains twelve particular factors. Consequently, a morphism f on $\{a, b\}$ is cube-free if and only if $f(aabbababbabbaabaababaabb)$ is cube-free (length 24 is optimal). Another consequence is an unpublished result of Leconte: A binary morphism is cube-free if and only if the images of all cube-free words of length 7 are cube-free.
We also prove that, given an alphabet A containing at least two letters, the monoid of cube-free endomorphisms on A is not finitely generated.

1 Introduction

At the beginning of the century, Thue [19,20] (see also [3,4]) worked on repetitions in words. In particular, he showed the existence of a square-free infinite word over a three-letter alphabet, and the existence of an overlap-free (and thus cube-free) infinite word over a binary alphabet. Since these works, many other results on repetitions in words have been achieved (see [7] for a recent survey, and [13] for related works), and Thue's results have been rediscovered in several instances (see for example [13]).

Thue obtained an infinite overlap-free word over a two-letter alphabet (called Thue-Morse word since the works of Morse [15]) by iterating a morphism μ ($\mu(a) = ab$ and $\mu(b) = ba$). Morphisms are widely used to generate infinite words. To obtain an infinite word with some property P, one very often uses P-preserving-morphisms, called P-morphisms. Naturally, some studies concern such morphisms: Sturmian morphisms (see [14] for a recent survey), power-free morphisms [12], square-free morphisms [2,8]...

Our paper is concerned with cube-free morphisms. The close problem of infinite cube-free words generated by morphism has already been studied for instance in [9,16]. Necessary conditions or sufficient conditions for cube-freeness of a morphism can be found in the studies of the general case of kth power-free morphisms [1,10,11]. But characterizations of cube-freeness exist only for

H. Reichel and S. Tison (Eds.): STACS 2000, LNCS 1770, pp. 99–109, 2000.

morphisms from a two-letter or a three-letter alphabet [11]. In the case of a two-letter alphabet, an unpublished result of Leconte is: A morphism on a binary alphabet is cube-free if and only if the images of all cube-free words of length at most seven are cube-free.

In this paper, we consider test-sets for cube-freeness of morphisms on an alphabet A, that is, subsets T of A^* such that given any morphism f defined on A, f is cube-free if and only if $f(T)$ is cube-free. The result of Leconte can be rephrased: The set of cube-free words over $\{a, b\}$ of length at most 7 is a test-set for morphisms on $\{a, b\}$. Similar results have been obtained for other property-free morphisms. Among which, let us mention that on a three-letter alphabet, a morphism is square-free if and only if the images of all square-free words of length at most 5 are square-free (Crochemore [8]). In [5], Berstel and Séébold show that an endomorphism f on $\{a, b\}$ is overlap-free if and only if the images of all overlap-free words of length at most 3 are overlap-free or, equivalently, if $f(abbabaab)$ is overlap-free. In [17], Richomme and Séébold improve this result showing that an endomorphism f on $\{a, b\}$ is overlap-free if and only if $f(bbabaa)$ is overlap-free. More precisely, they characterize all the finite test-sets for overlap-freeness of binary endomorphisms, that is, each set S such that a morphism f is overlap-free if and only if given any word w in S, $f(w)$ is overlap-free.

In Section 4, we give our main result which is a characterization of test-sets for cube-free morphisms on a two-letter alphabet: The set of factors of such a test-set just has to contain a particular finite subset. As one of the consequences of this characterization, we show that a morphism f defined on $\{a, b\}$ is cube-free if and only if the word $f(aabbababbabbaabaababaabb)$ is cube-free. Length 24 of this test-word is optimal: No word of length 23 or less can be used to test the cube-freeness of a morphism on a binary alphabet.

In Section 4, we also show that any test-set for cube-free morphisms defined on an alphabet containing at least three letters is infinite. Thus, test-sets give no general effective way to determine whether a morphism is cube-free.

The set of cube-free morphisms forms a monoid. When a monoid of morphisms is finitely generated, we have a natural way to determine if a morphism belongs to this monoid. Such a situation is known for instance for overlap-free morphisms [18,20]. In Section 3, we show that this is no longer true for cube-free morphisms. The monoid of cube-free endomorphisms on an alphabet A containing at least two letters is never finitely generated.

2 Preliminaries

In this section, we recall and introduce some basic notions on words and morphisms.

Let A be an *alphabet*, that is a finite non-empty set of abstract symbols called letters. The *Cardinal* of A, i.e., the number of elements of A, is denoted by Card(A). A *word* over A is a finite sequence of letters from A. The set of the words over A equipped with the concatenation of words and completed with a neutral element ε called *empty word* is a free monoid denoted by A^*.

Let $u = a_1a_2.....a_n$ be a word over A, with $a_i \in A$ $(1 \le i \le n)$. The number n of letters of u is called its *length* and is denoted by $|u|$. Observe $|\varepsilon| = 0$. When $n \ge 1$ the *mirror image* of u, denoted by $\tilde{u}$, is the word $u = a_n....a_2a_1$. In the particular case of the empty word, $\tilde{\varepsilon} = \varepsilon$. Given a set of words X, we denote by $\tilde{X}$ the set of all the words $\tilde{w}$ with w in X.

A word u is a *factor* of a word v if $v = v_1uv_2$ for some words v_1, v_2. If $v_1 = \varepsilon$, u is a *prefix* of v. If $v_2 = \varepsilon$, u is a *suffix* of v. Given a set of words X, $\text{Fact}(X)$ denotes the set of all the factors of the words in X.

Let us consider a non-empty word w and its letter-decomposition $x_1 \dots x_n$. For any integers i, j, $1 \le i \le j \le n$, we denote by $w_{[i..j]}$ the factor $x_i \dots x_j$ of w. We extend this notation when $i > j$: In this case, $w_{[i..j]} = \varepsilon$. We abbreviate $w_{[i..i]}$ in $w_{[i]}$. This notation denotes the ith letter of w.

For an integer $n \ge 2$, we denote by u^n the concatenation of n occurrences of the word u, $u^0 = \varepsilon$ and $u^1 = u$. In particular, a *cube* is a word of the form u^3 with $u \ne \varepsilon$. A word w contains a cube if at least one of its factors is a cube. A word is called *cube-free*, if it does not contain any cube as a factor. A set of cube-free words is said *cube-free*.

A *morphism* f from an alphabet A to another alphabet B is a mapping from A^* to B^* such that given any words u and v over A, we have $f(uv) = f(u)f(v)$. When $B = A$, f is called an *endomorphism* on A. When B has no importance, we say that f is defined on A or that f is a morphism on A (this does not mean that f is an endomorphism). Observe that for a morphism f on A, we necessary get $f(\varepsilon) = \varepsilon$, and f is uniquely defined by the values of $f(x)$ for all x in A. The *Identity* endomorphism (resp. the *Empty* morphism) on A, denoted Id (resp. ϵ) is defined by $Id(x) = x$ (resp. $\epsilon(x) = \varepsilon$), for all x in A. When A is a binary alphabet $\{a, b\}$, the *Exchange* endomorphism E is defined by $E(a) = b$ and $E(b) = a$. If X is a set of words, $f(X)$ denotes the set of all the images of the words in X.

A morphism f on A is called *cube-free* if for every cube-free word w over A, $f(w)$ is cube-free. The morphisms Id, ϵ and (on a binary alphabet) E are obviously cube-free. Let us remark that if two composable morphisms f and g are cube-free then $f \circ g$ is cube-free (where $\circ$ denotes the composition of functions). Thus the set of cube-free endomorphisms on a given alphabet is a monoid. It is also easy to verify that a morphism f on $\{a, b\}$ is cube-free if and only if $f \circ E$ is cube-free. Given a morphism f on A, the *mirror morphism* $\tilde{f}$ of f is defined for all words w over A, by $\tilde{f}(w) = \widetilde{f(\tilde{w})}$. Since $\tilde{\tilde{f}} = f$, f is cube-free if and only if $\tilde{f}$ is cube-free. One can see that for a non-empty cube-free morphism f, given two letters x and y, we cannot have $f(x)$ prefix nor suffix of $f(y)$ (such a morphism f is called *biprefix*) else $f(xxyx)$ or $f(xyxx)$ contains the cube $(f(x))^3$. In particular, for any cube-free morphism different from the empty morphism, $f(x) \ne \varepsilon$ for each letter x. We will use the following theorem:

Theorem 1. [1,6] *Given two alphabets A and B with $card(A) \ge 2$, $card(B) \ge 2$, there exist some cube-free morphisms from A to B.*

If the image of some word by a morphism contains a cube, we often want to consider the exact factor whose image contains this cube. Given a morphism f, we say that the image of a non-empty word w is a *minimal cover* of a cube u^3, if $f(w_{[1]})$ (resp. $f(w_{[|w|]})$ has a prefix $p \neq f(w_{[1]})$ (resp. a suffix $s \neq f(w_{[|w|]})$) such that $f(w) = puuus$. The image of a word contains a cube if and only if the image of one of its factors is a minimal cover of this cube.

3 About Monoids of Cube-Free Endomorphisms

The set of cube-free endomorphisms on a given alphabet is a monoid. The aim of this section is to prove the following result:

Theorem 2. *The monoid of cube-free endomorphisms on an alphabet A containing at least two letters is not finitely generated.*

Given an alphabet A, we denote by CF_A the monoid of cube-free endomorphisms on A different from the empty morphism ϵ. A set of generators of CF_A is a subset G of CF_A such that any morphism f in CF_A has a finite decomposition over G. Theorem 2 says that CF_A has no finite set of generators. In order to prove it, we first observe:

Lemma 1. *Given a morphism $f : \{a,b\}^* \to \{a,b\}^*$ such that $|f(a)| = 1$, f is cube-free if and only if $f = Id$ or $f = E$.*

Proof. The identity and exchange morphisms are cube-free. Let f be a cube-free morphism with $|f(a)| = 1$, and assume f is different from the identity and the exchange morphisms. Since a morphism f is cube-free if and only if $E \circ f$ is cube-free, without loosing generality, we can assume $f(a) = a$. We have $f(b) \neq \varepsilon$ since otherwise $f(aaba) = aaa$. Since $f \neq Id$, $f(b) \neq b$. Observe that $f(b)$ can not start nor end with a. Otherwise, $f(aab)$ or $f(baa)$ respectively contains aaa as prefix or suffix. Let $f(b) = bub$. The word bu cannot end with b otherwise $f(bb)$ contains bbb. In the same way, ub does not start with b. The word bub does not start with bab, otherwise $f(bab)$ contains $ababab$. In the same way bub does not end with bab. Thus bub starts and ends with $baab$. This is a final contradiction since in this case $f(baab)$ contains the cube $baabaabaa$. □

We now define a particular family of cube-free morphisms. Considering a word V over $\{b,c\}$ such that $cVcb$ is cube-free we define the morphism

$$\begin{cases} f_V(a) = a \\ f_V(b) = bacVc \end{cases}$$

One can verify that

Lemma 2. *The morphism f_V is cube-free.*

We are now able to sketch the proof of Theorem 2.

First, we consider the case $\text{Card}(A) = 2$. Let $A = \{a, b\}$ and assume $CF_{\{a,b\}}$ is finitely generated. Let G be a finite set of generators of this monoid. From Lemma 1, any morphism f in G different from E and Id verifies $|f(a)| \geq 2$ and $|f(b)| \geq 2$. Any cube-free morphism different from E and Id can be decomposed over G as $f = g_1 \circ g_2 \circ \ldots g_n$ with $g_i \in G \setminus \{Id\}$ in such a way that $g_i = E$ implies $g_{i+1} \neq E$. Thus, if $n \geq 6$, $|f(a)| \geq 8$. Consequently, since G is finite, there is a finite number of cube-free morphisms with $|f(a)| = 6$.

Since there exist infinite cube-free words over a two-letter alphabet $\{b, c\}$ (see, for instance [19,20]), and since in such a word, the factor cb occurs infinitely often, there exist arbitrarily large words V over $\{b, c\}$ such that $c\text{V}cb$ is a cube-free word. Thus, using Lemma 2, we have a contradiction. Indeed, there exist an infinity of cube-free morphisms (from $\{a, b\}^*$ onto itself) $g \circ f_{\text{V}}$ with $|g \circ f_{\text{V}}(a)| = 6$, where g is the cube-free morphism (see [6])

$$\begin{cases} g(a) = aababb, \\ g(b) = aabbab, \\ g(c) = abbaab. \end{cases}$$

Let us now consider the case $\text{Card}(A) \geq 3$. Let a and b be two particular letters of A. We consider the submonoid S of CF_A of the cube-free endomorphisms f on A such that $f(a), f(b)$ are in $\{a, b\}^*$ and for all letters x in $A \setminus \{a, b\}$, $f(x) = x$. Any morphism f of $CF_{\{a,b\}}$ can be extended into a morphism of S taking $f(x) = x$ for x in $A \setminus \{a, b\}$. Conversely, any morphism of S can be projected on a morphism of $CF_{\{a,b\}}$. Consequently, one can see that S finitely generated implies $CF_{\{a,b\}}$ finitely generated. Thus S is not finitely generated.

We give now the main ideas to prove that if CF_A is finitely generated then S is also finitely generated. This ends the proof of Theorem 2.

Let us recall that a permutation of the alphabet A is a bijective endomorphism p on A such that for all x in A, $|p(x)| = 1$. We denote by p^{-1} the inverse permutation (such that $p \circ p^{-1} = p^{-1} \circ p = Id$). Observe that a permutation is a cube-free morphism. We can prove:

Fact. Given a morphism f in S, and two morphisms g and h in CF_A, if $f = g \circ h$, then there exists a permutation p of A such that $g \circ p$ and $p^{-1} \circ h$ belong to S.

Now assume that G is a finite set of generators of CF_A. Given a morphism f in S, since $S \subset CF_A$, there exist some morphisms $g_1, \ldots, g_n$ in G such that $f = g_1 \circ g_2 \circ \ldots \circ g_n$. From the previous fact, there exist some permutations $p_1, p_2, \ldots, p_n, p_{n+1}$ with $p_1 = p_{n+1} = Id$ such that for all i $(1 \leq i \leq n)$, $p_i^{-1} \circ g_i \circ p_{i+1} \in S$. We have:

$$f = (p_1^{-1} \circ g_1 \circ p_2) \circ (p_2^{-1} \circ g_2 \circ p_3) \circ \ldots \circ (p_n^{-1} \circ g_n \circ p_{n+1}).$$

Since there is a finite number of permutations of A, S is finitely generated by the morphisms of S of the form $p \circ g \circ q$ with p, q two permutations of A and $g \in G$. $\square$

4 Test-Sets

In this section, we are interested in test-sets for cube-free morphisms. Let A, B be two alphabets. If $\text{Card}(B) = 1$, the only cube-free morphism from A to B is the empty morphism ϵ (and Id if $\text{Card}(A) = 1$). In the rest of the paper, we assume that $\text{Card}(B) \geq 2$.

A set of words T is a *test-set for cube-free morphisms* from A to B if, given any morphism f from A to B, f is cube-free if and only if $f(T)$ is cube-free.

We first examine the case $\text{Card}(A) \geq 3$.

Theorem 3. *Given two alphabets A and B with Card(A) ≥ 3 and Card(B) ≥ 2, there is no finite test-set for cube-free morphisms from A to B.*

Theorem 3 is a corollary of Proposition 1 below. Let A be an alphabet containing at least two letters. We consider one particular letter a in A, and three different letters x,y,z which do not belong to $A\setminus\{a\}$. Let $C = A\setminus\{a\}\cup\{x, y, z\}$, and let u and v be two cube-free words over $A\setminus\{a\}$ non simultaneously empty. We define the morphism $f_{u,v} : A^* \to C^*$ by:

$$\begin{cases} f_{u,v}(a) = xzyuxyvxzyuxyvxzy \\ f_{u,v}(b) = b \qquad \text{for all } b \text{ in } A \setminus \{a\} \end{cases}$$

We have (see Section 2 for the definition of cover):

Proposition 1. *Let w be a cube-free word over A. $f_{u,v}(w)$ is a minimal cover of a cube if and only if $w = avaua$.*

By lack of place, we leave it to the reader to verify this proposition.

Proof of Theorem 3. Consider a particular letter a in A. Let $C = A \setminus \{a\} \cup \{x, y, z\}$ where x,y,z are three different letters which do not belong to $A \setminus \{a\}$. Since $\text{Card}(A) \geq 3$ and $\text{Card}(B) \geq 2$, using Theorem 1, there exists a cube-free morphism g from C to B. From Proposition 1, given two words u and v over $A \setminus \{a\}$ non simultaneously empty, we know that $g \circ f_{u,v}(w)$ is a minimal cover of a cube if and only if $w = avaua$. Thus $avaua$ belongs to the set of factors of any test-set for cube-free morphisms from A to B. Since there is an infinity of cube-free words over $A \setminus \{a\}$, any test-set for cube-free morphisms from A to B is infinite.

□

From now, we will assume $\text{Card}(A) = 2$, for instance $A = \{a, b\}$. Cube-free words u and v over $\{b\}$ can take only three values ε, b and bb. So, using similar techniques as in the proof of Theorem 3, and considering $f_{u,v}$ and $f_{u,v} \circ E$ for (u, v) in $\{(bb, bb), (b, bb), (bb, b), (\varepsilon, bb), (bb, \varepsilon), (b, b)\}$, we get, as another consequence of Proposition 1, that for any test-set T for cube-free morphisms from $\{a, b\}$ to an alphabet of cardinal at least 2, the set of factors of T contains:

$$T_{\min} = \begin{Bmatrix} abbabba, \ baabaab, \ ababba, \ babaab, \ abbaba, \ baabab, \\ aabba, \ bbaab, \ abbaa, \ baabb, \ ababa, \ babab \end{Bmatrix}$$

The converse also holds and gives a characterization of test-sets for cube-free morphisms on a two-letter alphabet:

Theorem 4. *A subset T of $\{a,b\}^*$ is a test-set for cube-free morphisms from $\{a,b\}$ to an alphabet of cardinal at least two if and only if T is cube-free and $T_{\min} \subset \mathrm{Fact}(T)$.*

Before giving some ideas on the proof of this theorem, let us examine some particular test-sets. Obviously, the set $T_{\min}$ is one of them. Since it contains twelve elements, one can ask for a test-set of minimal cardinality. There exist some test-words (that is a test-set of cardinal 1). For instance, one can verify that the cube-free word *aabbababbabbaabaababaabb* is one of the 56 words of length 24 that fulfills the conditions of Theorem 4, and thus, is a test-set for cube-free morphisms on $\{a,b\}$. The length of this word is optimal: No cube-free word of length 23 contains all the words of $T_{\min}$ as factors.

Another direct corollary of Theorem 4 is the following unpublished result of Leconte [11]:

Corollary 1. *Given a morphism f on a binary alphabet, the following assertions are equivalent:*

1. *f is cube-free.*
2. *The images of all cube-free words of length 7 are cube-free.*
3. *The images of all cube-free words of length at most 7 are cube-free.*

About the works of Leconte [11], let us also mention that he used the morphism $f_{bb,bb}$ to show the optimality of the bound 7 in Corollary 1. To prove this corollary from Theorem 4, one just has to observe that each word of $T_{\min}$ is a factor of a cube-free word of length 7.

Since the Identity morphism is cube-free, any test-set for cube-free morphisms is necessarily cube-free. So to prove Theorem 4, it is enough to prove that if $T_{\min} \subset \mathrm{Fact}(T)$ for a set T of cube-free words, then T is a test-set for cube-free morphisms. Denoting by G_8 (resp. L_7) the set of all cube-free words over $\{a,b\}$ of length at least 8 (resp. at most 7), we cut the proof of Theorem 4 into two parts (in the two following propositions f is a morphism defined on $\{a,b\}$):

Proposition 2. *Given any word w of G_8, if $f(T_{\min})$ is cube-free then $f(w)$ is not a minimal cover of a cube.*

Proposition 3. *If $f(T_{\min})$ is cube-free then $f(L_7)$ is cube-free.*

Proposition 2 means that if $f(T_{\min})$ is cube-free, and if, for a cube-free word w, $f(w)$ contains a cube, a factor of w which is in $L_7 \setminus \mathrm{Fact}(T_{\min})$ is necessarily a minimal cover of this cube. Proposition 3 then proves that such a situation is impossible. Since $\{a,b\}^* = L_7 \cup G_8$, from these two propositions, $f(T_{\min})$ cube-free implies f cube-free. Thus if T is cube-free and $T_{\min} \subset \mathrm{Fact}(T)$, T is a test-set for cube-free morphisms. This ends the proof of Theorem 4.

In the next two subsections, we give the main ideas of the proofs of Propositions 2 and 3.

4.1 Ideas of the Proof of Proposition 2

Here, we only give the scheme of the proof (rather technical) of Proposition 2. By contradiction, we prove that if the image of a word of G_8 is a minimal cover of a cube then at least one word of $f(T_{\min})$ contains a cube. For this, we need to study more precisely the decomposition of a cube u^3 when $f(w)$ is a minimal cover of it, for some morphism f on $\{a, b\}$ and some cube-free word w.

Note that up to the end of the section, we denote by n the length of w.

By definition, the image of a word w is a minimal cover of a cube uuu if and only if $f(w) = p_1 uuus_n$ with $f(w_{[1]}) = p_1 s_1$, $f(w_{[n]}) = p_n s_n$, for some words p_1, $s_1 \neq \varepsilon$, $p_n \neq \varepsilon$ and s_n. In this case there exist two integers i and j between 1 and n such that $|f(w_{[1..i-1]})| < |p_1 u| \leq |f(w_{[1..i]})|$ and $|f(w_{[1..j-1]})| < |p_1 uu| \leq |f(w_{[1..j]})|$ (remember $w_{[1..0]} = \varepsilon$).

In the general case, we may have $i = 1, i = j$ or $i = n$. In the proof of Proposition 2, we will see that when the image of a cube-free word of length at least eight is a minimal cover of a cube, then we necessarily have $1 < i < j < n$. In this case, there exist some words p_i, p_j, s_i, s_j such that $f(w_{[i]}) = p_i s_i$, $f(w_{[j]}) = p_j s_j$, $u = s_1 f(w_{[2..i-1]}) p_i = s_i f(w_{[i+1..j-1]}) p_j = s_j f(w_{[j+1..n-1]}) p_n$. Since $|f(w_{[1..i-1]})| < |p_1 u|$ and $|f(w_{[1..j-1]})| < |p_1 uu|$, one can observe that p_i and p_j are non-empty words. But we may have $w_{[2..i-1]} = \varepsilon$, $w_{[i+1..j-1]} = \varepsilon$ or $w_{[j+1..n-1]} = \varepsilon$, i.e., $i = 2$, $j = i + 1$ or $n = j + 1$. The previous situation can be summed up by Figure 1.

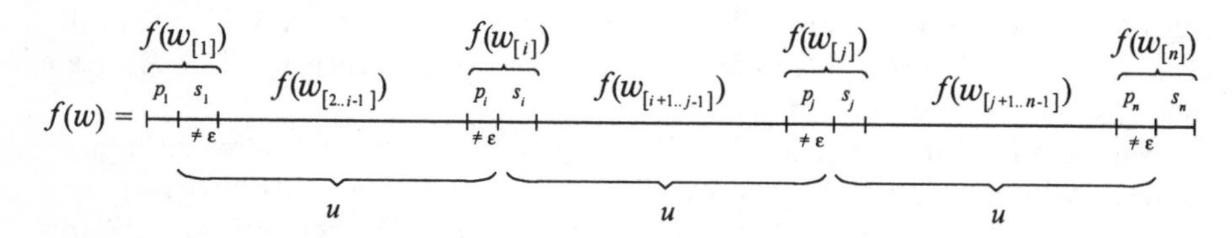

Fig. 1. Decomposition of a cube

In each situation of the following lemma, we place ourself in the situation of Figure 1. Adding some hypotheses, we deduce that $f(T_{\min})$ is not cube-free.

Lemma 3. *Consider f an injective morphism on $\{a, b\}$, w a cube-free word of length $n \geq 6$ and integers i, j such that the situation described in Figure 1 is verified. In each of the following cases, $f(T_{\min})$ contains a cube:*

1. $w_{[j+1]} = w_{[i]} \neq w_{[j]} = w_{[i+1]}$, *with* $1 < i < j - 1 < n - 2$.
2. $w_{[j-1]} = w_{[i]} \neq w_{[j]} = w_{[i-1]}$, *with* $2 < i < j - 1 < n - 1$.
3. $w_{[i]} = w_{[j]}$ *and* $1 < i < j < n$.

The proof of Lemma 3 can be done using the following property which has its own interest:

Property 1. Consider a morphism f on $\{a,b\}$, a letter c in $\{a,b\}$, and two words x and y respectively prefix and suffix of the images of two cube-free words over $\{a,b\}$ by f. If $f(c)x = yf(c)$ with $0 < |x| < |f(c)|$ then $f(T_{\min})$ is not cube-free.

The proof of Proposition 2 now consists in seven successive steps (by lack of place, we do not go further on details):

Step 1. We recall the hypotheses. We consider a morphism f on $\{a,b\}$, a non-empty word u and a cube-free word w such that the length n of w is at least 8, and $f(w)$ is a minimal cover of the cube uuu. We define the words $p_1, s_1 \neq \varepsilon, p_n, s_n \neq \varepsilon$ such that $f(w) = p_1 uuu s_n$, $f(w_{[1]}) = p_1 s_1$, $f(w_{[n]}) = p_n s_n$. We also define the integers i and j such that $1 \leq i \leq j \leq n$, $|f(w_{[1..i-1]})| < |p_1 u| \leq |f(w_{[1..i]})|$ and $|f(w_{[1..j-1]})| < |p_1 uu| \leq |f(w_{[1..j]})|$. We have to prove that $f(T_{\min})$ contains a cube.
We may assume that f is injective otherwise $f(aabaa)$ or $f(bbabb)$ is not cube-free, or $f(a) = f(b) = \varepsilon$ and f is cube-free.

Step 2. We prove that we are in the situation of Figure 1, that is $1 < i < j < n$. We use length reason to prove $i \neq 1$, $i \neq j$ and $j \neq n$. For instance, if $i = 1$, we get $|u| \leq |p_1 u| \leq |f(w_{[1]})|$ and $|uuf(w_{[n]})| > |uus_n| \geq |f(w_{[2..n-1]})f(w_{[n]})|$ which implies $|uu| > |f(w_{[2..n-1]})|$. We have $|w_{[2..n-1]}| \geq 6$ and $w_{[2..n-1]}$ cube-free. Such a word contains at least two as and two bs. It follows that $|u| > |f(a)| + |f(b)|$: A contradiction with $|u| \leq |f(w_{[1]})|$.

Step 3. Since the case $w_{[i]} = w_{[j]}$ is treated by Lemma 3 ($f(T_{\min})$ contains a cube), we assume $w_{[i]} \neq w_{[j]}$.

Step 4. Using length reason, we prove $i \neq j - 1$.

Step 5. We prove that we cannot have simultaneously $i = 2$ and $j = n - 1$.

Step 6. In case $i = 2$ (similarly $j = n - 1$), we prove that, in cases there is no contradiction, $f(T_{\min})$ contains a cube.

Step 7. In case $i \neq 2$ and $j \neq n - 1$, we show also that $f(T_{\min})$ is not cube-free.

4.2 Ideas of the Proof of Proposition 3

To prove Proposition 3, we again use Lemma 3. Moreover, we consider one new set:

$$S_{\text{comp}} = \{aabaaba,\ aabaabb,\ aababaa,\ aababba,\ aabbaab,\ aabbaba,\ abaabab,\ abaabba,\ ababaab,\ aababa,\ aababb,\ aabbaa,\ aabbab,\ abaaba,\ abbaab\}.$$

The main interest of S_{comp} is that: $L_7 \setminus \text{Fact}(T_{\min}) = S_{\text{comp}} \cup \tilde{S}_{\text{comp}} \cup E(S_{\text{comp}}) \cup E(\tilde{S}_{\text{comp}})$. To prove Proposition 3, we have to show that if the image by a morphism f of a word in $L_7 \setminus \text{Fact}(T_{\min})$ is a minimal cover of a cube then at least one word in $f(T_{\min})$ contains a cube. But, for any word w in $S_{\text{comp}} \cup \tilde{S}_{\text{comp}} \cup E(S_{\text{comp}}) \cup E(\tilde{S}_{\text{comp}})$, one of the words w, $\tilde{w}$, $E(w)$ or $E(\tilde{w})$ is in S_{comp}, and if $f(w)$ is a minimal cover of a cube, then $\tilde{f}(\tilde{w})$, $(f \circ E)(E(w))$ and $(\tilde{f} \circ E)(E(\tilde{w}))$ also are minimal covers of a cube. But $T_{\min} = E(T_{\min}) = \tilde{T}_{\min}$. It is thus sufficient to prove that for any word w in S_{comp}, $f(w)$ is a minimal cover of a cube implies $f(T_{\min})$ non cube-free.

Let w be a word in S_{comp} and let f be a morphism on $\{a,b\}$. Assume that $f(w)$ is a minimal cover of a cube u^3, i.e., $f(w) = p_1uuus_n$ with $f(w_{[1]}) = p_1s_1$, $f(w_{[n]}) = p_ns_n$, for some words p_1, $s_1 \neq \varepsilon$, $p_n \neq \varepsilon$ and s_n. And let i, j the integers such that $|f(w_{[1..i-1]})| < |p_1u| \leq |f(w_{[1..i]})|$ and $|f(w_{[1..j-1]})| < |p_1uu| \leq |f(w_{[1..j]})|$.

If $i = 1$, we have $|u| \leq |f(w_{[1]})|$. Moreover $2|u| > |f(w_{[2..n-1]})|$. But, for each word in S_{comp}, $w_{[2..n-1]}$ contains at least two $w_{[1]}$. Thus we can not have $i = 1$. In the same way, we get $j \neq n$. One can also prove $i \neq j$.

From now on, we are in the situation of Figure 1. We adopt its notations, that is, $f(w_{[i]}) = p_is_i$, $f(w_{[j]}) = p_js_j$ and $u = s_1f(w_{[2..i-1]})p_i = s_if(w_{[i+1..j-1]})p_j = s_jf(w_{[j+1..n-1]})p_n$ where s_1, p_i, p_j and p_n are non-empty words. We consider all the 3-uples (w, i, j) with w in S_{comp} and i, j some integers such that $1 < i < j < |w|$. We show that each configuration is impossible or implies that $f(T_{\min})$ is not cube-free.

Some configurations lead to immediate results. In particular, this is true for configurations that verify one case of Lemma 3. One configuration such that $i = 2$, $w_{[1]} \neq w_{[2]}$, $w_{[3..j-1]}$ or $w_{[j+1..n-1]}$ contains at least one a and one b is impossible. Indeed, in this case, $|u| \leq |f(w_{[1..2]})| = |f(ab)|$ and $|u| > \max\{|f(w_{[3..j-1]})|, |f(w_{[j+1..n-1]})|\} \geq |f(ab)|$. For the same reason, we cannot have "$i = j-1$, $w_{[i]} \neq w_{[j]}$, $w_{[2..i-1]}$ or $w_{[j+1..n-1]}$ contains at least one a and one b" and "$j = n-1$, $w_{[j]} \neq x_n$, $w_{[2..i-1]}$ or $w_{[i+1..j-1]}$ contains at least one a and one b".

After the elimination of the 3-uples for which one of the previous cases is verified, it keeps ten configurations to study (among 126): (*aabaabb*, 2, 6), (*aababb*, 2, 5), (*aababba*, 2, 5), (*aababba*, 3, 4), (*ababaab*, 3, 4), (*abbab*, 2, 4), (*abbaab*, 3, 5), (*aababa*, 3, 4), (*aababb*, 3, 4) and (*abbaab*, 3, 4). In each of these cases, we obtain by similar techniques as described in Part 4.1 a contradiction or the fact that $f(T_{\min})$ is cube-free.

Acknowledgments. The authors would like to thank P. Séébold for his encouragements during these works, and for all his helpful remarks.

References

1. D.R. Bean, A. Ehrenfeucht and G. McNulty, *Avoidable patterns in string of symbols*, Pacific J. Math. 95, p261-294, 1979.
2. J. Berstel, *Mots sans carré et morphismes itérés*, Discrete Mathematics 29, p235-244, 1980.
3. J. Berstel, *Axel Thue's Work on repetitions in words*, (4th Conf. on Formal Power Series and Algebraic Combinatorics, Montréal 1992) LITP Technical Report 92.70, 1992.
4. J. Berstel, *Axel Thue's papers on repetitions in words: a translation*, Publications of LaCIM 20, University of Québec at Montréal.
5. J. Berstel and P. Séébold, *A characterization of overlap-free morphisms*, Discrete Applied Mathematics 46, p275-281, 1993.

6. F.-J. Brandenburg, *Uniformly Growing k-th Power-Free Homomorphisms*, Theoretical Computer Science 23, p69-82, 1983.
7. C. Choffrut and J. Karhumäki, *chapter: Combinatorics of Words* in *Handbook of Formal Languages* vol.1 (G. Rozenberg and A. Salomaa Eds), Springer, 1997.
8. M. Crochemore, *Sharp Characterization of Squarefree Morphisms*, Theoretical Computer Science 18, p221-22, 1982.
9. J. Karhumäki, *On cube-free ω -words generated by binary morphisms*, Discrete Applied Mathematics 5, p279-297, 1983.
10. V. Keränen, *On the k-freeness of morphisms on free monoids*, Annales Academia Scientiarum Fennicae, 1986.
11. M. Leconte, *Codes sans répétition,* Thèse de 3ème cycle, LITP Université P. et M. Curie, 1985.
12. M. Leconte, *A characterization of power-free morphisms*, Theoretical Computer Science 38, p117-122, 1985.
13. M. Lothaire, *Combinatorics on words*, Encyclopedia of Mathematics, Vol. 17, Addison-Wesley; reprinted in 1997 by Cambridge University Press in the Cambridge Mathematical Library.
14. M. Lothaire, *Algebraic combinatorics on words*, Cambridge University Press, to appear.
15. M. Morse, *Recurrent geodesics on a surface of negative curvature*, Transactions Amer. Math. Soc. 22, p84-100, 1921.
16. F. Mignosi and P. Séébold, *If a D0L language is k-power free then it is circular*, ICALP'93, LNCS 700, Springer-Verlag, 1993, p507-518.
17. G. Richomme and P. Séébold, *Characterization of test-sets for overlap-free morphisms*, LaRIA Internal report 9724, 1997, to appear in Discrete Applied Mathematics.
18. P. Séébold, *Sequences generated by infinitely iterated morphisms*, Discrete Applied Mathematics 11, p255-264, 1985.
19. A. Thue, *Über unendliche Zeichenreihen*, Videnskapsselskapets Skrifter, I. Mat.-naturv. Klasse, Kristiania, p1-22, 1906.
20. A. Thue, *Über die gegenseitige Lage gleigher Teile gewisser Zeichenreihen*, Videnskapsselskapets Skrfter, I. Mat.-naturv. Klasse, Kristiania, p1-67, 1912.

Linear Cellular Automata with Multiple State Variables

Jarkko Kari*

Department of Computer Science, MLH 14,
University of Iowa, IA 52242, USA,
jjkari@cs.uiowa.edu,
http://www.cs.uiowa.edu/~jjkari/

Abstract. We investigate a finite state analog of subband coding, based on linear Cellular Automata with multiple state variables. We show that such a CA is injective (surjective) if and only if the determinant of its transition matrix is an injective (surjective, respectively) single variable automaton. We prove that in the one-dimensional case every injective automaton can be factored into a sequence of elementary automata, defined by elementary transition matrices. Finally, we investigate the factoring problem in higher dimensional spaces.

1 Introduction

Consider the frequently encountered task of encoding a discrete time signal $\ldots x_{-1}, x_0, x_1, \ldots$ where the sample values x_i are real numbers. Let us transform the samples x_i by applying on each site i a local linear function f, i.e., the transformed signal will be $\ldots y_{-1}, y_0, y_1, \ldots$ where for every $i \in \mathbb{Z}$

$$y_i = f(x_{i-r}, \ldots, f_{i+r})$$

for some positive integer r and a linear function $f : \mathbb{R}^{2r+1} \longrightarrow \mathbb{R}$. The local function f is a called a finite impulse response (FIR) filter, and the same filter f is used on all sites i. In order to use the transformed signal as an encoding of the original signal the global transformation has to be a one-to-one mapping. In signal processing literature this is known as the perfect reconstruction condition. It is easy to see that the only choices of linear f that have the perfect reconstruction property are functions that shift and/or scale the signal by a multiplicative constant.

In order to obtain non-trivial transforms one can relax the requirement that the same local function f is applied everywhere. Instead, let us use two linear functions f and g and let us apply f on even sites x_{2i} and g on odd sites x_{2i+1}. Then the transformed signal $\ldots y_{-1}, y_0, y_1, \ldots$ satisfies

$$y_i = \begin{cases} f(x_{i-r}, \ldots, f_{i+r}), & \text{for even } i, \text{ and} \\ g(x_{i-r}, \ldots, f_{i+r}), & \text{for odd } i. \end{cases}$$

* Research supported by NSF Grant CCR 97-33101

H. Reichel and S. Tison (Eds.): STACS 2000, LNCS 1770, pp. 110–121, 2000.

This approach is called subband coding. Even and odd samples form two subbands of the original signal. Suitable choices of f and g lead to perfect reconstruction transformations. Simple conditions on f and g are known for the perfect reconstruction condition, as well as for other desired properties such as the orthonormality of the global transformation. In compression applications the functions f and g are designed in such a way that if the input signal is "smooth" then most of the information and signal energy is packed in the even subband, while the odd subband contains little information and can be heavily compressed using entropy coding. The subband transformation can be repeated over the even subband, and further iterated using the even subband of the previous level as the input to the next level.

Digital signal processing literature contains detailed studies of subband coding of speech and images. State of the art wavelet compression algorithms are based on clever encodings of the subbands. The inherent multiresolution representation allows embedded encodings of signals, i.e., encodings where every prefix of a single compressed bit stream provides an approximation of the signal, and the quality of the approximation improves as the length of the prefix increases.

In this work we consider the analogous problem of encoding an infinite sequence $\ldots x_{-1}, x_0, x_1, \ldots$ using locally defined linear functions. However, we are interested in coding binary sequences, or more generally, sequences over a finite alphabet. For the sake of linearity the alphabet is assumed to have the algebraic structure of a commutative ring with identity. Potential applications include compression of bilevel images, or graphics containing a few colors. Transformations that apply the same local rule at all sites are known to computer scientists as cellular automata (CA). The perfect reconstruction condition is equivalent to the reversibility condition of CA. Reversible linear cellular automata have been studied in the past [7,10,17] and — analogously to the real valued case — they are of little use in compression applications. In the special case of linear CA over a finite field the only reversible rules are shifts and/or multiplications by a non-zero constant.

Following the example set by subband coding we generalize the notion of linear CA by allowing different local functions on odd and even lattice points, or more generally, m different local functions applied on m subbands. The same definition is more elegantly captured in the notion of a vector valued cellular automaton where each cell contains m components, updated according to a linear local rule.

Definitions of linear CA and multiband linear CA are given in Sections 2 and 3. In Section 3 we establish necessary and sufficient conditions for the perfect reconstruction (that is, the injectivity) property, as well as for the surjectivity property. In Section 4 we investigate the problem of factoring injective rules into elementary components. This is important in order to be able to efficiently find injective rules with desired properties. The elementary rules are also natural from the compression point of view as they correspond to computing predictions/prediction errors between different sublattices. In Section 5 we discuss the problems that arise in higher dimensional cellular spaces.

2 Linear Cellular Automata

Cellular Automata (CA) are discrete time dynamical systems consisting of an infinite regular lattice of cells. Cells are finite state machines that change their states synchronously according to a local rule f that specifies the new state of a cell as a function of the old states of some neighboring cells.

More precisely, let us consider a D-dimensional CA, with a finite state set Q. The cells are positioned at the integer lattice points of the D-dimensional Euclidean space, indexed by $\mathbb{Z}^D$. The global state – or the configuration – of the system at any given time is a function $c : \mathbb{Z}^D \longrightarrow Q$ that provides the states of all cells. Let C_Q^D denote the set of all D-dimensional configurations over state set Q.

The neighborhood vector $N = (\overline{v}_1, \overline{v}_2, \ldots, \overline{v}_n)$ of the CA specifies the relative locations of the neighbors of the cells: each cell $\overline{x} \in \mathbb{Z}^D$ has n neighbors, in positions $\overline{x} + \overline{v}_i$ for $i = 1, 2, \ldots, n$. The local rule $f : Q^n \longrightarrow Q$ determines the global dynamics $F : C_Q^D \longrightarrow C_Q^D$ as follows: For every $c \in C_Q^D$ and $\overline{x} \in \mathbb{Z}^D$ we have

$$F(c)(\overline{x}) = f\left[c(\overline{x} + \overline{v}_1), c(\overline{x} + \overline{v}_2), \ldots c(\overline{x} + \overline{v}_n)\right].$$

Function F is called a (global) CA function.

A Cellular Automaton is called linear (or additive) if its state set Q is a finite commutative ring with identity, usually the ring $\mathbb{Z}_m$ of integers modulo m, and the local rule f is a linear function $f(q_1, q_2, \ldots, q_n) = a_1 q_1 + a_2 q_2 + \ldots + a_n q_n$ where $a_1, a_2, \ldots, a_n$ are some constants from the ring Q. Linearity simplifies the analysis of the global function F, and many properties that are undecidable for general CA have polynomial time algorithms in the linear case. Throughout this paper, when Q is said to be a ring, we automatically assume that Q is a finite commutative ring with identity element 1.

A useful representation of linear local rules uses Laurent polynomials. For simplicity, consider the one-dimensional case $D = 1$ first – the higher dimensional cases are considered later in Section 5. Local rule

$$f(q_1, q_2, \ldots, q_n) = a_1 q_1 + a_2 q_2 + \ldots + a_n q_n$$

with neighborhood $N = (v_1, v_2, \ldots, v_n)$ is represented as the Laurent polynomial

$$p(x) = a_1 x^{-v_1} + a_2 x^{-v_2} + \ldots + a_n x^{-v_n}.$$

The polynomial is Laurent because both negative and positive powers of x are possible. Notice that if Laurent polynomials $p(x)$ and $q(x)$ over ring Q define global functions F and G, respectively, then the product $p(x)q(x)$ represents the composition $F \circ G$. Consequently, for any $k \geq 1$, $p^k(x)$ represents F^k. Also, it is useful to represent configurations as Laurent power series: Let one-dimensional configuration c correspond to the formal power series

$$s(x) = \sum_{i=-\infty}^{\infty} c(i)x^i.$$

Then the product $p(x)s(x)$ represents the configuration $F(c)$, and $p^k(x)s(x)$ represents $F^k(c)$.

Let us denote the set of Laurent polynomials over Q by $Q[x, x^{-1}]$. The set $Q[x, x^{-1}]$ itself is an (infinite) commutative ring. Let $Q[[x, x^{-1}]]$ denote the set of Laurent series over Q. Notice that elements of $Q[[x, x^{-1}]]$ can be added but not multiplied with each other. The product of a Laurent series and a Laurent polynomial is well defined.

It is easy to see that a CA function F is linear, i.e., defined by some Laurent polynomial $p(x)$, if and only if it is a linear transformation of $Q[[x, x^{-1}]]$, i.e., if and only if $F(c+d) = F(c) + F(d)$ and $F(a \cdot c) = aF(c)$ for all configurations c and d, and all $a \in Q$ [2].

Classical results concerning Cellular Automata address the injectivity and surjectivity problems. A CA in called injective if its global function F is one-to-one and surjective if the global function is onto. The following basic properties are valid in any dimension D, and they are valid for unrestricted CA, that is, not only in the linear case:

1. Every injective CA is also surjective [6,16].
2. If F is an injective CA function then its inverse F^{-1} is also a CA function [6,16], computed by the inverse cellular automaton. Sometimes injective CA are called invertible or reversible.
3. A CA function F is surjective if and only if F is injective on the set of finite configurations [13,14]. (A configuration c is called finite w.r.t. state q if only a finite number of cells are in states different from q. The statement is valid for any choice of $q \in Q$.)

In the unrestricted case it is difficult to characterize local rules that make the CA injective or surjective. If the space is at least two-dimensional then it is even undecidable whether a given local rule defines an injective or surjective dynamics [8]. (In the one-dimensional case decision algorithms exist [1].)

Linearity simplifies the analysis. The inverse of a linear injective function is also a linear function, so $p(x)$ defines an injective CA if and only if there exists a Laurent polynomial $q(x)$ such that $p(x)q(x) = 1$. Such $q(x)$ is the local rule of the inverse automaton. And based on property 3, $p(x)$ defines a surjective CA if and only if there does not exist a Laurent polynomial $q(x) \neq 0$ such that $p(x)q(x) = 0$. Such $q(x)$ would namely represent a finite configuration c such that $F(c) = 0 = F(0)$. In other words, the linear CA represented by Laurent polynomial $p(x)$ is injective (non-surjective) if and only if $p(x)$ is a unit (a zero divisor, respectively) of the ring $Q[x, x^{-1}]$.

It turns out that inclusions of the coefficients of $p(x)$ in the maximal ideals of ring Q determines the injectivity and surjectivity status of the CA, as proved in [17]:

Proposition 1. (Sato 1993) *The linear CA represented by polynomial $p(x)$ over ring Q is*

(a) surjective if and only if no maximal ideal of Q contains all coefficients of $p(x)$.
(b) injective if and only if for every maximal ideal exactly one coefficient of $p(x)$ is outside the ideal.

In the special case $Q = \mathbb{Z}_m$ we obtain the following well-known result [7]:

Corollary 1. (Ito et. al. 1983) *The linear CA represented by polynomial $a_1x^{v_1} + a_2x^{v_2} + \ldots + a_nx^{v_n}$ over ring $\mathbb{Z}_m$ is*

(a) surjective if and only $\gcd(m, a_1, a_2, \ldots, a_n) = 1$.
(b) injective if and only if every prime factor p of m divides all but exactly one coefficient $a_1, a_2, \ldots, a_n$.

The conditions of Proposition 1 can be rephrased in the following easy-to-check characterizations of injective and surjective linear rules:

Proposition 2. *The linear CA represented by polynomial $p(x)$ over ring Q is*

(a) surjective if and only if $a \cdot p(x) \neq 0$ for every $a \in Q \setminus \{0\}$,
(b) injective if and only if for every $a \in Q \setminus \{0\}$ there exists $b \in Q$ such that $ab \cdot p(x)$ is a monomial.

3 Multiband Linear Rules

In the previous section we considered linear Cellular Automata over ring Q, where each cell used the same linear local rule. Let us generalize the situation by allowing different linear rules on even and odd numbered cells. In this section we analyze such CA — and the more general case of m different local rules applied in different positions modulo m — and we provide algorithms to determine the injectivity and the surjectivity status of any such automaton.

Let $p_1(x)$ and $p_2(x)$ be the Laurent polynomials representing the local rules on even and odd cells, respectively. Let us separate the even and odd powers of x in the polynomials, i.e., let us find Laurent polynomials $p_{11}(x)$, $p_{12}(x)$, $p_{21}(x)$ and $p_{22}(x)$ such that

$$p_1(x) = p_{11}(x^2) + x \cdot p_{12}(x^2), \text{ and}$$
$$p_2(x) = x \cdot p_{21}(x^2) + p_{22}(x^2).$$

Notice that $p_{11}(x)$ and $p_{12}(x)$ represent the contributions of even and odd cells to even cells, respectively, and $p_{21}(x)$ and $p_{22}(x)$ represent the contributions of even and odd cells to odd cells, respectively. So if $c_1(x)$ and $c_2(x)$ are Laurent power series representing the states of even and odd cells at any given time, then series

$$p_{11}(x)c_1(x) + p_{12}(x)c_2(x), \text{ and}$$
$$p_{21}(x)c_1(x) + p_{22}(x)c_2(x)$$

represent the states of even and odd cells one time step later. Using matrix notation, the new configuration is represented by

$$\begin{pmatrix} p_{11}(x) & p_{12}(x) \\ p_{21}(x) & p_{22}(x) \end{pmatrix} \begin{pmatrix} c_1(x) \\ c_2(x) \end{pmatrix}.$$

This leads to the following definition of linear CA with multiple state variables. Let us combine blocks of two cells into single "super" cells. Then each cell contains two elements of the ring Q, and both elements are updated according to some local linear rules. The two elements may have different rules, but all "super" cells are identical.

More generally, let us allow every cell to store m elements of the ring Q, so the states are m-tuples $(q_1, q_2, \ldots, q_m) \in Q^m$. Let us extend the Laurent power series notation for configurations. A configuration c consists of m Laurent power series $c_i(x)$ over ring Q, organized as a column vector of size m:

$$c(x) = \begin{pmatrix} c_1(x) \\ c_2(x) \\ \vdots \\ c_m(x) \end{pmatrix}$$

Series $c_i(x)$ gives the states of the i'th state variables of all cells, and we call it the i'th band of the configuration.

The local rule is a linear function. It is defined by an $m \times m$ square matrix of Laurent polynomials:

$$A(x) = \begin{pmatrix} p_{11}(x) & p_{12}(x) & \ldots & p_{1m}(x) \\ p_{21}(x) & p_{22}(x) & \ldots & p_{2m}(x) \\ \vdots & \vdots & \ddots & \vdots \\ p_{m1}(x) & p_{m2}(x) & \ldots & p_{mm}(x) \end{pmatrix}$$

We call this the transition matrix of the automaton. Laurent polynomial $p_{ij}(x)$ gives the contribution of the j'th band to the i'th band. Analogously to the single variable case, the result of applying CA $A(x)$ to configuration $c(x)$ is configuration $A(x)c(x)$. Composition of two CA is then given by the product of their transition matrices, and the k'th iterate of CA $A(x)$ on initial configuration $c(x)$ is $A^k(x) \cdot c(x)$. All products are standard matrix products over matrices whose elements are Laurent polynomials and series. We call such CA linear m-band CA over ring Q, and we use the abbreviation m-band LCA. In the single band case $m = 1$ the definition is identical to the normal linear CA over Q.

Notice that a CA function F is an m-band LCA if and only if it is a linear function on the set $Q[[x, x^{-1}]]^m$ of configurations: $F(c + d) = F(c) + F(d)$ and $F(a \cdot c) = aF(c)$ for all configurations c and d and ring element $a \in Q$. Consequently, the inverse CA of any injective m-band LCA is also an m-band LCA.

The following proposition reduces the injectivity and surjectivity questions of m-band CA into the well understood single band case:

Proposition 3. *An m-band LCA over ring Q is injective (surjective) if and only if the determinant of its transition matrix is an injective (surjective, respectively) one-band LCA over Q.*

Proof. Let $A(x)$ be the transition matrix of an m-band LCA.
1. Injectivity: The CA defined by $A(x)$ is injective if and only if $A(x)$ is an invertible matrix, that is, if and only if there exits a transition matrix $B(x)$ (of the inverse automaton) such that $A(x)B(x) = I$. A matrix over a commutative ring is invertible if and only if its determinant is a unit of the ring (see for example [11], Theorem 50). Notice that the elements of transition matrices are elements of the commutative ring $Q[x, x^{-1}]$. So the CA is invertible if and only if $\det A(x)$ is an invertible element of $Q[x, x^{-1}]$, that is, if and only if the one-band LCA defined by $\det A(x)$ is injective.

2. Surjectivity: The CA defined by $A(x)$ is surjective if and only if $A(x)B(x) \neq 0$ for every transition matrix $B(x) \neq 0$. This is equivalent to $A(x)$ not being a zero divisor because any square matrix over a commutative ring is a left zero divisor if and only if it is a right zero divisor. On the other hand, a square matrix is a zero divisor if and and only if its determinant is a zero divisor (McCoy [11], Theorem 51), so $A(x)$ defines a surjective CA if and only if $\det A(x)$ is not a zero divisor of $Q[x, x^{-1}]$, i.e., if and only if $\det A(x)$ defines a surjective one-band LCA. □

According to the proposition $\det A(x)$ is a single band LCA that has the same injectivity and surjectivity status as the multiband CA $A(x)$.

4 Factoring Injective Rules into Elementary Components

Proposition 3 gives a characterization of injective m-band linear CA. However, it does not provide a simple method of constructing injective automata, apart from trying different matrices one-by-one and checking whether their determinants satisfy the condition of Proposition 2(b). In this section we consider simple CA rules, called elementary rules, that are trivially invertible and easy to construct. Then we show how any injective m-band LCA can be factored into a composition of such elementary CA.

For any Laurent polynomial $p(x) \in Q[x, x^{-1}]$ and any $i \neq j$ we define an elementary m-band LCA with transition matrix $E_{ij}(p(x))$ as follows: All diagonal elements of $E_{ij}(p(x))$ are 1, element (i, j) is $p(x)$, and all other elements are 0. In other words, the elementary automaton $E_{ij}(p(x))$ adds to band i the result of applying $p(x)$ to band j. This is known as an elementary row operation in linear algebra. The automaton is injective, with determinant 1, and the inverse of $E_{ij}(p(x))$ is $E_{ij}(-p(x))$.

Because $\det AB = \det A \det B$, the determinant of any composition of elementary CA is 1. We are interested in the opposite direction: Is every matrix with determinant 1 a product of elementary matrices ? It turns out that this claim is true. However, unlike other results we have seen so far, this claim does not easily generalize to higher dimensional cellular spaces. This point will be discussed in Section 5.

Let us use the following standard notations:

- (general linear group) $\mathrm{GL}_m(R)$ is the set of invertible $m \times m$ matrices over the ring R. These are the matrices over R whose determinant is a unit of R. In our case $R = Q[x, x^{-1}]$ the elements of $\mathrm{GL}_m(R)$ are exactly the injective m-band LCA.
- (special linear group) $\mathrm{SL}_m(R)$ is the set of $m \times m$ matrices over ring R whose determinant is 1.
- $\mathrm{E}_m(R)$ are the $m \times m$ matrices over R that are products of elementary matrices.

We are interested in determining whether $\mathrm{E}_m(R) = \mathrm{SL}_m(R)$ in the case of $R = Q[x, x^{-1}]$.

Clearly, if $A \in \mathrm{GL}_m(R)$ then, for any matrix M whose determinant is $(\det A)^{-1}$, we have $M^{-1}A \in \mathrm{SL}_m(R)$. Matrix M can be as simple as the diagonal matrix whose diagonal entries are 1 except for the first elements which is $(\det A)^{-1}$. Therefore, a factorization of matrices in $\mathrm{SL}_m(R)$ into elementary matrices provides a factorization of matrices in $\mathrm{GL}_m(R)$ into products of a diagonal matrix and a sequence of elementary matrices.

Proposition 4. *Let $R = Q[x, x^{-1}]$ where Q is a finite commutative ring with identity. Then, for every $m \geq 1$, $SL_m(R) = E_m(R)$.*

Proof. Without loss of generality we may assume that Q is a local ring. Every finite commutative ring is namely a direct sum of local rings [12].

Let us briefly review some properties of local rings. See [12] for proofs and more details. A local ring Q has a unique maximal ideal M, and the quotient ring Q/M is a field. Ideal M consists of the zero-divisors of Q. Every element outside M is a unit. Set M is nilpotent, which means that there exists an integer n such that the product of any n elements of M is 0.

Let us consider Laurent polynomials $p(x)$ over local ring Q, with maximal ideal M. We know that $p(x)$ is invertible iff exactly one coefficient is not in M. Let us define the degree of a Laurent polynomial in the following, slightly non-standard way. Let us separate the unit and non-unit terms of of the polynomial:

$$p(x) = (a_1 x^{v_1} + a_2 x^{v_2} + \ldots + a_n x^{v_n}) + (b_1 x^{u_1} + b_2 x^{u_2} + \ldots + b_k x^{u_k})$$

where $v_1 < v_2 < \ldots < v_n$, $u_1 < u_2 < \ldots < u_k$, $v_i \neq u_j$ for all i and j, every a_i is a unit and every b_i is a non-unit, i.e. an element of M. The degree $\deg p(x)$ is then $v_n - v_1$ if $n \geq 1$, and $-\infty$ if $n = 0$. In other words, the non-unit terms are ignored in the calculation of the degree. We have the following easily verifiable properties: For all $p(x)$ and $q(x)$, $\deg p(x)q(x) = \deg p(x) + \deg q(x)$, and if $\deg p(x) = -\infty$ then $\deg[p(x) + q(x)] = \deg q(x)$. Laurent polynomial $p(x)$ is invertible iff its degree is 0.

The existence of a division algorithm is crucial: Let $f(x)$ and $g(x)$ be Laurent polynomials such that $\deg g(x) \geq 0$. Then there exist Laurent polynomials $q(x)$ and $r(x)$ such that

$$f(x) = g(x)q(x) + r(x)$$

and $\deg r(x) < \deg g(x)$.

To prove Proposition 4 we show that any member of $\mathrm{SL}_m(R)$ can be reduced to the identity matrix using elementary row and column operations. Because elementary row and column operations correspond to multiplying the matrix by an elementary matrix from the left and right, respectively, and because the inverse of an elementary matrix is elementary, this proves the proposition.

Let us use mathematical induction on m, the size of the matrices. If $m = 1$ the claim is trivial. Assume then the claim has been proved for matrices of size $(m-1) \times (m-1)$ and consider an $m \times m$ matrix $A(x) = (p_{ij}(x))$ whose determinant is 1. So the degree of the determinant is 0. The determinant is a linear expression

$$\det A(x) = p_{11}(x)q_1(x) + p_{21}(x)q_2(x) + \ldots + p_{m1}(x)q_m(x)$$

of the first column where the coefficients $q_i(x)$ are cofactors of the matrix. If all the elements $p_{i1}(x)$ of the first column would have degree $-\infty$ then we would have $\deg[\det A(x)] = -\infty$. So at least one $p_{i1}(x)$ has degree ≥ 0.

If more than one element on the first column has degree ≥ 0 we can use the division algorithm to reduce the degree of one of them: If $\deg p_{i1}(x) \geq \deg p_{j1}(x) \geq 0$ then there exist $q(x)$ and $p'_{i1}(x)$ such that $p_{i1}(x) = q(x)p_{j1}(x) + p'_{i1}(x)$ and $\deg p'_{i1}(x) < \deg p_{i1}(x)$. So the degree of the element $(i, 1)$ can be reduced using the elementary row operation $E_{ij}(-q(x))$. This process can be repeated until only one element of the first column has degree ≥ 0.

Once only one element, say $p_{i1}(x)$, of the first column has degree ≥ 0, then

$$\begin{aligned} 0 &= \deg[\det A(x)] \\ &= \deg\left[p_{11}(x)q_1(x) + p_{21}(x)q_2(x) + \ldots + p_{m1}(x)q_m(x)\right] \\ &= \deg p_{i1}(x) + \deg q_i(x). \end{aligned}$$

We must have $\deg p_{i1}(x) = 0$, i.e., $p_{i1}(x)$ is invertible. If $i \neq 2$ we add row i to row 2 to obtain an invertible element in position $(2, 1)$. Then we can add suitable multiples of row 2 to other rows to obtain 1 in position $(1, 1)$, and 0 in positions $(j, 1)$ for all $j > 2$. Finally, using the elementary row operation $E_{21}(-p_{21}(x))$ we reduce also element $(2, 1)$ of the matrix to 0.

Using elementary row operations we were able to transform $A(x)$ into a matrix whose first column is the first column of the identity matrix. Using $m-1$ elementary column operations we can then make the elements $(1, i)$, $i \geq 2$, of the first row equal to 0. We end up with matrix

$$\left(\begin{array}{c|ccc} 1 & 0 & \ldots & 0 \\ \hline 0 & & & \\ \vdots & & A'(x) & \\ 0 & & & \end{array}\right)$$

where $A'(x)$ is an $(m-1) \times (m-1)$ matrix with determinant 1. According to the inductive hypothesis $A'(x)$ can be reduced to the identity matrix using elementary row and column operations. □

5 Higher Dimensional Cellular Spaces

The results of Sections 2 and 3 can readily be generalized to higher dimensional cellular spaces. Linear rules on D-dimensional cellular spaces are represented as Laurent polynomials over D variables $x_1, x_2, \ldots, x_D$. We will abbreviate $\overline{x} = (x_1, x_2, \ldots, x_D)$, and denote $p(\overline{x})$ if p is a Laurent polynomial of variables $x_1, x_2, \ldots, x_D$. Let $Q[\overline{x}, \overline{x}^{-1}]$ be the ring of Laurent polynomials over $x_1, x_2, \ldots, x_D$.

Configurations are Laurent power series over the same variables. Let $Q[[\overline{x}, \overline{x}^{-1}]]$ denote the set of Laurent power series over $x_1, x_2, \ldots, x_D$. Product $p(\overline{x})c(\overline{x})$ represents the configuration $F(c)$ if $p(\overline{x})$ represents F and power series $c(\overline{x})$ represents configuration c. As in the one-dimensional case, set $Q[\overline{x}, \overline{x}^{-1}]$ is a commutative ring.

A D-dimensional m-band LCA is defined analogously to the one-dimensional case. It is specified by an $m \times m$ matrix of Laurent polynomials over $x_1, x_2, \ldots, x_D$. Propositions 1, 2 and 3 remain valid in the D-dimensional case: the proofs are analogous and use only the fact that $Q[\overline{x}, \overline{x}^{-1}]$ is a commutative ring.

However, factoring injective CA into elementary components is considerably harder in the higher dimensional spaces. The proof of Proposition 4 in the previous section does not work if $D \geq 2$ because the division algorithm is no longer available. The proof was essentially the Euclidean algorithm where the degree of the polynomials on the first column was reduced using the division algorithm. In the higher dimensional case there is no natural notion of the degree of a polynomial, and the Euclidean algorithm cannot be used. Nevertheless, factorization is possible if the number of bands is at least three, but other techniques need to be used.

The problem of factoring matrices of the special linear group $\mathrm{SL}_m(R)$ into products of elementary matrices has been investigated for various commutative rings R. A classic result by Suslin [19] states that $\mathrm{SL}_m(R) = \mathrm{E}_m(R)$ if $m \geq 3$ and $R = k[\overline{x}]$ is the ring of polynomials over a field k. In contrast, case $m = 2$ is different [4]: $\mathrm{SL}_2(R) \neq \mathrm{E}_2(R)$ for polynomial rings $R = k[\overline{x}]$. Note that the results concern polynomials, not Laurent polynomials. But in [15] H. Park introduced a technique to transform Laurent polynomials into non-Laurent polynomials in such a way that Suslin's theorem can be extended to Laurent polynomials over a field.

Because the quotient Q/M of a local ring Q and its unique maximal ideal M is a field, Park's result proves that one can use elementary operations to reduce any $A(\overline{x}) \in \mathrm{SL}_m(Q[\overline{x}, \overline{x}^{-1}])$, for $m \geq 3$ and local ring Q, into $I + B(\overline{x})$, where I is the identity matrix and $B(\overline{x})$ is a matrix of Laurent polynomials with coefficients in M. It is then straightforward to reduce $I + B(\overline{x})$ into I. Because every finite commutative ring is a direct sum of local rings, we have

Proposition 5. *Let $R = Q[\overline{x}, \overline{x}^{-1}]$ where Q is a finite commutative ring. Then $SL_m(R) = E_m(R)$ for every $m \geq 3$.*

The case $m = 2$ remains open. Cohn's counter example

$$\begin{pmatrix} x^2 & xy+1 \\ xy-1 & y^2 \end{pmatrix}$$

for polynomials over any field [4] is not a counter example for Laurent polynomials because the diagonal elements are invertible as Laurent polynomials.

6 Conclusions

We have introduced and investigated linear Cellular Automata with m state variables over a commutative ring Q. Especially we studied automata that are invertible and proved that, in the one-dimensional case, such automata can be factored into elementary components. In higher dimensional spaces with two bands it remains an outstanding open problem whether such a factorization exists. One should note that our factorization corresponds to so-called ladder decompositions in the theory of subband coding of signals [3].

A motivation for this study came from potential applications to compression of binary images and signals. The idea, analogous to subband coding, is to divide the signal into odd and even samples and use a 2-band CA to pack most of the information into the even samples. The process can then be repeated on the even band, and iterated over several levels. Using zerotree coding [18] one can encode the odd samples on different levels into a compact and fully embedded representation of the original signal.

In this work we restricted the study to linear CA. Unfortunately linearity is a constraint that severely limits the compression results one can hope to get. Non-linear CA allow more flexibility, but they are also much harder to analyze. One possible next step would be to extend the factorization result to the non-linear case. It is straightforward to generalize the notion of an elementary operation to non-linear CA over m bands. A non-linear elementary operation changes variables of one band only. The variables are changed according to some permutations π of Q. Which permutation π is used in any given cell is determined by the variables on the other $m-1$ bands in the neighborhood. Such elementary step is trivially invertible because the inverse permutation π^{-1} restores the original states. It would be interesting to investigate which injective CA can be obtained by combining such non-linear elementary operations. Notice that each elementary step can be viewed as making a prediction for the value of a variable based on known values on the other $m-1$ bands in the vicinity and storing the prediction error in the variable. If the prediction is good then the prediction error will have low information content, and will be compressable.

References

1. S. Amoroso and Y. Patt. Decision Procedures for Surjectivity and Injectivity of Parallel Maps for Tessellation Structures. *Journal of Computer and System Sciences* **6** (1972): 448-464.
2. H. Aso and N. Honda. Dynamical Characteristics of Linear Cellular Automata. *Journal of Computer and System Sciences* **30** (1985), 291–317.
3. A. Bruckers and A. van den Enken. New networks for perfect inversion and perfect reconstruction. *IEEE J. Select. Areas Commun.* **10** (1992), 130–137.
4. P.M. Cohn. On the structure of the GL_2 of a ring. *Inst. Hautes Études Sci. Publ. Math.* **30** (1966), 365–413.
5. K. Culik II. On Invertible Cellular Automata. *Complex Systems* **1** (1987): 1035-1044.
6. G. Hedlund. Endomorphisms and automorphisms of shift dynamical systems, *Mathematical Systems Theory* **3** (1969), 320–375.
7. M. Ito, N. Osato and M. Nasu. Linear Cellular Automata over $\mathbb{Z}_m$. *Journal of Computer and System Sciences* **27** (1983), 125–140.
8. J. Kari. Reversibility and surjectivity problems of cellular automata, *Journal of Computer and System Sciences* **48**, 149–182, 1994.
9. J. Kari. Representation of reversible cellular automata with block permutations. *Mathematical Systems Theory* **29**, 47–61, 1996.
10. G. Manzini and L. Margara. Invertible Linear Cellular Automata over $\mathbb{Z}_m$: Algorithmic and Dynamical Aspects. *Journal of Computer and System Sciences* **56** (1998), 60–67.
11. N. McCoy. *Rings and Ideals*, 1948.
12. B. McDonald. *Finite Rings with Identity*, 1974.
13. E.F. Moore. Machine Models of Self-reproduction. *Proc. Sympos. Appl. Math.* **14** (1962): 17-33.
14. J. Myhill. The Converse to Moore's Garden-of-Eden Theorem. *Proc. Amer. Math. Soc.* **14** (1963): 685-686.
15. H.A. Park. *A Computational Theory of Laurent Polynomial Rings and Multidimensional FIR Systems*, Ph.D. thesis, UC Berkeley, 1995.
16. D. Richardson. Tessellation with local transformations, *Journal of Computer and System Sciences* **6** (1972), 373–388.
17. T. Sato. Decidability of some problems of linear cellular automata over finite commutative rings. *Information Processing Letters* **46** (1993), 151–155.
18. J.M.Shapiro. Embedded image coding using zerotrees of wavelet coefficients. *IEEE Transactions on Signal Processing, Special Issue on Wavelets and Signal Processing* **41(12)** (1993), 3445–3462.
19. A.A. Suslin. On the structure of the special linear group over polynomial rings. *Math. USSR Izv.* **11** (1977), 221–238.
20. T. Toffoli and N. Margolus. *Cellular Automata Machines*, MIT Press, Cambridge, Massachusetts, 1987.
21. T. Toffoli and N. Margolus. Invertible Cellular Automata: A Review, *Physica D*, **45** (1990), 229–253.
22. L. Tolhuizen H. Hollmann and T. Kalker. On the Realizability of Biorthogonal, m-Dimensional Two-Band Filter Banks. *IEEE Transactions on Signal Processing* **43** (1995), 640–648.

Two-Variable Word Equations
(Extended Abstract)

Lucian Ilie[1*] and Wojciech Plandowski[1,2**]

[1] Turku Centre for Computer Science TUCS
FIN-20520 Turku, Finland
lucili@cs.utu.fi

[2] Institute of Informatics, Warsaw University
Banacha 2, 02-097 Warsaw, Poland
wojtekpl@mimuw.edu.pl

Abstract. We consider languages expressed by word equations in two variables and give a complete characterization for their complexity functions, that is, the functions that give the number of words of a given length. Specifically, we prove that there are only five types of complexities: constant, linear, exponential, and two in between constant and linear. For the latter two, we give precise characterizations in terms of the number of solutions of Diophantine equations of certain types. There are several consequences of our study. First, we show that the linear upper bound on the non-exponential complexities by Karhumäki et al., cf. [KMP], is optimal. Second, we derive that both of the sets of all finite Sturmian words and of all finite Standard words are expressible by word equations. Third, we characterize the languages of non-exponential complexity which are expressible by two-variable word equations as finite unions of several simple parametric formulae and solutions of a two-variable word equation with a finite graph. Fourth, we find optimal upper bounds on the solutions of (solvable) two-variable word equations, namely, linear bound for one variable and quadratric for the other. From this, we obtain an $\mathcal{O}(n^6)$ algorithm for testing the solvability of two-variable word equations.

Keywords: word equation, expressible language, complexity function, minimal solution, solvability

1 Introduction

Word equations constitute one of the basic parts of combinatorics on words. The fundamental result in word equations is Makanin's algorithm, cf. [Ma], which decides whether or not a word equation has a solution. The algorithm is one of the most complicated ones existing in the literature. The structure of solutions

* Research supported by the Academy of Finland, Project 137358. On leave of absence from Faculty of Mathematics, University of Bucharest, Str. Academiei 14, R-70109, Bucharest, Romania.

** Supported by KBN grant 8 T11C 039 15.

H. Reichel and S. Tison (Eds.): STACS 2000, LNCS 1770, pp. 122–132, 2000.

of word equations is not well understood; see [Hm, Raz1, Raz2]. A new light on that topic has been led recently by [KMP] where the languages which are defined by solutions of word equations are studied.

The structure of languages which are defined by equations with one variable is very simple. The infinite languages which are defined by one-variable word equations consist of a finite part and an infinite part which is of the form $A^n A'$ for A' a prefix of A. The structure of the finite part is not completely known [OGM]. Our analysis deals with languages which are defined by two-variable word equations. We prove that the complexity of those languages, which is measured by the number of words of a given length, belongs to one of five classes: constant, $\mathcal{D}_1$-type, $\mathcal{D}_2$-type, linear and exponential. The complexities $\mathcal{D}_1$-type and $\mathcal{D}_2$-type are in between linear and constant and they are related to the number of solutions of certain Diophantine equations. As a side effect of our considerations we prove that the linear upper bound given in [KMP] for languages which do not contain a pattern language is optimal. An interesting related result is that the sets of Sturmian and Standard words are expressible by simple word equations. As another consequence, we characterize the languages of non-exponential complexity which are expressible by two-variable word equations as finite unions of several simple parametric formulae and solutions of simple two-variable word equations.

Based on our analysis, we find optimal upper bounds on the solutions of (solvable) two-variable word equations, namely, linear bound for one variable and quadratric for the other. From this, we obtain an $\mathcal{O}(n^6)$ algorithm for testing the solvability of two-variable word equations. We recall that the only polynomial-time algorithm known for this problem is the one given by Charatonik and Pacholski [ChPa]. Its complexity, as computed in [ChPa], is $\mathcal{O}(n^{100})$. It should be added that they did not take very much care of the complexity. They mainly intended to prove that the problem can be solved in polynomial time.

Due to space limitations we remove all proofs in particular several lemmas which are used to prove our main theorem Theorem 3.

2 Expressible Languages

In this section we give basic definitions we need later on, as well as recalling some previous results. For an alphabet Σ, we denote by card(Σ) the number of elements of Σ; Σ^* is the set of words over Σ with 1 the empty word. For $w \in \Sigma^*$, $|w|$ is the length of w; for $a \in \Sigma$, $|w|_a$ is the number of occurrences of a in w. By $\rho(w)$ we denote the primitive root of w. If $w = uv$, then we denote $u^{-1}w = v$ and $wv^{-1} = u$. For any notions and results of combinatorics on words, we refer to [Lo] and [ChKa].

Consider two disjoint alphabets, of constants, Σ, and of variables, Ξ. A *word equation* e is a pair of words $\varphi, \psi \in (\Sigma \cup \Xi)^*$, denoted $e : \varphi = \psi$. The *size* of e, denoted $|e|$, is the sum of the lengths of φ and ψ. The equation e is said to be *reduced* if φ and ψ start with different letters and end with different letters, as words over $\Sigma \cup \Xi$. Throughout the paper, all equations we consider are assumed to be reduced.

A *solution* of e is a morphism $h : (\Sigma \cup \Xi)^* \to \Sigma^*$ such that $h(a) = a$, for any $a \in \Sigma$, and $h(\varphi) = h(\psi)$. The set of solutions of e is denoted by $\mathrm{Sol}(e)$.

Notice that a solution can be given also as an ordered tuple of words, each component of the tuple corresponding to a variable of the equation. Therefore, we may take, for a variable $X \in \Xi$, the X-component of all solutions of e, that is,

$$L_X(e) = \{x \in \Sigma^* \mid \text{ there is a solution } h \text{ of } e \text{ such that } h(X) = x\}.$$

The set $L_X(e)$ is called *the language expressed by X in e*. A language $L \subseteq \Sigma^*$ is *expressible* if there is a word equation e and a variable X such that $L = L_X(e)$. Notice that, if X does not appear in e, then $L_X(e) = \Sigma^*$ as soon as $\mathrm{Sol}(e) \neq \emptyset$. Also, if $\mathrm{card}(\Sigma) = 1$, that is, there is only one constant letter, then all expressible languages are trivially regular, as we work here with numbers. Therefore, we shall assume that always $\mathrm{card}(\Sigma) \geq 2$. The *complexity function* of a language $L \subseteq \Sigma^*$, is the natural function $\#_L : N \to N$ defined by $\#_L(n) = \mathrm{card}\{w \in L \mid |w| = n\}$.

Example 1. Consider the equation $e : XX = Y$. The complexity of its solutions with respect to Y is

$$\#_{L_Y(e)}(n) = \begin{cases} \mathrm{card}(\Sigma)^{\frac{n}{2}}, & \text{if } n \text{ is even,} \\ 0, & \text{if } n \text{ is odd.} \end{cases}$$

Since the function $\#_L$ can be very irregular, as can be seen from the above example, we use in our considerations a function $\bar{\#}_L$, which is defined by $\bar{\#}_L(n) = \max_{1 \leq i \leq n} \#_L(i)$. We say that a function f is *constant* if $f(n) = \Theta(1)$, is *linear* if $f(n) = \Theta(n)$, and is *exponential* if $f(n) = 2^{\Theta(n)}$.

We make the following conventions concerning notations:

– $a, b, \ldots \in \Sigma$ are constant letters,

– $A, B, \ldots \in \Sigma^*$ are (fixed) constant words,

– $X, Y, \ldots \in \Xi$ are variables,

– $x, y, \ldots \in \Sigma^*$ may denote some arbitrary constant words but may also stand for images of variables by some morphisms from $(\Sigma \cup \Xi)^*$ to Σ^*, that is, $x = h(X), y = h(Y)$, etc.,

– $\varphi, \psi, \ldots \in (\Sigma \cup \Xi)^*$ are mixed words, which *may* (but need not) contain both constants and variables.

We shall use also the following notation (due to Hmelevskii, cf. [Hm]); for $\alpha_i \in (\Sigma \cup \Xi)^*, 1 \leq i \leq n$, we denote $[\alpha_i]_{i=1}^n = \alpha_1 \alpha_2 \cdots \alpha_n$.

Example 2. For a fixed word $A \in \Sigma^*$, the language $L_1 = \{A^n \mid n \geq 0\}$ is expressed by the variable Y in the two-variable word equation $e_1 : XAY = AXX^t$ where t is such that $A = \rho(A)^t$. Then $\bar{\#}_{L_Y(e)} = \bar{\#}_{L_1}$ is constant.

We shall prove that the following result from [KMP] gives an optimal bound on non-exponential complexity of languages expressible by two variable word equations.

Theorem 1. (Karhumäki et al. [KMP]) *Any non-exponential complexity function of languages expressible by two-variable word equations is at most linear.*

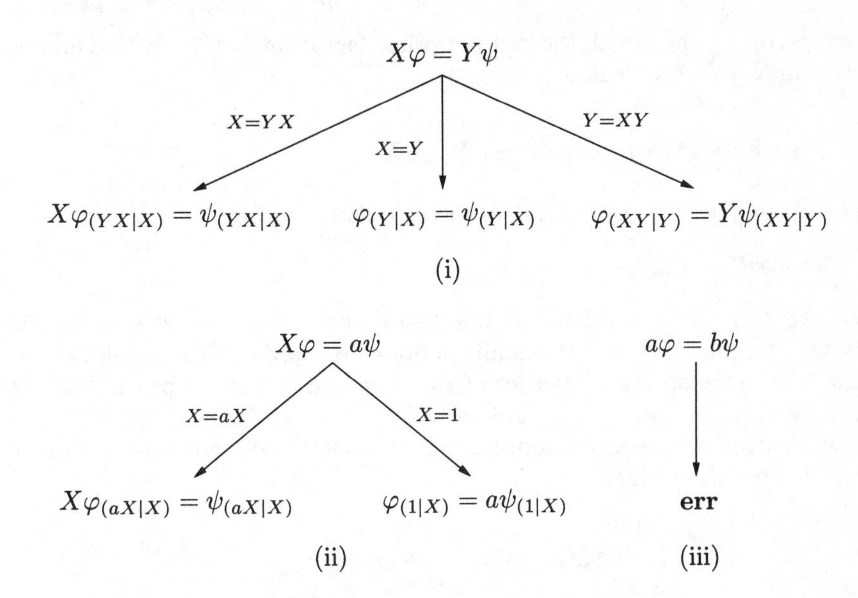

Fig. 1. The graph associated with a word equation

We shall need also the *graph* associated with an equation $e : \varphi = \psi$, see [Lo]. It is constructed by applying exhaustively the so-called Levi's lemma which states that if $uv = wt$, for some words u, v, w, t, then either u is a proper prefix of w or $u = w$ or w is a proper prefix of u. The vertices of the graph are different equations (including e) and the directed edges are put as follows. We start with e and draw the graph by considering iteratively the following three cases, depicted in Fig. 1: (i) both sides of e start with variables (which are different since the equation is assumed to be reduced), (ii) one side starts with a constant and the other starts with a variable, and (iii) the two sides start with constants which are different. Clearly, in the last case the equation has no solution, which is marked by an error node. In Fig. 1, we denote by $\varphi_{(\alpha|\beta)}$ the word obtained from φ by replacing all occurrences of β by α.

Thus, we start by processing e and then process all unprocessed vertices. When we find an equation already obtained, we do not create a new vertex but direct the corresponding edge to the old one.

We notice that the graph associated with a word equation may be infinite but, if it is finite, then all solutions of the equation are obtained starting from a vertex with no outgoing edges and different from **err** and going in the oposite direction of the edges to the root; at the same time, the corresponding operations on the values of the variables are performed.

We recall that the *Euler's totient function* $\phi : N \to N$ is defined by

$$\phi(n) = \text{card}\{k|k \text{ is coprime with } n\} = n(1-\frac{1}{p_1})(1-\frac{1}{p_2})\dots(1-\frac{1}{p_k})$$

where $p_1, p_2, \dots, p_k$ are all the distinct prime factors of n. Clearly, the function $\bar{\phi}(n) = \max_{1 \le i \le n} \phi(i)$ is linear.

3 The Equation $XbaY = YabX$

The starting point of our analysis is the equation

$$\mathbf{e_0} : XbaY = YabX$$

which we study in this section. We show first that there is a very close connection between solutions of $\mathbf{e_0}$ and the family of Standard words which we define below. Using then some strong properties of the Standard words, we prove that both functions $\bar{\#}_{L_X(\mathbf{e_0})}$ and $\bar{\#}_{L_Y(\mathbf{e_0})}$ are linear.

Let us consider the set of solutions of our equation $\mathbf{e_0}$. For this we draw its associated graph in Fig. 2.

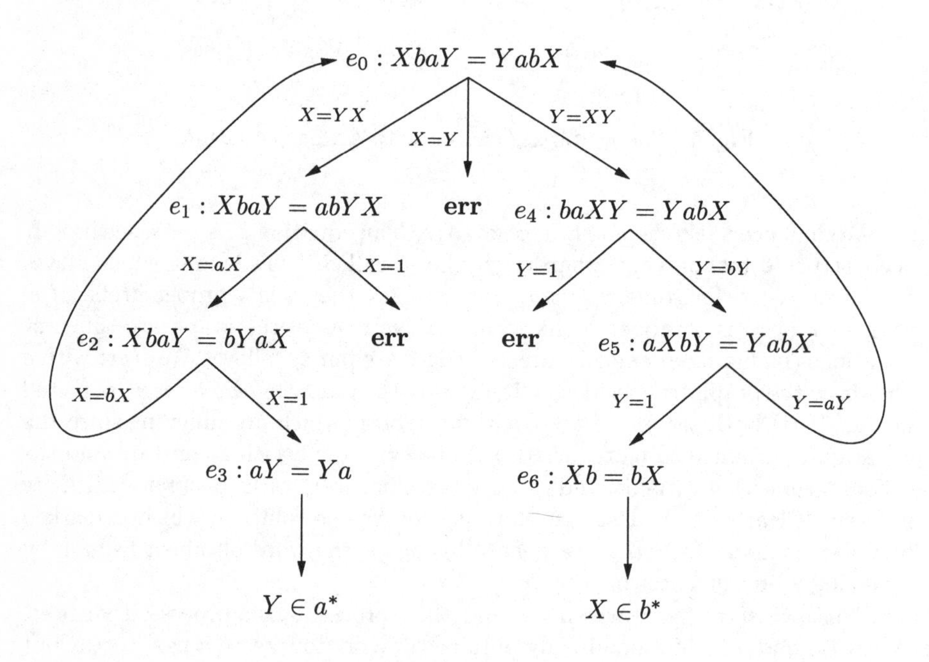

Fig. 2. The graph of $\mathbf{e_0}$

Consider the following two mappings

$$\alpha_1, \alpha_2 : \{a,b\}^* \times \{a,b\}^* \to \{a,b\}^* \times \{a,b\}^*$$

defined by

$$\alpha_1(u,v) = (u, ubav), \alpha_2(u,v) = (vabu, v),$$

for any $u, v \in \{a,b\}^*$. Using these two mappings, we give the following result which characterizes the set of solutions of $\mathbf{e_0}$.

Lemma 1. *The solutions of* $\mathbf{e_0}$ *are precisely the pairs of words obtained by: (i) starting with a pair* (u,v) *of words in the set*

$$\{(a^{n+1}, a^n) \mid n \geq 0\} \cup \{(b^n, b^{n+1}) \mid n \geq 0\},$$

(ii) applying to (u,v) *a finite (possible empty) sequence* $\alpha_{i_1}, \alpha_{i_2}, \ldots, \alpha_{i_k}$, *for some* $k \geq 0, 1 \leq i_j \leq 2$, *for any* $1 \leq j \leq k$.

We define next the Standard words. The set $\mathcal{R}$ of *Standard pairs* (as defined by Rauzy, cf. [Ra]) is the minimal set included in $\{a,b\}^* \times \{a,b\}^*$ such that:
(i) $(a,b) \in \mathcal{R}$ and
(ii) $\mathcal{R}$ is closed under the two mappings

$$\beta_1, \beta_2 : \{a,b\}^* \times \{a,b\}^* \to \{a,b\}^* \times \{a,b\}^*$$

defined by $\beta_1(u,v) = (u, uv), \beta_2(u,v) = (vu, v)$, for any $u, v \in \{a,b\}^*$. The set $\mathcal{S}$ of *Standard words* is defined by

$$\mathcal{S} = \{u \in \{a,b\}^* \mid \text{ there is } v \in \{a,b\}^* \text{ such that either } (u,v) \in \mathcal{R} \text{ or } (v,u) \in \mathcal{R}\}.$$

We shall use the following strong properties of Standard words, proved by de Luca and Mignosi, cf. [dLMi].

Lemma 2. (de Luca, Mignosi [dLMi]) *The set of Standard words satisfies the formula* $\mathcal{S} = \{a,b\} \cup \Pi\{ab, ba\}$ *where*

$$\Pi = \{w \in \{a,b\}^* \mid w \text{ has two periods } p, q \text{ which are coprime and } |w| = p + q - 2\}.$$

Lemma 3. (de Luca, Mignosi [dLMi]) $\#_\Pi(n) = \phi(n+2)$, *for any* $n \geq 0$, *where* ϕ *is Euler's totient function.*

We now establish a connection between the solutions of $\mathbf{e_0}$ and the set of Standard pairs $\mathcal{R}$.

Lemma 4. $\mathrm{Sol}(\mathbf{e_0}) = \{(u,v) \mid (uba, vab) \in \mathcal{R}\}$.

Theorem 2. *1.* $\#_{L_X(\mathbf{e_0})}(n) = \#_{L_Y(\mathbf{e_0})}(n) = \phi(n)$, *for any* $n \geq 1$.
2. $L_X(\mathbf{e_0}) = L_Y(\mathbf{e_0}) = \Pi$

As a corollary of Theorem 2 we obtain that the upper bound of Karhumäki et al. in Theorem 1 is optimal.

Corollary 1. *The linear upper bound for the non-exponential complexities of languages expressible by two-variable word equations is optimal.*

Furthermore, using the above considerations, we prove that both sets, of Standard and of Sturmian (finite) words are expressible by word equations.

Example 3. **Standard words.** The set of Standard words $\mathcal{S}$ is expressed by the variable Z in the following system:

$$\begin{cases} XbaY = YabX \\ Z = Xba \text{ or } Z = Yab \text{ or } Z = a \text{ or } Z = b \end{cases}$$

First, by Lemma 4, it is clear that the Z-components of all solutions of the system above are precisely the Standard words. Second, from the above system we can derive a single equation, as well known; see, e.g., [KMP]. Hence, the set of Standard words is expressible.

Example 4. **Sturmian words.** There are many definitions of the finite Sturmian words (see, e.g., [Be] and references therein). We use here only the fact that the set $\mathcal{S}t$ of finite Sturmian words is the set of factors of Π, cf. [dLMi]. Therefore, the set $\mathcal{S}t$ is expressed by the variable Z in the system

$$\begin{cases} XbaY = YabX \\ X = WZT \end{cases}$$

Again, this can be expressed using a single equation.

4 Special Equations

We study in this section equations in two variables such that the first from the left variable appearing in both sides of the equation is the same. We show that in this case there are three possible types of complexity for the language expressed by the other component: constant, exponential, and $\mathcal{D}_1$-type. The last type lies in between constant and linear and is defined in terms of the number of solutions of certain Diophantine equations.

Before giving the definition of the $\mathcal{D}_1$-type, we give an example showing how it arises naturally.

Example 5. Consider the equation

$$e : aXXbY = XaYbX.$$

Clearly, the set of solutions of e is $\mathrm{Sol}(e) = \{(a^n, (a^n b)^m a^n) \mid n \geq 0\}$. Here $\#_{L_X(e)}$ is constant but $\#_{L_Y(e)}$ is not. Indeed, for any $p \geq 0$, $\#_{L_Y(e)}(p)$ is the number of solutions of the Diophantine equation in unknowns n and m, $(n+1)m + n = p$.

We now define the $\mathcal{D}_1$-type precisely. A function $f : N \to N$ is of *divisor-type* if there are some non-negative integers $c_i, 1 \le i \le 4$, such that $c_1 \ge c_3, c_2 \ge c_4$ and, $f(k)$ is the number of solutions of the Diophantine equation in unknowns n and m, $(c_1 n + c_2)m + c_3 n + c_4 = k$. A function f is $\mathcal{D}_1$*-type* if there is a divisor-type function g such that $f = \Theta(\bar{g})$ where $\bar{g}(n) = \max_{1 \le i \le n} g(i)$.

Lemma 5. *If $e : \varphi = \psi$ is an equation over $\Xi = \{X, Y\}$ such that the first variable appearing in each of φ and ψ is X, then $\bar{\#}_{L_X(e)}$ is constant and $\bar{\#}_{L_Y(e)}$ is either constant, or $\mathcal{D}_1$-type, or else exponential.*

We now give some examples in order to see that all situations in Lemma 5 are indeed possible.

Example 6. (i) Consider first the equation $e_1 : aXaY = XaYa$. Then, clearly, both $\bar{\#}_{L_X(e_1)}$ and $\bar{\#}_{L_Y(e_1)}$ are constant.

(ii) For the equation in Example 5, we have that $\bar{\#}_{L_Y(e)}$ is $\mathcal{D}_1$-type.

(iii) Our last equation is

$$e_2 : aXYXa = XaYaX.$$

Then, clearly, $\mathrm{Sol}(e_2) = \{(a^n, w) \mid n \ge 0, w \in \Sigma^*\}$, hence $\#_{L_Y(e_2)}$ is exponential as soon as $\mathrm{card}(\Sigma) \ge 2$.

5 The General Form

We start by defining $\mathcal{D}_2$-type complexity functions. We say that a natural function $f : N \to N$ is of *divisor2-type* if there are some integers $c_i, 1 \le i \le 8$, such that, $f(k)$ is the number of solutions of the Diophantine equation in unknowns n, m, and p

$$((c_1 n + c_2)m + c_3 n + c_4)p + (c_5 n + c_6)m + c_7 n + c_8 = k.$$

Note here that each divisor-type function is also divisor2-type function We say that a function f is of $\mathcal{D}_2$*-type* complexity if there is a divisor2-type function g such that $f = \Theta(\bar{g})$ where $\bar{g}(n) = \max_{1 \le i \le n} g(i)$.

We give next an example in which the $\mathcal{D}_2$-type complexity is reached.

Example 7. Consider the equation

$$e : XabcXcbabcY = YcbaXcbabcX,$$

which we solve completely in what follows. Consider a solution $(x, y) \in \mathrm{Sol}(e)$. Then $xabcxcbabc$ and $cbaxcbabcx$ are conjugated by y, that is, $xabcxcbabc = \mathrm{cycle}^t(cbaxcbabcx)$, for some $0 \le t \le 2|x| + 7$. It is not difficult to see that the only possibilities for t are (i) $t = 1$, (ii) $t = |x| + 4$, and (iii) $|x| + 9 \le t \le 2|x| + 7$. (The cases $t \in \{0, 2, 3, |x| + 3\}$ and $|x| + 5 \le t \le |x| + 8$ are immediately ruled out; the cases $4 \le t \le |x| + 2$, are ruled out using a reasoning which is in the proof of Theorem 3.)

In case (i) we have $x = bab, y = (ba(babc)^3)^n ba(babc)^2 bab, n \geq 0$, and in case (ii) we obtain $x = b, y = (ba(bc)^2 babc)^n babcb, n \geq 0$. Thus, in both cases, we have a constant contribution to either of $\#_{L_X(e)}$ and $\#_{L_Y(e)}$.

The interesting case is (iii). Applying the reasoning in the proof of Theorem 3, we obtain

$$x = (bc)^n b(a(bc)^{n+1}b)^m a(bc)^n b, y = (xabxaab)^p uv(cba)^{-1} \tag{1}$$

for any $n, m, p \geq 0$ where $u = (bc)^n b$ and $v = (a(bc)^{n+1}b)^m a$. Hence, for any $k \geq 0$, $\#_{L_Y(e)}$ differs by at most 2 from the number of solutions of the Diophantine equation (in unknowns n, m, and p)

$$((4n+8)m + 8n + 11)p + (2n+4)m + 2n - 1 = k.$$

We have used also the fact that, if we denote y in (1) by $y_{m,n,p}$, then $(n_1, m_1, p_1) \neq (n_2, m_2, p_2)$ implies $y_{n_1,m_1,p_1} \neq y_{n_2,m_2,p_2}$. Consequently, $\#_{L_Y(e)}$ is of $\mathcal{D}_2$-type.

We now study two-variable word equations of the general form. We show that one cannot obtain as complexities of their expressed languages anything but the five types we have identified so far, namely constant, $\mathcal{D}_1$-type, $\mathcal{D}_2$-type, linear, and exponential.

Theorem 3. *Let e be an equation with two variables X, Y. Then*

$$\bar{\#}_{L_X(e)}, \bar{\#}_{L_Y(e)} \in \{constant, \mathcal{D}_1\text{-}type, \mathcal{D}_2\text{-}type, linear, exponential\}.$$

As another consequence of our study, we can give the general forms of the languages expressible by two-variable word equations.

First, we need the notion of pattern language, from [An], cf. also [JSSY]. A *pattern* is a word over the alphabet $\Sigma \cup \Xi$. A *pattern language* generated by a pattern α, denoted $L(\alpha)$ is the set of all morphic images of α under morphisms $h : (\Sigma \cup \Xi)^* \to \Sigma^*$ satisfying $h(a) = a$, for any $a \in \Sigma$.

By Theorem 13 in [KMP], we know that, for any language L which is expressible by a two-variable word equation, if $\bar{\#}_L$ is exponential, then there exists a pattern α containing occurrences of one variable only such that $L(\alpha) \subseteq L$.

We have then the following theorem which characterizes the languages expressible by two-variable word equations.

Theorem 4. *For any language L which is expressible by a two-variable word equation, we have*
(i) if $\bar{\#}_L$ is exponential, then L contains a pattern language,
(ii) if $\bar{\#}_L$ is not exponential, then L is a union of
(a) a finite language,
(b) finitely many parametric formulae of the forms

- $A^n B$,
- $(A^n B)^m A^n C$,
- $([A^n B_i]_{i=1}^k)^m \operatorname{pref}([A^n B_i]_{i=1}^k)$,
- $([(A^n B)^m A^n C D_i]_{i=1}^k)^p \operatorname{pref}([(A^n B)^m A^n C D_i]_{i=1}^k)$, *and*

(c) solutions of an equation $XAY = YBX$; these solutions can be expressed as compositions of finite number of substitutions which can be computed on the basis of the graph for $XAY = YBX$.

6 Minimal Solutions and Solvability

We consider here the lengths of solutions for two-variable word equations.

Example 8. Consider the equation

$$e: aXa^n bX^n = XaXbY.$$

The equation e has a unique solution which is $(x, y) = (a^n, a^{n^2})$. As $|e| = 2n+8$, we have that $|x| = \Theta(|e|)$ and $|y| = \Theta(|e|^2)$.

Using our analysis of two-variable word equations one can prove that the bounds for the lengths of words in solutions in Example 8 are optimal.

Theorem 5. *If e is a solvable two-variable word equation over $\Xi = \{X, Y\}$, then e has a solution (x, y) such that $|x| \leq 2|e|$, $|y| \leq 2|e|^2$.*

Given a two-variable word equation $e: \varphi = \psi$, and two non-negative numbers l_x, l_y, it is clear that we can check in time $|e| + |\varphi\psi|_X(l_x - 1) + |\varphi\psi|_Y(l_y - 1)$ whether e has a solution (x, y) for some x, y with $|x| = l_x$, $|y| = l_y$. Therefore, we get immediately from Theorem 5 the following result.

Theorem 6. *The solvability of two-variable word equations can be tested in time $\mathcal{O}(n^6)$.*

Another consequence of Theorem 5 concerns the complexity of languages expressible by three-variable word equations. The following result can be proved as Theorem 13 in [KMP].

Theorem 7. *Let L be a language expressible by a three-variable word equation. Then either there is a one-variable pattern α such that $L(\alpha) \subseteq L$ or $\#_L(n) = \mathcal{O}(n^3)$.*

Acknowledgements

The authors would like to thank Dr. Filippo Mignosi (University of Palermo, Italy) for useful comments on the paper.

References

[An] Angluin, D., Finding patterns common to a set of strings, *J. Comput. System Sci.* **21**(1) (1980) 46 – 62.

[Be] Berstel, J., Recent results in Sturmian words, in J. Dassow, G. Rozenberg, A. Salomaa, eds., *Developments in Language Theory II*, 13 – 24, World Sci. Publishing, 1996.

[ChPa] Charatonik, W. and Pacholski, L., Word equations with two variables, *Proc. of IWWERT'91*, H. Abdulrab, J. P. Pecuchet, eds., 43 – 57, LNCS 667, Springer, Berlin, 1991.

[ChKa] Choffrut, C. and Karhumäki, J., Combinatorics of words, in G. Rozenberg, A. Salomaa, eds., *Handbook of Formal Languages*, 329 – 438, Springer, Berlin, 1997.

[dLMi] de Luca, A. and Mignosi, F., Some combinatorial properties of sturmian words, *Theoret. Comput. Sci.* **136** 361 – 385, 1994.

[OGM] Eyono Obono, S., Goralcik, P., and Maksimenko, M., Efficient solving of the word equations in one variable, in *Proc. of MFCS'94*, 336 – 341, LNCS 841, Springer, Berlin, 1994.

[Hm] Hmelevskii, Yu. I., Equations in free semigroups, *Trudy Mat. Inst. Steklov* **107** 1971. English transl. *Proc Steklov Inst. of Mathematics* **107** (1971), Amer. Math. Soc., 1976.

[JSSY] Jiang, T., Salomaa, A., Salomaa, K., and Yu, S., Decision problems for patterns, *J. Comput. System Sci.* **50**(1) (1995) 53 – 63.

[KMP] Karhumäki, J., Mignosi, F., and Plandowski, W., The expressibility of languages and relations by word equations, in *Proc. of ICALP'97*, 98 – 109, LNCS 1256, Springer, Berlin, 1997.

[KoPa] Koscielski, A. and Pacholski, L., Complexity of Makanin's algorithm, *Journal of the ACM*, **43**(4), 670-684, 1996.

[Lo] Lothaire, M., *Combinatorics on Words*, Addison-Wesley, Reading, Mass., 1983.

[Ma] Makanin, G.S., The problem of solvability of equations in a free semigroup, *Mat. Sb.* **103** (145), 147-233, 1977. English transl. in *Math. U.S.S.R. Sb.* **32**, 1977.

[Ra] Rauzy, G., Mots infinis en arithmetique, in M. Nivat, D. Perrin, eds., *Automata on infinite words*, LNCS 192, Springer, Berlin, 1984.

[Raz1] Razborov, A., On systems of equations in a free group, *Math. USSR Izvestija* **25**(1), 115 – 162, 1985.

[Raz2] Razborov, A., On systems of equations in a free group, Ph.D. Thesis, Moscow State University, 1987.

[Si] Sierpinski, W., *Elementary Theory of Numbers*, Elseviers Science Publishers B.V., Amsterdam, and PWN - Polish Scientific Publishers, Warszawa, 1988.

Average-Case Quantum Query Complexity

Andris Ambainis[1*] and Ronald de Wolf[2,3]

[1] Computer Science Department, University of California, Berkeley CA 94720,
ambainis@cs.berkeley.edu
[2] CWI, P.O. Box 94079, 1090 GB Amsterdam, The Netherlands,
rdewolf@cwi.nl
[3] ILLC, University of Amsterdam

Abstract. We compare classical and quantum query complexities of total Boolean functions. It is known that for *worst-case* complexity, the gap between quantum and classical can be at most polynomial [3]. We show that for *average-case* complexity under the uniform distribution, quantum algorithms can be exponentially faster than classical algorithms. Under non-uniform distributions the gap can even be super-exponential. We also prove some general bounds for average-case complexity and show that the average-case quantum complexity of MAJORITY under the uniform distribution is nearly quadratically better than the classical complexity.

1 Introduction

The field of quantum computation studies the power of computers based on quantum mechanical principles. So far, most quantum algorithms—and *all* physically implemented ones—have operated in the so-called *black-box* setting. Examples are [9,18,11,7,8]; even period-finding, which is the core of Shor's factoring algorithm [17], can be viewed as a black-box problem. Here the input of the function f that we want to compute can only be accessed by means of queries to a "black-box". This returns the ith bit of the input when queried on i. The complexity of computing f is measured by the required number of queries. In this setting we want quantum algorithm that use significantly fewer queries than the best classical algorithms.

We restrict attention to computing total Boolean functions f on N variables. The query complexity of f depends on the kind of errors one allows. For example, we can distinguish between exact computation, zero-error computation (a.k.a. Las Vegas), and bounded-error computation (Monte Carlo). In each of these models, *worst-case* complexity is usually considered: the complexity is the number of queries required for the "hardest" input. Let $D(f)$, $R(f)$ and $Q(f)$ denote the worst-case query complexity of computing f for classical deterministic algorithms, classical randomized bounded-error algorithms, and quantum bounded-error algorithms, respectively. Clearly $Q(f) \leq R(f) \leq D(f)$.

* Part of this work was done when visiting Microsoft Research.

H. Reichel and S. Tison (Eds.): STACS 2000, LNCS 1770, pp. 133–144, 2000.

The main quantum success here is Grover's algorithm [11]. It can compute the OR-function with bounded-error using $\Theta(\sqrt{N})$ queries (this is optimal [4,5,20]). Thus $Q(\text{OR}) \in \Theta(\sqrt{N})$, whereas $D(\text{OR}) = N$ and $R(\text{OR}) \in \Theta(N)$. This is the biggest gap known between quantum and classical worst-case complexities for total functions. (In contrast, for *partial* Boolean functions the gap can be much bigger [9,18].) A recent result is that the gap between $D(f)$ and $Q(f)$ is at most polynomial for *every* total f: $D(f) \in O(Q(f)^6)$ [3]. This is similar to the best-known relation between classical deterministic and randomized algorithms: $D(f) \in O(R(f)^3)$ [16].

Given some probability distribution μ on the set of inputs $\{0,1\}^N$ one may also consider *average-case* complexity instead of worst-case complexity. Average-case complexity concerns the expected number of queries needed when the input is distributed according to μ. If the hard inputs receive little μ-probability, then average-case complexity can be significantly smaller than worst-case complexity. Let $D^\mu(f)$, $R^\mu(f)$, and $Q^\mu(f)$ denote the average-case analogues of $D(f)$, $R(f)$, and $Q(f)$, respectively. Again $Q^\mu(f) \leq R^\mu(f) \leq D^\mu(f)$. The objective of this paper is to compare these measures and to investigate the possible gaps between them. Our main results are:

- Under uniform μ, $Q^\mu(f)$ and $R^\mu(f)$ can be super-exponentially smaller than $D^\mu(f)$.
- Under uniform μ, $Q^\mu(f)$ can be exponentially smaller than $R^\mu(f)$. Thus the [3]-result for worst-case quantum complexity does not carry over to the average-case setting.
- Under non-uniform μ the gap can be even larger: we give distributions μ where $Q^\mu(\text{OR})$ is constant, whereas $R^\mu(\text{OR})$ is almost $\sqrt{N}$. (Both this gap and the previous one still remains if we require the quantum algorithm to work with zero-error instead of bounded-error.)
- For every f and μ, $R^\mu(f)$ is lower bounded by the expected *block sensitivity* $E_\mu[bs(f)]$ and $Q^\mu(f)$ is lower bounded by $E_\mu[\sqrt{bs(f)}]$.
- For the MAJORITY-function under uniform μ, we have $Q^\mu(f) \in O(N^{1/2+\varepsilon})$ for every $\varepsilon > 0$, and $Q^\mu(f) \in \Omega(N^{1/2})$. In contrast, $R^\mu(f) \in \Omega(N)$.
- For the PARITY-function, the gap between Q^μ and R^μ can be quadratic, but not more. Under uniform μ, PARITY has $Q^\mu(f) \in \Omega(N)$.

2 Definitions

Let $f : \{0,1\}^N \to \{0,1\}$ be a Boolean function. It is *symmetric* if $f(X)$ only depends on $|X|$, the Hamming weight (number of 1s) of X. $\overline{0}$ denotes the input with weight 0. We will in particular consider the following functions: $\text{OR}(X) = 1$ iff $|X| \geq 1$; $\text{MAJ}(X) = 1$ iff $|X| > N/2$; $\text{PARITY}(X) = 1$ iff $|X|$ is odd. If $X \in \{0,1\}^N$ is an input and S a set of (indices of) variables, we use X^S to denote the input obtained by flipping the values of the S-variables in X. The *block sensitivity* $bs_X(f)$ of f on input X is the maximal number b for which there are b disjoint sets of variables $S_1, \ldots, S_b$ such that $f(X) \neq f(X^{S_i})$ for all $1 \leq i \leq b$. The block sensitivity $bs(f)$ of f is $\max_X bs_X(f)$.

We focus on three kinds of algorithms for computing f: classical *deterministic*, classical *randomized* bounded-error, and *quantum* bounded-error algorithms. If A is an algorithm (quantum or classical) and $b \in \{0,1\}$, we use $\Pr[A(X)=b]$ to denote the probability that A answers b on input X. We use $T_A(X)$ for the expected number of queries that A uses on input X.[1] Note that this only depends on A and X, not on the input distribution μ. For deterministic A, $\Pr[A(X)=b] \in \{0,1\}$ and the expected number of queries $T_A(X)$ is the same as the actual number of queries.

Let $\mathcal{D}(f)$ denote the set of classical *deterministic* algorithms that compute f. Let $\mathcal{R}(f) = \{\text{classical } A \mid \forall X \in \{0,1\}^N : \Pr[A(X)=f(X)] \geq 2/3\}$ be the set of classical *randomized* algorithms that compute f with bounded error probability. Similarly let $\mathcal{Q}(f)$ be the set of *quantum* algorithms that compute f with bounded-error. We define the following *worst-case complexities*:

$$D(f) = \min_{A \in \mathcal{D}(f)} \max_{X \in \{0,1\}^N} T_A(X)$$

$$R(f) = \min_{A \in \mathcal{R}(f)} \max_{X \in \{0,1\}^N} T_A(X)$$

$$Q(f) = \min_{A \in \mathcal{Q}(f)} \max_{X \in \{0,1\}^N} T_A(X)$$

$D(f)$ is also known as the *decision tree complexity* of f and $R(f)$ as the *bounded-error* decision tree complexity of f. Since quantum generalizes randomized and randomized generalizes deterministic computation, we have $Q(f) \leq R(f) \leq D(f)$ for all f. The three worst-case complexities are polynomially related: $D(f) \in O(R(f)^3)$ [16] and $D(f) \in O(Q(f)^6)$ [3] for all total f.

Let $\mu : \{0,1\}^N \to [0,1]$ be a probability distribution. We define the *average-case complexity* of an algorithm A with respect to a distribution μ as:

$$T_A^\mu = \sum_{X \in \{0,1\}^N} \mu(X) T_A(X).$$

The average-case deterministic, randomized, and quantum complexities of f with respect to μ are

$$D^\mu(f) = \min_{A \in \mathcal{D}(f)} T_A^\mu$$

$$R^\mu(f) = \min_{A \in \mathcal{R}(f)} T_A^\mu$$

$$Q^\mu(f) = \min_{A \in \mathcal{Q}(f)} T_A^\mu$$

Note that the algorithms still have to output the correct answer on *all* inputs, even on X that have $\mu(X) = 0$. Clearly $Q^\mu(f) \leq R^\mu(f) \leq D^\mu(f)$ for all μ and

[1] See [3] for definitions and references for the quantum circuit model. A satisfactory formal definition of *expected* number of queries $T_A(X)$ for a quantum algorithm A is a hairy issue, involving the notion of a stopping criterion. We will not give such a definition here, since in the bounded-error case, expected and worst-case number of queries can be made the same up to a small constant factor.

f. Our goal is to examine how large the gaps between these measures can be, in particular for the uniform distribution $unif(X) = 2^{-N}$.

The above treatment of average-case complexity is the standard one used in average-case analysis of algorithms [19]. One counter-intuitive consequence of these definitions, however, is that the average-case performance of polynomially related algorithms can be superpolynomially apart (we will see this happen in Section 5). This seemingly paradoxical effect makes these definitions unsuitable for dealing with polynomial-time reducibilities and average-case complexity classes, which is what led Levin to his alternative definition of "polynomial time on average" [13].[2] Nevertheless, we feel the above definitions are the appropriate ones for our query complexity setting: they just *are* the average number of queries that one needs when the input is drawn according to distribution μ.

3 Super-Exponential Gap between $D^{unif}(f)$ and $Q^{unif}(f)$

Here we show that $D^{unif}(f)$ can be much larger then $R^{unif}(f)$ and $Q^{unif}(f)$:

Theorem 1. *Define f on N variables such that $f(X) = 1$ iff $|X| \geq N/10$. Then $Q^{unif}(f)$ and $R^{unif}(f)$ are $O(1)$ and $D^{unif}(f) \in \Omega(N)$.*

Proof. Suppose we randomly sample k bits of the input. Let $a = |X|/N$ denote the fraction of 1s in the input and $\tilde{a}$ the fraction of 1s in the sample. Standard Chernoff bounds imply that there is a constant $c > 0$ such that

$$\Pr[\tilde{a} < 2/10 \mid a \geq 3/10] \leq 2^{-ck}.$$

Now consider the following randomized algorithm for f:

1. Let $i = 1$.
2. Sample $k_i = i/c$ bits. If the fraction $\tilde{a}_i$ of 1s is $\geq 2/10$, output 1 and stop.
3. If $i < \log N$, increase i by 1 and repeat step 2.
4. If $i \geq \log N$, count N exactly using N queries and output the correct answer.

It is easily seen that this is a bounded-error algorithm for f. Let us bound its average-case complexity under the uniform distribution.

If $a \geq 3/10$, the expected number of queries for step 2 is

$$\sum_{i=1}^{\log N} \Pr[\tilde{a}_1 \leq 2/10, \ldots, \tilde{a}_{i-1} \leq 2/10 \mid a > 3/10] \cdot \frac{i}{c} \leq$$

$$\sum_{i=1}^{\log N} \Pr[\tilde{a}_{i-1} \leq 2/10 \mid a > 3/10] \cdot \frac{i}{c} \leq \sum_{i=1}^{\log N} 2^{-(i-1)} \cdot \frac{i}{c} \in O(1).$$

The probability that step 4 is needed (given $a \geq 3/10$) is at most $2^{-c \log N/c} = 1/N$. This adds $\frac{1}{N}N = 1$ to the expected number of queries.

[2] We thank Umesh Vazirani for drawing our attention to this.

The probability of $a < 3/10$ is $2^{-c'N}$ for some constant c'. This case contributes at most $2^{-c'N}(N + (\log N)^2) \in o(1)$ to the expected number of queries. Thus in total the algorithm uses $O(1)$ queries on average, hence $R^{unif}(f) \in O(1)$.

It is easy to see that any deterministic classical algorithm for f must make at least $N/10$ queries on every input, hence $D^{unif}(f) \geq N/10$. □

Accordingly, we can have huge gaps between $D^{unif}(f)$ and $Q^{unif}(f)$. However, this example tells us nothing about the gaps between quantum and classical bounded-error algorithms. In the next section we exhibit an f where $Q^{unif}(f)$ is exponentially smaller than $R^{unif}(f)$.

4 Exponential Gap between $R^{unif}(f)$ and $Q^{unif}(f)$

4.1 The Function

We use the following modification of Simon's problem [18]:[3]

Input: $X = (x_1, \ldots, x_{2^n})$, where each $x_i \in \{0,1\}^n$.
Output: $f(X) = 1$ iff there is a non-zero $k \in \{0,1\}^n$ such that $x_{i \oplus k} = x_i \; \forall i$.

Here we treat $i \in \{0,1\}^n$ both as an n-bit string and as a number, and $\oplus$ denotes bitwise XOR. Note that this function is total (unlike Simon's). Formally, f is not a Boolean function because the variables are $\{0,1\}^n$-valued. However, we can replace every variable x_i by n Boolean variables and then f becomes a Boolean function of $N = n2^n$ variables. The number of queries needed to compute the Boolean function is at least the number of queries needed to compute the function with $\{0,1\}^n$-valued variables (because we can simulate a query to the Boolean oracle with a query to the $\{0,1\}^n$-valued oracle by just throwing away the rest of the information) and at most n times the number of queries to the $\{0,1\}^n$-valued oracle (because one $\{0,1\}^n$-valued query can be simulated using n Boolean queries). As the numbers of queries are so closely related, it does not make a big difference whether we use the $\{0,1\}^n$-valued oracle or the Boolean oracle. For simplicity we count queries to the $\{0,1\}^n$-valued oracle.

The main result is the following exponential gap:

Theorem 2. *For f as above, $Q^{unif}(f) \leq 22n + 1$ and $R^{unif}(f) \in \Omega(2^{n/2})$.*

4.2 Quantum Upper Bound

The quantum algorithm is similar to Simon's. Start with the 2-register superposition $\sum_{i \in \{0,1\}^n} |i\rangle|0\rangle$ (for convenience we ignore normalizing factors). Apply the oracle once to obtain

$$\sum_{i \in \{0,1\}^n} |i\rangle|x_i\rangle.$$

[3] The recent preprint [12] proves a related but incomparable result about another modification of Simon's problem.

Measuring the second register gives some j and collapses the first register to

$$\sum_{i:x_i=j} |i\rangle.$$

Applying a Hadamard transform H to each qubit of the first register gives

$$\sum_{i:x_i=j} \sum_{i'\in\{0,1\}^n} (-1)^{(i,i')}|i'\rangle. \tag{1}$$

(a, b) denotes inner product mod 2; if $(a, b) = 0$ we say a and b are orthogonal.

If $f(X) = 1$, then there is a non-zero k such that $x_i = x_{i\oplus k}$ for all i. In particular, $x_i = j$ iff $x_{i\oplus k} = j$. Then the final state (1) can be rewritten as

$$\sum_{i'\in\{0,1\}^n} \sum_{i:x_i=j} (-1)^{(i,i')}|i'\rangle = \sum_{i'\in\{0,1\}^n} \left(\sum_{i:x_i=j} \frac{1}{2}((-1)^{(i,i')} + (-1)^{(i\oplus k,i')}) \right) |i'\rangle$$
$$= \sum_{i'\in\{0,1\}^n} \left(\sum_{i:x_i=j} \frac{(-1)^{(i,i')}}{2}(1 + (-1)^{(k,i')}) \right) |i'\rangle.$$

Notice that $|i'\rangle$ has non-zero amplitude only if $(k, i') = 0$. Hence if $f(X) = 1$, then measuring the final state gives some i' orthogonal to the unknown k.

To decide if $f(X) = 1$, we repeat the above process $m = 22n$ times. Let $i_1, \ldots, i_m \in \{0,1\}^n$ be the results of the m measurements. If $f(X) = 1$, there must be a non-zero k that is orthogonal to all i_r. Compute the subspace $S \subseteq \{0,1\}^n$ that is generated by $i_1, \ldots, i_m$ (i.e. S is the set of binary vectors obtained by taking linear combinations of $i_1, \ldots, i_m$ over $GF(2)$). If $S = \{0,1\}^n$, then the only k that is orthogonal to all i_r is $k = 0^n$, so then we know that $f(X) = 0$. If $S \neq \{0,1\}^n$, we just query all 2^n values $x_{0\ldots0}, \ldots, x_{1\ldots1}$ and then compute $f(X)$. This latter step is of course very expensive, but it is needed only rarely:

Lemma 1. *Assume that* $X = (x_{0\ldots0}, \ldots, x_{1\ldots1})$ *is chosen uniformly at random from* $\{0,1\}^N$. *Then, with probability at least* $1-2^{-n}$, $f(X) = 0$ *and the measured* $i_1, \ldots, i_m$ *generate* $\{0,1\}^n$.

Proof. It can be shown by a small modification of [1, Theorem 5.1, p.91] that with probability at least $1 - 2^{-c2^n}$ $(c > 0)$, there are at least $2^n/8$ values j such that $x_i = j$ for exactly one $i \in \{0,1\}^n$. We assume that this is the case.

If $i_1, \ldots, i_m$ generate a proper subspace of $\{0,1\}^n$, then there is a non-zero $k \in \{0,1\}^n$ that is orthogonal to this subspace. We estimate the probability that this happens. Consider some fixed non-zero vector $k \in \{0,1\}^n$. The probability that i_1 and k are orthogonal is at most $\frac{15}{16}$, as follows. With probability at least 1/8, the measurement of the second register gives j such that $f(i) = j$ for a unique i. In this case, the measurement of the final superposition (1) gives a uniformly random i'. The probability that a uniformly random i' has $(k, i') \neq 0$ is 1/2. Therefore, the probability that $(k, i_1) = 0$ is at most $1 - \frac{1}{8} \cdot \frac{1}{2} = \frac{15}{16}$.

The vectors $i_1, \ldots, i_m$ are chosen independently. Therefore, the probability that k is orthogonal to each of them is at most $(\frac{15}{16})^{22n} < 2^{-2n}$. There are $2^n - 1$ possible non-zero k, so the probability that there is a k which is orthogonal to each of $i_1, \ldots, i_m$, is at most $(2^n - 1)2^{-2n} < 2^{-n}$. □

Note that this algorithm is actually a *zero-error* algorithm: it always outputs the correct answer. Its expected number of queries on a uniformly random input is at most $m = 22n$ for generating $i_1, \ldots, i_m$ and at most $\frac{1}{2^n}2^n = 1$ for querying all the x_i if the first step does not give $i_1, \ldots, i_m$ that generate $\{0,1\}^n$. This completes the proof of the first part of Theorem 2.

4.3 Classical Lower Bound

Let D_1 be the uniform distribution over all inputs $X \in \{0,1\}^N$ and D_2 be the uniform distribution over all X for which there is a unique $k \neq 0$ such that $x_i = x_{i \oplus k}$ (and hence $f(X) = 1$). We say an algorithm A *distinguishes* between D_1 and D_2 if the average probability that A outputs 0 is $\geq 3/4$ under D_1 and the average probability that A outputs 1 is $\geq 3/4$ under D_2.

Lemma 2. *If there is a bounded-error algorithm A that computes f with $m = T_A^{unif}$ queries on average, then there is an algorithm that distinguishes between D_1 and D_2 and uses $O(m)$ queries on all inputs.*

Proof. We run A until it stops or makes $4m$ queries. The average probability (under D_1) that it stops is at least $3/4$, for otherwise the average number of queries would be more than $\frac{1}{4}(4m) = m$. Under D_1, the probability that A outputs $f(X) = 1$ is at most $1/4 + o(1)$ ($1/4$ is the maximum probability of error on an input with $f(X) = 0$ and $o(1)$ is the probability of getting an input with $f(X) = 1$). Therefore, the probability under D_1 that A outputs 0 after at most $4m$ queries, is at least $3/4 - (1/4 + o(1)) = 1/2 - o(1)$.

In contrast, the D_2-probability that A outputs 0 is $\leq 1/4$ because $f(X) = 1$ for any input X from D_2. We can use this to distinguish D_1 from D_2. □

Lemma 3. *No classical randomized algorithm A that makes $m \in o(2^{n/2})$ queries can distinguish between D_1 and D_2.*

Proof. For a random input from D_1, the probability that all answers to m queries are different is

$$1 \cdot (1 - 1/2^n) \cdots (1 - (m-1)/2^n) \geq (1 - m/2^n)^m \to e^{-m^2/2^n} = 1 - o(1).$$

For a random input from D_2, the probability that there is an i s.t. A queries both x_i and $x_{i \oplus k}$ (k is the hidden vector) is $\leq \binom{m}{2}/(2^n - 1) \in o(1)$, since:

1. for every pair of distinct i, j, the probability that $i = j \oplus k$ is $1/(2^n - 1)$
2. since A queries only m of the x_i, it queries only $\binom{m}{2}$ distinct pairs i, j

If no pair x_i, $x_{i\oplus k}$ is queried, the probability that all answers are different is

$$1 \cdot (1 - 1/2^{n-1}) \cdots (1 - (m-1)/2^{n-1}) = 1 - o(1).$$

It is easy to see that all sequences of m different answers are equally likely. Therefore, for both distributions D_1 and D_2, we get a uniformly random sequence of m different values with probability $1-o(1)$ and something else with probability $o(1)$. Thus A cannot "see" the difference between D_1 and D_2 with sufficient probability to distinguish between them. □

The second part of Theorem 2 now follows: a classical algorithm that computes f with an average number of m queries can be used to distinguish between D_1 and D_2 with $O(m)$ queries (Lemma 2), but then $O(m) \in \Omega(2^{n/2})$ (Lemma 3).

5 Super-Exponential Gap for Non-uniform μ

The last section gave an exponential gap between Q^μ and R^μ under uniform μ. Here we show that the gap can be even larger for non-uniform μ. Consider the average-case complexity of the OR-function. It is easy to see that $D^{unif}(\text{OR})$, $R^{unif}(\text{OR})$, and $Q^{unif}(\text{OR})$ are all $O(1)$, since the average input will have many 1s under the uniform distribution. Now we give some examples of non-uniform distributions μ where $Q^\mu(\text{OR})$ is super-exponentially smaller than $R^\mu(\text{OR})$:

Theorem 3. *If $\alpha \in (0, 1/2)$ and $\mu(X) = c/\binom{N}{|X|}(|X|+1)^\alpha(N+1)^{1-\alpha}$ ($c \approx 1-\alpha$ is a normalizing constant), then $R^\mu(\text{OR}) \in \Theta(N^\alpha)$ and $Q^\mu(\text{OR}) \in \Theta(1)$.*

Proof. Any classical algorithm for OR requires $\Theta(N/(|X|+1))$ queries on input X. The upper bound follows from random sampling, the lower bound from a block-sensitivity argument [16]. Hence (omitting the intermediate Θs):

$$R^\mu(\text{OR}) = \sum_X \mu(X) \frac{N}{|X|+1} = \sum_{t=0}^{N} \frac{cN^\alpha}{(t+1)^{\alpha+1}} \in \Theta(N^\alpha).$$

Similarly, for a quantum algorithm $\Theta(\sqrt{N/(|X|+1)}$ queries are necessary and sufficient on input X [11,5], so

$$Q^\mu(\text{OR}) = \sum_X \mu(X) \sqrt{\frac{N}{|X|+1}} = \sum_{t=0}^{N} \frac{cN^{\alpha-1/2}}{(t+1)^{\alpha+1/2}} \in \Theta(1).$$ □

In particular, for $\alpha = 1/2 - \varepsilon$ we have the huge gap $O(1)$ quantum versus $\Omega(N^{1/2-\varepsilon})$ classical. Note that we obtain this super-exponential gap by weighing the complexity of two algorithms (classical and quantum OR-algorithms) which are only quadratically apart on each input X.

In fact, a small modification of μ gives the same big gap even if the quantum algorithm is forced to output the correct answer always. We omit the details.

6 General Bounds for Average-Case Complexity

In this section we prove some general bounds. First we make precise the intuitively obvious fact that if an algorithm A is faster on every input than another algorithm B, then it is also much faster on average under any distribution:

Theorem 4. *If $\phi : \mathbf{R} \to \mathbf{R}$ is a concave function and $T_A(X) \leq \phi(T_B(X))$ for all X, then $T_A^\mu \leq \phi(T_B^\mu)$ for every μ.*

Proof. By Jensen's inequality, if ϕ is concave then $E_\mu[\phi(T)] \leq \phi(E_\mu[T])$, hence

$$T_A^\mu \leq \sum_{X\in\{0,1\}^N} \mu(X)\phi(T_B(X)) \leq \phi\left(\sum_{X\in\{0,1\}^N} \mu(X)T_B(X)\right) = \phi\left(T_B^\mu\right). \qquad \square$$

In words: taking the average cannot make the complexity-gap between two algorithms smaller. For instance, if $T_A(X) \leq \sqrt{T_B(X)}$ (say, A is Grover's algorithm and B is a classical algorithm for OR), then $T_A^\mu \leq \sqrt{T_B^\mu}$. On the other hand, taking the average *can* make the gap much larger, as we saw in Theorem 3: the quantum algorithm for OR runs only quadratically faster than any classical algorithm on each input, but the *average-case* gap between quantum and classical can be much bigger than quadratic.

We now prove a general lower bound on R^μ and Q^μ. Using an argument from [16] for the classical case and an argument from [3] for the quantum case, we can show:

Lemma 4. *Let A be a bounded-error algorithm for some function f. If A is classical then $T_A(X) \in \Omega(bs_X(f))$, and if A is quantum then $T_A(X) \in \Omega(\sqrt{bs_X(f)})$.*

A lower bound in terms of the μ-expected block sensitivity follows:

Theorem 5. *For all f, μ: $R^\mu(f) \in \Omega(E_\mu[bs_X(f)])$ and $Q^\mu(f) \in \Omega(E_\mu[\sqrt{bs_X(f)}])$.*

7 Average-Case Complexity of MAJORITY

Here we examine the average-case complexity of the MAJORITY-function. The hard inputs for majority occur when $t = |X| \approx N/2$. Any quantum algorithm needs $\Omega(N)$ queries for such inputs [3]. Since the uniform distribution puts most probability on the set of X with $|X|$ close to $N/2$, we might expect an $\Omega(N)$ average-case complexity. However we will prove that the complexity is nearly $\sqrt{N}$. For this we need the following result about approximate quantum counting, which follows from [8, Theorem 5] (see also [14] or [15, Theorem 1.10]):

Theorem 6 (Brassard, Høyer, Tapp; Mosca). *Let $\alpha \in [0,1]$. There is a quantum algorithm with worst-case $O(N^\alpha)$ queries that outputs an estimate $\tilde{t}$ of the weight $t = |X|$ of its input, such that $|\tilde{t} - t| \leq N^{1-\alpha}$ with probability $\geq 2/3$.*

Theorem 7. *For every $\varepsilon > 0$, $Q^{unif}(\mathrm{MAJ}) \in O(N^{1/2+\varepsilon})$.*

Proof. Consider the following algorithm, with input X, and $\alpha \in [0,1]$ to be determined later.

1. Estimate $t = |X|$ by $\tilde{t}$ using $O(N^\alpha)$ queries.
2. If $\tilde{t} < N/2 - N^{1-\alpha}$ then output 0; if $\tilde{t} > N/2 + N^{1-\alpha}$ then output 1.
3. Otherwise use N queries to classically count t and output its majority.

It is easy to see that this is a bounded-error algorithm for MAJ. We determine its average complexity. The third step of the algorithm will be invoked iff $|\tilde{t} - N/2| \leq N^{1-\alpha}$. Denote this event by "$\tilde{t} \approx N/2$". For $0 \leq k \leq N^\alpha/2$, let D_k denote the event that $kN^{1-\alpha} \leq |t - N/2| < (k+1)N^{1-\alpha}$. Under the uniform distribution the probability that $|X| = t$ is $\binom{N}{t}2^{-N}$. By Stirling's formula this is $O(1/\sqrt{N})$, so the probability of the event D_k is $O(N^{1/2-\alpha})$. In the quantum counting algorithm, $\Pr[kN^{1-a} \leq |\tilde{t} - t| < (k+1)N^{1-a}] \in O(1/(k+1))$ (this follows from [6], the upcoming journal version of [8] and [14]). Hence also $\Pr[\tilde{t} \approx N/2 \mid D_k] \in O(1/(k+1))$. The probability that the second counting stage is needed is $\Pr[\tilde{t} \approx N/2]$, which we bound by

$$\sum_{k=0}^{N^\alpha/2} \Pr[\tilde{t} \approx N/2 \mid D_k] \cdot \Pr[D_k] = \sum_{k=0}^{N^\alpha/2} O(\frac{1}{k+1}) \cdot O(N^{1/2-\alpha}) = O(N^{1/2-\alpha} \log N).$$

Thus we can bound the average-case query complexity of our algorithm by

$$O(N^\alpha) + \Pr[\tilde{t} \approx N/2] \cdot N = O(N^\alpha) + O(N^{3/2-\alpha} \log N).$$

Choosing $\alpha = 3/4$, we obtain an $O(N^{3/4} \log N)$ algorithm.

However, we can reiterate this scheme: instead of using N queries in step 3 we could count using $O(N^{\alpha_2})$ instead of N queries, output an answer if there is a clear majority (i.e. $|\tilde{t} - N/2| > N^{1-\alpha_2}$), otherwise count again using $O(N^{\alpha_3})$ queries etc. If after k stages we still have no clear majority, we count using N queries. For any fixed k, we can make the error probability of each stage sufficiently small using only a constant number of repetitions. This gives a bounded-error algorithm for MAJORITY. (The above algorithm is the case $k = 1$.)

It remains to bound the complexity of the algorithm by choosing appropriate values for k and for the α_i (put $\alpha_1 = \alpha$). Let p_i denote the probability under *unif* that the ith counting-stage will be needed, i.e. that all previous counts gave results close to $N/2$. Then $p_{i+1} \in O(N^{1/2-\alpha_i} \log N)$ (as above). The average query complexity is now bounded by:

$$O(N^{\alpha_1}) + p_2 \cdot O(N^{\alpha_2}) + \cdots + p_k \cdot O(N^{\alpha_k}) + p_{k+1} \cdot N =$$

$$O(N^{\alpha_1}) + O(N^{1/2-\alpha_1+\alpha_2} \log N) + \cdots + O(N^{1/2-\alpha_{k-1}+\alpha_k} \log N) + O(N^{3/2-\alpha_k} \log N).$$

Clearly the asymptotically minimal complexity is achieved when all exponents in this expression are equal. This induces $k-1$ equations $\alpha_1 = 1/2 - \alpha_i + \alpha_{i+1}$, $1 \leq i < k$, and a kth equation $\alpha_1 = 3/2 - \alpha_k$. Adding up these k equations we obtain $k\alpha_1 = -\alpha_1 + (k-1)/2 + 3/2$, which implies $\alpha_1 = 1/2 + 1/(2k+2)$. Thus we have average query complexity $O(N^{1/2+1/(2k+2)} \log N)$. Choosing k sufficiently large, this becomes $O(N^{1/2+\varepsilon})$. □

The nearly matching lower bound is:

Theorem 8. $Q^{unif}(\text{MAJ}) \in \Omega(N^{1/2})$.

Proof. Let A be a bounded-error quantum algorithm for MAJORITY. It follows from the worst-case results of [3] that A uses $\Omega(N)$ queries on the hardest inputs, which are the X with $|X| = N/2 \pm 1$. Since the uniform distribution puts $\Omega(1/\sqrt{N})$ probability on the set of such X, the average-case complexity of A is at least $\Omega(1/\sqrt{N})\Omega(N) = \Omega(\sqrt{N})$. □

What about the *classical* average-case complexity? Alonso, Reingold, and Schott [2] prove that $D^{unif}(\text{MAJ}) = 2N/3 - \sqrt{8N/9\pi} + O(\log N)$. We can also prove that $R^{unif}(\text{MAJ}) \in \Omega(N)$ (for reasons of space we omit the details), so quantum is almost quadratically better than classical for this problem.

8 Average-Case Complexity of PARITY

Finally we prove some results for the average-case complexity of PARITY. This is in many ways the hardest Boolean function. Firstly, $bs_X(f) = N$ for all X, hence by Theorem 5:

Corollary 1. *For every* μ, $R^{\mu}(\text{PARITY}) \in \Omega(N)$ *and* $Q^{\mu}(\text{PARITY}) \in \Omega(\sqrt{N})$.

We can bounded-error quantum count $|X|$ exactly, using $O(\sqrt{(|X|+1)N})$ queries [8]. Combining this with a μ that puts $O(1/\sqrt{N})$ probability on the set of all X with $|X| > 1$, we obtain $Q^{\mu}(\text{PARITY}) \in O(\sqrt{N})$.

We can prove $Q^{\mu}(\text{PARITY}) \leq N/6$ for any μ by the following algorithm: with probability $1/3$ output 1, with probability $1/3$ output 0, and with probability $1/3$ run the exact quantum algorithm for PARITY, which has worst-case complexity $N/2$ [3,10]. This algorithm has success probability $2/3$ on every input and has expected number of queries equal to $N/6$.

More than a linear speed-up on average is not possible if μ is uniform:

Theorem 9. $Q^{unif}(\text{PARITY}) \in \Omega(N)$.

Proof. Let A be a bounded-error quantum algorithm for PARITY. Let B be an algorithm that flips each bit of its input X with probability $1/2$, records the number b of actual bitflips, runs A on the changed input Y, and outputs $A(Y) \oplus b$. It is easy to see that B is a bounded-error algorithm for PARITY and that it uses an expected number of T_A^{μ} queries on *every* input. Using standard techniques, we can turn this into an algorithm for PARITY with *worst-case* $O(T_A^{\mu})$ queries. Since the worst-case lower bound for PARITY is $N/2$ [3,10], the theorem follows. □

Acknowledgments

We thank Harry Buhrman for suggesting this topic, and him, Lance Fortnow, Lane Hemaspaandra, Hein Röhrig, Alain Tapp, and Umesh Vazirani for helpful discussions. Also thanks to Alain for sending a draft of [6].

References

1. N. Alon and J. H. Spencer. *The Probabilistic Method.* Wiley-Interscience, 1992.
2. L. Alonso, E. M. Reingold, and R. Schott. The average-case complexity of determining the majority. *SIAM Journal on Computing*, 26(1):1–14, 1997.
3. R. Beals, H. Buhrman, R. Cleve, M. Mosca, and R. de Wolf. Quantum lower bounds by polynomials. In *Proceedings of 39th FOCS*, pages 352–361, 1998. http://xxx.lanl.gov/abs/quant-ph/9802049.
4. C. H. Bennett, E. Bernstein, G. Brassard, and U. Vazirani. Strengths and weaknesses of quantum computing. *SIAM Journal on Computing*, 26(5):1510–1523, 1997. quant-ph/9701001.
5. M. Boyer, G. Brassard, P. Høyer, and A. Tapp. Tight bounds on quantum searching. *Fortschritte der Physik*, 46(4–5):493–505, 1998. Earlier version in Physcomp'96. quant-ph/9605034.
6. G. Brassard, P. Høyer, M. Mosca, and A. Tapp. Quantum amplitude amplification and estimation. Forthcoming.
7. G. Brassard, P. Høyer, and A. Tapp. Quantum algorithm for the collision problem. *ACM SIGACT News (Cryptology Column)*, 28:14–19, 1997. quant-ph/9705002.
8. G. Brassard, P. Høyer, and A. Tapp. Quantum counting. In *Proceedings of 25th ICALP*, volume 1443 of *Lecture Notes in Computer Science*, pages 820–831. Springer, 1998. quant-ph/9805082.
9. D. Deutsch and R. Jozsa. Rapid solution of problems by quantum computation. In *Proceedings of the Royal Society of London*, volume A439, pages 553–558, 1992.
10. E. Farhi, J. Goldstone, S. Gutmann, and M. Sipser. A limit on the speed of quantum computation in determining parity. quant-ph/9802045, 16 Feb 1998.
11. L. K. Grover. A fast quantum mechanical algorithm for database search. In *Proceedings of 28th STOC*, pages 212–219, 1996. quant-ph/9605043.
12. E. Hemaspaandra, L. A. Hemaspaandra, and M. Zimand. Almost-everywhere superiority for quantum polynomial time. quant-ph/9910033, 8 Oct 1999.
13. L. A. Levin. Average case complete problems. *SIAM Journal on Computing*, 15(1):285–286, 1986. Earlier version in STOC'84.
14. M. Mosca. Quantum searching, counting and amplitude amplification by eigenvector analysis. In *MFCS'98 workshop on Randomized Algorithms*, 1998.
15. A. Nayak and F. Wu. The quantum query complexity of approximating the median and related statistics. In *Proceedings of 31th STOC*, pages 384–393, 1999. quant-ph/9804066.
16. N. Nisan. CREW PRAMs and decision trees. *SIAM Journal on Computing*, 20(6):999–1007, 1991. Earlier version in STOC'89.
17. P. W. Shor. Polynomial-time algorithms for prime factorization and discrete logarithms on a quantum computer. *SIAM Journal on Computing*, 26(5):1484–1509, 1997. Earlier version in FOCS'94. quant-ph/9508027.
18. D. Simon. On the power of quantum computation. *SIAM Journal on Computing*, 26(5):1474–1483, 1997. Earlier version in FOCS'94.
19. J. S. Vitter and Ph. Flajolet. Average-case analysis of algorithms and data structures. In J. van Leeuwen, editor, *Handbook of Theoretical Computer Science. Volume A: Algorithms and Complexity*, pages 431–524. MIT Press, Cambridge, MA, 1990.
20. Ch. Zalka. Grover's quantum searching algorithm is optimal. *Physical Review A*, 60:2746–2751, 1999. quant-ph/9711070.

Tradeoffs between Nondeterminism and Complexity for Communication Protocols and Branching Programs*
(Extended Abstract)

Juraj Hromkovič[1] and Martin Sauerhoff[2]

[1] Department of Computer Science I (Algorithms and Complexity)
Technological University of Aachen, Ahornstraße 55, 52074 Aachen, Germany
jh@i1.informatik.rwth-aachen.de

[2] Department of Computer Science, LS 2
University of Dortmund, 44221 Dortmund, Germany
sauerhof@ls2.cs.uni-dortmund.de

Abstract. In this paper, lower bound and tradeoff results relating the computational power of determinism, nondeterminism, and randomness for communication protocols and branching programs are presented. The main results can be divided into the following three groups.

(i) One of the few major open problems concerning nondeterministic communication complexity is to prove an asymptotically exact tradeoff between complexity and the number of available advice bits. This problem is solved here for the case of *one-way* communication.

(ii) *Multipartition protocols* are introduced as a new type of communication protocols using a restricted form of non-obliviousness. In order to be able to study methods for proving lower bounds on multilective and/or non-oblivious computation, these protocols are allowed to either deterministically or nondeterministically choose between different partitions of the input. Here, the first results showing the potential increase of the computational power by non-obliviousness as well as boundaries on this power are derived.

(iii) The above results (and others) are applied to obtain several new exponential lower bounds for different types of oblivious branching programs, which also yields new insights into the power of nondeterminism and randomness for the considered models. The proofs rely on a general technique described here which allows to prove explicit lower bounds on the size of oblivious branching programs in an easy and transparent way.

1 Introduction and Definitions

The communication complexity of two-party protocols has been introduced by Abelson [1] and Yao [29]. The initial goal was to develop a method for proving lower bounds on the complexity of distributed and parallel computations.

* This work has been supported by DFG grants HR 14/3-2 and We 1066/8-2.

H. Reichel and S. Tison (Eds.): STACS 2000, LNCS 1770, pp. 145–156, 2000.

Let $f\colon \{0,1\}^n \to \{0,1\}$ be a Boolean function defined on a set X of n Boolean variables, and let $\Pi = (X_1, X_2)$ ($X_1 \cup X_2 = X$, $X_1 \cap X_2 = \emptyset$) be a partition of X. A *(communication) protocol P computing f according to Π* consists of two computers C_{I} and C_{II} with unbounded computational power. At the beginning of the computation, C_{I} obtains an input $x\colon X_1 \to \{0,1\}$ and C_{II} obtains an input $y\colon X_2 \to \{0,1\}$. Then C_{I} and C_{II} communicate according to the protocol by exchanging binary coded messages until one of them knows the result $f(x,y)$. The cost of the computation of P on an input (x,y) is the sum of the lengths of exchanged messages. The cost of the protocol P, $\mathrm{cc}(P)$, is the maximum of the cost over all inputs x, y. The *communication complexity of f according to Π*, $\mathrm{cc}(f,\Pi)$, is the cost of the best protocol computing f according to Π. There are several ways to define the communication complexity of f, the choice depending on the application considered. Usually, the *communication complexity of f*, $\mathrm{cc}(f)$, is defined as the minimum of $\mathrm{cc}(f,\Pi)$ over all balanced partitions of the set of input variables. Analogously, $\mathrm{ncc}(f)$ stands for the *nondeterministic communication complexity*[1] *of f*. Finally, for a communication complexity measure $x(f)$ used in this paper, $x_k(f)$ denotes the k-round version of $x(f)$, and $\mathrm{n}x^r(f)$ denotes the corresponding nondeterministic complexity with the bound r on the number of advice bits.

In the two decades of its existence, communication complexity has established itself as a well defined subarea of complexity theory (see [13,21] for a thorough introduction). The reason for this is the success in the following two main streams of research:

I. *Comparison of the power of different modes of computation.* Communication protocols are one of the few models of computation where the relative power of deterministic, nondeterministic, and randomized computation could be characterized, leading to a better understanding of these modes of computation.

II. *Proving lower bounds.* Communication complexity has considerably contributed to proving lower bounds on the amount of resources required to solve concrete problems in several fundamental sequential and parallel models of computation, e. g., circuits, Turing machines, and branching programs.

In this paper, we contribute to both of these streams of research. In the first part of the paper, new results concerning the power of nondeterminism for communication protocols are presented. The second part deals with the application these results (and others) to prove lower bounds for branching programs.

Communication with Restricted Nondeterminism. With respect to the first point from above, we may not only be interested in the question whether nondeterminism or randomness helps at all to compute a given function, but we may also ask the following, more sophisticated questions:

- How many advice or random bits are needed to achieve the full power of the respective model of computation?

[1] We always talk about *private* nondeterminism, where advice bits have to be explicitly communicated if the other computer has to know them. In an unrestricted public model, each function would have complexity at most 1 (see [13]).

- Can all nondeterministic or random guesses be moved to the very beginning of the computation without increasing the complexity and the required number of advice or random bits, resp.?

For communication protocols and the resource randomness, the following answers to these questions have been given. Newman [23] has proven that at most $O(\log n)$ random bits evaluated at the beginning of the computation are sufficient to get the full power out of randomized protocols with bounded error or zero-error. Canetti and Goldreich [7] and Fleischer, Jung, and Mehlhorn [11] have proven lower bounds on the number of required random bits in terms of the communication complexity for various models. These bounds are asymptotically optimal.

The dependence of the nondeterministic communication complexity of two-way protocols on the available number of advice bits has been analyzed by Hromkovič and Schnitger [14]. They have shown that for every fixed number of advice bits $r(n) = O(\log^c n)$, $c \geq 1$ an arbitrary constant, there is a function $f_{r(n)}\colon \{0,1\}^{2n} \to \{0,1\}$ which has nondeterministic communication complexity $O(\log^c n)$ if at least $r(n)$ advice bits are available; but $\Omega(n/\log n)$, if only $o(r(n)/\log n)$ advice bits may be used. Up to now, it is open to prove an asymptotically exact tradeoff between nondeterministic communication complexity and the number of advice bits.

We present a partial solution to this problem by proving such a tradeoff for *one-way* communication, where the second computer has to output the result after receiving a single message from the first computer. We prove that the conjunction of s copies of the well-known "pointer" or "index" function IND_n has nondeterministic one-way complexity $\Theta(sn \cdot 2^{-r/s} + r)$, where r is the number of advice bits (the input size is $\Theta(s(n + \log n))$).

As our second main contribution, we introduce and investigate a new type of communication protocols. Usual protocols are oblivious in the sense that they work only with a fixed partition of the input variables. *Multipartition protocols* introduced here may work with different partitions depending on the input and allow a controlled degree of non-obliviousness. In order to study the dependence of the complexity on the available amount of nondeterminism, we allow nondeterministic multipartition protocols to guess a partition from a given collection.

Definition 1. *Let k be a positive integer, and let f be a Boolean function defined on a set X of input variables. A* (deterministic) k-partition protocol P for f *consists of $k+1$ two-party communication protocols $(P_0, \Pi_0), (P_1, \Pi_1), \ldots, (P_k, \Pi_k)$, where Π_i is a balanced partition for $i = 0, 1, \ldots, k$. For an arbitrary input $x\colon X \to \{0,1\}$, the protocol P works as follows.*

(i) The protocol (P_0, Π_0) computes a value from $\{1, 2, \ldots, k\}$ for the input x partitioned according to Π_0.

(ii) If $P_0(x) = i$, then protocol P_i is executed on the input x partitioned according to Π_i, and its output is $P_i(x) = f(x)$.

The communication complexity of P is

$$\mathrm{cc}\,(P_0, \Pi_0) + \max\{\mathrm{cc}\,(P_i, \Pi_i) \mid i = 1, \ldots, k\}.$$

The k-partition communication complexity *of* f, k-pcc(f), *is the minimum over the communication complexities of all k-partition protocols computing f. The* multipartition communication complexity *of f is* $\text{pcc}(f) := \min_k\{k\text{-pcc}(f)\}$.

A nondeterministic k-partition protocol P for f *is a collection of k deterministic protocols $(P_1, \Pi_1), \ldots, (P_k, \Pi_k)$, where Π_i is a balanced partition of X for $i = 1, \ldots, k$. For an arbitrary input x, the protocol P works as follows:*

(i) If $f(x) = 0$, then $P_i(x) = 0$ for all $i = 1, \ldots, k$; and

(ii) if $f(x) = 1$, then there exists an $i \in \{1, \ldots, k\}$ such that $P_i(x) = 1$.

The communication complexity of P is $\lceil \log k \rceil + \max\{\text{cc}(P_i, \Pi_i) \mid i = 1, \ldots, k\}$. The nondeterministic k-partition communication complexity *of f,* k-pncc(f), *is the cost of the best nondeterministic k-partition protocol for f. The* nondeterministic multipartition communication complexity *of f is defined as* $\text{pncc}(f) := \min_k\{k\text{-pncc}(f)\}$.

Note that it is important to add $\lceil \log k \rceil$ to the nondeterministic k-partition communication complexity because the computers have to agree on the partition they use. If this agreement were for free, then our model would be as powerful as public advice nondeterministic communication protocols.

In order to apply communication complexity for proving lower bounds, it is often convenient to consider the *uniform* version of protocols introduced and applied in [10,13]. Informally, a uniform protocol decides whether a given input $w \in \Sigma^*$ belongs to a language $L \subseteq \Sigma^*$ for *arbitrary* partitions $w = xy$, where $x, y \in \Sigma^*$. For Boolean functions, we use the following definition.

Definition 2. *Let f be a Boolean function defined on the variables $x_1, \ldots, x_n$, and let π be a permutation of the set $\{1, \ldots, n\}$, here called a* variable ordering. *A* (deterministic) uniform communication protocol for f with variable ordering π, *denoted by (P, π), is a collection of deterministic* one-way *communication protocols $(P_1, \Pi_1), \ldots, (P_{n-1}, \Pi_{n-1})$, $\Pi_i := (\{x_{\pi(1)}, \ldots, x_{\pi(i)}\}, \{x_{\pi(i+1)}, \ldots, x_{\pi(n)}\})$ for $i = 1, \ldots, n-1$. The* uniform communication complexity of f according to π *is* $\text{ucc}(f, \pi) := \max\{\text{cc}_1(f, \Pi_i) \mid i = 1, \ldots, n-1\}$.

A nondeterministic uniform k-partition protocol P with variable orderings $\pi_1, \ldots, \pi_k$ *for a Boolean function f on n variables is a collection of k uniform protocols $(P_1, \pi_1), \ldots, (P_k, \pi_k)$, where for an arbitrary input x:*

(i) if $f(x) = 0$, then $P_i(x) = 0$ for all $i = 1, \ldots, k$; and

(ii) if $f(x) = 1$, there exists an $i \in \{1, \ldots, k\}$ such that $P_i(x) = 1$.

The cost of P is $\lceil \log k \rceil + \max\{\text{ucc}(P_i, \pi_i) \mid i = 1, \ldots, k\}$. The nondeterministic uniform k-partition communication complexity *of f,* k-pnucc(f), *is the cost of the best nondeterministic uniform k-partition protocol for f.*

We now present our results for multipartition communication complexity. Our goal is to compare usual nondeterministic protocols with nondeterministic multipartition protocols.

On the one hand, it turns out that already between the deterministic two-partition model and the usual nondeterministic model without restrictions we have the maximal possible gap, i. e., constant versus linear complexity. Hence, multipartition protocols appear to be quite powerful.

It is much harder to obtain also a result in the opposite direction. As the second main result of the paper, we will show that a function which is easy for one-way nondeterministic protocols using at most $\log n$ advice bits has large complexity for nondeterministic uniform multipartition protocols which may use at most logarithmically many partitions:

(i) For every $n = m^2 + 1$, $m \in \mathbb{N}$, there is a Boolean function f_n on n variables such that 2-pcc$(f_n) = O(1)$, but ncc$(f_n) = \Omega(\sqrt{n})$.

(ii) For every $n = 6 \cdot 3^{2\lceil \log m \rceil} \cdot m$, $m \in \mathbb{N}$, there is a Boolean function g_n on n variables such that $\text{ncc}_1^{\lceil \log n \rceil}(g_n) = O(\log n)$; but k-pnucc$(g_n) = \Omega(n^{1/5}/k)$ for all $k \leq \log n/5$.

These are results for explicitly defined functions. For the complexity measure k-pncc we finally show the following non-constructive result: There are functions with complexity at most $m + k$ for usual deterministic one-way protocols, but with complexity larger than m for nondeterministic 2^{k-2}-partition protocols. Hence, one needs exponentially many partitions in the nondeterministic multipartition model only to halve the complexity compared to the usual deterministic one-way model.

Branching Program Complexity. Branching programs (BPs) are one of the standard nonuniform models of computation and an especially interesting "testing field" for lower bound techniques, see, e. g., [27,28] for an introduction. A branching program is a graph representing a single Boolean function. The complexity of a branching program, called *branching program size*, is the number of its nodes. Nondeterministic and randomized variants of branching programs are defined in a straightforward way by specifying a subset of the input variables whose values are chosen by coin tosses or nondeterministic guesses, resp. (see, e. g., [25] for details). Such variables are called *probabilistic* or *nondeterministic variables* of the branching program, and we require that each such variable may appear at most once on each path from the source to a sink of the branching program.

An *oblivious branching program* is a branching program with an associated sequence s of input variables (which may contain duplicates). For each path from the source to a sink of the BP, the sequence of variables on this path has to be a subsequence of s. The length of a path corresponds to the time of computation for this path, and oblivious BPs have originally been introduced to study time-space tradeoffs. A special time-restricted model are *oblivious read-k-times branching programs* which have the property that on every path from the source to one of the sinks each input variable appears at most k times. For $k = 1$, we obtain oblivious read-once BPs, better known as *OBDDs (ordered binary decision diagrams)*. In this case, the variable sequence s is simply a permutation, called the *variable ordering* of the OBDD. Lower bounds for oblivious BPs have been proven in [4,5,12,17,18,19]. Some of these papers also contain (or imply) lower bounds for nondeterministic oblivious BPs, and also a few results for the randomized case are known [2,3,24,25].

A variant of the standard model of nondeterminism for oblivious BPs are partitioned BDDs which have been originally invented for application purposes [16]. A *k-partitioned BDD* (k-PBDD) is a branching program with a tree of nodes labeled by nondeterministic variables at the top by which one of k OBDDs is chosen, where each of these OBDDs may use a different variable ordering. A *partitioned BDD* is a k-PBDD for some k. This type of nondeterministic BPs is interesting for complexity theory because it allows a fine control of the available amount of nondeterminism as well as a bounded non-oblivious access to the input variables. Theoretical results for partitioned BDDs have been proven by Bollig and Wegener [6]. Among other results, they have shown that the classes of functions with polynomial size k-PBDDs form a proper hierarchy with respect to k. It is an open problem to compare this type of nondeterminism with usual nondeterministic OBDDs.

We will apply the results on communication complexity proven here (together with known results) to attack some open questions concerning the power of nondeterminism for the above types of BPs. As a tool, we use so-called *overlapping communication complexity* introduced in [15], which allows to prove lower bounds for oblivious read-k-times BPs in a clean and transparent way superior to the previously used techniques.

We first deal with the dependence of the size of nondeterministic OBDDs on the available amount of nondeterminism. For randomized OBDDs, it is known that $O(\log n)$ random bits are sufficient to exploit the full power of randomness for n-input functions [24]. We show here that imposing a logarithmic bound on the number of available nondeterministic variables may increase the OBDD size from $(n/\log n)^{O(\log n)}$ to $2^{\Omega(n)}$. (Recently, it has been proven by a different technique that even an increase from polynomial to exponential size is possible [26].)

Furthermore, we partially solve the open problem to compare nondeterministic OBDDs and k-PBDDs by showing that these two kinds of nondeterminism are incomparable if k is logarithmically bounded.

Finally, we compare the power of the deterministic and the Las Vegas variant of oblivious read-k-times BPs. It has already been proven that the size measures for deterministic OBDDs and Las Vegas OBDDs are polynomially related [10]. On the other hand, for general (non-oblivious) read-once BPs an exponential gap could be established [25]. Here we shed new light on this surprisingly different behavior by showing that the size of oblivious (deterministic) read-k-times BPs, where $2 \leq k \leq \log n/5$ and n is the input size, may be superpolynomial in the size of oblivious Las Vegas read-2-times BPs.

This extended abstract is organized as follows. In Section 2, we present and discuss our results. The common technique behind the proofs of the lower bounds for uniform multipartition communication complexity and oblivious BPs is sketched in Section 3. For the full proofs of these lower bounds, as well as the involved constructions required for the upper bounds, we have to refer to the journal version of this paper.

2 Results

2.1 Communication Complexity

The first main result of the paper is the asymptotically exact tradeoff between nondeterministic one-way communication complexity and the number of allowed advice bits. Such a result is easy to prove if a logarithmic number of advice bits is sufficient to obtain small nondeterministic complexity (e. g., it holds that $\mathrm{ncc}^r(\mathrm{NE}_n) = \Theta(n/2^r + r)$ for the string-nonequality function NE_n, see [14]). Here we are concerned with the much harder case where superlogarithmically many advice bits are required.

We consider the well-known function $\mathrm{IND}_n\colon \{0,1\}^n \times \{1,\ldots,n\} \to \{0,1\}$ defined by $\mathrm{IND}_n(x,y) = x_y$. The conjunction of s copies of this function, $\mathrm{IND}_{s,n}\colon \{0,1\}^{sn} \times \{1,\ldots,n\}^s \to \{0,1\}$, is defined by

$$\mathrm{IND}_{s,n}((x^1,\ldots,x^s),(y^1,\ldots,y^s)) := \mathrm{IND}_n(x^1,y^1) \wedge \cdots \wedge \mathrm{IND}_n(x^s,y^s),$$

where $x^1,\ldots,x^s \in \{0,1\}^n$, $y^1,\ldots,y^s \in \{1,\ldots,n\}$. Our analysis yields the following nearly exact bounds for this function:

Theorem 1. *For every $n, r, s \in \mathbb{N}$,*
(i) $\mathrm{ncc}_1^r(\mathrm{IND}_{s,n}) \le \alpha \cdot s \cdot n \cdot 2^{-r/s} + s + r$, *where* $\alpha := 2/(e \cdot \ln 2) < 1.062$; *and*
(ii) $\mathrm{ncc}_1^r(\mathrm{IND}_{s,n}) \ge \frac{1}{2} \cdot s \cdot n \cdot 2^{-r/s} + r$.

Our next two results deal with the relation between multipartition communication complexity and usual deterministic and nondeterministic complexity.

Consider the function $\mathrm{MRC}_n\colon \{0,1\}^{n^2+1} \to \{0,1\}$ ("monochromatic rows or columns") defined on a Boolean $n \times n$-matrix $X = (x_{i,j})$ and an additional variable z by

$$\mathrm{MRC}_n(X,z) = \Bigl(z \wedge \bigwedge_{1\le i\le n} (x_{i,1} \equiv \cdots \equiv x_{i,n})\Bigr) \vee \Bigl(\overline{z} \wedge \bigwedge_{1\le i\le n} (x_{1,i} \equiv \cdots \equiv x_{n,i})\Bigr).$$

Theorem 2.
(i) $2\text{-}\mathrm{pcc}(\mathrm{MRC}_n) \le 3$; *but*
(ii) $\mathrm{ncc}(\mathrm{MRC}_n) \ge \lfloor n/\sqrt{2} \rfloor$.

Theorem 2 shows that already the possibility of a deterministic choice of one out of two partitions may be much more powerful than unrestricted nondeterminism. Our next results lie in the opposite direction, i. e., we are going to prove limits on the power of multipartition protocols. We first describe a general construction technique which will allow us to derive variants of standard functions which are "hard" for multipartition protocols (and also for oblivious BPs).

Definition 3. *Let a function $f_n\colon \{0,1\}^n \times \{0,1\}^n \to \{0,1\}$, $n \in \mathbb{N}$, be given, and let $m \ge 2$ be arbitrarily chosen. We define the function m-Masked-$f_n\colon \{0,1\}^{3mn} \to \{0,1\}$ on Boolean vectors $s = (s_1,\ldots,s_{mn})$, $t = (t_1,\ldots,t_{mn})$, and $z = (z_1,\ldots,z_{mn})$ as follows. If either s or t do not contain exactly n ones, then we set m-Masked-$f_n(s,t,z) := 0$. Otherwise, let $i_1 < \cdots < i_n$ and $j_1 < \cdots < j_n$ be the positions of ones in s and t, resp., and define m-Masked-$f_n(s,t,z) := f(z_{i_1},\ldots,z_{i_n},z_{j_1},\ldots,z_{j_n})$.*

We apply this construction to the *string-nonequality function* $\mathrm{NE}_n\colon \{0,1\}^n \times \{0,1\}^n \to \{0,1\}$, defined by $\mathrm{NE}_n(x,y) = 1$ iff $x \ne y$. For arbitrary $m \ge 2$, we obtain a function m-Masked-NE_n on $3mn$ variables.

Theorem 3. *Let* $m := 2 \cdot 3^{2\lceil \log n \rceil}$ *and* $N := 6 \cdot 3^{2\lceil \log n \rceil} n$ *(the input size of* m-Masked-NE$_n$*). Then*

(i) $\mathrm{ncc}_1^{\lceil \log N \rceil}(m\text{-Masked-NE}_n) = O(\log N)$*; but*

(ii) $k\text{-pnucc}(m\text{-Masked-NE}_n) = \Omega(N^{1/5}/k)$ *for all* $k \le \log N/5$.

For the multipartition model with uniform protocols, logarithmically many partitions may thus be exponentially weaker than logarithmically many usual advice bits. What happens if we replace uniform protocols by usual protocols? Although our present technique does not yield a result similar to Theorem 3 in this case, we can show that there must exist hard functions also for the general multipartition model by counting arguments. By choosing $k = m = n/2$ in the following theorem, we obtain that $2^{n/2-2}$ partitions may not be sufficient to halve the communication complexity compared to deterministic one-way protocols.

Theorem 4. *For all positive integers,* $k, m, n \in \mathbb{N}$ *with* $k + m \le n$*, there exists a Boolean function* $F_n^{k,m} \colon \{0,1\}^{2n} \to \{0,1\}$ *such that*

(i) $\mathrm{cc}_1(F_n^{k,m}) \le m + k$*; and*

(ii) $2^{k-2}\text{-pncc}(F_n^{k,m}) > m$.

2.2 Branching Programs

We start by an analysis of the dependence of the size of nondeterministic OBDDs (nondeterministic oblivious read-once BPs) on the resource nondeterminism.

Let $\mathrm{UIND}_{s,n} \colon \{0,1\}^{sn} \times \{0,1\}^{sn} \to \{0,1\}$ be defined as the variant of the function $\mathrm{IND}_{s,n}$ from above where we use a unary encoding for the "pointers" instead of a binary one. We consider the function 2-Masked-UIND$_{s,n}$ on $N = 6sn$ variables obtained by applying Definition 3 with $m := 2$. Let $\text{N-OBDD}^r(f)$ be the minimum size of a nondeterministic OBDD for f that uses at most r nondeterministic variables.

Theorem 5.

(i) $\text{N-OBDD}(\text{2-Masked-UIND}_{s,n}) = O(n^{s+3} \cdot 2^s \cdot s^2)$*; but*

(ii) $\text{N-OBDD}^r(\text{2-Masked-UIND}_{s,n}) = 2^{\Omega(sn \cdot 2^{-r/s})}$ *for all* $r \in \mathbb{N}$.

Choosing $s := \lceil \log n \rceil$, we obtain the following gap.

Corollary 1. *For* $N = 6\lceil \log n \rceil n$*, the input size of* 2-Masked-UIND$_{\lceil \log n \rceil, n}$*,*

(i) $\text{N-OBDD}(\text{2-Masked-UIND}_{\lceil \log n \rceil, n}) = (N/\log N)^{O(\log N)}$*; but*

(ii) $\text{N-OBDD}^r(\text{2-Masked-UIND}_{\lceil \log n \rceil, n}) = 2^{\Omega(N)}$ *if* $r = O(\log N)$.

The next problem which we consider is the comparison of the usual form of nondeterminism in nondeterministic OBDDs with the nondeterministic choice of variable orderings in partitioned BDDs. Up to now, it has been open to find a concrete example for which k-partitioned OBDDs are superior to the usual type of nondeterminism for *any* k. Here we show that even 2-partitioned BDDs may be exponentially smaller than usual nondeterministic OBDDs. Let k-PBDD(f) denote the minimal size of a k-partitioned BDD for a function f.

Theorem 6.
(i) 2-PBDD(MRC$_n$) $= O(n^2)$; *but*
(ii) N-OBDD(MRC$_n$) $= 2^{\Omega(n)}$.

For the result in the opposite direction, we again consider the "masked variant" of the string-nonequality function from Theorem 3.

Theorem 7. *Let* $m := 2 \cdot 3^{2\lceil \log n \rceil}$ *and* $N := 6 \cdot 3^{2\lceil \log n \rceil} n$ *(the input size of* m-Masked-NE$_n$*). Then*
(i) N-OBDD(m-Masked-NE$_n$) $= O(N^2)$; *but*
(ii) k-PBDD(m-Masked-NE$_n$) $= 2^{\Omega(N^{1/5}/k)}$ *for every* $k \le \log N/5$.

Finally, we show a superpolynomial gap between Las Vegas and determinism for oblivious read-k-times BPs. Let NE$_{n,n}$ be the function obtained by taking the conjunction of n disjoint string-nonequality functions NE$_n$ (analogous to the definition of IND$_{s,n}$). We again apply Definition 3 with parameter $m := 2 \cdot 3^{2\lceil \log n \rceil}$, obtaining functions m-Masked-NE$_{n,n}$ on $N = 6 \cdot 3^{2\lceil \log n \rceil} \cdot n^2$ input variables. Let kOBP(f) (LV-kOBP(f)) denote the minimal size of oblivious deterministic (Las Vegas, resp.) read-k-times BPs for a function f.

Theorem 8. *Let* $m := 2 \cdot 3^{2\lceil \log n \rceil}$ *and* $N := 6 \cdot 3^{2\lceil \log n \rceil} \cdot n^2$ *(the input size of* m-Masked-NE$_{n,n}$*). Let* $\alpha := 1/(2\log 3 + 2)$.
(i) LV-2OBP(m-Masked-NE$_{n,n}$) $= 2^{O(N^{\alpha} \log N)}$; *but*
(ii) kOBP(m-Masked-NE$_{n,n}$) $= 2^{\Omega(N^{2\alpha}/k)}$ *for every* $k \le \log N/5$.
Hence, there is a sequence of functions $f_N\colon \{0,1\}^N \to \{0,1\}$ *which is (explicitly) defined for infinitely many* N *such that, for every* $k \le \log N/5$,

$$k\mathrm{OBP}(f_N) = 2^{\Omega\left(\log^2(\text{LV-2OBP}(f_N))/(k\cdot\log^2 N)\right)}.$$

3 On the Proofs of the Lower Bounds

In this section, we comment on the general technique used for proving the lower bounds on uniform multipartition communication complexity and on the size of oblivious BPs.

It is well-known how results on communication complexity can be applied to prove lower bounds for models of computation which are *read-once* and *oblivious*. Several attempts have been made to extend this approach also to models with multiple read access to the input variables and with a limited degree of non-obliviousness; in the case of branching programs, e. g., in [4,5,18]. Techniques which explicitly use the language of communication complexity theory are found in [13,15], where the notion of *overlapping communication complexity* has been introduced, and in the monograph of Kushilevitz and Nisan [21].

The idea common to these extended approaches is that partitions of the input variables are replaced by *covers*. Let $\Gamma = (X_1, X_2)$ be a cover of the input variables from X, i. e., $X_1 \cup X_2 = X$ and X_1, X_2 need *not* be disjoint.

Considering the communication problem where the computers C_{I} and C_{II} obtain inputs $x\colon X_1 \to \{0,1\}$ and $y\colon X_2 \to \{0,1\}$ such that $x|_{X_1 \cap X_2} = y|_{X_1 \cap X_2}$, one can define an *overlapping protocol according to Γ* analogous to the classical scenario where $X_1 \cap X_2 = \emptyset$. The complexity of such a protocol is defined as the maximum of all valid input assignments, and the *overlapping communication complexity of f with respect to Γ*, $\mathrm{occ}(f, \Gamma)$, is the minimum complexity of an overlapping protocol for f according to Γ.

Unfortunately, there are only a few functions for which lower bounds on the overlapping complexity could be proven so far, and the respective proofs often hide the fact that these bounds rely on lower bounds for usual communication complexity. The approach followed for the results in this paper (based on [13,15,21]) allows to apply known results for usual communication complexity to derive new lower bounds on overlapping communication complexity in a simple and straightforward way.

We can only briefly sketch the ideas behind our technique here:

(1) As an extension of the notion of rectangular reductions from communication complexity theory, we define *generalized rectangular reductions* in such a way that, if a generalized rectangular reduction from f to g exists, then lower bounds on the usual communication complexity of f yield lower bounds for the overlapping communication complexity of g.

(2) For each considered model of computation (e. g., uniform multipartition protocols, oblivious branching programs), we show how overlapping communication complexity yields lower bounds on the respective complexity in the model. It turns out that it is sufficient to consider overlapping complexity with respect to a special type of covers, called *alternating covers* (which are also implicitly considered in the paper of Alon and Maass [4]).

(3) It remains to prove lower bounds on overlapping communication complexity with respect to alternating covers. At this point, our scheme for constructing "masked versions" of functions (Definition 3) comes into play. Together with the reductions from (1) and a "Ramsey-like" combinatorial lemma, this allows us to exploit the whole collection of results for usual communication complexity to obtain the required lower bounds on overlapping complexity.

Here we only consider oblivious BPs as an example for the application of these ideas. Given an arbitrary sequence s of variables from X (possibly with duplicates), partition this sequence into contiguous segments and number them, say from 1 to ℓ. Then the alternating cover $\Gamma = (X_1, X_2)$ of X with respect to s is defined by putting all segments with odd number into X_1 and all segments with even number into X_2.

To carry out part (2) of the above plan, we construct an $(\ell - 1)$-round overlapping protocol from any given oblivious BP G with variable sequence s, using ideas from the papers [17,18]. Communication rounds correspond to sets of "cut nodes" in G, and thus it can be shown that $\lceil \log |G| \rceil \geq \mathrm{occ}_{\ell-1}(g, \Gamma)/(\ell - 1)$, where g is the function represented by G. It remains to show that g has "high" overlapping complexity, which is done as described in (3) above.

Putting all this together, we arrive at the following theorem which summarizes the proof technique for the special case of oblivious read-k-times BPs.

Theorem 9. *Let f be an arbitrary Boolean function defined on $2n$ variables, and let Π be an arbitrary balanced partition of the variable set of f. Let $k, p \in \mathbb{N}$ with $k \leq p$ and define $m := 2 \cdot 3^{2p}$. Let G be an oblivious read-k-times BP for m-Masked-f with respect to an arbitrary variable sequence. Then*

$$\lceil \log |G| \rceil \geq \mathrm{cc}_{2k-1}(f, \Pi)/(2k-1).$$

Analogous assertions hold for nondeterministic and randomized variants of oblivious read-k-times BPs.

Using this theorem, it is easy to obtain several new lower bounds for oblivious read-k-times BPs using the known results from communication complexity. The lower bounds in Theorem 7 and Theorem 8 are proven in this way. For Theorem 7, we need the additional observation that $k\text{-PBDD}(f) \geq k\text{OBP}(f)$ for all Boolean functions f due to Bollig and Wegener [6]. The lower bound for nondeterministic uniform multipartition protocols in Theorem 3 is proven by adapting the proof for partitioned BDDs to communication protocols and hence also relies on the same technique.

Acknowledgement. Thanks to Ingo Wegener for the idea to look at the "Las Vegas versus determinism" problem for oblivious read-k-times BPs and to Detlef Sieling for helpful comments on an early version of the proof of Theorem 1.

References

1. H. Abelson. Lower bounds on information transfer in distributed computations. In *Proc. of 19th IEEE Symp. on Foundations of Computer Science (FOCS)*, 151–158, 1978.
2. F. Ablayev. Randomization and nondeterminism are incomparable for polynomial ordered binary decision diagrams. In *Proc. of 24th Int. Coll. on Automata, Languages, and Programming (ICALP), LNCS 1256*, 195–202. Springer, 1997.
3. F. Ablayev and M. Karpinski. On the power of randomized branching programs. In *Proc. of 23rd Int. Coll. on Automata, Languages, and Programming (ICALP), LNCS 1099*, 348–356. Springer, 1996.
4. N. Alon and W. Maass. Meanders and their applications in lower bounds arguments. *Journal of Computer and System Sciences*, 37:118–129, 1988.
5. L. Babai, N. Nisan, and M. Szegedy. Multiparty protocols, pseudorandom generators for logspace and time-space trade-offs. *Journal of Computer and System Sciences*, 45:204–232, 1992.
6. B. Bollig and I. Wegener. Complexity theoretical results on partitioned (nondeterministic) binary decision diagrams. *Theory of Computing Systems*, 32:487–503, 1999. (Earlier version in *Proc. of 22nd Int. Symp. on Mathematical Foundations of Computer Science (MFCS), LNCS 1295*, 159–168. Springer, 1997.)
7. R. Canetti and O. Goldreich. Bounds on tradeoffs between randomness and communication complexity. *Computational Complexity*, 3:141 - 167, 1993.
8. P. Ďuriš and Z. Galil. On the power of multiple reads in a chip. *Information and Computation*, 104:277–287, 1993.
9. P. Ďuriš, Z. Galil, and G. Schnitger. Lower bounds on communication complexity. In *Proc. of 16th Ann. ACM Symp. on Theory of Computing (STOC)*, 81–91, 1984.

10. P. Ďuriš, J. Hromkovič, J. D. P. Rolim, and G. Schnitger. Las Vegas versus determinism for one-way communication complexity, finite automata, and polynomial-time computations. In *Proc. of 14th Ann. Symp. on Theoretical Aspects of Computer Science (STACS), LNCS 1200*, 117–128. Springer, 1997. To appear in *Information and Computation.*
11. R. Fleischer, H. Jung, and K. Mehlhorn. A communication-randomness tradeoff for two-processor systems. *Information and Computation*, 116:155–161, 1995.
12. J. Gergov. Time-space tradeoffs for integer multiplication on various types of input oblivious sequential machines. *Information Processing Letters*, 51:265 – 269, 1994.
13. J. Hromkovič. *Communication Complexity and Parallel Computing.* EATCS Texts in Theoretical Computer Science. Springer, Berlin, 1997.
14. J. Hromkovič and G. Schnitger. Nondeterministic communication with a limited number of advice bits. In *Proc. of 28th Ann. ACM Symp. on Theory of Computing (STOC)*, 551 – 560, 1996.
15. J. Hromkovič. Communication complexity and lower bounds on multilective computations. *Theoretical Informatics and Applications (RAIRO)*, 33:193–212, 1999.
16. J. Jain, J. Bitner, J. A. Abraham, and D. S. Fussell. Functional partitioning for verification and related problems. In T. Knight and J. Savage, editors, *Advanced Research in VLSI and Parallel Systems: Proceedings of the 1992 Brown/MIT Conference*, 210–226, 1992.
17. S. P. Jukna. Lower bounds on communication complexity. *Mathematical Logic and Its Applications*, 5:22–30, 1987.
18. M. Krause. Lower bounds for depth-restricted branching programs. *Information and Computation*, 91(1):1–14, Mar. 1991.
19. M. Krause and S. Waack. On oblivious branching programs of linear length. *Information and Computation*, 94:232–249, 1991.
20. K. Kriegel and S. Waack. Lower bounds on the complexity of real-time branching programs. *Theoretical Informatics and Applications (RAIRO)*, 22:447–459, 1988.
21. E. Kushilevitz and N. Nisan. *Communication Complexity.* Cambridge University Press, Cambridge, 1997.
22. K. Mehlhorn and E. Schmidt. Las-Vegas is better than determinism in VLSI and distributed computing. In *Proc. of 14th Ann. ACM Symp. on Theory of Computing (STOC)*, 330 – 337, 1982.
23. I. Newman. Private vs. common random bits in communication complexity. *Information Processing Letters*, 39:67 – 71, 1991.
24. M. Sauerhoff. *Complexity Theoretical Results for Randomized Branching Programs.* PhD thesis, Univ. of Dortmund. Shaker, 1999.
25. M. Sauerhoff. On the size of randomized OBDDs and read-once branching programs for k-stable functions. In *Proc. of 16th Ann. Symp. on Theoretical Aspects of Computer Science (STACS), LNCS 1563*, 488–499. Springer, 1999.
26. M. Sauerhoff. Computing with restricted nondeterminism: The dependence of the OBDD size on the number of nondeterministic variables. To appear in *Proc. of FST & TCS.*
27. I. Wegener. *The Complexity of Boolean Functions.* Wiley-Teubner, 1987.
28. I. Wegener. *Branching Programs and Binary Decision Diagrams—Theory and Applications.* Monographs on Discrete and Applied Mathematics. SIAM, 1999. To appear.
29. A. C. Yao. Some complexity questions related to distributive computing. In *Proc. of 11th Ann. ACM Symp. on Theory of Computing (STOC)*, 209 – 213, 1979.

The Boolean Hierarchy of NP-Partitions
(Extended Abstract)

Sven Kosub and Klaus W. Wagner

Theoretische Informatik, Julius-Maximilians-Universität Würzburg
Am Hubland, D-97074 Würzburg, Germany
{kosub,wagner}@informatik.uni-wuerzburg.de

Abstract. We introduce the boolean hierarchy of k-partitions over NP for $k \geq 3$ as a generalization of the boolean hierarchy of sets (i.e., 2-partitions) over NP. Whereas the structure of the latter hierarchy is rather simple the structure of the boolean hierarchy of k-partitions over NP for $k \geq 3$ turns out to be much more complicated. We establish the Embedding Conjecture which enables us to get a complete idea of this structure. This conjecture is supported by several partial results.

1 Introduction

To divide the real world into two parts like big and small, black and white, or good and bad usually oversimplifies things. In most cases a partition into many parts is more appropriate. For example, take marks in school, scores for papers submitted to a conference, salary groups, or classes of risk. In mathematics, k-valued logic is just a language for dealing with k-valent objects, and in the computer science field of artificial intelligence, this language has become a powerful tool for reasoning about incomplete knowledge. In computational complexity for instance, proper partitions, although not mentioned explicitely, emerge in connection with locally definable acceptance types (cf. [5]).

Nevertheless, complexity theoreticians mainly investigate the complexity of sets, i.e., partitions into two parts, or, the other extreme, the complexity of functions, i.e., partitions into usually infinitely many parts. But what about partitions into $3, 4, 5, \ldots$ parts?

This paper studies, as a first step in this direction, complexity classes of k-partitions which correspond to the classes of the boolean hierarchy of sets (i.e., 2-partitions). This investigation is justified by the fact that in the cases $k \geq 3$ there are interesting new phenomena which cannot be treated appropriately when encoding k-partitions by sets. On the other hand, with the boolean hierarchy of sets we have a well-studied reference structure.

The most general way to define the boolean hierarchy of sets over NP is as follows (see [8]): For a boolean function $f : \{0,1\}^m \rightarrow \{0,1\}$ and sets $B_1, \ldots, B_m$ define the set $f(B_1, \ldots, B_m)$ by $c_{f(B_1,\ldots,B_m)}(x) =_{\text{def}} f(c_{B_1}(x), \ldots, c_{B_m}(x))$. The class NP($f$) consists of all sets $f(B_1, \ldots, B_m)$ when varying the sets B_i over NP. The boolean hierarchy (of sets) over NP consists of the classes NP(f). It was

H. Reichel and S. Tison (Eds.): STACS 2000, LNCS 1770, pp. 157–168, 2000.

proved in [8] that every class NP(f) coincides with one of the classes NP(i) or coNP(i), where NP(i) is the class of all sets which are the symmetric difference of i NP-sets.

This approach is generalized in Sect. 3 to the case of k-partitions. The characteristic function of a k-partition $A = (A_1, \ldots, A_k)$ is defined by $c_A(x) = i \iff x \in A_i$. For a function $f : \{1,2\}^m \to \{1,2,\ldots,k\}$ and sets $B_1, \ldots, B_m$, taken as the 2-partitions $(B_i, \overline{B_i})$ for every $i \in \{1,\ldots,m\}$, define a k-partition $A = f(B_1,\ldots,B_m)$ by $c_A(x) =_{\text{def}} f(c_{B_1}(x),\ldots,c_{B_m}(x))$. The boolean hierarchy of k-partitions over NP consists of the classes $\text{NP}(f) =_{\text{def}} \{\, f(B_1,\ldots,B_m) \mid B_1,\ldots, B_m \in \text{NP}\,\}$. The boolean hierarchy of sets over NP now appears as the special case $k = 2$.

Whereas the boolean hierarchy of sets over NP has a very simple structure (note that $\text{NP}(i) \cup \text{coNP}(i) \subseteq \text{NP}(i+1) \cap \text{coNP}(i+1)$ for all $i \geq 1$), the situation is much more complicated for the boolean hierarchy of k-partitions in the case $k \geq 3$. The main question is: Can we get an overview over the structure of this hierarchy? This question is not answered completely so far, but in the remaining sections we give partial answers, and we establish a conjecture.

A function $f : \{1,2\}^m \to \{1,2,\ldots,k\}$ which defines the class NP(f) of k-partitions corresponds to the finite boolean lattice $(\{1,2\}^m, \leq)$ with the labeling function f. Generalizing this idea we define for every finite lattice G with labeling function $f : G \to \{1,2,\ldots,k\}$ (for short: the k-lattice (G,f)) a class NP(G,f) of k-partitions. This does not result in more classes: In Sect. 4 we state that for every k-lattice (G,f) there exists a finite function f' such that $\text{NP}(G,f) = \text{NP}(f')$. However, the use of arbitrary lattices instead of only boolean lattices simplifies many considerations.

To get an idea of the structure of the boolean hierarchy of k-partitions over NP it is very important to have a criterion for $\text{NP}(G,f) \subseteq \text{NP}(G',f')$ for k-lattices (G,f) and (G',f'). In Sect. 5 we define a relation $\leq$ as follows: $(G,f) \leq (G',f')$ if and only if there is a monotonic mapping $\varphi : G \to G'$ such that $f(x) = f'(\varphi(x))$. We prove the Embedding Lemma which says that $(G,f) \leq (G',f')$ implies $\text{NP}(G,f) \subseteq \text{NP}(G',f')$, and we establish the Embedding Conjecture which says that the converse is also true unless the polynomial-time hierarchy collapses.

In Sect. 6 we collect evidence for our Embedding Conjecture. For $k = 2$ we confirm this conjecture to be true. Moreover, we give a theorem which enables us to verify the Embedding Conjecture for $k \geq 3$ for a large class of k-lattices including all k-chains. The proof of this theorem uses Kadin's easy-hard-technique (cf. [6]).

Assume the Embedding Conjecture is true. Then the set inclusion structure of the boolean hierarchy of k-partitions is isomorphic to the partial order of $\leq$-equivalence classes of k-lattices with respect to $\leq$. In Sect. 7 we present the partial order of all 132 equivalence classes which contain boolean 3-lattices of the form $(\{1,2\}^3, f)$. Furthermore, the partial order of equivalence classes of 3-lattices does not have bounded width. This gives an impression on the complexity of the (conjectured) structure of the boolean hierarchy of 3-partitions over NP.

2 Preliminaries

For the classes $\mathcal{K}$ and $\mathcal{K}'$ of subsets of a set M, define co$\mathcal{K} = \{\,\overline{L} \mid L \in \mathcal{K}\,\}$, $\mathcal{K} \wedge \mathcal{K}' =_{\mathrm{def}} \{\, A \cap B \mid A \in \mathcal{K}, B \in \mathcal{K}'\,\}$, and $\mathcal{K} \oplus \mathcal{K}' =_{\mathrm{def}} \{\, A \triangle B \mid A \in \mathcal{K}, B \in \mathcal{K}'\,\}$ where $A \triangle B$ denotes the symmetric difference of A and B. The classes $\mathcal{K}(i)$ and co$\mathcal{K}(i)$ defined by $\mathcal{K}(1) =_{\mathrm{def}} \mathcal{K}$ and $\mathcal{K}(i+1) =_{\mathrm{def}} \mathcal{K}(i) \oplus \mathcal{K}$ for $i \geq 1$ build the *boolean hierarchy over* $\mathcal{K}$ that has many equivalent definitions (see [8,2,7,1]). Since $\mathcal{K}(i) \cup \mathrm{co}\mathcal{K}(i) \subseteq \mathcal{K}(i+1) \cap \mathrm{co}\mathcal{K}(i+1)$ for all $i \geq 1$ and $\mathcal{K}$ with $M \in \mathcal{K}$, the boolean hierarchy has a very clear structure. The class BC($\mathcal{K}$) is the boolean closure of $\mathcal{K}$, that is the smallest class which contains $\mathcal{K}$ and which is closed under intersection, union, and complementation.

Further we need some notions from lattice theory and order theory (see e.g., [3]). A finite poset $(G, \leq)$ is a lattice if for all $x, y \in G$ there exist (a) exactly one maximal element $z \in G$ such that $z \leq x$ and $z \leq y$ (which will be denoted by $x \wedge y$), and (b) exactly one minimal element $z \in G$ such that $z \geq x$ and $z \geq y$ (which will be denoted by $x \vee y$). For a lattice G we denote by 1_G the unique element greater than or equal to all $x \in G$ and by 0_G the unique element less than or equal to all $x \in G$. An element $x \neq 1_G$ is said to be *meet-irreducible* iff $x = a \wedge b$ implies $x = a$ or $x = b$ for all $a, b \in G$.

For symbols $a_1, a_2, \ldots, a_m$ from an alphabet Σ, we identify the m-tuple $(a_1, a_2, \ldots, a_m)$ with the word $a_1 a_2 \ldots a_m \in \Sigma^m$. If there is an order $\leq$ on Σ, we assume Σ^m to be partially ordered by the vector-ordering, that is $a_1 a_2 \ldots a_m \leq b_1 b_2 \ldots b_m$ if and only if $a_i \leq b_i$ for all $i \in \{1, 2, \ldots, m\}$.

Finally, let us make some notational conventions about partitions. For any set M, a k-tuple $A = (A_1, \ldots, A_k)$ is said to be a *k-partition of* M if and only if $A_1 \cup A_2 \cup \cdots \cup A_k = M$ and $A_i \cap A_j = \varnothing$ if $i \neq j$. The set A_i is said to be the *i-th component of* A. Let $c_A : M \to \{1, 2, \ldots, k\}$ be the characteristic function of a k-partition $A = (A_1, \ldots, A_k)$ of M, i.e., $c_A(x) = i$ if and only if $x \in A_i$ for every $x \in M$ and $1 \leq i \leq k$. For classes $\mathcal{K}_1, \ldots, \mathcal{K}_k \subseteq \mathcal{P}(M)$, we define

$$(\mathcal{K}_1, \ldots, \mathcal{K}_k) =_{\mathrm{def}} \{\, A \mid A \text{ is a } k\text{-partition of } M \text{ and } A_i \in \mathcal{K}_i \text{ for } 1 \leq i \leq k \,\}$$

and for $1 \leq i \leq k$,

$$(\mathcal{K}_1, \ldots, \mathcal{K}_{i-1}, \cdot, \mathcal{K}_{i+1}, \ldots, \mathcal{K}_k) =_{\mathrm{def}} (\mathcal{K}_1, \ldots, \mathcal{K}_{i-1}, \mathcal{P}(M), \mathcal{K}_{i+1}, \ldots, \mathcal{K}_k).$$

For a class $\mathcal{K}$ of k-partitions, let $\mathcal{K}_i =_{\mathrm{def}} \{\, A_i \mid A \in \mathcal{K}\,\}$ be the *i-th projection of* $\mathcal{K}$. Obviously, $\mathcal{K} \subseteq (\mathcal{K}_1, \ldots, \mathcal{K}_k)$. In what follows we identify a set A with the 2-partition $(A, \overline{A})$, and we identify a class $\mathcal{K}$ of sets with the class $(\mathcal{K}, \mathrm{co}\mathcal{K}) = (\mathcal{K}, \cdot) = (\cdot, \mathrm{co}\mathcal{K})$ of 2-partitions.

3 Partition Classes Defined by Finite Functions

Let $\mathcal{K}$ be a class of subsets of M such that $\varnothing, M \in \mathcal{K}$ and $\mathcal{K}$ is closed under intersection and union. In the literature, one way to define the classes of the boolean hierarchy of sets over $\mathcal{K}$ is as follows (see [8]). Let $f : \{1, 2\}^m \to$

$\{1,2\}$ be a boolean function. For $B_1, \dots, B_m \in \mathcal{K}$ the set $f(B_1, \dots, B_m)$ is defined by $c_{f(B_1,\dots,B_m)}(x) = f(c_{B_1}(x), \dots, c_{B_m}(x))$. Then the classes $\mathcal{K}(f) =_{\text{def}} \{ f(B_1, \dots, B_m) \mid B_1, \dots, B_m \in \mathcal{K} \}$ form the boolean hierarchy over $\mathcal{K}$. Using finite functions $f : \{1,2\}^m \to \{1, 2, \dots, k\}$ we generalize this definition (remember in which sense sets are 2-partitions) to obtain the classes of the boolean hierarchy of k-partitions over $\mathcal{K}$ as follows.

Definition 1. *Let $k \geq 2$.*

1. *For $f : \{1,2\}^m \to \{1, 2, \dots, k\}$ and for sets $B_1, \dots, B_m \in \mathcal{K}$, the k-partition $f(B_1, \dots, B_m)$ is defined by $c_{f(B_1,\dots,B_m)}(x) = f(c_{B_1}(x), \dots, c_{B_m}(x))$.*
2. *For $f : \{1,2\}^m \to \{1, 2, \dots, k\}$, the* class of k-partitions over $\mathcal{K}$ defined by f *is given by the class $\mathcal{K}(f) =_{\text{def}} \{ f(B_1, \dots, B_m) \mid B_1, \dots, B_m \in \mathcal{K} \}$.*
3. *The family $\mathrm{BH}_k(\mathcal{K}) =_{\text{def}} \{ \mathcal{K}(f) \mid f : \{1,2\}^m \to \{1, 2, \dots, k\}$ and $m \geq 1 \}$ is the* boolean hierarchy of k-partitions over $\mathcal{K}$.
4. $\mathrm{BC}_k(\mathcal{K}) =_{\text{def}} \bigcup \mathrm{BH}_k(\mathcal{K})$.

Obviously, if $i \in \{1, \dots, k\}$ is not a value of $f : \{1,2\}^m \to \{1, 2, \dots, k\}$ then $\mathcal{K}(f)_i = \{\varnothing\}$, i.e., $\mathcal{K}(f)$ does not really have an i-th component. Therefore we assume in what follows that f is surjective.

The following proposition shows that every partition in $\mathcal{K}(f)$ consists of sets from the boolean hierarchy over $\mathcal{K}$. This also justifies the use of the term *boolean* in the above definition.

Proposition 1. *Let $f : \{1,2\}^m \to \{1, 2, \dots, k\}$, $k \geq 2$.*

1. $(\mathcal{K}, \dots, \mathcal{K}) \subseteq \mathcal{K}(f) \subseteq (\mathrm{BC}(\mathcal{K}), \dots, \mathrm{BC}(\mathcal{K}))$.
2. *If $\mathcal{K}$ is closed under complementation then $\mathcal{K}(f) = (\mathcal{K}, \dots, \mathcal{K})$.*
3. $\mathrm{BC}_k(\mathcal{K}) = (\mathrm{BC}(\mathcal{K}), \dots, \mathrm{BC}(\mathcal{K}))$.

For $k = 2$ the classes $\mathcal{K}(f)$ of the boolean hierarchy $\mathrm{BH}_2(\mathcal{K})$ of sets (2-partitions) over $\mathcal{K}$ have been completely characterized. For $f : \{1,2\}^m \to \{1,2\}$ let $\mu(f)$ be the maximum number of alternations of f-labels which can occur in a $\leq$-chain in $(\{1,2\}^m, \leq)$.

Theorem 1. [8] *For $f : \{1,2\}^m \to \{1,2\}$,*

$$\mathcal{K}(f) = \begin{cases} \mathcal{K}(\mu(f)), & \text{if } f(2^m) = 2, \\ \mathrm{co}\mathcal{K}(\mu(f)), & \text{if } f(2^m) = 1. \end{cases}$$

Consequently, $\mathrm{BH}_2(\mathcal{K}) = \{ \mathcal{K}(n) \mid n \geq 1 \} \cup \{ \mathrm{co}\mathcal{K}(n) \mid n \geq 1 \}$, and given a function $f : \{1,2\}^m \to \{1,2\}$ it is easy to determine the class $\mathcal{K}(n)$ or $\mathrm{co}\mathcal{K}(n)$ which coincides with $\mathcal{K}(f)$. As mentioned above, the classes of $\mathrm{BH}_2(\mathcal{K})$ form a simple structure with respect to set inclusion. There do not exist three classes in $\mathrm{BH}_2(\mathcal{K})$ which are incomparable in this sense.

It is the goal of this paper to get insights into the structure of the boolean hierarchy $\mathrm{BH}_k(\mathrm{NP})$ of k-partitions over NP for $k \geq 3$. What we can say at this point is, that already for $k = 3$ the structure of $\mathrm{BH}_k(\mathrm{NP})$ with respect to set inclusion is not as simple as for $k = 2$ (unless $\mathrm{NP} = \mathrm{coNP}$). This is shown by the following example.

Example 1. For a, b, c such that $\{a, b, c\} = \{1, 2, 3\}$ define the function $f_{abc} : \{1, 2\}^2 \to \{1, 2, 3\}$ by $f_{abc}(11) = a$, $f_{abc}(12) = f_{abc}(21) = b$, and $f_{abc}(22) = c$. Obviously, $\mathrm{NP}(f_{abc})_a = \mathrm{NP}$, $\mathrm{NP}(f_{abc})_b = \mathrm{NP}(2)$, and $\mathrm{NP}(f_{abc})_c = \mathrm{coNP}$. Now let $abc \neq a'b'c'$. If $\mathrm{NP}(f_{abc}) = \mathrm{NP}(f_{a'b'c'})$ then $\mathrm{NP} = \mathrm{NP}(2)$ or $\mathrm{NP} = \mathrm{coNP}$, or $\mathrm{NP}(2) = \mathrm{coNP}$. In each of this cases we obtain $\mathrm{NP} = \mathrm{coNP}$. Consequently, if $\mathrm{NP} \neq \mathrm{coNP}$ then the six classes $\mathrm{NP}(f_{abc})$ are pairwise incomparable with respect to set inclusion.

4 Partition Classes Defined by Lattices

It turns out that, for $f : \{1, 2\}^m \to \{1, 2, \ldots, k\}$, a k-partition $f(B_1, \ldots, B_m)$ has a very natural equivalent lattice-theoretical definition. Consider the boolean lattice $\{1, 2\}^m$ with the partial vector-ordering $\leq$, and consider the function $S : \{1, 2\}^m \to \mathcal{K}$ defined by $S(a_1, \ldots, a_m) =_{\mathrm{def}} \bigcap_{a_i = 1} B_i$, where we define an intersection over an empty index set to be M. For an example see Fig. 1. Note that $S(2, \ldots, 2) = M$ and $S(a \wedge b) = S(a) \cap S(b)$ for all $a, b \in \{1, 2\}^m$. Defining $T_S(a) =_{\mathrm{def}} S(a) \setminus \bigcup_{b < a} S(b)$ we obtain the i-th component of $f(B_1, \ldots, B_m)$ as $f(B_1, \ldots, B_m)_i = \bigcup_{f(a) = i} T_S(a)$, i.e., $f(B_1, \ldots, B_m)$ can also be given by the function $S : \{1, 2\}^m \to \mathcal{K}$.

On the other side, if we have any function $S : \{1, 2\}^m \to \mathcal{K}$ such that $S(2, \ldots, 2) = M$ and $S(a \wedge b) = S(a) \cap S(b)$ for all $a, b \in \{1, 2\}^m$ we can define $B_j =_{\mathrm{def}} S(2^{j-1} 1 2^{m-j})$ for $j \in \{1, 2, \ldots, m\}$, and we obtain $f(B_1, \ldots, B_m)_i = \bigcup_{f(a) = i} T_S(i)$ for $i \in \{1, 2, \ldots, k\}$. In this manner the class $\mathcal{K}(f)$ of k-partitions is completely characterized by the labeled boolean lattice $((\{1, 2\}^m, \leq), f)$.

In this section we will see that classes of k-partitions can also be defined by weaker structures than boolean algebras. Again we always suppose $\mathcal{K}$ to be a class such that $\emptyset, M \in \mathcal{K}$ and which is closed under intersection and union.

Definition 2. *Let G be a lattice.*

1. *A mapping $S : G \to \mathcal{K}$ is said to be a $\mathcal{K}$-*homomorphism on *G if and only if $S(1_G) = M$ and $S(a \wedge b) = S(a) \cap S(b)$ for all $a, b \in G$.*
2. *For a $\mathcal{K}$-homomorphism S on G, let $T_S(a) =_{\mathrm{def}} S(a) \setminus \bigcup_{b < a} S(b)$ for $a \in G$.*

Lemma 1. *Let G be a lattice, and let S be a $\mathcal{K}$-homomorphism on G.*

1. *$T_S(a) \in \mathcal{K} \wedge \mathrm{co}\mathcal{K}$ for every $a \in G$.*
2. *$S(a) = \bigcup_{b \leq a} T_S(b)$ for every $a \in G$.*
3. *The set of all $T_S(a)$ for $a \in G$ yields a partition of M.*
4. *S is completely determined by its values for the meet-irreducible elements. That is, if S and S' are two $\mathcal{K}$-homomorphisms on G such that $S(a) = S'(a)$ for all meet-irreducible $a \in G$ then $S(a) = S'(a)$ for all $a \in G$.*

Any pair (G, f) of an arbitrary finite poset G and a function $f : G \to \{1, 2, \ldots, k\}$ is called a *k-poset.* A k-poset which is a lattice (boolean lattice) is called a *k-lattice* (*boolean k-lattice*, resp.).

Lemma 1 provides the soundness of the following definition.

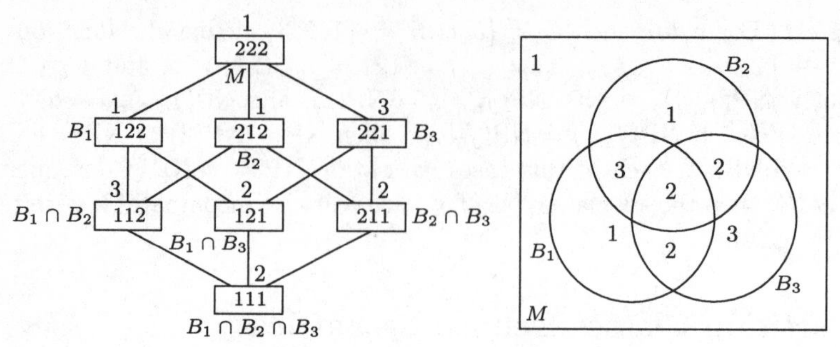

Fig. 1. Partition defined by a boolean 3-lattice.

Definition 3. *Let (G, f) be a k-lattice, $k \geq 2$.*

1. *For a $\mathcal{K}$-homomorphism S on G, the* k-partition defined by (G, f) and S *is given by* $(G, f, S) =_{\text{def}} \left(\bigcup_{f(a)=1} T_S(a), \ldots, \bigcup_{f(a)=k} T_S(a) \right)$.
2. $\mathcal{K}(G, f) =_{\text{def}} \left\{ (G, f, S) \,\middle|\, S \text{ is } \mathcal{K}\text{-homomorphism on } G \right\}$ *is the* class of k-partitions defined by (G, f).

Example 2. Consider the 3-lattice (G, f) in Fig. 2. The meet-irreducible elements of G are a, b, and c. By point 4 of Lemma 1 every $\mathcal{K}$-homomorphism $S : G \to \mathcal{K}$ is determined by fixing $S(a) = A$, $S(b) = B$, and $S(c) = C$. By the definition of $\mathcal{K}$-homomorphisms we get $S(1) = M$, $S(d) = S(a \wedge b) = S(a) \cap S(b) = A \cap B$, and $S(0) = S(d \wedge c) = S(d) \cap S(c) = A \cap B \cap C$. Furthermore, $C = S(c) = S(c \wedge b) = S(c) \cap S(b) = C \cap B$, i.e., $C \subseteq B$. We obtain

$$\begin{aligned}
T_S(1) &= M \setminus (A \cup B) &&= \overline{A} \cap \overline{B}, \\
T_S(a) &= A \setminus (A \cap B) &&= A \cap \overline{B}, \\
T_S(b) &= B \setminus ((A \cap B) \cup C) &&= \overline{A} \cap B \cap \overline{B}, \\
T_S(c) &= C \setminus (A \cap B \cap C) &&= \overline{A} \cap C, \\
T_S(d) &= (A \cap B) \setminus (A \cap B \cap C) &&= A \cap B \cap \overline{C}, \\
T_S(0) &= A \cap B \cap C &&= A \cap C.
\end{aligned}$$

Hence

$$\begin{aligned}
(G, f, S) &= (T_S(a) \cup T_S(0), T_S(1) \cup T_S(c), T_S(b) \cup T_S(d)) \\
&= (A \cap (\overline{B} \cup C), \overline{A} \cap (\overline{B} \cup C), B \cap \overline{C}),
\end{aligned}$$

and

$$\begin{aligned}
\mathcal{K}(G, f) &= \left\{ (A \cap (\overline{B} \cup C), \overline{A} \cap (\overline{B} \cup C), B \cap \overline{C}) \,\middle|\, A, B, C \in \mathcal{K} \text{ and } C \subseteq B \right\} \\
&\subseteq (\mathcal{K}(3), \text{co}\mathcal{K}(3), \mathcal{K}(2)).
\end{aligned}$$

The discussion at the beginning of the section yields the following proposition.

Proposition 2. $\mathcal{K}(f) = \mathcal{K}((\{1,2\}^m, \leq), f)$ *for every* $f : \{1,2\}^m \to \{1, 2, \ldots, k\}$.

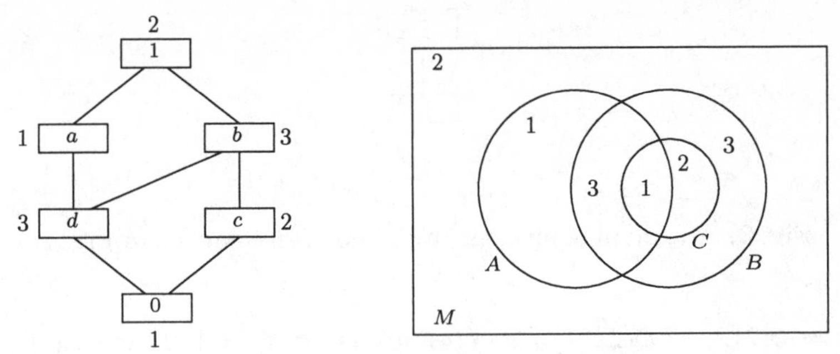

Fig. 2. Partition defined by a 3-lattice.

So, if (G, f) is a boolean k-lattice then $\mathcal{K}(G, f) = \mathcal{K}(f)$. But if (G, f) is an arbitrary k-lattice, is $\mathcal{K}(G, f)$ also of the form $\mathcal{K}(f')$ for a suitable function f'? The following theorem says that this is generally true. This turns out to be very important for the further study of the structure of the boolean hierarchy of k-partitions because instead of large boolean k-lattices one can handle with usually much smaller equivalent k-lattices.

Theorem 2. *For every k-lattice (G, f) there is an $f' : \{1,2\}^m \to \{1, 2, \ldots, k\}$ with $\mathcal{K}(G, f) = \mathcal{K}(f')$, where m is the number of meet-irreducible elements of G.*

The proof of this theorem is an application of the Embedding Lemma below. In fact, it is enough to construct to any k-lattice a boolean k-lattice which is equivalent in the sense explained right in the next section.

5 Comparing Partition Classes: The Embedding Conjecture

To study the structure of the boolean hierarchy of k-partitions over NP it would be important to have a criterion to decide whether $\mathrm{NP}(G, f) \subseteq \mathrm{NP}(G', f')$ for any two k-lattices (G, f) and (G', f'). To this end we establish a relation $\leq$ between k-lattices. For k-lattices (G, f) and (G', f') we write $(G, f) \leq (G', f')$ if and only if there is a monotonic mapping $\varphi : G \to G'$ such that $f(x) = f'(\varphi(x))$ for every $x \in G$. We write $(G, f) \equiv (G', f')$ and we say that (G, f) and (G', f') are *equivalent* if $(G, f) \leq (G', f')$ and $(G', f') \leq (G, f)$.

The following lemma gives a sufficient condition for $\mathrm{NP}(G, f) \subseteq \mathrm{NP}(G', f')$.

Lemma 2. (Embedding Lemma.) *Let $\mathcal{K}$ be a class with $M \in \mathcal{K}$ and which is closed under finite union. Let (G, f) and (G', f') be k-lattices. If $(G, f) \leq (G', f')$, then $\mathcal{K}(G, f) \subseteq \mathcal{K}(G', f')$.*

Example 3. The 3-lattice (G, f) shown in Fig. 1 and the 3-lattice (G', f') shown in Fig. 3 are equivalent. This can be seen as follows: Define the functions $\varphi : G \to G'$ and $\psi : G' \to G$ by $\varphi(111) = \varphi(121) = \varphi(211) = a$, $\varphi(112) = \varphi(221) = b$,

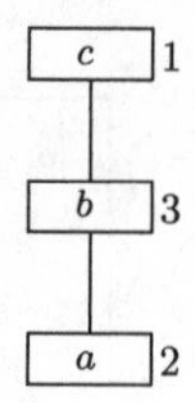

Fig. 3. A 3-chain equivalent to the boolean 3-lattice in Fig. 1.

$\varphi(122) = \varphi(212) = \varphi(222) = c$, $\psi(a) = 111$, $\psi(b) = 112$, and $\psi(c) = 222$. It is easy to see that φ and ψ are monotonic, $f(x) = f'(\varphi(x))$ for all $x \in G$, and $f'(x) = f(\psi(x))$ for all $x \in G'$. By the Embedding Lemma we obtain $\mathcal{K}(G,f) = \mathcal{K}(G',f')$ for all $\mathcal{K}$. Obviously, $\mathcal{K}(G',f') = \{\, (\overline{B}, A, B \setminus A) \mid A, B \in \mathcal{K} \text{ and } A \subseteq B \,\} = (\mathrm{co}\mathcal{K}, \mathcal{K}, \cdot) = (\mathrm{co}\mathcal{K}, \mathcal{K}, \mathcal{K}(2))$.

Let us come back to the Embedding Lemma which shows that $(G,f) \leq (G',f')$ implies $\mathcal{K}(G,f) \subseteq \mathcal{K}(G',f')$. Because of Proposition 1.2, we cannot hope to convert this implication without an additional assumption to $\mathcal{K}$. Even an infinite boolean hierarchy of *sets* over $\mathcal{K}$ is not sufficient to redirect the implication. To see this consider the class IS $=_{\mathrm{def}} \{\, \{1, 2, \ldots, n\} \mid n \in \mathbb{N} \text{ or } n = \infty \,\}$. Obviously, IS is closed under intersection and union. Moreover, it is an easy exercise to confirm that $\mathrm{BH}_2(\mathrm{IS})$ is strict. For $\mathrm{BH}_3(\mathrm{IS})$ this is not true.

Proposition 3. *Let (G,f) and (G',f') be the 3-lattices shown in Fig. 4. Then* $\mathrm{IS}(G,f) = \mathrm{IS}(G',f')$ *but* $(G,f) \not\leq (G',f')$.

Up to this proposition, all results so far hold for arbitrary classes with some simple closure properties. The forthcoming now makes use of the very nature of the class NP. Since the collapse of the boolean hierarchy over NP implies the collapse of the polynomial-time hierarchy (cf. [6]) the following conjecture seems to be reasonable.

Embedding Conjecture. Assume the polynomial-time hierarchy does not collapse. Let (G,f) and (G',f') be k-lattices. Then $\mathrm{NP}(G,f) \subseteq \mathrm{NP}(G',f')$ if and only if $(G,f) \leq (G',f')$.

To provide evidence for the Embedding Conjecture we formulate in the next section a theorem (Theorem 3) which shows that the Embedding Conjecture is true for a large subclass of all k-lattices including all 2-lattices (Corollary 2) and all k-chains (Corollary 1). Furthermore, the 3-lattices in Fig. 4 turn out to be no counterexample.

6 Evidence for the Embedding Conjecture

We establish a theorem which shows that the Embedding Conjecture is true for a large subclass of k-lattices. Proving this theorem, we detect some normal forms of

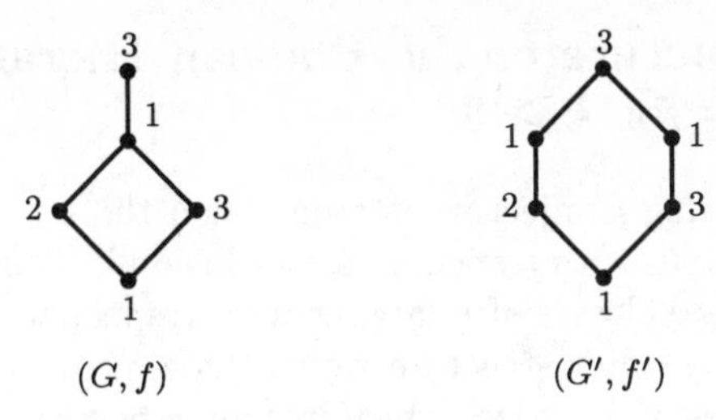

Fig. 4. Critical 3-lattices.

(hypothetical) inclusions between partition classes enabling us a generalization of the easy-hard-arguments developed by Kadin (cf. [6]) to the context of partition classes. A k-chain is called *repetition-free* iff neighbored elements have different labels.

Theorem 3. *Assume that the polynomial-time hierarchy does not collapse. Let* (G, f) *and* (G', f') *be* k*-lattices. If* $\mathrm{NP}(G, f) \subseteq \mathrm{NP}(G', f')$ *then every repetition-free* k*-subchain of* (G, f) *occurs as a* k*-subchain of* (G', f').

Theorem 3 easily gives that the 3-lattices in Fig. 2 and Fig. 3 define incomparable partition classes over NP, unless the polynomial-time hierarchy collapses. The following corollary shows that the Embedding Conjecture is true for k-chains.

Corollary 1. *Assume the polynomial-time hierarchy does not collapse. For* k*-chains* (G, f) *and* (G', f') *it holds that* $\mathrm{NP}(G, f) \subseteq \mathrm{NP}(G', f')$ *if and only if* $(G, f) \leq (G', f')$.

Furthermore, the Embedding Conjecture is generally true for 2-lattices. This is a consequence of Theorem 3 and the following simple proposition.

Proposition 4. *Every* 2*-lattice is equivalent to its longest chain with alternating labels* 1 *and* 2.

Corollary 2. *Assume the polynomial-time hierarchy does not collapse. For* 2*-lattices* (G, f) *and* (G', f') *it holds that* $\mathrm{NP}(G, f) \subseteq \mathrm{NP}(G', f')$ *if and only if* $(G, f) \leq (G', f')$.

Assume the polynomial-time hierarchy does not collapse. By Theorem 3, if the k-lattice (G, f) has a repetition-free k-subchain which is not a k-subchain of the k-lattice (G', f') then $\mathrm{NP}(G, f) \not\subseteq \mathrm{NP}(G', f')$. But what about k-lattices which have the same repetition-free k-subchains? For example, take the 3-lattices (G, f) and (G', f') presented in Fig. 4. Since $(G, f) \not\leq (G', f')$ the Embedding Conjecture says that $\mathrm{NP}(G, f) \not\subseteq \mathrm{NP}(G', f')$, but Theorem 3 does not help to prove this. However, this can be proved by a clever exploitation of the situation in order to simplify the self-reduction tree of the satisfiability problem. The proof is inspired by a work of Hemaspaandra *et al.* [4].

Theorem 4. *Let* (G, f) *and* (G', f') *be the* 3*-lattices shown in Fig. 4. Then it holds that* $\mathrm{NP}(G, f) \subseteq \mathrm{NP}(G', f')$ *if and only if* $\mathrm{NP} = \mathrm{coNP}$.

7 On the Structure of the Boolean Hierarchy of k-Partitions for $k \geq 3$

Assume the Embedding Conjecture is true. Then the structure of the boolean hierarchy of k-partitions with respect to set inclusion is identical with the partial order of $\leq$-equivalence classes of k-lattices with respect to $\leq$. To get an idea of the complexity of the latter structure we will now present the partial order of all equivalence classes of 3-lattices which include a boolean 3-lattice of the form $(\{1,2\}^3, f)$ with surjective f (for non-surjective f these k-lattices do not really define 3-partitions). The 5796 different boolean 3-lattices of the form $(\{1,2\}^3, f)$ with surjective f are in 132 different equivalence classes.

Figure 5 shows the partial order of the 44 equivalence classes which contain boolean 3-lattices of the form $(\{1,2\}^3, f)$ such that $f(1,1,1) = 1$. The cases $f(1,1,1) = 2$ and $f(1,1,1) = 3$ yield isomorphic partial orders. A line from equivalence class $\mathcal{G}$ up to equivalence class $\mathcal{G}'$ means that $(G,f) < (G',f')$ for every $(G,f) \in \mathcal{G}$ and $(G',f') \in \mathcal{G}'$. We emphasize that such a study would be intractable without the possibility to present boolean k-lattices by equivalent k-lattices.

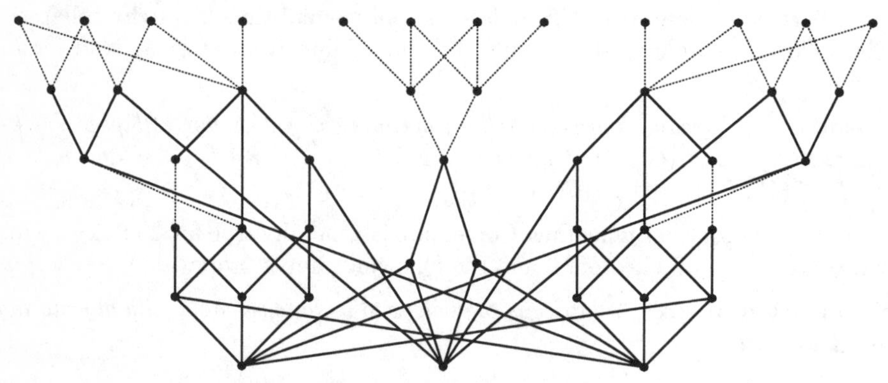

Fig. 5. Scheme of all boolean 3-lattices of the form $(\{1,2\}^3, f)$.

Figure 6 shows the right part of the partial order in Fig. 5 where each equivalence class is represented by a minimal equivalent 3-lattice. The left part of the partial order in Fig. 2 is symmetric to the right part where the labels 2 and 3 change their role.

Theorem 5. *Assume the polynomial-time hierarchy does not collapse. If there is a solid line from class $\mathcal{G}$ up to class $\mathcal{G}'$ in Figure 6 then* $\mathrm{NP}(G,f) \subset \mathrm{NP}(G',f')$ *for every $(G,f) \in \mathcal{G}$ and $(G',f') \in \mathcal{G}'$.*

Every "solid line" in this theorem is an application of Theorem 3 besides the one marked by $*$ which is just Theorem 4.

At the end of this section we mention that the partial order of equivalence classes of 3-lattices does not have "bounded width".

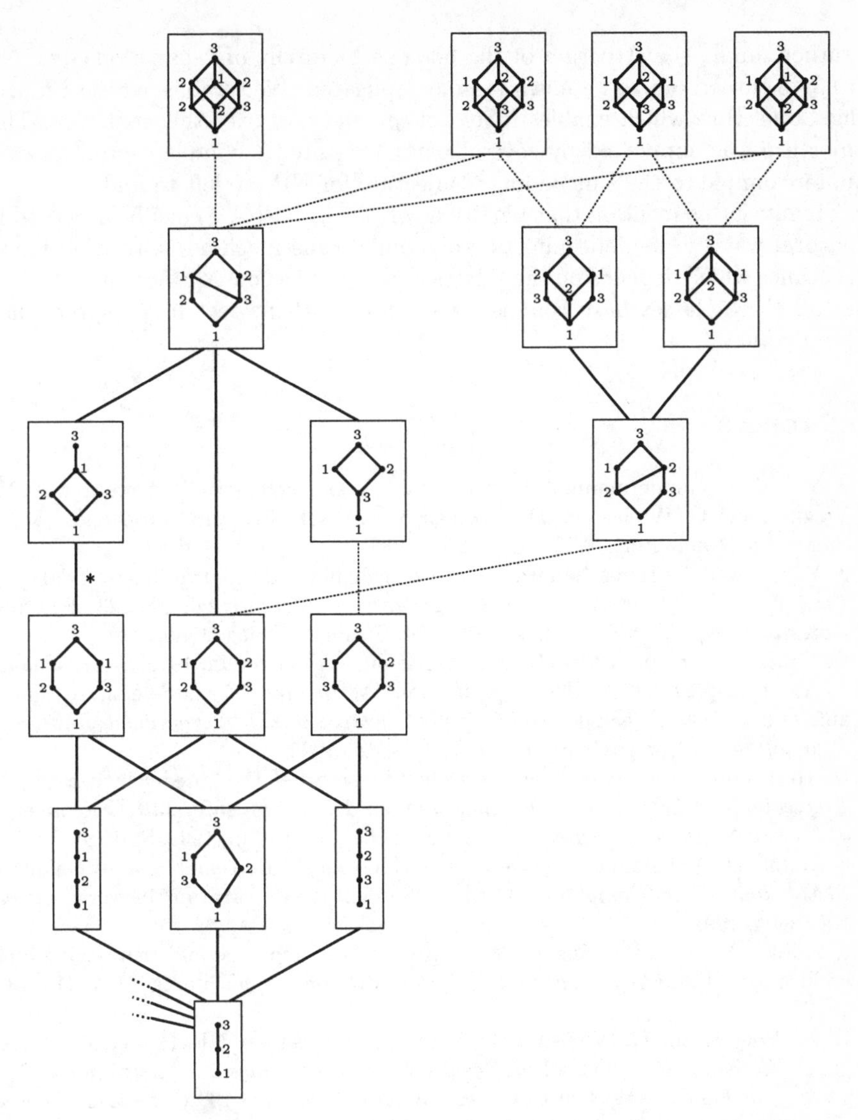

Fig. 6. Magnification of the right part of the scheme in Fig. 5.

Proposition 5. *For every* $m \in \mathbb{N}$ *there exist at least* m *3-lattices that are incomparable with respect to* $\leq$.

8 Conclusion

In the preceding sections, we have investigated the boolean hierarchy of k-partitions over NP for $k \geq 3$ as a generalization of the boolean hierarchy of sets (i.e., 2-partitions) over NP. Whereas the structure of the latter hierarchy is rather simple the structure of the boolean hierarchy of k-partitions over NP for $k \geq 3$ turned out to be much more complicated. We established the Embedding Conjecture which enables us to get an overview over this structure. This conjecture was supported by several partial results. A complete proof of or a counterexample to the Embedding Conjecture for NP are left to find.

Finally, let us mention that partitions of classes $\mathrm{NP}(G, f)$ can be accepted in a natural way by nondeterministic polynomial-time machines with a notion of acceptance which depends on the k-lattice (G, f). As a consequence one can show that all these classes have complete partitions with respect to an appropriate $\leq_m^p$-reduction.

References

1. J.-Y. Cai, T. Gundermann, J. Hartmanis, L. A. Hemachandra, V. Sewelson, K. W. Wagner, and G. Wechsung. The boolean hierarchy I: Structural properties. *SIAM Journal on Computing*, 17(6):1232–1252, 1988.
2. J.-Y. Cai and L. Hemachandra. The Boolean hierarchy: hardware over NP. In *Proceedings 1st Structure in Complexity Theory Conference*, volume 223 of *Lecture Notes in Computer Science*, pages 105–124. Springer-Verlag, 1986.
3. G. Grätzer. *General Lattice Theory*. Akademie-Verlag, Berlin, 1978.
4. L. A. Hemaspaandra, A. Hoene, A. V. Naik, M. Ogihara, A. L. Selman, T. Thierauf, and J. Wang. Nondeterministically selective sets. *International Journal of Foundations of Computer Science*, 6(4):403–416, 1995.
5. U. Hertrampf. Locally definable acceptance types — the three-valued case. In *Proceedings 1st Latin American Symposium on Theoretical Informatics*, volume 583 of *Lecture Notes in Computer Science*, pages 262–271, Berlin, 1992. Springer-Verlag.
6. J. Kadin. The polynomial time hierarchy collapses if the Boolean hierarchy collapses. *SIAM Journal on Computing*, 17(6):1263–1282, 1988. Erratum in same journal 20(2):404, 1991.
7. J. Köbler, U. Schöning, and K. W. Wagner. The difference and truth-table hierarchies for NP. *RAIRO Theoretical Informatics and Applications*, 21(4):419–435, 1987.
8. K. W. Wagner and G. Wechsung. On the boolean closure of NP. Extended abstract as: G. Wechsung. On the boolean closure of NP. *Proceedings 5th International Conference on Fundamentals in Computation Theory,* volume 199 of *Lecture Notes in Computer Science*, pages 485-493, Berlin, 1985.

Binary Exponential Backoff Is Stable for High Arrival Rates

Hesham Al-Ammal[1], Leslie Ann Goldberg[2], and Phil MacKenzie[3]

[1] Department of Computer Science, University of Warwick, Coventry, CV4 7AL, United Kingdom,
heshamQdcs.warwick.ac.uk
[2] Department of Computer Science, University of Warwick, Coventry, CV4 7AL, United Kingdom. This work was partially supported by EPSRC grant GR/L6098,
leslie@dcs.warwick.ac.uk, http://www.dcs.warwick.ac.uk/~leslie/
[3] Information Sciences Center, Bell Laboratories, Lucent Technologies, 600 Mountain Avenue, Murray Hill, NJ 07974–0636,
philmac@research.bell-labs.com

Abstract. Goodman, Greenberg, Madras and March gave a lower bound of $n^{-\Omega(\log n)}$ for the maximum arrival rate for which the n-user binary exponential backoff protocol is stable. Thus, they showed that the protocol is stable as long as the arrival rate is at most $n^{-\Omega(\log n)}$. We improve the lower bound, showing that the protocol is stable for arrival rates up to $O(n^{-.9})$.

1 Introduction

A *multiple-access channel* is a broadcast channel that allows multiple users to communicate with each other by sending messages onto the channel. If two or more users simultaneously send messages, then the messages interfere with each other (collide), and the messages are not transmitted successfully. The channel is not centrally controlled. Instead, the users use a contention-resolution protocol to resolve collisions. Thus, after a collision, each user involved in the collision waits a random amount of time (which is determined by the protocol) before re-sending. Perhaps the best-known contention-resolution protocol is the *Ethernet* protocol of Metcalfe and Boggs [9]. The Ethernet protocol is based on the following simple *binary exponential backoff protocol.* Time is divided into discrete units called steps. If the i'th user has a message to send during a given step, then it sends this message with probability 2^{-b_i}, where b_i denotes the number of collisions that this message has already had. With probability $1 - 2^{-b_i}$, user i does not send during the step. The Ethernet protocol is based on binary exponential backoff, but some modifications are made to make it easier to build. See [6,9] for details.

Håstad, Leighton and Rogoff [6] have studied the performance of the binary exponential backoff protocol in the following natural model. The system consists of n users. Each user maintains a queue of messages that it wishes to send. At the beginning of the t'th time step, the length of the queue of the i'th user is denoted $q_i(t)$ and the number of times that the message at the head of its

H. Reichel and S. Tison (Eds.): STACS 2000, LNCS 1770, pp. 169–180, 2000.

queue has collided is denoted $b_i(t)$. At the beginning of the t'th step, each queue receives 0 or 1 new messages. In particular, a new message is added to the end of each queue independently with probability λ/n, where λ is the *arrival rate* of the system. After the new messages are added to the queues, each user makes an independent decision about whether or not to send the message at the head of its queue, using the binary exponential backoff protocol. (If the message at the head of the i'th queue has never been sent before then $b_i = 0$, so it is now sent. Otherwise, $b_i = b_i(t)$, so it is sent independently with probability $2^{-b_i(t)}$.) If exactly one message is sent (so there are no collisions), then this message is delivered successfully, and it leaves its queue. Otherwise, the messages that are sent collide and no messages are delivered successfully.

Since the arrivals are modelled by a stochastic process, the evolution of the whole system over time can be viewed as a Markov chain in which the state just before step t is $X(t) = ((q_1(t), \ldots, q_n(t)), (b_1(t), \ldots, b_n(t)))$ and the next state is $X(t+1)$. One measure of the performance of the system is the expectation of the random variable T_{ret}, which is the number of steps required for the system to return to the start state $X(0) = ((0, \ldots, 0), (0, \ldots, 0))$. Håstad et al. [6] proved that if the arrival rate is too high, then the system is unstable, in the sense that the expected recurrence time is infinite.

Theorem 1 (Håstad, Leighton, and Rogoff). *Suppose that for some positive ϵ, $\lambda \geq \frac{1}{2} + \epsilon$. Suppose that n is sufficiently large (as a function of ϵ). Then $E[T_{ret}] = \infty$.*

On the other hand, Goodman, Greenberg, Madras and March [5] showed that if the arrival rate is sufficiently low, then the system is stable.

Theorem 2 (Goodman, Greenberg, Madras and March). *There is a positive constant α such that $E[T_{ret}]$ is finite for the n-user system, provided that $\lambda < \frac{1}{n^{\alpha \log n}}$.*

While Goodman, Greenberg, Madras, and March's result is the only known stability result for the finitely-many-users binary-exponential-backoff protocol, their upper bound ($\lambda < \frac{1}{n^{\alpha \log n}}$) is very small. In this paper, we narrow the gap between the two results. In particular, we prove the following theorem.

Theorem 3. *There is a positive constant α such that, as long as n is sufficiently large and $\lambda < \frac{1}{\alpha n^{.9}}$ then $E[T_{ret}]$ is finite for the n-user system.*

The point of Theorem 3 is to show that n-user Binary Exponential Backoff is stable for arrival rates which grow faster asymptotically than $1/n$. That is, the purpose of the result is to show that, for positive constants α and η, $\lambda < \frac{1}{\alpha n^{1-\eta}}$ guarantees stability. We have chosen $\eta = .1$ for concreteness. We believe that the same methods could be used for slightly larger values of η, but an interesting (and difficult) question raised by this work is whether the same result would be true for $\eta = 1$. That is, is there a constant α such that the n-user system is stable whenever $\lambda < \frac{1}{\alpha}$?

The organisation of the paper is as follows. In Section 2 we summarise other related work. In Section 3 we give the proof of Theorem 3.

2 Related Work

We now summarize some other related work. We start by observing that the results in Theorem 1 and 2 can be extended to more general models. For example, the result of Goodman et al. can be extended to a more general model of stochastic arrivals in which the expected number of arrivals at user i at time t (conditioned on all events up to time t) is a quantity, λ_i, and $\sum_i \lambda_i$ is required to be equal to λ. The result of Håstad et al. can be extended to small values of n, provided that $\lambda > .568 + 1/(4n-2)$. The instability result of Håstad et al. implies that, when λ is sufficiently large, the expected average waiting time of messages is infinite.

Next, we mention that the binary exponential backoff protocol is known to be unstable in the infinitely-many-users Poisson-arrivals model. Kelly and MacPhee [7,8] showed this for $\lambda < \ln 2$ and Aldous [1] showed that it holds for all positive λ [1].

Finally, we mention that, while the goal of this paper is to understand the binary-exponential backoff protocol, on which Ethernet is based, there are other acknowledgement-based protocols which are known to be stable in the same model for larger arrival rates. In particular, Håstad et al. have shown that *polynomial-backoff* protocols are stable as long as $\lambda < 1$. The expected waiting time of messages is high in polynomial-backoff protocols, but Raghavan and Upfal [10] have given a protocol that is stable for $\lambda < 1/10$, in which the expected waiting time of every message is $O(\log n)$, provided that the users are given a reasonably good estimate of $\log n$. Finally, Goldberg, MacKenzie, Paterson and Srinivasan [4] have given a protocol that is stable for $\lambda < 1/e$, in which the expected average message waiting-time is $O(1)$, provided that the users are given an upper bound on n.

We conclude by observing that the technique of Goldberg and MacKenzie [3] can be used to extend Theorem 3 so that it applies to a non-geometric version of binary-exponential backoff, which is closer to the version used in the Ethernet. (Instead of deciding whether to send on each step independently with probability 2^{-b_i}, the user simply chooses the number of steps to wait before sending uniformly at random from $[1, \ldots, 2^{b_i}]$.) The ideas are the same as those used in the proof that follows, but the details are messier. Our result can also be extended along the lines of [6] to show that, when λ is sufficiently low, the expected average message waiting time is finite.

[1] Note that it can be misleading to view the infinitely-many-users model as the limit (as n tends to infinity) of the n-users model. For example, the "polynomial backoff" protocol is known to be *unstable* (for any positive λ) in the infinitely-many-users Poisson-arrivals model [7,8], but it is *stable* (for any $\lambda < 1$) in the n-users model [6]. Thus, Aldous's result does not rule out the possibility that there is a positive constant λ^* such that the n-user binary exponential backoff protocol is stable whenever $\lambda < \lambda^*$.

3 The Stability Proof

In order to prove Theorem 3, let $\lambda = \frac{1}{\alpha' n^{.9}}$, where $\alpha' \geq \alpha$. We will now define the relevant potential function. Let $f(X(t))$ be the following function of the state just before step t.

$$f(X(t)) = \alpha n^{1.8} \sum_{i=1}^{n} q_i(t) + \sum_{i=1}^{n} 2^{b_i(t)}.$$

We will use the following generalisation of Foster's theorem [2]. Note that the Markov chain X satisfies the initial conditions of the theorem. That is, it is time-homogeneous, irreducible, and aperiodic and has a countable state space.

Theorem 4 (Foster; Fayolle, Malyshev, Menshikov). *A time-homogeneous irreducible aperiodic Markov chain X with a countable state space $\mathcal{A}$ is positive recurrent iff there exists a positive function $f(\rho)$, $\rho \in \mathcal{A}$, a number $\epsilon > 0$, a positive integer-valued function $k(\rho)$, $\rho \in \mathcal{A}$, and a finite set $C \subseteq \mathcal{A}$, such that the following inequalities hold.*

$$E[f(X(t + k(X(t)))) - f(X(t)) \mid X(t) = \rho] \leq -\epsilon k(\rho), \rho \notin C \quad (1)$$

$$E[f(X(t + k(X(t)))) \mid X(t) = \rho)] < \infty, \rho \in C. \quad (2)$$

We use the following notation, where $\beta = 3$. For a state $X(t)$, let $m(X(t))$ denote the number of users i with $q_i(t) > 0$ and $b_i(t) < \lg \beta + \lg n$, and let $m'(X(t))$ denote the number of users i with $q_i(t) > 0$ and $b_i(t) < .8 \lg n + 1$. Note that $m'(X(t)) \leq m(X(t))$. We will take ϵ to be $1 - 2/\alpha$ and C to be the set consisting of the single state $((0, \ldots, 0), (0, \ldots, 0))$. We define $k(((0, \ldots, 0), (0, \ldots, 0))) = 1$, so Equation 2 is satisfied. For every state $\rho \notin C$, we will define $k(\rho)$ in such a way that Equation 1 is also satisfied. We give the details in three cases.

3.1 Case 1: $m'(X(t)) = 0$ and $m(X(t)) < n^{.8}$.

For every state ρ such that $m'(\rho) = 0$ and $m(\rho) < n^{.8}$ we define $k(\rho) = 1$. We wish to show that, if $\rho \neq ((0, \ldots, 0), (0, \ldots, 0))$ and $X(t) = \rho$, then $E[f(X(t + 1) - f(X(t))] \leq -\epsilon$. Our general approach is the same as the approach used in the proof of Lemma 5.7 of [6]. For convenience, we use m as shorthand for $m(X(t))$ and we use ℓ to denote the number of users i with $q_i(t) > 0$. Without loss of generality, we assume that these are users $1, \ldots, \ell$. We use p_i to denote the probability that user i sends on step t. (So $p_i = 2^{-b_i(t)}$ if $i \in [1, \ldots, \ell]$ and $p_i = \lambda/n$ otherwise.) We let T denote $\prod_{i=1}^{n}(1 - p_i)$ and we let S denote $\sum_{i=1}^{n} \frac{p_i}{1-p_i}$. Note that the expected number of successes at step t is ST. Let $I_{a,i,t}$ be the 0/1 indicator random variable which is 1 iff there is an arrival at user i during step t and let $I_{s,i,t}$ be the 0/1 indicator random variable which is 1 iff user i succeeds in sending a message at step t. Then

$$
\begin{aligned}
&E[f(X(t+1)) - f(X(t))] \\
&= \alpha n^{1.8} \sum_{i=1}^{n} (E[I_{a,i,t} - I_{s,i,t}]) + \sum_{i=1}^{n} E[2^{b_i(t+1)}] - 2^{b_i(t)}, \\
&= \alpha n^{1.8}(\lambda - ST) + \sum_{i=1}^{n} \left(2^{b_i(t)}\sigma_i - (2^{b_i(t)} - 1)\pi_i \right), \qquad (3) \\
&= \alpha n^{1.8}(\lambda - ST) + \sum_{i=1}^{n} \left(2^{b_i(t)} p_i (1 - \frac{T}{1-p_i}) - (2^{b_i(t)} - 1) p_i \frac{T}{1-p_i} \right), \\
&= \alpha n^{1.8}(\lambda - ST) + \sum_{i=1}^{\ell} (1 - \frac{T}{1-p_i}) + \sum_{i=\ell+1}^{n} \frac{\lambda}{n}(1 - \frac{T}{1-p_i}) - \ell T, \\
&= \alpha n^{1.8}(\lambda - ST) + \ell - \ell T + \frac{(n-\ell)\lambda}{n} - T\left(\sum_{i=1}^{\ell} \frac{1}{1-p_i} + \sum_{i=\ell+1}^{n} \frac{p_i}{1-p_i} \right), \\
&= \alpha n^{1.8}(\lambda - ST) + \ell - \ell T + \frac{(n-\ell)\lambda}{n} - ST - \ell T, \\
&= \alpha n^{1.8}\lambda + \ell + \frac{(n-\ell)\lambda}{n} - T((\alpha n^{1.8} + 1)S + 2\ell), \qquad (4)
\end{aligned}
$$

where σ_i in Equality 3 denotes the probability that user i collides at step t and π_i denotes the probability that user i sends successfully at step t. (To see why Equality 3 holds, note that with probability σ_i, $b_i(t+1) = b_i(t) + 1$, with probability π_i, $b_i(t+1) = 0$, and otherwise, $b_i(t+1) = b_i(t)$.) We now find lower bounds for S and T. First,

$$
\begin{aligned}
S &= \sum_{i=1}^{n} \frac{p_i}{1-p_i} \\
&= \sum_{i=1}^{\ell} \left(\frac{2^{-b_i(t)}}{1 - 2^{-b_i(t)}} \right) + \frac{\lambda(n-\ell)}{n-\lambda} \\
&\geq \sum_{i=1}^{m} \left(\frac{1}{\beta n - 1} \right) + \frac{\lambda(n-\ell)}{n-\lambda} \\
&= \frac{m}{\beta n - 1} + \frac{\lambda(n-\ell)}{n-\lambda}. \qquad (5)
\end{aligned}
$$

Next,

$$
\begin{aligned}
T &= \prod_{i=1}^{n} (1 - p_i) \\
&\geq (1 - \frac{1}{2n^{.8}})^{m} (1 - \frac{1}{\beta n})^{\ell - m} (1 - \frac{\lambda}{n})^{n-\ell} \\
&\geq 1 - \frac{m}{2n^{.8}} - \frac{\ell - m}{\beta n} - \frac{\lambda(n-\ell)}{n} \qquad (6)
\end{aligned}
$$

Combining Equations 4, 5 and 6, we get the following equation.

$$E[f(X(t+1)-f(X(t))] \leq \alpha n^{1.8}\lambda + \ell + \frac{(n-\ell)\lambda}{n} - \quad (7)$$
$$\left(1 - \frac{m}{2n^{.8}} - \frac{\ell - m}{\beta n} - \frac{\lambda(n-\ell)}{n}\right)\left((\alpha n^{1.8}+1)\left(\frac{m}{\beta n - 1} + \frac{\lambda(n-\ell)}{n-\lambda}\right) + 2\ell\right).$$

We will let $g(m,\ell)$ be the quantity in Equation 7 plus ϵ and we will show that $g(m,\ell)$ is negative for all values of $0 \leq m < n^{.8}$ and all $\ell \geq m$. In particular, for every fixed positive value of m, we will show that

1. $g(m,m)$ is negative,
2. $g(m,n)$ is negative, and
3. $\frac{\partial^2}{\partial \ell^2} g(m,\ell) > 0$. ($g(m,\ell)$ is concave up as a function of ℓ for the fixed value of m so $g(m,\ell)$ is negative for all $\ell \in [m,n]$.)

We will handle the case $m=0$ similarly except that $m=\ell=0$ corresponds to the start state, so we will replace Item 1 with the following for $m=0$.

1'. $g(0,1)$ is negative.

The details of the proof are now merely calculations.

1. $g(m,m)$ **is negative:** $g(m,m) \times 2\alpha' n^{1.9}(\beta n - 1)(\alpha' n^{1.9} - 1)$ is equal to the following.

$$\begin{aligned}&-2m-4m^2+2n+6mn+2\beta mn+6\beta m^2n+2\alpha' m^2n^{1.1}+2\alpha'\epsilon n^{1.9}+2\alpha' m^2n^{1.9}\\&-2n^2-2\beta n^2-8\beta mn^2-\alpha' mn^{2.1}-3\alpha'\beta m^2n^{2.1}+2\alpha n^{2.8}+2\alpha mn^{2.8}+2\alpha\beta m^2n^{2.8}\\&-2\alpha'\beta\epsilon\, n^{2.9}-2\alpha' mn^{2.9}+2\alpha'\beta mn^{2.9}-4\alpha'\beta m^2n^{2.9}+2\beta n^3-\alpha'^2m^2n^3+\alpha'\beta mn^{3.1}\\&-2\alpha\alpha' m^2n^{3.7}-2\alpha n^{3.8}-2\alpha\beta n^{3.8}-2\alpha'^2\epsilon n^{3.8}-4\,\alpha\beta mn^{3.8}-\alpha\alpha' mn^{3.9}+4\alpha'\beta mn^{3.9}\\&-\alpha\alpha'\beta m^2n^{3.9}+2\alpha'^2\beta m^2n^4+2\alpha\alpha' mn^{4.7}+2\alpha\alpha'\beta mn^{4.7}+2\alpha\beta n^{4.8}+2\alpha'^2\beta\epsilon n^{4.8}\\&-2\alpha'^2\beta mn^{4.8}+\alpha\alpha'^2m^2n^{4.8}+\alpha\alpha'\beta mn^{4.9}-2\alpha\alpha'^2mn^{5.6}\end{aligned}$$

The dominant term is $-2\alpha\alpha'^2 mn^{5.6}$. Note that there is a positive term ($\alpha\alpha'^2m^2n^{4.8}$) which could be half this big if m is as big as $n^{.8}$ (the upper bound for Case 1), but all other terms are asymptotically smaller.

2. $g(m,n)$ **is negative:** $g(m,n) \times 2\alpha'\beta n(\beta n - 1)$ is equal to the following.

$$\begin{aligned}&-2\alpha' m^2+\alpha'\beta m^2n^{.2}-2\alpha'\beta\epsilon n+6\alpha' mn-2\alpha'\beta mn-2\alpha'\beta mn^{1.2}\\&-2\alpha\alpha' m^2n^{1.8}-2\alpha\beta n^{1.9}-4\alpha' n^2+2\alpha'\beta n^2+2\alpha'\beta^2\epsilon n^2\\&-4\alpha'\beta mn^2+\alpha\alpha'\beta m^2n^2+2\alpha'\beta^2mn^{2.2}+2\alpha\alpha' mn^{2.8}-2\alpha\alpha'\beta mn^{2.8}+2\alpha\beta^2n^{2.9}\\&+4\alpha'\beta n^3-2\alpha'\beta^2n^3\end{aligned}$$

Since $\beta > 2$, the term $-2\alpha'\beta^2n^3$ dominates $+4\alpha'\beta n^3$. For the same reason, the term $-2\alpha\alpha'\beta mn^{2.8}$ dominates the two terms $+2\alpha\alpha' mn^{2.8}$ and $+\alpha\alpha'\beta m^2n^2$. The other terms are asymptotically smaller.

3. $\frac{\partial^2}{\partial \ell^2} g(m, \ell) > 0$:

$$\frac{\partial^2}{\partial \ell^2} g(m, \ell) = 2\left(\frac{1}{\beta n} - \frac{\lambda}{n}\right)\left(2 - \frac{(\alpha n^{1.8} + 1)\lambda}{n - \lambda}\right).$$

1'. **$g(0,1)$ is negative:** $g(0,1) \times \alpha' \beta n^{1.9}(\alpha' n^{1.9} - 1)$ is equal to the following.

$$\begin{aligned}&+ 4\beta - 3\alpha' n^{.9} - 5\beta n + \alpha\beta n^{1.8} + \alpha' n^{1.9} - \alpha'\beta m^{1/9} - \alpha'\beta\epsilon n^{1.9}\\ &+ \beta n^2 - \alpha\alpha' n^{2.7} + 2\alpha'^2 n^{2.8} - 3\alpha\beta n^{2.8} + 2\alpha'\beta n^{2.9} + \alpha\alpha' n^{3.7}\\ &+ \alpha\alpha'\beta n^{3.7} + \alpha\beta n^{3.8} - \alpha'^2\beta n^{3.8} + \alpha'^2\beta\epsilon n^{3.8}\end{aligned}$$

Since $\alpha'(1-\epsilon) \geq \alpha(1-\epsilon) > 1$, the term $-\alpha'^2\beta(1-\epsilon)n^{3.8}$ dominates the term $+\alpha\beta n^{3.8}$. The other terms are asymptotically smaller.

3.2 Case 2: $m(X(t)) \geq n^{.8}$ or $m'(X(t)) > n^{.4}$.

For every state ρ such that $m(\rho) \geq n^{.8}$ or $m'(\rho) > n^{.4}$, we will define an integer k (which depends upon ρ) and we will show that, if $X(t) = \rho$, then $E[f(X(t+k) - f(X(t))] \leq -\epsilon k$, where $\epsilon = 1 - 2/\alpha$.

For convenience, we will use m as shorthand for $m(X(t))$ and m' as shorthand for $m'(X(t))$. If $m \geq n^{.8}$ then we will define $r = m$, $W = m^{1/4}\lceil \lg r \rceil 2^{-8}$, $A = W$, $b = \lg \beta + \lg n$ and $v = n$. Otherwise, we will define $r = m'$, $W = \lceil \lg r \rceil 2^{-8}$, $A = 0$, $b = .8 \lg n + 1$, and $v = 2\lceil n^{.8} \rceil$. In either case, we will define $k = 4(r+v)\lceil \lg r \rceil$. Let τ be the set of steps $\{t, \ldots, t+k-1\}$ and let $\mathcal{S}$ be the random variable which denotes the number of successes that the system has during τ. Let p denote $\Pr(\mathcal{S} \geq W)$. Then we have

$$\begin{aligned}E[f(X(t+k) - f(X(t))] &\leq \alpha n^{1.8}\lambda k - \alpha n^{1.8} E[\mathcal{S}] + \sum_{i=1}^{n} \sum_{t'=t+1}^{t+k} E[2^{b_i(t')} - 2^{b_i(t'-1)}]\\ &\leq \alpha n^{1.8}\lambda k - \alpha n^{1.8} W p + kn\\ &\leq -\epsilon k,\end{aligned}$$

where the final inequality holds as long as $\alpha p \geq 2^{13}$ and n is sufficiently big (see the Appendix). Thus, it suffices to find a positive lower bound for p which is independent of n. We do this with plenty to spare. In particular, we show that $p \geq 1 - 5 \times 10^{-5}$.

We start with a technical lemma, which describes the behaviour of a single user.

Lemma 1. *Let j be a positive integer, and let δ be a positive integer which is at least 2. Suppose that $q_i(t) > 0$. Then, with probability at least $1 - \frac{\lceil \lg j \rceil}{j^{\delta/(2\ln 2)}}$, either user i succeeds in steps $[t, \ldots, t + \delta j \lceil \lg j \rceil - 1]$, or $b_i(t + \delta j \lceil \lg j \rceil) \geq \lceil \lg j \rceil$.*

Proof. Suppose that user i is running in an externally-jammed channel (so every send results in a collision). Let X_z denote the number of steps $t' \in [t, \ldots, t + \lceil \delta j \lg(j) \rceil]$ with $b_i(t') = z$. We claim that $\Pr(X_z > \delta \lceil \lg j \rceil 2^{z-1}) < j^{-\delta/(2 \ln 2)}$. This proves the lemma since $\sum_{z=0}^{\lceil \lg j \rceil - 1} \delta \lceil \lg j \rceil 2^{z-1} \leq \delta j \lceil \lg j \rceil$. To prove the claim, note that $X_0 \leq 1$, so $\Pr(X_0 > \delta \lceil \lg j \rceil 2^{-1}) = 0 < j^{-\delta/(2 \ln 2)}$. For $z > 0$, note that

$$\Pr(X_z > \delta \lceil \lg j \rceil 2^{z-1}) \leq (1 - 2^{-z})^{\delta \lceil \lg j \rceil 2^{z-1}} < j^{-\delta/(2 \ln 2)}.$$

Next, we define some events. We will show that the events are likely to occur, and, if they do occur, then $\mathcal{S}$ is likely to be at least W. This will allow us to conclude that $p \geq 1 - 5 \times 10^{-5}$, which will finish Case 2. We start by defining $B = \lceil W \rceil + \lceil A \rceil$, $k' = 4r \lceil \lg r \rceil$, $k'' = 4B \lceil \lg B \rceil$ and $\tau_0 = \{t, \ldots, t + k' - 1\}$. Let $\tau'(i)$ be the set of all $t' \in \tau$ such that $b_i(t') = 0$ and either (1) $q_i(t') > 0$ or (2) there is an arrival at user i at t'. Let τ_2 be the set of all $t' \in \tau$ such that $|\{(t'', i) \mid t'' \in \tau'(i) \text{ and } t'' < t'\}| \geq B$. Finally, let τ_1 be the set of all $t' \in \tau - \tau_0 - \tau_2$ such that, for some i, $\tau'(i) \cap [t' - k'' + 1, t'] \neq \emptyset$. We can now define the events E1–E4.

E1. There are at most A arrivals during τ.
E2. Every station with $q_i(t) > 0$ and $b_i(t) < b$ either sends successfully during τ_0 or has $b_i(t + k') \geq \lceil \lg r \rceil$.
E3. Every station with $q_i(t) > 0$ and $b_i(t) < b$ has $b_i(t') \leq b + \lg(r)/2 + 3$ for all $t' \in \tau$.
E4. For all $t' \in \tau'(i)$ and all $t'' > t'$ such that $t'' \in \tau - \tau_1 - \tau_2$, $b_i(t') \geq \lceil \lg B \rceil$.

Next, we show that E1–E4 are likely to occur.

Lemma 2. *If n is sufficiently large then* $\Pr(\overline{E1}) \leq 10^{-5}$.

Proof. The expected number of arrivals in τ is λk. If $m \geq n^{.8}$, then $A = m^{1/4} \lceil \lg r \rceil 2^{-8} \geq 2\lambda k$. By a Chernoff bound, the probability that there are this many arrivals is at most $e^{-\lambda k/3} \leq 10^{-5}$. Otherwise, $A = 0$ and $\lambda k = o(1)$. Thus, $\Pr(\text{E1}) \geq (1 - \lambda/n)^{nk} \geq 1 - \lambda k \geq 1 - 10^{-5}$.

Lemma 3. *If n is sufficiently large then* $\Pr(\overline{E2}) \leq 10^{-5}$.

Proof. Apply Lemma 1 to each of the r users with $\delta = 4$ and $j = r$. Then $\Pr(\overline{\text{E2}}) \leq r \frac{\lceil \lg r \rceil}{r^{2/(\ln 2)}} \leq 10^{-5}$.

Lemma 4. *If n is sufficiently large then* $\Pr(\overline{E3}) \leq 10^{-5}$.

Proof. Let $y = \left\lceil \frac{\lceil \lg r \rceil}{4} \right\rceil$. Note that $2y \leq \frac{\lg r}{2} + 3$. Suppose that user i has $b_i(t') > b + 2y$. Then this user sent when its backoff counter was $\lceil b + z \rceil$ for all $z \in \{y, \ldots, 2y - 1\}$. The probability of such a send on any particular step is at most $\frac{1}{2^b 2^y}$. Thus, the probability that it makes all y of the sends is at most

$$\binom{k}{y} \left(\frac{1}{2^b 2^y} \right)^y \leq \left(\frac{ke}{2^b y 2^y} \right)^y \leq 10^{-5}/r.$$

Thus, the probability that any of the r users obtains such a big backoff counter is at most 10^{-5}.

Lemma 5. *If n is sufficiently large then* $\Pr(\overline{E4}) \le 10^{-5}$.

Proof. We can apply Lemma 1 separately to each of the (up to B) pairs (t', i) with $\delta = 4$ and $j = B$. The probability that there is a failure is at most $\frac{B\lceil \lg B \rceil}{B^{2/(\ln 2)}} \le 10^{-5}$.

We now wish to show that $\Pr(\mathcal{S} < W \mid \mathrm{E1} \wedge \mathrm{E2} \wedge \mathrm{E3} \wedge \mathrm{E4}) \le 10^{-5}$. We begin with the following lemma.

Lemma 6. *Given any fixed sequence of states $X(t), \ldots, X(t+z)$ which does not violate E2 or E4, and satisfies $t + z \in \tau - \tau_0 - \tau_1 - \tau_2$, $q_i(t+z) > 0$, and $b_i(t+z) \le b + \lg(r)/2 + 3$, the probability that user i succeeds at step $t+z$ is at least $\frac{1}{2^{10}2^b r^{1/2}}$.*

Proof. The conditions in the lemma imply the following.

- There are no users j with $b_j(t+z) < \lceil \lg B \rceil$.
- There are at most B users j with $b_j(t+z) < \lceil \lg r \rceil$.
- There are at most $r + B$ users j with $b_j(t+z) < b$.
- There are at most $m + B$ users j with $b_j(t+z) < \lg \beta + \lg n$.

Thus, the probability that user i succeeds is at least

$$\begin{aligned} & 2^{-(b+\lg(r)/2+3)} \left(1 - \frac{1}{B}\right)^B \left(1 - \frac{1}{r}\right)^r \left(1 - \frac{1}{2^b}\right)^{m-r} \left(1 - \frac{1}{\beta n}\right)^{n-m-B} \\ & \ge \frac{1}{2^b r^{1/2} 2^3} \left(\frac{1}{4}\right) \left(\frac{1}{4}\right) \left(\frac{1}{4}\right) \left(1 - \frac{n-m-B}{\beta n}\right) \\ & \ge \frac{1}{2^{10} 2^b r^{1/2}}. \end{aligned}$$

Corollary 1. *Given any fixed sequence of states $X(t), \ldots, X(t+z)$ which does not violate E2, E3, or E4, and satisfies $t + z \in \tau - \tau_0 - \tau_1 - \tau_2$, the probability that some user succeeds at step $t+z$ is at least $\frac{r-B}{2^{10}2^b r^{1/2}} \ge \frac{1}{2^{13} n^{.6}}$.*

Proof. Since $t + z \notin \tau_2$, at least $r - B$ of the users i with $q_i(t) > 0$ and $b_i(t) < b$ have not succeeded before step $t+z$. Since E3 holds, each of these has $b_i(t+z) \le b + \lg(r)/2 + 3$. For all i and i', the event that user i succeeds at step $t+z$ is disjoint with the event that user i' succeeds at step $t+z$.

Lemma 7. *If n is sufficiently large then* $\Pr(\mathcal{S} < W \mid E1 \wedge E2 \wedge E3 \wedge E4) \le 10^{-5}$.

Proof. If E1 is satisfied then τ_2 does not start until there have been at least W successes. Since $|\tau - \tau_0 - \tau_1| \ge k - k' - Bk'' \ge v\lceil \lg r \rceil/2$, Corollary 1 shows that the probability of having fewer than W successes is at most the probability of having fewer than W successes in $v\lceil \lg r \rceil/2$ Bernoulli trials with success probability $\frac{1}{2^{13} n^{.6}}$. Since W is at most half of the expected number of successes, a Chernoff bound shows that the probability of having fewer than W successes is at most $\exp(-\frac{v\lceil \lg r \rceil}{2^{17} n^{.6}}) \le 10^{-5}$.

We conclude Case 2 by observing that p is at least $1 - \Pr(\overline{\mathrm{E1}}) - \Pr(\overline{\mathrm{E2}}) - \Pr(\overline{\mathrm{E3}}) - \Pr(\overline{\mathrm{E4}}) - \Pr(\mathcal{S} < W \mid \mathrm{E1} \wedge \mathrm{E2} \wedge \mathrm{E3} \wedge \mathrm{E4})$. By Lemmas 2, 3, 4, 5, and 7, this is at least $1 - 5 \times 10^{-5}$.

3.3 Case 3: $0 < m'(X(t)) \leq n^{.4}$ and $m(X(t)) < n^{.8}$.

For every state ρ such that $0 < m'(\rho) \leq n^{.4}$ and $m(\rho) < n^{.8}$, we will define $k = 32m'(\rho)\lceil \lg m'(\rho) \rceil + \lceil n^{.8} \rceil$. We will show that, if $X(t) = \rho$, then $E[f(X(t+k) - f(X(t))] \leq -\epsilon k$. Once again, we will use m as shorthand for $m(X(T))$ and m' as shorthand for $m'(X(t))$. Let $\tau = \{t, \ldots, t+k-1\}$, let $\mathcal{S}$ be the number of successes that the system has in τ. Let p denote $\Pr(\mathcal{S} \geq 1)$. As in Case 2, $E[f(X(t+k)-f(X(t))] \leq \alpha n^{1.8}\lambda k - \alpha n^{1.8} p + kn$, and this is at most $-\epsilon k$ as long as $\alpha p > 9$. Thus, we will finish by finding a positive lower bound for p which is independent of n.

Since $m' > 0$, there is a user γ such that $b_\gamma(t) < .8 \lg n + 1$. Let $k' = 32m' \lceil \lg m' \rceil$ and $\tau_0 = \{t, \ldots, t+k'-1\}$. We will now define some events, as in Case 2.

E1. There are no arrivals during τ.
E2. Every station with $q_i(t) > 0$ and $b_i(t) < .8 \lg n + 1$ either sends successfully during τ_0 or has $b_i(t+k') \geq \lceil \lg m' \rceil$.
E3. $b_\gamma(t') < .8 \lg n + 7$ for all $t' \in \tau$.

Lemma 8. *If n is sufficiently large then $Pr(\overline{E1}) \leq 10^{-5}$.*

Proof. As in the proof of Lemma 2,

$$\Pr(\text{E1}) \geq \left(1 - \frac{\lambda}{n}\right)^{nk} \geq 1 - \lambda k \geq 1 - 10^{-5}.$$

Lemma 9. $Pr(\overline{E2}) \leq 10^{-5}$.

Proof. We use lemma 5 with $\delta = 32$ and $j = m'$ to get

$$Pr(\overline{E3}) \leq m' \cdot \frac{\lceil \lg m' \rceil}{(m')^{16/\ln(2)}} \leq 10^{-5}. \tag{8}$$

Lemma 10. *If n is sufficiently large then $Pr(\overline{E3}) \leq 10^{-5}$.*

Proof. Let $y = 6$, and suppose that user γ sends with backoff $b_\gamma = \lceil .8 \lg n + r \rceil$ for $r \in \{1, \ldots, 6\}$. The probability of this happening is

$$\begin{aligned} Pr(\overline{E3}) &\leq \binom{k}{6} \prod_{r=1}^{6} 2^{-\lceil .8 \lg n \rceil - r} \\ &\leq \left(\frac{ke}{6}\right)^6 \left(\frac{1}{n^{.8}}\right) 2^{-\sum_{r=1}^{6} r} \\ &\leq \left(\frac{2en^{.8}}{6n^{.8} 2^3}\right)^6 \\ &\leq 10^{-5}. \end{aligned}$$

Lemma 11. *Given any fixed sequence of states $X(t), \ldots, X(t+z)$ which does not violate E1, E2, or E3 such that $t+z \in \tau - \tau_0$ and there are no successes during steps $[t, \ldots, t+z-1]$, the probability that user γ succeeds at step $t+z$ is at least $\frac{1}{2^{12} n^{.8}}$.*

Proof. The conditions in the statement of the lemma imply the following.

- $q_\gamma(t+z) > 0$ and $b_\gamma(t+z) < .8 \lg n + 7$.
- There are no users j with $b_j(t+z) < \lceil \lg m' \rceil$.
- There are at most m' users j with $b_j(t+z) < .8 \lg n + 1$.
- There are at most m users j with $b_j(t+z) < \lg \beta + \lg n$.
- There will be no arrivals on step $t+z$.

The probability of success for user γ is at least

$$\begin{aligned} & 2^{-(.8 \lg n + 7)} \left(1 - \frac{1}{m'}\right)^{m'} \left(1 - \frac{1}{2n^{.8}}\right)^{m-m'} \left(1 - \frac{1}{\beta n}\right)^{n-m} \\ & \geq \frac{1}{2^7 n^{.8}} \left(\frac{1}{4}\right) \left(\frac{1}{4}\right) \left(\frac{1}{2}\right) \\ & \geq \frac{1}{2^{12} n^{.8}}. \end{aligned}$$

Lemma 12. *If n is sufficiently large then $Pr(\mathcal{S} < 1 \mid E1 \wedge E2 \wedge E3) \leq e^{-1/2^{12}}$.*

Proof. Lemma 11 implies that the probability of having no successes is at most the probability of having no successes in $|\tau - \tau_0|$ Bernoulli trials, each with success probability $\frac{1}{2^{12} n^{.8}}$. Since $|\tau - \tau_0| \geq n^{.8}$, this probability is at most

$$\left(1 - \frac{1}{2^{12} n^{.8}}\right)^{n^{.8}} \leq e^{-1/2^{12}}.$$

We conclude Case 3 by observing that p is at least $1 - \Pr(\overline{\mathrm{E1}}) - \Pr(\overline{\mathrm{E2}}) - \Pr(\overline{\mathrm{E3}}) - \Pr(\mathcal{S} < 1 \mid \mathrm{E1} \wedge \mathrm{E2} \wedge \mathrm{E3})$. By Lemmas 8, 9, 10, and 12, this is at least $1 - 3 \times 10^{-5} - e^{-1/2^{12}} \geq .0002$.

References

1. D. Aldous, Ultimate instability of exponential back-off protocol for acknowledgement-based transmission control of random access communication channels, *IEEE Trans. Inf. Theory* **IT-33(2)** (1987) 219–233.
2. G. Fayolle, V.A. Malyshev and M.V. Menshikov, *Topics in the Constructive Theory of Countable Markov Chains,* (Cambridge Univ. Press, 1995)
3. L.A. Goldberg and P.D. MacKenzie, Analysis of practical backoff protocols for contention resolution with multiple servers, *Journal of Computer and Systems Sciences,* **58** (1999) 232–258.

4. L.A. Goldberg, P.D. MacKenzie, M. Paterson and A. Srinivasan, Contention resolution with constant expected delay, Pre-print (1999) available at `http://www.dcs.warwick.ac.uk/~leslie/pub.html`. (Extends a paper by the first two authors in *Proc. of the Symposium on Foundations of Computer Science (IEEE)* 1997 and a paper by the second two authors in *Proc. of the Symposium on Foundations of Computer Science (IEEE)* 1995.)
5. J. Goodman, A.G. Greenberg, N. Madras and P. March, Stability of binary exponential backoff, *J. of the ACM,* **35(3)** (1988) 579–602.
6. J. Håstad, T. Leighton and B. Rogoff, Analysis of backoff protocols for multiple access channels, *SIAM Journal on Computing* **25(4)** (1996) 740-774.
7. F.P. Kelly, Stochastic models of computer communication systems, *J.R. Statist. Soc. B* **47(3)** (1985) 379–395.
8. F.P. Kelly and I.M. MacPhee, The number of packets transmitted by collision detect random access schemes, *The Annals of Probability,* **15(4)** (1987) 1557–1568.
9. R.M. Metcalfe and D.R. Boggs, Ethernet: Distributed packet switching for local computer networks. *Commun. ACM,* **19** (1976) 395–404.
10. P. Raghavan and E. Upfal, Contention resolution with bounded delay, *Proc. of the ACM Symposium on the Theory of Computing* **24** (1995) 229–237.

Appendix : Supplementary Calculations for Case 2

Here we show the inequality $\alpha n^{1.8}\lambda k - \alpha n^{1.8}Wp + kn \leq -\epsilon k$ holds when $\alpha p \geq 2^{13}$ and n is sufficiently large.

Case A: ($m \geq n^{.8}$) Since $k \leq 8n\lceil \log m \rceil$ for large n,

$$\begin{aligned}
&\alpha n^{1.8}\lambda k - \alpha n^{1.8}Wp + kn \\
&\leq \alpha n^{1.8}(\alpha' n^{.9})^{-1}k - 2^{13}n^{1.8}W + kn \\
&\leq (\alpha/\alpha')n^{.9}k - 2^{13}n^{1.8}m^{1/4}\lceil \log m \rceil 2^{-8} + kn \\
&\leq (\alpha/\alpha')n^{.9}(4(m+n)\lceil \log m \rceil) - 2^5 n^2 \lceil \log m \rceil + 4(m+n)\lceil \log m \rceil n \\
&\leq 8n^{1.9}\lceil \log m \rceil - 32n^2\lceil \log m \rceil + 8n^2\lceil \log m \rceil \\
&\leq -16n^2\lceil \log m \rceil \leq -2nk \leq -\epsilon k.
\end{aligned}$$

Case B: ($m < n^{.8}$, $m' > n^{.4}$) Since $k \leq 12\lceil n^{.8}\rceil \lceil \log m' \rceil$ for large n,

$$\begin{aligned}
&\alpha n^{1.8}\lambda k - \alpha n^{1.8}Wp + kn \\
&\leq \alpha n^{1.8}(\alpha' n^{.9})^{-1}k - 2^{13}n^{1.8}W + kn \\
&\leq (\alpha/\alpha')n^{.9}k - 2^{13}n^{1.8}\lceil \log m' \rceil 2^{-8} + kn \\
&\leq (\alpha/\alpha')n^{.9}(4(m' + 2\lceil n^{.8}\rceil)\lceil \log m' \rceil) - 2^5 n^{1.8}\lceil \log m' \rceil \\
&\quad +4(m + 2\lceil n^{.8}\rceil)\lceil \log m' \rceil n \\
&\leq 12n^{.9}\lceil n^{.8}\rceil \lceil \log m' \rceil) - 2^5 n^{1.8}\lceil \log m' \rceil + 12n\lceil n^{.8}\rceil \lceil \log m' \rceil \\
&\leq 12n^{.9}\lceil n^{.8}\rceil \lceil \log m' \rceil) - 32n^{1.8}\lceil \log m' \rceil + 13n^{1.8}\lceil \log m' \rceil \\
&\leq -18n^{1.8}\lceil \log m' \rceil \leq -nk \leq -\epsilon k.
\end{aligned}$$

The Data Broadcast Problem with Preemption

Nicolas Schabanel

LIP, ENS Lyon, 46, allée d'Italie, F-69364 Lyon Cedex 07, France

Abstract. The data-broadcast problem consists in finding an infinite schedule to broadcast a given set of messages so as to minimize the average response time to clients requesting messages, and the cost of the broadcast. This is an efficient means of disseminating data to clients, designed for environments, such as satellites, cable TV, mobile phones, where there is a much larger capacity from the information source to the clients than in the reverse direction.
Previous work concentrated on scheduling indivisible messages. Here, we studied a generalization of the model where the messages can be preempted. We show that this problem is *NP*-hard, even in the simple setting where the broadcast costs are zero, and give some practical 2-approximation algorithms for broadcasting messages. We also show that preemption can improve the quality of the broadcast by an arbitrary factor.

1 Introduction

1.1 Motivation

Data-broadcast is an efficient means of disseminating data to clients in wireless communication environment, where there is a much larger capacity from the information source to the recipients than in the reverse direction, such as happens when mobile clients (*e.g.* car navigation systems) retrieve information (*e.g.* traffic information) from base-station (*e.g.* the emitter) through a wireless medium. In a broadcasting protocol, items are broadcast according to an infinite horizon schedule and clients do not explicity send a request for an item to the server, but connect to the broadcast channels (shared by all the clients) and wait until the requested item is broadcast. These system are therefore known as *pseudo-interactive* or *push-based*: the server "pushes" the items, or messages, to the clients (even if disconnected) according to a schedule which is oblivious to the effective requests; as opposed to the "traditional" *pull-based* model, where the clients send a request to "pull" the required item from the server when they need it. The quality of the broadcast schedule is measured by the expected service time of the addressed requests. Furthermore, as each message has a cost for broadcasting (*e.g.* a weather broadcast and a news broadcast may have different costs for the emitter), the server also tries to minimize the resulting cost of service. The server has then to minimize the expected service response time of the requests (quality of service) and the broadcast cost of the resulting schedule

H. Reichel and S. Tison (Eds.): STACS 2000, LNCS 1770, pp. 181–192, 2000.

(cost of service). The server designs the broadcast schedule from the *profile* of the users: given the messages $M_1, \ldots, M_m$, the profile consists of the *popularities* of the different messages, that is to say the probabilities $(p_i)_{1 \leqslant i \leqslant m}$, that Message M_i is requested by a random user. [17] proposes some techniques to gauge user profiles in push-based environment.

With the impressive growth of the wireless, satellite and cable network, the data dissemination protocols have a number applications in research and commercial frameworks. One of the earliest applications was the Boston Community Information System (BCIS, 1982) developed at the MIT to deliver news and information to clients equipped personally with radio receivers in metropolitan Boston. It was also introduced in early 1980's in the context of Teletext and Videotex [8, 3]. It is now used by applications that require dissemination among a huge number of clients. The Advanced Traffic Information System (ATIS) [14], which provides traffic and route planning information to cars specially equipped with computers, may have to serve over 100,000 clients in a large metropolitan city during the rush hours. The news delivery systems on the Internet, such as Pointcast inc. (1997), or Airmedia inc. (1997), require efficient information dissemination system. A comparison of the push-based system to the traditional pull-based approach for those problems can be found in [1].

Note that the data-broadcast problem also models the maintenance scheduling problem and the multi-item replenishment problem [5, 6, 10].

While previous work made the assumption that messages transmission cannot be preempted, we focus in this paper on the case where the messages do not have uniform transmission times and can be split.

1.2 Background

Since the early 1980's, many authors [8, 3, 4, 5, 6, 11] have studied the data-broadcast problem in the restrictive setting where all messages have the same length, the broadcast is done on a single channel, and time is discrete (this restricted problem is also known as Broadcast disks problem or Dissemination-based systems). In particular, Ammar and Wong [3, 4] give an algebraic expression of the expected service time of periodic schedules, provide a lower bound, and prove the existence of an optimal schedule which is periodic. Our Lemmas 2, 3 and Proposition 1 are generalizations of these results to our setting. Bar-Noy, Bhatia, Naor and Schieber [6] prove that the problem with broadcast costs is *NP*-hard, and after a sequence of papers giving constant factor approximations [5, 6], Kenyon, Schabanel and Young [11] design a PTAS for the problem. The papers [2, 1, 9, 15, 12, 13] study related questions pertaining to prefetching, to caching and to indexing.

As can be seen from the example of broadcasting weather and news reports, in many applications, it does not make sense to assume that all messages have the same transmission time; thus a couple of recent papers have explored the case of non-uniform transmission times. In [16] Vaidya and Hameed report some experimental results for heuristics on one or two channels. In [10] Kenyon and

Schabanel show that the case where the messages do not have the same transmission time, the data-broadcast problem is *NP*-hard, even if message have zero broadcast cost, and does not always admit an periodic optimal schedule. They show that the natural extension of the lower bound given in [3, 6] is arbitrarily far from the optimal when the messages have very different length. The main difficulty is due to the fact that, while a long message is being broadcast, all requests for shorter and more popular messages have to be put on hold. But in that case, it seems reasonable to allow a occasional interruption of a long "boring" message transmission so as to broadcast a short popular message. In other word, one should allow preemption. This is the main motivation to the preemptive model introduced and studied in this paper.

1.3 Our Contribution

This paper introduces and studies the model where the messages to be broadcast have non uniform transmission time and where their transmission can be preempted. One of the most interesting contribution from the practical point of view is that our algorithms (Section 4) generate preemptive schedules whose costs can be arbitrarily smaller than the optimal cost of any non-preemptive schedule on some inputs (See Note 1 in Section 4). Thus there is an *infinite* gap between the preemptive and non-preemptive problem.

We adopt the following model. The input consists of m messages $M_1, \ldots, M_m$ and an user profile determined by the *probabilities* $(p_i)_{1 \leqslant i \leqslant m}$ that a user requests Message M_i ($p_1 + \cdots + p_m = 1$). Each message M_i, $i = 1..m$, is composed of ℓ_i *packets* with transmission time 1 and each broadcast of a packet costs $c_i \geqslant 0$. The packets of the messages are broadcast over W identical and synchronized broadcast channels split into *time slots* of length 1 (time slot t is the period of time $[t-1, t]$). Given a schedule S of the packets into the slots, over the W channels, a client requesting Message M_i, starts monitoring all the channels at some (continuous) point, downloads the different packets of M_i *one at a time* when they are broadcast on some channel, and is served as soon as it has downloaded all the ℓ_i packets of Message M_i. The order in which the client has received the packet of M_i is irrelevant, as in TCP/IP.

The problem is to design a sequence S to schedule the packets over time, so as to minimize the sum of the expected service time of Poisson requests and of the average broadcast cost, i.e. so as to minimize $\limsup_{T \to \infty} (\mathrm{EST}(S, [0, T]) + \mathrm{BC}(S, [0, T]))$; here, $\mathrm{EST}(S, [0, T])$ denotes the expected service time of a request which is generated at a random uniform (continuous) instant between 0 and T, requests Message M_i with probability p_i, and must wait until the ℓ_i packets of M_i have been broadcast and downloaded; and $\mathrm{BC}(S, [0, T])$ is the average broadcast cost of the packets whose broadcast starts between 0 and T. Note that this definition agrees with the one in the literature (*e.g.* [6]), in the uniform-length case where the messages are composed of a single packet.

The results presented in this paper are obtained thanks to the simple but crucial observation made in Lemma 1: for all i, *an optimal schedule broadcasts* *the packets of Message M_i in Round Robin order.* We can thus restrict our search to Round Robin schedules. From this observation, we get an tractable algebraic expression for the cost of such a schedule in Lemma 2, from which we derive the lower bound in Lemma 3. This lower bound is the key to the two main results of the papers: 1) the problem is strongly *NP*-hard, even if no broadcast cost are assumed, in Theorem 1 (note that the *NP*-hardness proof given in [6] for the uniform length case requires non-zero broadcast cost); 2) there exists polynomial algorithm which constructs a periodic schedule with cost at most twice the optimal, in Section 4.

The lower bound also reveals some important structural differences between our model and the previous models. First, surprisingly, as opposed to *all* the previous studies, the lower bound *cannot* be realized by *scheduling the packets regularly* but by gluing them together (see Lemma 3): from the individual point of view of a request for a given message, the message should not be preempted. This allows to derive some results from the non-preemptive case studied in [10]. But, whereas non-preemptive strategies cannot approach this lower bound, we obtain, all the same, efficient approximation scheme within a factor of 2 by broadcasting the packets of each message regularly. Second, although the lower bound specializes to the one designed in [6] when all messages are composed of a single packet, deriving the lower bound is no longer a straight forward relaxation on the constraints on the schedule and requires a finer study of the "ideal" schedules. Moreover, its objective function is no longer convex and its resolution (in particular the unicity of its solution) needs a careful adaptation, presented Section 4.5, of the methods introduced in [6, 10].

Note that our preemptive setting models also the case where users do request single messages but batches of messages. We can indeed consider the packets of a message as messages of a batch. The preemptive case studied here is the case where the batches are all disjoint. In that sense the paper is an extension of some results in [7].

1.4 The Cost Function

We are interested in minimizing the *cost* of the schedule S, which is a combination of two quantities on S. The first one, denoted by $\mathrm{EST}(S)$, is the *expected service time* of a random request (where the average is taken over the moments when requests occur, and the type M_i of message requested). If we define by $\mathrm{EST}(S, I)$, the expected service time of a random request arrived in time interval I, $\mathrm{EST}(S)$ is: $\mathrm{EST}(S) = \limsup_{T\to\infty} \mathrm{EST}(S, [T_0, T])$, for any T_0. If we denote by $\mathrm{ST}(S, M_i, t)$, the service time of a request for M_i arrived at time t, and by $\mathrm{EST}(S, M_i, I)$ the expected service time of a request for M_i arriving in time interval I, we get: $\mathrm{EST}(S, M_i, I) = \frac{1}{|I|} \int_I \mathrm{ST}(S, M_i, t)\, dt$, and $\mathrm{EST}(S, I) = \sum_{i=1}^{m} p_i \, \mathrm{EST}(S, M_i, I)$.

The second quantity is the *broadcast cost* $\mathrm{BC}(S)$ of the messages, defined as the asymptotic value of the broadcast cost $\mathrm{BC}(S, I)$ over a time interval I: $\mathrm{BC}(S) = \limsup_{T\to\infty} \mathrm{BC}(S, [T_0, T])$, for any T_0. By definition, each broadcast of a packet of M_i costs c_i. For a time interval I, $\mathrm{BC}(S, I)$ is the sum of the cost of all the packets whose broadcast begins in I, divided by the length of I. The quantity which we want to minimize is then: $\mathrm{COST}(S) = \mathrm{EST}(S) + \mathrm{BC}(S)$. Note that up to scaling the costs c_i, any linear combinaison of EST and BC can be considered.

2 Preliminary Results

2.1 Structural Properties

The following lemma is a crucial observation that will allow to deal with the dependencies in a tractable way. From this observation, we derive an algebraic expression for the cost of periodic schedule. In the next section, we show that this expression yields to a lower bound on the cost of any schedule. The lower bound will be used in Section 4 to design efficient approximation algorithm.

Definition 1. *A schedule S is said* Round Robin *if at most one packet of each message M_i is broadcast in any time slot according to S, and if S schedules the packets of each message in Round Robin order (i.e. according to a cyclic order).*

Lemma 1 (Round Robin). *For any schedule S, consider the Round Robin schedule S' constructed from S by rescheduling in Round Robin order the packets of each message M_i within the slots reserved in S to broadcasting a packet of M_i. Then:* $\mathrm{COST}(S') \leqslant \mathrm{COST}(S)$.

Moreover, if S is periodic and is not Round Robin, then S' is periodic and: $\mathrm{COST}(S') < \mathrm{COST}(S)$.

Proof. First, S and S' have the same broadcast cost. Second, consider a request for M_i arriving at time t in S, and the ℓ_i first time slots where a packet of M_i is broadcast in S after time t. The service time of the request is minimized *iff* the ℓ_i packets of M_i are broadcast in those slots. Thus the expected service time in S' is at most as large as in S. Moreover, if S is periodic with period T and is not Round Robin, then S' is periodic with period $\leqslant T \prod_{i=1}^{m} \ell_i$ and its expected service time is smaller than S's. □

W.l.o.g. we will now only consider Round Robin schedules.

Lemma 2 (Cost). *Consider a periodic schedule S with period T. For each i, n_i is the number of broadcasts of message M_i in a period, and $(t_j^i)_{1 \leqslant j \leqslant n_i \ell_i}$ the time elapsed between the beginnings of the jth and the $(j+1)^{th}$ broadcasts of a packet of Message M_i. Then:*

$$\mathrm{EST}(S) = 1 + \sum_{i=1}^{m} p_i \sum_{j=1}^{n_i \ell_i} \frac{t_j^i}{T} \left\{ \frac{t_j^i}{2} + \left(t_{j+1}^i + \cdots + t_{j+\ell_i-1}^i\right) \right\}$$

and $\mathrm{BC}(S) = \dfrac{1}{T} \displaystyle\sum_{i=1}^{m} c_i n_i \ell_i$*, where the indices are considered modulo $n_i \ell_i$.*

Proof. Consider i in $\{1, \dots, m\}$. Message M_i is broadcast n_i times per period, its contribution to the broadcast cost is then $n_i \ell_i c_i / T$. A request is for Message M_i with probability p_i and arrives between the jth and the $(j+1)^{\text{th}}$ broadcasts of a packet of M_i with probability t_j^i / T. It starts then downloading the first packet after $t_j^i/2$ time on expectation and ends downloading the last packet after $t_{j+1}^i + \cdots + t_{j+\ell_i-1}^i + 1$ other time slots.

Remark 1 (Trapezoids representation). Note that we can represent the cumulated response time to request for a given message over a period of time by the sum of the areas of trapezoids as shown Figure 1; the black arrows are two example of requests, their waits are highlight in black, and the extra cost for downloading the last packet is in grey.

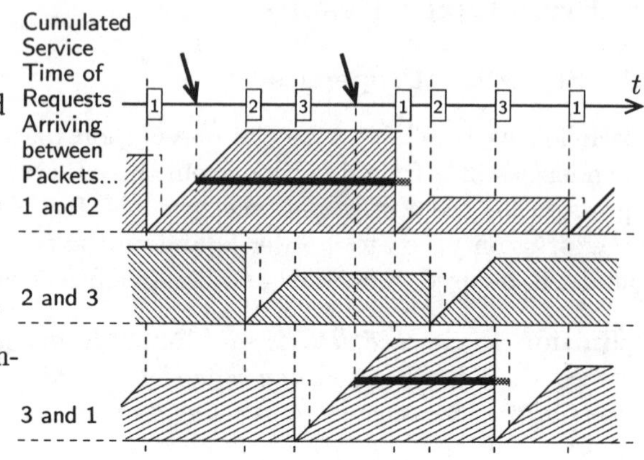

Fig. 1: The expected service time.

2.2 Optimality Results

Theorem 1 (NP-Hardness). *Finding the optimal schedule is strongly NP-hard on a single channel and with zero cost messages.*

Proof sketch. (Omitted) The proof is derived from the *NP*-hardness proof of the non-preemptive case given in [10]: we show that deciding whether the lower bound in Lemma 3 is realized is at least as hard as N-partition. □

Remark 2. Note that [6] yields an other *NP*-hardness proof by stating that the uniform length case with non zero cost is already *NP*-hard; however the present proof does not use costs.

Proposition 1 (Optimal periodic). *There exists an optimal schedule which is periodic. It can be computed in exponential time.*

Proof sketch. (Omitted) The proof is based on the search of a minimum cost cycle in a finite graph, and the lemmas are broadly inspired from [4, 5, 10] but their proofs need to be widely adapted in order to take into account the segmentation of the messages into packets. □

3 A Lower Bound

Finding a good lower bound is a key point to designing and proving efficient approximation algorithms for this problem. An algorithm to compute the value of the following lower bound, will be given Section 4.5.

Lemma 3 (Lower bound). *The following minimization problem is a lower bound to the cost of any schedule of the packets of* $M_1, \ldots, M_m$ *on* W *channels:*

$$\mathrm{LB}(M) \begin{cases} \min\limits_{\tau>0} \sum\limits_{i=1}^{m} p_i \left(\dfrac{\tau_i \ell_i}{2} + \ell_i - \dfrac{\ell_i - 1}{2\tau_i} \right) + \dfrac{c_i}{\tau_i} \\ \text{Subject to:} \quad (i) \quad \forall i,\ \tau_i \geqslant 1 \ \text{ and } (ii) \quad \sum\limits_{i=1}^{m} \dfrac{1}{\tau_i} \leqslant W \end{cases}$$

This minimization problem admits a unique solution τ^**.* $\mathrm{LB}(M)$ *is realized if and only if one can broadcast all the packets of each* M_i *consecutively periodically exactly every* $(\tau_i^* \cdot \ell_i)$*.*

Proof sketch. According to Lemma 1, let S be a periodic Round Robin schedule of the packets of messages $M_1, \ldots, M_m$ on W channels with period T. We use the same notations (n_i) and (t_j^i) as in Lemma 2. Given that Message M_i is broadcast n_i times per period, we seek for the optimal value of the (t_j^i) for each message independently. We relax the constraints on the schedule by authorizing messages to overlap and to be scheduled outside the slots. The proof works in three steps:
1. If the expected service time for M_i with $\ell_i \geqslant 2$ is minimized, then for any pair of consecutive broadcasts of the same packet of M_i at time t_1 and t_2 $(t_1 < t_2)$, a packet of M_i is broadcast at time $(t_1 + 1)$ or $(t_2 - 1)$.
2. If the expected service time of M_i is minimized, the packets of M_i are broadcast within blocks of ℓ_i consecutive time slots.
3. The blocks are optimally scheduled periodically every T/n_i.

Step 1. Consider M_i with $\ell_i \geqslant 2$ and two consecutive packets of M_i (w.l.o.g. packets 1 and 2). For $1 \leqslant k \leqslant n_i$, let I_k, J_k, and K_k be the intervals delimited by the end of the kth broadcast of packet 1, the beginning and the end of the kth broadcast of packet 2, and the beginning of the next broadcast of a packet of M_i as illustrated below (Note that $|J_k| = 1$).

(S) — x — 1 2 — 2 — x — 1 2 — 2 — x → t

I_k J_k K_k I_{k+1} J_{k+1} K_{k+1}

Let S' be the schedule that schedules the packets of M_i as in S except that packet 2 is always scheduled next to packet 1. A request for M_i that raises outside intervals I_k, J_k and K_k has the same service time in S and in S'. A request that raises in I_k is served one time unit later in S' than in S. But a request that raises in $J_k \cup K_k$ is served $|I_{k+1}|$ earlier in S' than in S. The expected service time varies then from S to S' by:

$$\sum_{i=1}^{n_i} (|I_k| \times 1 - (1 + |K_k|) \times |I_{k+1}|) = -\sum_{i=1}^{n_i} |K_k| \cdot |I_{k+1}| \leqslant 0$$

Thus, the expected service time in S' is at most as big as in S and smaller if there exists in S a pair of consecutive broadcasts of packet 2 occuring at time t_1 and t_2 $(t_1 < t_2)$ so that no packet of M_i is broadcast at time $(t_1 + 1)$ $(|K_k| \neq 0)$ and $(t_2 - 1)$ $(|I_{k+1}| \neq 0)$.

Step 2 is obtained by contradiction using the transformation in Step 1.

Step 3. We are thus left with n_i blocks of ℓ_i packets of M_i. Let t_k be the time elapsed between the beginning of the kth and the $(k+1)^{\text{th}}$ block. Lemma 2 yields that the expected service time for M_i is:

$$\sum_{k=1}^{n_i} \left(\frac{{t_k}^2}{2T}\right) + \ell_i - \frac{n_i \ell_i(\ell_i - 1)}{2T}$$

which is minimized under the constraint $\sum_{k=1}^{n_i} t_k = T$, when for all k, $t_k = T/n_i$. Define $\tau_i = T/(n_i \ell_i)$. The cost of S is thus bounded from below by:

$$\text{COST}(S) \geqslant \sum_{k=1}^{n_i} \left\{ p_i \left(\frac{\tau_i \ell_i}{2} + \ell_i - \frac{\ell_i - 1}{2\tau_i} \right) + \frac{c_i}{\tau_i} \right\}$$

Finally $n_i \ell_i \leqslant T$ and $\sum_{i=1}^{m} n_i \ell_i \leqslant WT$ imply: (*i*) $\tau_i \geqslant 1$ and (*ii*) $\sum_{i=1}^{m} 1/\tau_i \leqslant W$. Minimizing over those constraints yields the lower bound on the cost of any schedule. The unicity of the solution τ^* to the minimization problem will be proved Section 4.5.
Moreover by construction, the lower bound is realized *iff* there exists a periodic Round Robin schedule that broadcast the ℓ_i packets of each M_i in consecutive slots, periodically every $\tau_i^* \ell_i$. □

Remark 3. One can derive a trivial lower bound close up to an additive term $\sum_{i=1}^{m} p_i \ell_i$ to ours by simply optimizing the time needed to download a given packet for each message. If this later lower bound is sufficient to analyze our heurisitics, it is never realized and cannot be used to yield our *NP*-hardness result.

4 Constant Factor Approximation Algorithms

Note 1. The optimal ficticious schedule suggested by the lower bound LB(M) is not realizable in general. Actually, as shown in [10], if no preemption are used, the optimal cost of a schedule can be arbitrary far from the lower bound LB(M). Consider the problem of scheduling $W+1$ messages $M_1, \ldots, M_{W+1}$ on W channels, where M_i counts $\ell_i = L^{i-1}$ packets, cost $c_i = 0$, and request probability $p_i = \alpha/L^{i-1}$, where α is such that $p_1 + \cdots + p_m = 1$. In that case, one can show by induction on W that when L goes to infinity, the optimal schedule without preemption has a cost $\text{OPT}_{\text{whithout preemption}} = \Theta(L^{1/2^W})$, but LB($M$) $= \Theta(1)$.

In order to minimize the cost of the schedule, we won't follow exactly the ficticious schedule suggested by the lower bound in Lemma 3. In fact, remark that if we spread regularly the packets of each message M_i, every τ_i, in this ficticious schedule, the expected service time to a random request increases by less than a factor of 2. This will be helpful in order to design an efficient approximation algorithm for the preemptive case.

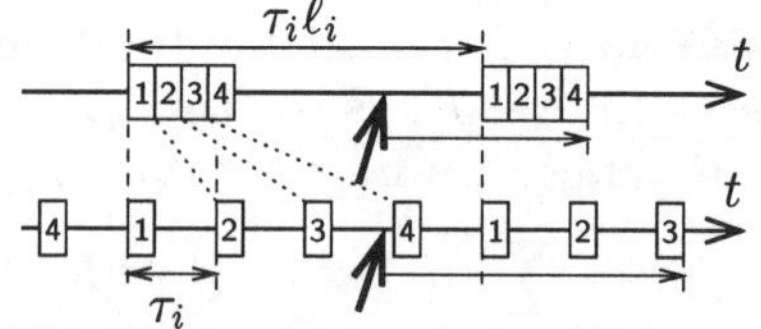

Fig. 2: Spreading the packets regularly.

Algorithm 1 Randomized algorithm

Input: Some positive numbers $\tau_1, \dots, \tau_m$, verifying $\sum_{i=1}^m 1/\tau_i \leqslant 1$.
Let $\tau_0 > 0$ so that:
$$1/\tau_0 = 1 - \sum_{i=1}^m 1/\tau_i$$
Output:
for $t = 1..\infty$ **do**
Draw $i \in \{0, 1, \dots, m\}$ with probability $1/\tau_i$. Schedule during slot t, the next packet of Message M_i in the Round Robin order, if $i \geqslant 1$; and Idle during slot t, otherwise.

Algorithm 2 Greedy algorithm

Input: Some positive numbers $\tau_1, \dots, \tau_m$, verifying $\sum_{i=1}^m 1/\tau_i \leqslant 1$.
Let $c_0 = p_0 = 0$ and $\tau_0 > 0$ so that:
$$1/\tau_0 = 1 - \sum_{i=1}^m 1/\tau_i$$
Output:
for $t = 1..\infty$ **do**
Select $i \in \{0, 1, \dots, m\}$ which minimizes $(c_i - p_i\tau_i \sum_{j=1}^{\ell_i} s_{i,j}^{t-1})$.
Schedule during slot t, the next packet of M_i in the Round Robin order, if $i \geqslant 1$; and Idle during slot t, otherwise.

We will first present algorithms that construct efficient schedules on a single channel in Sections 4.1, 4.2 and 4.3; then Section 4.4 shows how to extend these algorithms to the multichannel case, using a result of [6].

4.1 A Randomized Algorithm

Theorem 2. *Given m messages $M_1, \dots, M_m$, the expected cost of the one-channel schedule S generated by the randomized algorithm 1, is:*
$$\mathbb{E}[\mathrm{COST}(S)] = \frac{1}{2} + \sum_{i=1}^{m} \left(p_i \tau_i \ell_i + \frac{c_i}{\tau_i} \right)$$
Thus if $\tau = \tau^*$ realizes $\mathrm{LB}(M)$*:* $\quad \mathbb{E}[\mathrm{COST}(S)] \leqslant 2 \cdot \mathrm{LB}(M) - 3/2$.

Proof. A packet of M_i is broadcast with probability $1/\tau_i$ in S. The expected frequency of M_i is then $1/\tau_i$ and $\mathbb{E}[\mathrm{BC}(S)] = \sum_{i=1}^m c_i/\tau_i$. A request for M_i is served after ℓ_i downloads of a packet of M_i: it waits on expectation $1/2$ until the end of the current time-slot and $\tau_i\ell_i$ upto the end of the download of the last packet of M_i. Then, $\mathbb{E}[\mathrm{EST}(S)] = 1/2 + \sum_{i=1}^m p_i\ell_i\tau_i$.
Finally $\tau_i^* \geqslant 1$ and $\ell_i \geqslant 1$ imply: $2\,\mathrm{LB}(M) \geqslant \sum_{i=1}^m (p_i\tau_i^*\ell_i + c_i/\tau_i^*) + 2$, which yields the last statement.

4.2 A Greedy Approximation

We present in this section a derandomized version of the randomized algorithm above.

As shown Figure 3, we define the *state* of the schedule at time slot t as a vector $\boldsymbol{s}^t$, such that: for any i and $1 \leqslant j \leqslant \ell_i$, the jth of the ℓ_i last broadcasts of a packet of M_i before time t starts at time $(t - (s_{i,j}^t + \dots + s_{i,\ell_i})$. Since no request arrive before $t = 0$, we equivalently assume that all the packets of all messages are fictively broadcast at time $t = 0$, and initially, at time $t = 0$: for all i and j, $s_{i,j}^0 = 0$;.

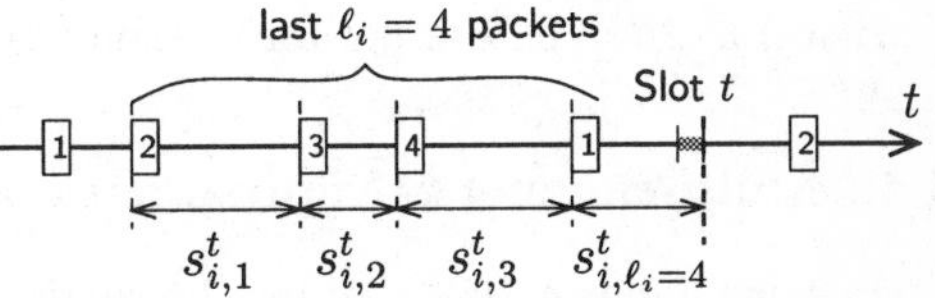

Fig. 3: The state $(s_{i,j}^t)$ at time t.

Theorem 3. *Given m messages $M_1, \dots, M_m$, the cost of the one-channel schedule S generated by the greedy algorithm 2, is:*

$$\mathrm{COST}(S) \leqslant \frac{1}{2} + \sum_{i=1}^{m} \left(p_i \tau_i \ell_i + \frac{c_i}{\tau_i} \right)$$

Thus if $\tau = \tau^*$ realizes $\mathrm{LB}(M)$*:* $\mathrm{COST}(S) \leqslant 2 \cdot \mathrm{LB}(M) - 3/2$.

Proof sketch. (Omitted) The greedy algorithm is a derandomized version of the algorithm above. The greedy choice ensures that at any time t, the choice made in time slot t minimizes the expected cost of the already allocated slots $1, \dots, t-1$, if the schedule would continue with the randomized scheme. Its cost is then, at any time, bounded from above by the expected cost of the randomized schedule. □

4.3 A Deterministic Periodic Approximation

It is sometimes required to have a fixed schedule instead of generating it on the fly. For instance, it helps to design caching strategies [1]. The next result shows that one can construct an efficient periodic schedule with polynomial period. Note that this allow also to *guarantee* a bound (the period) on the service time of any request.

Theorem 4. *One can construct in polynomial time, a periodic schedule with cost* $\leqslant 2 \cdot \mathrm{LB}(M)$ *and period polynomial in the total length and cost of the messages* $(\frac{14}{3}(\sum_{i=1}^m \ell_i)^2 + 2\sum_{i=1}^m c_i \ell_i)$.

Proof sketch. (Omitted) The schedule is constructed as shown Figure 4: 1) First, schedule all the packets of each message during the first $\mathcal{L} =_{\mathrm{def}} \sum_i \ell_i$ time slots; 2) Second, executes T steps of the greedy algorithm above; 3) Third, sort the set $X = \{k\tau_i^* : 1 \leqslant k \leqslant \ell_i\}$ in increasing order and schedule during the next $\mathcal{L}$ time slots, the kth packets of the messages M_i in order of increasing $k\tau_i^*$; 4) Finally, complete with some packets of the messages in order to ensure that for all i, the number of broadcasts of a packet of M_i in a period is a multiple of ℓ_i, and thus guaranty the Round Robin property. One can show that the cost of the resulting schedule is at most $2\,\mathrm{LB}(M)$ as soon as the period is bigger than $\frac{14}{3}(\sum_i \ell_i)^2 + 2\sum_i c_i \ell_i$. □

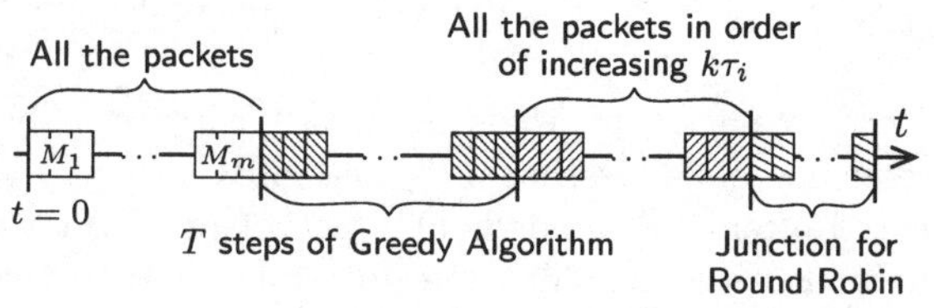

Fig. 4: A periodic approximation.

4.4 Multi-channel 2-Approximations

The performance ratio proof for the randomized algorithm given above only rely on the fact that we know how to broadcast the packets of each M_i every τ_i on expectation. In order to extend the result to the multi-channel case, we only need to manage to broadcast the packets of each M_i with probability $1/\tau_i$, while ensuring that two packets of the same messages are not broadcast during the same time slot. A straight forward application of the method designed in [6], to extend the single channel randomized algorithm to the multi-channel, yields then the result.

The multi-channel greedy algorithm is again obtained by derandomizing the schedule, and by extending the greedy choice as in [6]. Finally the extension of the periodic approximation is then constructed exactly as in Section 4.3, except that one uses the multi-channel greedy algorithm instead of the single channel one.

4.5 Solving the Lower Bound

The aim of this last section is to solve the following generic non-linear program (A), defined by:

$$(A) \quad \min_{\tau>0} \sum_{i=1}^{m} a_i\tau_i + \frac{b_i}{\tau_i} \quad \text{Subject to: } (i)\ \forall i,\ \tau_i \geqslant 1 \text{ and } (ii)\ \sum_{i=1}^{m} \frac{1}{\tau_i} \leqslant W$$

where W is a positive integer, $a_1, \dots, a_m$ are positive numbers, and $b_1, \dots, b_m$ are *arbitrary* numbers.

We present essentially an extension of the method designed in [6] for the special case where for all i, $b_i \geqslant 0$. The results presented are basically the same but the proofs need to be adapted. As in [6], we introduce a relaxed minimization problem (A'), which do not require the constraint (i), and which can be solved algebraically. The solution to the relaxed problem will allow to construct and prove the unicity of the solution to (A).

Lemma 4 (Relaxation). *Given some positive numbers $a_1, \dots, a_m$, a positive integer W and some numbers $b_1, \dots, b_m$, the following minimization problem:*

$$(A') \quad \min_{\tau>0} \sum_{i=1}^{m} a_i\tau_i + \frac{b_i}{\tau_i} \quad \text{Subject to: } \sum_{i=1}^{m} \frac{1}{\tau_i} \leqslant W$$

admits a unique solution τ' verifying: $\tau'_i = \sqrt{(b_i + \lambda')/a_i}$, for a certain $\lambda' \geqslant 0$. If, for all i, $b_i \geqslant 0$ and $\sum_i \sqrt{a_i/b_i} \leqslant W$, then $\lambda' = 0$; else λ' is the unique solution to: $\sum_i \sqrt{a_i/(b_i + \lambda')} = W$.

Proof sketch. (Omitted) Solved by carefull use of Lagrangian relaxation. □

Lemma 5. *Consider the two non-linear minimization problems (A) and (A'), a solution τ^* to (A) and the solution τ' to (A'). Then, for all i, if $\tau'_i < 1$, then $\tau^*_i = 1$.*

Proof. The proof given in [6] is only based on the unimodularity (and not on the convexity) of the terms $a_i\tau_i + b_i/\tau_i$. Their proof then naturally extends to the case where some b_i may be negative. □

Corollary 1 (Unicity). *The minimization problem (A) admits a unique solution τ^* which can be computed in polynomial time.*

Proof. Consider a solution τ^* to (A). We compute the solution τ' to (A'). If for some i_0, $\tau'_{i_0} < 1$, then $\tau^*_{i_0} = 1$. Thus, we remove this variable from Problem (A) by fixing its value to 1, and iterate. If for all i, $\tau'_i \geqslant 1$, τ' is also solution of (A), which is thus unique: $\tau^* = \tau'$. □

Acknowledgment. We'd like to thank Neal E. Young and Claire Kenyon, for useful comments and careful reading of the paper.
The full version of the paper is available at //www.ens-lyon.fr/~nschaban.

References

[1] ACHARYA, S. *Broadcast Disks: Dissemination-based Management for Assymmetric Communication Environments.* PhD thesis, Brown University, 1998.

[2] AMMAR, M. H. Response time in a teletext system: An individual user's perspective. *IEEE Transactions on Communications COM-35,11* (Nov. 1987), 1159–1170.

[3] AMMAR, M. H., AND WONG, J. W. The design of teletext broadcast cycles. In *Performance Evaluation* (1985), vol. 5(4), pp. 235–242.

[4] AMMAR, M. H., AND WONG, J. W. On the optimality of cyclic transmission in teletext systems. In *IEEE Trans. on Comm.* (1987), vol. COM-35(11), pp. 1159–1170.

[5] ANILY, S., GLASS, C. A., AND HASSIN, R. The scheduling of maintenance service. To appear (http://www.math.tau.ac.il/~hassin).

[6] BAR-NOY, A., BHATIA, R., NAOR, J. S., AND SCHIEBER, B. Minimizing service and operation costs of periodic scheduling. In *Proc. of the 9th Annual ACM-SIAM Symp. on Discrete Algorithms (SODA)* (1998), pp. 11–20.

[7] BAR-NOY, A., AND SHILO, Y. Optimal broadcasting of two files over an asymmetric channel. In *Proc. of Infocom* (1999).

[8] GECSEI, J. The architecture of videotex systems. *Prentice Hall, Englewood Cliffs, N.J.* (1983).

[9] IMIELINSKI, T., VISWANATHAN, S., AND BADRINATH, B. Energy efficient indexing on air. In *SIGMOD* (May 1994).

[10] KENYON, C., AND SCHABANEL, N. The data broadcast problem with non-uniform transimission times. In *Proc. of the 10th SODA* (1999), pp. 547–556.

[11] KENYON, C., SCHABANEL, N., AND YOUNG, N. E. Polynomial time approximation scheme for data broadcast. Submitted, Oct 1999.

[12] KHANNA, S., AND LIBERATORE, V. On broadcast disk paging. In *Proceedings of 30th STOC* (1998), vol. 30, pp. 634–643.

[13] KHANNA, S., AND ZHOU, S. On indexed data broadcast. In *Proceedings of 30th STOC* (1998), vol. 30, pp. 463–472.

[14] SHEKHAR, S., AND LIU, D. Genesis: An approach to data dissemination in Advanced Traveller Information Systems (ATIS). In *IEEE Data Engineering Bulletin* (Sept. 1996), vol. 19(3).

[15] TAN, K., AND XU, J. Energy efficient filtering of nonuniform broadcast. In *Proc. of the 16th Int. Conf. in Distributed Computing System* (1996), pp. 520–527.

[16] VAIDYA, N., AND HAMEED, S. Log time algorithms for scheduling single and multiple channel data broadcast. In *Proc. of the 3rd ACM/IEEE Conf. on Mobile Computing and Networking (MOBICOM)* (Sep. 1997).

[17] VISHWANATH, S. *Publishing in wireless and wireline environments.* PhD thesis, Rutgers University, 1994.

An Approximate L^p-Difference Algorithm for Massive Data Streams*
(Extended Abstract)**

Jessica H. Fong[1] and Martin J. Strauss[2]

[1] Princeton University, 35 Olden St., Princeton NJ 08544, USA
jfong@cs.princeton.edu,
http://www.cs.princeton.edu/~jfong/
[2] AT&T Labs—Research, 180 Park Ave., Florham Park, NJ, 07932 USA
mstrauss@research.att.com,
http://www.research.att.com/~mstrauss/

Abstract. Several recent papers have shown how to approximate the difference $\sum_i |a_i - b_i|$ or $\sum |a_i - b_i|^2$ between two functions, when the function values a_i and b_i are given in a data stream, and their order is chosen by an adversary. These algorithms use little space (much less than would be needed to store the entire stream) and little time to process each item in the stream and give approximations with small relative error. Using different techniques, we show how to approximate the L^p-difference $\sum_i |a_i - b_i|^p$ for any rational-valued $p \in (0, 2]$, with comparable efficiency and error. We also show how to approximate $\sum_i |a_i - b_i|^p$ for larger values of p but with a worse error guarantee. These results can be used to assess the difference between two chronologically or physically separated massive data sets, making one quick pass over each data set, without buffering the data or requiring the data source to pause.

1 Introduction

[Some of the following material is excerpted from [7], with the authors' permission. Readers familiar with [7] may skip to Section 1.1.]

Massive data sets are increasingly important in a wide range of applications, including observational sciences, product marketing, and monitoring and operations of large systems. In network operations, raw data typically arrive in *streams*, and decisions must be made by algorithms that make one pass over each stream, throw much of the raw data away, and produce "synopses" or "sketches" for further processing. Moreover, network-generated massive data sets are often *distributed*: Several different, physically separated network elements may receive or generate data streams that, together, comprise one logical data set. To be of use in operations, the streams must be analyzed locally and their synopses

* Part of this work was done while the first author was visiting AT&T Labs.
** An expanded version of this paper is available in preprint form at
http://www.research.att.com/~mstrauss/pubs/lp.ps

H. Reichel and S. Tison (Eds.): STACS 2000, LNCS 1770, pp. 193–204, 2000.

sent to a central operations facility. The enormous scale, distributed nature, and one-pass processing requirement on the data sets of interest must be addressed with new algorithmic techniques.

In [2,11,1,7], the authors presented a new technique: a space-efficient, one-pass algorithm for approximating the L^1 difference $\sum_i |a_i - b_i|$ or L^2 difference[1] $\left(\sum_i |a_i - b_i|^2\right)^{1/2}$ between two functions, when the function values a_i and b_i are given as data streams, and their order is chosen by an adversary. Here we continue that work by showing how to compute $\sum_i |a_i - b_i|^p$ for any rational-valued $p \in (0, 2]$. These algorithms fit naturally into a toolkit for Internet-traffic monitoring. For example, Cisco routers can now be instrumented with the NetFlow feature [5]. As packets travel through the router, the NetFlow software produces summary statistics on each *flow*.[2] Three of the fields in the flow records are source IP-address, destination IP-address, and total number of bytes of data in the flow. At the end of a day (or a week, or an hour, depending on what the appropriate monitoring interval is and how much local storage is available), the router (or, more accurately, a computer that has been "hooked up" to the router for monitoring purposes) can assemble a set of values $(x, f_t(x))$, where x is a source-destination pair, and $f_t(x)$ is the total number of bytes sent from the source to the destination during a time interval t. The L^p difference between two such functions assembled during different intervals or at different routers is a good indication of the extent to which traffic patterns differ.

Our algorithm allows the routers and a central control and storage facility to compute L^p differences efficiently under a variety of constraints. First, a router may want the L^p difference between f_t and f_{t+1}. The router can store a small "sketch" of f_t, throw out all other information about f_t, and still be able to approximate $\|f_t - f_{t+1}\|_p$ from the sketch of f_t and (a sketch of) f_{t+1}.

The functions $f_t^{(i)}$ assembled at each of several remote routers R_i at time t may be sent to a central tape-storage facility C. As the data are written to tape, C may want to compute the L^p difference between $f_t^{(1)}$ and $f_t^{(2)}$, but this computation presents several challenges. First, each router R_i should transmit its statistical data when R_i's load is low and the R_i-C paths have extra capacity; therefore, the data may arrive at C from the R_i's in an arbitrarily interleaved manner. Also, typically the x's for which $f(x) \neq 0$ constitute a small fraction of all x's; thus, R_i should only transmit $(x, f_t^{(i)}(x))$ when $f_t^{(i)}(x) \neq 0$. The set of transmitted x's is not predictable by C. Finally, because of the huge

---

[1] Approximating the L^p difference, $\|\langle a_i\rangle - \langle b_i\rangle\|_p = \left(\sum |a_i - b_i|^p\right)^{1/p}$, is computationally equivalent to approximating the easier-to-read expression $\sum |a_i - b_i|^p$. We will use these interchangeably when discussing computational issues.

[2] Roughly speaking, a "flow" is a semantically coherent sequence of packets sent by the source and reassembled and interpreted at the destination. Any precise definition of "flow" would have to depend on the application(s) that the source and destination processes were using to produce and interpret the packets. From the router's point of view, a flow is just a set of packets with the same source and destination IP-addresses whose arrival times at the routers are close enough, for a tunable definition of "close."

size of these streams,[3] the central facility will not want to buffer them in the course of writing them to tape (and cannot read from one part of the tape while writing to another), and telling R_i to pause is not always possible. Nevertheless, our algorithm supports approximating the L^p difference between $f_t^{(1)}$ and $f_t^{(2)}$ at C, because it requires little workspace, requires little time to process each incoming item, and can process in one pass all the values of both functions $\{(x, f_t^{(1)}(x))\} \cup \{(x, f_t^{(2)}(x))\}$ in any permutation.

Our L^p-difference algorithm achieves the following performance for rational $p \in (0, 2]$:

> Consider two data streams of length at most n, each representing the non-zero points on the graph of an integer-valued function on a domain of size n. Assume that the maximum value of either function on this domain is M. Then a one-pass streaming algorithm can compute with probability $1 - \delta$ an approximation A to the L^p-difference B of the two functions, such that $|A - B| \leq \epsilon B$, using total space and per-item processing time $(\log(M) \log(n) \log(1/\delta)/\epsilon)^{O(1)}$. The input streams may be interleaved in an arbitrary (adversarial) order.

1.1 L^p-Differences for p Other than 1 or 2

Our results fill in gaps left by recent work. While the L^1- and L^2- differences are important, the L^p-differences for other p, say the $L^{1.5}$-difference, provide additional information. In particular, there are $\langle a_i \rangle$, $\langle b_i \rangle$, $\langle a'_i \rangle$, and $\langle b'_i \rangle$ such that $\sum |a_i - b_i| = \sum |a'_i - b'_i|$ and $\sum |a_i - b_i|^2 = \sum |a'_i - b'_i|^2$ but $\sum |a_i - b_i|^{1.5}$ and $\sum |a'_i - b'_i|^{1.5}$ are different.

By showing how to compute the L^p difference for varing p, we provide an approximate difference algorithm that is precisely tunable for the application at hand.

We also give an algorithm for $p > 2$, though with an error guarantee somewhat worse than the guarantee available for the $p \leq 2$ cases. Still, that result is a randomized algorithm with the correct mean, which is an advantage in some situtations.

1.2 Organization

The rest of this paper is organized as follows. In Section 2, we describe precisely our model of computation and its complexity measure. We present our main technical results in Section 3. We discuss the relationship of our algorithm to other recent work and present some open problems, in Section 4. In this extended abstract, all proofs are omitted.

[3] In 1999, a WorldNet gateway router generated more that 10Gb of NetFlow data each day.

2 Background

We describe the details of our algorithm in terms of the streaming model used in [7]. This model is closely related to that of [10]. It is immediate to adapt our algorithm to the sketch model used in [7,4]; we give only brief comments.

2.1 The Streaming Model

A *data stream* is a sequence of data items $\sigma_1, \sigma_2, \ldots, \sigma_n$ such that, on each *pass* through the stream, the items are read once in increasing order of their indices. We assume the items σ_i come from a set of size M, so that each σ_i has size $\log M$. In the computational model, we assume that the input is one or more data streams. We focus on two resources—the *workspace* required in words and the *time to process* an item in the stream, but disregard pre- and post-processing time.

Definition 1. *The complexity class* $\mathrm{PASST}(s(\delta, \epsilon, n, M), t(\delta, \epsilon, n, M))$ *(read as "probably approximately correct streaming space complexity* $s(\delta, \epsilon, n, M)$ *and time complexity* $t(\delta, \epsilon, n, M)$*") contains those functions* f *for which one can output a random variable* X *such that* $|X - f| < \epsilon f$ *with probability at least* $1 - \delta$ *and computation of* X *can be done by making a single pass over the data, using workspace at most* $s(\delta, \epsilon, n, M)$ *and taking time at most* $t(\delta, \epsilon, n, M)$ *to process each of the* n *items, each of which is in the range* 0 *to* $M - 1$.

If $s = t$, *we also write* $\mathrm{PASST}(s)$ *for* $\mathrm{PASST}(s, t)$.

2.2 The Sketch Model

Sketches were used in [4] to check whether two documents are nearly duplicates. A sketch can also be regarded as a *synopsis data structure* [9].

Definition 2. *The complexity class* $\mathrm{PAS}(s(\delta, \epsilon, n, M)))$ *(to be read as "probably approximately correct sketch complexity* $s(\delta, \epsilon, n, M)$*") contains those functions* $f : X \times X \to Z$ *of two inputs for which there exists a set* S *of size* 2^s, *a randomized* sketch function $h : X \to S$, *and a randomized* reconstruction function $\rho : S \times S \to Z$ *such that, for all* $x_1, x_2 \in X$, *with probability at least* $1 - \delta$, $|\rho(h(x_1), h(x_2)) - f(x_1, x_2)| < \epsilon f(x_1, x_2)$.

By "randomized function" of k inputs, we mean a function of $k+1$ variables. The first input is distinguished as the source of randomness. It is not necessary that, for all settings of the last k inputs, for most settings of the first input, the function outputs the same value.

2.3 Medians and Means of Unbiased Estimators

We now recall a general technique of randomized approximation schemes.

Lemma 1. *Let* X *be a real-valued random variable such that, for some* c, *we have* $E[X^2] \le c \cdot \mathrm{var}[X]$. *Then, for any* $\epsilon, \delta > 0$, *there exists a random variable* Z *such that* $\Pr(|Z - E[X]| \ge \epsilon E[X]) \le \delta$. *Furthermore,* Z *is a function of* $O(\log(1/\delta)/\epsilon^2)$ *independent samples of* X.

3 The Algorithm

In this section we prove our main theorem:

Theorem 1. *For rational $p \in (0,2]$, the L^p-difference of two functions $\langle a_i \rangle$ and $\langle b_i \rangle$ is in*

$$\text{PASST}\left((\log(n)\log(M)\log(1/\delta)/\epsilon)^{O(1)}\right) , \tag{1}$$

when the stream items consist of values a_i or b_i, presented in arbitrary order. This L^p-difference is also in

$$\text{PAS}\left((\log(n)\log(M)\log(1/\delta)/\epsilon)^{O(1)}\right) . \tag{2}$$

3.1 Intuition

We first give an intuitive overview of the algorithm. Our goal is to approximate $L_p = \sum |a_i - b_i|^p$, where the values a_i, $b_i < M$ are presented in a stream in any order, and the index i runs up to n. We are given tolerance ϵ and maximum error probability δ. We wish to output a random variable Z such that $\Pr(|Z - L_p| > \epsilon L_p) < \delta$, using total space and per-item processing time polynomial in $(\log(n)\log(M)\log(1/\delta)/\epsilon)$. The input is a stream consisting of tuples of the form (i, c, θ), where $i \in [0, n)$, $c \in [0, M)$, and $\theta \in \{\pm 1\}$. The tuple (i, c, θ) denotes that $a_i = c$ if $\theta = +1$ and $b_i = c$ if $\theta = -1$.

Below, the reader may consider f to be a deterministic function with $(f(b) - f(a))^2 = |b - a|^p$. (In the next few sections, we will construct a randomized function, $f(r, x)$, such that $E\left[(f(r,b) - f(r,a))^2\right] \approx |b - a|^p$.) The algorithm proceeds as in Figure 1.

To see how the algorithm works, first focus on single values for k and ℓ. Note that $Z = Z_{k,\ell} = \sum_i \sigma_i(f(a_i) - f(b_i))$. Separating the diagonal and off-diagonal terms of Z^2,

$$E\left[Z^2\right] \tag{3}$$

$$= E\left[\sum_i \sigma_i^2 (f(a_i) - f(b_i))^2 + \sum_{i \neq i'} \pm \sigma_i \sigma_{i'} (f(a_i) - f(b_i))(f(a_{i'}) - f(b_{i'}))\right] \tag{4}$$

$$\approx E\left[\sum_i |a_i - b_i|^p + \sum_{i \neq i'} \pm \sigma_i \sigma_{i'} (f(a_i) - f(b_i))(f(a_{i'}) - f(b_{i'}))\right] \tag{5}$$

$$= \sum_i |a_i - b_i|^p . \tag{6}$$

In the last line, we used the fact that $E[\sigma_i] = 0$ and that σ_i and $\sigma_{i'}$ are independent for $i \neq i'$. A similar calculation shows that $\text{var}(Z^2) \leq O(E^2[Z^2])$. We can therefore apply Lemma 1 and take a median of means of independent copies of Z^2 to get the desired result.

It is straightforward to check that cost bounds are met. The analysis is omitted in this extended abstract.

```
Algorithm L1(⟨(i, c, θ)⟩)

Initialize:
        For k = 1 to O(log(1/δ)) do
            For ℓ = 1 to O(1/ε²) do
                Z_{k,ℓ} = 0;
                pick sample points for random variable
                families {σ_i} and {R_i};
                //σ_i = ±1 and R_i is described below
Stream processing:
        For each tuple (i, c, θ) in the input stream do
            For k = 1 to O(log(1/δ)) do
                For ℓ = 1 to O(1/ε²) do
                    Z_{k,ℓ}  +=  σ_i θ f(R_i, c);
Report:
        Output median_k avg_ℓ Z²_{k,ℓ};
```

Fig. 1. Main algorithm, intuition

3.2 Construction of f, Overview

Construction of f is the main technical content of this paper. We construct a function $f : \mathbb{Z} \to \mathbb{Z}$ such that

$$E\left[(f(b) - f(a))^2\right] = (1 \pm \epsilon)|b - a|^p . \quad (7)$$

To do this, we will first define a function $d(a, b)$ such that

- $|d(a, b)| \in O(|b - a|^{p/2})$ for all a and b,
- $|d(a, b)| \in \Omega(|b - a|^{p/2})$ for a significant fraction of a and b, and
- $d(r, b) - d(r, a) = d(a, b)$ for all r.

Next, we define a family $\{T_{\boldsymbol{R}}\}$ of transformations on the reals, with corresponding *inverse scale factors* $\phi(\boldsymbol{R})$ such that:

- the transformation is an approximate isometry, *i.e.*, $|a-b|^p \approx \phi^2(\boldsymbol{R})|T_{\boldsymbol{R}}(b) - T_{\boldsymbol{R}}(a)|^p$, and,
- for random $\boldsymbol{R}$, the distribution of $|\phi(\boldsymbol{R})d(T_{\boldsymbol{R}}(a), T_{\boldsymbol{R}}(b))|/|b-a|^{p/2}$ is approximately γ_0, for γ_0 independent of a and b.

We then put $f(x) = c\phi(\boldsymbol{R})d(T_{\boldsymbol{R}}(0), T_{\boldsymbol{R}}(x))$ (rounding appropriately from reals to integers). We have

$$\begin{aligned} E_{\boldsymbol{R}}\left[(f(b) - f(a))^2\right] &= E_{\boldsymbol{R}}\left[c^2\phi^2(\boldsymbol{R})(d(T_{\boldsymbol{R}}(0), T_{\boldsymbol{R}}(b)) - d(T_{\boldsymbol{R}}(0), T_{\boldsymbol{R}}(a)))^2\right] && (8) \\ &= E_{\boldsymbol{R}}\left[c^2\phi^2(\boldsymbol{R})(d(T_{\boldsymbol{R}}(a), T_{\boldsymbol{R}}(b)))^2\right] && (9) \\ &\in O\left(E_{\boldsymbol{R}}\left[c^2\phi^2(\boldsymbol{R})|T_{\boldsymbol{R}}(b) - T_{\boldsymbol{R}}(a)|^p\right]\right) && (10) \\ &\approx O(|b - a|^p). && (11) \end{aligned}$$

Because $|d(\alpha, \beta)| \in \Omega\left(|\beta - \alpha|^{p/2}\right)$ for a significant fraction of α, β (according to the distribution $(\alpha, \beta) = (T_{\boldsymbol{R}}(a), T_{\boldsymbol{R}}(b))$), from the Markov inequality we conclude that

$$E_{\boldsymbol{R}}\left[d(T_{\boldsymbol{R}}(a), T_{\boldsymbol{R}}(b))^2\right] \in \Omega\left(|T_{\boldsymbol{R}}(b) - T_{\boldsymbol{R}}(a)|^p\right) . \tag{12}$$

We have

$$E_{\boldsymbol{R}}\left[(f(b) - f(a))^2\right] = E_{\boldsymbol{R}}\left[c^2\phi^2(\boldsymbol{R})(d(T_{\boldsymbol{R}}(a), T_{\boldsymbol{R}}(b)))^2\right] \tag{13}$$

$$\in \Omega\left(E_{\boldsymbol{R}}\left[c^2\phi^2(\boldsymbol{R})|T_{\boldsymbol{R}}(b) - T_{\boldsymbol{R}}(a)|^p\right]\right) \tag{14}$$

$$\approx \Omega(|b - a|^p) . \tag{15}$$

It follows that $E_{\boldsymbol{R}}\left[(f(b) - f(a))^2\right] \in \Theta\left(|b - a|^p\right)$. Because the distribution on $|d(T_{\boldsymbol{R}}(a), T_{\boldsymbol{r}}(b))/|b - a|^{p/2}$ is approximately independent of a and b, it follows that $E_{\boldsymbol{R}}\left[(f(b) - f(a))^2\right] \approx c'(1 \pm \epsilon)|b - a|^p$, for c' independent of a and b. By choosing c appropriately, we can arrange that $c' = 1$.

We now proceed with a detailed construction of f.

3.3 Construction of f

The function $d(a, b)$ takes the form $d(a, b) = \sum_{a \leq i < b} \pi_j$, where π_j is a ± 1-valued function of j, related to a function described in [7].

Lemma 2. *For all rational $p \in (0, 2]$, there exist integers u and v such that $\frac{\log(v-u)}{\log(v+u)} = p/2$ and $v - u \geq 17$.*

Proof. If $p/2 = \alpha/\beta$ for $\alpha \geq 5$, put $v = 2^{\beta-1} + 2^{\alpha-1}$ and $u = 2^{\beta-1} - 2^{\alpha-1}$.

Now, we define a sequence π of $+1$'s and -1's, as follows. Let $\pi = \lim_{i \to \infty} \pi_{(i)}$ where $\pi_{(i)}$ is defined recursively, for $i \geq 1$, as

$$\pi_{(1)} = (+1)^u(-1)^v \tag{16}$$

$$\pi_{(i+1)} = \pi_{(i)}^u \overline{\pi_{(i)}^v} , \tag{17}$$

and $\overline{\pi_{(i)}}$ denotes $\pi_{(i)}$ with all $+1$'s replaced by -1's and all -1's replaced by $+1$'s. Note that $\pi_{(i)}$ is a prefix of $\pi_{(i+1)}$. For example, a graph of π with $u = 1$ and $v = 3$ is given in Figure 2 (Figure 2 also describes sets $S_{s,t}$, to be defined later).

Let π_j (as opposed to $\pi_{(j)}$) denote the j'th symbol of π.

Definition 3. *Let $d(a, b) = \sum_{j=a}^{b-1} \pi_j$ be the discrepancy of π between $+1$'s and -1's in the interval $[a, b)$.*

Note that d and π depend on u and v. We only consider one set of values for u, v at a time and drop them from the notation.

We will need the following notation:

Definition 4. *The* **randomized rounding** $[x]_\rho$ *of a real number* x *by a (random) real* $\rho \in [0,1]$ *is defined by*

$$[x]_\rho = \begin{cases} \lceil x \rceil \,, x \geq \rho + \lfloor x \rfloor \\ \lfloor x \rfloor \,, \textit{otherwise} \end{cases} \tag{18}$$

We now define the transformation $T_{\boldsymbol{R}}()$ and the inverse scale factor $\phi(\boldsymbol{R})$.

Definition 5. *Let* r *be* $(u+v)^s$, *where* s *is chosen uniformly at random from the real interval* $[N_1, N_2]$. *Let* r' *be an integer chosen uniformly at random from* $[0, N_3)$. *Finally, let* ρ *be a uniformly chosen real number in* $[0,1]$. *For* $\boldsymbol{R} = (r, r', \rho)$, *put* $T_{\boldsymbol{R}}(a) = [ra]_\rho + r'$ *and put* $\phi(\boldsymbol{R}) = r^{-p/2}$.

In this extended abstract, we don't give N_1, N_2, and N_3 precisely. The following is true:

$$N_1 = \log(8)/\log(u+v) \tag{19}$$
$$N_2 = N_1 + O(\log(M)/\epsilon) \tag{20}$$
$$N_3 = O(M^{1+1/\epsilon}) \tag{21}$$

Let $\hat{d}(a,b)$ denote $\phi(\boldsymbol{R})d(T_{\boldsymbol{R}}(a), T_{\boldsymbol{r}}(b))$, *i.e.*, d acting in the transformed domain.

We apply the below properties about π, d, $T_{\boldsymbol{R}}$, and $\phi(\boldsymbol{R})$ in our proof of the main theorem. Some of the following assume that $v - u \geq 17$. The constants c_1, c_2, c below may depend on p, M, and ϵ, but are bounded uniformly in M and ϵ.

$$d((u+v)a, (u+v)b) \;=\; -(v-u)d(a,b)\;. \tag{22}$$

For all r, all $a \leq b < (u+v)^r$ and all x,

$$|d(a,b)| = |d(a + x(u+v)^r, b + x(u+v)^r)|\;. \tag{23}$$

$$\text{For some } c_2,\ |d(a,b)| \;\leq\; c_2(b-a)^{p/2}\;. \tag{24}$$

$$\text{For some } c_1 > 0 \text{ and some } \eta > 0, \Pr_{\boldsymbol{R}}\left(\left|\hat{d}(a,b)\right| \;\geq\; c_1|b-a|^{p/2}\right) > \eta\;. \tag{25}$$

$$\left.\begin{array}{ll} \text{For some } \gamma_0 > 0, & E_{\boldsymbol{r}}\left[\left|\hat{d}(a,b)\right|\right] = \gamma_0|(b-a)|^{p/2}(1\pm\epsilon)\;. \\ \text{For some } \gamma_1 > 0, & E_{\boldsymbol{r}}\left[\hat{d}^2(a,b)\right] = \gamma_1|(b-a)|^{p}(1\pm\epsilon)\;. \\ \text{For some } \gamma_2 > 0, & E_{\boldsymbol{r}}\left[\hat{d}^4(a,b)\right] = \gamma_2|(b-a)|^{2p}(1\pm\epsilon)\;. \end{array}\right\} \tag{26}$$

We omit the proofs of (22) (homogeneity), (23) (periodicity), (24) (upper bound), and (26) (average) altogether. The proof of (25) (averaged lower bound) consists of two lemmas, that we present without proof.

Proof (of (25), averaged lower bound).
We identify a set S of (a,b) values. We then show in Lemma 3 that $|d(a,b)|$ is large on S and we show in Lemma 4 that the set S itself is big. The result follows.

Definition 6. *Fix u and v. We define a set S of (a,b) values as follows. For each integer s and t, let $S_{s,t}$ consist of the pairs (a,b) such that*

$$\begin{cases} t(u+v)^{s+1} + u(u+v)^s & \leq a < t(u+v)^{s+1} + (u+1)(u+v)^s \\ t(u+v)^{s+1} + (u+v-1)(u+v)^s < b \leq (t+1)(u+v)^{s+1} \end{cases} \tag{27}$$

Let S be the (disjoint) union of all $S_{s,t}$.

One can view this definition geometrically. (See Figure 2.)

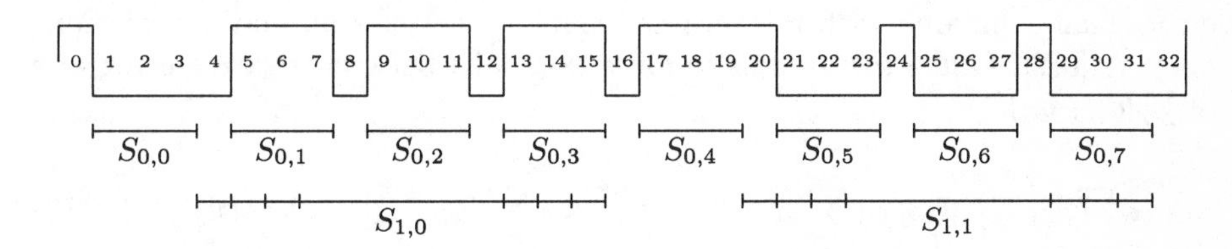

Fig. 2. Geometric view of π (continuous polygonal curve) for $u = 1$ and $v = 3$. The sets $S_{s,t}$ are indicated by segments with vertical ticks. Each element of $S_{s,t}$ is a pair (α, β), indicated in the diagram by two of the vertical ticks near opposite ends of the interval labeled $S_{s,t}$. The discrepancy of π is relatively high over intervals with endpoints in $S_{s,t}$. Note that the pattern π and sets $S_{s,t}$ are self-similar (loosely defined). Elements of $S_{s,0}$ are close to analogs of $S_{0,0} = \{(u, u+v)\}$, scaled-up (by $(u+v)^s$). Elements of $S_{s,t}$ are analogs of elements of $S_{s,0}$, translated (by $t(u+v)^s$)

Lemma 3. *For some constant c_1, for each $(a,b) \in S$, we have*

$$|d(a,b)| \geq c_1 |b-a|^{p/2} . \tag{28}$$

Lemma 4. *Let $a, b < M$ be arbitrary. Then for any u, v there exists η such that* $\Pr\left((T_{\mathbf{R}}(a), T_{\mathbf{R}}(b)) \in S\right) \geq \eta$.

3.4 Algorithm in the Sketch Model

We can use this algorithm in the sketching model, also. Perform $O(\log(1/d)/\epsilon^2)$ parallel repetitions of the following. Given one function $\langle a_i \rangle$, construct the small sketch $\sum_i \sigma_i f(a_i)$. Later, given the two sketches $A = \sum_i \sigma_i f(a_i)$ and $B = \sum_i \sigma_i f(b_i)$, one can reconstruct the L^p difference by outputting a median of means of independent copies of $|A - B|^{2/p}$.

3.5 Toplevel Algorithm, $2 < p \leq 4$

For rational $p \in (2,4)$, a similar algorithm (with similar analysis) approximates

$$A \sum_i |a_i - b_i|^p + B \left(\sum_i |a_i - b_i|^{p/2} \right)^2 \tag{29}$$

with small relative error, whence one can approximate $\sum_i |a_i - b_i|^p$, for $2 < p \leq 4$, with error small compared with $\left(\sum_i |a_i - b_i|^{p/2}\right)^2$. We omit the details.

We now analyze this error guarantee compared with error guarantees of related work. In previous work, there are three types of error guarantees given. In typical sampling arguments, an additive error is produced which is small compared with n or even Mn. The techniques of [4] can be used to approximate $\sum |a_i - b_i|$ when $a_i, b_i \in \{0,1\}$. In this case, the error is guaranteed to be small compared with $\sum |a_i + b_i|$—already a substantial improvement over additive error and the best possible in the original context of [4]. Relative error, *i.e.*, error small compared with the returned value, is better still, and is achievable for $p \in [0,2]$. Our error guarantee for $p > 2$ falls between relative error and $O(\sum |a_i + b_i|)$:

$$\epsilon \sum |a_i - b_i|^p \leq \epsilon \left(\sum |a_i - b_i|^{p/2}\right)^2 \leq \epsilon \sum |a_i + b_i| \leq O(\epsilon M n) . \tag{30}$$

In particular, our error in approximating $\sum |a_i - b_i|^p$ is small compared with $(\sum |a_i - b_i|^p)^2$, so our error gets small as the returned value $\sum |a_i - b_i|^p$ gets small—this is not true for an error bound of, say, $\sum |a_i + b_i|^p$ or $\sum |a_i + b_i|$.

Since $\sum |a_i - b_i|^{p/2} \leq \sum |a_i - b_i|^p \leq \left(\sum |a_i - b_i|^{p/2}\right)^2$, one could also approximate $\sum |a_i - b_i|^p$ by

$$\left(\sum |a_i - b_i|^{p/2}\right)^{3/2} . \tag{31}$$

This will be correct to within the factor $\left(\sum |a_i - b_i|^{p/2}\right)^{1/2}$. Note that our algorithm is an unbiased estimator, *i.e.*, it has the correct mean—an advantage in some contexts; this is not true of $\left(\sum |a_i - b_i|^{p/2}\right)^{3/2}$. Furthermore, our algorithm provides a smooth trade-off between guaranteed error and cost, which is not directly possible with the trivial solution. We hope that, in some applications, our approximation to $\sum |a_i - b_i|^p$ provides information not contained in $\sum |a_i - b_i|^{p/2}$.

4 Discussion

4.1 Relationship with Previous Work

We give an approximation algorithm for, among other cases, $p = 1$ and $p = 2$. The $p = 1$ case was first solved in [7], using different techniques. Our algorithm is less efficient in time and space, though by no more than a power. The case $p = 2$ was first solved in [2,1], and it is easily seen that our algorithm for the case $p = 2$ coincides with the algorithm of [2,1]. Our algorithm is similar to [2,7] at the top level, using the strategy proposed by [2].

4.2 Random-Self-Reducibility

Our proof technique can be regarded as exploitation of a random-self-reduction [6] of the L^p difference. Roughly, a function $f(x)$ is random-self-reducible via (σ, ϕ) if, for random r, $f(x) = \phi(r, f(\sigma(r,x)))$, where the distribution $\sigma(\cdot, x)$ does not depend on x. The function L^p of two streams is random-self-reducible in the sense that $|a_i - b_i| = \frac{1}{r}\left|(ra_i + r') - (rb_i + r')\right|$, where the distribution $(ra_i + r', rb_i + r')$ is only weakly dependent on (a_i, b_i). We omit further discussion due to space considerations.

4.3 Determination of Constants

Our function f involves some constant c, such that $E[(f(a) - f(b))^2] \approx |b-a|^p$, which we do not explicitly provide. This needs to be investigated further. We give a few comments here.

One can approximate c using a randomized experiment. Due to our fairly tight upper and lower bounds for c, we can, using Lemma 1, estimate c reliably as $\left|\hat{d}(a,b)\right| \cdot |b-a|^{-p/2}$. This occurs once for each p, M, n, and ϵ. It is not necessary to do this once for each item or even once for each stream, and one can fix generously large M and n and generously small ϵ to avoid repeating the estimation of c for changes in these values.

In some practical cases, not knowing c may not be a drawback. In practice, as in [4], one may use the measure $\sum |a_i - b_i|^p$ to quantify the difference between two web pages, where a_i is the number of occurrences of feature i in page A and b_i is the number of occurrences of feature i in page B. For example, one may want to keep a list of non-duplicate web pages, where two web pages that are close enough may be deemed to be duplicates. According to this model, there are sociological empirical constants $\hat{c}$ and $\hat{p}$ such that web pages with small value of $\hat{c} \sum |a_i - b_i|^{\hat{p}}$ are considered to be duplicates. To apply this model, one must estimate the parameters $\hat{c}$ and $\hat{p}$ by doing sociological experiments, *e.g.*, by asking human subjects whether they think pairs of webpages, with varying measures of $\hat{c} \sum |a_i - b_i|^{\hat{p}}$ for various values of $\hat{c}$ and $\hat{p}$, are or are not duplicates. If one does not know c, one can simply estimate $\hat{c}/c$ at once by the same sociological experiment.

4.4 Non-grouped Input Representation

Often, in practice, one wants to compare $\langle a_i \rangle$ and $\langle b_i \rangle$ when the values a_i and b_i are represented differently. For example, suppose there are two grocery stores, A and B, that sell the same type of items. Each time either store sells an item it sends a record of this to headquarters in an ongoing stream. Suppose item i sells a_i times in store A and b_i times in store B. Then headquarters is presented with two streams, A and B, such that i appears a_i times in A and b_i times in B; $\sum |a_i - b_i|^p$ measures the extent to which sales differ in the two stores. Unfortunately, we don't see how to apply our algorithm in this situation. Apparently, in order to use our algorithm, each store would have to aggregate sales data and present a_i or b_i, rather than present a_i or b_i non-grouped occurrences of i. The algorithm of [2,1] solves the $p = 2$ case in the non-grouped case, but the problem for other p is important and remains open.

We have recently learned of a possible solution the non-grouped problem. Note that, in general, a solution A in the non-grouped representation yields a solution in the function-value representation, since, on input a_i, an algorithm can simulate A on a_i occurrences of i; this simulation takes time exponential in the size of a_i to process a_i. The proposed solution, however, appears to be of efficiency comparable to ours in the function-value representation, at least in theory, but there may be implementation-related reasons to prefer our algorithm in the grouped case.

References

1. N. Alon, P. Gibbons, Y. Matias, and M. Szegedy. Tracking Join and Self-Join Sizes in Limited Storage. In *Proc. of the 18'th Symp. on Principles of Database Systems*, ACM Press, New York, pages 10–20, 1999.
2. N. Alon, Y. Matias, and M. Szegedy. The space complexity of approximating the frequency moments. In *Proc. of 28'th STOC*, pages 20–29, 1996. To appear in Journal of Computing and System Sciences.
3. N. Alon and J. Spencer. *The Probabilistic Method.* Wiley, 1992.
4. A. Broder, M. Charikar, A. Frieze, and M. Mitzenmacher. Min-wise independent permutations. In *Proc. of the 30'th STOC*, pages 327–336, 1998.
5. Cisco NetFlow, 1998. `http://www.cisco.com/warp/public/732/netflow/`.
6. J. Feigenbaum. Locally random reductions in interactive complexity theory. *DIMACS Series in Discrete Mathematics and Theoretical Computer Science*, vol. 13, pages 73–98. American Mathematical Society, Providence, 1993.
7. J. Feigenbaum, S. Kannan, M. Strauss, and M. Viswanathan. An Approximate L^1-Difference Algorithm for Massive Data Streams. To appear in *Proc. of the 40'th IEEE Symposium on Foundataions of Computer Science*, 1999.
8. J. Feigenbaum and M. Strauss. An Information-Theoretic Treatment of Random-Self-Reducibility. *Proc. of the 14'th Symposium on Theoretical Aspects of Computer Science*, pages 523–534. Lecture Notes in Computer Science, vol. 1200, Springer-Verlag, New York, 1997.
9. P. Gibbons and Y. Matias. Synopsis Data Structures for Massive Data Sets. To appear in *Proc. 1998 DIMACS Workshop on External Memory Algorithms.* DIMACS Series in Discrete Mathematics and Theoretical Computer Science, American Mathematical Society, Providence. Abstract in *Proc. Tenth Symposium on Discrete Algorithms*, ACM Press, New York and Society for Industrial and Applied Mathematics, Philadelphia, pages S909–910, 1999.
10. M. Rauch Henzinger, P. Raghavan, and S. Rajagopalan. Computing on data streams. Technical Report 1998-011, Digital Equipment Corporation Systems Research Center, May 1998.
11. E. Kushilevitz, R. Ostrovsky, Y. Rabani. Efficient Search for Approximate Nearest Neighbor in High Dimensional Spaces. *Proc. of The 30's ACM Symposium on Theory of Computing*, ACM Press, New York, pages 514-523.

Succinct Representations of Model Based Belief Revision*
(Extended Abstract)

Paolo Penna

Dipartimento di Matematica, Università di Roma "Tor Vergata",
penna@mat.uniroma2.it

Abstract. In this paper, following the approach of Gocic, Kautz, Papadimitriou and Selman (1995), we consider the ability of belief revision operators to succinctly represent a certain set of models. In particular, we show that some of these operators are more efficient than others, even though they have the sane model checking complexity. We show that these operators are partially ordered, i.e. some of them are not comparable. We also strengthen some of the results by Cadoli, Donini, Liberatore and Shaerf (1995) by showing that for some of the so called "model based" operators, a polynomial size representation does not exist even if we allow the new knowledge base to have a non polynomial time model checking (namely, either in NP or in co-NP). Finally, we show that Dalal's and Weber's operators can be compiled one into the other via a formalism whose model checking is in NP. All of our results also hold when iterated revision, for one or more of the operators, is considered.

1 Introduction

Several formalisms for knowledge representation and nonmonotonic reasoning have been proposed and studied in the literature. Such formalisms often give rise to intractable problems, even when propositional versions of such formalisms are considered (see [7] for a survey).

Knowledge compilation aims to avoid these difficulties through an off-line process where a given knowledge base is compiled into an equivalent one that supports queries more efficiently. The feasibility of the above approach has been deeply investigated depending on several factors such as: the formalism used for the original and resulting knowledge base, the kind of equivalence we require, and so on (see [4] for a survey). For example, let us consider the propositional version of *circumscription* ($\mathcal{CIRC}$), a well known form of nonmonotonic reasoning introduced in the AI literature in [16, 17]. Informally, $\mathcal{CIRC}(T)$ denotes those truth assignments that satisfy T and that have a "minimal" set of variables mapped into 1. The idea behind minimality is to assume that a fact is false whenever possible. In particular, we represent a truth assignment as a subset m of variables of

* Part of this work has been done while the author was visiting the research center of INRIA Sophia Antipolis (SLOOP project).

H. Reichel and S. Tison (Eds.): STACS 2000, LNCS 1770, pp. 205–216, 2000.

T (those mapped into 1) and we say that m is a *model* if the corresponding truth assignment satisfies T. Then, $\mathcal{CIRC}(T)$ contains only the models of T that are minimal w.r.t. set inclusion (see Sect. 1.2 for a formal definition). Although it is possible to explicitly represent all the models in $\mathcal{CIRC}(T)$, this representation in general has size exponential in the size of T. So, a shorter (implicit) representation is given by the propositional formula T. However, representing the set of models $\mathcal{CIRC}(T)$ simply as T yields an overhead from the computational point of view. For instance, given T and a subset m of its variables, deciding whether $m \in \mathcal{CIRC}(T)$ (*model checking*) is an co-NP-complete problem [2]. Notice that in the classical propositional logic ($\mathcal{PL}$) a formula F simply represents all of its models, so model checking for $\mathcal{PL}$ is clearly in P. Similarly, deciding whether a formula logically follows from $\mathcal{CIRC}(T)$ (*inference*) is a $\prod_2^p$-complete problem [9], while inference for $\mathcal{PL}$ is co-NP-complete. A natural question is therefore: is it possible to "translate" $\mathcal{CIRC}(T)$ into a *propositional formula* F and then use F (instead of T) to solve the model checking problem in time polynomial in $|T|$? Clearly such translation cannot be performed in polynomial time unless P = NP (that is why we need to do it off-line). Additionally, a necessary condition is the size of F to be polynomially bounded in the size of T. A negative answer to this question has been given in [6] where the authors proved that, in general, $|F|$ is not polynomially bounded in $|T|$. Informally, this is due to the fact that $\mathcal{CIRC}$ allows for representations of the information (i.e. a set of models) that are much more "succinct" than any equivalent representation in $\mathcal{PL}$ (see [6] for more formal definitions of what 'equivalent' means).

The above idea of compiling one formalism into another has been extended in [13] where the relative *succinctness* – also known as *compactness* or *space efficiency* – of several propositional logical formalisms has been investigated. The way two formalisms can be compared is the following. A formalism $\mathcal{F}_1$ is more efficient than a formalism $\mathcal{F}_2$ if: (a) $\mathcal{F}_1$ can be compiled into $\mathcal{F}_2$ and (b) $\mathcal{F}_2$ cannot be compiled into $\mathcal{F}_1$, where the compilation requires the new knowledge base being model equivalent and having size polynomial w.r.t. the original one. It is worth observing that, by one hand, *succincteness implies non-compactability.* By the other hand, the converse does not always hold since it might be the case that $\mathcal{F}_1$ and $\mathcal{F}_2$ cannot be compiled one into the other, i.e. they are *not comparable.* A somehow surprising result of [13] is that formalisms having the same model checking time complexity are instead totally ordered in terms of succinctness. In this case, succinctness becomes crucial in choosing one formalism instead of another to represent the knowledge.

Another important aspect of nonmonotonic reasoning is that we have to deal with uncertain and/or incomplete information. Several criteria for updating and/or revising a knowledge base have been proposed [1, 11, 12, 18, 21, 22, 24]. Suppose we have a knowledge base T and a new piece of information, represented by a formula P, is given. It might be the case that T and P are not consistent. In this case the *revision* of T with P, denoted as $T \circ P$, contains those models of P defined by means of a belief revision operator '$\circ$'. The so called *model based* operators define the set of models of $T \circ P$ as those models of

P that are "close" to the models of T. To different definitions of closeness correspond different revision operators. *Syntax based* approaches are instead defined in terms of syntactic operations on the knowledge base T. In general, model based approaches are preferred to syntax based ones because of their *syntax irrelevance*, i.e. revising two logical equivalent knowledge bases T and T' with a formula P always yields the same set of models. Also in this case model checking and inference become harder than in $\mathcal{PL}$ [15, 8] (see also Table 1). This is a first motivation for investigating compilability of belief revision into $\mathcal{PL}$. Moreover, since we are dealing with revision of knowledge, it is often required to explicitly compute a propositional formula T' equivalent to $T \circ P$, that is the revised knowledge base. In such a case it might be desirable not to have an exponential increase in the size of the original knowledge base. Unfortunately, for several revision operators non-compactability results have been proved in [5]. In the same paper also a weaker kind of equivalence has been considered: *query equivalence*. In this case the compilation does not preserve the set of models but just the set of formulas that logically follow. So, it can be used for inference but not for model checking. In Table 1 we summarize both the complexity and the compactability results proved for several belief revision operators, both model and syntax based (Ginsberg's and WIDTIO).

It is interesting to observe that some revision operators and $\mathcal{CIRC}$ have similar properties. For instance, Ginsberg's operator and $\mathcal{CIRC}$ have the same time complexity and the same compactability properties (see [2, 9, 6] for the results on $\mathcal{CIRC}$). It is therefore natural to ask whether this is a chance or not. A first study of relationships between belief revision and $\mathcal{CIRC}$ has been done in [23] where the author remarked similarities between $\mathcal{CIRC}$ and her operator. Subsequently, in [14] the authors pointed out interesting connections between $\mathcal{CIRC}$ and several belief revision operators, thus extending the result of [23]. In particular, they proved that $\mathcal{CIRC}$ can be compactly represented by means of several belief revision operators, i.e. given a propositional formula F, two formulas T and P, of size polynomial w.r.t. $|F|$, exist such that $T \circ P$ is logically equivalent to $\mathcal{CIRC}(F)$. As remarked in [14], this allows to import results from one field into the other. For example, the above mentioned result combined with the non-compactability results of $\mathcal{CIRC}$ can be used to prove several of the negative results in [5]. Also inverse reductions have been investigated, i.e. compiling belief revision into $\mathcal{CIRC}$, but in this case query equivalence (instead of model equivalence) is considered. In [3], among other results, a precise characterizations of compactability properties of Ginsberg's operators is given. In fact, it can be compiled in $\mathcal{CIRC}$ and vice versa. Additionally, such result also holds for the case of iterated revision, i.e. when a polynomial number of revision steps is considered, by making use of the fact that also in this case the model checking is in co-NP [10].

Finally, we remark that all of the non compilability results in [5, 6, 3] and some of those in [13] relies on the standard hypothesis that the polynomial hierarchy does not collapse. Moreover, the results in [6] hold if and only if this hypothesis is true.

Operator	Complexity		Compactability	
	Model Checking [15]	**Inference** [8]	**Model** [5]	**Query** [5]
Ginsberg	co-NP-complete	$\prod_2^p$-complete	No	No
Winslett, Borgida, Forbus, Satoh.	$\sum_2^p$-complete	$\prod_2^p$-complete	No	No
Dalal	$\mathrm{P}^{\mathrm{NP}[O(\log n)]}$-comp.	$\mathrm{P}^{\mathrm{NP}[O(\log n)]}$-comp.	No	Yes
Weber	$\sum_2^p$-complete	$\prod_2^p$-comp. [19] & [8]	No	Yes
WIDTIO	$\sum_2^p$-complete	$\prod_2^p$-comp. [19] & [8]	Yes	Yes

Table 1. Previous results: complexity and compactability of belief revision operators.

1.1 Results of the Paper

In this paper we give a better characterization of (non) compactability properties of belief revision and we provide important connections between such operators and $\mathcal{CIRC}$. We consider the model based revision operators in Table 1 and we compare their space efficiency with that of $\mathcal{CIRC}$, as well as their relative compactness. In particular, we show that, for some model based operators (Winslett's, Borgida's, Forbus's, and Satoh's) belief revision is more difficult to be compiled than $\mathcal{CIRC}$, Ginsberg's, Dalal's or Weber's revision. For the latter two operators we give a precise characterization of their compactability properties. Our results significantly strengthen several non-compactability results in [5] and the results of [15]. Moreover, they provide an intuitive explanation of the (non) compactability results when query equivalence is considered. The results are obtained under the assumption that the polynomial hierarchy does not collapse.

To this aim, we introduce a formalism, denoted as $\overline{\mathcal{CIRC}}$, whose model checking is in NP and is not comparable to $\mathcal{CIRC}$. Roughly speaking, $\overline{\mathcal{CIRC}}$ can be seen as the "complement" of $\mathcal{CIRC}$, i.e. $\overline{\mathcal{CIRC}}$ corresponds to the set of *non minimal models* of a propositional formula. In Fig. 1 we show relationships among revision operators and their space efficiency with respect to $\mathcal{PL}$, $\mathcal{CIRC}$ and $\overline{\mathcal{CIRC}}$, where one way arrows represent the fact that one operator is strictly more succinct then another. The results are consequences of previously known results combined with the following two:

- $\overline{\mathcal{CIRC}}$ can be compiled into model based operators;
- Dalal's and Weber's operators can be compiled into $\overline{\mathcal{CIRC}}$.

As a consequence we have that Winslett's, Borgida's, Forbus's and Satoh's operators are more succinct than all the other operators, and Ginsberg's operator is not comparable to Dalal's or Weber's one. Moreover, Dalal's and Weber's can be reduced each other via $\overline{\mathcal{CIRC}}$. This yields a *precise characterization* of their space efficiency w.r.t. $\mathcal{CIRC}$ and the other belief revision operators. Additionally, the fact that Dalal's and Weber's operators are equivalent to $\overline{\mathcal{CIRC}}$ gives an intuitive explanation of their query compactability properties [5] (see Table 2).

Motivated by the non-compactability results of [5], we attempt to find a trade-off between compactability and the complexity of model checking of the

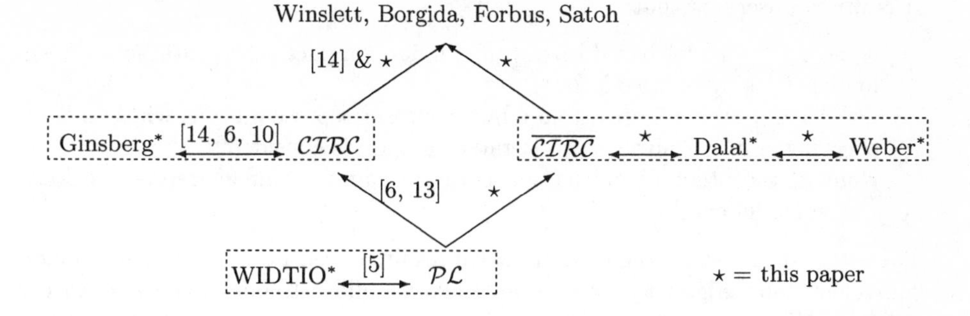

Fig. 1. Results of the paper: the relative succinctness of belief revision operators, where '$*$' means that the result also holds for iterated revision.

knowledge base in which the original one is compiled. In particular, we consider the following question:

> Can we succinctly represent a revised knowledge base by means of $\mathcal{CIRC}$? More generally, can it be compiled into a knowledge base whose model checking is either in NP or in co-NP?

Since model checking for model based operators is harder than any problem in NP or in co-NP, a positive answer to the above question can be used to make model checking easier through an off-line preprocessing of compilation. In Table 2 we summarized the obtained results, which follow from properties of $\mathcal{CIRC}$ and $\overline{\mathcal{CIRC}}$.

Operator	**Compactable into a knowledge base whose model checking is in**	
	NP	co-NP
Ginsberg	No Corollary 3 & [14]	Yes, also iterated [15, 6, 10]
Winslett	No Corollary 3 & [14]	No Corollary 2
Borgida	No Corollary 3 & [14]	No Corollary 2
Satoh	No Corollary 3 & [14]	No Corollary 2
Dalal	Yes, also iterated [5], also Theorem 4	No Corollary 2
Weber	Yes, also iterated [5], also Theorem 4	No Corollary 2

Table 2. Results of the paper: compactability w.r.t. model checking time complexity of the new knowledge base.

It is worth observing that:

- None of the model based operators admits compact representations whose model checking is in co-NP.
- Dalal's and Weber's operators admit compact representations whose model checking is in NP, even when iterated revision is considered.
- None of the other model based operators admits compact representations whose model checking is in NP.

The latter result strengthen the negative result proved in [5] in that no model equivalent knowledge base exists even when we allow its model checking to be either in NP or in co-NP. Additionally, it is not possible to compile a model based revision operator into $\mathcal{CIRC}$, thus implying that the result of [15] (which holds in the case of query equivalence) cannot be extended to model equivalence.

We emphasize that the compactness of belief revision operators, in general, does not seem to depend on either the complexity of inference and model checking or the previously known compactability results. For instance, Winslett's and Weber's ones have the same complexity (see Table 1) while they are ordered in terms of space efficiency (see Fig. 1). Additionally, Dalal's and Weber's, that have different complexity, can be compiled one into the other, instead. The reducibility of those two operators to $\overline{\mathcal{CIRC}}$ also gives an intuitive explanation of their compactability properties (see Table 1) which, actually, are the same as $\overline{\mathcal{CIRC}}$

Due to lack of space some of the proofs of the above results will be omitted or only sketched in this extended abstract.

1.2 Preliminaries

Given a propositional formula F and given a truth assignment m to the variables of F, we say that m is a *model* of F if m satisfies F. Models will be denoted as sets of variables (those mapped into 1). We denote by $\mathcal{M}(F)$ the set of models of F. A *theory* is a set T of propositional formulas. The set of models $\mathcal{M}(T)$ of the theory T is the set of models that satisfy all of the formulas in T. If $\mathcal{M}(F) \neq \emptyset$ then the formula is satisfiable. Similarly, a theory T is consistent if $\mathcal{M}(T) \neq \emptyset$. We use $a \rightarrow b$ and $a \leftrightarrow b$ as a shorthand for $\neg a \vee b$ and $(a \wedge b) \vee (\neg a \wedge \neg b)$, respectively. Given two models m and n, we denote by $m \Delta n$ their symmetric difference. Given a set of sets $\mathcal{S}$, we denote by $\min_{\subseteq} \mathcal{S}$ (respectively, $\max_{\subseteq} \mathcal{S}$), the minimal (respectively, maximal) subset of $\mathcal{S}$ w.r.t. set inclusion. The *circumscription* of a propositional formula F (denoted by $\mathcal{CIRC}(F)$) is defined as

$$\mathcal{CIRC}(F) \doteq \{m \in \mathcal{M}(F) \mid \forall m' \subset m, m' \notin \mathcal{M}(F)\} = \min_{\subseteq} \mathcal{M}(F).$$

Given a theory T and a propositional formula P, we denote by $T \circ P$ the theory T revised with P according to some belief revision operator $\circ$. We distinguish the following belief revision operators:

Ginsberg. Let $\mathcal{W}(T, P) \doteq \max_{\subseteq} \{T' \subseteq T \mid T' \cup \{P\} \not\models \perp\}$. Then
$T \circ_G P \doteq \{T' \cup \{P\} \mid T' \in \mathcal{W}(T, P)\}$.

Winslett. Let $\mu(m, P) \doteq \min_{\subseteq} \{m \Delta n \mid n \in \mathcal{M}(P)\}$. Then,
$\mathcal{M}(T \circ_{Win} P) \doteq \{n \in \mathcal{M}(P) \mid \exists m \in \mathcal{M}(T) : m \Delta n \in \mu(m, P)\}$.

Borgida. It is defined as $\circ_{Win}$ if $T \cup \{P\}$ is not consistent, and it is defined as $T \cup \{P\}$ otherwise.

Forbus. For any two models m_1 and m_2 let $d(m_1, m_2) = |m_1 \Delta m_2|$. Also let $k_{m,P} = \min\{d(m, n) : n \in \mathcal{M}(P)\}$. Then,
$\mathcal{M}(T \circ_F P) \doteq \{n \in \mathcal{M}(P) \mid \exists m \in \mathcal{M}(T) : d(m, n) = k_{m,P}\}$.

Satoh. Let $\delta(T, P) \doteq \min_{\subseteq}\{\bigcup_{m \in M(P)} \mu(m, P)\}$. Then,
$\mathcal{M}(T \circ_S P) \doteq \{n \in \mathcal{M}(P) \mid \exists m \in \mathcal{M}(T) : m \Delta n \in \delta(T, P)\}$.

Dalal. Let $k_{T,P} = \min\{k_{m,P} \mid m \in \mathcal{M}(T)\}$. Then,
$\mathcal{M}(T \circ_D P) \doteq \{n \in \mathcal{M}(P) \mid \exists m \in \mathcal{M}(T) : d(m, n) = k_{T,P}\}$.

Weber. Let $\Omega = \bigcup \delta(T, P)$, i.e. Ω contains all of the variables appearing on a minimal difference between models of T and model of P. Then,
$\mathcal{M}(T \circ_{Web} P) \doteq \{n \in \mathcal{M}(P) \mid \exists m \in \mathcal{M}(T) : m \Delta n \subseteq \Omega\}$.

An *advise taking Turing machine* is a Turing machine that can access an advice $a(n)$, i.e. an "oracle" whose output depends *only* on the size n of the input. The class NP/poly is the class of those languages that are accepted by a nondeterministic Turing machine with an advice of polynomial size (see [20] for a formal definition). The class co-NP/poly is similarly defined. In [25] non-uniform classes such as NP/poly and the polynomial hierarchy have been related. In particular, it has been proved that if NP $\subseteq$ co-NP/poly or co-NP $\subseteq$ NP/poly then the polynomial hierarchy (denoted by PH) collapses at the third level, i.e. PH $= \sum_3^p$ (see [20] for a formal definition of those concepts), which is considered *very unlikely* in the complexity community.

2 The Complemented Circumscription and Its Properties

In this section we introduce $\overline{\mathcal{CIRC}}$ and state its basic (non) compactability properties that will be used in the rest of the paper. To this aim we first introduce a model equivalence preserving reduction used in [3] and we assume that a knowledge base K in a formalism $\mathcal{F}$ represents a set of models $\mathcal{F}(K)$.

Definition 1 ([3]). *Given two logical formalisms $\mathcal{F}_1$ and $\mathcal{F}_2$, $\mathcal{F}_1 \mapsto \mathcal{F}_2$ if the following holds: for each knowledge base K_1 in $\mathcal{F}_1$, there exists a knowledge base K_2 in $\mathcal{F}_2$ and a polynomial time computable function g_{K_1} such that (i) for any set of variables m_1, $m_1 \in \mathcal{F}_1(K_1) \Leftrightarrow g_{K_1}(m_1) \in \mathcal{F}_2(K_2)$; (ii) $|K_2|$ is polynomially bounded in $|K_1|$.*

The above definition implies that, once we have computed (off-line) the formula K_2, we can decide whether m_1 is a model of K_1, by checking if $g_{K_1}(m_1)$ is a model of K_2. Additionally, $g_{K_1}(m_1)$ can be computed in polynomial time. Finally, the '$\mapsto$' relation is transitive.

Definition 2. *We denote by $\overline{\mathcal{CIRC}}(F)$ the set of non minimal models of a propositional formula F, that is*

$$\overline{\mathcal{CIRC}}(F) = \mathcal{M}(F) \setminus \mathcal{CIRC}(F) = \{m \in \mathcal{M}(F) \mid \exists m' \in \mathcal{M}(F) : m' \subset m\}.$$

Lemma 1 ([6, 13]). *For any n a propositional formula F_n (of size polynomial in n) exists such that for every n-variables 3CNF propositional formula f there exists a model m_f (computable in polynomial time) such that*

$$f \text{ is unsatisfiable} \Leftrightarrow m_f \in \mathcal{CIRC}(F_n).$$

The above lemma states that the circumscription of a formula F_n is able to capture all of the unsatisfiable 3CNF formulas with n variables (notice that F_n depends only on n). As a consequence we obtain the following result, whose proof is similar to non compilability proofs given in [6, 13].

Theorem 1. *The following hold: (i) $\mathcal{CIRC} \mapsto \overline{\mathcal{CIRC}} \Rightarrow$ co-NP $\subseteq$ NP/poly; (ii) $\overline{\mathcal{CIRC}} \mapsto \mathcal{CIRC} \Rightarrow$ NP $\subseteq$ co-NP/poly.*

The above theorem can be easily generalized to any two formalisms whose model checking is in co-NP and NP, respectively. Thus, the following corollary holds.

Corollary 1. *Let $\mathcal{F}_{\text{co-NP}}$ and $\mathcal{F}_{\text{NP}}$ be any two formalism whose model checking is in co-NP and NP, respectively. Unless the polynomial hierarchy collapses at the third level, the following two hold: (i) $\mathcal{CIRC} \not\mapsto \mathcal{F}_{\text{NP}}$; (ii) $\overline{\mathcal{CIRC}} \not\mapsto \mathcal{F}_{\text{co-NP}}$.*

In the rest of the paper we will make use of the above result and thus we will always assume PH $\neq \sum_3^p$.

3 Reducing $\overline{\mathcal{CIRC}}$ to Belief Revision

In this section we provide some reductions from $\overline{\mathcal{CIRC}}$ to any of the model based belief revision operators. As a consequence we have that none of such operators can be compactly represented by $\mathcal{CIRC}$ or by $\circ_G$.

Theorem 2. *$\overline{\mathcal{CIRC}} \mapsto \circ_{Win}$, $\overline{\mathcal{CIRC}} \mapsto \circ_B$, $\overline{\mathcal{CIRC}} \mapsto \circ_F$.*

Proof. We will prove the theorem only for the $\circ_{Win}$ operator, since the proof can be easily adapted to the other two operators. Let F be a propositional formula over the variables $x_1, \ldots, x_n$. We show that two formulas T and P of polynomial size exist such that $\mathcal{M}(T \circ_{Win} P) = \overline{\mathcal{CIRC}}(F)$.

Let $y_1, \ldots, y_n$ be a set of new variables in correspondence one-to-one with $x_1, \ldots, x_n$ and let m^y be the set $\{y_i | x_i \in m\}$. We construct two formulas T and P over the set of variables $x_1, \ldots, x_n, y_1, \ldots, y_n$ such that

$$\mathcal{M}(T) = \{m_1 \cup m_2^y \mid m_1, m_2 \in \mathcal{M}(F), m_2 \subset m_1\}$$

and $\mathcal{M}(P) = \mathcal{M}(F)$. Let us observe that if no two models $m_1, m_2 \in \mathcal{M}(F)$ exist such that $m_2 \subset m_1$, then T is not satisfiable. We will see in the sequel how to deal with that case. Thus, let us suppose $\mathcal{M}(T) \neq \emptyset$ and let

$$T = F \wedge F[x_i/y_i] \wedge \underbrace{\neg \left(\bigwedge_{i=1}^{n} x_i \rightarrow y_i \right)}_{m_1 \not\subseteq m_2} \wedge \underbrace{\left(\bigwedge_{i=1}^{n} y_i \rightarrow x_i \right)}_{m_2 \subseteq m_1}$$

and $P = F \wedge \bigwedge_{i=1}^{n} \neg y_i$. We now prove that $m \in \mathcal{M}(T \circ_{Win} P) \Leftrightarrow m \in \overline{\mathcal{CIRC}}(F)$.

($\Rightarrow$) By the definition of $\circ_{Win}$ we have that a model $m^T \in \mathcal{M}(T)$ exists such that $m \in \mu(m^T, P)$. Let $m^T = m_1 \cup m_2^y$, where $m_1, m_2 \in \mathcal{M}(F)$ and $m_2 \subset m_1$. Let us first observe that, since m does not contain any variable y_i,

$$m\Delta m^T = (m\Delta m_1) \cup m_2^y.$$

We now prove that $m = m_1$. Suppose, by contradiction, that $m\Delta m_1 \neq \emptyset$. Then, $m_1 \Delta m^T = m_2^y \subset m\Delta m^T$, which implies that $m \notin \mu(m^T, P)$, thus a contradiction. So, $m_2 \subset m_1 = m$, that is $m \in \overline{\mathcal{CIRC}}(F)$.

($\Leftarrow$) There exists $m_2 \in \mathcal{M}(F)$ such that $m_2 \subset m$. Let $m^T = m \cup m_2^y$. Clearly $m^T \in \mathcal{M}(T)$. Suppose by contradiction that $m \notin \mu(m^T, P)$. Thus, an $m_1 \in \mathcal{M}(P)$ exists such that $m_1 \Delta m^T = (m_1 \Delta m) \cup m_2^y \subset m\Delta m^T = m_2^y$, thus a contradiction.

We now consider the case in which no two models $m_1, m_2 \in \mathcal{M}(F)$ exist such that $m_2 \subset m_1$. To this aim, we have to slightly modify the above construction and consider the formula $F' = F \vee \bigwedge_{i=1}^{n+1} x_i$, where x_{n+1} is a new variable. Let T' and P' be the formulas obtained by replacing F with F' in the definition of T and P, respectively. Let $\overline{x} = \{x_1, \ldots, x_{n+1}\}$. We then have that

$$\mathcal{M}(T' \circ_{Win} P') = \overline{\mathcal{CIRC}}(F') = \overline{\mathcal{CIRC}}(F) \cup \{\overline{x}\}.$$

Finally, the above reduction also apply to $\circ_F$, while it can be easily adapted for $\circ_B$ (it suffices to guarantee that $T' \wedge P'$ is not consistent). Hence, the theorem follows.

Theorem 3. $\overline{\mathcal{CIRC}} \mapsto \circ_D$, $\overline{\mathcal{CIRC}} \mapsto \circ_{Web}$, $\overline{\mathcal{CIRC}} \mapsto \circ_S$.

Proof. (*sketch of*) First of all we slightly modify the formulas T and P of Theorem 2 as follows:

$$T = F' \wedge F'[x_i/y_i] \wedge \underbrace{\neg\left(\bigwedge_{i=1}^{n+1} x_i \rightarrow y_i\right)}_{m_1 \not\subseteq m_2} \wedge \underbrace{\left(\bigwedge_{i=1}^{n+1} y_i \rightarrow x_i\right)}_{m_2 \subseteq m_1} \wedge \left(\bigwedge_{i=1}^{n+1} \neg y_i \leftrightarrow z_i\right),$$

and $P = F' \wedge \bigwedge_{i=1}^{n+1} \neg y_i \bigwedge_{i=1}^{n+1} \neg z_i$, where F' is defined as in the proof of Theorem 2.
In the case of $\circ_D$, the proof is a consequence of the following claims:

Claim 1: $k_{T,P} = n + 1$.
Claim 2: *For all* $m \in \overline{\mathcal{CIRC}}(F)$, *there exists* $m^T \in \mathcal{M}(T)$ *such that* $d(m, m^T) = n + 1$.
Claim 3: *For any* $m \in \mathcal{CIRC}(F)$ *and for all* $m^T \in \mathcal{M}(T)$, $d(m, m^T) > n + 1$.

As far as $\circ_{Web}$ and $\circ_S$ concerns, we first observe that $\delta(T, P)$ does not contain any variable x_i. Moreover, it is easy to see that, for any $n \in \mathcal{CIRC}(F)$, and for any $m^T \in \mathcal{M}(T)$, $n\Delta m^T$ contains at least one variable x_i. This proves the theorem.

4 (Non) Compactability of Model Based Revision

In this section we consider the problem of compiling the revised knowledge base into a model equivalent one that has model checking either in NP or co-NP. To this aim we will denote by $\mathcal{F}_{\text{NP}}$ and $\mathcal{F}_{\text{co-NP}}$ any two formalisms[1] whose model checking is in NP and co-NP, respectively.

Let us first observe that an immediate consequence of the reductions given in Sect. 3 and of Corollary 1 is the following fact.

Corollary 2. *For any* $\circ \in \{\circ_{Win}, \circ_B, \circ_F, \circ_S, \circ_D, \circ_{Web}\}$, $\circ \not\mapsto \mathcal{F}_{\text{co-NP}}$.

The above result implies that such operators cannot be represented by means of $\mathcal{CIRC}$. Motivated by this fact we ask whether it is possible to obtain compact representations of $\mathcal{M}(T \circ P)$ by means of F_{NP}. In this case, we show that the situation is more tangled.

We first consider Dalal's and Weber's revision and show that they admit a compact representation by means of $\overline{\mathcal{CIRC}}$.

The main idea of the reductions is that both $k_{T,P}$ and Ω can be represented in polynomial space ($k_{T,P}$ is an integer and $|\Omega| \leq n$). Moreover, once those two entities have been computed (off-line), then the problem of deciding $m \in \mathcal{M}(T \circ P)$ is in NP for both the operators.

Theorem 4. $\circ_D \mapsto \overline{\mathcal{CIRC}}$, $\circ_{Web} \mapsto \overline{\mathcal{CIRC}}$.

The above result can be easily extended to the case of a polynomial number of revision steps. Notice that a different proof can be derived by making use of the fact that such two operators are query compactable [5].

We now consider the other revision operators. To this aim we combine the results proved in Sect. 3 with the results given in [14]. In particular we exploit the fact that such operators can be used to represent $\mathcal{CIRC}$.

Theorem 5 ([14]). *For any* $\circ \in \{\circ_{Win}, \circ_B, \circ_F, \circ_S\}$, $\mathcal{CIRC} \mapsto \circ$

The above theorem combined with Corollary 1 yields the following result.

Corollary 3. *For any* $\circ \in \{\circ_{Win}, \circ_B, \circ_F, \circ_S\}$, $\circ \not\mapsto \mathcal{F}_{\text{NP}}$.

5 Succinctness of Belief Revision

We compare the space efficiency of the belief revision operators and consider the problem of compiling one operator into another. By combining our results with previously known results we will obtain the partial ordering shown in Fig. 1.

To this aim we first introduce the following notation.

[1] In this case the term formalism is quite general, since it refers to any representation of a set of models.

Definition 3. *For any two logical formalisms* $\mathcal{F}_1$ *and* $\mathcal{F}_2$*: (i)* $\mathcal{F}_1 \prec \mathcal{F}_2$ *if both* $\mathcal{F}_1 \mapsto \mathcal{F}_2$ *and* $\mathcal{F}_2 \not\mapsto \mathcal{F}_1$*; (ii)* $\mathcal{F}_1 \approx \mathcal{F}_2$ *if* $\mathcal{F}_1 \mapsto \mathcal{F}_2$ *and* $\mathcal{F}_2 \mapsto \mathcal{F}_1$*; (iii)* $\mathcal{F}_1 \not\approx \mathcal{F}_2$ *if both* $\mathcal{F}_1 \not\mapsto \mathcal{F}_2$ *and* $\mathcal{F}_2 \not\mapsto \mathcal{F}_1$.

All of the following results are easy consequences of Theorem 3, Theorem 4 and Corollary 3. We first compare $\circ_D$ and $\circ_{Web}$ operators with the other model based ones. All of the results also hold for a polynomial number of revision steps of $\circ_D$ or $\circ_{Web}$.

Corollary 4. *For any* $\circ \in \{\circ_{Win}, \circ_B, \circ_F, \circ_S\}$, $\circ_D \approx \circ_{Web} \approx \overline{\mathcal{CIRC}} \prec \circ$.

Corollary 5. *For any* $\circ \in \{\circ_D, \circ_{Web}\}$, $\mathcal{CIRC} \not\approx \circ$ *and* $\circ_G \not\approx \circ$.

6 Conclusions and Open Problems

We have shown that belief revision operators with the same model checking and inference complexity have different behaviours in terms of compilability and space efficiency. We precisely characterized the space efficiency of $\circ_D$ and $\circ_{Web}$ which, following the definitions of [3], are model-NP-complete. Moreover, our results combined with those in [14] imply that $\circ_{Win}$, $\circ_B$, $\circ_F$ and $\circ_S$ are both model-NP-hard and model-co-NP-hard.

The first problem left open is that of finding similar characterizations for the latter operators, as well as that of understanding their relative space efficiency. More generally, it could be interesting to investigate relationships with other formalisms considered in [13, 3] such as default logic, model preference and autoepistemic logic. Furthermore, compactability results for the case of iterated revision are not known. Do these operators became even harder to be compacted when more than one step of revision is considered?

It is interesting to observe that a different situation occurs when query equivalence is considered. Indeed, in [14] the authors proved that in this case $\circ_G$ and $\circ_S$ can be reduced one to the other and $\circ_{Win}$ can be reduced to both.

Acknowledgments. I am grateful to Marco Cadoli for introducing me to the area of knowledge compilation and for several useful discussions. I am also grateful to Riccardo Silvestri for his valuable comments on a preliminary version of this work.

References

[1] A. Borgida. Language features for flexible handling of exceptions in information systems. *ACM Transactions on Database Systems*, 10:563–603, 1981.
[2] M. Cadoli. The complexity of model checking for circumscriptive formulae. *Information Processing Letters*, 44:113–118, 1992.
[3] M. Cadoli, F. M. Donini, P. Liberatore, and M. Schaerf. Comparing space efficiency of propositional knowledge representation formalisms. In *Proc. of KR-96*, pages 100–109, 1996.

[4] M. Cadoli and F.M. Donini. A survey on knowledge compilation. *AI Communications-The European Journal for Artificial Intelligence*, 10:137–150, 1998.

[5] M. Cadoli, F.M. Donini, P. Liberatore, and M. Shaerf. The size of a revised knowledge base. In *Proc. of PODS-95*, pages 151–162, 1995.

[6] M. Cadoli, F.M. Donini, M. Shaerf, and R. Silvestri. On compact representations of propositional circumscription. *Theoretical Computer Science*, 182:183–202, 1995.

[7] M. Cadoli and M. Shaerf. A survey on complexity results for nonmonotonic logics. *Journal of Logic Programming*, 17:127–160, 1993.

[8] T. Eiter and G. Gottlob. On the complexity of propositional knowledge base revision, updates and counterfactuals. *Artificial Intelligence*, 57:227–270, 1992.

[9] T. Eiter and G. Gottlob. Propositional circumscription and extended closed world reasoning are $\prod_2^p$-complete. *Theoretical Computer Science*, 114:231–245, 1993.

[10] T. Eiter and G. Gottlob. The complexity of nested counterfactuals and iterated knowledge base revisions. *Journal of Computer and System Sciences*, 53:497–512, 1996.

[11] R. Fagin, J.D. Ullman, and M.Y. Vardi. On the semantics of updates in databases. In *Proc. of PODS-83*, pages 352–365, 1983.

[12] M.L. Ginsberg. Counterfactuals. *Artificial Intelligence*, 30:35–79, 1986.

[13] G. Gogic, H. Kautz, C. Papadimitriou, and B. Selman. The comparative linguistics of knowledge representation. In *Proc. of IJCAI-95*, pages 862–869, 1995.

[14] P. Liberatore and M. Shaerf. Relating belief revision and circumscription. In *Proc. of IJCAI-95*, pages 1557–1563, 1995.

[15] P. Liberatore and M. Shaerf. The complexity of model checking for belief revision and update. In *Proc. of AAAI-96*, pages 556–561, 1996.

[16] J. McCarthy. Circumscription - a form of non-monotonic reasoning. *Artificial Intelligence Journal*, 13:27–39, 1980.

[17] J. McCarthy. Applications of circumscription to formalizing common sense knowledge. *Artificial Intelligence Journal*, 28:89–116, 1986.

[18] B. Nebel. Belief revision and default reasoning: Syntax-based approaches. In *Proc. of KR-91*, pages 417–428, 1991.

[19] B. Nebel. How hard is it to revise a belief base? In *Handbook of Defeasible Reasoning and Uncertainty Management Systems, Vol. 3: Belief Change*, pages 77–145, 1998.

[20] C.H. Papadimitriou. *Computational complexity.* Addison Wesley, 1993.

[21] K. Satoh. Nonmonotonic reasoning by minimal belief revision. In *Proc. of FGCS-88*, pages 455–462, 1988.

[22] A. Weber. Updating propositional formulas. In *Proc. of the First Conference on Expert Database Systems*, pages 487–500, 1986.

[23] M. Winslett. Sometimes updates are circumscription. In *Proc. of IJCAI-89*, pages 859–863, 1989.

[24] M. Winslett. *Updating logical databases.* Cambridge University Press, 1990.

[25] H.P. Yap. Some consequences of non-uniform conditions on uniform classes. *Theoretical Computer Science*, 26:287–300, 1983.

Logics Capturing Local Properties

Leonid Libkin*

Bell Laboratories, 600 Mountain Avenue, Murray Hill, NJ 07974, USA.
libkin@research.bell-labs.com

Abstract. Well-known theorems of Hanf's and Gaifman's establishing locality of first-order definable properties have been used in many applications. These theorems were recently generalized to other logics, which led to new applications in descriptive complexity and database theory. However, a logical characterization of local properties that correspond to Hanf's and Gaifman's theorems, is still lacking. Such a characterization only exists for structures of bounded valence.
In this paper, we give logical characterizations of local properties behind Hanf's and Gaifman's theorems. We first deal with an infinitary logic with counting terms and quantifiers, that is known to capture Hanf-locality on structures of bounded valence. We show that testing isomorphism of neighborhoods can be added to it without violating Hanf-locality, while increasing its expressive power. We then show that adding local second-order quantification to it captures precisely all Hanf-local properties. To capture Gaifman-locality, one must also add a (potentially infinite) case statement. We further show that the hierarchy based on the number of variants in the case statement is strict.

1 Introduction

It is well known that first-order logic (FO) only expresses local properties. Two best known formal results stating locality of FO are Hanf's and Gaifman's theorems [12,8]. They both found numerous applications in computer science, due to the fact that they are among relatively few results in first-order model theory that extend to *finite* structures. Gaifman's theorem itself works for both finite and infinite structures, while for Hanf's theorem an extension to finite structures was formulated by Fagin, Stockmeyer, and Vardi [7].

More recently, the statements underlying Hanf's and Gaifman's theorems have been abstracted from the statements of the theorems, and used in their own right. In essence, Hanf's theorem states that two structures cannot be distinguished by sentences of quantifier rank k whenever they realize the same multiset of d-neighborhoods of points; here d depends only on k. Gaifman's theorem states that in a given structure, two tuples cannot be distinguished by formulae of quantifier rank k whenever d-neighborhoods of these tuples are isomorphic; again d is determined by k.

* Part of this work was done while visiting INRIA.

H. Reichel and S. Tison (Eds.): STACS 2000, LNCS 1770, pp. 217–229, 2000.

It was shown that Hanf's theorem is strictly stronger than Gaifman's, and that both apply to a variety of logics that extend FO with counting mechanisms and limited infinitary connectives [11,14,15,19,22]. Since the complexity class TC^0 (with the appropriate notion of uniformity) can be captured by FO with counting quantifiers [1], these results found applications in descriptive complexity, where they were used to prove lower bounds for logics coming very close to capturing TC^0 [6,21]. They were also applied in database theory, where they were used to prove expressivity bounds for relational query languages with aggregation [4,15] that correspond to practical query languages such as SQL. For applications to automata, see [24].

The abstract notions of locality were themselves characterized only on finite structures of bounded valence (e.g., for graphs of fixed maximum degree). The characterization for Hanf-locality uses a logic $\mathcal{L}^*_{\infty\omega}(\mathbf{C})$ introduced in [19] as a counterpart of a finite variable logic $\mathcal{L}^\omega_{\infty\omega}$. While $\mathcal{L}^\omega_{\infty\omega}$ subsumes a number of fixpoint logics and is easier to study, $\mathcal{L}^*_{\infty\omega}(\mathbf{C})$ subsumes a number of counting extensions of FO (such as FO with counting quantifiers [17], FO with unary generalized quantifiers [13,18], FO with unary counters [2]) and is quite easy to deal with. A result in [14] states that Hanf-local properties on structures of bounded valence are precisely those definable in $\mathcal{L}^*_{\infty\omega}(\mathbf{C})$.

The question naturally arises whether this continues to hold for arbitrary finite structures. We show in this paper that this is not the case. We do so by first finding a simple direct proof of Hanf-locality of $\mathcal{L}^*_{\infty\omega}(\mathbf{C})$, and then using it to show that adding new atomic formulae testing isomorphism of neighborhoods of a fixed radius does not violate Hanf-locality, while strictly increasing the expressive power. We next define a logic that captures precisely the Hanf-local properties. It is obtained by adding *local second-order* quantification to $\mathcal{L}^*_{\infty\omega}(\mathbf{C})$. That is, second-order quantifiers bind predicates that are only allowed to range over fixed radius neighborhoods of free first-order variables. We will also show that this amounts to adding arbitrarily powerful computations to $\mathcal{L}^*_{\infty\omega}(\mathbf{C})$ as long as they are bound to some neighborhoods.

For Gaifman-locality, a characterization theorem in [14] stated that it is equivalent, over structures of bounded valence, to first-order definition by cases. That is, there are $m > 0$ classes of structures and m FO formulae φ_i such that over the ith class, the given property is described by φ_i. Again, this falls short of a general characterization. We show that over the class of all finite structures (no restriction on valence), Gaifman-locality is equivalent to definition by cases, where the number of classes can be infinite. Furthermore, the hierarchy given by the number of those classes (that is, the number of cases) is strict.

Organization. Section 2 introduces notations and notions of locality. Section 3 gives a new simple proof of Hanf-locality of $\mathcal{L}^*_{\infty\omega}(\mathbf{C})$ which is then used to show that adding tests for neighborhood isomorphism preserves locality. Section 4 characterizes Hanf-local properties as those definable in $\mathcal{L}^*_{\infty\omega}(\mathbf{C})$ with local second-order quantification. Section 5 characterizes Gaifman-local properties as those definable by (finite or infinite) case statements, and shows the strictness of the hierarchy. All proofs can be found in the full version [20].

2 Notation

Finite structures and neighborhoods All structures are assumed to be *finite.* A relational signature σ is a set of relation symbols $\{R_1, ..., R_l\}$, with associated arities $p_i > 0$. A σ-structure is $\mathcal{A} = \langle A, R_1^{\mathcal{A}}, \ldots, R_l^{\mathcal{A}}\rangle$, where A is a finite set, and $R_i^{\mathcal{A}} \subseteq A^{p_i}$ interprets R_i. The class of finite σ-structures is denoted by STRUCT[σ]. When there is no confusion, we write R_i in place of $R_i^{\mathcal{A}}$. Isomorphism is denoted by $\cong$. The carrier of a structure $\mathcal{A}$ is always denoted by A and the carrier of $\mathcal{B}$ is denoted by B.

Given a structure $\mathcal{A}$, its *Gaifman graph* [5,8,7] $\mathcal{G}(\mathcal{A})$ is defined as $\langle A, E\rangle$ where (a, b) is in E iff there is a tuple $\vec{c} \in R_i^{\mathcal{A}}$ for some i such that both a and b are in $\vec{c}$. The distance $d(a, b)$ is defined as the length of the shortest path from a to b in $\mathcal{G}(\mathcal{A})$; we assume $d(a, a) = 0$. If $\vec{a} = (a_1, \ldots, a_n)$ and $\vec{b} = (b_1, \ldots, b_m)$, then $d(\vec{a}, \vec{b}) = \min_{ij} d(a_i, b_j)$. Given $\vec{a}$ over A, its *r-sphere* $S_r^{\mathcal{A}}(\vec{a})$ is $\{b \in A \mid d(\vec{a}, b) \leq r\}$. Its *r-neighborhood* $N_r^{\mathcal{A}}(\vec{a})$ is defined as a structure in the signature that extends σ with n new constant symbols:

$$\langle S_r^{\mathcal{A}}(\vec{a}), R_1^{\mathcal{A}} \cap S_r^{\mathcal{A}}(\vec{a})^{p_1}, \ldots, R_l^{\mathcal{A}} \cap S_r^{\mathcal{A}}(\vec{a})^{p_l}, a_1, \ldots, a_n\rangle$$

That is, the carrier of $N_r^{\mathcal{A}}(\vec{a})$ is $S_r^{\mathcal{A}}(\vec{a})$, the interpretation of the σ-relations is inherited from $\mathcal{A}$, and the n extra constants are the elements of $\vec{a}$. If $\mathcal{A}$ is understood, we write $S_r(\vec{a})$ and $N_r(\vec{a})$.

If $\mathcal{A}, \mathcal{B} \in$ STRUCT[σ], and there is an isomorphism $N_r^{\mathcal{A}}(\vec{a}) \to N_r^{\mathcal{B}}(\vec{b})$ (that sends $\vec{a}$ to $\vec{b}$), we write $\vec{a} \approx_r^{\mathcal{A},\mathcal{B}} \vec{b}$. If $\mathcal{A} = \mathcal{B}$, we write $\vec{a} \approx_r^{\mathcal{A}} \vec{b}$.

Given a tuple $\vec{a} = (a_1, \ldots, a_n)$, we write $\vec{a}c$ for the tuple $(a_1, \ldots, a_n, c)$.

The quantifier rank of a formula is denoted by $\mathsf{qr}(\cdot)$.

Hanf's and Gaifman's theorems An *m-ary query* on σ-structures, Q, is a mapping that associates to each $\mathcal{A} \in$ STRUCT[σ] a structure $\langle A, S\rangle$, where $S \subseteq A^m$. We always assume that queries are invariant under isomorphisms. We write $\vec{a} \in Q(\mathcal{A})$ if $\vec{a} \in S$, where $\langle A, S\rangle = Q(\mathcal{A})$. A query Q is definable in a logic $\mathcal{L}$ if there exists an $\mathcal{L}$ formula $\varphi(x_1, \ldots, x_m)$ such that $Q(\mathcal{A}) = \langle A, \{\vec{a} \mid \mathcal{A} \models \varphi(\vec{a})\}\rangle$. If $m = 0$, then Q is naturally associated with a subclass of STRUCT[σ] and definability means definability by a sentence of $\mathcal{L}$.

Definition 1. (cf. [4,14]) *An m-ary query Q, $m \geq 1$, is called* Gaifman-local *if there exists a number $r \geq 0$ such that, for any structure $\mathcal{A}$ and any $\vec{a}, \vec{b} \in A^m$*

$$\vec{a} \approx_r^{\mathcal{A}} \vec{b} \quad \textit{implies} \quad \vec{a} \in Q(\mathcal{A}) \;\; \textit{iff} \;\; \vec{b} \in Q(\mathcal{A}).$$

The minimum such r is called the locality rank *of Q, and is denoted by* $\mathsf{lr}(Q)$. □

Theorem 1 (Gaifman). *Every FO formula $\varphi(x_1, \ldots, x_m)$ defines a Gaifman-local query Q with $\mathsf{lr}(Q) \leq (7^{\mathsf{qr}(\varphi)} - 1)/2$.*

The statement of Gaifman's theorem actually provides more information about FO definable properties; it states that every formula is a Boolean combination of sentences of a special form and open formulae in which quantifiers are restricted to certain neighborhoods. However, it is the above statement that is used in most applications for proving expressivity bounds, and it also extends beyond FO. Note also that better bounds of the order $O(2^{\mathsf{qr}(\varphi)})$ are known for $\mathsf{lr}(Q)$, see [19].

For $\mathcal{A}, \mathcal{B} \in \text{STRUCT}[\sigma]$, we write $\mathcal{A} \leftrightarrows_d \mathcal{B}$ if the multisets of isomorphism types of d-neighborhoods of points are the same in $\mathcal{A}$ and $\mathcal{B}$. That is, $\mathcal{A} \leftrightarrows_d \mathcal{B}$ if there exists a bijection $f : A \to B$ such that $N_d^{\mathcal{A}}(a) \cong N_d^{\mathcal{B}}(f(a))$ for every $a \in A$. We also write $(\mathcal{A}, \vec{a}) \leftrightarrows_d (\mathcal{B}, \vec{b})$ if there is a bijection $f : A \to B$ such that $N_d^{\mathcal{A}}(\vec{a}c) \cong N_d^{\mathcal{B}}(\vec{b}f(c))$ for every $c \in A$.

Definition 2 (Hanf-locality). (see [12,7,14]) *An m-ary query Q, $m \geq 0$, is called* Hanf-local *if there exist a number $d \geq 0$ such that for any two structures $\mathcal{A}, \mathcal{B}$ and any $\vec{a} \in A^m, \vec{b} \in B^m$,*

$$(\mathcal{A}, \vec{a}) \leftrightarrows_d (\mathcal{B}, \vec{b}) \quad \textit{implies} \quad \vec{a} \in Q(\mathcal{A}) \ \textit{ iff } \ \vec{b} \in Q(\mathcal{B}).$$

The minimum d for which this holds is called Hanf locality rank *of Q, and is denoted by $\mathsf{hlr}(Q)$.*

For a Boolean query Q ($m = 0$) this means that Q cannot distinguish two structures $\mathcal{A}$ and $\mathcal{B}$ whenever $\mathcal{A} \leftrightarrows_d \mathcal{B}$.

Theorem 2 (Hanf, Fagin-Stockmeyer-Vardi). *Every FO sentence Φ defines a Hanf-local Boolean query Q with $\mathsf{hlr}(Q) \leq 3^{\mathsf{qr}(\Phi)}$.* □

An extension to open formulae, although easily derivable from the proof of [7], was probably first explicitly stated in [14]: every FO formula $\varphi(\vec{x})$ defines a Hanf-local query. Better bounds of the order $O(2^{\mathsf{qr}(\varphi)})$ are also known for Hanf-locality [16,19].

It was shown in [14] that every Hanf-local m-ary query, $m \geq 1$, is Gaifman-local.

Logic $\mathcal{L}^*_{\infty\omega}(\mathbf{C})$ The logic $\mathcal{L}^*_{\infty\omega}(\mathbf{C})$ subsumes a number of counting extensions of FO, such as FO with counting quantifiers [6,17], unary quantifiers [13], and unary counters [2]. (When we speak of counting extensions of FO, we mean extensions that only add a counting mechanism, as opposed to those – extensively studied in the literature, see [3,23] – that add both counting and fixpoint.) It is a two-sorted logic, with one sort being the universe of a finite structure, and the other sort being $\mathbb{N}$, and it uses counting terms that produce constants of the second sort, similarly to the logics studied in [10]. The formal definition is as follows.

We denote the infinitary logic by $\mathcal{L}_{\infty\omega}$; it extends FO by allowing infinite conjunctions $\bigwedge$ and disjunctions $\bigvee$. Then $\mathcal{L}_{\infty\omega}(\mathbf{C})$ is a two-sorted logic, that extends $\mathcal{L}_{\infty\omega}$. Its structures are of the form $(\mathcal{A}, \mathbb{N})$, where $\mathcal{A}$ is a finite relational structure, and $\mathbb{N}$ is a copy of natural numbers. We shall use $\vec{x}, \vec{y}$, etc., for variables ranging over the first (non-numerical) sort, and $\vec{\imath}, \vec{\jmath}$, etc., for variables ranging over the second (numerical) sort. Assume that every constant $n \in \mathbb{N}$ is a second-sort term. To $\mathcal{L}_{\infty\omega}$, add *counting quantifiers* $\exists ix$ for every $i \in \mathbb{N}$, and *counting terms:* If φ is a formula and $\vec{x}$ is a tuple of free first-sort variables in φ, then $\#\vec{x}.\varphi$ is a term of the second sort, and its free variables are those in φ except $\vec{x}$. Its interpretation is the number of $\vec{a}$ over the finite first-sort universe that satisfy φ. That is, given a structure $\mathcal{A}$, a formula $\varphi(\vec{x}, \vec{y}; \vec{\jmath})$, $\vec{b} \subseteq A$, and $\vec{\jmath}_0 \subset \mathbb{N}$, the value of the term $\#\vec{x}.\varphi(\vec{x}, \vec{b}; \vec{\jmath}_0)$ is the cardinality of the (finite) set $\{\vec{a} \subseteq A \mid \mathcal{A} \models \varphi(\vec{a}, \vec{b}; \vec{\jmath}_0)\}$. For example, the interpretation of $\#x.E(x, y)$ is the in-degree of node y in a graph with the edge-relation E. The interpretation of $\exists ix\varphi$ is $\#x.\varphi \geq i$.

As this logic is too powerful (it expresses every property of finite structures), we restrict it by means of the *rank* of formulae and terms, denoted by rk. It is defined as quantifier rank (that is, it is 0 for variables and constants $n \in \mathbb{N}$, $\mathsf{rk}(\bigvee_i \varphi_i) = \max_i \mathsf{rk}(\varphi_i), \mathsf{rk}(\neg\varphi) = \mathsf{rk}(\varphi), \mathsf{rk}(\exists x\varphi) = \mathsf{rk}(\exists ix\varphi) = \mathsf{rk}(\varphi) + 1)$ but does not take into account quantification over $\mathbb{N}$: $\mathsf{rk}(\exists i\varphi) = \mathsf{rk}(\varphi)$. Furthermore, $\mathsf{rk}(\#\vec{x}.\psi) = \mathsf{rk}(\psi) + \mid \vec{x} \mid$, and the rank of an atomic formula is the maximum rank of a term in it.

Definition 3. (see [19]) *The logic $\mathcal{L}^*_{\infty\omega}(\mathbf{C})$ is defined to be the restriction of $\mathcal{L}_{\infty\omega}(\mathbf{C})$ to terms and formulae of finite rank.*

It is known [19] that $\mathcal{L}^*_{\infty\omega}(\mathbf{C})$ is closed under finitary Boolean connectives and all quantification, and that every predicate on $\mathbb{N} \times \ldots \times \mathbb{N}$ is definable by a $\mathcal{L}^*_{\infty\omega}(\mathbf{C})$ formula of rank 0. Thus, we assume that $+, *, -, \leq$, and in fact *every* predicate on $\mathbb{N}$ is available. Furthermore, counting terms can be eliminated in $\mathcal{L}^*_{\infty\omega}(\mathbf{C})$ without increasing the rank (that is, counting quantifiers suffice, although expressing properties with just counting quantifiers is often quite awkward).

Fact 3 *(see [15,19]) Queries expressed by $\mathcal{L}^*_{\infty\omega}(\mathbf{C})$ formulae without free variables of the second-sort are Hanf-local and Gaifman-local.* □

Gaifman-locality of $\mathcal{L}^*_{\infty\omega}(\mathbf{C})$ was proved by a simple direct argument in [19]; Hanf-locality was shown in [15] using *bijective Ehrenfeuct-Fraïssé games* of [13].

Structures of bounded valence (degree) If $\mathcal{A} \in \mathrm{STRUCT}[\sigma]$, and R_i is of arity p_i, then $degree_j(R_i^{\mathcal{A}}, a)$ for $1 \leq j \leq p_i$ is the number of tuples $\vec{a}$ in $R_i^{\mathcal{A}}$ having a in the jth position. In the case of directed graphs, this gives us the usual notions of in- and out-degree. By $deg\_set(\mathcal{A})$ we mean the set of all degrees realized in $\mathcal{A}$. We use the notation $\mathrm{STRUCT}_k[\sigma]$ for $\{\mathcal{A} \in \mathrm{STRUCT}[\sigma] \mid deg\_set(\mathcal{A}) \subseteq \{0, 1, \ldots, k\}\}$.

Fact 4 *(see [14]) For any fixed k, a query Q on $\mathrm{STRUCT}_k[\sigma]$ is Hanf-local iff it is expressed by a formula of $\mathcal{L}^*_{\infty\omega}(\mathbf{C})$ (without free second-sort variables).* □

An m-ary query Q on a class $\mathcal{C} \subseteq \text{STRUCT}[\sigma]$ is given by a *first-order definition by cases* if there exists a number p, a partition $\mathcal{C} = \mathcal{C}_1 \cup \mathcal{C}_2 \cup \ldots \cup \mathcal{C}_p$ and first order formulae $\alpha_1(x_1, \ldots, x_m), \ldots, \alpha_p(x_1, \ldots, x_m)$ in the language σ such that on all structures $\mathcal{A} \in \mathcal{C}_i$, Q is definable by α_i. That is, for all $1 \leq i \leq p$ and $\mathcal{A} \in \mathcal{C}_i$, $\vec{a} \in Q(\mathcal{A})$ iff $\mathcal{A} \models \alpha_i(\vec{a})$.

Fact 5 *(see [14]) For any fixed k, a query Q on* $\text{STRUCT}_k[\sigma]$ *is Gaifman-local iff it is given by a first-order definition by cases.* □

3 Isomorphism of Neighborhoods and $\mathcal{L}^*_{\infty\omega}(\mathbf{C})$

We start with a slightly modified definition of locality that makes it convenient to work with two-sorted logics, like $\mathcal{L}^*_{\infty\omega}(\mathbf{C})$. We say that such a logic expresses Hanf-local (or Gaifman-local) queries if for every formula $\varphi(\vec{x}, \vec{\imath})$ there exists a number d such that for every $\vec{\imath}_0 \subset \mathbb{N}$, the formula $\varphi_{\vec{\imath}_0}(\vec{x}) = \varphi(\vec{x}, \vec{\imath}_0)$ (without free second-sort variables) expresses a query Q with $\mathsf{hlr}(Q) \leq d$ ($\mathsf{lr}(Q) \leq d$, respectively).

Consider a set θ of relation symbols, disjoint from σ, and define $\mathcal{L}^*_{\infty\omega}(\mathbf{C}) + \theta$ by allowing for each k-ary $U \in \theta$ and a k-tuple $\vec{x}$ of variables of the first sort, $U(\vec{x})$ to be a new atomic formula. The rank of this formula is 0. Assume that we fix a semantics of predicates from θ. We then say that θ is Hanf-local if there exists a number d such that each predicate in θ defines a Hanf-local query Q with $\mathsf{hlr}(Q) \leq d$.

Theorem 6. *Let θ be Hanf-local. Then $\mathcal{L}^*_{\infty\omega}(\mathbf{C}) + \theta$ expresses only Hanf-local queries.*

Proof sketch. Let d witness Hanf-locality of θ. We show that every $\mathcal{L}^*_{\infty\omega}(\mathbf{C}) + \theta$ formula of rank m defines a Hanf-local query Q with $\mathsf{hlr}(Q) \leq 3^m \cdot d + (3^m - 1)/2$ (for all instantiations of free variables of the second sort).

The proof is by induction on a formula. The atomic case follows from the assumption that θ is Hanf-local. The cases of Boolean and infinitary connectives, as well as negation and quantification over the numerical sort are simple. It remains to consider the case of $\psi(\vec{x}, \vec{\imath}) \equiv \exists iy(\varphi(y, \vec{x}, \vec{\imath}))$ (as counting terms can be eliminated without increasing the rank [19]) and to show that if φ defines a query of Hanf locality rank r for every $\vec{\imath}_0$, then ψ defines a query Q with $\mathsf{hlr}(Q) \leq 3r+1$. For this, we need the following result from [14]: if $(\mathcal{A}, \vec{a}) \leftrightarrows_{3r+1} (\mathcal{B}, \vec{b})$, then there exists a bijection $f : A \to B$ such that $(\mathcal{A}, \vec{a}c) \leftrightarrows_r (\mathcal{B}, \vec{b}f(c))$ for all $c \in A$. We then fix $\vec{\imath}_0$ and assume $(\mathcal{A}, \vec{a}) \leftrightarrows_{3r+1} (\mathcal{B}, \vec{b})$. Then, for f as above, it is the case that $\mathcal{A} \models \varphi(c, \vec{a}, \vec{\imath})$ iff $\mathcal{B} \models \varphi(f(c), \vec{b}, \vec{\imath})$, due to Hanf-locality of φ, and thus $\mathcal{A} \models \psi(\vec{a}, \vec{\imath})$ iff $\mathcal{B} \models \psi(\vec{b}, \vec{\imath})$, as the number of elements satisfying $\varphi(\cdot, \vec{a}, \vec{\imath})$ and $\varphi(\cdot, \vec{b}, \vec{\imath})$ is the same. This completes the proof. □

We now consider the following example. For each d, k, define a $2k$-ary predicate $I^k_d(x_1, \ldots, x_k, y_1, \ldots, y_k)$ to be interpreted as follows: $\mathcal{A} \models I^k_d(\vec{a}, \vec{b})$ iff $N^{\mathcal{A}}_d(\vec{a}) \cong N^{\mathcal{A}}_d(\vec{b})$. Clearly, $(\mathcal{A}, \vec{a}_1\vec{a}_2) \leftrightarrows_d (\mathcal{B}, \vec{b}_1\vec{b}_2)$ implies $N^{\mathcal{A}}_d(\vec{a}_1\vec{a}_2) \cong N^{\mathcal{B}}_d(\vec{b}_1\vec{b}_2)$, and thus $\vec{a}_1 \approx^{\mathcal{A}}_d \vec{a}_2$ iff $\vec{b}_1 \approx^{\mathcal{B}}_d \vec{b}_2$. This shows Hanf-locality of I^k_d and gives us

Corollary 1. *For any fixed d,* $\mathcal{L}^*_{\infty\omega}(\mathbf{C}) + \{I^k_d \mid k > 0\}$ *only expresses Hanf-local properties.* □

We next show that this gives us an increase in expressive power. The result below is proved using bijective games.

Proposition 1. *For any* $d, k > 0$, $\mathcal{L}^*_{\infty\omega}(\mathbf{C}) + I^k_d$ *is strictly more expressive than* $\mathcal{L}^*_{\infty\omega}(\mathbf{C})$. □

Corollary 2. *The logic* $\mathcal{L}^*_{\infty\omega}(\mathbf{C})$ *fails to capture Hanf-local properties over arbitrary finite structures.* □

Note that we only used I^k_ds as atomic formulae. A natural extension would be to use them as generalized quantifiers. In this case we extend the definition of the logic by a rule that if $\varphi_1(\vec{v}_1, \vec{z}), \ldots, \varphi_l(\vec{v}_l, \vec{z})$ are formulae with $\vec{v}_i$ being an m_i-tuple of first-sort variables, then $\psi(\vec{x}, \vec{y}, \vec{z}) \equiv \mathbf{I}^k_d[m_1, \ldots, m_l](\vec{v}_1, \ldots, \vec{v}_l)(\varphi_1(\vec{v}_1, \vec{z}), \ldots, \varphi_l(\vec{v}_l, \vec{z}))$ is a formula with $\vec{x}$ and $\vec{y}$ being k-tuples of fresh free variables of the first sort. The semantics is that for each $\mathcal{A}$ and $\vec{c}$, one defines a new structure on A in which the ith predicate of arity m_i is interpreted as $\{\vec{u} \in A^{m_i} \mid \mathcal{A} \models \varphi_i(\vec{u}, \vec{c})\}$. Then $\mathcal{A} \models \psi(\vec{a}, \vec{b}, \vec{c})$ if in this structure the d-neighborhoods of $\vec{a}$ and $\vec{b}$ are isomorphic. However, this generalization does not preserve locality.

Proposition 2. *Adding* $\mathbf{I}^k_d[m_1, \ldots, m_l]$ *to* $\mathcal{L}^*_{\infty\omega}(\mathbf{C})$ *violates Hanf-locality. In fact, with addition of* $\mathbf{I}^1_1[2]$ *to FO one can define properties that are neither Hanf-local nor Gaifman-local.* □

4 Characterizing Hanf-Local Properties

We have seen that the logic $\mathcal{L}^*_{\infty\omega}(\mathbf{C})$ fails to capture Hanf-local properties over arbitrary finite structures. To fill the gap between $\mathcal{L}^*_{\infty\omega}(\mathbf{C})$ and Hanf-locality, we introduce the notion of *local second-order quantification.* The idea is similar to local first-order quantification which restricts quantified variables to fixed radius neighborhoods of free variables. This kind of quantification was used in Gaifman's locality theorem [8] as well as in translations of various modal logics into fragments of FO [9,25].

Definition 4. *Fix* $r \geq 0$ *and a relational signature* σ. *Suppose that we have, for every arity* $k > 0$, *a countably infinite set of k-ary relational symbols* T^i_k, $i \in \mathbb{N}$, *disjoint from* σ. *Define a set of formulae* $\mathcal{F}$ *by starting with* $\mathcal{L}^*_{\infty\omega}(\mathbf{C})$ *atomic formulae involving symbols from* σ *as well as* T^i_k*s, and closing under the formation rules of* $\mathcal{L}^*_{\infty\omega}(\mathbf{C})$ *and the following rule: If* $\varphi(\vec{x}, \vec{\imath})$ *is a formula,* $\vec{y}$ *is a subtuple of* $\vec{x}$ *and* $d \leq r$, *then*

$$\psi_1(\vec{x}, \vec{\imath}) \equiv \exists T^i_k \sqsubseteq S_d(\vec{y})\ \varphi(\vec{x}, \vec{\imath}) \quad \textit{and} \quad \psi_2(\vec{x}, \vec{\imath}) \equiv \forall T^i_k \sqsubseteq S_d(\vec{y})\ \varphi(\vec{x}, \vec{\imath})$$

are formulae of rank $\mathsf{rk}(\varphi) + 1$. *We say that the symbol* T^i_k *is bound in these formulae.*

We then define $\mathcal{LSO}^r_{\infty\omega}(\mathbf{C})$ *over* STRUCT$[\sigma]$ *as the set of all formulae in* $\mathcal{F}$ *of finite rank in which all occurrences of the symbols* T^i_k*s are bound. The logic* $\mathcal{LSO}^*_{\infty\omega}(\mathbf{C})$ *(*local second-order with counting*) is defined as* $\bigcup_{r\geq 0} \mathcal{LSO}^r_{\infty\omega}(\mathbf{C})$.

The semantics of the new construct is as follows. Given a σ*-structure* $\mathcal{A}$ *and an interpretation* $\mathcal{T}$ *for all the symbols* T^i_k*s occurring freely in* ψ_1*, we have* $(\mathcal{A}, \mathcal{T}) \models \psi_1(\vec{a}, \vec{\imath})$ *iff there exists a set* $T \subseteq S_d(\vec{b})^k$*, where* $\vec{b}$ *is the subtuple of* $\vec{a}$ *corresponding to* $\vec{y}$*, such that* $(\mathcal{A}, \mathcal{T}, T) \models \varphi(\vec{a}, \vec{\imath})$*. For* ψ_2*, one replaces 'exists' by 'for all'.* □

For example, the formula

$$\exists x \exists T \sqsubseteq S_r(x) \exists T' \sqsubseteq S_r(x) \left(\begin{array}{r} \forall y \in S_r(x)\ (T(y) \wedge \neg T'(y)) \vee (\neg T(y) \wedge T'(y)) \\ \wedge\, \forall z, v\ (T(z) \wedge E(z,v) \rightarrow \qquad\qquad\qquad\qquad \\ T'(v)) \wedge (T'(z) \wedge E(z,v) \rightarrow T(v)) \end{array} \right)$$

tests if there is a 2-colorable r-neighborhood of a node in a graph. Note that local first-order quantification $\forall y \in S_r(x)$ is definable in FO for every fixed r.

Our main result can now be stated as follows.

Theorem 7. *An m-ary query Q,* $m \geq 0$*, is Hanf-local iff it is definable by a formula of* $\mathcal{LSO}^*_{\infty\omega}(\mathbf{C})$ *(without free second-sort variables).*

Proof sketch. We first show that queries definable in $\mathcal{LSO}^*_{\infty\omega}(\mathbf{C})$ are Hanf-local. The same argument as in [19] shows that counting terms can be eliminated from $\mathcal{LSO}^r_{\infty\omega}(\mathbf{C})$ without increasing the rank of a formula. Suppose we are given a signature σ' disjoint from σ. If $\mathcal{A} \in$ STRUCT$[\sigma]$, $\vec{a}$ is a k-tuple of elements of A, and $\vec{C}$ is an interpretation of σ' predicates as relations of appropriate arity over A, we write $(\mathcal{A}, \vec{C}, \vec{a})$ for the corresponding structure in the language of $\sigma \cup \sigma'$ union constants for elements of $\vec{a}$. By *adom*$(\vec{C})$ we mean the active domain of $\vec{C}$, that is, the set of all elements of A that occur in relations from $\vec{C}$. We then write, for $d \geq r$,

$$(\mathcal{A}, \vec{C}, \vec{a}) \ \ \sim^r_d \ \ (\mathcal{B}, \vec{D}, \vec{b})$$

if $\vec{D}$ interprets σ' over B, $\vec{a}$, $\vec{b}$ are of the same length, and the following three conditions hold: (1) $(\mathcal{A}, \vec{a}) \leftrightarrows_d (\mathcal{B}, \vec{b})$; (2) *adom*$(\vec{C}) \subseteq S^{\mathcal{A}}_r(\vec{a})$ and *adom*$(\vec{D}) \subseteq S^{\mathcal{B}}_r(\vec{b})$; and (3) there exists an isomorphism $h : N^{\mathcal{A}}_d(\vec{a}) \rightarrow N^{\mathcal{B}}_d(\vec{b})$ such that $h(\vec{C}) = \vec{D}$. The *if* direction is now implied by the lemma below, simply by taking σ' to be empty.

Lemma 1. *Let* $\varphi(\vec{x}, \vec{\imath}, \vec{X})$ *be a* $\mathcal{LSO}^r_{\infty\omega}(\mathbf{C})$ *formula. Then there exists a number* $d \geq r$ *such that, for every interpretation* $\vec{\imath}_0$ *of* $\vec{\imath}$*, it is the case that* $(\mathcal{A}, \vec{a}, \vec{C}) \sim^r_d (\mathcal{B}, \vec{b}, \vec{D})$ *implies*

$$\mathcal{A} \models \varphi(\vec{a}, \vec{\imath}_0, \vec{C}) \quad \textit{iff} \quad \mathcal{B} \models \varphi(\vec{b}, \vec{\imath}_0, \vec{D}).$$

Proof of the lemma is by induction on formulae. Let $\mathsf{rk}_0(\varphi)$ be defined as $\mathsf{rk}(\varphi)$ but without taking into account second-order quantification (in particular, $\mathsf{rk}_0(\varphi) \leq \mathsf{rk}(\varphi)$). We show that d can be taken to be $9^m r + \frac{9^m-1}{2}$ where $m = \mathsf{rk}_0(\varphi)$.

The case requiring most work is that of counting quantifiers; that is, of a formula $\psi(\vec{x}, \vec{\imath}, \vec{X}) \equiv \exists iz\ \varphi(\vec{x}, z, \vec{\imath}, \vec{X})$. Applying the hypothesis to φ, we obtain a number $d \geq r$ such that for every $\vec{\imath}_0$, $(\mathcal{A}, \vec{a}, c, \vec{C}) \sim_d^r (\mathcal{B}, \vec{b}, e, \vec{D})$ implies that $\mathcal{A} \models \varphi(\vec{a}, c, \vec{\imath}_0, \vec{C})$ iff $\mathcal{B} \models \varphi(\vec{b}, e, \vec{\imath}_0, \vec{D})$. To conclude, we must prove that $(\mathcal{A}, \vec{a}, \vec{C}) \sim_{9d+4}^r (\mathcal{B}, \vec{b}, \vec{D})$ implies that $\mathcal{A} \models \psi(\vec{a}, \vec{\imath}_0, \vec{C})$ iff $\mathcal{B} \models \psi(\vec{b}, \vec{\imath}_0, \vec{D})$. For this, it suffices to establish a bijection $f : A \to B$ such that for every c, $(\mathcal{A}, \vec{a}, c, \vec{C}) \sim_d^r (\mathcal{B}, \vec{b}, f(c), \vec{D})$ – then clearly the number of elements satisfying φ will be preserved. This proof of this is based on the following combinatorial lemma: Assume that $(\mathcal{A}, \vec{a}) \leftrightarrows_{9d+4} (\mathcal{B}, \vec{b})$, and h is an arbitrary isomorphism $N_{9d+4}^{\mathcal{A}}(\vec{a}) \to N_{9d+4}^{\mathcal{B}}(\vec{b})$. Then there exists a bijection $f : A \to B$ such that on $S_{6d+3}(\vec{a})$ it coincides with h, and $(\mathcal{A}, \vec{a}c) \leftrightarrows_d (\mathcal{B}, \vec{b}f(c))$ for every $c \in A$.

To prove the *only if* part, we show that with local second-order quantification, one can define local orderings on neighborhoods, and then the counting power of $\mathcal{L}^*_{\infty\omega}(\mathbf{C})$ allows one to code neighborhoods with numbers. The construction can be carried out in such a way that the entire multiset of isomorphism types of neighborhoods in a structure is coded by a formula whose rank is only determined by the radius of neighborhoods and the signature σ. Using this, one can express any Hanf-local query in $\mathcal{LSO}^*_{\infty\omega}(\mathbf{C})$. □

There are several corollaries to the proof. First notice that if we defined $\mathcal{LSO}^*_{\infty\omega}(\mathbf{C})$ without increasing the rank of a formula for every second-order local quantifier, the proof would go through verbatim. We can also define a logic $\mathbb{L}^r_{\infty\omega}(\mathbf{C})$ just as $\mathcal{LSO}^r_{\infty\omega}(\mathbf{C})$ except that first-order local quantification $\exists z \in S_r(\vec{x})$ and $\forall z \in S_r(\vec{x})$ is used in place of second-order local quantifiers, and those local quantifiers do not increase the rank (in particular, the depth of their nesting can be infinite, which allows one to define arbitrary computations on those neighborhoods). Let then $\mathbb{L}^*_{\infty\omega}(\mathbf{C})$ be $\bigcup_r \mathbb{L}^r_{\infty\omega}(\mathbf{C})$. The proof of Hanf-locality of $\mathbb{L}^*_{\infty\omega}(\mathbf{C})$ goes through as before, and proving that every Hanf-local query is definable in $\mathbb{L}^*_{\infty\omega}(\mathbf{C})$ is very similar to that of $\mathcal{LSO}^*_{\infty\omega}(\mathbf{C})$ as with infinitely many local first-order quantifiers we can write out diagrams of neighborhoods. We thus obtain:

Corollary 3. *The following have the same expressive power as $\mathcal{LSO}^*_{\infty\omega}(\mathbf{C})$ (and thus capture Hanf-local properties):*

- *the logic obtained from $\mathcal{LSO}^*_{\infty\omega}(\mathbf{C})$ by allowing the depth of nesting of local quantifiers to be infinite, and*
- *the logic $\mathbb{L}^*_{\infty\omega}(\mathbf{C})$.* □

Analyzing the proof of Theorem 7, we also obtain the following normal form for $\mathcal{LSO}^*_{\infty\omega}(\mathbf{C})$ formulae, which shows that the depth of nesting of local second-order quantifiers need not exceed 1.

Corollary 4. *Every $\mathcal{LSO}^*_{\infty\omega}(\mathbf{C})$ formula $\varphi(\vec{x})$ is equivalent to a formula in the form*

$$\bigvee_i \bigwedge_j (n_{ij} = \#y.(\exists S \sqsubseteq S_d(\vec{x})\ \psi_{ij}(\vec{x}, y, S)))$$

where the conjunctions are finite, S is binary, and each ψ_{ij} is a $\mathcal{L}^*_{\infty\omega}(\mathbf{C})$ formula.

As a final remark, we note that $\mathcal{LSO}^*_{\infty\omega}(\mathbf{C})$ is strictly more expressive than $\mathcal{L}^*_{\infty\omega}(\mathbf{C})$ extended with tests for neighborhood isomorphisms.

Proposition 3. $\bigcup_{d>0}(\mathcal{L}^*_{\infty\omega}(\mathbf{C}) + \{I^k_d \mid k > 0\}) \subsetneq \mathcal{LSO}^*_{\infty\omega}(\mathbf{C})$. □

5 Characterizing Gaifman-Local Properties

We now turn to Gaifman's notion of locality, which states that a query Q is local with $\mathsf{lr}(Q) \leq r$ if $N^{\mathcal{A}}_r(\vec{a}_1) \cong N^{\mathcal{A}}_r(\vec{a}_2)$ implies that $\vec{a}_1 \in Q(\mathcal{A})$ iff $\vec{a}_2 \in Q(\mathcal{A})$. For structures of bounded valence, this notion was characterized by first-order definition by cases. An extended version of this notion captures Gaifman-locality in the general case.

Definition 5. *An m-ary query, $m > 0$, on* STRUCT$[\sigma]$ *is given by a* Hanf-local definition by cases *if there exists a finite or countable partition of* STRUCT$[\sigma]$ *into classes $\mathcal{C}_i$, $i \in \mathbb{N}$, a number $d \geq 0$, and Hanf-local queries Q_i, $i \in \mathbb{N}$, with $\mathsf{hlr}(Q_i) \leq d$, such that for every i and every $\mathcal{A} \in \mathcal{C}_i$, it is the case that $Q(\mathcal{A}) = Q_i(\mathcal{A})$.*

Theorem 8. *A query is Gaifman-local iff it is given by a Hanf-local definition by cases.*

Proof sketch. Assume that Q is given by a Hanf-local definition by cases. Let d be an upper bound on $\mathsf{hlr}(Q_i)$. Then Q is Gaifman-local and $\mathsf{lr}(Q) \leq 3d+1$. Fix $\mathcal{A}$, and assume $\mathcal{A} \in \mathcal{C}_i$. Let $\vec{a}_1 \approx^{\mathcal{A}}_{3d+1} \vec{a}_2$. Then by [14], $(\mathcal{A}, \vec{a}_1) \leftrightarrows_d (\mathcal{A}, \vec{a}_2)$, and Hanf-locality of Q_i implies $\vec{a}_1 \in Q_i(\mathcal{A}) = Q(\mathcal{A})$ iff $\vec{a}_2 \in Q_i(\mathcal{A}) = Q(\mathcal{A})$. Conversely, let a Gaifman-local Q be given, with $\mathsf{lr}(Q) = d$. Let $\tau_1, \tau_2 \ldots$ be an enumeration of isomorphism types of finite σ-structures. Let $\mathcal{C}_i$ be the class of structures of type τ_i. We define Q_i as follows: $\vec{b} \in Q_i(\mathcal{B})$ iff there exists $\mathcal{A}$ of type τ_i and $\vec{a} \in A^m$ such that $(\mathcal{B}, \vec{b}) \leftrightarrows_d (\mathcal{A}, \vec{a})$ and $\vec{a} \in Q(\mathcal{A})$. One then shows that each Q_i is Hanf-local, with $\mathsf{hlr}(Q_i) \leq d$, and for every $\mathcal{A}$ of type τ_i, $Q(\mathcal{A}) = Q_i(\mathcal{A})$. □

Unlike in Fact 5, the number of cases in a Hanf-local definition by cases can be infinite. A natural question to ask is whether a finite number of cases is sufficient (in particular, whether the statement of Fact 5 holds for arbitrary finite structures). We now show that the infinite number of cases is unavoidable. In fact, we show a stronger result.

Definition 6. *For $k > 0$, let* LOCAL$_k$ *be the class of queries given by a Hanf-local definition by cases, where the number of cases is at most k. Let* LOCAL* *be* $\bigcup_{k>0}$ LOCAL$_k$*, and* G_LOCAL *be the class of all Gaifman-local queries.*

Note that LOCAL_1 is precisely the class of Hanf-local queries.

Theorem 9. *The hierarchy*

$$\text{LOCAL}_1 \subset \text{LOCAL}_2 \subset \ldots \subset \text{LOCAL}^* \subset \text{G\_LOCAL}$$

is strict.

Proof sketch. We first exhibit a query $Q \in \text{LOCAL}_{l+1} - \text{LOCAL}_l$. Intuitively, a query from LOCAL_l cannot make $l+1$ choices, and thus is different from every query in LOCAL_{l+1} on some class of the partition. More precisely, we define a class $\mathcal{C}_i^{l+1}$, $1 \leq i \leq l+1$ to be the class of graphs with the number of connected components being $i-1$ modulo $l+1$. Let Q_i^{l+1} be a FO-definable query returning the set of nodes reachable by a path of length $i-1$ from a node of indegree 0. Form the query Q that coincides with Q_i^{l+1} on $\mathcal{C}_i^{l+1}$. (Note that Q is not FO, as the classes $\mathcal{C}_i^{l+1}$ are not FO-definable.) From Theorem 8, this is a Gaifman-local query, and it belongs to LOCAL_{l+1}. Suppose Q is in LOCAL_l; that is, there is a partition of the class of all finite graphs into l classes $\mathcal{C}'_1, \ldots, \mathcal{C}'_l$ and Hanf-local queries Q'_i such that on $\mathcal{C}'_i$, Q coincides with Q'_i, $i = 1, \ldots, l$. Let $d = 1 + \max \mathsf{hlr}(Q'_i)$. Let G_0 be a successor relation on $l+1$ nodes. Define a graph H_i^{l+1} as the union of i cycles with $\frac{(l+1)!(2d+1)}{i}$ nodes each, $i = 1, \ldots, l+1$. As the total number of nodes in each H_i^{l+1} is $(l+1)!(2d+1)$ and all d-neighborhoods are isomorphic, we have $H_i^{l+1} \leftrightarrows_d H_j^{l+1}$ for all $i, j \leq l+1$. Let now G_i^{l+1} be the disjoint union of G_0 and H_i^{l+1}, $i = 1, \ldots, l+1$. By pigeonhole, there exists a class $\mathcal{C}'_k$ and $i \neq j, i, j \leq l+1$ such that $G_i^{l+1}, G_j^{l+1} \in \mathcal{C}'_k$. We then show that Q cannot give correct results on both G_i^{l+1} and G_j^{l+1}. The separation G_LOCAL from LOCAL^* is proved by a minor modification of the construction above. □

Thus, similarly to the case of Hanf-local queries, the characterization for structures of bounded valence fails to extend to the class of all finite structures.

Corollary 5. *There exist Gaifman-local queries that cannot be given by first-order definition by cases.* □

6 Conclusion

Notions of locality have been used in logic numerous times. The local nature of first-order logic is particularly transparent when one deals with fragments corresponding to various modal logics; in general, Gaifman's and Hanf's theorems state that FO can only express local properties. These theorems were generalized, and, being applicable to finite structures, they found applications in areas such as complexity and databases.

However, while more and more powerful logics were proved to be local, there was no clear understanding of what kind of mechanisms can be added to logics while preserving locality. Here we answered this question by providing logical characterizations of local properties on finite structures. For Hanf-locality, arbitrary counting power and arbitrary computations over small neighborhoods and can be added to first-order logic while retaining locality; moreover, with a limited form of infinitary connectives, such a logic captures all Hanf-local properties. For Gaifman-locality, one can in addition permit definition by cases, and the number of cases be either finite or infinite.

References

1. D.A.M. Barrington, N. Immerman, H. Straubing. On uniformity within NC^1. *JCSS*, 41:274–306,1990.
2. M. Benedikt, H.J. Keisler. Expressive power of unary counters. *Proc. Int. Conf. on Database Theory (ICDT'97)*, Springer LNCS 1186, January 1997, pages 291–305.
3. J. Cai, M. Fürer and N. Immerman. On optimal lower bound on the number of variables for graph identification. *Combinatorica*, 12 (1992), 389–410.
4. G. Dong, L. Libkin and L. Wong. Local properties of query languages. *Theoretical Computer Science*, to appear. Extended abstract in *ICDT'97*, pages 140–154.
5. H.-D. Ebbinghaus and J. Flum. *Finite Model Theory*. Springer Verlag, 1995.
6. K. Etessami. Counting quantifiers, successor relations, and logarithmic space, *JCSS*, 54 (1997), 400–411.
7. R. Fagin, L. Stockmeyer and M. Vardi, On monadic NP vs monadic co-NP, *Information and Computation*, 120 (1995), 78–92.
8. H. Gaifman. On local and non-local properties, *Proceedings of the Herbrand Symposium, Logic Colloquium '81*, North Holland, 1982.
9. E. Grädel. On the restraining power of guards. *J. Symb. Logic*, to appear.
10. E. Grädel and Y. Gurevich. Metafinite model theory. *Information and Computation* 140 (1998), 26–81.
11. M. Grohe and T. Schwentick. Locality of order-invariant first-order formulas. In *MFCS'98*, pages 437–445.
12. W. Hanf. Model-theoretic methods in the study of elementary logic. In J.W. Addison et al, eds, *The Theory of Models*, North Holland, 1965, pages 132–145.
13. L. Hella. Logical hierarchies in PTIME. *Information and Computation*, 129 (1996), 1–19.
14. L. Hella, L. Libkin and J. Nurmonen. Notions of locality and their logical characterizations over finite models. *J. Symb. Logic*, to appear. Extended abstract in *LICS'97*, pages 204–215 (paper by the 2nd author).
15. L. Hella, L. Libkin, J. Nurmonen and L. Wong. Logics with aggregate operators. In *LICS'99*, pages 35–44.
16. N. Immerman. *Descriptive Complexity*. Springer Verlag, 1999.
17. N. Immerman and E. Lander. Describing graphs: A first order approach to graph canonization. In *"Complexity Theory Retrospective"*, Springer Verlag, Berlin, 1990.
18. Ph. Kolaitis and J. Väänänen. Generalized quantifiers and pebble games on finite structures. *Annals of Pure and Applied Logic*, 74 (1995), 23–75.
19. L. Libkin. On counting logics and local properties. In *LICS'98*, pages 501–512.
20. L. Libkin. Logics capturing local properties. Bell Labs Technical Memo, 1999.
21. L. Libkin and L. Wong. Unary quantifiers, transitive closure, and relations of large degree. In *STACS'98*, Springer LNCS 1377, pages 183–193.
22. J. Nurmonen. On winning strategies with unary quantifiers. *J. Logic and Computation*, 6 (1996), 779–798.
23. M. Otto. *Bounded Variable Logics and Counting: A Study in Finite Models*. Springer Verlag, 1997.

24. T. Schwentick and K. Barthelmann. Local normal forms for first-order logic with applications to games and automata. In *STACS'98*, Springer LNCS 1377, 1998, pages 444–454.
25. M. Vardi. Why is monadic logic so robustly decidable? In *Proc. DIMACS Workshop on Descriptive Complexity and Finite Models*, AMS 1997.

The Complexity of Poor Man's Logic

Edith Hemaspaandra*

Dept. of Comp. Sci., Rochester Institute of Technology, Rochester, NY 14623, USA.
eh@cs.rit.edu

Abstract. Motivated by description logics, we investigate what happens to the complexity of modal satisfiability problems if we only allow formulas built from literals, $\wedge$, $\Diamond$, and $\Box$. Previously, the only known result was that the complexity of the satisfiability problem for K dropped from PSPACE-complete to coNP-complete (Schmidt-Schauss and Smolka [8] and Donini et al. [3]). In this paper we show that not all modal logics behave like K. In particular, we show that the complexity of the satisfiability problem with respect to frames in which each world has at least one successor drops from PSPACE-complete to P, but that in contrast the satisfiability problem with respect to the class of frames in which each world has at most two successors remains PSPACE-complete. As a corollary of the latter result, we also solve the open problem from Donini et al.'s complexity classification of description logics [2]. In the last section, we classify the complexity of the satisfiability problem for K for all other restrictions on the set of operators.

1 Introduction

Since consistent normal modal logics contain propositional logic, the satisfiability problems for all these logics are automatically NP-hard. In fact, as shown by Ladner [6], many of them are even PSPACE-hard.

But we don't always need all of propositional logic. For example, in some applications we may use only a finite set of propositional variables. Propositional satisfiability thus restricted is in P, and, as shown by Halpern [4], the complexity of satisfiability problems for some modal logics restricted in the same way also decreases. For example, the complexity of S5 satisfiability drops from NP-complete to P. On the other hand, K satisfiability remains PSPACE-complete. The same restriction for linear temporal logics was studied in Demri and Schnoebelen [1].

Restricting the number of propositional variables is not the only propositional restriction on modal logics that occurs in the literature. For example, the description logic $\mathcal{ALE}$ can be viewed as multi-modal K where the formulas are built from literals, $\wedge$, $\Diamond$s, and $\Box$s.

As in the case of a fixed number of propositional variables, satisfiability for propositional logic for formulas built from literals and $\wedge$ is easily seen to be in P. After all, in that case every propositional formula is the conjunction of literals.

* Supported in part by grant NSF-INT-9815095. Work done in part while visiting the University of Amsterdam.

H. Reichel and S. Tison (Eds.): STACS 2000, LNCS 1770, pp. 230–241, 2000.

Such a formula is satisfiable if and only if there is no propositional variable p such that both p and $\overline{p}$ are conjuncts of the formula.

Hence, satisfiability for modal logics for formulas built from literals, $\wedge$, $\Box$, and $\Diamond$ is not automatically NP-hard. Of course, it does not necessarily follow that the complexity of modal satisfiability problems will drop significantly. The only result that was previously known is that the complexity of K satisfiability (i.e., satisfiability with respect to the class of all frames) drops from PSPACE-complete to coNP-complete. The upper bound was shown by Schmidt-Schauss and Smolka [8], and the lower bound by Donini et al. [3]. It should be noted that these results were shown in the context of description logics (a.k.a. concept languages), so that the notation in these papers is quite different from ours.[1] In addition, their language contains the constants *true* and *false*. However, it is easy to simulate these constants by propositional variables. See the full version of this paper for details [5].

In this paper we investigate if it is always the case that the complexity of the satisfiability problem decreases if we only look at formulas that are built from literals, $\wedge$, $\Diamond$, and $\Box$, and if so, if there are upper or lower bounds on the amount that the complexity drops.

We will show that not all logics behave like K. Far from it, by looking at simple restrictions on the number of successors that are allowed for each world in a frame, we obtain different levels of complexity, making apparent a subtle interplay between frames and operators. In particular, we will show that

1. The complexity of the satisfiability problem with respect to linear frames drops from NP-complete to P.
2. The complexity of the satisfiability problem with respect to  remains NP-complete.
3. The complexity of the satisfiability problem with respect to frames in which every world has at least one successor drops from PSPACE-complete to P.
4. The complexity of the satisfiability problem with respect to frames in which every world has at most two successors remains PSPACE-complete.

As a corollary of the last result, we also solve the open problem from Donini et al.'s complexity classification of description logics [2].

In the last section, we completely classify the complexity of the satisfiability problem (with respect to the class of all frames) for all possible restrictions on the set of operators allowed, to gain more insight in the sources of complexity for modal logics. It turns out that the restriction studied in this paper, which we will call poor man's logic, is the only fragment whose satisfiability problem

[1] Certain description logics can be viewed as syntactic variations of modal logics in the following way: the universal concept corresponds to *true*, the empty concept corresponds to *false*, atomic concepts correspond to propositional variables, atomic negation corresponds to propositional negation, complementation corresponds to negation, intersection corresponds to conjunction, union corresponds to disjunction, universal role quantifications correspond to $\Box$ operators, and existential role quantifications correspond to $\Diamond$ operators [7].

is so unusual. For all other restrictions, the satisfiability problems are PSPACE-complete, NP-complete, or in P. These are exactly the complexity classes that one would expect to show up in this context.

2 Definitions

We will first briefly review syntax, Kripke semantics, and some basic terminology for modal logic.

Syntax

The set of $\mathcal{L}$ formulas is inductively defined as follows. (As usual, we assume that we have a countably infinite set of propositional variables.)

- p and $\overline{p}$ are $\mathcal{L}$ formulas for every propositional variable p,
- if ϕ and ψ are $\mathcal{L}$ formulas, then so are $\phi \wedge \psi$ and $\phi \vee \psi$, and
- if ϕ is an $\mathcal{L}$ formula, then $\Box\phi$ and $\Diamond\phi$ are $\mathcal{L}$ formulas.

We will identify $\overline{\overline{p}}$ with p.

The *modal depth* of a formula ϕ (denoted by $md(\phi)$) is the depth of nesting of the modal operators $\Box$ and $\Diamond$.

Semantics

A *frame* is a tuple $F = <W, R>$ where W is a non-empty set of possible worlds, and R is a binary relation on W called the accessibility relation.

A *model* is of the form $M = <W, R, \pi>$ such that $<W, R>$ is a frame (we say that M is *based on* this frame), and π is a function from the set of propositional variables to $Pow(W)$: a valuation, i.e., $\pi(p)$ is the set of worlds in which p is true. For ϕ an $\mathcal{L}$ formula, we will write $M, w \models \phi$ for ϕ is *true /satisfied at* w *in* M. The truth relation $\models$ is defined with induction on ϕ in the following way.

- $M, w \models p$ iff $w \in \pi(p)$ for p a propositional variable.
- $M, w \models \overline{p}$ iff $w \notin \pi(p)$ for p a propositional variable.
- $M, w \models \phi \wedge \psi$ iff $M, w \models \phi$ and $M, w \models \psi$.
- $M, w \models \phi \vee \psi$ iff $M, w \models \phi$ or $M, w \models \psi$.
- $M, w \models \Box\phi$ iff $\forall w' \in W[wRw' \Rightarrow M, w' \models \phi]$.
- $M, w \models \Diamond\phi$ iff $\exists w' \in W[wRw'$ and $M, w' \models \phi]$.

The *size* of a model or frame is the number of worlds in the model or frame.

The notion of satisfiability can be extended to models, frames, and classes of frames in the following way. ϕ is satisfied in model M if $M, w \models \phi$ for some world w in M, ϕ is satisfiable in frame F (F satisfiable) if ϕ is satisfied in M for some model M based on F, and ϕ is satisfiable with respect to class of frames $\mathcal{F}$ ($\mathcal{F}$ satisfiable) if ϕ is satisfiable in some frame $F \in \mathcal{F}$.

As is usual, we will look at satisfiability with respect to classes of frames. For a class of frames $\mathcal{F}$, the satisfiability problem with respect to $\mathcal{F}$ is the problem of determining, given an $\mathcal{L}$ formula ϕ, whether ϕ is $\mathcal{F}$ satisfiable. For a complete logic L, we will sometimes view L as the class of frames where L is valid. For example, we will speak of K satisfiability when we mean satisfiability with respect to all frames. Likewise, we will on occasion identify a class of frames with its logic, i.e., with the set of formulas valid on this class of frames.

Poor Man's Logic

The set of poor man's formulas is the set of $\mathcal{L}$ formulas that do not contain $\vee$. The poor man's satisfiability problem with respect to $\mathcal{F}$ is the problem of determining, given a poor man's formula ϕ, whether ϕ is $\mathcal{F}$ satisfiable.

In poor man's language, we will view $\wedge$ as a multi-arity operator, and we will assume that all conjunctions are "flattened," that is, a conjunct will not be a conjunction. Thus, a formula ϕ in this language is of the following form: $\phi = \Box\psi_1 \wedge \cdots \wedge \Box\psi_k \wedge \Diamond\xi_1 \wedge \cdots \wedge \Diamond\xi_m \wedge \ell_1 \wedge \cdots \wedge \ell_s$, where the ℓ_is are literals.

In all but the last section of this paper, we will compare the complexity of satisfiability to the complexity of poor man's satisfiability with respect to the same class of frames. We are interested in simple restrictions on the number of successor worlds that are allowed. Let $\mathcal{F}_{\leq 1}$, $\mathcal{F}_{\leq 2}$, $\mathcal{F}_{\geq 1}$ be the classes of frames in which every world has at most one, at most two, and at least one successor, respectively.

3 Poor Man's Versions of NP-Complete Satisfiability Problems

We already know that the poor man's version of an NP-complete modal satisfiability problem can be in P. Look for example at satisfiability with respect to the class of frames where no world has a successor. This is plain propositional logic in disguise, and it inherits the complexity behavior of propositional logic. As mentioned in the introduction, the complexity of satisfiability drops from NP-complete to P.

In this section, we will give an example of a non-trivial modal logic with the same behavior. We will show that the poor man's version of satisfiability with respect to linear frames is in P. In contrast, we will also give a very simple example of a modal logic where the complexity of poor man's satisfiability remains NP-complete.

Theorem 1. *Satisfiability with respect to $\mathcal{F}_{\leq 1}$ is NP-complete and poor man's satisfiability with respect to $\mathcal{F}_{\leq 1}$ is in P.*

Proof. Clearly, $\mathcal{F}_{\leq 1}$ satisfiability is in NP (and thus NP-complete), since every satisfiable formula is satisfiable on a linear frame with $\leq$ *md*(ϕ) worlds, where *md*(ϕ) is the modal depth of ϕ. This immediately gives the following NP algorithm for $\mathcal{F}_{\leq 1}$ satisfiability: Guess a linear frame of size $\leq$ *md*(ϕ), and for every world in the frame, guess a valuation on the propositional variables that occur in ϕ. Accept if and only if the guessed model satisfies ϕ.

It is easy to see that the following polynomial-time algorithm decides poor man's satisfiability with respect to $\mathcal{F}_{\leq 1}$. Let $\phi = \Box\psi_1 \wedge \cdots \wedge \Box\psi_k \wedge \Diamond\xi_1 \wedge \cdots \wedge \Diamond\xi_m \wedge \ell_1 \wedge \cdots \wedge \ell_s$, where the ℓ_is are literals. ϕ is $\mathcal{F}_{\leq 1}$ satisfiable if and only if

- $\ell_1 \wedge \cdots \wedge \ell_s$ is satisfiable (that is, for all i and j, $\ell_i \neq \overline{\ell_j}$), and
-
  - $m = 0$, (that is, ϕ does not contain conjuncts of the form $\Diamond\xi$, in which case the formula is satisfied in a world with no successors), or
  - $\bigwedge_{i=1}^{k} \psi_i \wedge \bigwedge_{i=1}^{m} \xi_i$ is $\mathcal{F}_{\leq 1}$ satisfiable (the world has exactly one successor).

□

From the previous example, you might think that the poor man's versions of logics with the poly-size frame property are in P, or even that the poor man's versions of all NP-complete satisfiability problems are in P. Not so. The following theorem gives a very simple counterexample.

Theorem 2. *Satisfiability and poor man's satisfiability with respect to the frame* *are NP-complete.*

Proof. Because the frame is finite, both satisfiability problems are in NP. Thus it suffices to show that poor man's satisfiability with respect to is NP-hard.

Since we are working with a fragment of propositional modal logic, it is extremely tempting to try to reduce from an NP-complete propositional satisfiability problem. However, because poor man's logics contain only a fragment of propositional logic, these logics don't behave like propositional logic at all. Because of this, propositional satisfiability problems are not the best choice of problems to reduce from. In fact, they are particularly confusing.

It turns out that it is much easier to reduce a partitioning problem to our poor man's satisfiability problem. We will reduce from the following well-known NP-complete problem.

> GRAPH 3-COLORABILITY: Given an undirected graph G, can you color every vertex of the graph using only three colors in such a way that vertices connected by an edge have different colors?

Suppose $G = (V, E)$ where $V = \{1, 2, \ldots, n\}$. We introduce a propositional variable p_e for every edge e. The three leaves of will correspond to the three colors. To ensure that adjacent vertices in the graph end up in different leaves, we will make sure that the smaller endpoint of e satisfies p_e and that the larger endpoint of e satisfies $\overline{p_e}$.

The requirements for vertex i are given by the following formula:

$$\psi_i = \bigwedge\{p_e \mid e = \{i, j\} \text{ and } i < j\} \wedge \bigwedge\{\overline{p_e} \mid e = \{i, j\} \text{ and } i > j\}.$$

Define $f(G) = \bigwedge_{i=1}^{n} \Diamond\psi_i$.

f is clearly computable in polynomial-time. To show that f is indeed a reduction from GRAPH 3-COLORABILITY to poor man's satisfiability with respect to  , first note that it is easy to see that for every set $V' \subseteq V$, the following holds: $\bigwedge_{i \in V'} \psi_i$ is satisfiable if and only if no two vertices in V' are connected by an edge.

It follows that $f(G) = \bigwedge_{i=1}^{n} \Diamond\psi_i$ is satisfiable on  if and only if there exist sets of vertices V_1, V_2, and V_3 such that $V = V_1 \cup V_2 \cup V_3$ and $\bigwedge_{i \in V_j} \psi_i$ is satisfiable for $j \in \{1, 2, 3\}$. This holds if and only if there exist sets of vertices V_1, V_2, and V_3 such that $V = V_1 \cup V_2 \cup V_3$ and no two vertices in V_j are adjacent for $j \in \{1, 2, 3\}$, which is the case if and only if G is 3-colorable. (We obtain a coloring by coloring each vertex v by the smallest j such that $v \in V_j$.)

□

4 Poor Man's Versions of PSPACE-Complete Satisfiability Problems

It is well-known that the satisfiability problems for many modal logics including K are PSPACE-complete [6]. We also know that poor man's satisfiability for K is coNP-complete [8,3]. That is, in that particular case the complexity of the satisfiability problem drops from PSPACE-complete to coNP-complete. Is this the general pattern? We will show that this is not the case. We will give an example of a logic where the complexity of the satisfiability problem drops from PSPACE-complete all the way down to P, and another example in which the complexity of both the satisfiability and the poor man's satisfiability problems are PSPACE-complete. Both examples are really close to K; they are satisfiability with respect to $\mathcal{F}_{\geq 1}$ and $\mathcal{F}_{\leq 2}$, respectively.

We will first consider $\mathcal{F}_{\geq 1}$. This logic is very close to K and it should come as no surprise that the complexity of $\mathcal{F}_{\geq 1}$ satisfiability and K satisfiability are the same. It may come as a surprise to learn that poor man's satisfiability with respect to $\mathcal{F}_{\geq 1}$ is in P. It is easy to show that poor man's satisfiability with respect to $\mathcal{F}_{\geq 1}$ is in coNP, because the following function f reduces the poor man's satisfiability problem with respect to $\mathcal{F}_{\geq 1}$ to the poor man's satisfiability problem for K.

$$f(\phi) = \phi \wedge \bigwedge_{i=0}^{md(\phi)} \Box^i \Diamond q,$$

where q is a propositional variable not in ϕ. The formula ensures that every world in the relevant part of the K frame has at least one successor.

It is very surprising that poor man's satisfiability with respect to $\mathcal{F}_{\geq 1}$ is in P, because the relevant part of the $\mathcal{F}_{\geq 1}$ frame may require an exponential number of worlds to satisfy a formula in poor man's language. For example, consider the following formula:

$$\Diamond\Box\Box p_1 \wedge \Diamond\Box\Box\overline{p_1} \wedge \Box(\Diamond\Box p_2 \wedge \Diamond\Box\overline{p_2}) \wedge \Box\Box(\Diamond p_3 \wedge \Diamond\overline{p_3}).$$

If this formula is satisfiable in world w, then for every assignment to p_1, p_2, and p_3, there exists a world reachable in three steps from w that satisfies that assignment.

In its general version, the formula becomes

$$\phi_{asg} = \bigwedge_{i=1}^{n} \Box^{i-1}(\Diamond\Box^{n-i}p_i \wedge \Diamond\Box^{n-i}\overline{p_i}).$$

The formula is of length polynomial in n and forces the relevant part of the model to be of exponential size.

Now that we have seen how surprising it is that poor man's satisfiability with respect to $\mathcal{F}_{\geq 1}$ is in P, let's prove it.

Theorem 3. *Satisfiability with respect to $\mathcal{F}_{\geq 1}$ is PSPACE-complete and poor man's satisfiability with respect to $\mathcal{F}_{\geq 1}$ is in P.*

Proof. The proof that satisfiability with respect to $\mathcal{F}_{\geq 1}$ is PSPACE-complete is very close to the proof that K satisfiability is PSPACE-complete [6] and therefore omitted.

For the poor man's satisfiability problem, note that a simplified version of Ladner's PSPACE upper bound construction for K can be used to show the following.

Let $\phi = \Box\psi_1 \wedge \cdots \wedge \Box\psi_k \wedge \Diamond\xi_1 \wedge \cdots \wedge \Diamond\xi_m \wedge \ell_1 \wedge \cdots \wedge \ell_s$, where the ℓ_is are literals. ϕ is $\mathcal{F}_{\geq 1}$ satisfiable if and only if

1. $\ell_1 \wedge \cdots \wedge \ell_s$ is satisfiable,
2. for all j, $\psi_1 \wedge \cdots \wedge \psi_k \wedge \xi_j$ is $\mathcal{F}_{\geq 1}$ satisfiable, and
3. $\psi_1 \wedge \cdots \wedge \psi_k$ is $\mathcal{F}_{\geq 1}$ satisfiable. (only relevant when $m = 0$.)

Note that this algorithm takes exponential time and polynomial space. Of course, we already know that poor man's satisfiability with respect to $\mathcal{F}_{\geq 1}$ is in PSPACE, since satisfiability with respect to $\mathcal{F}_{\geq 1}$ is in PSPACE. How can this PSPACE algorithm help to prove that poor man's satisfiability with respect to $\mathcal{F}_{\geq 1}$ is in P?

Something really surprising happens here. We will prove that for every poor man's formula ϕ, ϕ is $\mathcal{F}_{\geq 1}$ satisfiable if and only if (the conjunction of) every pair of (not necessary different) conjuncts of ϕ is $\mathcal{F}_{\geq 1}$ satisfiable. Using dynamic programming, we can compute all pairs of subformulas of ϕ that are $\mathcal{F}_{\geq 1}$ satisfiable in polynomial-time. This proves the theorem. It remains to show that for every poor man's formula ϕ, ϕ is $\mathcal{F}_{\geq 1}$ satisfiable if and only if every pair of conjuncts of ϕ is $\mathcal{F}_{\geq 1}$ satisfiable. We will prove this claim by induction on $md(\phi)$, the modal depth of ϕ. In the proof, we will write "satisfiable" for "satisfiable with respect to $\mathcal{F}_{\geq 1}$."

If $md(\phi) = 0$, ϕ is a conjunction of literals. In that case ϕ is not satisfiable if and only if there exist i and j such that $\ell_i = \overline{\ell_j}$. This immediately implies our claim.

For the induction step, suppose $\phi = \Box\psi_1 \wedge \cdots \wedge \Box\psi_k \wedge \Diamond\xi_1 \wedge \cdots \wedge \Diamond\xi_m \wedge \ell_1 \wedge \cdots \wedge \ell_s$ (where the ℓ_is are literals), $md(\phi) \geq 1$, and suppose that our claim holds for all formulas of modal depth $< md(\phi)$. Suppose for a contradiction that ϕ is not satisfiable, though every pair of conjuncts of ϕ is satisfiable. Then, by the Ladner-like construction given above, we are in one of the following three cases:

1. $\ell_1 \wedge \cdots \wedge \ell_s$ is not satisfiable,
2. for some j, $\psi_1 \wedge \cdots \wedge \psi_k \wedge \xi_j$ is not satisfiable, or
3. $\psi_1 \wedge \cdots \wedge \psi_k$ is not satisfiable.

By induction, it follows immediately that we are in one of the following four cases:

1. There exist i, i' such that $\ell_i \wedge \ell_{i'}$ is not satisfiable,
2. there exist i, i' such that $\psi_i \wedge \psi_{i'}$ is not satisfiable,
3. there exist i, j such that $\psi_i \wedge \xi_j$ is not satisfiable, or
4. there exists a j such that $\xi_j \wedge \xi_j$ is not satisfiable.

If we are in case 2, $\Box\psi_i \wedge \Box\psi_{i'}$ is not satisfiable. In case 3, $\Box\psi_i \wedge \Diamond\xi_j$ is not satisfiable. In case 4, $\Diamond\xi_j \wedge \Diamond\xi_j$ is not satisfiable. So in each case we have found a pair of conjuncts of ϕ that is not satisfiable, which contradicts the assumption. □

Why doesn't the same construction work for K? It is easy enough to come up with a counterexample. For example, $\{\Box p, \Box \overline{p}, \Diamond q\}$ is not satisfiable, even though every pair is satisfiable. The deeper reason is that we have some freedom in K that we don't have in $\mathcal{F}_{\geq 1}$. Namely, on a K frame a world can have successors or no successors. This little bit of extra freedom is enough to encode coNP in poor man's language.

Theorem 2 showed that poor man's satisfiability can be as hard as satisfiability for NP-complete logics. In light of the fact that poor man's satisfiability for K is coNP-complete and poor man's satisfiability with respect to $\mathcal{F}_{\geq 1}$ is even in P, you might wonder if the complexity of PSPACE-complete logics always decreases.

To try to keep the complexity as high as possible, it makes sense to look at frames in which each world has a restricted number of successors, as in the construction of Theorem 2. Because we want the logic to be PSPACE-complete, we also need to make sure that the frames can simulate binary trees. The obvious class of frames to look at is $\mathcal{F}_{\leq 2}$ – the class of frames in which each world has at most two successors. This gives us the desired example.

Theorem 4. *Satisfiability and poor man's satisfiability with respect to* $\mathcal{F}_{\leq 2}$ *are PSPACE-complete.*

Proof. Satisfiability with respect to $\mathcal{F}_{\leq 2}$ is PSPACE-complete by pretty much the same proof as the PSPACE-completeness proof for K [6]. To show that the poor man's version remains PSPACE-complete, first note that a formula is $\mathcal{F}_{\leq 2}$ satisfiable if and only if it is satisfiable in the root of a binary tree. Stockmeyer [9] showed that the set of true quantified 3CNF formulas is PSPACE-complete. Using padding, it is immediate that the following variation of this set is also PSPACE-complete.

QUANTIFIED 3SAT: Given a quantified Boolean formula $\exists p_1 \forall p_2 \exists p_3 \cdots \exists p_{n-1} \forall p_n \phi$, where ϕ is a propositional formula over $p_1, \ldots, p_n$ in 3CNF (that is, a formula in conjunctive normal form with exactly 3 literals per clause), is the formula true?

We will reduce QUANTIFIED 3SAT to poor man's satisfiability with respect to binary trees. To simulate the quantifiers, we need to go back to the formula that forces models to be of exponential size.

$$\phi_{asg} = \bigwedge_{i=1}^{n} \Box^{i-1}(\Diamond \Box^{n-i} p_i \wedge \Diamond \Box^{n-i} \overline{p_i}).$$

ϕ_{asg} is clearly satisfiable in the root of a binary tree and if ϕ_{asg} is satisfied in the root of a binary tree, the worlds of depth $\leq n$ form a complete binary tree of depth n and every assignment to $p_1, \ldots, p_n$ occurs exactly once in a world at depth n. We will call the worlds at depth n the assignment-worlds.

The assignment-worlds in a subtree rooted at a world at distance $i \leq n$ from the root are constant with respect to the value of p_i. It follows that $\exists p_1 \forall p_2 \exists p_3 \cdots \exists p_{n-1} \forall p_n \phi \in$ QUANTIFIED 3SAT if and only if $\phi_{asg} \wedge (\Diamond \Box)^{n/2} \phi$ is satisfiable with respect to binary trees.

This proves that satisfiability for $\mathcal{F}_{\leq 2}$ is PSPACE-hard, but it does *not* prove that the poor man's version is PSPACE-hard. Recall that ϕ is in 3CNF and thus not a poor man's formula.

Below, we will show how to label all assignment-worlds where ϕ does not hold by f (for *false*). It then suffices to add the conjunct $(\Diamond \Box)^{n/2} \overline{f}$ to obtain a reduction.

How can we label all assignment-worlds where ϕ does not hold by f? Let k be such that $\phi = \psi_1 \wedge \psi_2 \wedge \cdots \wedge \psi_k$, where each ψ_i is the disjunction of exactly 3 literals: $\psi_i = \ell_{i1} \vee \ell_{i2} \vee \ell_{i3}$. We assume without loss of generality that n is even and that each ψ_i contains 3 different propositional variables.

For every i, we will label all assignment-worlds where ψ_i does not hold by f. Since $\psi_i = \ell_{i1} \vee \ell_{i2} \vee \ell_{i3}$, this implies that we have to label all assignment-worlds where $\overline{\ell_{i1}} \wedge \overline{\ell_{i2}} \wedge \overline{\ell_{i3}}$ holds by f. In general, this cannot be done in poor man's logic, but in this special case we are able to do it, because the relevant part of the model is completely fixed by ϕ_{asg}.

As a warm-up, first consider how you would label all assignment-worlds where $\overline{p_3}$ holds by f. This is easy; add the conjunct

$$\Box \Box \Diamond \Box^{n-3}(\overline{p_3} \wedge f).$$

You can label all assignment-worlds where $\overline{p_3} \wedge p_5$ holds as follows:

$$\Box \Box \Diamond \Box \Diamond \Box^{n-5}(\overline{p_3} \wedge p_5 \wedge f).$$

This can easily be generalized to a labeling for $\overline{p_3} \wedge p_5 \wedge \overline{p_8}$:

$$\Box \Box \Diamond \Box \Diamond \Box \Box \Diamond \Box^{n-8}(\overline{p_3} \wedge p_5 \wedge \overline{p_8} \wedge f).$$

Note that we can write the previous formula in the following suggestive way:

$$\Box^{3-1}\Diamond\Box^{5-3-1}\Diamond\Box^{8-5-1}\Diamond\Box^{n-8}(\overline{p_3} \wedge p_5 \wedge \overline{p_8} \wedge f).$$

In general, suppose you want to label all assignment-worlds where $\ell_1 \wedge \ell_2 \wedge \ell_3$ hold by f, where ℓ_1, ℓ_2, and ℓ_3 are literals. Suppose that ℓ_1, ℓ_2, and ℓ_3's propositional variables are p_a, p_b, and p_c, respectively. Also suppose that $a < b < c$. The labeling formula *label_false*$(\ell_1 \wedge \ell_2 \wedge \ell_3)$ is defined as follows.

$$\mathit{label\_false}(\ell_1 \wedge \ell_2 \wedge \ell_3) = \Box^{a-1}\Diamond\Box^{b-a-1}\Diamond\Box^{c-b-1}\Diamond\Box^{n-c}(\ell_1 \wedge \ell_2 \wedge \ell_3 \wedge f).$$

If *label_false*$(\ell_1 \wedge \ell_2 \wedge \ell_3)$ is satisfied in the root of a complete binary tree, then there exist at least 2^{n-3} worlds at depth n such that $(\ell_1 \wedge \ell_2 \wedge \ell_3 \wedge f)$ holds.

If ϕ_{asg} is satisfied in the root of a binary tree, then the worlds of depth $\leq n$ form a complete binary tree and there are exactly 2^{n-3} assignment-worlds such that $(\ell_1 \wedge \ell_2 \wedge \ell_3)$ holds.

It follows that if ϕ_{asg} is satisfied in the root of a binary tree, then *label_false*$(\ell_1 \wedge \ell_2 \wedge \ell_3)$ is satisfied in the root if and only if f holds in every assignment-world where $(\ell_1 \wedge \ell_2 \wedge \ell_3)$ holds.

Thus, the following function g is a reduction from QUANTIFIED 3SAT to poor man's satisfiability with respect to $\mathcal{F}_{\leq 2}$.

$$g(\exists p_1 \forall p_2 \exists p_3 \cdots \exists p_{n-1} \forall p_n \phi) = \phi_{asg} \wedge \bigwedge_{i=1}^{k} \mathit{label\_false}(\overline{\ell_{i1}} \wedge \overline{\ell_{i2}} \wedge \overline{\ell_{i3}}) \wedge (\Diamond\Box)^{n/2}\overline{f}.$$

□

Why doesn't the construction of Theorem 4 work for K? A formula that is satisfiable in a world with exactly two successors is also satisfiable in a world with more than two successors. Because of this, the *label_false* formula will not necessarily label all assignment-worlds where ϕ does not hold by f. For a very simple example, consider the formula

$$\Diamond p \wedge \Diamond\overline{p} \wedge \Diamond(p \wedge f) \wedge \Diamond(\overline{p} \wedge f) \wedge \Diamond\overline{f}.$$

This formula is not $\mathcal{F}_{\leq 2}$ satisfiable, since both the p successor and the $\overline{p}$ successor are labeled f. However, this formula is satisfiable in a world with three successors, satisfying $p \wedge f$, $\overline{p} \wedge f$, and $\overline{f}$, respectively.

5 $\mathcal{ALEN}$ Satisfiability Is PSPACE-Complete

In the introduction, we mentioned that poor man's logic is closely related to certain description logics. Donini et al. [2] almost completely characterize the complexity of the most common description logics. The only language they couldn't completely characterize is $\mathcal{ALEN}$. $\mathcal{ALEN}$ is $\mathcal{ALE}$ (the poor man's version of multi-modal K) with number restrictions. Number restrictions are of the form

($\leq n$) and ($\geq n$). ($\leq n$) is true if and only if a world has $\leq n$ successors and ($\geq n$) is true if and only if a world has $\geq n$ successors.

In [2], it was shown that $\mathcal{ALEN}$ satisfiability is in PSPACE, assuming that the number restrictions are given in unary. The best lower bound for satisfiability was the coNP lower bound that is immediate from the fact that this is an extension of $\mathcal{ALE}$.

We will use Theorem 4 to prove PSPACE-hardness for a very restricted version of $\mathcal{ALEN}$.

Theorem 5. *Satisfiability for the poor man's version of K extended with the number restriction* (≤ 2) *is PSPACE-hard.*

Proof. The reduction from poor man's satisfiability with respect to $\mathcal{F}_{\leq 2}$ is obvious. It suffices to use the number restriction (≤ 2) to make sure that every world in the relevant part of the model has at most two successors. Let $md(\phi)$ be the modal depth of ϕ. All worlds that are of importance to the satisfiability of ϕ are at most $md(\phi)$ steps away from the root. The reduction is as follows:

$$f(\phi) = \phi \wedge \bigwedge_{i=0}^{md(\phi)} \Box^i (\leq 2)$$

□

Combining this with the PSPACE upper bound from [2] completely characterizes the complexity of $\mathcal{ALEN}$ satisfiability.

Corollary 1. *$\mathcal{ALEN}$ satisfiability is PSPACE-complete.*

6 Other Restrictions on the Set of Operators

As mentioned in the introduction, restricting the modal language in the way that we have, i.e., looking at formulas built from literals, $\wedge$, $\Box$, and $\Diamond$, was motivated by the fact that this restriction occurs in description logics and also by the rather bizarre complexity behavior of this fragment.

From a more technical point of view however, we might well wonder what happens to other restrictions on the set of operators allowed. After all, who is to say which sublanguages will be useful in the future? Also, we might hope to gain more insight in the sources of complexity for modal logics by looking at different sublanguages.

For $S \subseteq \{\neg, ^-, \wedge, \vee, \Box, \Diamond, \mathit{true}, \mathit{false}\}$, let $\mathcal{L}(S)$ denote the modal language whose formulas are built from an infinite set of propositional variables and operators from S. We will write $^-$ for propositional negation, and $\neg$ for general negation. So, our "old" language $\mathcal{L}$ will be denoted by $\mathcal{L}(\{^-, \wedge, \vee, \Box, \Diamond\})$, and poor man's language by $\mathcal{L}(\{^-, \wedge, \Box, \Diamond\})$.

Completely characterizing the complexity of $\mathcal{L}(S)$ satisfiability (with respect to the class of all frames) for every $S \subseteq \{\neg, ^-, \wedge, \vee, \Box, \Diamond, \mathit{true}, \mathit{false}\}$ may seem to be a daunting task, since there are 2^8 subsets to consider. But it turns out that there are only four possibilities for the complexity of these satisfiability problems: P, NP-complete, coNP-complete, and PSPACE-complete. Also, there are not many surprises: Languages that contain a complete basis for modal logic obviously have PSPACE-complete satisfiability problems, languages that contain a complete basis for propositional logic, but not for modal logic have NP-complete satisfiability problems, and poor man's logic (with or without constants) is the only coNP-complete case. All other cases are in P, except for the one surprise that $\mathcal{L}(\{\wedge, \vee, \Box, \Diamond, \mathit{false}\})$ satisfiability is PSPACE-complete.

Due to space limitations, we refer the reader to the full version of this paper for the proofs [5].

Acknowledgments

I would like to thank Johan van Benthem for suggesting this topic, Johan van Benthem, Hans de Nivelle, and Maarten de Rijke for helpful conversations and suggestions, and the anonymous STACS referees for useful comments and suggestions.

References

1. S. Demri and Ph. Schnoebelen. The Complexity of Propositional Linear Temporal Logics in Simple Cases (Extended Abstract). In *15th Annual Symposium on Theoretical Aspects of Computer Science*, pp. 61–72, 1998.
2. F. Donini, M. Lenzerini, D. Nardi, and W. Nutt. The complexity of concept languages. *Information and Computation*, 134, pp. 1-58, 1997.
3. F. Donini, B. Hollunder, M. Lenzerini, D. Nardi, W. Nutt, and A. Spaccamela. The complexity of existential quantification in concept languages. *Artificial Intelligence*, 53, pp. 309–327, 1992.
4. J. Y. Halpern. The effect of bounding the number of primitive propositions and the depth of nesting on the complexity of modal logic. *Artificial Intelligence*, 75(2), pp. 361-372, 1995.
5. E. Hemaspaandra. The complexity of poor man's logic. ACM Computing Research Repository Technical Report cs.LO/9911014, 1999.
6. R. Ladner. The computational complexity of provability in systems of modal propositional logic. *SIAM Journal on Computing*, 6(3), pp. 467–480, 1977.
7. K. Schild. A correspondence theory for terminological logics: preliminary report. In *Proceedings of the 12th International Joint Conference on Artificial Intelligence*, pp. 466–471, 1991.
8. M. Schmidt-Schauss and G. Smolka. Attributive concept descriptions with complements. *Artificial Intelligence*, 48, pp. 1–26, 1991.
9. L. Stockmeyer. The polynomial-time hierarchy. *Theoretical Computer Science*, 3, pp. 1–22, 1977.

Fast Integer Sorting in Linear Space

Yijie Han

Computer Science Telecommunications Program
University of Missouri - Kansas City
5100 Rockhill Road
Kansas City, Missouri 64110, USA
han@cstp.umkc.edu
http://welcome.to/yijiehan

Abstract. We present a fast deterministic algorithm for integer sorting in linear space. Our algorithm sorts n integers in linear space in $O(n(\log\log n)^{1.5})$ time. This improves the $O(n(\log\log n)^2)$ time bound given in [11]. This result is obtained by combining our new technique with that of Thorup's[11]. The approach and technique we provide are totally different from previous approaches and techniques for the problem. As a consequence our technique can be extended to apply to nonconservative sorting and parallel sorting. Our nonconservative sorting algorithm sorts n integers in $\{0, 1, ..., m-1\}$ in time $O(n(\log\log n)^2/(\log k+\log\log\log n))$ using word length $k\log(m+n)$, where $k \leq \log n$. Our EREW parallel algorithm sorts n integers in $\{0, 1, ..., m-1\}$ in $O((\log n)^2)$ time and $O(n(\log\log n)^2/\log\log\log n)$ operations provided $\log m = \Omega((\log n)^2)$.

1 Introduction

Sorting is a classical problem which has been studied by many researchers. Although the complexity for comparison sorting is now well understood, the picture for integer sorting is still not clear. The only known lower bound for integer sorting is the trivial $\Omega(n)$ bound. Recent advances in the design of algorithms for integers sorting have resulted in fast algorithms[3][6][12]. However, these algorithms use randomization or superlinear space. For sorting integers in $\{0, 1, ..., m-1\}$ $O(nm^{\epsilon})$ space is used in the algorithms reported in [3][6]. When m is large (say $m = \Omega(2^n)$) the space used is excessive. Integer sorting using linear space is therefore extensively studied by researchers. An earlier work by Fredman and Willard[4] shows that n integers can be sorted in $O(n\log n/\log\log n)$ time in linear space. Raman showed that sorting can be done in $O(n\sqrt{\log n\log\log n})$ time in linear space[10]. Later Andersson improved the time bound to $O(n\sqrt{\log n})$[2]. Then Thorup improved the time bound to $O(n(\log\log n)^2)$ [11]. In this paper we further improve upon previous results. We show that n integers can be sorted in $O(n(\log\log n)^{1.5})$ time in linear space.

Unlike previous techniques [2][4][10][11], our technique can be extended to apply to nonconservative sorting and parallel sorting. Conservative sorting is to sort n integers in $\{0, 1, ...m-1\}$ with word length (the number of bits in a word)

H. Reichel and S. Tison (Eds.): STACS 2000, LNCS 1770, pp. 242–253, 2000.

$O(\log(m + n))$ [8]. Nonconservative sorting is to sort with word length larger than $O(\log(m + n))$. We show that n integers in $\{0, 1, ..., m - 1\}$ can be sorted in time $O(n(\log\log n)^2/(\log k + \log\log\log n))$ with word length $k\log(m + n)$ where $k \leq \log n$. Thus if $k = (\log n)^\epsilon$, $0 < \epsilon < 1$, the sorting can be done in linear space and $O(n \log\log n)$ time. Andersson[2] and Thorup[11] did not show how to extend their linear space sorting algorithm to nonconservative sorting. Thorup [11] used an algorithm to insert a batch of n integers into the search tree in $O(n \log\log n)$ time. When using word length $k\log(m+n)$ this time complexity can be reduced to $O(n(\log\log n - \log k))$, thus yielding an $O(n \log\log n(\log\log n - \log k))$ time algorithm for linear space sorting, which is considerably worse than our algorithm.

Also note that previous results [2][4][11] do not readily extend to parallel sorting. Our technique can be applied to obtain a more efficient parallel algorithm for integer sorting. In this regard the best previous result on the EREW PRAM is due to Han and Shen[7] which sorts n integers in $\{0, 1, ..., m-1\}$ in $O(\log n)$ time and $O(n\sqrt{\log n})$ operations(time processor product). We show when $\log m = \Omega((\log n)^2)$ we can sort in $O((\log n)^2)$ time with $O(n(\log\log n)^2/\log\log\log n)$ operations on the EREW PRAM. Thus for large integers our new algorithm is more efficient than the best previous algorithm.

2 Preparation

Word length is the number of bits in a word. For sorting n integers in the range $\{0, 1, 2, ..., m - 1\}$ we assume that the word length used in our conservative algorithm is $O(\log(m + n))$. The same assumption is made in previous designs[2][4][10][11]. In integer sorting we often pack several small integers into one word. We always assume that all the integers packed in a word use the same number of bits. Suppose k integers each having l bits are packed into one word. By using the test bit technique[1][3] we can do a pairwise comparison of the corresponding integers in two words and extract the larger integers into one word and smaller integers into another word in constant time. Therefore by adapting well-known selection algorithms (e.g. select median for every 5 elements, find the median a among the selected elements, use a to eliminate 1/4 of the elements and then recurse), we immediately have the following lemma:

Lemma 1: Selecting the s-th largest integer among the n integers packed into n/k words can be done in $O(n \log k/k)$ time and $O(n/k)$ space. In particular the median can be found in $O(n \log k/k)$ time and $O(n/k)$ space.

The factor $\log k$ in Lemma 1 comes from the fact that after a constant number of integers are eliminated we have to pack the integers into fewer number of words. This packing incurs the factor $\log k$ in the time complexity.

Now consider sorting small integers. Let k integers be packed in one word. We say that the nk integers in n words are sorted if ki-th to $(k(i + 1) - 1)$-th smallest integers are sorted and packed in the i-th word, $0 \leq i < n$. We have the following lemma:

Lemma 2: If $k = 2^t$ integers using a total of $(\log n)/2$ bits are packed into one word, then the nk integers in n words can be sorted in $O(nt) = O(n \log k)$ time and $O(n)$ space.

Proof: Because only $(\log n)/2$ bits are used in each word to store k integers we can use bucket sorting to sort all words by treating each word as one integer and this takes $O(n)$ time and space. Because only $(\log n)/2$ bits are used in each word there are only $\sqrt{n}$ patterns for all the words. We then put $k < (\log n)/2$ words with the same pattern into one group. For each pattern there are at most $k-1$ words left which cannot form a group. Therefore at most $\sqrt{n} \cdot (k-1)$ words cannot form groups. For each group we use the following algorithm to move the i-th integer in all k words into one word.

Algorithm Transpose($s, A_0, A_1, ..., A_{2g-1}$)
/* $s = (g \log n)/(2k)$. Let A_i be the i-th word, $0 \leq i < 2g$. Let B_i and C_i, $0 \leq i < k$ be words used for temporary storage. Let D_i be the constant which is $(0^i 1^i)^{(\log n)/(4i)}$ when represented as a binary string. Let E_i be the constant which is 1's complement of D_i, that is $(1^i 0^i)^{(\log n)/(4i)}$ when represented as a binary string. *AND* and *OR* are bit-wise AND and OR operations. */
for $i = 0$ **to** $g - 1$ **do**
  begin
    $B_i = A_i \; AND \; D_s$;
    $B_{i+g} = A_{i+g} \; AND \; D_s$;
    $B_i = B_i * 2^s$;
    $B_i = B_i \; OR \; B_{i+g}$;
    $C_i = A_i \; AND \; E_s$;
    $C_{i+g} = A_{i+g} \; AND \; E_s$;
    $C_i = C_i / 2^s$;
    $C_i = C_i \; OR \; C_{i+g}$;
    $A_i = B_i$;
    $A_{i+g} = C_i$;
  end
if $g = 1$ **return**;
Call Transpose($s/2, A_0, A_1, ..., A_{g-1}$);
Call Transpose($s/2, A_g, A_{g+1}, ..., A_{2g-1}$);

We invoke Transpose($(\log n)/4, A_0, A_1, ..., A_{k-1}$) to transpose the integers in a group. This takes $O(k \log k)$ time and $O(k)$ space for each group. Therefore for all groups it takes $O(n \log k)$ time and $O(n)$ space. For the words not in a group (there are at most $\sqrt{n} \cdot (k-1)$ of them) we simply disassemble the words and then reassemble the words. This will take no more than $O(n)$ time and space. After all these are done we then use bucket sorting again to sort the n words. This will have all the integers sorted. □

Note that when $k = O(\log n)$ we are sorting $O(n \log n)$ integers packed in n words in $O(n \log \log n)$ time and $O(n)$ space. Therefore the saving is considerable.

Lemma 3: Assume that each word has $\log m > \log n$ bits, that k integers each having $(\log m)/k$ bits are packed into one word, that each integer has a label containing $(\log n)/(2k)$ bits, and that the k labels are packed into one word the same way as integers are packed into words (that is, if integer a is packed as the s-th integer in the t-th word then the label for a is packed as the s-th label in the t-th word for labels), then n integers in n/k words can be sorted *by their labels* in $O((n \log\log n)/k)$ time and $O(n/k)$ space.

Proof: The words for labels can be sorted by bucket sorting because each word uses $(\log n)/2$ bits. The sorting will group words for integers into groups as in Lemma 2. We can then call Transpose on each group of words for integers. □

Note also that the sorting algorithm given in Lemma 2 and Lemma 3 are not stable. As will be seen later we will use these algorithms to sort arbitrarily large integers. Even though we do not know how to make the algorithm in Lemma 2 stable, as will be seen that our sorting algorithm for sorting large integers can be made stable by using the well known method of appending the address bits to each input integer.

If we have larger word length the sorting can be done faster as shown in the following lemma.

Lemma 4: Assume that each word has $\log m \log\log n > \log n$ bits, that k integers each having $(\log m)/k$ bits are packed into one word, that each integer has a label containing $(\log n)/(2k)$ bits, and that the k labels are packed into one word the same way as integers are packed into words, then n integers in n/k words can be sorted by their labels in $O(n/k)$ time and $O(n/k)$ space.

Proof: Note that although word length is $\log m \log\log n$ only $\log m$ bits are used for storing packed integers. As in Lemmas 2 and 3 we sort the words containing packed labels by bucket sorting. Instead of putting k words into one group we put $k \log\log n$ words into one group. To transpose the integers in a group containing $k \log\log n$ words we first further pack $k \log\log n$ words into k words by packing $\log\log n$ words into one word. We then do transpose on the k words. Thus transpose takes only $O(k \log\log n)$ time for each group and $O(n/k)$ time for all integers. After finishing transpose we then unpack the integers in the k words into $k \log\log n$ words. □

Note also if the word length is $\log m \log\log n$ and only $\log m$ bits are used to pack $k \le \log n$ integers into one word. Then the selection in Lemma 1 can be done in $O(n/k)$ time and space.

3 The Approach and the Technique

Consider the problem of sorting n integers in $\{0, 1, ..., m-1\}$. We assume that each word has $\log m$ bits and that $\log m \ge \log n \log\log n$. Otherwise we can use radix sorting to sort in $O(n \log\log n)$ time and linear space. We divide the $\log m$ bits used for representing each integer into $\log n$ blocks. Each block thus contains at least $\log\log n$ bits. The i-th block containing $(i \log m/\log n)$-th to $((i+1)\log m/\log n - 1)$-th bits. Bits are counted from the least significant bit starting at 0. We sort from high order bits to low order bits. We now propose a $2\log n$ stage algorithm which works as follows.

In each stage we work on one block of bits. We call these blocks small integers because each small integer now contains only $\log m/\log n$ bits. Each integer is represented by and corresponds to a small integer which we are working on. Consider the 0-th stage which works on the most significant block (the $(\log n-1)$-th block). Assume that the bits in these small integers are packed into $n/\log n$ words with $\log n$ small integers packed into one word. For the moment we ignore the time needed for packing these small integers into $n/\log n$ words and assume that this is done for free. By Lemma 1 we can find the median of these n small integers in $O(n\log\log n/\log n)$ time and $O(n/\log n)$ space. Let a be the median found. Then n small integers can be divided into at most three sets S_1, S_2, and S_3. S_1 contains small integers which are less than a. S_2 contains small integers which are equal to a. S_3 contains small integers which are greater than a. We also have $|S_1| \leq n/2$ and $|S_3| \leq n/2$. Although $|S_2|$ could be larger than $n/2$ all small integers in S_2 are equal. Let S_2' be the set of integers whose most significant block is in S_2. Then we can eliminate $\log m/\log n$ bits (the most significant block) from each integer in S_2' from further consideration. Thus after one stage each integer is either in a set whose size is at most half of the size of the set at the beginning of the stage, or one block of bits ($\log m/\log n$ bits) of the integer can be eliminated from further computation. Because there are only $\log n$ blocks in each integer, each integer takes at most $\log n$ stages to eliminate blocks of bits. An integer can be put in a half sized set for at most $\log n$ times. Therefore after $2\log n$ stages all integers are sorted. Because in each stage we are dealing with only $n/\log n$ words, if we ignore the time needed for packing small integers into words and for moving small integers to the right set then the remaining time and space complexity will be $O(n\log\log n)$ because there are only $2\log n$ stages.

The subtle part of the algorithm is how to move small integers into the set where the corresponding integer belongs after previous set dividing operations of our algorithm. Suppose that n integers have already been divided into k sets. Also assume that $(\log n)/(2\log k)$ small integers each containing $\log k$ continuous blocks of an integer are packed into one word. For each small integer we use a label of $\log k$ bits indicating which set it belongs. Assume that the labels are also packed into words the same way as the small integers are packed into words with $(\log n)/(2\log k)$ labels packed into one words. Thus if small integer a is packed as the s-th small integer in the t-th word then the label for a is packed as the s-th label in the t-th word for labels. Note that we cannot disassemble the small integers from the words and then move them because this will incur $O(n)$ time. Because each word for labels contains $(\log n)/(2\log k)$ labels therefore only $(\log n)/2$ bits are used for each such word. Thus Lemma 3 can be applied here to move the small integers into the sets they belong to. Because only $O((n\log k)/\log n)$ words are used the time complexity for moving small integers to their sets is $O((n\log\log n\log k)/\log n)$.

Note that $O(\log k)$ blocks for each small integer is the most number of bits we can move in applying Lemma 3 because each word has $\log m$ bits. Note also that the moving process is not stable as the sorting algorithm in Lemma 3 is not stable.

With such a moving scheme we immediately face the following problem. If integer a is the k-th member of a set S. That is, a block of a (call it a') is listed as the k-th (small) integer in S. When we use the above scheme to move the next several blocks of a (call it a'') into S, a'' is merely moved into a position in set S, but not necessarily to the k-th position (the position where a' locates). If the value of the block for a' is identical for all integers in S that does not create problem because that block is identical no matter which position in S a'' is moved to. If the value of the block for a' is not identical for all integers in S then we have problem continue the sorting process. What we do is the following. At each stage the integers in one set works on a common block which is called the current block of the set. The blocks which proceed the current block contain more significant bits of the integer and are identical for all integers in the set. When we are moving more bits into the set we move the following blocks together with the current block into the set. That is, in the above moving process we assume the most significant block among the $\log k$ continuous blocks is the current block. Thus after we move these $\log k$ blocks into the set we delete the original current block because we know that the $\log k$ blocks are moved into the correct set and that where the original current block locates is not important because that current block is contained in the $\log k$ blocks.

Another problem we have to pay attention is that the size of the sets after several stages of dividing will become small. The scheme of Lemmas 2 and 3 relies on the fact that the size of the set is not very small. We cope with this problem in this way. If the size of the set is larger than $\sqrt{n}$ we keep dividing the set. In this case each word for packing the labels can use at least $(\log n)/4$ bits. When the size of the set is no larger than $\sqrt{n}$ we then use a recursion to sort the set. In each next level of recursion each word for packing the labels uses less number of bits. The recursion has $O(\log\log n)$ levels.

Below is our sorting algorithm which is used to sort integers into sets of size no larger than $\sqrt{n}$. This algorithm uses yet another recursion (do not confuse this recursion with the recursion mentioned in the above paragraph).

Algorithm Sort($level, a_0, a_1, ..., a_t$)
/* a_i's are the input integers in a set to be sorted. *level* is the recursion level. */
1. **if** *level* $= 1$ **then** examine the size of the set (i.e. t). If the size of the set is less than or equal to $\sqrt{n}$ then return. Otherwise use the current block to divide the set into at most three sets by using Lemma 1 to find the median and then using Lemma 3 to sort. For the set all of its elements are equal to the median eliminate the current block and note the next block to become the current block. Create a label which is the set number (0, 1 or 2 because the set is divided into at most three sets) for each integers. Then reverse the computation to route the label for each integer back to the position where the integer located in the input to the procedure call. Also route a number (a 2 bit number) for each integer indicating the current block back to the location of the integer. This is possible because we can assume each block has at least $\log\log n$ bits. Return.

2. Cut the bits in each integer a_i into equal two segments a_i^{High} (high order bits) and a_i^{Low} (low order bits). Pack a_i^{High}'s into half the number of words. Call Sort($level-1, a_0^{High}, a_1^{High}, ..., a_t^{High}$). /*When the algorithm returns from this recursive call the label for each integer indicating the set the integer belongs is already routed back to the position where the integer locates in the input of the procedure call. */
3. For each integer a_i extract out a_i^{Low} which has half the number of bits as in a_i and is a continuous segment with the most significant block being the current block of a_i. Pack a_i^{Low}'s into half the number of words as in the input. Route a_i^{Low}'s to their sets by using Lemma 3.
4. For each set $S = \{a_{i_0}, a_{i_1}, ..., a_{i_s}\}$ call Sort($level-1, a_{i0}^{Low}, a_{i_1}^{Low}, ..., a_{i_s}^{Low}$).
5. Route the label which is the set number for each integers back to the position where the integer located in the input to the procedure call. Also route a number (a $2(level+1)$ bit number) for each integer indicating the current block back to the location of the integer. This step is the reverse of the routing in Step 3.

In Step 3 of algorithm Sort we need to extract a_i^{Low}'s and to pack them. The extraction requires a mask. This mask can be computed in $O(\log\log n)$ time for each word. Suppose k small integers each containing $(\log n)/(4k)$ blocks are packed in a word. We start with a constant which is $(0^{(t\log n)/(8k)}1^{(t\log n)/(8k)})^k$ when represented as a binary string, where t is the number of bits in a block. Because a $2(level+1)$ bit number a is used to note the current block we can check 1 bit of a in a step for all a's packed in a word (there are k of them). This can determine whether we need to shift the $1^{(t\log n)/(8k)}$ for each small integer to the left or not. Thus using $O(\log\log n)$ time we can produce the mask for each word. Suppose the current block is the $((\log n)/(8k)+g)$-th block then the resulting mask corresponding to this small integer will be $0^{t((\log n)/(8k)-g)}1^{(t\log n)/(8k)}0^{tg}$. Packing is to pack $s \le \log n$ blocks to consecutive locations in a word. This can be done in $O(\log\log n)$ time for each word by using the packing algorithm in [9](Section 3.4.3).

We let a block contain $(4\log m)/\log n$ bits. Then if we call Sort($\log((\log n)/4)$, $a_0, a_1, ..., a_{n-1}$) where a_i's are the input integers, then $(\log n)/4$ calls to the level 1 procedure will be executed. This could split the input set into $3^{(\log n)/4}$ sets. And therefore we need $\log 3^{(\log n)/4}$ bits to represent/index each set. When the procedure returns the number of eliminated bits in different sets could be different. Therefore we need modify our procedure a little bit. At level j we form a_i^{High} by extract out the 2^{j-1} continuous blocks with the most significant block being the current block from a_i. After this modification we call Sort six times as below:

Algorithm IterateSort
Call Sort($\log((\log n)/4), a_0, a_1, ..., a_{n-1}$);
for $j = 1$ **to** 5 **do**
  begin
    Move a_i to its set by bucket sorting because there are only about $\sqrt{n}$ sets;
    For each set $S = \{a_{i_0}, a_{i_1}, ..., a_{i_t}\}$ if $t > \sqrt{n}$ then

call Sort($\log((\log n)/4), a_{i_0}, a_{i_1}, ..., a_{i_t}$);

end

Then $(3/2)\log n$ calls to the level 1 procedure are executed. Blocks can be eliminated at most $\log n$ times. The other $(1/2)\log n$ calls are sufficient to partition the input set of size n into sets of size no larger than $\sqrt{n}$.

At level j we use only $n/2^{\log((\log n)/4)-j}$ words to store small integers. Each call to the Sort procedure involves a sorting on labels and a transposition of packed integers (use Lemma 3) and therefore involves a factor of $\log\log n$ in time complexity. Thus the time complexity of algorithm Sort is:

$$T(level) = 2T(level-1) + cn\log\log n/2^{\log((\log n)/4)-level}; \quad (1)$$
$$T(0) = 0.$$

where c is a constant. Thus $T(\log((\log n)/4)) = O(n(\log\log n)^2)$. Algorithm IterateSort only sorts sets into sizes less than $\sqrt{n}$. We need another recursion to sort sets of size less than $\sqrt{n}$. This recursion has $O(\log\log n)$ levels. Thus the time complexity to have the input integers sorted is $O(n(\log\log n)^3)$.

The sorting process is not stable. Since we are sorting arbitrarily large integers we can append the address bits to each input integer to stabilize the sorting. Although this requires that each word contains $\log m + \log n$ bits, when $m \geq n$ the number of bits for each word can be kept at $\log m$ by using the idea of radix sorting.

The space used for each next level of recursion in Sort uses half the size of the space. After recursion returns the space can be reclaimed. Thus the space used is linear, i.e. $O(n)$.

Theorem 1: n integers can be sorted in linear space in time $O(n(\log\log n)^3)$. □

4 An Algorithm with Time Complexity $O(n(\log\log n)^2)$

We first note the following Lemma.

Lemma 5: If the word length used in the algorithm is $\log m \log\log n$. then n integers in $\{0, 1, ..., m-1\}$ can be sorted into sets of size no larger than $\sqrt{n}$ in linear space in time $O(n\log\log n)$.

Proof: In this case the median finding takes linear time and we can use Lemma 4 to sort packed small integers. Also it takes $O(\log\log n)$ time to extract out a_i^{High}'s and a_i^{Low}'s for $\log\log n$ words (including computing mask and packing) because we can pack $\log\log n$ words further into one word. Therefore formula (1) becomes:

$$T(level) = 2T(level-1) + cn/2^{\log((\log n)/4)-level}; \quad (2)$$
$$T(0) = 0.$$

Therefore $T(\log((\log n)/4)) = O(n\log\log n)$. That is, the time complexity for dividing the input set to sets of size no larger than $\sqrt{n}$ is $O(n\log\log n)$. □

We apply the following technique to improve the time complexity of our algorithm further.

We divide $\log m$ bits of an integers into $\log n \log\log n$ blocks with each block containing $(\log m)/(\log n \log\log n)$ bits. Note that each block has at least $\log\log n$ bits because we can assume that $\log m \geq \log n(\log\log n)^2$ for otherwise we can use radix sort to sort the integers. We execute the following algorithm:

Algorithm SpeedSort

while there is a set S which has size $> \sqrt{n}$ **do**

begin

1. for each integer $a_i \in S$ extract out a'_i which contains $\log n$ continuous blocks of a_i with the most significant block being the current block, put all a'_i's in S';
2. Call IterateSort on set S';

end

Now since during the sorting process each word stores only $(\log n)/4$ blocks therefore only $\log m/\log\log n$ bits are used. By Lemma 5 one iteration of the while loop in SpeedSort takes $O(\log\log n)$ time for each integer. We account the time for each integer in the sorting process by two variables D and E. If an integer a has gone through t iterations of the while loop of SpeedSort then $(t-1)\log n$ blocks of a has been eliminated we add $O((t-1)\log\log n)$ to variable E indicating that that much time has been expended to eliminate $(t-1)\log n$ blocks. We also add $O(\log\log n)$ time to variable D indicating that that much time has been expended to divide the set in SpeedSort. Because we can eliminate at most $\log\log n \log n$ blocks therefore the value of E is upbounded by $O((\log\log n)^2)$ throughout the integer sorting process including the call to SpeedSort to dividing integers into sets of size $\leq \sqrt{n}$ and recursive calls to SpeedSort to finish dividing the resulting sets into singleton sets. The value of variable D is also upbounded by $O((\log\log n)^2)$ because there are $\log\log n$ levels of recursion (one level divides set of size $n^{1/2^i}$ to sets of size $n^{1/2^{i+1}}$) to divide integers into singleton sets. Therefore we have

Theorem 2: n integers can be sorted in linear space in $O(n(\log\log n)^2)$ time.

□

5 Nonconservative Sorting and Parallel Sorting

When the word length is $k\log(m+n)$ for sorting integers in $\{0, 1, ..., m-1\}$ we modify algorithm Sort in section 4. Here whenever we sort t bits in the integer we can move tk bits in step 3 of Sort. Thus in step 2 of Sort we can divide a_i into equal k segments. Subsequently we can invoke recursion k times. Each time we sort on a segment. Immediately upon the finish of each recursion we move a_i to its sorted set. We can move the whole a_i instead of a segment of a_i because we have the advantage of the nonconservatism. Therefore algorithm Sort can be done in $O(n(\log\log n)^2/\log k)$ time if each integer has only $O(\log m/k)$ bits. Here we assume that transposition is done in $O(n\log\log n)$ time for n words. If we apply the technique in section 5 then in each pass we are sorting only $O(\log m/(k \log\log n))$ bits for each integer. And therefore we can assume that the transposition can be done in $O(n)$ time for n words. Therefore the time complexity for algorithm Sort becomes $O(n \log\log n/\log k)$. Since there are $O(\log\log n)$ calls to Sort which are are made in the whole sorting process, the time complexity of our nonconservative sorting algorithm to sort n integers is $O(n(\log\log n)^2/\log k)$.

Theorem 3: n integers in $\{0, 1, ..., m-1\}$ can be sorted in $O(n(\log\log n)^2/\log k)$ time and linear space with word length $k\log(m+n)$, where $1 \leq k \leq \log n$.

Concerning parallel integer sorting we note that on the EREW PRAM we can have the following lemma to replace Lemma 1.

Lemma 6: An integer a among the n integers packed into n/k words can be computed on the EREW PRAM in $O(\log n)$ time and $O(n/k)$ operations using $O(n/k)$ space such that a is ranked at least $n/4$ and at most $3n/4$ among the n integers.

The proof of Lemma 6 can be obtained by applying Cole's parallel selection algorithm[5]. Note that we can do without packing and therefore the factor $\log k$ does not show up in the time complexity.

Currently Lemma 2 cannot be parallelized satisfactorily. On the EREW PRAM the currently best result[7] sorts in $O(\log n)$ time and $O(n\sqrt{\log n})$ operations. To replace Lemma 2 for parallel sorting we resort to nonconservatism.

Lemma 7: If $k = 2^t$ integers using a total of $(\log n)/2$ bits are packed into one word, then the nk integers in n words can be sorted in $O(\log n)$ time and $O(n)$ operations on the EREW PRAM using $O(n)$ space, provided that the word length is $\Omega((\log n)^2)$.

The sorting of words in Lemma 7 is done with the nonconservative sorting algorithm in [7]. The transposition can also be done in $O(n)$ operations because of nonconservatism.

For Lemma 3 we have to assume that $\log m = \Omega((\log n)^2)$. Then we can sort the n integers in n/k words by their labels in $O(\log n)$ time and $O((n \log\log n)/k)$ opearations on the EREW PRAM using $O(n/k)$ space. Note here that labels are themselves being sorted by nonconservative sorting algorithm in Lemma 7. Note also that the transposition here incurs a factor of $\log\log n$ in the operation complexity.

Lemma 4 and Section 5 say how do we remove the factor $\log\log n$ from the time complexity incurred in transposition with nonconservatism. This applies to parallel sorting as well to reduce the factor of $\log\log n$ from the operation complexity.

Because algorithm Sort uses algorithms in Lemmas 1 to 3 $O(\log n)$ times and because we can now replace Lemmas 1 to 3 with corresponding Lemmas for parallel computation, algorithm Sort is in effect converted into a parallel EREW PRAM algorithm with time complexity $O((\log n)^2)$ and operation complexity $O(n(\log\log n)^2)$. The technique in section 5 applies to parallel sorting. Therefore we have

Theorem 4: n integers in $\{0, 1, ..., m-1\}$ can be sorted in $O((\log n)^2)$ time and $O(n(\log\log n)^2)$ operations provided that $\log m = \Omega((\log n)^2)$.

Note that although algorithm Sort takes $O((\log n)^2)$ time, the whole sorting algorithm takes $O((\log n)^2)$ time as well because subsequent calls to Sort takes geometrically decreasing time.

6 Improving the Complexity

The results given in the previous sections can be improved further by using our approach and technique alone. In particular, $O(n(\log\log n)^2/\log\log\log n)$ time can be achieved for sequential linear space conservative sorting. $O(n(\log\log n)^2/\log\log\log n)$ operations and $O((\log n)^2)$ time can be achived for parallel linear space sorting on EREW PRAM. And $O(n(\log\log n)^2/(\log k + \log\log\log n))$ time can be achieved for nonconservative linear space sorting with word length $k\log(m+n)$, where $k \leq \log n$. The techniques for these improvements are too involved and warrants too much space for presentation. Therefore the details of these algorithms are omitted here and will be given in the full version of the paper. We instead show how to combine our algorithm with that of Thorup's[11] to obtain an algorithm with time complexity $O(n(\log\log n)^{1.5})$.

In [11] Thorup builds an exponential search tree and associates buffer $B(v)$ with each node v of the tree. He defines that a buffer $B(v)$ is over-full if $|B(v)| > d(v)$, where $d(v)$ is the number of children of v. Our modification on Thorup's approch is that we define $B(v)$ to be over-full if $|B(v)| > (d(v))^2$. Other aspects of Thorup's algorithm are not modified. Since a buffer is flushed (see Thorup's definition [11]) only when it is over-full, using our modification we can show that the time for flush can be reduced to $|B(v)|\sqrt{\log\log n}$. This will give the $O(n(\log\log n)^{1.5})$ time for sorting by Thorup's analysis[11].

The flush can be done in theory by sorting the elements in $B(v)$ together with the set $D(v)$ of keys at v's children. In our algorithm this theoretical sorting is done as follows. First, for each integer in $B(v)$, execute $\sqrt{\log\log n}$ steps of the binary search on the dictionary built in Section 3 of [11]. After that we have converted the original theoretical sorting problem into the problem of sorting $|B(v)|$ integers (come from $B(v)$ and denoted by $B'(v)$) of $\log m/2^{\sqrt{\log\log n}}$ bits with $d(v)$ integers (coming from $D(v)$ and denoted by $D'(v)$) of $\log m/2^{\sqrt{\log\log n}}$ bits. Note that here a word has $\log m$ bits. Also note that $|B(v)| > (d(v))^2$ and what we needed is to partition $B'(v)$ by the $d(v)$ integers in $D'(v)$ and therefore sorting all integers in $B'(v) \cup D'(v)$ is not necessary. By using the nonconservative version of the algorithm Sort we can then partition integers in $B'(v)$ into sets such that the cardinality of a set is either $< \sqrt{d(v)}$ or all integers in the set are equal. The partition maintains that for any two sets all integers in one set is larger than all integers in anther set. Because we used the nonconservative version of Sort the time complexity is $O(|B(v)|\sqrt{\log\log n})$. Then each integer in $D(v)$ can find out which set it falls in. Since each set has cardinality no larger than $\sqrt{d(v)}$ the integers in $D(v)$ can then further partition the sets they fall in and therefore partitioning $B(v)$ in an additional $O(|B(v)|)$ time. Overall the flush thus takes $O(|B(v)|\sqrt{\log\log n})$ time. By the analysis in [11] the time complexity for sorting in linear space is thus $O(n(\log\log n)^{1.5})$.

Theorem 5: n integers in $\{0, 1, ..., m-1\}$ can be sorted in $O(n(\log\log n)^{1.5})$ time and linear space.

7 Conclusions

The complexity of our algorithm could be improved further if we could sort better in Lemma 2, i.e. either to sort stably with the same complexity or to sort more bits in a word instead of $(\log n)/2$ bits. It is not clear whether $O(n(\log\log n)^{1.5})$ time is the lower bound for sorting integers in linear space. Note that our bound is very close to the current bound for sorting integer in nonlinear space (which is $O(n\log\log n)$)[3][6]. Also note that Thorup[12] showed that $O(n\log\log n)$ time and linear space can be achieve with randomization. It would be interesting to see whether the current $O(n\log\log n)$ time complexity for nonlinear space and randomization can be achieved in linear space deterministically.

References

1. S. Albers and T. Hagerup, *Improved parallel integer sorting without concurrent writing*, Information and Computation, **136**, 25-51(1997).
2. A. Andersson, *Fast deterministic sorting and searching in linear space*, Proc. 1996 IEEE Symp. on Foundations of Computer Science, 135-141(1996).
3. A. Andersson, T. Hagerup, S. Nilsson, R. Raman, *Sorting in linear time?* Proc. 1995 Symposium on Theory of Computing, 427-436(1995).
4. M.L. Fredman, D.E. Willard, *Surpassing the information theoretic bound with fusion trees*, J. Comput. System Sci. 47, 424-436(1994).
5. R. Cole, *An optimally efficient selection algorithm*, Information Processing Letters, **26**, 295-299(1987/88).
6. Y. Han, X. Shen, *Conservative algorithms for parallel and sequential integer sorting*, Proc. 1995 International Computing and Combinatorics Conference, Lecture Notes in Computer Science **959**, 324-333(August, 1995).
7. Y. Han, X. Shen, *Parallel integer sorting is more efficient than parallel comparison sorting on exclusive write PRAMs*. Proc. 1999 Tenth Annual ACM-SIAM Symposium on Discrete Algorithms (SODA'99), Baltimore, Maryland, 419-428(January 1999).
8. D. Kirkpatrick and S. Reisch, *Upper bounds for sorting integers on random access machines*, Theoretical Computer Science **28**, 263-276(1984).
9. F. T. Leighton, *Introduction to Parallel Algorithms and Architectures: Arrays, Trees, Hypercubes*, Morgan Kaufmann Publ., San Mateo, CA. 1992.
10. R. Raman, *Priority queues: small, monotone and trans-dichotomous*, Proc. 1996 European Symp. on Algorithms, Lecture Notes in Computer Science 1136, 121-137(1996).
11. M. Thorup. *Fast deterministic sorting and priority queues in linear space*, Proc. 1998 ACM-SIAM Symp. on Discrete Algorithms (SODA'98), 550-555(1998).
12. M. Thorup. *Randomized sorting in $O(n\log\log n)$ time and linear space using addition, shift, and bit-wise boolean operations*, Proc. 8th ACM-SIAM Symp. on Discrete Algorithms (SODA'97), 352-359(1997).

On the Performance of *WEAK-HEAPSORT*

Stefan Edelkamp[1] and Ingo Wegener[2]

[1] Institut für Informatik, Albert-Ludwigs-Universität,
Am Flughafen 17, D-79110 Freiburg;
edelkamp@informatik.uni-freiburg.de
[2] Lehrstuhl 2, Fachbereich Informatik
Universität Dortmund, D-44221 Dortmund
wegener@ls2.cs.uni-dortmund.de

Abstract. Dutton (1993) presents a further *HEAPSORT* variant called *WEAK-HEAPSORT*, which also contains a new data structure for priority queues. The sorting algorithm and the underlying data structure are analyzed showing that *WEAK-HEAPSORT* is the best *HEAPSORT* variant and that it has a lot of nice properties.
It is shown that the worst case number of comparisons is $n\lceil\log n\rceil - 2^{\lceil\log n\rceil} + n - \lceil\log n\rceil \le n\log n + 0.1n$ and weak heaps can be generated with $n-1$ comparisons. A double-ended priority queue based on weak-heaps can be generated in $n + \lceil n/2\rceil - 2$ comparisons.
Moreover, examples for the worst and the best case of *WEAK-HEAPSORT* are presented, the number of *Weak-Heaps* on $\{1,\dots,n\}$ is determined, and experiments on the average case are reported.

1 Introduction

General sequential sorting algorithms require at least $\lceil\log(n!)\rceil = n\log n - n\log e + \Theta(\log n) \approx n\log n - 1.4427n$ key comparisons in the worst case and $\lceil\log(n!)\rceil - \frac{1}{n!}2^{\lceil\log(n!)\rceil} \le \lceil\log(n!)\rceil - 1$ comparisons in the average case. We assume that the time for all other operations should be small compared to the time of a key comparison. Therefore, in order to compare different sorting algorithms the following six criteria are desirable:

1. The sorting algorithm should be general, i.e., objects of any totally ordered set should be sorted.
2. The implementation of the sorting algorithm should be easy.
3. The sorting algorithm should allow internal sorting, i.e., beside the space consumption for the input array only limited extra space is available.
4. For a small constant c the average case on key comparisons should be less than $n\log n + cn$.
5. For a small constant c' the worst case on key comparisons should be less than $n\log n + c'n$.
6. The number of all other operations such as exchanges, assignments and other comparisons should exceed the number of key comparisons by at most a constant factor.

H. Reichel and S. Tison (Eds.): STACS 2000, LNCS 1770, pp. 254–266, 2000.

BUCKETSORT and its *RADIXSORT* variants (Nilsson (1996)) are not general as required in Property 1.

Given $n = 2^k$ traditional *MERGESORT* performs at most $n \log n - n + 1$ key comparisons, but requires $O(n)$ extra space for the objects, which violates Property 3. Current work of J. Katajainen, T. A. Pasanen, and J. Teuhola (1996), J. Katajainen and T. A. Pasanen (1999), and K. Reinhardt (1992) shows that *MERGESORT* can be designed to be *in-place* and to achieve promising results, i.g. $n \log n - 1.3n + O(\log n)$ comparisons in the worst case. However, for practical purposes these algorithms tend to be too complicated and too slow.

INSERTIONSORT (Steinhaus (1958)) invokes less than $\sum_{i=1}^{n-1} \lceil \log(i+1) \rceil = \log(n!) + n - 1$ key comparisons, but even in the average case the number of exchanges is in $\Theta(n^2)$ violating Property 6.

SHELLSORT (Shell (1959)) requires as an additional input a decreasing integer sequence $h_1, \ldots, h_t = 1$. According to these distances of array indices traditional *INSERTIONSORT* is invoked. The proper choice of the distances is important for the running time of *SHELLSORT*. In the following we summarize the worst case number of operations (comparisons and exchanges). For $n = 2^k$ Shell's original sequence $(2^k, \ldots, 2, 1)$, leads to a quadratic running time. A suggested improvement of Hibbard (1963) achieves $O(n^{3/2})$ with the analysis given by Papernov and Stasevich (1965). Pratt (1979) provided a sequence of length $\Theta(\log^2 n)$ that led to $\Theta(n \log^2 n)$ operations. Sedgewick (1986) improves the $O(n^{3/2})$ bound for sequences of maximal length $O(\log n)$ to $O(n^{4/3})$ and in a joint work with Incerpi (1985) he further improves this to $O(n^{1+\epsilon/\sqrt{\log n}})$ for a given $\epsilon > 0$. Based on incompressibiliy results in Kolmogorov complexity, a very recent result of Jiang, Li and Vitányi (1999) states that the average number of operations in (so-called p pass) *SHELLSORT* for any incremental sequence is in $\Omega(pn^{1+1/p})$. Therefore, *SHELLSORT* violates Properties 4. and 5.

QUICKSORT (Hoare (1962)) consumes $\Theta(n^2)$ comparisons in the worst case. For the average case number of comparisons $V(n)$ we get the following recurrence equation $V(n) = n - 1 + \frac{1}{n} \sum_{k=1}^{n} (V(k-1) + V(n-k))$. This sum can be simplified to $V(n) = 2(n+1)H_n - 4n$, with $H_n = \sum_{i=1}^{n} 1/i$, and to the approximation $V(n) \approx 1.386 n \log n - 2.846 n + O(\log n)$.

Hoare also proposed *CLEVER-QUICKSORT*, the median-of-tree variant of *QUICKSORT*. In the worst case we still have $\Theta(n^2)$ key comparisons but the average case number can be significantly reduced. A case study reveals that the median of three objects can be found in $8/3$ comparisons on the average. Therefore, we have $n - 3 + 8/3 = n - 1/3$ comparisons in the divide step leading to the following recurrence for the average case $V(n) = n - 1/3 + \binom{n}{3}^{-1} \sum_{k=1}^{n} (k-1)(n-k)(V(k-1) + V(n-k))$. This sum simplifies to $V(n) = \frac{12}{7}(n+1)H_{n-1} - \frac{477}{147}n + \frac{223}{147} + \frac{252}{147n} \approx 1.188 n \log n - 2.255 n + O(\log n)$ (Sedgewick (1977)). No variant of *QUICKSORT* is known with $n \log n + o(n \log n)$ comparisons on average (cf. van Emden (1970)). Hence, it violates Properties 4. and 5.

The worst case case number on key comparisons in *HEAPSORT* independently invented by Floyd (1964) and Williams (1964) is bounded by $2n \log n + O(n)$. For the generating phase less than $2n - 2$ comparisons are required.

BOT-TOM-UP-HEAPSORT (Wegener (1993)) is a variant of *HEAPSORT* with $1.5n \log n + O(n)$ key comparisons in the worst case. The idea is to search the path to the leaf independently to the place for the root element to sink. Since the expected depth is high this path is traversed bottom-up. Fleischer (1991) as well as Schaffer and Sedgewick (1993) give worst case examples for which *BOT-TOM-UP-HEAPSORT* requires at least $1.5n \log n - o(n \log n)$ comparisons. Based on the idea of Ian Munro (cf. Li and Vitányi (1992)) one can infer that the average number of comparisons in this variant is bounded by $n \log n + O(n)$.

MDR-HEAPSORT proposed by McDiarmid and Reed (1989) performs less than $n \log n + cn$ comparisons in the worst case and extends *BOT-TOM-UP--HEAPSORT* by using one bit to encode on which branch the smaller element can be found and another one to mark if this information is unknown. The analysis that bounds c in *MDR-HEAPSORT* to 1.1 is given by Wegener (1993). *WEAK-HEAPSORT* is more elegant and faster. Instead of two bits per element *WEAK-HEAPSORT* uses only one and the constant c is less than 0.1.

In order to ensure the upper bound $n \log n + O(n)$ on the number of comparisons, *ULTIMATE-HEAPSORT* proposed by Katajainen (1998) avoids the worst case examples for *BOT-TOM-UP-HEAPSORT* by restricting the set of heaps to two layer heaps. It is more difficult to guarantee this restricted form. Katajainen obtains an improved bound for the worst case number of comparisons but the average case number of comparisons is larger as for *BOT-TOM-UP-HEAPSORT*. Here we allow a larger class of heaps that are easier to handle and lead to an improvement for the worst case *and* the average case.

There is one remaining question: How expensive is the extra space consumption of one bit per element? Since we assume objects with time-costly key comparisons we can conclude that their structure is more complete than an integer, which on current machines consumes 64 bits to encode the interval $[-2^{63} - 1, 2^{63} - 1]$. Investing one bit per element only halves this interval.

The paper is structured as follows. Firstly, we concisely present the design, implementation and correctness of the *WEAK-HEAPSORT* algorithm with side remarks extending the work of Dutton (1993). Secondly, we determine the number of *Weak-Heaps* according to different representation schemas. Afterwards we prove that there are both worst and best case examples that exactly meet the given bounds. Finally, we turn to the use of *Weak-Heaps* as a priority queue.

2 The *WEAK-HEAPSORT* Algorithm

2.1 Definition *Weak-Heap* and Array-Representation

A (Max-) *Weak-Heap* is established by relaxing the heap condition as follows:

1. Every key in the right subtree of each node is smaller than or equal to the key at the node itself.
2. The root has no left child.
3. Leaves are found on the last two levels of the tree only.

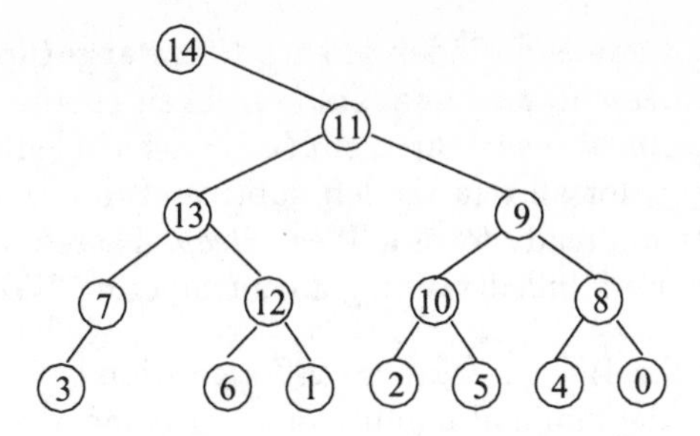

Fig. 1. Example of a *Weak-Heap*.

An example of a *Weak-Heap* is given in Figure 1. The underlying structure to describe the tree structure is a combination of two arrays. First, we have the array a in which the objects are found and second, the array of so-called *reverse* bits represents whether the tree associated with a node is rotated or not.

For the array representation we define the left child of index i as $2i + r_i$ and the right child as $2i + 1 - r_i$, $0 \leq i \leq n - 1$. Thus by flipping r_i we exchange the indices of the right and left children. The subtree of i is therefore rotated. For example, one array representation according to Fig. 1 is $a = [14, 11, 9, 13, 8, 10, 12, 7, 0, 4, 5, 2, 1, 6, 3]$ and $r = [0, 1, 1, 1, 1, 1, 1, 0, 1, 1, 1, 1, 1, 1, 1]$.

The successors of a node on the left branch of the right subtree are called grandchildren. For the example of Figure 1 the grandchildren of the root are labeled by 11, 13, 7 and 3. The inverse function `Gparent(x)` (for *grandparent*) is defined as `Gparent(Parent(x))` in case x is a left child and `Parent(x)` if x is a right one. `Gparent(x)` can be calculated by the following pseudo-code `while(odd(`x`) = `$r_{x/2}$`)` $x \leftarrow x/2$ followed by `return` $x/2$ with an obvious interpretation of `odd`.

2.2 Generating Phase

Let y be the index of the root of a tree T and x be the index of a node with $a_x \geq a_z$ for all z in the left subtree of T and let the right subtree of T and y itself be a *Weak-Heap. Merging* x and y gives a new *Weak-Heap* according to the following case study. If $a_x \geq a_y$ then the tree with root x and right child T is a *Weak-Heap*. If, however, $a_y > a_x$ we swap a_y with a_x and rotate the subtrees in T. By the definition of a *Weak-Heap* it is easy to see that if the leaves in T are located only on the last two levels merging x with y results in a *Weak-Heap*. The pseudo-code according to `merge` is given by `if (`$a_x < a_y$`) swap(`a_x`,`a_y`);` $r_y \leftarrow 1 - r_y$.

In the generating phase all nodes at index i for decreasing $i = n-1, \ldots, 1$ are merged to their grandparents. The pseudo-code for the so-called `WeakHeapify` procedure can be specified as: `for` $i \in \{n - 1, \ldots, 1\}$ `Merge(Gparent(i),i)`.

Theorem 1. `WeakHeapify` *generates a* Weak-Heap *according to its definition.*

Proof. Assume that there is an index y, such that `Merge(Gparent(y),y)` does not return a *Weak-Heap* at x = `Gparent(y)`. Then choose y maximal in this sense. Since all nodes $w > y$ with `Gparent(w)` $= x$ have led to a correct *Weak-Heap*, we have $a_x \geq a_z$ for all z in the left subtree of root y. On the other hand y and its right subtree already form a *Weak-Heap*. Therefore, all preconditions of merging x with y are fulfilled yielding a contradicting *Weak-Heap* at root x.

One reason why the *WEAK-HEAPSORT* algorithm is fast is that the generating phase requires the minimal number of $n-1$ comparisons.

Note that the `Gparent` calculations in `WeakHeapify` lead to several shift operations. This number is linear with respect to the accumulated path length $L(n)$, which can recursively be fixed as $L(2) = 1$ and $L(2^k) = 2 \cdot L(2^{k-1}) + k$. For $n = 2^k$ this simplifies to $2n - \log n - 2$. Therefore, the additional computations in the generating phase are in $O(n)$.

2.3 Sorting Phase

Similar to *HEAPSORT* we successively swap the top element a_0 with the last element a_m in the array, $n - 1 \geq m \geq 2$, and restore the defining *Weak-Heap* conditions in the interval $[0...m-1]$ by calling an operation `MergeForest(m)`:

First of all we traverse the grandchildren of the root. More precisely, we set an index variable x to the value 1 and execute the following loop: `while` $(2x + r_x < m)$ $x \leftarrow 2x + r_x$. Then, in a bottom-up traversal, the *Weak-Heap* conditions are regained by a series of merge operations. This results in a second loop: `while(`$x > 0$`) Merge(`$0, x$`);` $x \leftarrow x/2$ with at most $\lceil \log(m+1) \rceil$ key comparisons.

Theorem 2. `MergeForest` *generates a* Weak-Heap *according to its definition.*

Proof. After traversing the grandchildren set of the root, x is the leftmost leaf in the *Weak-Heap*. Therefore, the preconditions to the first `Merge` operation are trivially fulfilled. Hence the root and the subtree at x form a *Weak-Heap*. Since the *Weak-Heap* definition is reflected in all substructures for all grandchildren y of the root we have that y and its right subtree form a *Weak-Heap*. Therefore, we correctly combine the *Weak-Heaps* at position 0 and y and continue in a bottom-up fashion.

2.4 The *WEAK-HEAPSORT* Algorithm

WEAK-HEAPSORT combines the generating and the sorting phase. It invokes `WeakHeapify` and loops on the two operations `swap(0,m)` and `MergeForest(m)`. Since the correctness has already been shown above, we now turn to the time complexity of the algorithm measured in the number of key comparisons.

Theorem 3. *Let* $k = \lceil \log n \rceil$. *The worst case number of key comparisons of* WEAK-HEAPSORT *is bounded by* $nk - 2^k + n - 1 \leq n \log n + 0.086013n$.

Proof. The calls `MergeForest(i)` perform at most $\sum_{i=2}^{n-1}\lceil\log(i+1)\rceil = nk - 2^k$ comparisons (and at least $\sum_{i=2}^{n-1}\lceil\log(i+1)\rceil - 1 = nk - 2^k - n + 2$ comparisons). Together with the $n-1$ comparisons to build the *Weak-Heap* we have $nk - 2^k + n - 1$ comparisons altogether. Utilizing basic calculus we deduce that for all n there is an x in $[0,1]$ with $nk - 2^k + n - 1 = n\log n + nx - n2^x + n - 1 = n\log n + n(x - 2^x + 1) - 1$ and that the function $f(x) = x - 2^x + 1$ takes it maximum at $x_0 = -\ln\ln 2/\ln 2$ and $f(x_0) = 0.086013$. Therefore, the number of key comparisons in *WEAK-HEAPSORT* is less than $n\log n + 0.086013n$.

3 The Number of *Weak-Heaps*

Let $W(n)$ be the set of roots of complete subtrees in the *Weak-Heap* of size n.

Theorem 4. *If the input of* WEAK-HEAPSORT *is a random permutation of the elements* $\{1,\ldots,n\}$*, then every possible and feasible* Weak-Heap *occurs with the same probability. Moreover, there are* $n!/2^{|W(n)|}$ *different* Weak-Heaps *represented as a binary tree.*

Instead of a formal proof (cf. [3]) in Fig. 2 we give a simple example illustrating the idea of the backward analysis: For $n = 5$ we find two roots of complete subtrees (these nodes are double-encircled). In the first step swapping the top two elements leads to a dead end, since the generated *Weak-Heap* becomes infeasible. No further operation can move the (deepest) leaf to the leftmost branch as required for a correct input. Swapping 2 and 5 in the second step analogously leads to an infeasible *Weak-Heap*. Only in the following steps (according to roots of complete subtrees) both successor *Weak-Heaps* are feasible.

On the other hand, the assignments to reserve bits uniquely determines which cases in `Merge` have been chosen in the generating phase.

Theorem 5. *There are* $n!$ *different array-embedded* Weak-Heaps.

4 The Best Case of *WEAK-HEAPSORT*

This section proves Dutton's conjecture (1992) that an increasing sequence of input elements leads to the minimal number of key comparisons.

For the best case it will be sufficient that in every invocation of *MergeForest* the traversed path P to the leaf node, *special path* for short, terminates at the last position of the array. Subsequently, by exchanging the current root with the element at this position the path is pruned by one element. Fig. 1 depicts an example for this situation.

Therefore, by successively traversing the $2i + r_i$ successors from index 1 onwards we end up at index $n-1$. Hence, r_i has to coincide with the binary encoding $(b_k \ldots b_0)_2$ of $n-1$. More precisely, if $r_{\lfloor \frac{n-1}{2^i} \rfloor} = b_{i-1}$ for $i \in \{1,\ldots,k\}$, then $n-1$ is the last element of the special path. In case of the input $a_i = i$ for $i \in \{0,\ldots,n-1\}$, `WeakHeapify` leads to $r_0 = 0$ and $r_j = 1$ for $j \notin P$. Moreover,

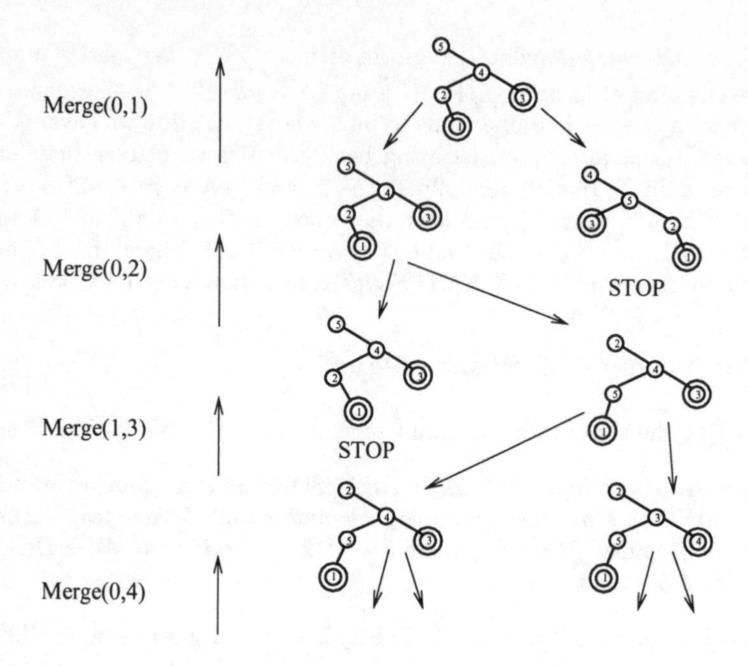

Fig. 2. Backward analysis of *WEAK-HEAPSORT*.

for r_j with $j \in P$ we get the binary representation of $n-1$ as required. In other words, $Heap(n)$ defined as

$$\begin{array}{l}
\tau^1_{Gparent(n-1),n-1} \circ \tau^1_{Gparent(n-2),n-2} \circ \ldots \circ \\
\tau^{b_0}_{Gparent(\lfloor \frac{n-1}{2^1} \rfloor),\lfloor \frac{n-1}{2^1} \rfloor} \circ \tau^1_{Gparent(\lfloor \frac{n-1}{2^1} \rfloor -1),\lfloor \frac{n-1}{2^1} \rfloor -1} \circ \ldots \circ \\
\tau^{b_1}_{Gparent(\lfloor \frac{n-1}{2^2} \rfloor),\lfloor \frac{n-1}{2^2} \rfloor} \circ \tau^1_{Gparent(\lfloor \frac{n-1}{2^2} \rfloor -1),\lfloor \frac{n-1}{2^2} \rfloor -1} \circ \ldots \circ \\
\vdots \\
\tau^{b_{k-2}}_{Gparent(\lfloor \frac{n-1}{2^{k-1}} \rfloor),\lfloor \frac{n-1}{2^{k-1}} \rfloor} \circ \tau^{b_{k-1}}_{0,1}
\end{array}$$

correctly determines the transpositions according to the generating phase, where $\tau^a_{i,j}$ is the transposition of i and j if a is odd and the identity, otherwise. As an example consider $Heap(15) = (14\ 3)^1(13\ 6)^1(12\ 1)^1(11\ 5)^1$ $(10\ 2)^1(9\ 4)^1(8\ 0)^1(7\ 3)^0(6\ 1)^1(5\ 2)^1(4\ 0)^1(3\ 1)^1(2\ 0)^1(1\ 0)^1$.

We now consider the second largest element $n-2$ and assume that $n-2$ has the binary encoding $(c_l \ldots c_0)_2$. Further, let $\oplus$ denote the exclusive or operation.

Lemma 1. *Let $a_i = i$ for $i \in \{0, \ldots, n-1\}$ be the input for* WEAK-HEAPSORT. *After the generating phase the element $n-2$ will be placed at position $\lfloor \frac{n-1}{2^{i^*}} \rfloor$ with $i^* = max\{i \mid b_{i-1} \oplus c_{i-1} = 1\}$. Moreover, for $j < i \leq i^*$ we have $a[\lfloor \frac{n-1}{2^j} \rfloor] < a[\lfloor \frac{n-1}{2^i} \rfloor]$.*

Proof. If $b_0 = 1$ then $n-1$ is a right child and $n-2$ is a left child. Hence, $Parent(n-2) = Gparent(n-1)$ and $Gparent(n-2) = Gparent(Gparent(n-1))$. Therefore, $n-2$ will be finally located at position $Gparent(n-1) = \lfloor \frac{n-1}{2^1} \rfloor$. Moreover, the key $n-2$ at $\lfloor \frac{n-1}{2^1} \rfloor$ is larger than $\lfloor \frac{n-1}{2^1} \rfloor + 1$ located at $n-1$. Therefore, for $j < i \le i^*$ we have $a_{\lfloor \frac{n-1}{2^j} \rfloor} < a_{\lfloor \frac{n-1}{2^i} \rfloor}$ as required.

For the other case $b_0 = 0$ we first consider $n-1 \neq 2^k$. The leaf $n-1$ is as long a left child as $n-2$ is a right child. Therefore, $n-2$ will be a left child at $\lfloor \frac{n-2}{2^{i^*-1}} \rfloor$ with $i^* = max\{i \mid b_{i-1} \oplus c_{i-1} = 1\}$. Since $Gparent(Gparent(n-1)) = Gparent(\lfloor \frac{n-1}{2^{i^*}} \rfloor) = Gparent(\lfloor \frac{n-2}{2^{i^*-1}} \rfloor)$, $n-2$ will finally be located at $\lfloor \frac{n-1}{2^{i^*}} \rfloor$. Now let $n-1 = 2^k$. In this case $Gparent(n-1)$ is the root. Since for all i we have $Parent(\lfloor \frac{n-2}{2^i} \rfloor) = Gparent(\lfloor \frac{n-2}{2^i} \rfloor)$, the element $n-2$ will eventually reach position $1 = \lfloor \frac{n-1}{2^{i^*}} \rfloor$.

To complete the proof the monotonicity criterion remains to be shown. An element can *escape* from a *Weak-Heap* subtree structure only via the associated root. The position $Gparent(n-1)$ will be occupied by $n-1$ and for all elements on P we know that the key at position $\lfloor \frac{n-1}{2^i} \rfloor$ is equal to $max\{\{\lfloor \frac{n-1}{2^i} \rfloor\} \cup \{k \mid k \in rT(\lfloor \frac{n-1}{2^i} \rfloor)\}\} = \lfloor \frac{n-1}{2} \rfloor + 2^{i-1}$, with $rT(x)$ denoting the right subtree of x. Therefore, for all $j < i < i^*$ we conclude that the key at $\lfloor \frac{n-1}{2^j} \rfloor$ is larger than the key at $\lfloor \frac{n-1}{2^i} \rfloor$. Since $a_{\lfloor \frac{n-1}{2^{i^*}} \rfloor} = n-2$, this condition also holds at $i = i^*$.

Lemma 2. *After the initial swap of position 0 and n-1* `MergeForest` *invokes the following set of transpositions* $CaseB(n) := \tau_{0,\lfloor \frac{n-1}{2^1} \rfloor}^{b_0 \oplus c_0} \circ \ldots \circ \tau_{0,1}^{b_{k-1} \oplus c_{k-1}}$.

Proof. Lemma 1 proves that all swaps of position $\lfloor \frac{n-1}{2^j} \rfloor$ with position 0 with $j < i^*$ are executed, since at the root we always find a smaller element of the currently considered one. We also showed that the maximum element $n-2$ is located at $\lfloor \frac{n-1}{2^{i^*}} \rfloor$. Therefore, no further element can reach the root. This corresponds to the observation that only for $j > i^*$ we have $b_{j-1} \oplus c_{j-1} = 0$.

For the example given above we have $CaseB(15) = (0\ \ 7)^1(0\ \ 3)^1(0\ \ 1)^0$.

The proof of the following two results

Lemma 3. $Heap(n) \circ (0\ \ n-1) = (Gparent(n-1)\ \ n-1) \circ Heap(n)$.

Lemma 4. $Heap(n-1)^{-1} \circ (Gparent(n-1)\ \ n-1) \circ Heap(n) = CaseB(n)^{-1}$.

is technically involved and can be found [3].

Lemma 5. $Heap(n) \circ Swap(0, n-1) \circ CaseB(n) \circ Heap(n-1)^{-1} = id_{\{0,\ldots,n-1\}}$.

Proof. By right and left multiplication with $Heap(n-1)$ and $Heap(n-1)^{-1}$ the equation $Heap(n) \circ Swap(0, n-1) \circ CaseB(n) \circ Heap(n-1)^{-1} = id_{\{0,\ldots,n-1\}}$ can be transformed into $Heap(n-1)^{-1} \circ Heap(n) \circ Swap(0, n-1) \circ CaseB(n) = id_{\{0,\ldots,n-1\}}$. By Lemma 3 this is equivalent to $Heap(n-1)^{-1} \circ (Gparent(n-1)\ \ n-1) \circ Heap(n) \circ CaseB(n) = id_{\{0,\ldots,n-1\}}$. Lemma 4 completes the proof.

Continuing our example with $n = 15$ we infer $n-1 = 14 = (b_3\ b_2\ b_1\ b_0)_2 = (1\ 1\ 1\ 0)_2$ and $n-2 = 13 = (c_3\ c_2\ c_1\ c_0)_2 = (1\ 1\ 0\ 1)_2$. Furthermore, $Heap(14)^{-1} = (1\ 0)^1(2\ 0)^1(3\ 1)^0(4\ 0)^1(5\ 2)^1(6\ 1)^1(7\ 3)^1(8\ 0)^1(9\ 4)^1(10\ 2)^1 (11\ 5)^1(12\ 1)^1(13\ 6)^1$. Therefore,

$$(1\ 0)(2\ 0)(4\ 0)(5\ 2)(6\ 1)(7\ 3)(8\ 0)(9\ 4)(10\ 2)(11\ 5)(12\ 1)(13\ 6)$$
$$(14\ 3)(13\ 6)(12\ 1)(11\ 5)(10\ 2)(9\ 4)(8\ 0)(6\ 1)(5\ 2)(4\ 0)(3\ 1)(2\ 0)(1\ 0)$$
$$(0\ 14)(0\ 7)(0\ 3) = id_{\{0,\ldots,14\}}.$$

Inductively, we get the following result

Theorem 6. *The best case of* WEAK-HEAPSORT *is met given an increasing ordering of the input elements.*

5 The Worst Case of *WEAK-HEAPSORT*

The worst case analysis, is based on the best case analysis. The main idea is that the special path misses the best case by one element. Therefore, the `Merge` calls in `MergeForest(m)` will contain the index $m-1$. This determines the assignment of the reverse bits on P: If $n-2 = (b_k \ldots b_0)_2$ and if $r_{\lfloor \frac{n-2}{2^i} \rfloor} = b_{i-1}$ for all $i \in \{1,\ldots,k\}$ then $n-2$ is the last element of P.

An appropriate example fulfilling this property, is the input $a_i = i+1$ with $i \in \{0,\ldots,n-2\}$ and $a_{n-1} = 0$. After termination of `WeakHeapify` we have $r_0 = 0$, $r_j = 1$ for $j \notin P$, $r_{n-1} = 0$, $r_{n-2} = 1$, and $r_{\lfloor \frac{n-2}{2^i} \rfloor} = b_{i-1}$ for $i \in \{1,\ldots,k\}$. The transpositions $Heap(n)$ of the *Weak-Heap* generation phase are:

$$\begin{array}{l}
\tau^1_{Gparent(n-2),n-2} \circ \tau^1_{Gparent(n-2),n-2} \circ \tau^1_{Gparent(n-2),n-2} \circ \ldots \circ \\
\tau^{b_0}_{Gparent(\lfloor \frac{n-2}{2^1} \rfloor),\lfloor \frac{n-2}{2^1} \rfloor} \circ \tau^1_{Gparent(\lfloor \frac{n-2}{2^1} \rfloor-1),\lfloor \frac{n-2}{2^1} \rfloor-1} \circ \ldots \circ \\
\tau^{b_1}_{Gparent(\lfloor \frac{n-2}{2^2} \rfloor),\lfloor \frac{n-2}{2^2} \rfloor} \circ \tau^1_{Gparent(\lfloor \frac{n-2}{2^2} \rfloor-1),\lfloor \frac{n-2}{2^2} \rfloor-1} \circ \ldots \circ \\
\vdots \\
\tau^{b_{k-2}}_{Gparent(\lfloor \frac{n-2}{2^{k-1}} \rfloor),\lfloor \frac{n-2}{2^{k-1}} \rfloor} \circ \tau^{b_{k-1}}_{0,1}
\end{array}$$

Unless once per level (when the binary tree rooted at position 1 is complete) we have $\lceil \log(n+1) \rceil$ instead of $\lceil \log(n+1) \rceil - 1$ comparisons. If $n-1$ is set to $n-2$ Lemma 1 remains valid according to the new input. Therefore, we conclude

Lemma 6. *The first invocation of* `MergeForest` *(with the above input) leads to following set of transpositions* $\text{CaseW}(n) = \tau^1_{0,n-2} \circ \tau^{b_0 \oplus c_0}_{0,\lfloor \frac{n-2}{2^1} \rfloor} \circ \ldots \circ \tau^{b_{k-1} \oplus c_{k-1}}_{0,\lfloor \frac{n-2}{2^k} \rfloor = 1}$.

The following two results are obtained by consulting the best case analysis.

Lemma 7. $Heap(n) \circ (0\ \ n-1) = (n-2\ \ n-1) \circ Heap(n)$.

Lemma 8. $CaseW(n) = Swap(0, n-2) \circ CaseB(n-1)$.

Since the definitions of $Heap(n)$ are different in the worst case and best case analysis we invent labels $Heap_b(n)$ for the best case and $Heap_w(n)$ for the worst case, respectively.

Lemma 9.

$$Heap_w(n) \circ Swap(0, n-1) \circ CaseW(n) \circ Heap_w(n-1)^{-1} = (n-2\ \ n-1).$$

Proof. According to Lemma 7 the stated equation is equivalent to $(n-2\ \ n-1) \circ Heap_w(n) \circ CaseW(n) \circ Heap_w(n-1)^{-1} = (n-2\ \ n-1)$.

The observation $Heap_w(n) = Heap_b(n-1)$ and Lemma 8 results in $(n-2\ \ n-1) \circ Heap_b(n-1) \circ Swap(0, n-2) \circ CaseB(n-1) \circ Heap_b(n-2)^{-1} = (n-2\ \ n-1)$, which is equivalent to Lemma 5 of the best case.

Inductively, we get

Theorem 7. *The worst case of* WEAK-HEAPSORT *is met with an input of the form* $a_{n-1} < a_i < a_{i+1}$, $i \in \{0, \ldots, n-3\}$.

As an example let $n = 16$, $n - 2 = 14 = (b_3\ \ b_2\ \ b_1\ \ b_0)_2 = (1\ \ 1\ \ 1\ \ 0)_2$ and $n - 3 = 13 = (c_3\ \ c_2\ \ c_1\ \ c_0)_2 = (1\ \ 1\ \ 0\ \ 1)_2$. Further let $Heap_w(16) = (14\ \ 3)^1(13\ \ 6)^1(12\ \ 1)^1(11\ \ 5)^1(10\ \ 2)^1(9\ \ 4)^1(8\ \ 0)^1(7\ \ 3)^0(6\ \ 1)^1(5\ \ 2)^1(4\ \ 0)^1 (3\ \ 1)^1(2\ \ 0)^1(1\ \ 0)^1$, $Swap(0, 16) = (0\ \ 15)^1$, $CaseW(16) = (0\ \ 14)^1(0\ \ 7)^1(0\ \ 3)^1$, and $Heap(15)^{-1} = (1\ \ 0)^1(2\ \ 0)^1(3\ \ 1)^0(4\ \ 0)^1(5\ \ 2)^1(6\ \ 1)^1(7\ \ 3)^1(8\ \ 0)^1(9\ \ 4)^1 (10\ \ 2)^1(11\ \ 5)^1(12\ \ 1)^1(13\ \ 6)^1$. Then $Heap_w(16) \circ Swap(0, 16) \circ CaseW(16) \circ Heap(15)^{-1} = (14\ \ 15)$.

6 The Average Case of *WEAK-HEAPSORT*

Let $d(n)$ be given such that $n \log n + d(n)n$ is the expected number of comparisons of *WEAK-HEAPSORT*. Then the following experimental data show that $d(n) \in [-0.47, -0.42]$. Moreover $d(n)$ is small for $n \approx 2^k$ and big for $n \approx 1.4 \cdot 2^k$.

n	1000	2000	3000	4000	5000	6000	7000	8000
$d(n)$	-0.462	-0.456	-0.437	-0.456	-0.445	-0.429	-0.436	-0.458

n	9000	10000	11000	12000	13000	14000	15000	16000
$d(n)$	-0.448	-0.437	-0.432	-0.430	-0.436	-0.443	-0.449	-0.458

n	17000	18000	19000	20000	21000	22000	23000	24000
$d(n)$	-0.458	-0.449	-0.443	-0.437	-0.433	-0.431	-0.436	-0.427

n	25000	26000	27000	28000	29000	30000
$d(n)$	-0.431	-0.437	-0.436	-0.440	-0.440	-0.447

There was no significant difference between the execution of one trial and the average of 20 trials. The reason is that the variance of the number of comparisons in *WEAK-HEAPSORT* is very small: At $n = 30000$ and 20 trials we achieved a best case 432657 and a worst case of 432816 comparisons.

According to published results of Wegener (1992) and own experiments *WEAK-HEAPSORT* requires approx. $0.81n$ less comparisons than *BOT-TOM-UP-HEAPSORT* and approx. $0.45n$ less comparisons than *MDR-HEAPSORT*.

7 The *Weak-Heap* Priority-Queue Data Structure

A priority queue provides the following operations on the set of items: *Insert* to enqueue an item and *DeleteMax* to extract the element with the largest key value. To insert an item v in a *Weak-Heap* we start with the last index x of the array a and put v in a_x. Then we climb up the grandparent relation until the *Weak-Heap* definition is fulfilled. Thus, we have the following pseudo-code: `while` $(x \neq 0)$ `and` $(a_{\texttt{Gparent}(x)} < a_x)$ `Swap(Gparent(`x`),`x`)`; $r_x = 1 - r_x$; $x \leftarrow$`Gparent(`x`)`. Since the expected path length of grandparents from a leaf node to a root is approximately half the depth of the tree, we expect at about $\log n/4$ comparisons in the average case. The argumentation is as follows. The sum of the length of the grandparent relation from all nodes to the root in a weak-heap of size $n = 2^k$ satisfy the following recurrence formula: $S(2^1) = 1$ and $S(2^k) = 2S(2^{k-1}) + 2^{k-1}$ with closed form of $nk/2 + n$ such that the average length is at about $k/2 + 1$.

A double-ended priority queue, *deque* for short, extends the priority queue operation by *DeleteMin* to extract the smallest key values. The transformation of a *Weak-Heap* into its dual in $\lfloor (n-1)/2 \rfloor$ comparisons is performed by the following pseudo-code:

`for` $i = \{size - 1, \ldots, \lfloor (size-1)/2 \rfloor + 1\}$ `Swap(Gparent(`i`),`i`)` followed by
`for` $i = \{\lfloor (size-1)/2 \rfloor, \ldots, 1\}$ `Merge(Gparent(`i`),`i`)`

By successively building the two heaps we have solved the well-known min-max-problem in the optimal number of $n + \lceil n/2 \rceil - 2$ comparisons.

Each operation in a general priority queue can be divided into several compare and exchange steps where only the second one changes the structure. We briefly sketch the implementation. Let M be a Max-*Weak-Heap* and M' be a Min-*Weak-Heap* on a set of n items a and a', respectively. We implicitly define the bijection ϕ by $a_i = a'_{\phi(i)}$. In analogy we might determine ϕ' for M'. The conditions $a_i = a'_{\phi(i)}$ and $a'_i = a_{\phi'(i)}$ are kept as an invariance. Swapping j and k leads to the following operations: Swap a_j and a_k, exchange $\phi(j)$ and $\phi(k)$, set $\phi'(\phi(j))$ to j and set $\phi'(\phi(k))$ to k. We see that the invariance is preserved. A similar result is obtained if a swap-operation on M' is considered.

8 Conclusion

Weak-Heaps are a very fast data structure for sorting in theory and practice. The worst case number of comparisons for sorting an array of size n is bounded by $n \log n + 0.1n$ and empirical studies show that the average case is at $n \log n + d(n)n$ with $d(n) \in [-0.47, -0.42]$. Let $k = \lceil \log n \rceil$. The exact worst case bound for *WEAK-HEAPSORT* is $nk - 2^k + n - k$ and appears if all but the last two elements are ordered whereas the exact best case bound of $nk - 2^k + 1$ is found if all elements are in ascending order. On the other hand the challenging algorithm *BOT-TOM-UP-HEAPSORT* is bounded by $1.5n \log n + O(n)$ in the worst case. Its *MDR-HEAPSORT* variant consumes at most $n \log n + 1.1n$ comparisons. Therefore, the sorting algorithm based on *Weak-Heaps* can be judged to be the fastest *HEAPSORT* variant and to compete fairly well with other algorithms.

References

[1] R. D. Dutton. The weak-heap data structure. Technical report, University of Central Florida, Orlando, FL 32816, 1992.

[2] R. D. Dutton. Weak-heap sort. *BIT*, 33:372–381, 1993.

[3] S. Edelkamp and I. Wegener. On the performance of *WEAK-HEAPSORT*. Technical Report TR99-028, Electronic Colloquium on Computational Complexity, 1999. ISSN 1433-8092, 6th Year.

[4] R. Fleischer. A tight lower bound for the worst case of Bottom-Up-Heapsort. *Algorithmica*, 11(2):104–115, 1994.

[5] R. W. Floyd. ACM algorithm 245: Treesort 3. *Communications of the ACM*, 7(12):701, 1964.

[6] T. N. Hibbard. A empirical study of minimal storage sorting. *Communications of the ACM*, 6(5):206–213, 1963.

[7] C. A. R. Hoare. Quicksort. *Computer Journal*, 5(1):10–15, 1962.

[8] J. Incerpi and R. Sedgewick. Improved upper bounds on shellsort. *Journal of Computer and System Sciences*, 31:210–224, 1985.

[9] T. Jiang, M. Li, and P. Vitányi. Average complexity of shellsort. In *ICALP'99*, volume 1644 of *LNCS*, pages 453–462, 1999.

[10] J. Katajainen. The ultimate heapsort. In *Proceedings of the Computing: the 4th Australasian Theory Symposium, Australian Computer Science Communications 20(3)*, pages 87–95, 1998.

[11] J. Katajainen, T. Pasanen, and J. Teuhola. Practical in-place mergesort. *Nordic Journal of Computing*, 3(1):27–40, 1996.

[12] J. Katajainen and T. A. Pasanen. In-place sorting with fewer moves. *Information Processing Letters*, 70(1):31–37, 1999.

[13] M. Li and P. Vitányi. *An Introduction to Kolmogorov Complexity and Its Applications.* Text and Monographs in Computer Science. Springer-Verlag, 1993.

[14] C. J. H. McDiarmid and B. A. Reed. Building heaps fast. *Journal of Algorithms*, 10:352–365, 1989.

[15] S. Nilsson. *Radix Sorting & Searching.* PhD thesis, Lund University, 1996.

[16] A. Papernov and G. Stasevich. The worst case in shellsort and related algorithms. *Problems Inform. Transmission*, 1(3):63–75, 1965.

[17] V. Pratt. *Shellsort and Sorting Networks.* PhD thesis, Stanford University, 1979.

[18] K. Reinhardt. Sorting in-place with a worst case complexity of $n \log n - 1.3n + O(\log n)$ comparisons and $\epsilon n \log n + O(1)$ transports. *Lecture Notes in Computer Science*, 650:489–499, 1992.

[19] W. Rudin. *Real and Complex Analysis.* McGraw–Hill, 1974.

[20] R. Schaffer and R. Sedgewick. The analysis of heapsort. *Journal of Algorithms*, 15(1):76–100, 1993.

[21] R. Sedgewick. The analysis of quicksort programs. *Acta Inform.*, 7:327–355, 1977.

[22] R. Sedgewick. A new upper bound for shellsort. *Journal of Algorithms*, 2:159–173, 1986.

[23] D. Shell. A high-speed sorting procedure. *Communications of the ACM*, 2(7):30–32, 1959.

[24] H. Steinhaus. *One hundred problems in elemantary mathematics (Problems 52,85).* Pergamon Press, London, 1958.

[25] M. H. van Emden. Increasing the efficiency of QUICKSORT. *Communications of the ACM*, 13:563–567, 1970.

[26] I. Wegener. The worst case complexity of McDiarmid and Reed's variant of BOTTOM-UP HEAPSORT is less than $n \log n + 1.1n$. *Information and Computation*, 97(1):86–96, 1992.
[27] I. Wegener. BOTTOM-UP-HEAPSORT, a new variant of HEAPSORT, beating, on an average, QUICKSORT (if n is not very small). *Theoretical Computer Science*, 118:81–98, 1993.
[28] J. W. J. Williams. ACM algorithm 232: Heapsort. *Communications of the ACM*, 7(6):347–348, 1964.

On the Two-Variable Fragment of the Equational Theory of the Max-Sum Algebra of the Natural Numbers

Luca Aceto[1], Zoltán Ésik[2*], and Anna Ingólfsdóttir[1**]

[1] BRICS[***], Department of Computer Science, Aalborg University, Fredrik Bajers Vej 7-E, DK-9220 Aalborg Ø, Denmark.
[2] Department of Computer Science, A. József University, Árpád tér 2, 6720 Szeged, Hungary.

Abstract. This paper shows that the collection of identities in two variables which hold in the algebra **N** of the natural numbers with constant zero, and binary operations of sum and maximum does not have a finite equational axiomatization. This gives an alternative proof of the non-existence of a finite basis for **N**—a result previously obtained by the authors.

1 Introduction

Since Birkhoff's original developments, equational logic has been one of the classic topics of study within universal algebra. (See, e.g., [9] for a survey of results in this area of research.) In particular, the research literature is, among other things, rich in results, both of a positive and negative nature, on the existence of finite bases for theories (i.e. finite sets of axioms for them).

In this paper, we contribute to the study of equational theories that are not finitely based by continuing our analysis of the equational theory of the algebra **N** of the natural numbers with constant zero, and binary operations of summation and maximum (written $\vee$ in infix form). Our investigations of this equational theory started in the companion paper [2]. In *op. cit.* we showed that the equational theory of **N** is not finitely based. Moreover, we proved that, for all $n \geq 0$, the collection of all the equations in at most n variables that hold in **N** does not form an equational basis.

The equational theory of **N** is surprisingly rich in non-trivial families of identities. For example, the following infinite schemas of equations also hold in **N**:

$$e_n : nx_1 \vee \ldots \vee nx_n \vee (x_1 + \ldots + x_n) = nx_1 \vee \ldots \vee nx_n$$
$$e'_n : nx \vee ny = n(x \vee y) \ ,$$

* Partially supported by grant no. T30511 from the National Foundation of Hungary for Scientific Research.
** Supported by a research grant from the Danish Research Council.
*** Basic Research in Computer Science.

H. Reichel and S. Tison (Eds.): STACS 2000, LNCS 1770, pp. 267–278, 2000.

where $n \in \mathbb{N}$, and nx denotes the n-fold sum of x with itself. By convention, nx stands for 0 when $n = 0$, and so does the empty sum.

Let $\mathrm{Eq}_2(\mathbf{N})$ denote the collection of equations that hold in $\mathbf{N}$ containing occurrences of two distinct variables. A natural question suggested by the family of equations e'_n above is the following: Is there a finite set E of equations that hold in $\mathbf{N}$ such that $E \vdash \mathrm{Eq}_2(\mathbf{N})$? This paper is devoted to proving that no finite axiomatization exists for $\mathrm{Eq}_2(\mathbf{N})$. Apart from its intrinsic mathematical interest, this result offers yet another view of the non-existence of a finite basis for the variety generated by $\mathbf{N}$ proven in [2].

The proof of our main technical result is model-theoretic in nature, and follows standard lines. The details are, however, rather challenging. More precisely, for every prime number p, we construct an algebra $\mathbf{A}_p$ in which all the equations that hold in $\mathbf{N}$ and whose "measure of complexity" is strictly smaller than p hold, but neither e_p nor e'_p hold in $\mathbf{A}_p$. As a consequence of this result, we obtain that not only the equational theory of $\mathbf{N}$ is not finitely based, but not even the collection of equations in two variables included in it is.

Although the proof of our main theorem uses results from [2], we have striven to make the paper self contained. The interested reader is referred to *op. cit.* and the textbook [4] for further background information. Full proofs of our results may be found in [1].

Related Work. The algebra $\mathbf{N}$, although itself not a semiring, is closely related to many instances of these structures whose addition operation (here, the maximum of two natural numbers) is idempotent. Interest in idempotent semirings arose in the 1950s through the observation that some problems in discrete optimization could be linearized over such structures (see, e.g., [12] for a survey). Since then, the study of idempotent semirings has forged productive connections with such diverse fields as, e.g., automata theory, discrete event systems and non-expansive mappings. The interested reader is referred to [5] for a survey of these more recent developments. Here we limit ourselves to mentioning that variations on the algebra $\mathbf{N}$, in the form of the so-called *tropical semirings*, have found deep applications in automata theory and the study of formal power series. The tropical semiring $(\mathbb{N} \cup \{+\infty\}, \min, +)$ was originally introduced by Simon in his solution to Brzozowski's celebrated finite power property problem [10]. It was also used by Hashiguchi in his independent solution to the aforementioned problem [6], and in his study of the star height of regular languages (see, e.g., [7]). Further examples of applications of the tropical semiring may be found in, e.g., [8,11].

2 The Max-Sum Algebra

Let $\mathbf{N} = (\mathbb{N}, \vee, +, 0)$ denote the algebra of the natural numbers equipped with the usual sum operation $+$, constant 0 and the operation $\vee$ for the maximum of two numbers, i.e.,

$$x \vee y = \max\{x, y\} \ .$$

We study the equational theory of the algebra **N**—that is, the collection Eq(**N**) of equations that hold in **N**. The reader will have no trouble in checking that the following axioms, that express expected properties of the operations of maximum and sum, hold in **N**:

$$\begin{array}{ll} \vee 1 \; x \vee y = y \vee x & +1 \; x + y = y + x \\ \vee 2 \; (x \vee y) \vee z = x \vee (y \vee z) & +2 \; (x + y) + z = x + (y + z) \\ \vee 3 \; x \vee 0 = x & +3 \; x + 0 = x \\ & +\vee \; (x \vee y) + z = (x + z) \vee (y + z) \end{array}$$

This set of equations will be denoted by Ax_1. Note that the equation $(x+y) \vee x = x + y$ is derivable from +3 and $+\vee$, and, using such an equation, it is a simple matter to derive the idempotency law for $\vee$, i.e.,

$$\vee 4 \; x \vee x = x \; .$$

We denote by Ax_0 the set consisting of the equations $\vee 1$, $\vee 2$, $\vee 4$, +1–+3 and $+\vee$. Moreover, we let $\mathbf{V}_0$ stand for the class of all models of Ax_0, and $\mathbf{V}_1$ for the class of all models of the equations Ax_1. Thus, both $\mathbf{V}_0$ and $\mathbf{V}_1$ are varieties and, by the above discussion, $\mathbf{V}_1$ is a subvariety of $\mathbf{V}_0$, i.e., $\mathbf{V}_1 \subseteq \mathbf{V}_0$.

Since the reduct $(A, \vee)$ of any algebra $A = (A, +, \vee, 0)$ in $\mathbf{V}_0$ is a semilattice, we can define a partial order $\leq$ on the set A by $a \leq b$ if and only if $a \vee b = b$, for all $a, b \in A$. This partial order is called the induced partial order. When A is in the variety $\mathbf{V}_1$, the constant 0 is the least element of A with respect to $\leq$. Moreover, for any $A \in \mathbf{V}_0$, the $\vee$ and + operations are monotonic with respect to the induced partial order.

The axiom system Ax_1 completely axiomatizes the collection of equations in at most one variable which hold in the algebra **N**. However, the interplay between the operations of maximum and sum generates some non-trivial collections of equations in two or more variables. For example, the infinite schemas of equations e_n and e'_n (defined in Sect. 1), which will play an important role in the technical developments of this paper, also hold in **N**. It is not too difficult to see that, for any n, the equation e_n is derivable from Ax_1 and e'_n.

Let $\mathrm{Eq}_2(\mathbf{N})$ denote the collection of equations that hold in **N** containing occurrences of two distinct variables. The remainder of this paper is devoted to proving that no finite axiomatization exists for $\mathrm{Eq}_2(\mathbf{N})$.

3 Explicit Description of the Free Algebras

In this section we give a brief review of some results on the equational theory of the algebra **N** that we obtained in [2]. We start by offering an explicit description of the free algebras in the variety **V** generated by **N**. Since **N** satisfies the equations in Ax_1, we have that **V** is a subvariety of $\mathbf{V}_1$, i.e., $\mathbf{V} \subseteq \mathbf{V}_1$.

For the sake of clarity, and for future reference, we shall describe the finitely generated free algebras in **V**. We recall that any infinitely generated free algebra is a directed union of the finitely generated free ones.

Let $n \geq 0$ denote a fixed integer. The set $\mathbb{N}^n$ is the collection of all n-dimensional vectors over $\mathbb{N}$. We use $P_f(\mathbb{N}^n)$ to denote the collection of all finite non-empty subsets of $\mathbb{N}^n$, and define the operations in the following way: for all $U, V \in P_f(\mathbb{N}^n)$,

$$\begin{aligned} U \vee V &:= U \cup V \\ U + V &:= \{\overline{u} + \overline{v} : \overline{u} \in U,\ \overline{v} \in V\} \\ 0 &:= \{\overline{0}\}\ , \end{aligned}$$

where $\overline{0}$ stands for the vector whose components are all 0. For each $i \in [n] = \{1, \ldots, n\}$, let $\overline{u}_i$ denote the ith unit vector in $\mathbb{N}^n$, i.e., the vector whose only non-zero component is a 1 in the ith position.

Proposition 1. *The algebra $P_f(\mathbb{N}^n)$ is freely generated in $\mathbf{V}_0$ by the n singleton sets $\{\overline{u}_i\}$, $i \in [n]$, containing the unit vectors.*

Note that the induced partial order on $P_f(\mathbb{N}^n)$ is given by set inclusion.

It is easy to see that any term t in the variables $x_1, \ldots, x_n$ $(n \geq 0)$ can be rewritten, using the equations in Ax_0, to the maximum of linear combinations of the variables $x_1, \ldots, x_n$, i.e., there are $m \geq 1$, and $c_i^j \in \mathbb{N}$ for $j \in [n]$ and $i \in [m]$ such that the equation

$$t = \bigvee_{i \in [m]} (\sum_{j \in [n]} c_i^j x_j)$$

holds in $\mathbf{V}_0$. We refer to such terms as *normal forms*. Thus we may assume that any equation which holds in a given subvariety of $\mathbf{V}_0$ is in normal form, i.e., of the form $t_1 = t_2$ where t_1 and t_2 are normal forms. Furthermore, an equation $t_1 \vee \ldots \vee t_m = t'_1 \vee \ldots \vee t'_{m'}$ holds in a subvariety of $\mathbf{V}_0$ if and only if, for all $i \in [m]$ and $j \in [m']$,

$$t_i \leq t'_1 \vee \ldots \vee t'_{m'} \text{ and } t'_j \leq t_1 \vee \ldots \vee t_m$$

hold in the subvariety. We refer to an inequation of the form $t \leq t_1 \vee \ldots \vee t_m$, where $t, t_1, \ldots, t_m$ are linear combinations of variables, as *simple inequations*.

A simple inequation $t \leq t_1 \vee \ldots \vee t_m$ that holds in $\mathbf{N}$ is *irredundant* if, for every $j \in [m]$,

$$\mathbf{N} \not\models t \leq t_1 \vee \ldots \vee t_{j-1} \vee t_{j+1} \vee \ldots \vee t_m\ .$$

By the discussion above, we may assume, without loss of generality, that every set of inequations that hold in $\mathbf{N}$ consists of simple, irredundant inequations only.

In order to give an explicit description of the finitely generated free algebras in $\mathbf{V}_1$, we need to take into account the effect of equation $\vee 3$. Let $\leq$ denote the pointwise partial order on $\mathbb{N}^n$. As usual, we say that a set $U \subseteq \mathbb{N}^n$ is an order ideal, if $\overline{u} \leq \overline{v}$ and $\overline{v} \in U$ jointly imply that $\overline{u} \in U$, for all vectors $\overline{u}, \overline{v} \in \mathbb{N}^n$. Each set $U \subseteq \mathbb{N}^n$ is contained in a least ideal $(U]_n$, the ideal generated by U. The relation that identifies two sets $U, V \in P_f(\mathbb{N}^n)$ if $(U]_n = (V]_n$ is a congruence relation on $P_f(\mathbb{N}^n)$, and the quotient with respect to this congruence is easily seen to be isomorphic to the subalgebra $I_f(\mathbb{N}^n)$ of $P_f(\mathbb{N}^n)$ generated by the finite non-empty ideals.

For each $i \in [n]$, let $(\overline{u}_i]_n$ denote the principal ideal generated by the unit vector $\overline{u}_i$, i.e., the ideal $(\{\overline{u}_i\}]_n$.

Proposition 2. *$I_f(\mathbb{N}^n)$ is freely generated in $\mathbf{V}_1$ by the n principal ideals $(\overline{u}_i]_n$.*

Again, the induced partial order on $I_f(\mathbb{N}^n)$ is the partial order determined by set inclusion.

We note that, if $n \geq 2$, then the equation e_n fails in $I_f(\mathbb{N}^n)$, and *a fortiori* in $\mathbf{V}_1$. Since for $n \geq 2$ the equation e_n holds in $\mathbf{N}$ but fails in $\mathbf{V}_1$, in order to obtain a concrete description of the free algebras in $\mathbf{V}$ we need to make further identifications of the ideals in $I_f(\mathbb{N}^n)$. Technically, we shall start with $P_f(\mathbb{N}^n)$.

Let $\overline{v}_1, \ldots, \overline{v}_k$ $(k \geq 1)$ be vectors in $\mathbb{N}^n$, and suppose that λ_i $(i \in [k])$ are non-negative real numbers with $\sum_{i\in[k]} \lambda_i = 1$. We call the vector of real numbers $\sum_{i\in[k]} \lambda_i \overline{v}_i$ a convex linear combination of the vectors $\overline{v}_i$ $(i \in [k])$.

Definition 1. *We call a set $U \subseteq P_f(\mathbb{N}^n)$ a* convex ideal *if for any convex linear combination $\sum_{i\in[k]} \lambda_i \overline{v}_i$, with $\overline{v}_i \in U$ for all $i \in [k]$, and for any $\overline{v} \in \mathbb{N}^n$, if*

$$\overline{v} \leq \sum_{i\in[k]} \lambda_i \overline{v}_i$$

in the pointwise order, then $\overline{v} \in U$.

Note that any convex ideal is an ideal. Moreover, the intersection of any number of convex ideals is a convex ideal. Thus, any subset U of $\mathbb{N}^n$ is contained in a least convex ideal, $[U]_n$. When U is finite, so is $[U]_n$. We let $\overline{c} \leq U$ mean that the simple inequation

$$\overline{c} \cdot \overline{x} \leq \bigvee_{\overline{d}\in U} \overline{d} \cdot \overline{x}$$

holds in $\mathbf{N}$. Then we have the following useful characterization of the simple inequations that hold in $\mathbf{N}$.

Lemma 1. *Suppose that $U \in P_f(\mathbb{N}^n)$ and $\overline{c} \in \mathbb{N}^n$. Then $\overline{c} \in [U]_n$ iff $\overline{c} \leq U$.*

As a corollary of Lemma 1, we obtain the following alternative characterization of simple equations which hold in the algebra $\mathbf{N}$.

Corollary 1. *Let $\overline{c}, \overline{d}_j$ $(j \in [m])$ be vectors in $\mathbb{N}^n$. Then $\overline{c} \leq \{\overline{d}_1, \ldots, \overline{d}_m\}$ iff there are $\lambda_1, \ldots, \lambda_m \geq 0$ such that $\lambda_1 + \ldots + \lambda_m = 1$ and $\overline{c} \leq \lambda_1 \overline{d}_1 + \ldots + \lambda_m \overline{d}_m$ with respect to the pointwise ordering. Moreover, if $\overline{c} \leq \{\overline{d}_1, \ldots, \overline{d}_m\}$ is irredundant, then $\lambda_1, \ldots, \lambda_m > 0$.*

The above result offers a geometric characterization of the simple inequations in Eq(**N**), viz. an inequation $\overline{c}\cdot\overline{x} \leq \overline{d}_1\cdot\overline{x} \vee \ldots \vee \overline{d}_m\cdot\overline{x}$ (where $\overline{x} = (x_1, \ldots, x_n)$ is a vector of variables) holds in **N** iff the vector $\overline{c}$ lies in the ideal generated by the convex hull of the vectors $\overline{d}_1, \ldots, \overline{d}_m$.

Let $\sim$ denote the congruence relation on $P_f(\mathbb{N}^n)$ that identifies two sets of vectors iff they generate the same least convex ideal. It is immediate to see that the quotient algebra $P_f(\mathbb{N}^n)/\sim$ is isomorphic to the following algebra $CI_f(\mathbb{N}^n) = (CI_f(\mathbb{N}^n), \vee, +, 0)$ of all non-empty finite convex ideals in $P_f(\mathbb{N}^n)$. For any two $I, J \in CI_f(\mathbb{N}^n)$,

$$\begin{aligned} I \vee J &:= [I \cup J]_n \\ I + J &:= [\{\overline{u} + \overline{v} : \overline{u} \in I,\ \overline{v} \in J\}]_n \\ 0 &:= \{\overline{0}\}\ . \end{aligned}$$

Indeed, an isomorphism $P_f(\mathbb{N}^n)/\sim \to CI_f(\mathbb{N}^n)$ is given by the mapping $U/\sim \mapsto [U]_n$.

Recall that, for each $i \in [n]$, $\overline{u}_i$ denotes the ith unit vector in $\mathbb{N}^n$. For each $i \in [n]$, the set $[\overline{u}_i]_n = (\overline{u}_i]_n = \{\overline{u}_i, \overline{0}\}$ is the least convex ideal containing $\overline{u}_i$.

Theorem 1. *$CI_f(\mathbb{N}^n)$ is freely generated by the n convex ideals $[\overline{u}_i]_n$ in the variety* **V**.

4 The Two-Variable Fragment is not Finitely Based

We now proceed to apply the results that we have recalled in the previous section to the study of the two variable fragment of the equational theory of the algebra **N**. The main aim of this paper is to prove the following result to the effect that the collection of equations $\mathrm{Eq}_2(\mathbf{N})$ cannot be deduced using any finite number of equations in Eq(**N**).

Theorem 2. *There is no finite set E of equations in* Eq(**N**) *such that* $E \vdash \mathrm{Eq}_2(\mathbf{N})$.

To prove Thm. 2 we shall define a sequence of algebras $\mathbf{A}_n$ $(n \geq 1)$ in $\mathbf{V}_1$ such that following holds: For any finite set E of equations which hold in **N**, there is an n such that

$$\mathbf{A}_n \models E \text{ but } \mathbf{A}_n \not\models e_n'\ .$$

Recalling that, for any n, the equation e_n is derivable from e_n' and Ax_1, it is sufficient to prove the statement above with e_n' replaced by e_n. Furthermore, in light of our previous analysis, the result we are aiming at in this section, viz. Thm. 2, may now be reformulated as follows.

Proposition 3. *Let E be a finite set of simple, irredundant equations such that $\mathbf{N} \models E$. Then there is an $n \in \mathbb{N}$ and an algebra $\mathbf{A}_n \in \mathbf{V}_1$ such that $\mathbf{A}_n \models E$ but $\mathbf{A}_n \not\models e_n$.*

The non-existence of a finite axiomatization of the two variable fragment of the equational theory of **N** follows easily from Propn. 3 and the preceding discussion. In fact, let E be any finite subset of Eq(**N**). Without loss of generality, we may assume that E includes Ax_1, and that $E \setminus Ax_1$ consists of simple, irredundant inequations. Then, by Propn. 3, there is an algebra $\mathbf{A}_n \in \mathbf{V}_1$ such that $\mathbf{A}_n \models E$ but $\mathbf{A}_n \not\models e_n$. Since e_n is derivable from Ax_1 and e'_n, it follows that $\mathbf{A}_n \not\models e'_n$. We may therefore conclude that $E \not\vdash \mathrm{Eq}_2(\mathbf{N})$, which was to be shown.

4.1 The Algebras $\mathbf{A}_n$

We let the weight of a vector $\overline{u} \in \mathbb{N}^n$, notation $|\overline{u}|$, be defined as the sum of its components. (Equivalently $|\overline{u}| = \overline{u} \cdot \overline{\delta}_n$ where $\overline{\delta}_n = (1, \ldots, 1)$.) To define the algebra $\mathbf{A}_n$, where $n \geq 1$ is a fixed integer, let us call a set $I \subseteq \mathbb{N}^n$ an n-convex ideal if it is an ideal and for any convex linear combination $\overline{v} = \lambda_1\overline{v}_1 + \ldots + \lambda_m\overline{v}_m$ and vector $\overline{u} \in \mathbb{N}^n$ of weight $|\overline{u}| < n$, if $\overline{v}_1, \ldots, \overline{v}_m \in I$ and $\overline{u} \leq \overline{v}$, where $\leq$ is the pointwise order, then $\overline{u} \in I$. It is clear that any convex ideal in $\mathbb{N}^n$ is n-convex. Any set $U \subseteq \mathbb{N}^n$ is contained in a least n-convex ideal, denoted $[\![U]\!]_n$. (The subscript n will often be omitted when it is clear from the context.) Call a vector $\overline{v} = (v_1, \ldots, v_n) \in \mathbb{N}^n$ n-ok, written $ok_n(\overline{v})$, if $|\overline{v}| \leq n$ and

$$|\overline{v}| = n \Rightarrow \exists i \in [n].\ v_i = n \ .$$

A set of vectors is n-ok if all of its elements are. Note that if U is a finite non-empty set consisting of n-ok vectors, then $[\![U]\!]$ is also finite and contains only n-ok vectors.

The algebra $\mathbf{A}_n$ consists of all non-empty (finite) n-convex ideals of n-ok vectors, as well as the element $\top$. The operations are defined as follows: for all $I, J \in \mathbf{A}_n$, $I, J \neq \top$, let $K = \{\overline{u} + \overline{v} : \overline{u} \in I,\ \overline{v} \in J\}$. Then,

$$I + J := \begin{cases} [\![K]\!] & \text{if } K \text{ contains only } n\text{-ok vectors} \\ \top & \text{otherwise.} \end{cases}$$
$$I \vee J := [\![I \cup J]\!]$$
$$0 := \{\overline{0}\} = [\![\overline{0}]\!] \ .$$

Moreover, we define $\top + I = \top \vee I = \top$, and symmetrically, for all $I \in \mathbf{A}_n$.

Proposition 4. *For each* $n \geq 1$, $\mathbf{A}_n \in \mathbf{V}_1$.

We shall now show that, for every $n \geq 2$, the algebra $\mathbf{A}_n$ is not in $\mathbf{V}$. In particular, if $n \geq 2$, then the equations e_n and e'_n do not hold in $\mathbf{A}_n$.

Lemma 2. *If* $n \geq 2$ *then* $\mathbf{A}_n \not\models e_n$ *and* $\mathbf{A}_n \not\models e'_n$.

Note that the induced partial order on $\mathbf{A}_n$ has $\top$ as its top element, and coincides with the inclusion order over the elements in $\mathbf{A}_n$ that are different from $\top$. For $K \in P_f(\mathbb{N}^n)$, by slightly abusing notation, we let $[\![K]\!]$ stand for $[\![K]\!]$ in the original sense if K contains only n-ok vectors and $\top$ otherwise. With this extension of the definition of $[\![\_]\!]$ we have that $[\![K]\!] + [\![L]\!] = [\![K + L]\!]$ for all $K, L \in P_f(\mathbb{N}^n)$. Using this notation, denoting the induced preorder on $\mathbf{A}_n$ by $\sqsubseteq_n$, we obtain the following alternative characterization of $\mathbf{A}_n$ as a quotient algebra of $P_f(\mathbb{N}^n)$, which will be used in the technical developments to follow.

Let $L, M \in P_f(\mathbb{N}^n)$. We write

$$L \leq M \quad \Leftrightarrow \quad \forall \overline{u} \in L.\, \overline{u} \leq M \ .$$

Now,

$$\mathbf{A}_n = \{[\![L]\!] \ : \ L \in P_f(\mathbb{N}^n)\}$$

and

$$[\![L]\!] \sqsubseteq_n [\![M]\!] \text{ iff } L \preceq_n M$$

where

$$L \preceq_n M \quad \Leftrightarrow \quad [ok_n(M) \quad \Rightarrow \quad (ok_n(L) \wedge L \leq M)] \ .$$

Similarly Lemma 1 provides us with the following characterization of $CI_f(\mathbb{N}^n)$:

$$CI_f(\mathbb{N}^n) = \{[L] \ : \ L \in P_f(\mathbb{N}^n)\}$$

where, by Lemma 1,

$$[L] \subseteq [M] \text{ iff } L \leq M \ .$$

In what follows we let the simple inequation

$$a^1x_1 + \ldots + a^mx_m \leq (c_1^1x_1 + \ldots + c_1^mx_m) \vee \ldots \vee (c_k^1x_1 + \ldots + c_k^mx_m)$$

be represented by $(a^i)^{i \leq m} \leq (c_j^i)_{j \leq k}^{i \leq m}$ (or sometimes simply by $\overline{a} \leq \overline{\overline{C}}$ if the meaning is clear from the context), where $(a^i)^{i \leq m}$ denotes the the row vector $(a^1, \cdots, a^m)$ and $(c_j^i)_{j \leq k}^{i \leq m}$ the $k \times m$ matrix $\begin{bmatrix} c_1^1 \ldots c_1^m \\ \vdots \quad\quad \vdots \\ c_k^1 \ldots c_k^m \end{bmatrix}$. We also let an instantiation of the variables $x_1, \ldots, x_m$ by the singleton sets $\{\overline{\eta}_1\}, \ldots, \{\overline{\eta}_m\}$ (or equivalence classes generated by these sets) be represented as

$$\overline{\overline{\eta}} = \begin{bmatrix} \overline{\eta}_1 \\ \vdots \\ \overline{\eta}_m \end{bmatrix} = \begin{bmatrix} \eta_1^1 \cdots \eta_1^n \\ \vdots \quad\quad \vdots \\ \eta_m^1 \cdots \eta_m^n \end{bmatrix} ,$$

i.e. the matrix with row vectors $\overline{\eta}_1, \ldots, \overline{\eta}_m$. We note that, by commutativity of $\vee$, the simple inequation $\overline{a} \leq \overline{\overline{B}}$, where $\overline{\overline{B}}$ is any matrix obtained by permuting the rows of $\overline{\overline{C}}$, represents exactly the same simple inequation as $\overline{a} \leq \overline{\overline{C}}$, viz. the inequation $\overline{a} \leq U$, where $U = \{\overline{c}_1, \ldots, \overline{c}_k\}$ and the $\overline{c}_i$ $(i \in [k])$ are the row vectors of $\overline{\overline{C}}$. Similarly any permutation of the column vectors of $\overline{\overline{C}}$ combined with a corresponding permutation of the entries of $\overline{a}$ yields a simple inequation that holds in $\mathbf{N}$ iff $\overline{a} \leq \overline{\overline{C}}$ does. (Any instantiation matrix should be similarly permuted as well.) The weight of $\overline{\overline{C}}$, notation $|\overline{\overline{C}}|$, is defined as the sum of its entries.

The following result will play a key role in the proof of Thm. 3 to follow.

Corollary 2. *Assume that $\overline{a} \leq \overline{\overline{C}}$ holds in* $\mathbf{N}$ *and is irredundant. Then* $\mathbf{A}_p \not\models \overline{a} \leq \overline{\overline{C}}$*, where* p *is a prime number with* $|\overline{a}| < p$ *and* $|\overline{\overline{C}}| < p$*, iff there is an instantiation matrix* $\overline{\overline{\eta}} \in \mathbb{N}^{m \times p}$*, such that*

1. $\overline{a} \cdot \overline{\overline{\eta}}$ *has weight* p *and at least two non-zero coefficients, and*
2. $\overline{c}_j \cdot \overline{\overline{\eta}}$ *has weight* p *and exactly one non-zero coefficient for* $j \in [k]$.

Note that if $\overline{\overline{\eta}}$ is an instantiation matrix satisfying points 1–2 in the above statement, then so does any matrix obtained by permuting the columns of $\overline{\overline{\eta}}$. This observation will play an important role later in the proof of Thm. 3.

Using the previous result, we are now in a position to show the following theorem, which is the key to the proof of Propn. 3 and of Thm. 2.

Theorem 3. *If* $\overline{a} \leq \overline{\overline{C}}$ *is irredundant and holds in* $\mathbf{N}$*, then* $\mathbf{A}_p \not\models \overline{a} \leq \overline{\overline{C}}$ *for at most one prime number* $p > \max\{|\overline{a}|, |\overline{\overline{C}}|\}$.

Proof. Assume that $\overline{a} = (a^1, \ldots, a^m) \in \mathbb{N}^{1 \times m}$ and $\overline{\overline{C}} = (c_j^i)_{j \leq k}^{i \leq m} \in \mathbb{N}^{k \times m}$ (i.e. the inequation $\overline{a} \leq \overline{\overline{C}}$ contains at most m variables) and that $\mathbf{N} \models \overline{a} \leq \overline{\overline{C}}$. Assume furthermore that the statement of the theorem does not hold and that m is the smallest number of variables for which it fails; we shall show that this leads to a contradiction. So suppose that $\mathbf{N} \models \overline{a} \leq \overline{\overline{C}}$ but that $\mathbf{A}_p \not\models \overline{a} \leq \overline{\overline{C}}$ and $\mathbf{A}_q \not\models \overline{a} \leq \overline{\overline{C}}$, where p and q are prime and $\max\{|\overline{a}|, |\overline{\overline{C}}|\} < p < q$. First we note that we may assume that $a^i > 0$ for $i \in [m]$, or else we could immediately reduce the number of variables in the equation under consideration. Since $\overline{a} \leq \overline{\overline{C}}$ holds in $\mathbf{N}$, this implies that each column vector of $\overline{\overline{C}}$ is non-zero. We now continue with the proof as follows: First we use the assumption $\mathbf{A}_p \not\models \overline{a} \leq \overline{\overline{C}}$ and Corollary 2 to analyze the structure of $\overline{a}$ and $\overline{\overline{C}}$. Then we argue that the result of this analysis contradicts our second assumption, viz. the failure of the inequation $\overline{a} \leq \overline{\overline{C}}$ in $\mathbf{A}_q$.

As $\mathbf{A}_p \not\models \overline{a} \leq \overline{\overline{C}}$, by Corollary 2 there is an instantiation matrix $\overline{\overline{\eta}} = (\eta_i^r)_{i \leq m}^{r \leq p} \in \mathbb{N}^{m,p}$, such that

- $\overline{a} \cdot \overline{\overline{\eta}}$ has weight p and at least two non-zero coefficients and
- $\overline{c}_j \cdot \overline{\overline{\eta}}$ has weight p and exactly one non-zero coefficient, for every $j \in [k]$.

By minimality of m, we may furthermore assume that each row of $\overline{\overline{\eta}}$ contains a non-zero entry.

By rearranging the order of the rows of $\overline{\overline{C}}$ (i.e. the order of the terms that occur on the right-hand side of the simple equation) and of the columns of $\overline{\overline{\eta}}$, we may assume that there is a $0 < k_0 \leq k$ such that

1. $\overline{c}_j \cdot \overline{\overline{\eta}} = (p, 0, \ldots, 0)$ for $j \in [k_0]$ and
2. $\overline{c}_j \cdot \overline{\overline{\eta}} = (0, l_j^2, \ldots, l_j^m)$ for $k_0 < j \leq k$, where exactly one of the l_j^i is non-zero for $2 \leq i \leq m$.

By rearranging the columns of $\overline{\overline{C}}$ and correspondingly the order of the entries of $\overline{a}$ and $\overline{\overline{\eta}}$ (i.e. the order of the variables that occur in the equation), we may assume that there is an $m_0 \in [m]$ such that if $j \leq k_0$ then $c^i_j = 0$ for all $i > m_0$. Therefore we may suppose that $\overline{\overline{C}}$ looks as follows:

$$\overline{\overline{C}} = \left[\begin{array}{c|c} \begin{matrix} c^1_1 \dots c^{m_0}_1 \\ \vdots \qquad \vdots \\ c^1_{k_0} \dots c^{m_0}_{k_0} \end{matrix} & \overline{\overline{O}}_1 \\ \hline \overline{\overline{O}}_2 & ? \end{array}\right] \tag{1}$$

where $\overline{\overline{O}}_1$ and $\overline{\overline{O}}_2$ are (possibly empty) 0-matrices (i.e., matrices whose entries are all 0) and ? just means that this part of the matrix is unknown. Recall that, as previously observed, some of the c^i_j $(i \in [m_0], j \in [k_0])$ may be 0, but no column in the upper left corner of $\overline{\overline{C}}$ is identically 0.

We claim that $k_0 < k$ and $m_0 < m$, and therefore that $\overline{\overline{O}}_1$ and $\overline{\overline{O}}_2$ are both non-trivial. To see that this claim holds, note that, by 1.-2. above, $\overline{\overline{\eta}}$ has the form

$$\overline{\overline{\eta}} = \left[\begin{array}{c|c} \begin{matrix} \eta^1_1 \\ \vdots \\ \eta^1_{m_0} \end{matrix} & \overline{\overline{O}} \\ \hline ? & ? \end{array}\right]$$

where $\overline{\overline{O}}$ is a 0-matrix. This follows because if some of the columns of the matrix to the right of the first column (the one that starts with $\begin{bmatrix} \eta^1_1 \\ \vdots \\ \eta^1_{m_0} \end{bmatrix}$) has a non-zero entry above the horizontal solid line, at least one of the $\overline{c}_j \cdot \overline{\overline{\eta}}$, $j \in [k_0]$, is going to have more than one non-zero entry. Since the rows of $\overline{\overline{\eta}}$ are non-zero, we have that $\eta^1_j \neq 0$ for every $j \in [m_0]$. Using this fact it is not difficult to see that, as claimed, the cases where either $k_0 = k$ or $m_0 = m$ cannot occur. Indeed, if $m_0 = m$ then $\overline{\overline{\eta}}$ has only one non-zero column and consequently $\overline{a} \cdot \overline{\overline{\eta}}$ cannot have two non-zero coefficients. Moreover, if $k_0 = k$ and $m_0 < m$ then the right hand side of the inequation does not contain the variables x_i, with $m_0 < i \leq m$, whereas, by assumption, the left hand side does—a contradiction to the fact that the inequation $\overline{a} \leq \overline{\overline{C}}$ holds in $\mathbf{N}$. We may therefore assume that $k_0 < k$ and $m_0 < m$.

Now we recall from Cor. 1 that, since $\overline{a} \leq \overline{\overline{C}}$ is an irredundant inequation that holds in $\mathbf{N}$, there is a sequence of real numbers λ_i, $i \in [k]$, such that

$$\overline{a} \leq (\lambda_1\overline{c}_1 + \ldots + \lambda_{k_0}\overline{c}_{k_0}) + (\lambda_{k_0+1}\overline{c}_{k_0+1} + \ldots + \lambda_k\overline{c}_k) \ ,$$

where $\lambda_i > 0$ for $i \in [k]$ and $\lambda_1 + \ldots + \lambda_k = 1$. Next let

$$\overline{a}_1 = (a^1, \ldots, a^{m_0}, 0, \ldots, 0) \qquad \text{and} \qquad \overline{a}_2 = (0, \ldots, 0, a^{m_0+1}, \ldots, a^m) \ .$$

Then

$$\begin{aligned} &\overline{a} = \overline{a}_1 + \overline{a}_2, \\ &\overline{a}_1 \leq \lambda_1\overline{c}_1 + \ldots + \lambda_{k_0}\overline{c}_{k_0} \text{ and} \\ &\overline{a}_2 \leq \lambda_{k_0+1}\overline{c}_{k_0+1} + \ldots + \lambda_k\overline{c}_k \ . \end{aligned}$$

Recalling that $|\overline{c}| = \overline{c} \cdot \overline{\delta}_p$, for every vector $\overline{c} \in \mathbb{N}^p$, this implies that

$$\begin{aligned} &p = |\overline{a} \cdot \overline{\overline{\eta}}| = |\overline{a}_1 \cdot \overline{\overline{\eta}}| + |\overline{a}_2 \cdot \overline{\overline{\eta}}|, \\ &|\overline{a}_1 \cdot \overline{\overline{\eta}}| \leq \lambda_1(|\overline{c}_1 \cdot \overline{\overline{\eta}}|) + \ldots + \lambda_{k_0}(|\overline{c}_{k_0} \cdot \overline{\overline{\eta}}|) = (\lambda_1 + \cdots + \lambda_{k_0})p \text{ and} \\ &|\overline{a}_2 \cdot \overline{\overline{\eta}}| \leq \lambda_{k_0+1}(|\overline{c}_{k_0+1} \cdot \overline{\overline{\eta}}|) + \ldots + \lambda_k(|\overline{c}_k \cdot \overline{\overline{\eta}}|) = (\lambda_{k_0+1} + \cdots + \lambda_k)p \ . \end{aligned}$$

In particular, as $\lambda_1 + \cdots + \lambda_k = 1$, the inequalities above must be equalities. Thus we have proven that

$$n_p = |\overline{a}_1 \cdot \overline{\overline{\eta}}| = \lambda p \ ,$$

where $\lambda = \lambda_1 + \ldots + \lambda_{k_0}$ and $n_p \in \mathbb{N}$. Note that, as $k_0 < k$ and $\lambda_i > 0$ ($i \in [k]$), it holds that $\lambda < 1$. Hence we have that $n_p < p$.

In a similar way, using the form of $\overline{\overline{C}}$ in (1) and the fact that $q > p$, $\mathbf{N} \models \overline{a} \leq \overline{\overline{C}}$ and $\mathbf{A}_q \not\models \overline{a} \leq \overline{\overline{C}}$, we may conclude that there is a $\overline{\overline{\gamma}} \in \mathbb{N}^{m \times q}$ such that

$$n_q = |\overline{a}_1 \cdot \overline{\overline{\gamma}}| = \lambda q \ .$$

This in turn implies that $\frac{n_p}{p} = \frac{n_q}{q}$ or equivalently $n_p \cdot q = n_q \cdot p$, contradicting our assumption that $n_p < p$, $n_q < q$ and p and q are different primes. We may therefore conclude that no such minimal number of variables m exists and consequently that the statement of the theorem holds. This completes the proof of the theorem. □

Propn. 3 follows immediately from the above result, completing the proof of the non-existence of a finite equational axiomatization for the two-variable fragment of the equational theory of the algebra $\mathbf{N}$.

Remark 1. Using our results, it is easy to show that the reduct $(\mathbb{N}, \vee, +)$ of $\mathbf{N}$ is also not finitely based, and that the two-variable fragment of its equational theory has no finite equational axiomatization.

As a further corollary of Thm. 2, we obtain that the equational theory of the algebra $(\mathbb{N}, \vee, +, 0, 1)$ is also not finitely based. To see this, note that whenever an equation holds in $(\mathbb{N}, \vee, +, 0, 1)$ and one side contains an occurrence of the symbol 1, then so does the other side. Let E be an axiom system for $(\mathbb{N}, \vee, +, 0, 1)$, and let E_0 denote the subset of E consisting of all the equations not containing occurrences of the constant 1. In light of the above observation, E_0 is an axiom system for the reduct $\mathbf{N}$ of $(\mathbb{N}, \vee, +, 0, 1)$. Thus the existence of a finite basis for the algebra $(\mathbb{N}, \vee, +, 0, 1)$ would contradict Thm. 2. In similar fashion, it is easy to prove that the two-variable fragment of the equational theory of the algebra $(\mathbb{N}, \vee, +, 0, 1)$ has no finite equational axiomatization.

Remark 2. In [1], we also provide an application of our main result to process algebra. More precisely, we show that trace equivalence has no finite ω-complete equational axiomatization for the language of Basic Process Algebra [3] over a singleton alphabet, with or without the empty process.

Acknowledgments: The anonymous referees provided useful comments and pointers to the large body of literature on the max-sum algebra and its applications.

References

1. L. Aceto, Z. Ésik, and A. Ingólfsdóttir, *On the two-variable fragment of the equational theory of the max-sum algebra of the natural numbers*, BRICS Report RS-99-22, August 1999.
2. L. Aceto, Z. Ésik, and A. Ingólfsdóttir, *The max-plus algebra of the natural numbers has no finite equational basis*, BRICS Report RS-99-33, October 1999. To appear in *Theoretical Computer Science.*
3. J. Baeten and W. Weijland, *Process Algebra*, Cambridge Tracts in Theoretical Computer Science 18, Cambridge University Press, 1990.
4. S. Burris and H. P. Sankappanavar, *A Course in Universal Algebra*, Springer-Verlag, New York, 1981.
5. J. Gunawardena (ed.), *Idempotency*, Publications of the Newton Institute 11, Cambridge University Press, 1998.
6. K. Hashiguchi, *Limitedness theorem on finite automata with distance functions*, J. Comput. System Sci., 24 (1982), no. 2, pp. 233–244.
7. K. Hashiguchi, *Relative star-height, star-height and finite automata with distance functions*, in Formal Properties of Finite Automata and Applications (J.E. Pin ed.), Lecture Notes in Computer Science 386, Springer-Verlag, 1989, pp. 74–88.
8. W. Kuich and A. Salomaa, *Semirings, Automata, Languages*, EATCS Monographs on Theoretical Computer Science 5, Springer-Verlag, 1986.
9. G. McNulty, *A field guide to equational logic*, J. Symbolic Computation, 14 (1992), pp. 371–397.
10. I. Simon, *Limited subsets of a free monoid*, in Proceedings of the 19th Annual Symposium on Foundations of Computer Science (Ann Arbor, Mich., 1978), IEEE Computer Society, 1978, pp. 143–150.
11. I. Simon, *Recognizable sets with multiplicities in the tropical semiring*, in Proceedings of Mathematical Foundations of Computer Science (Carlsbad, 1988), Lecture Notes in Computer Science 324, Springer-Verlag, 1988, pp. 107–120.
12. U. Zimmermann, *Linear and Combinatorial Optimization in Ordered Algebraic Structures*, Annals of Discrete Mathematics 10, North-Holland, 1981.

Real-Time Automata and the Kleene Algebra of Sets of Real Numbers

Cătălin Dima*

Bucharest University, Department of Fundamentals of Computer Science
str. Academiei 14, Bucharest, Romania, R0-70109,
cdima@funinf.math.unibuc.ro

Abstract. A commutative complemented Kleene algebra of sets of (positive) real numbers is introduced. For the subalgebra generated by finite unions of rational intervals a normal form is found. These are then applied to the complementation problem for real-time automata.

1 Introduction

Computing with intervals has sometimes been a by-product whenever a quantitative study of time and succession was of interest. Examples are constraint networks [DMP91], timed automata [AD94] and logics of real-time [AH96]. It is trivial that finite unions of intervals form a boolean algebra and in [CG98] it has been noticed that "periodic" unions of intervals form a boolean algebra, but the next and simple step of looking at a Kleene algebra of sets of positive reals has not yet (up to our knowledge) been an issue of interest.

We make this step here by defining star with the usual least fixpoint construction based on addition of *sets* of numbers, hence transforming sets of real numbers into a Kleene algebra. This algebra naturally arises in the study of the so-called *real-time automata* (RTA) [DW96], which are timed automata with a single clock which is reseted at each transition. In this model removing silent steps (a necessary step for transforming nondeterministic automata into deterministic ones and then "complementing" them) involves computing stars of unions of intervals. Hence even at this lowest level of generalization from the untimed finite automata removing silent transitions [BDGP98] is a problem.

We solve this problem here by studying the sub-Kleene algebra generated by finite unions of intervals with rational bounds. We prove a normal form theorem for the elements of this subalgebra, result which is based on properties of integer division. We then find a normal form for circuits with silent transitions which allows complementation. Actually this is done by allowing clock constraints of the form $x \in X$ where X is no longer a finite union of intervals but some "periodic" union of intervals.

We note that, though RTA cannot be used for modeling distributed real-time systems, they show sufficient theoretical interest as they are incomparable

* This research was done during the author's visit to TIFR, Bombay, supported by an extension of a UNU/IIST fellowship.

H. Reichel and S. Tison (Eds.): STACS 2000, LNCS 1770, pp. 279–289, 2000.

to event-clock automata[AFH94, HRS98] which are so far the largest determinizable subclass of timed automata. This situation occurs as RTA may accept languages consisting of signals whose lengths are natural numbers while event-clock automata may not.

The rest of the paper is divided as follows: in the next section we remind some definitions and state the problem of complementation of real-time automata. In the third section the Kleene algebra of sets of positive reals is introduced and the normal form theorem is proved. The fourth section is for the determinization construction and the last section is for short comments and directions of further study.

2 Real-Time Automata

A **signal** over a finite alphabet V is a function $\sigma : [0, e) \longrightarrow V$ where e is a nonnegative number, function which has finitely many discontinuities, all of them being left discontinuities. Hence the domain of a signal σ splits into finitely many intervals $[e_{i-1}, e_i)$ on which σ is constant. We denote by $dom(\sigma)$ the domain of σ, and $Sig(V)$ the set of signals over V.

For $\sigma_1, \sigma_2 \in Sig(V)$ with $dom(\sigma_i) = [0, e_i)$ $(i = 1, 2)$ define their *concatenation* $\sigma_1; \sigma_2 = \sigma$ as the signal with $dom(\sigma) = [0, e_1 + e_2)$ and such that for $t \in [0, e_1)$, $\sigma(t) = \sigma_1(t)$ and for $[e_1, e_1 + e_2)$, $\sigma(t) = \sigma_2(t - e_1)$. Hence $(Sig(V), \text{";"}, \sigma_\epsilon)$ becomes a noncommutative monoid whose unit is the signal σ_ϵ with $dom(\sigma_\epsilon) = [0, 0)$. Then concatenation can be extended for sets of signals and gives rise to star: for $\Sigma \subseteq Sig(V)$ put $\Sigma^* = \bigcup_{n \in \mathbb{N}} \Sigma^n$. Here $\Sigma^0 = \{\sigma_\epsilon\}$ and $\Sigma^{n+1} = \Sigma^n; \Sigma$.

For the sequel, $Int_{\mathbb{Q}}^{\geq 0}$ denotes the set of intervals with positive rational or infinite bounds. Its elements are called *rational intervals*.

Definition 1. *A* ***real-time automaton*** *(RTA) over the alphabet V is a tuple $\mathcal{A} = (Q, V, \lambda, \iota, \delta, Q_0, F)$ where Q is the (finite) set of* states, *$\delta \subseteq Q \times Q$ is the* transition relation, *$Q_0, F \subseteq Q$ are the sets of* initial, *resp.* final *states, $\lambda : Q \longrightarrow V$ is the* state labeling function *and $\iota : Q \longrightarrow Int_{\mathbb{Q}}^{\geq 0}$ is the* interval labeling function*; also call q an a-*state *iff $\lambda(q) = a$.*

RTA work over signals: a *run* of length n is a sequence of states $(q_i)_{i \in [n]}$ connected by δ, i.e. $(q_{i-1}, q_i) \in \delta, \forall i \in [n]$. A run is **accepting** iff it starts in Q_0 and ends in F. A run is **associated to** a signal σ with $dom(\sigma) = [0, e)$ iff there exist some "splitting" points $0 = e_1 \leq \ldots \leq e_{n+1} = e$ such that $e_{i+1} - e_i \in \iota(q_i)$ and $\sigma(t) = \lambda(q_i)$ for all $t \in (e_i, e_{i+1})$ and all $i \in [n]$. Note that the "splitting" points must contain all the discontinuities but the reverse does not hold.

The **language** of some RTA $\mathcal{A}$ is the set of signals associated to some *accepting run* of $\mathcal{A}$ and is denoted $L(\mathcal{A})$. Two RTA are *equivalent* iff they have the same language. Define then the class of **timed recognizable languages** as $TRec(V) = \{\Sigma \in Sig(V) \mid \exists \mathcal{A} \text{ s.t. } L(\mathcal{A}) = \Sigma\}$.

Proposition 1. *The emptiness problem for RTA is decidable.*

The proof relies on the algorithms for computing the accessible (or coaccessible) states which can be done in linear time (w.r.t. $card(Q)$).

Definition 2. *The set* **ETRE**(V) *of* ***elementary timed regular expressions*** *over V is defined by the rules $E ::= 0 \mid [a]_I \mid E+E \mid E;E \mid E^*$ where the* atoms $[a]_I$ *are all expressions with $a \in V$ and $I \in Int_{\mathbb{Q}}^{\geq 0}$.*

Semantics of elementary timed regular expressions is the following:

$$|[a]_I| = \{\sigma \in Sig(V) \mid dom(\sigma) = [0,e), e \in I \text{ and } \forall t \in [0,e)\ \sigma(t) = a\}$$
$$|E{+}F| = |E| \cup |F| \qquad |E;F| = |E|;|F| \qquad |E^*| = |E|^* \qquad |0| = \emptyset$$

Define the class of **timed regular languages** as $TReg(V) = \{\Sigma \in Sig(V) \mid \exists E \in ETRE(V)$ such that $|E| = \Sigma\}$.

Theorem 1 (Kleene theorem for RTA). $TRec(V) = TReg(V)$.

This theorem is proved in [Dim99]. Note that the regular expressions defined here are weaker than the ones of [ACM97]. A class of regular expressions equivalent[1] to [ACM97] involve using *sets* of letters in the atoms, e.g. $[A]_{(1,2)}$ where $A \subseteq V$, see [Dim99].

We want to show that the class of timed recognizable languages is closed under complementation, hence we need a subclass of RTA in which each word has *a single run*, for then complementation would be accomplished simply by complementation of the set of final states.

Definition 3. *A RTA $\mathcal{A}$ is* ***language deterministic*** *iff each signal in $L(\mathcal{A})$ is associated to a unique run. $\mathcal{A}$ is* ***stuttering-free*** *iff it does not have time labels which contain 0 and transitions (q,r) with $\lambda(q) = \lambda(r)$. $\mathcal{A}$ is* ***state-deterministic*** *iff initial states have disjoint labels and transitions starting in the same states have disjoint labels too, i.e. whenever $r \neq s$ and either $r, s \in Q_0$ or $(q,r),(q,s) \in \delta$ then $\lambda(r) \neq \lambda(s)$ or $\iota(r) \cap \iota(s) = \emptyset$. $\mathcal{A}$ is simply called* ***deterministic*** *iff it is both state-deterministic and stuttering-free.*

A first observation is that determinism implies language determinism while state-determinism itself does not. But a more important observation is that deterministic RTA are *strictly less expressive* than general RTA: Consider the language $L_{\mathbb{N}} = \{\sigma : [0,n) \longrightarrow \{a\} \mid n \in \mathbb{N}\}$ of constant signals with integer length; it is accepted by the RTA in the figure $1(a)$.

Proposition 2. *$L_{\mathbb{N}}$ cannot be accepted by any stuttering-free RTA.*

The proof is based on the intuition that a stuttering-free RTA for $L_{\mathbb{N}}$ would need an infinite number of states. Note that this proof works for event-clock automata too [AFH94, HRS98], hence RTA and event-clock automata are incomparable.

However there is no problem for building a RTA for the complement of $L_{\mathbb{N}}$, as we see from the next figure. Also in this figure we find out that some stuttering RTA can still be transformed into stuttering-free RTA.

[1] This is actually the reason for denoting the expressions introduced here as *elementary*

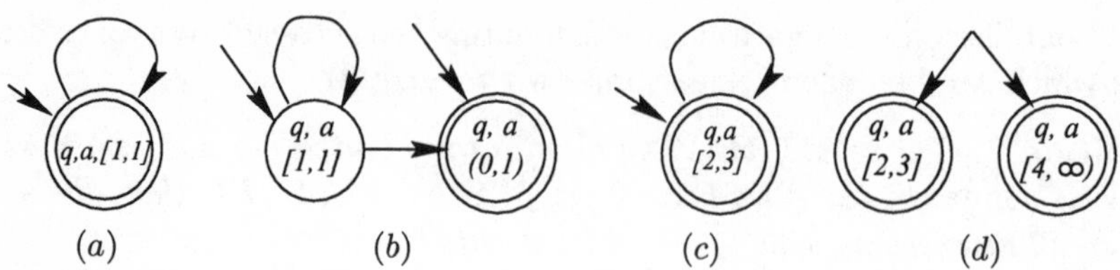

Fig. 1. The automaton at a accepts $L_{\mathbb{N}}$ while the complement of $L_{\mathbb{N}}$ is accepted by the RTA at (b). The stuttering RTA at (c) is equivalent to the stuttering-free one at (d).

Hence we discover the need of computing the "sum" of two intervals and the "star" of some interval, i.e. some operations that satisfy

$$\mathbb{R} \setminus \{1\}^* = \{1\}^* + (0,1) \quad \text{and} \quad [2,3]^* = [2,3] \cup [4,\infty)$$

relations which are suggested by figure 1.

3 Operations with Subsets of the Real Numbers

The powerset of the positive numbers $\mathcal{P}(\mathbb{R}_{\geq 0})$ is naturally endowed with an operation of "concatenation": it is addition extended over sets:

$$X + Y = \{x + y \mid x \in X, y \in Y\} \text{ for all } X, Y \subseteq \mathbb{R}$$

whose unit is $\mathbf{0} = \{0\}$.

Moreover we can define star via the usual least fixpoint construction $X^* = \bigcup_{n \in \mathbb{N}} nX$, where the multiples of X are defined as usual: $0X = \mathbf{0}$ and $(n+1)X = nX + X$.

The following theorem can then be easily verified:

Theorem 2. *The structure* $\mathcal{P}(\mathbb{R}_{\geq 0}) = (\mathcal{P}(\mathbb{R}_{\geq 0}), \cup, +, \cdot^*, \emptyset, \mathbf{0})$ *is a commutative Kleene algebra, i.e.* $(\mathcal{P}(\mathbb{R}_{\geq 0}), +, \mathbf{0})$ *is a commutative monoid,* $+$ *distributes over* $\cup$ *and* $\cdot^*$ *satisfies the following equations [Con71, Koz94]:*

$$X + Y \leq Y \Rightarrow X^* + Y \leq Y \tag{1}$$
$$\mathbf{0} \cup (X + X^*) \leq X^* \tag{2}$$
$$X^* + Y^* = (X + Y)^* + (X^* \cup Y^*) \tag{3}$$

where $X \leq Y$ *denotes* $X \cup Y = Y$.

Because a complement operation is available too: $\neg X = \mathbb{R}_{\geq 0} \setminus X$, we actually get a **commutative complemented Kleene algebra**, i.e. a boolean algebra which is also a commutative Kleene algebra.

Denote $\mathcal{K}(Int_{\mathbb{Q}}^{\geq 0})$ the sub-(commutative complemented Kleene) algebra generated by $Int_{\mathbb{Q}}^{\geq 0}$ in $\mathcal{P}(\mathbb{R}_{\geq 0})$.

Definition 4. *A set $X \in \mathcal{K}(Int_{\mathbb{Q}}^{\geq 0})$ can be written in* **normal form** *(NF) iff there exist finite unions of rational intervals $X_1, X_2 \in Int_{\mathbb{Q}}^{\geq 0}$ and some $k \in \mathbb{Q}_{\geq 0}$ such that*

$$X = X_1 \cup (X_2 + \{k\}^*) \tag{4}$$

with the requirement that there exists some $N \in \mathbb{N}$ such that $X_1 \subseteq [0, Nk)$ and $X_2 \subseteq [Nk, (N+1)k)$. We call this N the **bound** *of the NF.*

Normal forms are not unique: for the NF in the definition and some $p \in \mathbb{N}$, the following expression:

$$X = (X_1 \cup (X_2 + \{0, k, 2k, \ldots, (p-1)k\})) \cup (X_2 + \{pk\} + \{k\}^*)$$

is a NF too, but with bound $N + p$.

Clearly a finite union of rational intervals $X \in Int_{\mathbb{Q}}^{\geq 0}$ can always be put into NF. Also note that some NF $X = X_1 \cup (X_2 + \{k\}^*)$ has $X = \emptyset$ iff both X_1 and X_2 are empty.

Sometimes when applying different operations to NFs we might not be able to get very easily a new NF; instead, we might get a **weak normal form**, which is a decomposition like equation 4 but without the additional requirement on the existence of the bound. However we have the following:

Lemma 1. *Weak NFs can be transformed into NFs.*

The key result for NFs is the following:

Theorem 3. *Each $X \in \mathcal{K}(Int_{\mathbb{Q}}^{\geq 0})$ can be written in normal form.*

Proof. We must show that the result of any operation applied to NF can be put into NF. We first list some useful identities valid in $\mathcal{P}(\mathbb{R})$ [Con71]:

$$X^{**} = X^* \tag{5}$$

$$(X \cup Y)^* = X^* + Y^* \tag{6}$$

$$(X^* + Y)^* = \{0\} \cup (X^* + Y^* + Y) \tag{7}$$

Also note the following *ultimately periodicity property*:

> Given n distinct positive rationals $a_i \in \mathbb{Q}_{\geq 0}$ we have that $\{a_1, \ldots, a_n\}^*$ is *ultimately periodic*, i.e. there exist some *finite* set of rationals B and some rationals $q, r \in \mathbb{Q}_{\geq 0}$ such that
>
> $$\{a_1, \ldots, a_n\}^* = B \cup (\{q\} + \{r\}^*) \tag{8}$$

This result can be seen as a corollary of the normal form theorem of the regular languages over a one letter alphabet.

Fix now two NFs $X = X_1 \cup (X_2 + \{k\}^*)$ with bound M and $Y = Y_1 \cup (Y_2 + \{l\}^*)$ with bound N and denote $m = lcm(k, l)$. We then get the following form for $X \cup Y$:

$$X_1 \cup Y_1 \cup \left(\left(\bigcup_{i=0}^{m/k-1} (X_2 + \{ik\}) \cup \bigcup_{i=0}^{m/l-1} (Y_2 + \{il\}) \right) + \{m\}^* \right)$$

This is a WNF and Lemma 1 helps transforming it into NF.

For $X+Y$ distributivity of $+$ over $\cup$ transforms it into:

$$(X_1+Y_1)\cup(X_1+Y_2+\{l\}^*)\cup(X_2+Y_1+\{k\}^*)\cup(X_2+Y_2+\{k\}^*+\{l\}^*)$$

An instantiation of identity 6 gives $\{k\}^*+\{l\}^*=\{k,l\}^*$. The ultimately periodicity property 8 gives a NF for the last and thence we have above a union of (weak) NFs which we already know how to bring to NF.

For X^* we have two cases. The first one occurs when one of X_1 and X_2 contains a nonpoint interval. Then *the set X^* is a finite union of rational intervals*, so it is in NF. To prove this claim, note that for each nonpoint interval, say $(a,b]$ (that is $b-a>0$), denoting $m_0=\left\lceil\frac{a}{b-a}\right\rceil$, we have that $(a,b]^*=$ $\mathbf{0}\cup\bigcup_{i=1}^{m_0-1}(ia,ib]\cup(m_0a,\infty)$ since the choice of m_0 assures that $(m_0+1)a<m_0b$. Hence from the m_0-th iteration the intervals start to overlap. This observation can be easily adapted to prove our claim.

The second case for X^* is when both X_1 and X_2 consist of point intervals. Applying identity 6 we get that $X^*=X_1^*+(X_2+\{k\}^*)^*$. Then by the ultimately periodicity property 8 X_1 can be written into NF, so we concentrate on $(X_2+\{k\}^*)^*$. Therefore, by identities 7 and 6 we get

$$(X_2+\{k\}^*)^*=\mathbf{0}\cup(X_2+X_2^*+\{k\}^*)\;=\;\mathbf{0}\cup(X_2+(X_2\cup\{k\})^*)$$

Now the ultimately periodicity property 8 tells us that $(X_2\cup\{k\})^*$ can be put into NF, so we can also find a NF for X^* by the previous cases.

For $\neg$, note first that $\neg[a,b)=[0,a)\cup[b,\infty)$ (when $b<\infty$), and $\neg\{(a,\infty)\}=\{[0,a)\}$. Also, by De Morgan laws, $\neg(I_1\cup\ldots\cup I_n)=\neg I_1\cap\ldots\cap\neg I_n$ is a NF and, for any NF, $\neg X=\neg X_1\cap\neg(X_2+\{k\}^*)$. Then, by distributivity of $\cap$, we may restrict ourselves to the case $X_1=[a,b)$ with $a,b\in\mathbb{R}_{>0}$ and $k\neq 0$ (the cases with other parentheses are similar). But since $X_2\cup[0,Nk)\cup[Nk+k,\infty)$ is a finite union of rational intervals (hence we know how to write NFs for the complement of it) the following is a NF for $\neg X$:

$$[0,a)\cup[b,Nk)\cup(\neg(X_2\cup[0,Nk)\cup[Nk+k,\infty))+\{k\}^*)\qquad\square$$

Note that in [CG98] a weaker version of this theorem is proved, roughly saying that the set of finite unions of n-dimensional NFs forms a boolean algebra.

Though the theorem is based on the same technique that gives the normal form of regular languages over a one-letter alphabet it cannot be a simple corollary of that. Even if we restrict attention to the algebra generated by intervals with natural bounds, denote it $\mathbb{N}Int$, this has *two* generators: the point set $\{1\}$ and the nonpoint interval $(0,1)$. Neither of these generators alone may generate the whole $\mathbb{N}Int$: $\{1\}$ generates just sets with isolated points or complements of such sets (i.e. countable or co-countable sets) and hence does not generate $(0,1)$, while $(0,1)$ generates just finite unions of intervals (it cannot generate $(0,1)+\{1\}^*$).

One might also think that the result follows from Eilenberg's theory of automata with multiplicities [Ei74]. But this is not the case either since in that work star is defined via some formal power series and one cannot prove, unless defining some suitable equivalence on power series, that e.g. $[0,1)^* = [0,\infty)$.

Finally note the interesting relation which holds between the two generators of $\mathbb{N}Int$:

$$(0,1)^* = (\mathbf{0} \cup (0,1)) + \{1\}^*$$

At the end of this section we make a brief excursion into matrix theory. We construct, as in [Koz94] the Kleene algebra of matrices over $\mathcal{P}(\mathbb{R}_{\geq 0})$ whose operations are defined as follows:

$$(A \cup B)_{ij} = A_{ij} \cup B_{ij} \qquad (A+B)_{ij} = \bigcup\nolimits_{k=1}^{n} (A_{ik} + B_{kj})$$

$$A^* = \bigcup\nolimits_{n \in \mathbb{N}} nA$$

The star of matrix A can be computed by the usual Floyd-Warshall-Kleene algorithm [Con71, Ei74]: we recursively define a sequence of $n+1$ matrices $A(k)$ $(0 \leq k \leq n)$ with $A(0) = A$ and

$$A(k)_{ij} = A(k-1)_{ij} \cup (A(k-1)_{ik} + (A(k-1)_{kk})^* + A(k-1)_{kj}) \tag{9}$$

Proposition 3 ([Ei74]). *$A(n) = A^*$ for any matrix over $\mathcal{P}(\mathbb{R}_{\geq 0})$.*

Corollary 1. *If A is a matrix of normal forms then A^* is a matrix of normal forms too which can be computed by the above algorithm.*

4 Determinization and Complementation of RTA

Definition 5. *An **augmented real-time automaton** (ARTA for short) over V is a tuple $\mathcal{A} = (Q, V, \delta, \lambda, \iota, Q_0, F)$ where Q, Q_0, F, δ and λ are the same as for RTA while $\iota : Q \longrightarrow \mathcal{K}(Int_{\mathbb{Q}}^{\geq 0})$ (actually ι gives a NF).*

ARTA work similarly to RTA: *runs* have the same definition and a signal σ with $dom(\sigma) = [0,e)$ is *associated to* a run of length n iff there exist $0 = t_1 \leq \ldots \leq t_{n+1} = e$ with $t_{i+1} - t_i \in \iota(q_i)$ and $\sigma(t) = \lambda(q_i)$ for all $t \in [t_i, t_{i+1})$ and all $i \in [n]$. The emptiness problem is again decidable in linear time w.r.t. $card(Q)$. Note that we need a preliminary step in which states q whose $\iota(q) = \emptyset$ are removed. Also the different notions of determinism also apply to ARTA with the same definitions; hence we will speak of state-deterministic ARTA and stuttering-free ARTA.

The following theorem says that usage of NFs instead of just intervals do not add to the expressive power of RTA:

Theorem 4. *$TReg(V)$ equals the class of languages accepted by ARTA.*

The proof is very close to the one of Theorem 1 and is based on the following property of regular expressions:

$$\left|[a]_{(a,b)\cup((c,d)+\{k\}^*)}\right| = \left|[a]_{(a,b)} + [a]_{(c,d)}; [a]^*_{\{k\}}\right|$$

which can be easily extended to any atomic ETRE.

Hence we may focus on the problem of complementing ARTA, since ARTA and RTA are equally expressive by Theorem 4.

Theorem 5. *Each ARTA is equivalent to some deterministic ARTA.*

Proof. As a preliminary step, in the given ARTA we modify all states q in which $0 \in \iota(q)$ by removing 0 from $\iota(q)$ and adding some new transitions, the whole process resembling very closely to the removal of ϵ-transitions from finite automata. Also we assume that all states with empty time label have been removed.

We first achieve **stuttering freeness** by removing all stuttering transitions between a-states, for some $a \in V$, and repeating this for all the other letters in V. The idea is to find, for each pair of states (q, r) the set of positive numbers which are the duration of a signal that is associated to some run starting in q, ending in r and containing only a-states. For this we need to recursively add all the intervals of the states that may lie on such a run. This is one of the places where we apply the normal form theorem 3 and the algorithm for computing the star of a matrix of sets of positive numbers. The formalization is the following:

Start with some ARTA $\mathcal{A} = (Q, V, \delta, \lambda, \iota, Q_0, F)$ and number its transitions as $\delta = \{t_1, \ldots, t_p\}$ (denote $t_i = (in_i, out_i)$). Construct a matrix A whose elements are the sets of the a-labeled states: $A_{ij} = X$ iff $out_i = in_j$, $\lambda(out_i) = a$ and $\iota(out_i) = X$, otherwise $A_{ij} = \emptyset$. Then we add two more rows and columns to A (the $p+1$-th and the $p+2$-th) which intuitively record the time labels of initial, resp. final a-states: for $j \in [p]$ put $A_{p+1,j} = X$ for all j with $in_j \in Q_0$, $\lambda(in_j) = a$ and $\iota(in_j) = X$, otherwise $A_{p+1,j} = \emptyset$ and $A_{j,p+2} = X$ for all j with $out_j \in F$, $\lambda(out_j) = a$ and $\iota(out_j) = X\}$; moreover put $A_{p+1,p+2} = \bigcup\{X \mid \exists q \in Q_0 \cap F, \lambda(q) = a \text{ and } \iota(q) = X\}$.

Then A^* holds *the lengths of all signals associated to runs consisting of a-states only.* Hence $(A^*)_{ij}$ consists of the lengths of signals associated to runs starting in out_i, ending in in_j and consisting of a-states only. Also $(A^*)_{p+1,j}$ consists of the lengths of all runs that start in Q_0, end in in_i and consist of a-states only. Similarly for $(A^*)_{i,p+2}$ and $(A^*)_{p+1,p+2}$. Computation of A^* is done by the Floyd-Warshall-Kleene algorithm (9). Note here the importance of Corollary 1: the elements of A^* are still NFs, hence they may be used for labeling some new states of an ARTA.

Hence, while non-a-states will be preserved, the *nonempty* components of A^* will replace all a-states: their time label will be $(A^*)_{ij}$ and they will be connected only to non-a-states. Formally, denote $Q_{\neg a} = \{q \in A \mid \lambda(q) \neq a\}$ and $Q_a = \{(i, j) \mid (A^*)_{ij} \neq \emptyset\}$ and define $\mathcal{B} = (Q_{\neg a} \cup Q_a, V, \overline{\delta}, \overline{\lambda}, \overline{\iota}, T_0, T_f)$ where

$\overline{\lambda}(s) = \lambda(s)$ for $s \in Q_{\neg a}$ and $\overline{\lambda}((i,j)) = a$ for $(i,j) \in Q_a$;
$\overline{\iota}(s) = \iota(s)$ for $s \in Q_{\neg a}$ and $\overline{\iota}((i,j)) = (A^*)_{ij}$ for $(i,j) \in Q_a$;
$T_0 = (Q_{\neg a} \cap Q_0) \cup \{(p+1,j) \in Q_a\} \cup \{(p+1,p+2) \mid A^*_{p+1,p+2} \neq \emptyset\}$;
$T_f = (Q_{\neg a} \cap F) \cup \{(i,p+2) \in Q_a\} \cup \{(p+1,p+2) \mid A^*_{p+1,p+2} \neq \emptyset\}$;
$\overline{\delta} = \{(in_i,(i,j)) \mid \lambda(in_i) \neq a\} \cup \{((i,j),out_j) \mid \lambda(out_j) \neq a\} \cup$
$\cup (\delta \cap (Q_{\neg a} \times Q_{\neg a}))$

The proof that $L(\mathcal{B}) = L(\mathcal{A})$ relies on Theorem 3. Note that by construction no two a-states are directly connected and by the exclusion of zeroes from intervals in the beginning all components of A^* will not contain 0. Also, all transitions between non a-states are preserved, hence no stuttering transitions are added. This shows that after applying it for all letters in V we get a stuttering-free ARTA.

The **determinization** construction is an adaptation of the subset construction. Start with some ARTA $\mathcal{B} = (Q, V, \delta, \lambda, \iota, Q_0, F)$ assumed stuttering-free. If the time labels did not count then the states of the deterministic automaton were sets of *identically state-labeled states* and we would draw a transition from some S_1 with $\lambda(S_1) = \{a\}$ to some S_2 with $\lambda(S_2) = \{b\}$ iff $S_2 = \{r \in Q \mid \exists q \in S_1 \text{ s.t. } (q,r) \in \delta\}$. Taking into account the time labels is done by splitting S_2 into several "smaller" sets such that the time labels of these parts give a partition of $\mathbb{R}_{>0}$.

Therefore we start with $\mathcal{T}$, the set of triples (S, S', a) where $a \in V$, $S' \subseteq S \subseteq Q$ with $\lambda(S) = \{a\}$. and $\overline{\lambda}((S,S',a)) = a$. Intuitively the control passes through (S,S',a) iff in $\mathcal{B}$ the control may pass through some state in S' but not through any of the states in $S \setminus S'$. Formally we associate to each $U \subseteq Q$ with $\lambda(U) = \{a\}$ the set $Tl(U) = \{X \in \mathcal{K}(Int_{\mathbb{Q}}^{\geq 0}) \mid \exists q \in U \text{ s.t. } \iota(q) = X\}$ and define

$$\overline{\iota}((S,S',a)) = \mathbb{R}_{>0} \cap \left(\bigcap Tl(S')\right) \cap \neg \left(\bigcup Tl(S \setminus S')\right)$$

where the usual conventions $\bigcap \emptyset = \mathbb{R}_{\geq 0}$ and $\bigcup \emptyset = \emptyset$ apply. Note that we put $\mathbb{R}_{>0}$ in front of $\overline{\iota}((S,S',a))$ because otherwise we lose stuttering-freeness.

It is also important to note that it is *here* where we need the result that NFs are closed under complementation, because we need to put $\overline{\iota}((S,S',a))$ in NF and $\overline{\iota}((S,S',a))$ contains complementation.

Though $\mathcal{T}$ is not what we need: it might still happen that $\overline{\iota}(S,S',a) \cap \overline{\iota}(S,S'',a) \neq \emptyset$, but only when $\overline{\iota}(S,S',a) = \overline{\iota}(S,S'',a)$ Hence we can define an equivalence on $\mathcal{T}$ as $(S,S',a) \simeq (S,S'',a)$ iff $\overline{\iota}(S,S',a) = \overline{\iota}(S,S'',a)$. The quotient set $\mathcal{T}/\simeq$ is our desired set of states as we have that for any $S \subseteq Q$ the time labels of the classes $[S,S',a]$ for all $S' \subseteq S$ give a partition of $\mathbb{R}_{>0}$.

Hence we build $\mathcal{C} = (\mathcal{T}/\simeq, V, \tilde{\delta}, \tilde{\lambda}, \tilde{\iota}, \mathcal{T}_0, \mathcal{T}_f)$, where $\tilde{\lambda}([S,S',a]) = a$ and $\tilde{\iota}([S,S',a]) = \overline{\iota}((S,S',a))$. $\tilde{\delta}$ consists of transitions from $[S,S',a]$ to each $[U,U',b]$ where $U = \{q \in Q \mid \exists r \in S' \text{ s.t. } (r,q) \in \delta \text{ and } \lambda(q) = b\}$ and $U' \subseteq U$. Case $U' = \emptyset$ stands for the situation when the length of the current state in the signal is not in any of the sets from $Tl(U)$. (Note how states $[\emptyset,\emptyset,a]$ time-labeled with $\mathbb{R}_{>0}$ play the role of the trap states in dfa). Moreover put

$$\mathcal{T}_0 = \{[S, S', a] \in \overline{Q} \mid \emptyset \neq S \subseteq Q_0, a \in V\} \cup \{[\emptyset, \emptyset, a] \mid \forall q \in Q_0, \lambda(q) \neq a\}$$
$$\mathcal{T}_f = \{[S, S', a] \in \overline{Q} \mid S' \cap F \neq \emptyset, a \in V\}$$

The proof that $\mathcal{C}$ is equivalent to $\mathcal{B}$ follows the classical pattern. □

Theorem 6. *$TRec(V)$ is closed under complementation.*

Proof. This is a corollary of the above theorem: in the ARTA $\mathcal{C}$ constructed above, each signal *is associated to a unique run that starts in* $\mathcal{T}_0$. Hence, the ARTA that accepts $Sig(S) \setminus L(\mathcal{B})$ is the automaton obtained from $\mathcal{C}$ by complementing its set of final states. □

We actually have a normal form for RTAs resulting from this, namely that each RTA is equivalent to a RTA in which any circuit composed of stuttering transitions is a loop at a state labeled with a point interval, stuttering steps may only start from loop states and any chain of *different* stuttering steps has at most two transitions. An example for this is in figure $1(b)$.

5 Conclusions and Further Work

We have presented here a Kleene algebra of sets of positive numbers where elements have a finite representation and a class of real-time languages which is closed under complementation and is defined by some automata and by regular expressions.

We briefly note two problems for further study: the first one concerns the possible application of this theory to removing of circuits of silent transitions in timed automata [BDGP98], possibly using the Kleene theorem of [BP99]. The second is whether a normal form can be found for unions of intervals which have *any* bounds (not necessarily rational) and are generated from finite sets of intervals by $\cup$, complementation, concatenation and star.

Acknowledgments

I am grateful to Xu Qiwen, Paritosh Pandya and the anonymous reviewers for their helpful suggestions.

References

[AD94] R. Alur and D.L. Dill. A theory of timed automata, *Theoretical Computer Science*, 126:183-235, 1994.

[AFH94] R. Alur, L. Fix and T.A. Henzinger. A determinizable class of timed automata, in *Computer-Aided Verification*, LNCS 818:1-13, Springer Verlag, 1994.

[AH96] R. Alur and T. Henzinger. The benefits of relaxing punctuality, *Journal of ACM*, 43:116-146, 1996.

[ACM97] E. Asarin, P. Caspi and O. Maler, A Kleene Theorem for Timed Automata, in G. Winskel (Ed.) *Proc. LICS'97*, p.160-171, 1997.

[BDGP98] B. Bérard, V. Diekert, P. Gastin and A. Petit. Characterization of the expressive power of silent transitions in timed automata, *Fundamenta Informaticæ*, 36:145-182, 1998.

[BP99] P. Bouyer and A. Petit, Decomposition and composition of timed automata, *Proceedings of ICALP'99*, LNCS series, 1999.

[CG98] Ch. Choffrut and M. Goldwurm. Timed automata with periodic clock constraints, LIAFA report, 1998.

[Con71] J. H. Conway. *Regular Algebra and Finite Machines*, Chapman and Hall, 1971.

[DW96] Dang Van Hung and Wang Ji, On the design of Hybrid Control Systems Using Automata Models, in V. Chandru and V. Vinay (Eds.) *FSTTCS*, LNCS 1180:156-167, 1996.

[DMP91] R. Dechter, I. Meiri and J. Pearl, Temporal constraint networks, *Artificial Intelligence*, 49:61-95, 1991.

[Dim98] C. Dima. *Timed Regular Expressions, Real-Time Automata and Simple Duration Calculus*, draft of a UNU/IIST report, 1998. Available at `http://funinf.math.unibuc.ro/~cdima/work/report.ps.gz`.

[Dim99] C. Dima. Automata and regular expressions for real-time languages, *Proceedings of the AFL'99 workshop*, Vasszeczeny, Hungary, 9-13 august 1999.

[Ei74] S. Eilenberg. *Automata, Languages, and Machines, Vol. A*, Academic Press, 1974.

[HRS98] T.A. Henzinger, J.-F. Raskin and P.-Y. Schobbens. The regular real-time languages, in *Proc. 25-th ICALP*, LNCS series, Springer Verlag, 1998.

[HU] John E. Hopcroft and Jeffrey D. Ullman, *Introduction to Automata Theory, Languages and Computation*, Addison-Wesley/Narosa Publishing House, eighth edition, New Delhi, 1992.

[Koz94] D. Kozen. A completeness theorem for Kleene algebras and the algebra of regular events, *Information and Computation*, 110:366-390, 1994.

Small Progress Measures for Solving Parity Games

Marcin Jurdziński

BRICS*, Department of Computer Science, University of Aarhus,
Ny Munkegade, Building 540, 8000 Aarhus C, Denmark.
mju@brics.dk

Abstract. In this paper we develop a new algorithm for deciding the winner in parity games, and hence also for the modal μ-calculus model checking. The design and analysis of the algorithm is based on a notion of game progress measures: they are witnesses for winning strategies in parity games. We characterize game progress measures as pre-fixed points of certain monotone operators on a complete lattice. As a result we get the existence of the least game progress measures and a straightforward way to compute them. The worst-case running time of our algorithm matches the best worst-case running time bounds known so far for the problem, achieved by the algorithms due to Browne et al., and Seidl. Our algorithm has better space complexity: it works in small polynomial space; the other two algorithms have exponential worst-case space complexity.

1 Introduction

A parity game is an infinite path-forming game played by two players, player $\diamond$ and player $\square$, on a graph with integer priorities assigned to vertices. In order to determine the winner in an infinite play we check the parity of the lowest priority occurring infinitely often in the play: if it is even then player $\diamond$ wins, otherwise player $\square$ is the winner. The problem of deciding the winner in parity games is, given a parity game and an initial vertex, to decide whether player $\diamond$ has a winning strategy from the vertex.

There are at least two motivations for the study of the complexity of deciding the winner in parity games. One is that the problem is polynomial time equivalent to the modal μ-calculus model checking [3], hence developing better algorithms for parity games may lead to better model checking tools, which is a major objective in computer aided verification. The other is that the problem has an interesting status from the point of view of structural complexity theory. It is known to be in **NP** $\cap$ **co-NP** [3] (and even in **UP** $\cap$ **co-UP** [6]), and hence it is very unlikely to be **NP**-complete, but at the same time it is not known to be in **P**, despite substantial effort of the community (see [3, 1, 15, 20] and references therein).

* Basic Research in Computer Science,
Centre of the Danish National Research Foundation.

H. Reichel and S. Tison (Eds.): STACS 2000, LNCS 1770, pp. 290–301, 2000.

Progress measures [9] are decorations of graphs whose local consistency guarantees some global, often infinitary, properties of graphs. Progress measures have been used successfully for complementation of automata on infinite words and trees [7, 8]; they also underlie a translation of alternating parity automata on infinite words to weak alternating automata [10]. A similar notion, called a signature, occurs in the study of the modal μ-calculus [17]. Signatures have been used to prove memoryless determinacy of parity games [2, 18].

Our algorithm for parity games is based on the notion of game parity progress measures; Walukiewicz [18] calls them consistent signature assignments. Game parity progress measures are witnesses for winning strategies in parity games. We provide an upper bound on co-domains of progress measures; this reduces the search space of potential witnesses. Then we provide a characterization of game parity progress measures as pre-fixed points of certain monotone operators on a finite complete lattice. This characterization implies that the least game parity progress measures exist, and it also suggests an easy way to compute them.

The modal μ-calculus model checking problems is, given a formula φ of the modal μ-calculus and a Kripke structure K with a set of states S, to decide whether the formula is satisfied in the initial state of the Kripke structure. The problem has been studied by many researchers; see for example [5, 3, 1, 15, 11] and references therein. The algorithms with the best proven worst-case running time bounds so far are due to Browne et al. [1], and Seidl [15]. Their worst-case running times are roughly $O\big(m \cdot n^{\lceil d/2 \rceil}\big)$ and $O\big(m \cdot (n/d)^{\lceil d/2 \rceil}\big)$, respectively, where n and m are some numbers depending on φ and K, such that $n \leq |S| \cdot |\varphi|$, $m \leq |K| \cdot |\varphi|$, and d is the alternation depth of the formula φ.

In fact, number n above is the number of vertices in the parity game obtained from the formula and the Kripke structure via the standard reduction of the modal μ-calculus model checking to parity games, and m is the number of edges in the game graph; see for example [3, 16]. Moreover, the reduction can be done in such a way that the number of different priorities in the parity game is equal to the alternation depth d of the formula. Our algorithm has worst-case running time $O\big(m \cdot (n/\lfloor d/2 \rfloor)^{\lfloor d/2 \rfloor}\big)$, and it can be made to work in time $O\big(m \cdot (n/d)^{\lceil d/2 \rceil}\big)$, hence it matches the bounds of the other two algorithms. Moreover, it works in space $O(dn)$ while the other two algorithms have exponential worst-case space complexity. Our algorithm can be seen as a generic algorithm allowing many different evaluation policies; good heuristics can potentially improve performance of the algorithm. However, we show a family of examples for which worst-case running time occurs for all evaluation policies.

Among algorithms for parity games it is worthwhile to mention the algorithm of McNaughton [12] and its modification due to Zielonka [19]. In the extended version of this paper we show that Zielonka's algorithm can be implemented to work in time roughly $O\big(m \cdot (n/d)^d\big)$, and we also provide a family of examples for which the algorithm needs this time. Zielonka's algorithm works in fact for games with more general Muller winning conditions. By a careful analysis of the algorithm for games with Rabin (Streett) winning conditions we get a running time bound $O\big(m \cdot n^{2k}/(k/2)^k\big)$, where k is the number of pairs in the

Rabin (Streett) condition. The algorithm also works in small polynomial space. This compares favourably with other algorithms for the linear-time equivalent problem of checking non-emptiness of non-deterministic Rabin (Streett) tree automata [4, 14, 10], and makes it the best algorithm known for this **NP**-complete (co-**NP**-complete) [4] problem.

2 Parity Games

Notation: For all $n \in \mathbb{N}$, by $[n]$ we denote the set $\{0, 1, 2, \ldots, n-1\}$. If (V, E) is a directed graph and $W \subseteq V$, then by $(V, E) \restriction W$ we denote the subgraph (W, F) of (V, E), where $F = E \cap W^2$. [Notation] □

A *parity graph* $G = (V, E, p)$ consists of a directed graph (V, E) and a priority function $p : V \to [d]$, where $d \in \mathbb{N}$. A *parity game* $\Gamma = \big(V, E, p, (V_\Diamond, V_\Box)\big)$ consists of a parity graph $G = (V, E, p)$, called the *game graph* of Γ, and of a partition $(V_\Diamond, V_\Box)$ of the set of vertices V. For technical convenience we assume that all game graphs have the property that every vertex has at least one out-going edge. We also restrict ourselves throughout this paper to games with finite game graphs.

A parity game is played by two players: player $\Diamond$ and player $\Box$, who form an infinite path in the game graph by moving a token along edges. They start by placing the token on an initial vertex and then they take moves indefinitely in the following way. If the token is on a vertex in $V_\Diamond$ then player $\Diamond$ moves the token along one of the edges going out of the vertex. If the token is on a vertex in $V_\Box$ then player $\Box$ takes a move. In the result players form an infinite path $\pi = \langle v_1, v_2, v_3, \ldots \rangle$ in the game graph; for brevity we refer to such infinite paths as *plays*. The winner in a play is determined by referring to priorities of vertices in the play. Let $\mathrm{Inf}(\pi)$ denote the set of priorities occurring infinitely often in $\big\langle p(v_1), p(v_2), p(v_3), \ldots \big\rangle$. A play π is a *winning play* for player $\Diamond$ if $\min\big(\mathrm{Inf}(\pi)\big)$ is even, otherwise π is a winning play for player $\Box$.

A function $\sigma : V_\Diamond \to V$ is a *strategy* for player $\Diamond$ if $\big(v, \sigma(v)\big) \in E$ for all $v \in V_\Diamond$. A play $\pi = \langle v_1, v_2, v_3, \ldots \rangle$ is *consistent* with a strategy σ for player $\Diamond$ if $v_{\ell+1} = \sigma(v_\ell)$, for all $\ell \in \mathbb{N}$, such that $v_\ell \in V_\Diamond$. A strategy σ is a *winning strategy* for player $\Diamond$ from set $W \subseteq V$, if every play starting from a vertex in W and consistent with σ is winning for player $\Diamond$. Strategies and winning strategies are defined similarly for player $\Box$.

Theorem 1 (Memoryless Determinacy [2, 13])
For every parity game, there is a unique partition $(W_\Diamond, W_\Box)$ of the set of vertices of its game graph, such that there is a winning strategy for player $\Diamond$ from $W_\Diamond$, and a winning strategy for player $\Box$ from $W_\Box$.

We call the sets $W_\Diamond$ and $W_\Box$ the *winning sets* of player $\Diamond$ and player $\Box$, respectively. The problem of *deciding the winner* in parity games is, given a parity game and a vertex in the game graph, to determine whether the vertex is in the winning set of player $\Diamond$.

Before we proceed we mention a simple characterization of winning strategies for player $\diamond$ in terms of simple cycles in a subgraph of the game graph associated with the strategy. We say that a strategy σ for player $\diamond$ is *closed* on a set $W \subseteq V$ if for all $v \in W$, we have:

- if $v \in V_\diamond$ then $\sigma(v) \in W$, and
- if $v \in V_\square$ then $(v, w) \in E$ implies $w \in W$.

Note that if a strategy σ for player $\diamond$ is closed on W then every play starting from a vertex in W and consistent with σ stays within W.

If σ is a strategy for player $\diamond$ then by G_σ we denote the parity graph (V, E_σ, p) obtained from game graph $G = (V, E, p)$ by removing from E all edges (v, w) such that $v \in V_\diamond$ and $\sigma(v) \neq w$.

We say that a cycle in a parity graph is an *i-cycle* if i is the smallest priority of a vertex occurring in the cycle. A cycle is an even cycle if it is an i-cycle for some even i, otherwise it is an odd cycle. The following proposition is not hard to prove.

Proposition 2 Let σ be a strategy for player $\diamond$ closed on W. Then σ is a winning strategy for player $\diamond$ from W if and only if all simple cycles in $G_\sigma \restriction W$ are even.

3 Small Progress Measures

In this section we study a notion of progress measures. Progress measures play a key role in the design and analysis of our algorithm for solving parity games.

First we define parity progress measures for parity graphs, and we show that there is a parity progress measure for a parity graph if and only if all cycles in the graph are even. In other words, parity progress measures are witnesses for the property of parity graphs having only even cycles. The proof of the 'if' part also provides an upper bound on the size of the co-domain of a parity progress measure. Then we define game parity progress measures for parity games, we argue that they are witnesses for winning strategies for player $\diamond$, and we show that the above-mentioned upper bound holds also for game parity progress measures.

Notation: If $\alpha \in \mathbb{N}^d$ is a d-tuple of non-negative integers then we number its components from 0 to $d-1$, i.e., we have $\alpha = (\alpha_0, \alpha_1, \ldots, \alpha_{d-1})$. When applied to tuples of natural numbers, the comparison symbols $<, \leq, =, \neq, \geq$, and $>$ denote the lexicographic ordering. When subscripted with a number $i \in \mathbb{N}$ (e.g., $<_i, =_i, \geq_i$), they denote the lexicographic ordering on $\mathbb{N}^i$ applied to the arguments truncated to their first i components. For example, $(2,3,0,0) >_2 (2,2,4,1)$, but $(2,3,0,0) =_0 (2,2,4,1)$. [Notation] □

Definition 3 (Parity progress measure)
Let $G = \big(V, E, p : V \to [d]\big)$ be a parity graph. A function $\varrho : V \to \mathbb{N}^d$ is a parity progress measure for G if for all $(v, w) \in E$, we have $\varrho(v) \geq_{p(v)} \varrho(w)$, and the inequality is strict if $p(v)$ is odd. [Definition 3] □

Proposition 4 If there is a parity progress measure for a parity graph G then all cycles in G are even.

Proof: Let $\varrho : V \to \mathbb{N}^d$ be a parity progress measure for G. For the sake of contradiction suppose that there is an odd cycle $v_1, v_2, \ldots, v_\ell$ in G, and let $i = p(v_1)$ be the smallest priority on this cycle. Then by the definition of a progress measure we have $\varrho(v_1) >_i \varrho(v_2) \geq_i \varrho(v_2) \geq_i \cdots \geq_i \varrho(v_\ell) \geq_i \varrho(v_1)$, and hence $\varrho(v_1) >_i \varrho(v_1)$, a contradiction. [Proposition 4] ∎

If $G = \big(V, E, p : V \to [d]\big)$ is a parity graph then for every $i \in [d]$, we write V_i to denote the set $p^{-1}(i)$ of vertices with priority i in parity graph G. Let $n_i = |V_i|$, for all $i \in [d]$. Define M_G to be the following finite subset of $\mathbb{N}^d$: if d is even then

$$M_G = [1] \times [n_1 + 1] \times [1] \times [n_3 + 1] \times \cdots \times [1] \times [n_{d-1} + 1];$$

for odd d we have $\cdots \times [n_{d-2}+1] \times [1]$ at the end. In other words, M_G is the finite set of d-tuples of integers with only zeros on even positions, and non-negative integers bounded by $|V_i|$ on every odd position i.

Theorem 5 (Small parity progress measure)
If all cycles in a parity graph G are even then there is a parity progress measure $\varrho : V \to M_G$ for G.

Proof: The proof goes by induction on the number of vertices in $G = \big(V, E, p : V \to [d]\big)$. For the induction to go through we slightly strengthen the statement of the theorem: we additionally claim, that if $p(v)$ is odd then $\varrho(v) >_{p(v)} (0, \ldots, 0)$. The statement of the theorem holds trivially if G has only one vertex.

Without loss of generality we may assume that $V_0 \cup V_1 \neq \emptyset$; otherwise we can scale down the priority function of G by two, i.e., replace the priority function p by the function $p - 2$ defined by $(p-2)(v) = p(v) - 2$, for all $v \in V$. Suppose first that $V_0 \neq \emptyset$. By induction hypothesis there is a parity progress measure $\varrho : (V \setminus V_0) \to M_G$ for the subgraph $G \restriction (V \setminus V_0)$. Setting $\varrho(v) = (0, \ldots, 0) \in M_G$, for all $v \in V_0$, we get a parity progress measure for G.

Suppose that $V_0 = \emptyset$ then $V_1 \neq \emptyset$. We claim that there is a non-trivial partition (W_1, W_2) of the set of vertices V, such that there is no edge from W_1 to W_2 in G.

Let $u \in V_1$; define $U \subseteq V$ to be the set of vertices to which there is a non-trivial path from u in G. If $U = \emptyset$ then $W_1 = \{u\}$ and $W_2 = V \setminus \{u\}$ is a desired partition of V. If $U \neq \emptyset$ then $W_1 = U$ and $W_2 = V \setminus U$ is a desired partition. The partition is non-trivial (i.e., $V \setminus U \neq \emptyset$) since $u \notin U$: otherwise a non-trivial path from u to itself gives a 1-cycle because $V_0 = \emptyset$, contradicting the assumption that all cycles in G are even.

Let $G_1 = G \restriction W_1$, and $G_2 = G \restriction W_2$ be subgraphs of G. By induction hypothesis there are parity progress measures $\varrho_1 : W_1 \to M_{G_1}$ for G_1, and $\varrho_2 : W_2 \to M_{G_2}$ for G_2. Let $n'_i = |V_i \cap W_1|$, and let $n''_i = |V_i \cap W_2|$, for $i \in [d]$. Clearly $n_i = n'_i + n''_i$, for all $i \in [d]$. Recall that there are no edges from W_1 to W_2 in G. From this and our additional claim applied to ϱ_1 it follows that the function $\varrho = \varrho_1 \cup \big(\varrho_2 + (0, n_1', 0, n_3', \ldots)\big) : V \to M_G$ is a parity progress measure for G. [Theorem 5] ■

Let $\Gamma = \big(V, E, p, (V_\diamond, V_\square)\big)$ be a parity game and let $G = (V, E, p)$ be its game graph. We define $M_G^\top$ to be the set $M_G \cup \{\top\}$, where $\top$ is an extra element. We use the standard comparison symbols (e.g., $<, =, \geq$, etc.) to denote the order on $M_G^\top$ which extends the lexicographic order on M_G by taking $\top$ as the biggest element, i.e., we have $m < \top$, for all $m \in M_G$. Moreover, for all $m \in M_G$ and $i \in [d]$, we set $m <_i \top$, and $\top =_i \top$. If $\varrho : V \to M_G^\top$ and $(v, w) \in E$ then

> by $\mathrm{Prog}(\varrho, v, w)$ we denote the least $m \in M_G^\top$, such that $m \geq_{p(v)} \varrho(w)$, and if $p(v)$ is odd then either the inequality is strict, or $m = \varrho(w) = \top$.

Definition 6 (Game parity progress measure)
A function $\varrho : V \to M_G^\top$ is a game parity progress measure if for all $v \in V$, we have:

- if $v \in V_\diamond$ then $\varrho(v) \geq_{p(v)} \mathrm{Prog}(\varrho, v, w)$ for some $(v, w) \in E$, and
- if $v \in V_\square$ then $\varrho(v) \geq_{p(v)} \mathrm{Prog}(\varrho, v, w)$ for all $(v, w) \in E$;

by $\|\varrho\|$ we denote the set $\big\{\, v \in V \,:\, \varrho(v) \neq \top \,\big\}$. [Definition 6] □

For every game parity progress measure ϱ we define a strategy $\widetilde{\varrho} : V_\diamond \to V$ for player $\diamond$, by setting $\widetilde{\varrho}(v)$ to be a successor w of v, which minimizes $\varrho(w)$.

Corollary 7 If ϱ is a game parity progress measure then $\widetilde{\varrho}$ is a winning strategy for player $\diamond$ from $\|\varrho\|$.

Proof: Note first that ϱ restricted to $\|\varrho\|$ is a parity progress measure on $G_{\widetilde{\varrho}} \restriction \|\varrho\|$. Hence by Proposition 4 all simple cycles in $G_{\widetilde{\varrho}} \restriction \|\varrho\|$ are even.

It also follows easily from definition of a game parity progress measure that strategy $\widetilde{\varrho}$ is closed on $\|\varrho\|$. Therefore, by Proposition 2 we get that $\widetilde{\varrho}$ is a winning strategy for player $\diamond$ from $\|\varrho\|$. [Corollary 7] ■

Corollary 8 (Small game parity progress measure)
There is a game progress measure $\varrho : V \to M_G^\top$ such that $\|\varrho\|$ is the winning set of player $\diamond$.

Proof: It follows from Theorem 1 that there is a winning strategy σ for player $\diamond$ from her winning set $W_\diamond$, which is closed on $W_\diamond$. Therefore by Proposition 2 all cycles in parity graph $G_\sigma \restriction W_\diamond$ are even, hence by Theorem 5 there is a parity progress measure $\varrho : W_\diamond \to M_G$ for $G_\sigma \restriction W_\diamond$. It follows that setting $\varrho(v) = \top$ for all $v \in V \setminus W_\diamond$, makes ϱ a game parity progress measure. [Corollary 8] ■

4 The Algorithm

In this section we present a simple algorithm for solving parity games based on the notion of a game parity progress measure. We characterize game parity progress measures as (pre-)fixed points of certain monotone operators in a finite complete lattice. By Knaster-Tarski theorem it implies existence of the *least* game progress measure μ, and a simple way to compute it. It then follows from Corollaries 8 and 7 that $\|\mu\|$ is the winning set of player $\Diamond$.

Before we present the algorithm we define an ordering, and a family of $\mathrm{Lift}(\cdot, v)$ operators for all $v \in V$, on the set of functions $V \to M_G^\top$. Given two functions $\mu, \varrho : V \to M_G^\top$, we define $\mu \sqsubseteq \varrho$ to hold if $\mu(v) \leq \varrho(v)$ for all $v \in V$. The ordering relation $\sqsubseteq$ gives a complete lattice structure on the set of functions $V \to M_G^\top$. We write $\mu \sqsubset \varrho$ if $\mu \sqsubseteq \varrho$, and $\mu \neq \varrho$. Define $\mathrm{Lift}(\varrho, v)$ for $v \in V$ as follows:

$$\mathrm{Lift}\big(\varrho, v\big)(u) = \begin{cases} \varrho(u) & \text{if } u \neq v, \\ \min_{(v,w)\in E} \mathrm{Prog}(\varrho, v, w) & \text{if } u = v \in V_\Diamond, \\ \max_{(v,w)\in E} \mathrm{Prog}(\varrho, v, w) & \text{if } u = v \in V_\Box. \end{cases}$$

The following propositions follow immediately from definitions of a game parity progress measure, and of the $\mathrm{Lift}(\cdot, v)$ operators.

Proposition 9 For every $v \in V$, the operator $\mathrm{Lift}(\cdot, v)$ is $\sqsubseteq$-monotone.

Proposition 10 A function $\varrho : V \to M_G^\top$ is a game parity progress measure, if and only if is it is a simultaneous pre-fixed point of all $\mathrm{Lift}(\cdot, v)$ operators, i.e., if $\mathrm{Lift}(\varrho, v) \sqsubseteq \varrho$ for all $v \in V$.

From Knaster-Tarski theorem it follows that the $\sqsubseteq$-*least* game parity progress measure exists, and it can be obtained by running the following simple procedure computing the least simultaneous (pre-)fixed point of operators $\mathrm{Lift}(\cdot, v)$, for all $v \in V$.

```
ProgressMeasureLifting
    μ := λv ∈ V.(0,...,0)
    while μ ⊏ Lift(μ, v) for some v ∈ V do μ := Lift(μ, v)
```

Theorem 11 (The algorithm)
Given a parity game, procedure `ProgressMeasureLifting` *computes winning sets for both players and a winning strategy for player $\Diamond$ from her winning set; it works in space $O(dn)$, and its running time is*

$$O\Big(dm \cdot \Big(\frac{n}{\lfloor d/2 \rfloor}\Big)^{\lfloor d/2 \rfloor}\Big),$$

where n is the number of vertices, m is the number of edges, and d is the maximum priority in the parity game.

Proof: The result of running `ProgressMeasureLifting` on a parity game is the $\sqsubseteq$-least game progress measure μ. Let $W_\diamond$ be the winning set of player $\diamond$. By minimality of μ and by Corollary 8 it follows that $W_\diamond \subseteq \|\mu\|$. Moreover, Corollary 7 implies that $\widetilde{\mu}$ is a winning strategy for player $\diamond$ from $\|\mu\|$, and hence by Theorem 1 we get that $\|\mu\| \subseteq W_\diamond$, i.e., $\|\mu\| = W_\diamond$.

Procedure `ProgressMeasureLifting` algorithm works in space $O(dn)$ because it only needs to maintain a d-tuple of integers for every vertex in the game graph. The $\mathrm{Lift}(\cdot, v)$ operator, for every $v \in V$, can be implemented to work in time $O\big(d \cdot \text{out-deg}(v)\big)$, where out-deg$(v)$ is the out-degree of v. Every vertex can be "lifted" only $|M_G|$ many times, hence the running time of procedure `ProgressMeasureLifting` is bounded by

$$O\Big(\sum_{v \in V} d \cdot \text{out-deg}(v) \cdot |M_G|\Big) = O\big(dm \cdot |M_G|\big).$$

To get the claimed time bound it suffices to notice that

$$|M_G| = \prod_{i=1}^{\lfloor d/2 \rfloor} (n_{2i-1} + 1) \leq \left(\frac{n}{\lfloor d/2 \rfloor}\right)^{\lfloor d/2 \rfloor},$$

because $\sum_{i=1}^{\lfloor d/2 \rfloor}(n_{2i-1} + 1) \leq n$ if $n_i \neq 0$ for all $i \in [d]$, which we can assume without loss of generality; if $n_i = 0$ for some $i \in [d]$ then we can scale down the priorities bigger than i by two. [Theorem 11] ■

Remark: Our algorithm for solving parity games can be easily made to have

$$O\left(dm \cdot \left(\frac{n}{d}\right)^{\lceil d/2 \rceil}\right)$$

as its worst-case running time bound, which is better than $O\big(m \cdot (n/\lfloor d/2 \rfloor)\big)^{\lfloor d/2 \rfloor}$ for even d, and for odd d if $d \geq 2 \log n$. If $\sum_{i=1}^{\lfloor d/2 \rfloor}(n_{2i-1} + 1) \leq n/2$ then the above analysis gives the desired bound. Otherwise $\sum_{i=0}^{\lfloor d/2 \rfloor}(n_{2i} + 1) \leq n/2 + 1$. In this case it suffices to run procedure `ProgressMeasureLifting` on the dual game, i.e., the game obtained by scaling all priorities up by one, and swapping sets in the $(V_\diamond, V_\square)$ partition. The winning set of player $\diamond$ in the original game is the winning set of player $\square$ in the dual game. [Remark] □

Note that in order to make `ProgressMeasureLifting` a fully deterministic algorithm one has to fix a policy of choosing vertices at which the function μ is being "lifted". Hence it can be considered as a generic algorithm whose performance might possibly depend on supplying heuristics for choosing the vertices to lift. Unfortunately, as we show in the next section, there is a family of examples on which the worst case performance of the algorithm occurs for all vertex lifting policies.

5 Worst-Case Behaviour

Theorem 12 (Worst-case behaviour)
For all $d, n \in \mathbb{N}$ such that $d \leq n$, there is a game of size $O(n)$ with priorities not bigger that d, on which procedure `ProgressMeasureLifting` *performs at least $(\lceil n/d \rceil)^{\lceil d/2 \rceil}$ many lifts, for all lifting policies.*

Proof: We define the family of games $H_{\ell,b}$, for all $\ell, b \in \mathbb{N}$. The game graph of $H_{\ell,b}$ consists of ℓ "levels", each level contains b "blocks". There is one "odd" level, and $\ell - 1$ "even" levels.

The basic building block of the odd level is the following subgraph.

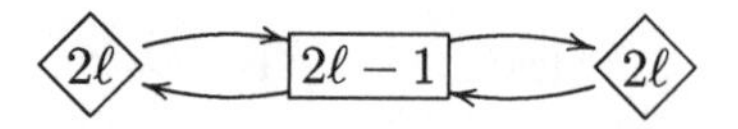

The numbers in vertices are their priorities. The odd level of $H_{\ell,b}$ consists of b copies of the above block assembled together by identifying the left-hand vertex with priority 2ℓ of the a-th block, for every $a \in \{1, 2, \ldots, b-1\}$, with the right-hand vertex with priority 2ℓ of the $(a+1)$-st block. For example the odd level of $H_{4,3}$ is the following.

In all our pictures vertices with a diamond-shaped frame are meant to belong to $V_\diamond$, i.e., they are vertices where player $\diamond$ moves; vertices with a box-shaped frame belong to $V_\square$. Some vertices have no frame; for concreteness let us assume that they belong to $V_\diamond$, but including them to $V_\square$ would not change our reasoning, because they all have only one successor in the game graph of $H_{\ell,b}$.

The basic building block of the k-th even level, for $k \in \{1, 2, \ldots, \ell - 1\}$, is the following subgraph.

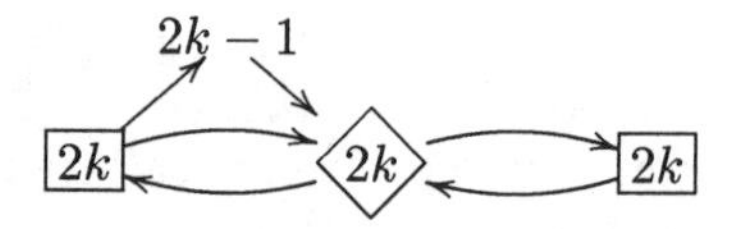

Every even level is built by putting b copies of the above block together in a similar way as for the odd level.

To assemble the game graph of $H_{\ell,b}$ we connect all $\ell - 1$ even levels to the odd level, by introducing edges in the following way. For every even level $k \in \{1, 2, \ldots, \ell - 1\}$, and for every block $a \in \{1, 2, \ldots, b\}$, we introduce edges in both directions between the box vertex with priority $2\ell - 1$ from the a-th block of the odd level, and the diamond vertex with priority $2k$ from the a-th block of the k-th even level. See Figure 1 for an example: the game $H_{4,3}$.

Claim 13 Every vertex with priority $2\ell - 1$ in game $H_{\ell,b}$ is lifted $(b+1)^\ell$ many times by procedure `ProgressMeasureLifting`.

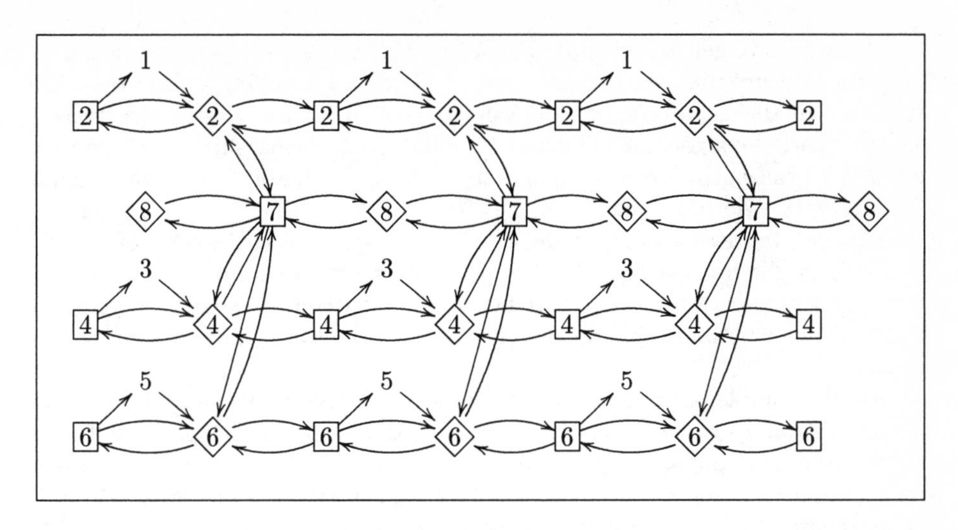

Fig. 1. The game $H_{4,3}$.

Proof: Note that in game $H_{\ell,b}$ player $\Diamond$ has a winning strategy from all vertices in even levels, and player $\Box$ has a winning strategy from all vertices in the odd level; see Figure 1. Therefore, the value of the least progress measure in all vertices with priority $2\ell - 1$ is $\top \in M^{\top}_{H_{\ell,b}}$. Hence it suffices to show that every vertex with priority $2\ell - 1$ can be lifted only to its immediate successor in the order on $M^{\top}_{H_{\ell,b}}$. Then it is lifted $|M_{H_{\ell,b}}| = (b+1)^{\ell}$ many times, because

$$M_{H_{\ell,b}} = \underbrace{[1] \times [b+1] \times [1] \times [b+1] \times \cdots \times [b+1] \times [1]}_{2\ell+1 \text{ components}}.$$

Let v be a vertex with priority $2\ell - 1$ in the odd level of $H_{\ell,b}$, and let w be a vertex, such that there is an edge from v to w in the game graph of $H_{\ell,b}$. Then there is also an edge from w to v in the game graph of $H_{\ell,b}$; see Figure 1. Therefore, function μ maintained by the algorithm satisfies $\mu(w) \leq \mu(v)$, because w is a diamond vertex with even priority, so $\text{Prog}(\mu, w, v) =_{p(w)} \mu(v)$, and $\left(\text{Prog}(\mu, w, v)\right)_i = 0$ for all $i > p(w)$. It follows that $\text{Lift}(\cdot, v)$ operator can only lift $\mu(v)$ to the immediate successor of $\mu(v)$ in the order on $M_{H_{\ell,b}}$, because the priority of v is $2\ell - 1$. [Claim 13] ■

Theorem 12 follows from the above claim by taking the game $H_{\lfloor d/2 \rfloor, \lceil n/d \rceil}$.
[Theorem 12] ■

6 Optimizations

Even though procedure `ProgressMeasureLifting` as presented above admits the worst-case performance, there is some room for improvements in its running time. Let us just mention here two proposals for optimizations, which should be considered when implementing the algorithm.

One way is to get better upper bounds on the values of the least game parity progress measure than the one provided by Corollary 8, taking into account the structure of the game graph. This would allow to further reduce the "search space" where the algorithm is looking for game progress measures. For example, let $G^{\geq i}$ be the parity graph obtained from the game graph G by removing all vertices with priorities smaller than i. One can show that if $v \in \|\mu\|$ for the least game progress measure μ then for odd i's the i-th component of $\mu(v)$ is bounded by the number of vertices of priority i reachable from v in graph $G^{\geq i}$. It requires further study to see whether one can get considerable improvements by pre-computing better bounds for the values of the least game parity progress measure.

Another simple but important optmization is to decompose game graphs into maximal strongly connected components. Note that every infinite play eventually stays within a strongly connected component, so it suffices to apply expensive procedure for solving parity games to the maximal strongly connected components separately. In fact, we need to proceed bottom up in the partial order of maximal strongly connected components. Each time one of the bottom components has been solved, we can also remove from the rest of the game the sets of vertices from which respective players have a strategy to force in a finite number of moves to their so far computed winning sets.

The above optimizations should considerably improve performance of our algorithm in practice, but they do not, as such, give any asymptotic worst-case improvement: see the examples $H_{\ell,b}$ from Section 5.

Acknowledgements

I am indebted to Mogens Nielsen, Damian Niwiński, Igor Walukiewicz, and Jens Vöge for numerous inspiring discussions on the subject. I thank anonymous referees for very helpful advice on improving the focus and presentation of the paper.

References

[1] A. Browne, E. M. Clarke, S. Jha, D. E. Long, and W. Marrero. An improved algorithm for the evaluation of fixpoint expressions. *Theoretical Computer Science*, 178(1–2):237–255, May 1997.

[2] E. A. Emerson and C. S. Jutla. Tree automata, mu-calculus and determinacy (Extended abstract). In *Proceedings of 32nd Annual Symposium on Foundations of Computer Science*, pages 368–377. IEEE Computer Society Press, 1991.

[3] E. A. Emerson, C. S. Jutla, and A. P. Sistla. On model-checking for fragments of μ-calculus. In Costas Courcoubetis, editor, *Computer Aided Verification, 5th International Conference, CAV'93*, volume 697 of *LNCS*, pages 385–396, Elounda, Greece, June/July 1993. Springer-Verlag.

[4] E. Allen Emerson and Charanjit S. Jutla. The complexity of tree automata and logics of programs. In *Proceedings of 29th Annual Symposium on Foundations of Computer Science*, pages 328–337, White Plains, New York, 24–26 October 1988. IEEE Computer Society Press.

[5] E. Allen Emerson and Chin-Laung Lei. Efficient model checking in fragments of the propositional mu-calculus (Extended abstract). In *Proceedings, Symposium on Logic in Computer Science*, pages 267–278, Cambridge, Massachusetts, 16–18 June 1986. IEEE.

[6] Marcin Jurdziński. Deciding the winner in parity games is in UP $\cap$ co-UP. *Information Processing Letters*, 68(3):119–124, November 1998.

[7] Nils Klarlund. Progress measures for complementation of ω-automata with applications to temporal logic. In *32nd Annual Symposium on Foundations of Computer Science*, pages 358–367, San Juan, Puerto Rico, 1–4 October 1991. IEEE.

[8] Nils Klarlund. Progress measures, immediate determinacy, and a subset construction for tree automata. *Annals of Pure and Applied Logic*, 69(2–3):243–268, 1994.

[9] Nils Klarlund and Dexter Kozen. Rabin measures and their applications to fairness and automata theory. In *Proceedings, Sixth Annual IEEE Symposium on Logic in Computer Science*, pages 256–265, Amsterdam, The Netherlands, 15–18 July 1991. IEEE Computer Society Press.

[10] Orna Kupferman and Moshe Y. Vardi. Weak alternating automata and tree automata emptiness. In *Proceedings of the Thirtieth Annual ACM Symposium on the Theory of Computing*, pages 224–233, Dallas, Texas, USA, 23–26 May 1998. ACM Press.

[11] Xinxin Liu, C. R. Ramakrishnan, and Scott A. Smolka. Fully local and efficient evaluation of alternating fixed points. In Bernhard Steffen, editor, *Tools and Algorithms for Construction and Analysis of Systems, 4th International Conference, TACAS '98*, volume 1384 of *LNCS*, pages 5–19, Lisbon, Portugal, 28 March–4 April 1998. Springer.

[12] Robert McNaughton. Infinite games played on finite graphs. *Annals of Pure and Applied Logic*, 65(2):149–184, 1993.

[13] A. W. Mostowski. Games with forbidden positions. Technical Report 78, University of Gdańsk, 1991.

[14] Amir Pnueli and Roni Rosner. On the synthesis of a reactive module. In *Conference Record of the 16th Annual ACM Symposium on Principles of Programming Languages (POPL '89)*, pages 179–190, Austin, Texas, January 1989. ACM Press.

[15] Helmut Seidl. Fast and simple nested fixpoints. *Information Processing Letters*, 59(6):303–308, September 1996.

[16] Colin Stirling. Local model checking games (Extended abstract). In Insup Lee and Scott A. Smolka, editors, *CONCUR'95: Concurrency Theory, 6th International Conference*, volume 962 of *LNCS*, pages 1–11, Philadelphia, Pennsylvania, 21–24 August 1995. Springer-Verlag.

[17] Robert S. Streett and E. Allen Emerson. An automata theoretic decision procedure for the propositional mu-calculus. *Information and Computation*, 81(3):249–264, 1989.

[18] Igor Walukiewicz. Pushdown processes: Games and model checking. In Thomas A. Henzinger and Rajeev Alur, editors, *Computer Aided Verification, 8th International Conference, CAV'96*, volume 1102 of *LNCS*, pages 62–74. Springer-Verlag, 1996. Full version available through `http://zls.mimuw.edu.pl/~igw`.

[19] Wiesław Zielonka. Infinite games on finitely coloured graphs with applications to automata on infinite trees. *Theoretical Computer Science*, 200:135–183, 1998.

[20] Uri Zwick and Mike Paterson. The complexity of mean payoff games on graphs. *Theoretical Computer Science*, 158:343–359, 1996.

Multi-linearity Self-Testing with Relative Error

Frédéric Magniez*

Université Paris–Sud, 91405 Orsay, France
magniez@lri.fr, http://www.lri.fr/~magniez

Abstract. We investigate self-testing programs with relative error by allowing error terms proportional to the function to be computed. Until now, in numerical computation, error terms were assumed to be either constant or proportional to the p-th power of the magnitude of the input, for $p \in [0, 1)$. We construct new self-testers with relative error for real-valued multi-linear functions defined over finite rational domains. The existence of such self-testers positively solves an open question in [KMS99]. Moreover, our self-testers are very efficient: they use few queries and simple operations.

Keywords — Program verification, approximation error, self–testing programs, robustness and stability of functional equations.

1 Introduction

It is not easy to write a program P to compute a real-valued function f. By definition of floating point computations, a program P can only compute an approximation of f. The succession of inaccuracies in computational operations could be significant. Moreover once P is implemented it is more difficult to verify its correctness, *i.e.* that $P(x)$ is a good approximation of $f(x)$ for all valid inputs x. In a good approximation one would like the significant figures to be correct. This leads us to the notion of relative error. If a is a real number and $\hat{a}$ is its approximation, then the quantity $\theta \stackrel{\text{def}}{=} |\hat{a} - a|/a$ is called the *relative error* of the approximation.

In recent years, several notions were developed to address the software correctness problem. Here we focus on the following scenario. First, the program to be tested is viewed as a black box, *i.e.* we can only query it on some inputs. Second, we want a very efficient testing procedure. In particular, a test should be more efficient than any known correct implementation. For exact computation, program checking [Blu88, BK95], self-testing programs [BLR93], and self-correcting programs [BLR93, Lip91] were developed in the early 90's. A *program checker* for f verifies whether the program P computes f on a particular input x; a *self-tester* for f verifies whether the program P is correct on most inputs; and a *self-correcting program* for f uses a program P, which is correct on most

* Partially supported by a CNRS–Conicyt'98 Project, ESPRIT Working Group RAND2 No. 21726, and Franco-Hungarian bilateral project Balaton No. 99013.

H. Reichel and S. Tison (Eds.): STACS 2000, LNCS 1770, pp. 302–313, 2000.

inputs, to compute f correctly everywhere with high probability. Let us insist that checkers, self-testers and self-correctors can only use the program P as a black box, and are required to be different and simpler than any known implementation of f (see [BK95] for a formal definition). In this context, results on testing linear functions and polynomials have theoretical implications for probabilistically checkable proofs [ALM+92, AS92] and in approximation theory. For a survey see [Bab93].

Let us recall the problem of linearity testing which has been fundamental in the development of testers [BLR93]. Given a program P which computes a function from one Abelian group G into another group, we want to verify that P computes a homomorphism on most inputs in G. The Blum-Luby-Rubinfeld linearity test is based on the linearity property $f(x+y) = f(x) + f(y)$, for all $x, y \in G$, which is satisfied when f is a homomorphism. The test consists in verifying the previous linearity equation on random instances. More precisely, it checks for random inputs $x, y \in G$ that $P(x+y) = P(x)+P(y)$. If the probability of failing the linearity test is small, then P computes a homomorphism except on a small fraction of inputs. This property of the linearity equation is usually called the *robustness* of the linearity equation. This term was defined in [RS96] and studied in [Rub99]. The analysis of the test is due to Coppersmith [Cop89]. It consists in correcting P by querying it on few queries. Let g be the function which takes at x the majority of the votes $(P(x+y) - P(y))$, for all $y \in G$. When the failure probability in the linearity test is small, majority turns out to quasi-unanimity, g equals P on a large fraction of inputs, and g is linear. This idea of *property testing* has been recently formalized and extended to testing graph properties in [GGR96, GR97].

These notions of testing were extended to approximate computation with absolute error for self-testers/correctors [GLR+91] and for checkers [ABCG93]. In [GLR+91] Gemmel et al. studied only functions defined over algebraically closed domains. Ergün, Ravi Kumar, and Rubinfeld [EKR96] initiated and solved the problem of self-testing with absolute error for linear functions, polynomials, and additive functions defined over rational domains. Rational domains were first considered by Lipton [Lip91] and these are the sets $\mathcal{D}_{n,s} \stackrel{\text{def}}{=} \{i/s : |i| \leq n, i \in \mathbb{Z}\}$, for some integer $n \geq 1$ and real $s > 0$. In these past works the absolute error of the approximation $\hat{a}$ of a is defined by $\varepsilon \stackrel{\text{def}}{=} |\hat{a} - a|$. In this approximate context the linearity testing problem consists now in verifying that a given program P computes approximately a real linear function over $\mathcal{D}_{n,s}$. To allow absolute error in the computation of P, the approximate linearity test consists in verifying that $|P(x+y) - P(x) - P(y)| \leq \varepsilon$, for random $x, y \in \mathcal{D}_{n,s}$ and some fixed $\varepsilon > 0$. Then the analysis is very similar to that of the exact case. Since the majority is not adapted to approximate computation, it is replaced by the median. Moreover both the closeness of g to P and the linearity of g are approximated. Therefore we need a second stage which consists in proving the *local stability* of the linearity equation for absolute error, that is, every function satisfying $|f(x+y) - f(x) - f(y)| \leq \varepsilon$, for all $x, y \in \mathcal{D}_{n,s}$, is close to a perfectly linear function. This part is a well-studied problem in mathematics for several kinds of error terms when x and

y describe a group like $\mathbb{Z}$. It corresponds to the study of Hyers-Ulam stability. The stability problem is due to Ulam and was first solved in the absolute error case in 1941 by Hyers [Hye41]. For a survey of Hyers-Ulam stability see [For95, HR92].

Using elegant techniques of Hyers-Ulam stability theory, Kiwi et al. extended a part of [EKR96]'s work for non-constant error terms [KMS99]. They considered error terms proportional in every input x to $|x|^p$, for any $0 \leq p < 1$, that is, they considered computations where inaccuracies depend on the size of the values involved in the calculations. This model corresponds to many practical situations. Among other things, they show how self-testing whether a program approximately computes a linear function for these errors terms. For this they proved the local stability of the linearity equation using its stability on the whole domain $\mathbb{Z}$ using techniques based on an argument due to Skof [Sko83]. The robustness part is similar to absolute error case, but the set of voters in the median defining $g(x)$ depends on x since big voters may induce big errors for small x. Since the linearity equation is unstable for the case $p = 1$ [HŠ92], their work did not lead to self-testers either for the case $p = 1$, which corresponds to linear error terms, or for relative error terms (*i.e.* proportional to the function to be computed) [KMS99, Sect. 5].

In this paper, we investigate the study of approximate self-testing with relative error. Relative error is one of the most important notions in numerical computation. Proving that a program is relatively close to its correct implementation is the challenge of many numerical analysts. We hope to contribute to make self-testers more adapted to numerical computation. In this setting self-testing consists in the following task:

Problem. *Given a class of real-valued functions $\mathcal{F}$ defined over a finite domain D, and some positive constants $c_1, c_2, \delta_1, \delta_2$, we want a simple and efficient probabilistic algorithm T such that, for any program $P : D \to \mathbb{R}$, which is an oracle for T:*

- *if for some $f \in \mathcal{F}$, $\mathbf{Pr}_{x \in D}\left[|P^T(x) - f(x)| > c_1|f(x)|\right] \leq \delta_1$, then T outputs* `PASS` *with high probability;*
- *if for all $f \in \mathcal{F}$, $\mathbf{Pr}_{x \in D}\left[|P^T(x) - f(x)| > c_2|f(x)|\right] > \delta_2$, then T outputs* `FAIL` *with high probability.*

We give a positive answer to this problem for the set of real-valued d-linear functions, for any integer $d \geq 1$. This is the first positive answer to this problem in the literature. In particular, we solve some problems in [KMS99] that were mentioned previously. For the sake of brevity and clarity we will consider functions defined over positive integer domains $\mathcal{D}_n^+ = \{i \in \mathbb{N} : 1 \leq i \leq n\}$, for some even integer $n \geq 1$. But all of our results remain valid for more general rational domains.

First we define in Theorem 2 a new probabilistic test for linear functions. It is constructed from a new functional equation for linearity which is robust (Theorem 3) and stable for linear error terms (Theorem 4). We use it to build an approximate self-tester for linear functions which allows linear error terms (Theorem 5). From it we are able to construct the first approximate self-tester with relative error in the sense of the stated problem (Theorem 6). This self-tester is generalized for multi-linear functions in Theorem 7 using an argument similar to that in [FHS94]. These self-testers are quite surprising since they only use comparisons, additions, and multiplications by powers of 2 (*i.e.* left or right shifts in binary representation). Moreover the number of queries and operations does not depend on n.

2 Linearity

The linearity test of [KMS99] is based on the linearity equation $f(x+y) = f(x) + f(y)$ which is robust and stable for error terms proportional to $|x|^p$, where $0 \leq p < 1$, but unstable when $p = 1$. More precisely they showed:

Theorem 1 ([KMS99, Theorem 2]) *Let $0 \leq \delta \leq 1$, $\theta \geq 0$, and $0 \leq p < 1$. If $P : \mathcal{D}^+_{8n} \to \mathbb{R}$ is such that*

$$\mathop{\mathbf{Pr}}_{x,y \in \mathcal{D}^+_{4n}} [|P(x+y) - P(x) - P(y)| > \theta \, \mathbf{Max}\{x^p, y^p\}] \leq \delta,$$

then there exists a linear function $l : \mathcal{D}^+_n \to \mathbb{R}$ such that for $C_p = (1+2^p)/(2-2^p)$,

$$\mathop{\mathbf{Pr}}_{x \in \mathcal{D}^+_n} [|P(x) - l(x)| > 17 C_p \theta x^p] \leq O(\sqrt{\delta}).$$

(If $p = 0$ then the latter inequality holds with $O(\delta)$ in its RHS.)

Remark. In this theorem and in the rest of the paper we only consider uniform probabilities.

For $p = 1$ the statement of this theorem does not hold anymore. Let $\theta > 0$ and $f(x) \stackrel{\text{def}}{=} \theta x \log_2(x+1)$, for all $x > 0$. In [HŠ92] it is shown that f satisfies $|f(x+y) - f(x) - f(y)| \leq 2\theta \, \mathbf{Max}\{x, y\}$, for all $x, y > 0$, but f is not close to any linear function. Hence either the test or the error term has to be modified, but both can not be kept. In [KMS99] the linearity test was unchanged and error terms proportional to x^p were considered, for some $0 \leq p < 1$. In this paper we change the test but keep a linear error term.

All results of this paper are based on the following theorem. It defines a probabilistic test such that the distance of any program to linear functions is upper bounded by a constant times its failure probability on it. Here the distance is not yet relative but it is defined for a linear error term. Let $\mathbf{Med}_{x \in X}(f(x))$ denote the median value of $f : X \to R$ when x ranges over X:

$$\mathop{\mathbf{Med}}_{x \in X}(f(x)) \stackrel{\text{def}}{=} \mathbf{Inf}\left\{a \in \mathbb{R} : \mathop{\mathbf{Pr}}_{x \in X}[f(x) \geq a] \leq 1/2\right\}.$$

For every integer $x \geq 1$, let k_x define the number:

$$k_x \stackrel{\text{def}}{=} \mathbf{Min}\left\{k \in \mathbb{N} : 2^k x \geq \frac{n}{2}\right\}.$$

Theorem 2 *Let $0 \leq \delta < 1/96$ and $\theta \geq 0$. If $P : \mathcal{D}^+_{8n} \to \mathbb{R}$ is such that*

$$\Pr_{x,y \in \mathcal{D}^+_{4n}} \left[|P(2^{k_x}x + y) - 2^{k_x}P(x) - P(y)| > \theta n \right] \leq \delta,$$

then the linear function $l : \mathcal{D}^+_n \to \mathbb{R}$, which is defined by

$$l(n) \stackrel{\text{def}}{=} \operatorname*{\mathbf{Med}}_{y \in \mathcal{D}^+_{2n}} (P(n + y) - P(y)),$$

satisfies

$$\Pr_{x \in \mathcal{D}^+_n} [|P(x) - l(x)| > 32\theta x] \leq 16\delta.$$

The proof of this theorem goes in two parts: the robustness (Theorem 3), and the stability (Theorem 4). Let us give the intuition for this test. When $x \geq n/2$, *i.e.* x is *large*, the test looks like the standard linearity test. But when $x < n/2$, *i.e.* x is *small*, we add a dilation term which amplifies small errors.

2.1 Robustness

This part consists in constructing, using P, a function g which is not linear, but approximately linear for large inputs, and perfectly homothetic for small inputs. In a sense g approximately corrects the program P.

The following theorem sates the existence of such a function g. The definition of g is based on the probability test and it consists in performing, for some $x \in \mathcal{D}^+_{2n}$, the median of votes $(P(2^{k_x}x + y) - P(y))/2^{k_x}$ for all $y \in \mathcal{D}^+_{2n}$. If the probability that P fails the test is small, then g satisfies the following theorem.

Theorem 3 (Robustness) *Let $0 \leq \delta < 1/96$ and $\theta \geq 0$. If $P : \mathcal{D}^+_{8n} \to \mathbb{R}$ is such that*

$$\Pr_{x,y \in \mathcal{D}^+_{4n}} \left[|P(2^{k_x}x + y) - 2^{k_x}P(x) - P(y)| > \theta n \right] \leq \delta,$$

then the function $g : \mathcal{D}^+_{2n} \to \mathbb{R}$ which is defined by

$$g(x) \stackrel{\text{def}}{=} \operatorname*{\mathbf{Med}}_{y \in \mathcal{D}^+_{2n}} (P(2^{k_x}x + y) - P(y))/2^{k_x},$$

satisfies

$$\Pr_{x \in \mathcal{D}^+_n} [|P(x) - g(x)| > 2\theta x] \leq 16\delta, \tag{1}$$

$$\forall x, y \in \{n/2, \ldots, n\}, \quad |g(x + y) - g(x) - g(y)| \leq 6\theta n, \tag{2}$$

$$\forall x \in \mathcal{D}^+_n, \quad g(x) = g(2^{k_x}x)/2^{k_x}. \tag{3}$$

Proof. The proof uses standard techniques developed in [BLR93, EKR96, KMS99]. Let us observe first that the function g satisfies

$$g(x) = \begin{cases} \mathbf{Med}_{y \in \mathcal{D}_{2n}^+}(P(x+y) - P(y)), & \text{if } x \geq n/2, \\ g(2^{k_x}x)/2^{k_x}, & \text{otherwise,} \end{cases}$$

and therefore g satisfies (3). Before proving that g also satisfies (1) and (2), we state a useful fact called the *halving principle* [KMS99].

Fact 1 (Halving principle) *Let Ω and S denote finite sets such that $S \subseteq \Omega$, and let ψ be a boolean function defined over Ω. Then for uniform probabilities,*

$$\mathop{\mathbf{Pr}}_{x \in S}[\psi(x)] \leq \frac{|\Omega|}{|S|} \mathop{\mathbf{Pr}}_{x \in \Omega}[\psi(x)].$$

First we show that g is close to P as defined in (1). To simplify notation, let $P_{x,y} = P(2^{k_x}x + y) - P(y) - 2^{k_x}P(x)$. By definition of g we get

$$\mathop{\mathbf{Pr}}_{x \in \mathcal{D}_n^+}[|g(x) - P(x)| > 2\theta x] = \mathop{\mathbf{Pr}}_{x \in \mathcal{D}_n^+}\left[|\mathop{\mathbf{Med}}_{y \in \mathcal{D}_{2n}^+}(P_{x,y})| > 2\theta 2^{k_x} x\right].$$

Notice that Markov's inequality gives a bound on the RHS of this equality:

$$\mathop{\mathbf{Pr}}_{x \in \mathcal{D}_n^+}\left[|\mathop{\mathbf{Med}}_{y \in \mathcal{D}_{2n}^+}(P_{x,y})| > 2\theta 2^{k_x} x\right] \leq 2 \mathop{\mathbf{Pr}}_{x \in \mathcal{D}_n^+, y \in \mathcal{D}_{2n}^+}\left[|P_{x,y}| > 2\theta 2^{k_x} x\right].$$

But $2^{k_x}x \geq n/2$, for all $x \in \mathcal{D}_n^+$, then using the halving principle we get

$$\mathop{\mathbf{Pr}}_{x \in \mathcal{D}_n^+, y \in \mathcal{D}_{2n}^+}\left[|P_{x,y}| > 2\theta 2^{k_x} x\right] \leq \frac{|\mathcal{D}_{4n}^+|^2}{|\mathcal{D}_n^+| \cdot |\mathcal{D}_{2n}^+|} \mathop{\mathbf{Pr}}_{x,y \in \mathcal{D}_{4n}^+}\left[|P_{x,y}| > \theta n\right].$$

Therefore g satisfies (1).

Now we prove that g satisfies (2). First we show that for all $c \in \{n/2, \ldots, 2n\}$ the median value $g(c)$ is close to any vote $(P(c+y) - P(y))$ with high probability:

$$\mathop{\mathbf{Pr}}_{y \in \mathcal{D}_{2n}^+}[|g(c) - (P(c+y) - P(y))| > 2\theta n] \leq 16\delta. \tag{4}$$

Note that $k_c = 0$, therefore Markov's inequality implies

$$\mathop{\mathbf{Pr}}_{y \in \mathcal{D}_{2n}^+}[|g(c) - (P(c+y) - P(y))| > 2\theta n] \leq 2 \mathop{\mathbf{Pr}}_{y,z \in \mathcal{D}_{2n}^+}[|P_{c+z,y} - P_{c+y,z}| > 2\theta n].$$

Then one can get inequality (4) using the union bound and the halving principle.

Now let a and b be two integers such that $\frac{n}{2} \leq a, b \leq n$. Let c take on the values a, b and $a+b$ in (4), and apply the halving principle to obtain:

$$\left.\begin{array}{r} \mathop{\mathbf{Pr}}_{y \in \mathcal{D}_n^+}[|g(a+b) - (P(a+b+y) - P(y))| > 2\theta n] \\ \mathop{\mathbf{Pr}}_{y \in \mathcal{D}_n^+}[|g(a) - (P(a+y) - P(y))| > 2\theta n] \\ \mathop{\mathbf{Pr}}_{y \in \mathcal{D}_n^+}[|g(b) - (P(b+(a+y)) - P(a+y))| > 2\theta n] \end{array}\right\} \leq 32\delta.$$

Therefore with probability at least $1 - 96\delta > 0$ there exists $y \in \mathcal{D}_n^+$ for which none of these inequalities are satisfied. Pick such a y to obtain inequality (2). □

2.2 Stability

In this section we prove that every function g satisfying the conditions of Theorem 3 is close to a perfectly linear one.

Theorem 4 (Stability) *Let $\theta' \geq 0$. If $g : \mathcal{D}_{2n}^{+} \to \mathbb{R}$ is such that*

$$\forall x, y \in \{n/2, \dots, n\}, \quad |g(x+y) - g(x) - g(y)| \leq \theta' n,$$

$$\text{and} \quad \forall x \in \mathcal{D}_n^{+}, \quad g(x) = g(2^{k_x} x)/2^{k_x},$$

then the linear function $l : \mathcal{D}_n^{+} \to \mathbb{R}$, which is defined by $l(n) \stackrel{\text{def}}{=} g(n)$, satisfies, for all $x \in \mathcal{D}_n^{+}$,

$$|g(x) - l(x)| \leq 5\theta' x.$$

Proof. Here we borrow a technique developed in [KMS99] that we apply to the function g where it is approximately linear. First we extend g restricted to $\{n/2, \dots, n\}$ to a function h defined over the whole semi-group $\{x \in \mathbb{N} : x \geq n/2\}$. The extension h is defined for all $x \geq n/2$ by

$$h(x) \stackrel{\text{def}}{=} \begin{cases} g(x) & \text{if } n/2 \leq x \leq n, \\ h(x - n/2) + g(n)/2 & \text{otherwise.} \end{cases}$$

One can verify that h satisfies the following doubling property, for all $x \geq \frac{n}{2}$,

$$|h(2x) - 2h(x)| \leq 5\theta' n/2.$$

Then we apply a result due to [KMS99] which is based on some techniques developed in [Hye41].

Lemma 1 ([KMS99, Lemma 3]) *Let E_1 be a semi-group and E_2 a Banach space. Let $\varepsilon \geq 0$ and $h : E_1 \to E_2$ be a mapping such that for all $x \in E_1$*

$$\|h(2x) - 2h(x)\| \leq \varepsilon.$$

If $f : E_1 \to E_2$ is such that $f(x) = \lim_{m \to \infty} h(2^m x)/2^m$ is a well defined mapping, then for all $x \in E_1$

$$\|h(x) - f(x)\| \leq \varepsilon.$$

Let f be this function. Then by definition of h we get that $f(x) = xg(n)/n$, for all $x \geq n/2$, therefore f is linear and $f = l$. We conclude the proof by recalling that g equals h on $\{n/2, \dots, n\}$, and when $1 \leq x < n/2$, $g(x) = g(2^{k_x} x)/2^{k_x}$. □

3 Testing with Relative Error

In this section we show how our results lead to approximate self-testers with linear error terms and with relative error. First let us define the relative distance. Let $\mathcal{F}$ be a collection of real functions defined over a finite domain D. For a real

$\theta \geq 0$ and functions $P, f : D \to \mathbb{R}$, we will define the *θ-relative distance* between P and f on D by

$$\theta\text{-}\mathbf{rdist}_D(P, f) \stackrel{\text{def}}{=} \mathbf{Pr}_{x \in D}\left[|P(x) - f(x)| > \theta|f(x)|\right],$$

and the θ-relative distance between P and $\mathcal{F}$ on D by

$$\theta\text{-}\mathbf{rdist}_D(P, \mathcal{F}) \stackrel{\text{def}}{=} \mathbf{Inf}_{f \in \mathcal{F}}\, \theta\text{-}\mathbf{rdist}_D(P, f).$$

Note that the relative distance between P and f is not symmetric in general. For example if $\theta > 1$, $P(x) = 0$, and $f(x) = \theta$, for all $x \in \mathcal{D}_n^+$, then $\theta\text{-}\mathbf{rdist}_{\mathcal{D}_n^+}(P, f) = 0$ and $\theta\text{-}\mathbf{rdist}_{\mathcal{D}_n^+}(f, P) = 1$. We will also need another distance which is symmetric but not relative. It is defined for any non negative error term $\beta : D \to \mathbb{R}_+$. The *$\beta$-distance* between P and f on D is

$$\beta\text{-}\mathbf{dist}_D(P, f) \stackrel{\text{def}}{=} \mathbf{Pr}_{x \in D}\left[|P(x) - f(x)| > \beta(x)\right],$$

and the β-distance between P and $\mathcal{F}$ on D is

$$\beta\text{-}\mathbf{dist}_D(P, \mathcal{F}) \stackrel{\text{def}}{=} \mathbf{Inf}_{f \in \mathcal{F}}\, \beta\text{-}\mathbf{dist}_D(P, f).$$

First we define the approximate self-tester for $\beta\text{-}\mathbf{dist}_D$ using the definition of [KMS99] which generalizes that of [EKR96, GLR+91].

Definition 1 *Let $\delta_1, \delta_2 \in [0, 1]$, $D_2 \subseteq D_1$, and $\mathcal{F}$ be a collection of real-valued functions defined over D_1. Let β_1 and β_2 be non negative real-valued functions also defined over D_1. A $(D_1, \beta_1, \delta_1; D_2, \beta_2, \delta_2)$-*self-tester *for $\mathcal{F}$ is a probabilistic oracle program T such that for any program $P : D_1 \to \mathbb{R}$:*

- *If $\beta_1\text{-}\mathbf{dist}_{D_1}(P, \mathcal{F}) \leq \delta_1$ then T^P outputs* PASS *with probability at least 2/3.*[1]
- *If $\beta_2\text{-}\mathbf{dist}_{D_2}(P, \mathcal{F}) > \delta_2$ then T^P outputs* FAIL *with probability at least 2/3.*[2]

Now we extend this definition for relative distance.

Definition 2 *Let $\delta_1, \delta_2 \in [0, 1]$, $D_2 \subseteq D_1$, and $\mathcal{F}$ be a collection of real-valued functions defined over D_1. Let θ_1 and θ_2 be non negative reals. A $(D_1, \theta_1, \delta_1; D_2, \theta_2, \delta_2)$-*self-tester with relative error *for $\mathcal{F}$ is a probabilistic oracle program T such that for any program $P : D_1 \to \mathbb{R}$:*

- *If $\theta_1\text{-}\mathbf{rdist}_{D_1}(P, \mathcal{F}) \leq \delta_1$ then T^P outputs* PASS *with probability at least 2/3.*
- *If $\theta_2\text{-}\mathbf{rdist}_{D_2}(P, \mathcal{F}) > \delta_2$ then T^P outputs* FAIL *with probability at least 2/3.*

---

[1] One can also want this probability to be greater than any confidence parameter $\gamma \in (0, 1)$. Here we simplify our discussion by fixing this parameter to 2/3.

[2] Same remark.

Usually one would like a self-tester to be different and simpler than any correct program. For example we can ask the self-tester to satisfy the *little-oh property* [BK95], *i.e.* its running time have to be asymptotically less than that of any known correct program. This property could be too restrictive for family testing. Here we simplify this condition. If T is a self-tester for d-linearity over $\mathcal{D}_n^+$ then T is required to use only comparisons, additions, and multiplications by powers of 2 (*i.e.* left or right shifts in binary representation). Moreover the number of queries and operations of T has to be independent of n.

A direct consequence of Theorem 2 is the existence of a self-tester for the set of linear functions, denoted by $\mathcal{L}$, where the distance is defined for a linear error term.

Theorem 5 *Let* $0 < \delta < 1/144$ *be a real,* $\theta \geq 0$ *a power of* 2, *and* $\beta(x) = \theta x$, *for all* x. *Then there exists a* $(\mathcal{D}_{8n}^+, \beta/16, \delta/12; \mathcal{D}_n^+, 32\beta, 24\delta)$*-self-tester for the set* $\mathcal{L}$. *Moreover it makes* $O(1/\delta)$ *queries to the program, and uses* $O(1/\delta)$ *comparisons, additions, and multiplications by powers of* 2.

Proof. Let $N \geq 1$ be an integer whose value will be fixed later. The self-tester T performs N independent rounds. Each round consists in performing the following experiment, where $\Theta \stackrel{\text{def}}{=} \theta n$:

Experiment linearity-test(P, Θ)
1. Randomly choose $x, y \in \mathcal{D}_{4n}^+$.
2. Check if $\lvert P(2^{k_x}x + y) - 2^{k_x}P(x) - P(y)\rvert \leq \Theta$.

A round *fails* if the inequality is not satisfied. Then T outputs `FAIL` if more than a δ fraction of the rounds fail, and `PASS` otherwise.

Let us define the failure probability of P in each round by

$$\mathbf{err}(P) \stackrel{\text{def}}{=} \Pr_{x,y \in \mathcal{D}_{4n}^+} \left[\lvert P(2^{k_x}x + y) - 2^{k_x}P(x) - P(y)\rvert > \theta n \right].$$

First suppose $(\beta/16)$-$\mathbf{dist}_{\mathcal{D}_{8n}^+}(P, \mathcal{L}) \leq \delta/12$. The halving principle and simple manipulations lead to $\mathbf{err}(P) \leq \delta/2$. Then a standard Chernoff bound argument yields that if $N = \Omega(1/\delta)$ then T^P outputs `PASS`, with probability at least 2/3.

Now if (32β)-$\mathbf{dist}_{\mathcal{D}_n^+}(P, \mathcal{L}) > 24\delta$, then, since $3\delta/2 < 1/96$, the contraposition of Theorem 2 implies $\mathbf{err}(P) > 3\delta/2$. Again, by a Chernoff bound argument if $N = \Omega(1/\delta)$ then T^P outputs `FAIL`, with probability at least 2/3. □

The previous self-tester has two main disadvantages. First, the error term is linear but not relative. Second, it needs to test the program on a bigger domain. The following theorem gets around these two problems.

Theorem 6 *Let* $0 < \delta < 1/144$ *be a real and* $0 \leq \theta \leq 16$ *a power of* 2. *Then there exists a* $(\mathcal{D}_n^+, \theta/64, \delta/12; \mathcal{D}_n^+, 32\theta, 24\delta)$*-self-tester with relative error for the set* $\mathcal{L}$. *Moreover it makes* $O(1/\delta)$ *queries to the program, and uses* $O(1/\delta)$ *comparisons, additions, and multiplications by powers of* 2.

Proof. Now the self-tester T performs $N = O(1/\delta)$ times the following experiment:

Experiment linearity-relative-test(P, θ)
1. Randomly choose $y \in \mathcal{D}_n^+$.
2. Compute $G = P(n - y) + P(y)$ (fix $P(0) = 0$).
3. Compute $\Theta = \theta|G|$.
4. Do **Experiment linearity-test(extension**(P, G),Θ**)**.

The function **extension** is easily computable using P, and it is defined by:

Function extension$(P, G)(x)$
1. $val = 0$.
2. While $x > n$ do $x = x - n$ and $val = val + G$.
3. Return $(val + P(x))$.

Again, T outputs FAIL if more than a δ fraction of the rounds fail, and PASS otherwise.

Fix $\Theta \stackrel{\text{def}}{=} \theta|G|$ and $\beta(x) \stackrel{\text{def}}{=} \theta|G|x/n$, for all x. Let $\widetilde{P} \stackrel{\text{def}}{=}$ **extension**(P, G), and denote the failure probability of one experiment by **rerr**(P).

First, suppose there exists a linear function l such that $(\theta/64)$-$\mathbf{rdist}_{\mathcal{D}_n^+}(P, l) \leq \delta/12$. Therefore $\mathbf{Pr}_{y \in \mathcal{D}_n^+}[|P(n - y) + P(y) - l(n)| > \theta|l(n)|/32] \leq \delta/6$. So $|G - l(n)| \leq \theta|l(n)|/32$ with probability at least $1 - \delta/6$. Suppose this last inequality is satisfied. Then one can verify that $(\theta/32)$-$\mathbf{rdist}_{\mathcal{D}_{8n}^+}(\widetilde{P}, l) \leq \delta/12$. Since $\theta/32 \leq 1/2$, we also obtain $|l(n)| \leq 2|G|$. Thus the combination of the two last inequalities gives $(\beta/16)$-$\mathbf{dist}_{\mathcal{D}_{8n}^+}(\widetilde{P}, l) \leq \delta/12$. In this case we previously proved that the failure probability of **linearity-test(extension**(P, G),Θ**)** is at most $\delta/2$. In conclusion, if $(\theta/64)$-$\mathbf{rdist}_{\mathcal{D}_n^+}(P, l) \leq \delta/12$ then $\mathbf{rerr}(P) \leq \delta/6 + \delta/2 = 2\delta/3$.

Suppose now that (32θ)-$\mathbf{rdist}_{\mathcal{D}_n^+}(P, \mathcal{L}) > 24\delta$. Then, for all real G, (32β)-$\mathbf{dist}_{\mathcal{D}_n^+}(\widetilde{P}, l) > 24\delta$, where $l \in \mathcal{L}$ is defined by $l(n) \stackrel{\text{def}}{=} G$. But $\widetilde{P}(n + y) = G + \widetilde{P}(y)$, so $\mathbf{Med}_{y \in \mathcal{D}_{2n}^+}(\widetilde{P}(n + y) - \widetilde{P}(y)) = G$. Therefore the contraposition of Theorem 2 implies that, for all real G, $\mathbf{rerr}(P) > 3\delta/2$.

To conclude the proof, apply a Chernoff bound argument. □

Now we can state our final result which extends the previous one to multilinear functions. It is quite surprising since it does not use multiplications but only comparisons, additions, and multiplications by powers of 2.

Theorem 7 *Let $d \geq 1$ be an integer. Let $0 < \delta \leq 1$ be a real and $0 \leq \theta \leq O(1/d^2)$ a power of 2. Then there exists a $((\mathcal{D}_n^+)^d, \theta, \delta; (\mathcal{D}_n^+)^d, O(d)\theta, O(d)\delta)$-self-tester with relative error for the set of real-valued d-linear functions defined on $(\mathcal{D}_n^+)^d$. Moreover it makes $O(1/\delta)$ queries to the program, and uses $O(1/\delta)$ comparisons, additions, and multiplications by powers of 2.*

*Proof (sketch).*We use some techniques from [FHS94] where a similar result for multi-variate polynomials in the context of exact computation was proven. Fact 2 and Lemma 2 lower and upper bound the distance between a d-variate function and d-linear functions by its successive distances from functions which are linear in one of their variables. Then, we estimate the latter quantity by repeating **Experiment linearity-relative-test**. More precisely, the self-tester T will repeat

$O(1/\delta)$ times the following experiment. Then T outputs FAIL if more than a δ fraction of the rounds fail, and PASS otherwise.

Experiment d-linearity-relative-test(P, θ)
1. Randomly choose $\vec{z} \in (\mathcal{D}_n^+)^d$.
2. Randomly choose $i \in \{1, \dots, d\}$.
3. Do **Experiment linearity-relative-test**$(\tilde{P}_{\vec{z}}^i, \theta)$.

The notation $\tilde{P}_{\vec{z}}^i$ denotes the function which takes at t the value $P(z_1, \dots, z_{i-1}, t, z_{i+1}, \dots, z_d)$.

Using Fact 2 and Lemma 2, one can conclude the proof using previous methods. □

The bounds involved in the previous proof are explicitly stated in the following, where $\mathcal{L}^d$ denote the set of d-linear functions defined over $(\mathcal{D}_n^+)^d$, and $\mathcal{L}_i^d$ the set of functions defined over $(\mathcal{D}_n^+)^d$ which are linear in their i-th variable. First let us state the easy one.

Fact 2 *Let $\theta \geq 0$ be a real. Then for all $f : (\mathcal{D}_n^+)^d \to \mathbb{R}$*

$$\frac{1}{d} \sum_{i=1}^{d} \theta\text{-}\mathbf{rdist}_{(\mathcal{D}_n^+)^d}(f, \mathcal{L}_i^d) \leq \theta\text{-}\mathbf{rdist}_{(\mathcal{D}_n^+)^d}(f, \mathcal{L}^d).$$

The other bound is more difficult and it can be proven by induction on d. Due to lack of place we omit the proof.

Lemma 2 *Let $0 \leq \theta \leq 1/(16d^2)$ be a real. Then for all $f : (\mathcal{D}_n^+)^d \to \mathbb{R}$*

$$(4d\theta)\text{-}\mathbf{rdist}_{(\mathcal{D}_n^+)^d}(f, \mathcal{L}^d) \leq 2 \sum_{i=1}^{d} \theta\text{-}\mathbf{rdist}_{(\mathcal{D}_n^+)^d}(f, \mathcal{L}_i^d).$$

Open Questions

In this paper we achieve the goal of approximate self-testing with relative error for multi-linear functions. We would like to extend this work for polynomials. More generally when we have no information *a priori* on the size of the function to be computed, constructing approximate self-testers with relative error is an interesting challenge.

Acknowledgments

We would like to thank Stéphane Boucheron, Marcos Kiwi, Sophie Laplante and Miklos Santha for useful discussions and assistance while writing this paper.

References

[ABCG93] S. Ar, M. Blum, B. Codenotti, and P. Gemmell. Checking approximate computations over the reals. In *Proc. 25th STOC*, pages 786–795, 1993.

[ALM+92] S. Arora, C. Lund, R. Motwani, M. Sudan, and M. Szegedy. Proof verification and intractibility of approximation problems. In *Proc. 33rd FOCS*, pages 14–23, 1992.

[AS92] S. Arora and S. Safra. Probabilistic checkable proofs: A new characterization of NP. In *Proc. 33rd FOCS*, pages 1–13, 1992.

[Bab93] L. Babai. Transparent (holographic) proofs. In *Proc. 10th STACS*, volume 665, pages 525–534. LNCS, 1993.

[BK95] M. Blum and S. Kannan. Designing programs that check their work. *J. ACM*, 42(1):269–291, 1995.

[BLR93] M. Blum, M. Luby, and R. Rubinfeld. Self-testing/correcting with applications to numerical problems. *J. Comp. and Syst. Sci.*, pages 549–595, 1993.

[Blu88] M. Blum. Designing programs to check their work. Technical Report 88-009, ICSI, 1988.

[Cop89] D. Coppersmith, December 1989. See discussion in [BLR93].

[EKR96] F. Ergün, S. Ravi Kumar, and R. Rubinfeld. Approximate checking of polynomials and functional equations. In *Proc. 37th FOCS*, pages 592–601, 1996.

[FHS94] K. Friedl, Z. Hátsági, and A. Shen. Low-degree tests. *Proc. 5th SODA*, pages 57–64, 1994.

[For95] G. L. Forti. Hyers-Ulam stability of functional equations in several variables. *Aeq. Mathematicae*, 50:143–190, 1995.

[GGR96] O. Goldreich, S. Goldwasser, and D. Ron. Property testing and its connection to learning and approximation. In *Proc. 37th FOCS*, pages 339–348, 1996.

[GLR+91] P. Gemmell, R. Lipton, R. Rubinfeld, M. Sudan, and A. Wigderson. Self-testing/correcting for polynomials and for approximate functions. In *Proc. 23rd STOC*, pages 32–42, 1991.

[GR97] O. Goldreich and D. Ron. Property testing in bounded degree graphs. In *Proc. 37th STOC*, pages 406–415, 1997.

[HR92] D. H. Hyers and T. M. Rassias. Approximate homomorphisms. *Aeq. Mathematicae*, 44:125–153, 1992.

[HŠ92] D. H. Hyers and P. Šemrl. On the behaviour of mappings which do not satisfy Hyers-Ulam stability. *Proc. AMS*, 144(4):989–993, April 1992.

[Hye41] D. H. Hyers. On the stability of the linear functional equation. *Proc. Nat. Acad. Sci., U.S.A.*, 27:222–224, 1941.

[KMS99] M. Kiwi, F. Magniez, and M. Santha. Approximate testing with relative error. In *Proc. 31st STOC*, pages 51–60, 1999.

[Lip91] R. Lipton. New directions in testing. *Series in Discrete Mathematics and Theoretical Computer Science*, 2:191–202, 1991.

[RS96] R. Rubinfeld and M. Sudan. Robust characterizations of polynomials with applications to program testing. *SIAM J. Comp.*, 25(2):23–32, April 1996.

[Rub99] R. Rubinfeld. On the robustness of functional equations. *SIAM J. Comp.*, 28(6):1972–1997, 1999.

[Sko83] F. Skof. Sull'approssimazione delle applicazioni localmente δ-additive. *Atti Acc. Sci. Torino*, 117:377–389, 1983. in Italian.

Nondeterministic Instance Complexity and Hard-to-Prove Tautologies

Vikraman Arvind[1], Johannes Köbler[2], Martin Mundhenk[3], and Jacobo Torán[4]

[1] Institute of Mathematical Sciences, C.I.T Campus, Madras 600113, India
[2] Humboldt-Universität zu Berlin, Institut für Informatik, D-10099 Berlin, Germany
[3] Universität Trier, Fachbereich IV – Informatik, D-54286 Trier, Germany
[4] Universität Ulm, Theoretische Informatik, D-89069 Ulm, Germany

Abstract. In this note we first formalize the notion of hard tautologies using a nondeterministic generalization of instance complexity. We then show, under reasonable complexity-theoretic assumptions, that there are infinitely many propositional tautologies that are hard to prove in any sound propositional proof system.

1 Introduction

In their seminal paper [4] Cook and Reckhow first formalized the study of lengths of proofs in propositional proof systems. They showed that NP = coNP if and only if there is a sound and complete propositional proof system S and polynomial p such that all tautologies F (i.e. all $F \in \mathit{TAUT}$) have proofs in S of size $p(|F|)$. The Cook-Reckhow paper led to an extensive and fruitful study of lengths of proofs for standard proof systems of increasing complexity and strength. A recent survey of the area by Beame and Pitassi is contained in [2]. A restatement of the Cook-Reckhow result is that if NP $\neq$ coNP then for every sound propositional proof system P and polynomial bound p there is an infinite collection of "hard" tautologies: i.e. the shortest proofs in P of each such tautology F is of length more than $p(|F|)$. A stronger notion of hard tautologies for a proof system is defined by Krajícek [8, Definition 14.2.1]. He defines a sequence of tautologies $\{\varphi_n\}$ to be hard for a proof system P if for every n the formula φ_n can be computed from 1^n in polynomial time, has size at least n, and there is no polynomial p so that $p(|\varphi_n|)$ bounds the size of the proof of φ_n in P for all n. Under a complexity assumption (slightly stronger than NP $\neq$ coNP) it is shown by Krajícek [8, Theorem 14.2.3] that for any non-optimal proof system P there is a sequence of hard tautologies for P. A more difficult question is whether there is a family of tautologies that is hard in the above sense for every sound proof system. Krajícek's construction does not yield this as the degree of the polynomial bounding the construction time of hard tautologies depends on the proof system and can become arbitrarily large. In a recent result Riis and Sitharam in [12] show under the assumption NEXP $\neq$ coNEXP how to construct a sequence of tautologies $\{\phi_n\}_{n>0}$ that do not have polynomial-size proofs in any propositional proof system. The property of $\{\phi_n\}_{n>0}$ here is that the sequence is, in a

H. Reichel and S. Tison (Eds.): STACS 2000, LNCS 1770, pp. 314–323, 2000.

formal logic sense, uniformly generated. Notice that this result is really about the hardness of the entire sequence and not about the individual tautologies ϕ_n. Indeed, the sequence $\{\phi_n\}_{n>0}$ might have an infinite subsequence that have polynomial-size proofs in some proof system.

Schöning [13] approaches the question of hard tautologies using the notion of complexity cores first defined in [10]. He shows in [13] that if NP $\neq$ coNP then there exist a constant ϵ and a collection $\mathcal{F}$ of tautologies of density at least $2^{\epsilon n}$ i.o. such that for *every* sound proof system for $TAUT$ and for every polynomial p, the shortest proof of F has length more than $p(|F|)$ for all but finitely many $F \in \mathcal{F}$. (The density part of this result is a consequence of the fact that $TAUT$ has a linear-time padding function.) Again, notice here that $\mathcal{F}$ could have individual tautologies that are easy for specific proof systems.

A natural question concerning hard tautologies that arises, which we address in this paper, is whether there exist individual tautologies which are "hard" for each sound proof system. Notice that none of the results mentioned above assert the existence of an infinite collection of tautologies in which every individual tautology is hard for every sound proof system. First, what do we mean by a tautology that is hard for all sound proof systems? We have to handle the following difficulty in the definition. For any $F \in TAUT$, we can always define a sound proof system S that includes F as an axiom. Clearly, F will have a trivial proof in the proof system S. The key point here is that F has to be somehow encoded into the proof system S, implying that the size of the description of S needs to be taken into account. We present a formal definition of hard tautology for a proof system based on the notion of instance complexity [7]. Intuitively, given a polynomial bound p for proof length, a tautology F is hard if for any given proof system P, either the shortest proof of F in P has size greater than $p(|F|)$ or else the size of P (considered as a string) is big enough to contain an encoding of F. We formalize the definition of hard tautology using the concepts of nondeterministic instance complexity and Kolmogorov complexity. We then prove, under reasonable complexity-theoretic assumptions, the existence of an infinite set of tautologies with the property that every single tautology in the set is hard (with respect to the discussed hardness notions) for every sound proof system.

2 Definitions

We fix the alphabet as $\Sigma = \{0, 1\}$. Let SAT denote the language of satisfiable propositional formulas suitably encoded as strings in Σ^*. Similarly, let $TAUT$ denote the language of propositional tautologies. For the definitions of standard complexity classes like P, NP, coNP, NEXP, the polynomial hierarchy PH etc. we refer the reader to a standard textbook, e.g. [1]. The class Θ_2^p (cf. [6]) – also described as $\mathrm{P}^{\mathrm{NP}}[O(\log n)]$ – consists of all sets decidable by a polynomial-time oracle machine that can make $O(\log n)$ queries to an NP oracle.

We start with a definition of proof systems that is equivalent to the original Cook-Reckhow formulation [4].

Definition 1. [2] *A (sound and complete) propositional proof system S is defined to be a polynomial-time predicate S such that for all F,*

$$F \in TAUT \Leftrightarrow \exists p : S(F,p).$$

In other words, a proof system can be identified with an efficient procedure for checking correctness of proofs. Thus, in complexity-theoretic terms we can also identify a propositional proof system S with a nondeterministic Turing machine M. The soundness and completeness properties of S can be expressed in terms of M as follows:

$$F \in L(M) \Rightarrow F \in TAUT \qquad \langle soundness \rangle$$

$$F \in TAUT \Rightarrow F \in L(M) \qquad \langle completeness \rangle$$

For the rest of the paper we use the characterization of propositional proof systems as nondeterministic Turing machines.

Nondeterministic Instance Complexity The notion of instance complexity was introduced by Orponen et al. in [7] as a measure of the complexity of individual instances. The motivation in that paper is to study whether the hardness of NP-complete problems, particularly SAT, can be pinpointed to individual hard instances in some formal sense.

We adapt the instance complexity idea to a nondeterministic setting that is suitable for defining hard individual tautologies. For a set $A \subseteq \Sigma^*$ we say that a nondeterministic machine M is A-consistent if $L(M) \subseteq A$.

Definition 2. *For a set A and a time bound t, the t-time-bounded* nondeterministic instance complexity *(nic for short) of x w.r.t. A is defined as:*

$$nic^t(x : A) = \min\{\ |M| \ \mid M \text{ is an } A\text{-consistent } t\text{-time-bounded nondeterministic machine, and } M \text{ decides correctly on } x\ \}.$$

As required for instance complexity [7] and Kolmogorov complexity [9], we formalize *nic* in terms of a fixed universal machine U that executes nondeterministic programs. For a time bound function t, and two strings $q, x \in \Sigma^*$, we denote by $U^t(q,x)$ the result of running in the universal machine U program q on input x for $t(|x|)$ steps. Notice that the *nic* measure is (up to an additive constant) essentially the same w.r.t. efficient universal machines.

We now turn to the formal definition of a set A having hard instances w.r.t. the *nic* measure. Intuitively, we consider x to be a hard instance if there is no easier way for an A-consistent nondeterministic program to decide x than to explicitly encode x into the program. To define this formally, we consider a nondeterministic version of Kolmogorov complexity first defined in [3]. More precisely, given a time bound t and a string $x \in \Sigma^*$,

$$CND^t(x) = \min\{\ |M| \ \mid M \text{ is a } t\text{-time-bounded nondeterministic Turing machine with } L(M) = \{x\}\ \}$$

is the nondeterministic t-time-bounded decision Kolmogorov complexity of x w.r.t. M. In the standard way (see [9]) we can consider the *CND* measure to be defined w.r.t. a fixed universal machine. Notice that the *CND* measure is a nondeterministic generalization of Sipser's *CD* measure [14]. We note in passing that there is no difference between the nondeterministic Kolmogorov complexity of checking and generating [3] (unlike in the deterministic case where the C and *CD* time-bounded measures appear to be different [14]). The *CND* measure gives an immediate upper bound to nondeterministic instance complexity.

The next proposition will provide an upper bound on the size of a sound proof system that for a given tautology has proofs of a certain length.

Proposition 1. *For any language A, and time bound t, there is a constant $c > 0$ such that $nic^t(x : A) \leq CND^t(x) + c$, for all $x \in \Sigma^*$.*

We now define hard instances for a language A w.r.t. the *nic* measure.

Definition 3. *A language A is said to have hard instances w.r.t. the nic measure if for every polynomial t there are a polynomial t' and a constant c such that for infinitely many x we have, $nic^t(x : A) > CND^{t'}(x) - c$.*

We observe next that hard instances of A, if any, must necessarily be in A. This is in contrast to the deterministic instance complexity definition [7] for which hard instances could be in A or $\overline{A}$.

Proposition 2. *For any set A there is a constant c such that for all time bounds t and for all $x \notin A$, $nic^t(x : A) \leq c$.*

Proof. It follows from the fact that any nondeterministic machine that accepts $\emptyset$ is consistent with A and decides correctly on all instances $x \notin A$. □

Observe that by Proposition 1, for every tautology F and time bound t, there is always a sound proof system P of size bounded by $CND^t(F) + c$ for some constant c, and in which F has proofs of size at most $t(|F|)$. Ideally we would like to obtain a matching lower bound for the *nic* measure, as stated in the following open question.

Problem 1. If NP $\neq$ coNP, then for every polynomial time bound t there are infinitely many tautologies F and a constant c such that $nic^t(F : TAUT) \geq CND^t(F) - c$ holds?

Because of the above observation, this result would be best possible. Although we cannot match the upper bound we prove somewhat weaker lower bounds using less restricted notions of Kolmogorov complexity instead of the *CND* measure. In particular, we consider the standard unbounded Kolmogorov complexity measure C [9] and a relativized time-bounded Kolmogorov complexity measure C^B: $C^{B,t}(x)$ is defined to be the t time-bounded Kolmogorov complexity of string x, where the universal machine has oracle access to oracle B. We summarize below the main results of this paper proved in Section 3.

- If PH does not collapse to Θ_2^p then for every polynomial t and constant $c > 0$ there are infinitely many tautologies F such that $nic^t(F : TAUT) \geq c \log |F|$.
- NP $\neq$ coNP if and only if for every polynomial t there exist a constant $c > 0$ and infinitely many tautologies F such that $nic^t(F : TAUT) > C(F) - c$.
- NP $\neq$ coNP if and only if for every polynomial t there exist a constant $c > 0$, a polynomial t', and infinitely many tautologies F such that $nic^t(F : TAUT) > C^{SAT_2,t'}(F) - c$, where SAT_2 denotes all true quantified boolean formulas with $\exists\forall$ as quantifier prefix.

Each of the above results can be interpreted as stating that, under a reasonable complexity-theoretic assumption, there exist infinitely many tautologies F such that any sound proof system S in which F has short proofs must have large description (the description size is some function of $|F|$).

Notice that the latter two results are weaker than the ideal possible result mentioned above because for all F and all polynomials t we have: $C(F) \leq C^{SAT_2,t}(F) \leq CND^t(F)$. Also, observe that the first result proving an $O(\log |F|)$ lower bound on the nondeterministic instance complexity of infinitely many tautologies F is incomparable to the other results (and also to the statement in Problem 1) because it does not refer to the Kolmogorov complexity of F.

Hard Instances and NP Cores We review NP cores in TAUT defined by Schöning [13] and briefly compare it with nondeterministic instance complexity. An NP core in TAUT is an infinite subset $C \subseteq TAUT$ such that for all polynomials p and for all proof systems S, at most finitely many $F \in C$ have proofs in system S of length bounded by $p(|F|)$. In [13] it is shown that NP $\neq$ coNP if and only if TAUT has an NP core (in fact, of density $2^{\epsilon n}$, for a constant fraction $\epsilon > 0$).

Theorem 1. [13] *Assuming* NP $\neq$ coNP, *there exist a constant* ϵ *and a collection* $\mathcal{F}$ *of tautologies of density at least* $2^{\epsilon n}$ *i.o. such that for every sound proof system for* $TAUT$ *and for every polynomial* p, *the shortest proof of* F *has length more than* $p(|F|)$ *for all but finitely many* $F \in \mathcal{F}$.

The above theorem does not really talk about tautologies that are hard for each proof system. We can only make the following easy connection to the nondeterministic instance complexity of tautologies in an NP core of $TAUT$.

Proposition 3. *A set* C *is an* NP *core of a recursive set* A *(and hence also* $TAUT$*) if and only if for every polynomial* t *and constant* c, $nic^t(x : A) > c$ *for all but finitely many* $x \in C$.

3 The Results

We prove two theorems which assert, in different ways, that there exist infinitely many hard tautologies assuming that the polynomial hierarchy does not collapse.

Theorem 2. *If* PH $\neq \Theta_2^p$ *then given any polynomial* t *and constant* c*, there exist infinitely many tautologies* F *such that* $nic^t(F : TAUT) > c \log |F|$.

Proof. Assume the contrary: Suppose there are a polynomial t and constant c such that for all $F \in TAUT : nic^t(F : TAUT) \leq c \log |F|$. In particular, for any n we have

$$\forall\, F \in TAUT^{\leq n} \;:\; nic^t(F : TAUT) \leq c \log n.$$

It suffices to show under this assumption that $\Sigma_2^p = \Theta_2^p$. Let M be an oracle-NP machine with TAUT as oracle. With no loss of generality, we can assume that on each computation path M makes precisely one positive oracle query to $TAUT$ and that too at the end of the path before it decides on acceptance. Let p denote a polynomial such that the running time of M on input x is bounded by $p(|x|)$, for all x. We will design a Θ_2^p machine N that accepts $L(M)$. Let x be a length m input to M. Let n denote $p(m)$ which bounds the size of $TAUT$ queries that $M(x)$ can make. Given the $O(\log n)$ bound on the nondeterministic instance complexity of tautologies in $TAUT^{\leq n}$, we only have to use the $TAUT$-consistent nondeterministic programs in the set $\Sigma^{\leq c \log n}$, whose size is bounded by some polynomial in n. But first we must get rid of the $TAUT$-inconsistent nondeterministic programs in $\Sigma^{\leq c \log n}$. We show that the inconsistent programs can be bundled out in an NP set, and by a census argument that can be carried out with a Θ_2^p computation we can exactly count the $TAUT$-inconsistent nondeterministic programs in $\Sigma^{\leq c \log n}$. After that a single NP computation can guess the $TAUT$-consistent subset of $\Sigma^{\leq c \log n}$ and use them to replace the $TAUT$ oracle.

$$BAD := \{\langle p, 0^n\rangle \mid p \in \Sigma^{\leq c \log n} \;:\; \exists\, F \in \Sigma^n - TAUT^{=n} \;:\; U^t(p, F) \text{ accepts } \}$$

Notice that BAD is in NP. In fact, BAD is a *sparse* NP set: $\|\{p \in \Sigma^{\leq c \log n} \mid \langle p, 0^n\rangle \in BAD\}\| = O(n^c)$. Let BAD_n denote the set $\{p \in \Sigma^{\leq c \log n} \mid \langle p, 0^n\rangle \in BAD\}$. We can use a standard census argument like the one in [6] to determine $\|BAD_n\|$ with a Θ_2^p computation. The Θ_2^p machine N will first compute $\|BAD_n\|$. Now it is easy to simulate $M^{TAUT}(x)$ with an NP computation: First guess the $\|BAD_n\|$ strings in BAD_n and verify with an NP computation that $\langle p, 0^n\rangle \in BAD$ for each guessed p. Having guessed and verified BAD_n, the remaining strings in $\Sigma^{\leq c \log n}$ are all $TAUT$-consistent programs. The NP computation now proceeds to simulate M^{TAUT}. When the simulation encounters a positive $TAUT$ query F, it accepts if and only if

$$\exists\, p \in (\Sigma^{\leq c \log n} - BAD_n) \;:\; U^t(p, F) \text{ accepts.}$$

To summarize, the Θ_2^p machine N on input x first computes $\|BAD_n\|$ with a Θ_2^p computation. What remains to simulate $M^{TAUT}(x)$ is just an NP computation. Hence PH collapses to Θ_2^p. □

We now turn to the next result which is closer in spirit to the definition of hard-to-prove tautologies.

Theorem 3. NP $\neq$ coNP *if and only if for every polynomial t there exist a constant* $c > 0$ *and infinitely many tautologies F such that* $nic^t(F : TAUT) > C(F) - c$.

Proof. We first prove the forward implication. Our proof is an easy modification of the Fortnow-Kummer proof technique in [5] for the instance complexity conjecture [7] for NP-hard sets under honest reductions (this technique is also used by Mundhenk [11] to prove a weaker form of the general instance complexity conjecture of [7]). The idea is to build a deterministic Turing machine M that on input 0^n searches for and outputs a tautology F such that $nic^t(F : TAUT) > cn$ for a constant c. The fact that M outputs F on input 0^n implies that $C(F) \leq \log n$, proving the theorem. The proof of correctness argues that if M never halts on some input 0^n then, in fact, $TAUT \in$ NP contradicting the assumption. We give the formal details.

Fix $n \in \mathcal{N}$. Let $I_n := \{q \in \Sigma^* \mid |q| \leq 2n\}$ be the candidate nondeterministic programs. We describe the machine M:

1. **input** 0^n; $m := 0$;
2. $I_n := \{q \in \Sigma^* \mid |q| \leq 2n\}$; $m := 0$;
3. *Stage* m:
   (a) Spend m steps in computing an initial segment σ of $TAUT$ using brute-force search;
   (b) Spend m steps in simulating $U^t(q, x)$ for $q \in I_n$ and where $x \in \sigma$ is picked
   in lexicographic order;
   (c) If q is found incompatible with σ then eliminate q from I_n;
   (d) Do a prefix search for $F \in \Sigma^m$ with the following property

$$F \in TAUT^{=m} \;:\; \text{all programs in } I_n \text{ reject } F$$

   (e) If the prefix search finds such an F then **output** F and stop. Else goto *Stage* $m + 1$.

Claim. $TAUT \in$ NP if $M(0^n)$ does not terminate for some n.

To see this, suppose $M(0^n)$ does not terminate. Steps (a), (b), and (c) ensure that after some stage $m = m_0$, all $TAUT$-inconsistent programs in I_n are eliminated. Let I denote the remaining TAUT-consistent programs. Now, if the prefix search in Step (d) fails for each $m > m_0$ it follows that for all $m > m_0$

$$F \in TAUT \Leftrightarrow U^t(q, F) \text{ accepts for some } q \in I$$

which implies that $TAUT \in$ NP.

By assumption, therefore, $M(0^n)$ halts for each n. Let F_n be the formula output by the computation of $M(0^n)$. The prefix search ensures that $nic^t(F_n : TAUT) > 2n$. The fact that $M(0^n)$ has output F_n implies that $C(F_n) \leq \log n$. Thus $nic^t(F_n : TAUT) > C(F_n) - c$, for a suitable constant c holds for all n, which completes the proof.

To see the reverse implication, notice that if NP = coNP then for some polynomial time bound t and some constant c we have $nic^t(F : TAUT) \leq c$. Now, since $C(F) > c$ for almost all F, it follows that $nic^t(F : TAUT) \leq C(F)$ for almost all F. □

Notice in the above proof that instead of $C(F)$ we can use polynomial-time bounded Kolmogorov complexity with a Σ_2^p oracle. This is due to the fact that with a Σ_2^p oracle the machine M can do the prefix search for F in polynomial time. Thus, we also have the following (stronger) result. Let SAT_2 denote the Σ_2^p-complete language consisting of all $\exists\forall$ quantified boolean formulas that are true.

Theorem 4. NP $\neq$ coNP *if and only if for every polynomial t there exist a constant $c > 0$, a polynomial t', and infinitely many tautologies F such that* $nic^t(F : TAUT) > C^{SAT_2,t'}(F) - c$.

4 Discussion

The interesting open question is to improve Theorems 3 and 4 by proving that if NP $\neq$ coNP then for every polynomial t there are infinitely many tautologies F, a polynomial t' and a constant c such that $nic^t(F : TAUT) \geq CND^{t'}(F) - c$ holds. The difficulty in adapting the Fortnow-Kummer proof technique to prove this is in carrying out the prefix search in Step 3(d) of the algorithm in the proof of Theorem 3. A sufficiently powerful oracle is required to carry out the prefix search for $F \in TAUT^{=n}$ in polynomial time. More accurately, the prefix search requires a leading existential quantifier followed by a universal quantifier (both for ensuring that $F \in TAUT^{=n}$ and checking that all nondeterministic programs in I_n reject F). Thus, we require a Σ_2^p oracle for the task; it is not clear if an NP oracle suffices.

This brings us to a related issue concerning the deterministic instance complexity result of [5]: In that paper a similar prefix search as mentioned above is carried out. Here the prefix search is for a hard instance $F \in \Sigma^n$ (not necessarily in SAT because, unlike for the *nic* measure, hard *ic* instances can be in SAT or $\overline{SAT}$). At first sight it appears that an NP oracle is required for the prefix search, but Fortnow and Kummer [5] apply the clever trick of using the programs in the set I to simulate this NP oracle and interpreting the program answers in some suitable way. However, the result of the prefix search could be a formula in SAT or $\overline{SAT}$, and there is no way of guaranteeing it to be in SAT, for instance. Thus, the proof in [5] cannot guarantee that there are infinitely many hard instances in SAT (although it certainly follows from their proof that either SAT or $\overline{SAT}$ has infinitely many hard instances). Thus, the following question concerning deterministic instance complexity of SAT remains open.

Problem 2. If P $\neq$ NP, then for each polynomial t there are a polynomial t', a constant c, and infinitely many $F \in SAT$, such that $ic^t(F : SAT) > C^{t'}(F) - c$?

However, guided by an NP oracle, a prefix search for a hard instance $F \in SAT^{=n}$ can be carried out in polynomial time (the additional Fortnow-Kummer trick [5] of replacing the NP oracle with the set of remaining programs does not appear possible if we need to search for $F \in SAT$). This gives us the following result that can be proved on the same lines as Theorems 3 and 4.

Theorem 5. P $\neq$ NP *if and only if for each polynomial t there are polynomial t', a constant c, and infinitely many $F \in SAT$, such that $ic^t(F : SAT) > C^{SAT,t'}(F) - c$.*

The above result is very similar to [11, Theorem 5] which proves a weaker form of the instance complexity conjecture of [7].

In order to put the above result in better perspective we define a one-sided version of deterministic instance complexity. For $A \subseteq \Sigma^*$ a deterministic machine M is *one-sided A-consistent* if $L(M) \subseteq A$.

Definition 4. *For $A \subseteq \Sigma^*$ and a time bound t, the t-time-bounded* one-sided deterministic *instance complexity (ic_+ for short) of x w.r.t. A is defined as:*

$$ic_+^t(x : A) = \min\{\ |M|\ \mid M \text{ is a one-sided } A\text{-consistent } t\text{-time-bounded deterministic Turing machine and } M \text{ decides correctly on } x\ \}.$$

As usual, ic_+ is defined w.r.t. a fixed universal machine. The one-sidedness of the definition implies, as in the case of *nic*, that hard instances w.r.t. the ic_+ measure can only be in A. It follows from the fact that any deterministic machine that accepts $\emptyset$ is consistent with A and decides correctly on all instances $x \notin A$.

Proposition 4. *For any set A there is a constant c such that for all time bounds t and and for all $x \notin A$, $ic_+^t(x : A) \leq c$.*

We say A has hard instances w.r.t. the ic_+ measure if for each polynomial t there are a constant c and polynomial t' such that for infinitely many x, $ic_+^t(x : A) > C^{t'}(x) - c$. The following result can also be proved exactly as Theorem 3.

Theorem 6. P $\neq$ NP *if and only if for each polynomial t there are a polynomial t', constant c, and infinitely many $F \in SAT$, such that $ic_+^t(F : SAT) > C^{SAT,t'}(F) - c$.*

Acknowledgments. The authors thank Uwe Schöning for interesting discussions. The first author is grateful to Uwe Schöning for hosting his visit to Ulm university during which he was partially supported by an Alexander von Humboldt fellowship.

References

1. J. L. Balcázar, J. Díaz, and J. Gabarró, *Structural Complexity I*, EATCS Monographs on Theoretical Computer Science. Springer-Verlag, second edition, 1995.

2. P. Beame and T. Pitassi, *Propositional proof complexity: past, present, and future*, ECCC Report TR98-067, 1998.
3. H. Buhrman and L. Fortnow, *Resource-bounded Kolmogorov complexity revisited*, In Proceedings of the 14th Symposium on Theoretical Aspects of Computer Science, LNCS, 1200: 105–116, Springer, 1997.
4. S. A. Cook and R. A. Reckhow, *The relative efficiency of propositional proof systems*, Journal of Symbolic Logic, 44(1):36–50, 1979.
5. L. Fortnow and M. Kummer, *On resource-bounded instance complexity*, Theoretical Computer Science A, 161:123–140, 1996.
6. J. Kadin, $\mathrm{P}^{\mathrm{NP}[\log n]}$ *and sparse Turing-complete sets for* NP, Journal of Computer and System Sciences, 39(3):282–298, 1989.
7. P. Orponen, K. Ko, U. Schöning, and O. Watanabe, *Instance complexity*, Journal of the ACM, 41:96–121, 1994.
8. J. Krajícek, *Bounded arithmetic, propositional logic, and complexity theory*, Cambridge University Press, 1995.
9. M. Li and P. Vitanyi, *Kolmogorov complexity and applications*, Springer Verlag, 1993.
10. N. Lynch, *On reducibility to complex or sparse sets*, Journal of the ACM, 22:341–345, 1975.
11. M. Mundhenk, *NP-hard sets have many hard instances*, In *Proc. MFCS*, LNCS 1295, 428–437, Springer Verlag, 1997.
12. S. Riis and M. Sitharam, *Generating hard tautologies using predicate logic and the symmetric group*, BRICS Report RS-98-19.
13. U. Schöning, *Complexity cores and hard to prove formulas*, In *Proc. CSL*, LNCS, Springer Verlag, 273–280, 1987.
14. M. Sipser, *A complexity theoretic approach to randomness*, In *Proc. ACM STOC*, 330–335, 1983.

Hard Instances of Hard Problems*

Jack H. Lutz, Vikram Mhetre, and Sridhar Srinivasan

Department of Computer Science
Iowa State University
Ames, Iowa 50011
U.S.A.

Abstract. This paper investigates the instance complexities of problems that are hard or weakly hard for exponential time under polynomial time, many-one reductions. It is shown that almost every instance of almost every problem in exponential time has essentially maximal instance complexity. It follows that every weakly hard problem has a dense set of such maximally hard instances. This extends the theorem, due to Orponen, Ko, Schöning and Watanabe (1994), that every hard problem for exponential time has a dense set of maximally hard instances. Complementing this, it is shown that every hard problem for exponential time also has a dense set of unusually easy instances.

1 Introduction

A problem that is computationally intractable in the worst case may or may not be intractable in the average case. In applications such as cryptography and derandomization, where intractability is a valuable resource, worst-case intractability seldom suffices, average-case intractability often suffices, and almost-everywhere intractability is sometimes required. Implicit in these distinctions is the truism that some instances of a computational problem may be hard while others are easy.

The complexity of an individual instance of a problem cannot be measured simply in terms of the running time required to solve that instance, because any algorithm for the problem can be modified to solve that instance quickly via a look-up table. Orponen, Ko, Schöning, and Watanabe [20] used ideas from algorithmic information theory to circumvent this difficulty, thereby introducing a precise formulation of the complexities of individual instances of computational problems.

Given a decision problem $A \subseteq \{0,1\}^*$, an instance $x \in \{0,1\}^*$, and a time bound $t : \mathbb{N} \to \mathbb{N}$, Orponen, Ko, Schöning, and Watanabe [20] defined the *t-time-bounded instance complexity of x relative to A*, written $ic^t(x : A)$, to be the number of bits in the shortest program π such that π decides x in at most $t(|x|)$ steps and π does not decide any string incorrectly for A. (See sections

* This research was supported in part by National Science Foundation Grant CCR-9610461.

H. Reichel and S. Tison (Eds.): STACS 2000, LNCS 1770, pp. 324–333, 2000.

2 and 3 below for complete definitions of this and other terms used in this introduction.) Instance complexity has now been investigated and applied in a number of papers, including [20, 10, 4, 7, 11, 3, 19], and is discussed at some length in the text [12].

In this paper we investigate the instance complexities of problems that are hard or weakly hard for exponential time under polynomial time, many-one reductions. Our most technical results establish the measure-theoretic abundance of problems for which almost all instances have essentially maximal instance complexities. From these results we derive our main results, which are lower bounds on the instance complexities of weakly hard problems, and we separately establish upper bounds on the instance complexities of hard problems. We now discuss these results in a little more detail.

The *t-time-bounded plain Kolmogorov complexity* of a string x, written $C^t(x)$, is the number of bits in the shortest program π that describes (i.e., prints) x in at most $t(|x|)$ steps. As observed in [20], it is easy to see that, for t' modestly larger than t, $ic^{t'}(x : A)$ cannot be much larger than $C^t(x)$, since a description of x contains all but one bit of the information required for a program to correctly decide whether $x \in A$ and decline to decide all other strings. An instance x thus has *essentially maximal* t-time-bounded instance complexity if $ic^t(x : A)$ is nearly as large as $C^{t'}(x)$, where t' is modestly larger than t. Orponen, Ko, Schöning, and Watanabe [20] established the existence of a problem $A \in \mathrm{E} = \mathrm{DTIME}(2^{\mathrm{linear}})$ for which all but finitely many instances x have instance complexities that are essentially maximal in the sense that $ic^{2^n}(x : A) > C^{t'}(x) - 2\log C^{t'}(x) - c$, where c is a constant and $t'(n) = cn2^{2n} + c$. In contrast with this existence result, we prove in this paper that *almost every* language $A \in \mathrm{E}$ has the property that all but finitely many instances x have essentially maximal instance complexities in the slightly weaker (but still very strong) sense that $ic^{2^n}(x : A) > (1 - \epsilon)C^{t'}(x)$, for any fixed real $\epsilon > 0$, where $t'(n) = 2^{3n}$. We also show that almost every $A \in \mathrm{E}_2 = \mathrm{DTIME}(2^{\mathrm{poly}})$ has the property that all but finitely many instances x satisfy the condition $ic^{2^n}(x : A) > C^{t'}(x) - C^{t'}(x)^{\epsilon}$, for any fixed real $\epsilon > 0$, where $t'(n) = 2^{n^2}$.

Naturally arising problems that are – or are presumed to be – intractable have usually turned out to be complete for NP or some natural complexity class containing NP. The complexities of such problems are thus of greater interest than the complexities of arbitrary problems. The instance complexities of problems that are complete (or just hard) for NP or exponential time under $\leq_m^{\mathrm{P}}$-reductions have consequently been a focus of investigation.

Regarding problems that are $\leq_m^{\mathrm{P}}$-hard for exponential time, Orponen, Ko, Schöning and Watanabe [20] have shown that every such problem H must have an exponentially dense set of instances x that are hard in the sense that for every polynomial t, $ic^t(x : H) > C^{t'}(x) - 2\log C^{t'}(x) - c$, where c is a constant and $t'(n) = cn2^{2n} + c$. Buhrman and Orponen [4] proved a related result stating that, if H is actually $\leq_m^{\mathrm{P}}$-complete for exponential time, then H has a dense set of instances x that are hard in the sense that for every polynomial $t(n) \geq n^2$, $ic^t(x : H) > C^t(x) - c$, where c is a constant.

The main results of this paper show that this phenomenon - a dense set of instances whose complexities are essentially maximal - holds not only for $\leq_m^{\mathrm{P}}$-hard problems for exponential time, but in fact for all *weakly* $\leq_m^{\mathrm{P}}$-hard problems for exponential time (with slight technical modifications in the instance complexity bounds). This is a significant extension of the earlier work because Ambos-Spies, Terwijn and Zheng [2] have shown that almost every problem in E is weakly $\leq_m^{\mathrm{P}}$-hard, but not $\leq_m^{\mathrm{P}}$-hard, for E, and similarly for E_2.

To be precise, we prove that for every weakly $\leq_m^{\mathrm{P}}$-hard language H for E_2 and every $\epsilon > 0$ there exists $\delta > 0$ such that the set of all instances x with $ic^{2^{n^\delta}}(x : H) > (1-\epsilon)C^{2^{4n}}(x)$ is dense, as is the set of all x for which $ic^{2^{n^\delta}}(x : H) > C^{2^{2n^2}}(x) - C^{2^{2n^2}}(x)^\epsilon$. Since Juedes and Lutz [9] have shown that every language that is weakly $\leq_m^{\mathrm{P}}$-hard for E is weakly $\leq_m^{\mathrm{P}}$-hard for E_2 (but not conversely, even for languages in E), our results hold *a fortiori* for problems that are weakly $\leq_m^{\mathrm{P}}$-hard for E.

Regarding problems that are NP-complete (of which we take SAT to be the canonical example), any nontrivial lower bound on instance complexity must be derived from some unproven hypothesis (or entail a proof that P $\neq$ NP) because languages in P have bounded instance complexities [20]. Assuming P $\neq$ NP, Orponen, Ko, Schöning and Watanabe [20] showed that for every polynomial t and constant c, the set $\{x | ic^t(x : SAT) \geq c \log|x|\}$ is infinite. Assuming the hypothesis that nonuniformly secure one-way functions exist (which implies P $\neq$ NP), Ko [10] proved that this set is nonsparse. Assuming E $\neq$ NE (which also implies P$\neq$NP), Orponen, Ko, Schöning and Watanabe [20] showed that SAT has an infinite set of instances of essentially maximal complexity in the sense that for every polynomial t there exist a polynomial t', a constant c, and infinitely many x such that $ic^t(x : SAT) > C^{t'}(SAT) - c$.

The hypothesis that NP does not have p-measure 0, written $\mu_{\mathrm{p}}(\mathrm{NP}) \neq 0$, has been proposed by Lutz. This hypothesis has been shown to imply reasonable answers to many complexity-theoretic questions not known to be resolvable using P $\neq$ NP or other "traditional" complexity-theoretic hypotheses. (Such results are discussed in the surveys [15, 1, 14, 5].) The $\mu_{\mathrm{p}}(\mathrm{NP}) \neq 0$ hypothesis implies the hypothesis E$\neq$NE [16] and is equivalent to the assertion that NP does not have measure 0 in E_2 [2]. Here we note that, if $\mu_{\mathrm{p}}(\mathrm{NP}) \neq 0$, then SAT is weakly $\leq_m^{\mathrm{P}}$-hard for E_2, whence our above-mentioned results imply that SAT has a *dense* set of instances of essentially maximal complexity. That is, if $\mu_{\mathrm{p}}(\mathrm{NP}) \neq 0$, then for every $\epsilon > 0$ there exists $\delta > 0$ such that the set of all x for which $ic^{2^{n^\delta}}(x : SAT) > (1-\epsilon)C^{2^{4n}}(x)$ is dense, as is the set of all x for which $ic^{2^{n^\delta}}(x : SAT) > C^{2^{2n^2}}(x) - C^{2^{2n^2}}(x)^\epsilon$.

In the course of this introduction, we have seen that almost every problem A in exponential time has both of the following properties.

1. All but finitely many instances of A have essentially maximal instance complexity (our abundance results).
2. A is weakly $\leq_m^{\mathrm{P}}$-hard for exponential time [2].

Thus weakly hard problems can have essentially maximal complexity at almost every instance. In contrast, we also show that every problem H that is actually $\leq_m^{\mathrm{P}}$-hard for exponential time must have a dense set of instances x that are unusually easy in the very strong sense that $ic^{2^{6n}}(x : H)$ is bounded above by a constant. Our proof of this fact is based largely on the proof by Juedes and Lutz [8] of an analogous result for complexity cores.

In section 3 below we present a complete definition of time-bounded instance complexity. Section 4 is the main section of this paper. In this section we prove our abundance theorems, derive our lower bounds on the instance complexities of weakly hard problems, and note the consequences for the complexity of SAT if $\mu_{\mathrm{p}}(\mathrm{NP}) \neq 0$. In section 5 we prove that every hard problem for exponential time has a dense set of unusually easy instances.

Due to space limitations, we have omitted all proofs from our paper in these proceedings. An expanded version of this paper, now available at http://www.cs.-iastate.edu/~lutz/papers.html, includes the proofs of our results, along with further discussion of relevant aspects of Kolmogorov complexity, resource-bounded measure, weak completeness, complexity cores, and instance complexity.

2 Notation and Terminology

A language $A \subseteq \{0,1\}^*$ is *sparse* if there exists a polynomial q such that $(\forall n)\ |A \cap \{0,1\}^{\leq n}| \leq q(n)$, and *exponentially dense* (or, simply, *dense*) if there exists a real number $\epsilon > 0$ such that $(\forall^\infty n)\ |A \cap \{0,1\}^{\leq n}| > 2^{n^\epsilon}$.

Our main results involve *resource-bounded measure*, which was developed by Lutz [13, 15]. We refer the interested reader to any of the surveys [14, 1, 5, 17] for discussion of this theory. Recall that a language H is $\leq_m^{\mathrm{P}}$*-hard* for a class $\mathcal{C}$ of languages if $A \leq_m^{\mathrm{P}} H$ for all $A \in \mathcal{C}$, and $\leq_m^{\mathrm{P}}$*-complete* for $\mathcal{C}$ if $H \in \mathcal{C}$ and H is $\leq_m^{\mathrm{P}}$-hard for $\mathcal{C}$. Resource-bounded measure allowed Lutz to generalize these notions as follows. (We write $\mathrm{P}_m(H) = \{A | A \leq_m^{\mathrm{P}} H\}$.)

Definition 2.1. *A language $H \subseteq \{0,1\}^*$ is weakly $\leq_m^{\mathrm{P}}$-hard for* E *(respectively, for* E_2*) if $\mu(\mathrm{P}_m(H)|\mathrm{E}) \neq 0$ (respectively, $\mu(\mathrm{P}_m(H)|\mathrm{E}_2) \neq 0$). A language $H \subseteq \{0,1\}^*$ is weakly $\leq_m^{\mathrm{P}}$-complete for E (respectively, for* E_2*) if $H \in \mathrm{E}$ (respectively, $H \in \mathrm{E}_2$) and H is weakly $\leq_m^{\mathrm{P}}$-hard for E (respectively, for* E_2*).*

It is clear that every $\leq_m^{\mathrm{P}}$-hard language for E is weakly $\leq_m^{\mathrm{P}}$-hard for E, and similarly for E_2.

3 Instance Complexity

Following [20], we define an *interpreter* to be a deterministic Turing machine with a read-only *program tape*, a read-only *input tape*, a write-only *output tape*, and an arbitrary number of read/write *work tapes*, all with alphabet $\{0, 1, \sqcup\}$, where $\sqcup$ is the blank symbol. Given a program $\pi \in \{0,1\}^*$ on the program tape and an input $x \in \{0,1\}^*$ on the input tape, an interpreter M may eventually halt in an *accepting* configuration, a *rejecting* configuration, an *undecided* configuration, or an *output* configuration, or it may fail to halt. If M halts in an accepting configuration, we say that π *accepts* x on M, and we write $M(\pi, x) = 1$. If M halts in a rejecting configuration, we say that π *rejects* x on M, and we write $M(\pi, x) = 0$. In either of these two cases, we say that π *decides* x on M. If M halts in an undecided configuration, or if M fails to halt, we say that π *fails to decide* x on M, and we write $M(\pi, x) = \perp$. If M halts in an output configuration with output $y \in \{0,1\}^*$ on the output tape, we write $M(\pi, x) = y$. (If y is 0 or 1, the context will always make it clear whether "$M(\pi, x) = y$" refers to a decision or an output.)

We write $time_M(\pi, x)$ for the running time of M with program π and input x. If $M(\pi, x) = \perp$, we stipulate that $time_M(\pi, x) = \infty$.

A program π is *consistent* with a language $A \subseteq \{0,1\}^*$ relative to an interpreter M if for all $x \in \{0,1\}^*$, $M(\pi, x) \in \{[\![x \in A]\!], \perp\}$, i.e., π either decides x correctly for A or else fails to decide x.

We now recall the definition of time-bounded instance complexity, which is the main topic of this paper.

Definition 3.1. *(Orponen, Ko, Schöning and Watanabe [20]) Let M be an interpreter, $t : \mathbb{N} \to \mathbb{N}$, $A \subseteq \{0,1\}^*$, and $x \in \{0,1\}^*$. The* t-time-bounded instance complexity *of x with respect to A given M is*

$$ic^t_M(x : A) = min\{|\pi| \mid \pi \text{ is consistent with } A \text{ relative to } M \text{ and } time_M(\pi, x) \le t(|x|)\},$$

where $min\ \phi = \infty$.

Thus $ic^t_M(x : A)$ is the minimum number of bits required for a program π to decide x correctly for A on M, subject to the constraints that π is consistent with A relative to M and $M(\pi, x)$ does not run for more than $t(|x|)$ steps.

Note. Our definition of $ic^t_M(x : A)$ differs from that in [20] in that we do not require $M(\pi, y)$ to halt within $t(|y|)$ steps – or even to halt at all – for $y \neq x$. In our complexity-theoretic setting, with time-constructible functions t, this difference is technical and minor (at most a constant number of bits and a $\log t$ factor in the time bound), and it simplifies some technical results. In other settings, such as that of the time-unbounded instance complexity conjecture [20], the halting behavior for $y \neq x$ is a more critical issue.

4 Hard Instances

In this section we prove our main results. We show that almost every instance of almost every problem in E has essentially maximal instance complexity, and similarly for E_2. Using this, we show that every problem that is weakly $\le^P_m$-hard for either of these classes has an exponentially dense set of such maximally hard instances. We begin with our abundance theorem in E. In contrast with Theorem 5.11 of Orponen, Ko, Schöning and Watanabe [20], which asserts the *existence* of a language in E with essentially maximal instance complexity, the following result says that *almost every* language in E has this property, albeit with a slightly weaker interpretation of "essentially maximal".

Theorem 4.1. *For all $c \in \mathbb{Z}^+$ and $\epsilon > 0$, the set*

$$X(c,\epsilon) = \{A|(\forall^\infty x) ic^{2^{cn}}(x : A) > (1-\epsilon) C^{2^{(c+2)n}}(x)\}$$

has p-measure 1, hence measure 1 in E.

Theorem 4.1 has the following analog in E_2.

Theorem 4.2. *For all $c \in \mathbb{Z}^+$ and $\epsilon > 0$, the set*

$$X_2(c,\epsilon) = \{A|(\forall^\infty x) ic^{2^{n^c}}(x : A) > C^{2^{n^{(c+1)}}}(x) - C^{2^{n^{(c+1)}}}(x)^\epsilon\}$$

has p_2-measure 1, hence measure 1 in E_2.

Before proceeding, we note that Theorems 4.1 and 4.2 imply the following known fact, which was proven independently by Juedes and Lutz [8] (as stated) and Mayordomo [18] (in terms of bi-immunity).

Corollary 4.3. *(Juedes and Lutz [8], Mayordomo [18]) Let $c \in \mathbb{Z}^+$.*

1. *Almost every language in E has $\{0,1\}^*$ as a DTIME(2^{cn})-complexity core.*
2. *Almost every language in E_2 has $\{0,1\}^*$as a DTIME(2^{n^c})-complexity core.*

Our next task is to use Theorems 4.1 and 4.2 to prove that every weakly $\leq^{\mathrm{P}}_m$-hard language for exponential time has a dense set of very hard instances. For this purpose we need a few basic facts about the behavior of polynomial-time reductions in connection with time-bounded Kolmogorov complexity, time-bounded instance complexity, and density.

The *data processing inequality* of classical information theory [6] says that the entropy (Shannon information content) of a source cannot be increased by performing a deterministic computation on its output. The analogous data processing inequality for plain Kolmogorov complexity [12] says that if f is a computable function, then $C(f(x))$, which is the algorithmic information content of $f(x)$, cannot exceed $C(x)$, the algorithmic information content of x, by more than a constant number of bits. The following lemma is a time-bounded version of this fact. It is essentially well-known, though perhaps not in precisely this form.

Lemma 4.4. *(data processing inequality) For each $f \in \mathrm{PF}$, there exist a polynomial q and a constant $c \in \mathbb{N}$ such that for all $x \in \{0,1\}^*$ and all nondecreasing $t : \mathbb{N} \to \mathbb{N}$,*

$$|f(x)| \geq |x| \Rightarrow C^{t''}(f(x)) \leq C^t(x) + c,$$

where $t''(n) = ct'(n)\log(t'(n)) + c$ and $t'(n) = t(n) + q(n)$.

Our next lemma is a straightforward extension of Proposition 3.5 of [20].

Lemma 4.5. *For each $f \in \mathrm{PF}$ there exist a polynomial q and a constant $c \in \mathbb{N}$ such that for all $A \subseteq \{0,1\}^*$, $x \in \{0,1\}^*$, and nondecreasing $t : \mathbb{N} \to \mathbb{N}$,*

$$ic^{t''}(x : f^{-1}(A)) \leq ic^{t}(f(x) : A) + c,$$

where $t''(n) = ct'(n)\log(t'(n)) + c$ and $t'(n) = q(n) + t(q(n))$.

The following consequence of Lemma 4.5 is especially useful here.

Corollary 4.6. *For each $f \in \mathrm{PF}$ there exist $\delta > 0$ and $c \in \mathbb{N}$ such that for all but finitely many $x \in \{0,1\}^*$, for all $A \subseteq \{0,1\}^*$,*

$$ic^{2^{n^\delta}}(f(x) : A) \geq ic^{2^n}(x : f^{-1}(A)) - c.$$

Juedes and Lutz [8] introduced the following useful notation. The *nonreduced image* of a language $S \subseteq \{0,1\}^*$ under a function $f : \{0,1\}^* \to \{0,1\}^*$ is the language

$$f^{\geq}(S) = \{f(x) | x \in S \text{ and } |f(x)| \geq |x|\}.$$

Lemma 4.7. *(Juedes and Lutz [8]) If $f \in \mathrm{PF}$ is one-to-one a.e. and $S \subseteq \{0,1\}^*$ is cofinite, then $f^{\geq}(S)$ is dense.*

We now prove that every weakly $\leq^{\mathrm{P}}_m$-hard language for exponential time has a dense set of very hard instances. Orponen, Ko, Schöning, and Watanabe [20] have shown that every $\leq^{\mathrm{P}}_m$-hard language for exponential time has a dense set of very hard instances, and Buhrman and Orponen [4] have proven a similar result with improved time bounds and density for languages that are $\leq^{\mathrm{P}}_m$-complete for exponential time. Theorems 4.8 and 4.9 below can be regarded as extending this phenomenon (with some modification in the precise bounds) to all *weakly* $\leq^{\mathrm{P}}_m$-hard languages for exponential time.

Juedes and Lutz [9] have proven that every weakly $\leq^{\mathrm{P}}_m$-hard language for E is weakly $\leq^{\mathrm{P}}_m$-hard for E_2, but that the converse fails, even for languages in E. We thus state our results in terms of weakly $\leq^{\mathrm{P}}_m$-hard languages for E_2, noting that they hold *a fortiori* for languages that are weakly $\leq^{\mathrm{P}}_m$-hard for E.

Theorem 4.8. *If H is weakly $\leq^{\mathrm{P}}_m$-hard for E_2, then for every $\epsilon > 0$ there exists $\delta > 0$ such that the set*

$$HI^{\epsilon,\delta}(H) = \{x | ic^{2^{n^\delta}}(x : H) > (1-\epsilon)C^{2^{4n}}(x)\}$$

is dense.

Using Theorem 4.1 in place of Theorem 4.2, we prove the following similar result.

Theorem 4.9. *If H is weakly $\leq_m^{\mathrm{P}}$-hard for E_2, then for every $\epsilon > 0$ there exists $\delta > 0$ such that the set*

$$HI_2^{\epsilon,\delta}(H) = \{x | ic^{2^{n^\delta}}(x : H) > C^{2^{2n^2}}(x) - C^{2^{2n^2}}(x)^\epsilon\}$$

is dense.

It can be shown that Theorem 4.9 implies (and is much stronger than) the following known result.

Corollary 4.10. *(Juedes and Lutz [8]). If H is weakly $\leq_m^{\mathrm{P}}$-hard for E_2, then H has a dense exponential complexity core.*

Theorems 4.8 and 4.9 are incomparable in strength because Theorem 4.8 gives a tighter time bound on the plain Kolmogorov complexity, while Theorem 4.9 gives a tighter bound on the closeness of the time-bounded instance complexity to the time-bounded plain Kolmogorov complexity. For most strings x, $C^t(x)$ and $C(x)$ are both very close to $|x|$, so the time bound on $C^t(x)$ is often of secondary significance. Thus for many purposes, the following simple consequence of Theorem 4.9 suffices.

Corollary 4.11. *If H is weakly $\leq_m^{\mathrm{P}}$-hard for E or E_2, then for every $\epsilon > 0$ there exists $\delta > 0$ such that the set*

$$HI_0^{\epsilon,\delta}(H) = \{x | ic^{2^{n^\delta}}(x : H) > C(x) - C(x)^\epsilon\}$$

is dense.

We conclude this section with a discussion of the instance complexities of NP-complete problems. For simplicity of exposition we focus on SAT, but the entire discussion extends routinely to other NP-complete problems.

We start with three known facts. The first says that the hypothesis $\mathrm{P} \neq \mathrm{NP}$ implies a lower bound on the instance complexity of SAT.

Theorem 4.12. *(Orponen, Ko, Schöning and Watanabe [20]) If $\mathrm{P} \neq \mathrm{NP}$, then for every polynomial t and constant $c \in \mathbb{N}$, the set*

$$\{x | ic^t(x : SAT) > c \log |x|\}$$

is infinite.

Each of the next two facts derives a stronger conclusion than Theorem 4.12 from a stronger hypothesis.

Theorem 4.13. *(Ko [10]) If nonuniformly secure one-way functions exist, then for every polynomial t and constant $c \in \mathbb{N}$, the set*

$$\{x | ic^t(x : SAT) > c \log |x|\}$$

is nonsparse.

Theorem 4.14. *(Orponen, Ko, Schöning and Watanabe [20]) If* $\mathrm{E} \neq \mathrm{NE}$*, then for every polynomial* t *there exist a polynomial* t' *and a constant* $c \in \mathbb{N}$ *such that the set*

$$\{x | ic^t(x : SAT) > C^{t'}(x) - c\}$$

is infinite.

The following theorem derives a strong lower bound on the instance complexity of SAT from the hypothesis that $\mu_{\mathrm{p}}(\mathrm{NP}) \neq 0$. This hypothesis, which was proposed by Lutz, has been proven to have many reasonable consequences [15, 1, 14, 5]. The $\mu_{\mathrm{p}}(\mathrm{NP}) \neq 0$ hypothesis implies $\mathrm{E} \neq \mathrm{NE}$ [16] and is equivalent to the assertion that NP does not have measure 0 in E_2 [2]. Its relationship to the hypothesis of Theorem 4.13 is an open question.

Theorem 4.15. *If* $\mu_{\mathrm{p}}(\mathrm{NP}) \neq 0$*, then for every* $\epsilon > 0$ *there exists* $\delta > 0$ *such that the sets*

$$HI_1^{\epsilon,\delta}(SAT) = \{x | ic^{2^{n^\delta}}(x : SAT) > (1-\epsilon)C^{2^{4n}}(x)\},$$
$$HI_2^{\epsilon,\delta}(SAT) = \{x | ic^{2^{n^\delta}}(x : SAT) > C^{2^{2n^2}}(x) - C^{2^{2n^2}}(x)^\epsilon\}$$

are dense.

5 Easy Instances

In this brief section, we note that languages that are $\leq_m^{\mathrm{P}}$-hard for exponential time have instance complexities that are *unusually low* in the sense that they obey an upper bound that is violated by almost every language in exponential time. Our proof is based on the following known result.

Theorem 5.1. *(Juedes and Lutz [8]) For every* $\leq_m^{\mathrm{P}}$*-hard language* H *for E, there exist* $B, D \in \mathrm{DTIME}(2^{4n})$ *such that* D *is dense and* $B = H \cap D$.

The following theorem gives an upper bound on the instance complexities of hard problems for exponential time. It says that every such problem has a *dense set of* (relatively) *easy instances.*

Theorem 5.2. *For every* $\leq_m^{\mathrm{P}}$*-hard language* H *for E there is a constant* $c \in \mathbb{N}$ *such that the set*

$$EI_c(H) = \{x | ic^{2^{6n}}(x : H) \leq c\}$$

is dense.

By Theorem 4.1, almost every language in exponential time violates the upper bound given by Theorem 5.2. Thus these two results together imply the known fact [8] that the set of $\leq_m^{\mathrm{P}}$-hard languages for exponential time has p-measure 0. It should also be noted that Ambos-Spies, Terwijn and Zheng [2] have shown that almost every language in E is weakly $\leq_m^{\mathrm{P}}$-hard for E. It follows by Theorem 4.1 that almost every language in E is weakly $\leq_m^{\mathrm{P}}$-hard for E and violates the instance complexity upper bound given by Theorem 5.2. Thus Theorem 5.2 cannot be extended to the weakly $\leq_m^{\mathrm{P}}$-hard problems for E.

References

[1] K. Ambos-Spies and E. Mayordomo. Resource-bounded measure and randomness. In A. Sorbi, editor, *Complexity, Logic and Recursion Theory*, Lecture Notes in Pure and Applied Mathematics, pages 1–47. Marcel Dekker, New York, N.Y., 1997.

[2] K. Ambos-Spies, S. A. Terwijn, and X. Zheng. Resource bounded randomness and weakly complete problems. *Theoretical Computer Science*, 172:195–207, 1997.

[3] H. Buhrman and E. Mayordomo. An excursion to the Kolmogorov random strings. *Journal of Computer and System Sciences*, 54:393–399, 1997.

[4] H. Buhrman and P. Orponen. Random strings make hard instances. *Journal of Computer and System Sciences*, 53:261–266, 1996.

[5] H. Buhrman and L. Torenvliet. Complete sets and structure in subrecursive classes. In *Proceedings of Logic Colloquium '96*, pages 45–78. Springer-Verlag, 1998.

[6] T. M. Cover and J. A. Thomas. *Elements of Information Theory.* John Wiley & Sons, Inc., New York, N.Y., 1991.

[7] L. Fortnow and M. Kummer. On resource-bounded instance complexity. *Theoretical Computer Science*, 161:123–140, 1996.

[8] D. W. Juedes and J. H. Lutz. The complexity and distribution of hard problems. *SIAM Journal on Computing*, 24(2):279–295, 1995.

[9] D. W. Juedes and J. H. Lutz. Weak completeness in E and E_2. *Theoretical Computer Science*, 143:149–158, 1995.

[10] K. Ko. A note on the instance complexity of pseudorandom sets. In *Proceedings of the Seventh Annual Structure in Complexity Theory Conference*, pages 327–337. IEEE Comput. Soc. Press, 1992.

[11] Martin Kummer. On the complexity of random strings. In *13th Annual Symposium on Theoretical Aspects of Computer Science*, pages 25–36. Springer, 1996.

[12] M. Li and P. M. B. Vitányi. *An Introduction to Kolmogorov Complexity and its Applications.* Springer-Verlag, Berlin, 1997. Second Edition.

[13] J. H. Lutz. Almost everywhere high nonuniform complexity. *Journal of Computer and System Sciences*, 44:220–258, 1992.

[14] J. H. Lutz. The quantitative structure of exponential time. In L.A. Hemaspaandra and A.L. Selman, editors, *Complexity Theory Retrospective II*, pages 225–254. Springer-Verlag, 1997.

[15] J. H. Lutz. Resource-bounded measure. In *Proceedings of the 13th IEEE Conference on Computational Complexity*, pages 236–248, New York, 1998. IEEE.

[16] J. H. Lutz and E. Mayordomo. Cook versus Karp-Levin: Separating completeness notions if NP is not small. *Theoretical Computer Science*, 164:141–163, 1996.

[17] J. H. Lutz and E. Mayordomo. Twelve problems in resource-bounded measure. *Bulletin of the European Association for Theoretical Computer Science*, 68:64–80, 1999.

[18] E. Mayordomo. Almost every set in exponential time is P-bi-immune. *Theoretical Computer Science*, 136(2):487–506, 1994.

[19] M. Mundhenk. NP-hard sets have many hard instances. In *Mathematical foundations of computer science 1997*, pages 428–437. Springer-Verlag, 1997.

[20] P. Orponen, K. Ko, U. Schöning, and O. Watanabe. Instance complexity. *Journal of the Association of Computing Machinery*, 41:96–121, 1994.

Simulation and Bisimulation over One-Counter Processes

Petr Jančar*[1], Antonín Kučera**[2], and Faron Moller***[3]

[1] Technical University Ostrava, Czech Republic (Petr.Jancar@vsb.cz)
[2] Faculty of Informatics MU, Czech Republic (tony@fi.muni.cz)
[3] Uppsala University, Sweden (fm@csd.uu.se)

Abstract. We show an effective construction of (a periodicity description of) the maximal simulation relation for a given one-counter net. Then we demonstrate how to reduce *simulation* problems over one-counter nets to analogous *bisimulation* problems over one-counter automata. We use this to demonstrate the decidability of various problems, specifically testing regularity and strong regularity of one-counter nets with respect to simulation equivalence, and testing simulation equivalence between a one-counter net and a deterministic pushdown automaton. Various obvious generalisations of these problems are known to be undecidable.

1 Introduction

In concurrency theory, a ***process*** is typically defined to be a state in a ***transition system***, which is a triple $T = \langle S, \Sigma, \rightarrow \rangle$ where S is a set of ***states***, Σ is a set of ***actions*** (assumed to be *finite* in this paper) and $\rightarrow \subseteq S \times \Sigma \times S$ is a ***transition relation***. We write $s \xrightarrow{a} t$ instead of $\langle s, a, t \rangle \in \rightarrow$, and we extend this notation in the natural way to elements of Σ^*. A state t is ***reachable*** from a state s iff $s \xrightarrow{w} t$ for some $w \in \Sigma^*$. T is ***image-finite*** iff for all $s \in S$ and $a \in \Sigma$ the set $\{t : s \xrightarrow{a} t\}$ is finite; T is ***deterministic*** if each such set is of size at most 1.

In this paper, we consider such processes generated by ***one-counter automata***, nondeterministic finite-state automata operating on a single counter variable ranging over the set $\mathbb{N}$ of nonnegative integers. Formally this is a tuple $M = \langle Q, \Sigma, \delta^=, \delta^> \rangle$ where Q is a finite set of ***control states***, Σ is a finite set of ***actions***, and $\delta^= : Q \times \Sigma \rightarrow \mathcal{P}(Q \times \{0, 1\})$, $\delta^> : Q \times \Sigma \rightarrow \mathcal{P}(Q \times \{-1, 0, 1\})$ are ***transition functions*** (where $\mathcal{P}(A)$ denotes the set of subsets of A). $\delta^=$ represents the transitions which are enabled when the counter value is zero, and $\delta^>$ represents the transitions which are enabled when the counter value is positive. M is a ***one-counter net*** iff $\forall q \in Q, \forall a \in \Sigma : \delta^=(q, a) \subseteq \delta^>(q, a)$. To M we associate the (image-finite) transition system $T_M = \langle S, \Sigma, \rightarrow \rangle$, where $S = \{ p(n) : p \in Q, n \in \mathbb{N}\}$ and $\rightarrow$ is defined as follows:

* Partially supported by the Grant Agency of the Czech Republic, grants No. 201/97/0456 and No. 201/00/0400.

** Supported by a Research Fellowship granted by the Alexander von Humboldt Foundation and by a Post-Doc grant GA ČR No. 201/98/P046.

*** Partially supported by TFR grant No. 221-98-103 and Np. 201/97/275.

H. Reichel and S. Tison (Eds.): STACS 2000, LNCS 1770, pp. 334–345, 2000.

$$p(n) \xrightarrow{a} p'(n+i) \quad \text{iff} \quad \begin{cases} n{=}0, \text{ and } (p',i) \in \delta^{=}(p,a); \text{ or} \\ n{>}0, \text{ and } (p',i) \in \delta^{>}(p,a). \end{cases}$$

Note that any transition increments, decrements, or leaves unchanged the counter value; and a decrementing transition is only possible if the counter value is positive. Also observe that when $n{>}0$ the transitions of $p(n)$ do not depend on the actual value of n. Finally, note that a one-counter *net* can in a sense test if its counter is nonzero (that is, it can perform some transitions only on the proviso that its counter is nonzero), but it cannot test in any sense if its counter is zero.

As an example, we might take $Q = \{p\}$, $\Sigma = \{a, z\}$, and take the only non-empty transition function values to be $\delta^{>}(p,a) = \{(p,+1),(p,-1)\}$, $\delta^{=}(p,a) = \{(p,+1)\}$, and $\delta^{=}(p,z) = \{(p,0)\}$. This one-counter automaton gives rise to the infinite-state transition system depicted in Fig. 1; if we eliminate the z-action, then this would be a one-counter net. The class of transition systems which are generated by one-counter nets is the same (up to isomorphism) as that generated by the class of labelled Petri nets with (at most) one unbounded place. The class of transition systems which are generated by one-counter automata is the same (up to isomorphism) as that generated by the class of realtime pushdown automata with a single stack symbol (apart from a special bottom-of-stack marker).

Given a transition system $T = \langle S, \Sigma, \rightarrow \rangle$, a ***simulation*** is a binary relation $\mathcal{R} \subseteq S \times S$ satisfying: whenever $\langle s,t \rangle \in \mathcal{R}$, if $s \xrightarrow{a} s'$ then $t \xrightarrow{a} t'$ for some t' with $\langle s',t' \rangle \in \mathcal{R}$. s is ***simulated*** by t, written $s \preccurlyeq t$, iff $\langle s,t \rangle \in \mathcal{R}$ for some simulation $\mathcal{R}$; and s and t are ***simulation equivalent***, written $s \preccurlyeq\succcurlyeq t$, iff $s \preccurlyeq t$ and $t \preccurlyeq s$. (The relation $\preccurlyeq$, being the union of all simulation relations, is in fact the maximal simulation relation.) A ***bisimulation*** is a symmetric simulation relation, and s and t are ***bisimulation equivalent***, or ***bisimilar***, written $s \sim t$, if they are related by a bisimulation. Simulations and bisimulations can also be used to relate states of *different* transition systems; formally, we can consider two transition systems to be a single one by taking the disjoint union of their state sets.

There are various other equivalences over processes which have been studied within the framework of concurrency theory; an overview and comparison of these is presented in [14]. Each has its specific advantages and disadvantages, and consequently none is universally accepted as the "best" one, although it seems that simulation and bisimulation equivalences are of particular importance as their accompanying theory has been intensively developed. Bisimilarity is especially mathematically tractable, having the best polynomial-time algorithms over finite-state transition systems (while all language-based equivalences by comparison are PSPACE-complete), and the only one which is decidable for various classes of infinite-state systems such as context-free processes and commutative context-free processes (see [12] for a survey of such results).

Let s be a state of a transition system T and $\approx$ be an equivalence over the class of all processes (that is, all states of all transition systems). s is ***$\approx$-regular***, or ***regular w.r.t.*** $\approx$, iff $s \approx f$ for some state f of a finite-state transition system; and s is ***strongly $\approx$-regular***, or ***strongly regular w.r.t.*** $\approx$, iff only finitely many states, up to $\approx$, are reachable from s. For bisimilarity, these two concepts coincide, but this is not true in general for other equivalences. For example, the state $p(0)$ of the infinite-state transition system depicted in Fig. 1 is $\preccurlyeq\succcurlyeq$-regular, being simulation equivalent to the state U of the

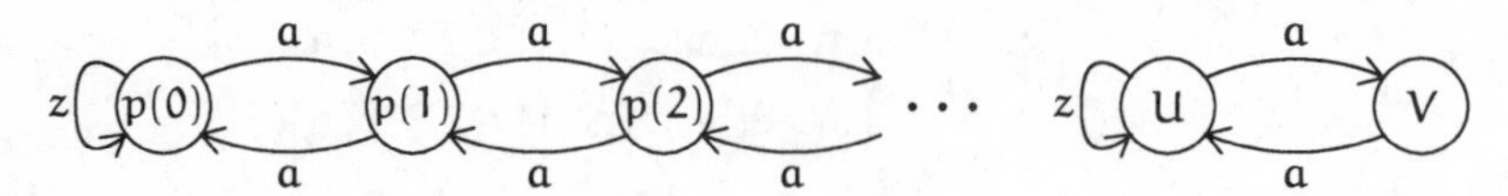

Fig. 1. A one-counter automata process and a simulation-equivalent finite-state process.

depicted finite-state system. However, it is not strongly $\preccurlyeq\succcurlyeq$-regular (nor $\sim$-regular) as $p(i) \not\preccurlyeq p(j)$ whenever $i < j$. The conditions of regularity and strong regularity say that a process can in some sense be finitely represented (up to the equivalence): in the first case there is an equivalent finite-state process; and in the second case the quotient of its state-space under the equivalence is finite. As all "reasonable" process equivalences are preserved under their respective quotients [8] (that is, each state is equivalent to its equivalence class in the automaton produced by collapsing equivalent states [2]), strong regularity in fact guarantees the existence of a finite-state process whose state-space is the same (up to the equivalence); this process provides a more robust description of the original process as it preserves strictly more logical properties than a process which is just equivalent [9].

Finite descriptions of infinite-state processes are important from the point of view of automatic formal verification. Verification tools typically work only for finite-state systems, and the types of systems which they analyze, such as protocols, are typically *semantically* finite-state. However, these systems are often expressed *syntactically* as infinite-state systems, for example maintaining a count of how many unacknowledged messages have been sent, so it is advantageous to develop algorithms which replace infinite-state processes with equivalent finite-state systems (when they exist). Examples of such algorithms appear in [2,4,5,8,11]

In Section 2 we show an effective construction of (a periodicity description of) the maximal simulation relation for a given one-counter net. Then, in Section 3, we study the connection between simulation and bisimulation relations, and demonstrate the decidability of the $\preccurlyeq\succcurlyeq$-regularity and strong $\preccurlyeq\succcurlyeq$-regularity problems for *one-counter nets*, a restricted form of Petri nets; the $\preccurlyeq\succcurlyeq$-regularity problem is reduced to the $\sim$-regularity problem for the more general class of *one-counter automata*, which is known to be decidable [3]. Note that the $\preccurlyeq\succcurlyeq$-regularity problem is known to be undecidable for general Petri nets [5] and an incomparable class of PA processes [10]. Finally, we demonstrate how to decide simulation equivalence between (a process related to) a one-counter net and (a process related to) a deterministic pushdown automaton. Here note that simulation equivalence between a (nondeterministic) one-counter automaton and a deterministic one-counter automaton (i.e., a special deterministic pushdown automaton) can be demonstrated to be undecidable [7].

2 Simulation on One-Counter Nets

In this section we fix a one-counter net with control state set Q, and present an algorithm which constructs a (simple) description of the set

$$\mathcal{S} \;=\; \{\langle p(m), q(n)\rangle \;:\; p, q \in Q,\; m, n \in \mathbb{N},\; p(m) \preccurlyeq q(n)\}$$

i.e., the maximal simulation relation on the transition system associated to the net. $\mathcal{S}$ can be viewed as a collection of $|Q|^2$ subsets of $\mathbb{N}\times\mathbb{N}$: to each $p, q \in Q$ we associate $\mathcal{S}_{\langle p,q\rangle} = \{\langle m,n\rangle : p(m) \preccurlyeq q(n)\}$. Observe that if $p(m) \preccurlyeq q(n)$ then $p(m') \preccurlyeq q(n')$ for all $m' \leq m$ and $n' \geq n$ since the set $\{\langle p(m'), q(n')\rangle : p(m) \preccurlyeq q(n)$ for some $m \geq m', n \leq n'\}$ is a simulation relation.

By a ***colouring*** we mean a function $\mathbb{C} : (Q\times Q) \to (\mathbb{N}\times\mathbb{N}) \to \{\text{black, white}\}$, where we write the function applications as $\mathbb{C}_{\langle p,q\rangle}(m,n)$. We further stipulate that a colouring must satisfy the following monotonicity condition: if $\mathbb{C}_{\langle p,q\rangle}(m,n)$=black then $\mathbb{C}_{\langle p,q\rangle}(m',n')$=black for all $m' \leq m$ and $n' \geq n$. With this proviso, each $\mathbb{C}_{\langle p,q\rangle}$ is determined by the ***frontier function*** $f^{\mathbb{C}}_{\langle p,q\rangle} : \mathbb{N} \to \mathbb{N} \cup \{\omega\}$ defined by: $f^{\mathbb{C}}_{\langle p,q\rangle}(n) = \min\{m : \mathbb{C}_{\langle p,q\rangle}(m,n)=\text{white}\}$; we put $f^{\mathbb{C}}_{\langle p,q\rangle}(n)=\omega$ if $\mathbb{C}_{\langle p,q\rangle}(m,n)$=black for all m. Note that this function is nondecreasing, i.e., each ***step*** $f^{\mathbb{C}}_{\langle p,q\rangle}(n+1) - f^{\mathbb{C}}_{\langle p,q\rangle}(n)$ is nonnegative. When $f_{\langle p,q\rangle}(n) \in \mathbb{N}$, we call the pair $\langle f_{\langle p,q\rangle}(n), n\rangle$ a ***frontier point*** and the set of all frontier points constitutes the ***frontier*** (in $\mathbb{C}_{\langle p,q\rangle}$).

We use $\mathbb{G}$ to denote the following distinguished colouring:

$$\mathbb{G}_{\langle p,q\rangle}(m,n) = \begin{cases} \text{black, if } p(m) \preccurlyeq q(n); \\ \text{white, if } p(m) \not\preccurlyeq q(n). \end{cases}$$

The observation about $\mathcal{S}$ from above confirms that this is a valid colouring, i.e., that the required monotonicity condition holds. We use $f_{\langle p,q\rangle}$ to denote the frontier function of $\mathbb{G}_{\langle p,q\rangle}$, and we understand the terms *frontier function* and *frontier* to be related to $\mathbb{G}$ when not specified otherwise.

The following "Belt Theorem" gives a crucial fact about frontiers; by a ***belt*** we mean the set of points of the (first quadrant of the) plane lying between two parallel lines.

Belt Theorem. Every frontier lies within a belt with nonnegative rational or infinite slope.

This theorem is central for the decidability of simulation over one-counter nets. It was proven in [6] by a combination of short and intuitive arguments; the theorem is also present (though not so explicitly) in [1] but the proof outlined there is formidable.

Note that if, for a frontier function f, $f(n)=\omega$ for some n then the respective frontier is finite and lies within a horizontal belt (i.e., with slope 0). Otherwise f (as a function $\mathbb{N} \to \mathbb{N}$) is almost linear, though its steps $\big(f(n+1)-f(n)\big)$ need not be constant. Nevertheless, we shall show that f is ***periodic***, i.e., from some n_0 a finite sequence of steps is repeated forever; and moreover, its ***periodicity description***—i.e., n_0, the sequence of steps to be repeated, and the values of $f(n)$ for all $n \leq n_0$—can be effectively computed, yielding the simple description of the set $\mathcal{S}$. (Note that the decision algorithms in both [1] and [6] only approximate the set $\mathcal{S}$, or equivalently the colouring $\mathbb{G}$, to a sufficient level to answer the relevant question; effective constructability of the functions $f_{\langle p,q\rangle}$ does not follow from there.)

We now show how the frontier functions $f_{\langle p,q\rangle}$ can be stepwise approximated. First we say that a point $\langle m,n\rangle$ (in $\mathbb{N}\times\mathbb{N}$) is ***locally correct in a colouring*** $\mathbb{C}$ iff the following holds for all $p, q \in Q$: if $\mathbb{C}_{\langle p,q\rangle}(m,n)$=black and $p(m) \xrightarrow{a} p'(m')$ then there is $q(n) \xrightarrow{a} q'(n')$ with $\mathbb{C}_{\langle p',q'\rangle}(m',n')$=black. Note that the local correctness of a point $\langle m, n\rangle$ depends only on the restriction of $\mathbb{C}$ to the ***neighbourhood*** of $\langle m, n\rangle$, i.e., to the set $\{\langle m', n'\rangle : |m'-m|\leq 1, |n'-n|\leq 1\}$; this follows from the fact that a transition in a one-counter net can change the counter value by at most 1. We say that $\mathbb{C}$ is k-***admissible***, where $k \in \mathbb{N} \cup \{\omega\}$, iff each point $\langle m, n\rangle$ with $m, n < k$ is locally correct in $\mathbb{C}$. In particular, note that $\mathbb{G}$ is ω-admissible.

The function $\mathbb{G}^k : (Q\times Q) \to (\mathbb{N}\times\mathbb{N})\to\{\text{black, white}\}$ defined by

$$\mathbb{G}^k_{\langle p,q\rangle}(m, n) = \text{black iff } \mathbb{C}_{\langle p,q\rangle}(m, n) = \text{black for some } k\text{-admissible colouring } \mathbb{C}$$

is easily seen to be a k-admissible colouring, and is in fact the ***maximal*** (i.e., *maximally-black*) k-***admissible*** colouring; furthermore, the maximal ω-admissible colouring $\mathbb{G}^\omega$ is clearly $\mathbb{G}$. For $k \in \mathbb{N}$, we denote the frontier function of $\mathbb{G}^k_{\langle p,q\rangle}$ by $f^k_{\langle p,q\rangle}$, and note that the range of $f^k_{\langle p,q\rangle}$ is $\{0, 1, \ldots, k-1\} \cup \{\omega\}$ and that $f^k_{\langle p,q\rangle}(n) = \omega$ for all $n\geq k$. The description of each function $f^k_{\langle p,q\rangle}$, i.e., (a table of) its values for $0, 1, \ldots, k-1$, is effectively computable, for example, by an exhaustive search. As $\mathbb{G}^k$ is i-admissible for any $i\leq k$, we have, for each p, q, $f^0_{\langle p,q\rangle}\geq f^1_{\langle p,q\rangle}\geq f^2_{\langle p,q\rangle}\geq \cdots \geq f_{\langle p,q\rangle}$ (where $f'\geq f''$ means $\forall n \in \mathbb{N} : f'(n) \geq f''(n)$). Therefore the function $g_{\langle p,q\rangle} = \lim_{n\to\infty} f^n_{\langle p,q\rangle}$ is well-defined, and $g_{\langle p,q\rangle}\geq f_{\langle p,q\rangle}$. But since the colouring defined by these limit functions $g_{\langle p,q\rangle}$ (as the frontier functions) is ω-admissible (recall the "locality" of the local correctness condition), and $\mathbb{G}$ is the *maximal* ω-admissible colouring, we have $g_{\langle p,q\rangle}\leq f_{\langle p,q\rangle}$. Thus $g_{\langle p,q\rangle}=f_{\langle p,q\rangle}$, and therefore we get the following.

Lemma 1. *For each* $n \in \mathbb{N}$ *there is* $k \geq n$ *such that each* $f^k_{\langle p,q\rangle}$ *coincides with* $f_{\langle p,q\rangle}$ *on the set* $\{0, 1, 2, \ldots, n\}$.

Our algorithm will construct $\mathbb{G}^k$ for $k = 0, 1, 2, \ldots$; Lemma 1 guarantees that larger and larger initial portions of (the graphs of) $\mathbb{G}_{\langle p,q\rangle}$ are appearing during the run of the algorithm (though we do not know the extent of the portion of $\mathbb{G}$ in $\mathbb{G}^k$). To show when our algorithm can terminate, recognizing an initial portion of $\mathbb{G}$ and providing a description of the whole $\mathbb{G}$, we now explore a certain "repeatable pattern" which is guaranteed to appear in $\mathbb{G}$.

By the Belt Theorem, we can fix a set of belts with nonnegative rational or infinite slopes such that each frontier is contained in one of them. We assume that the belts are "sufficiently" thick; thus we can, for instance, suppose that the belt slopes are pairwise distinct (merging parallel belts into one thicker).

Now we can choose $h_1, h_2, i \in \mathbb{N}$, where $0<h_1<h_2<i$, such that (see Fig. 2):

1. for each frontier function f with $f(h_2)<\omega$, all frontier points $\langle f(n), n\rangle$ between levels h_1 and h_2, (i.e., with $h_1\leq n\leq h_2$) lie in one of the belts (this follows trivially from our assumption; note that Fig. 2 depicts just one frontier in each belt, though in general there can be several frontiers in a single belt);
2. the belts are pairwise disjoint at and above level h_1-1 (i.e., we choose h_1 large enough so that at level h_1-1 each belt is to the right of any other belt with greater slope);
3. for each frontier function f: if $f(h_1-1)\leq 1$ then $f(h_2)=f(h_1-1)$; and if $f(h_2)=\omega$ then $f(h_1-1)=\omega$ (this is satisfied when h_1 and h_2 are chosen large enough);

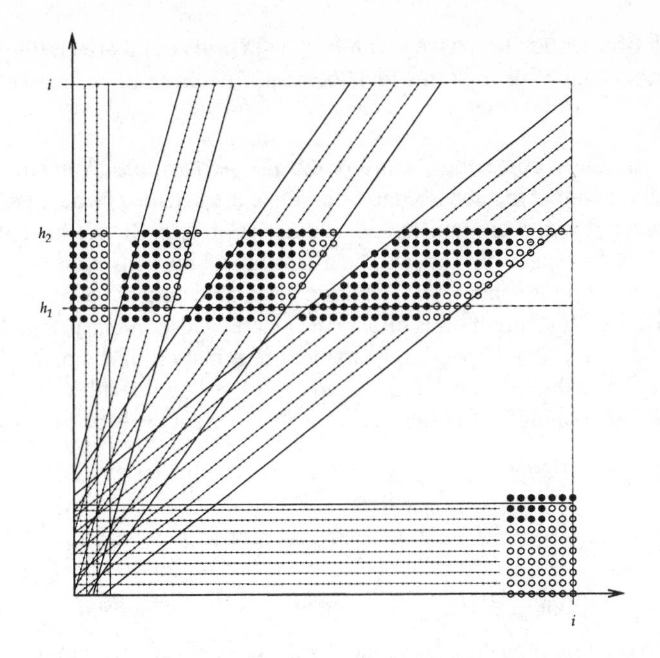

Fig. 2. Graphs of $\mathbb{G}_{\langle p,q\rangle}$ displaying a repeatable pattern, superimposed onto each other

4. for each frontier function f and each $n \leq h_2$: if $f(n) < \omega$ then $f(n) < i$ (this is satisfied by choosing i large enough after the choice of h_1 and h_2).

Each frontier point $\langle f(n), n\rangle$ has a certain (horizontal) *distance* to the left border line of the belt in which it lies. Since the slope of each belt is rational, it is clear that such distances range over finitely many possible values. So, by a straightforward use of the pigeonhole principle, we can additionally suppose (i.e., we could choose h_1, h_2, i so) that the frontier points of all frontiers inside a single belt have the same relative positions at levels h_2 and h_2-1 as at levels h_1 and h_1-1, respectively. More precisely:

5. for each frontier function f with $f(h_2) < \omega$, the slope of the belt in which the respective frontier appears between levels h_1 and h_2 is $(h_2-h_1)/(f(h_2)-f(h_1))$; moreover, $f(h_2)-f(h_2-1) = f(h_1)-f(h_1-1)$

The number of possible distances would allow us to calculate a bound b such that we can even suppose (i.e., choose so) that $h_2-h_1 \leq b$. Note that b does not depend on how thick the belts are chosen. In particular, we can assume each belt to be so thick that for each frontier point $\langle f(n), n\rangle$ in the belt, with $n \geq h_1$, the point $\langle f(n), n+b\rangle$ is still an ***interior point*** of the belt, i.e., its whole neighbourhood lies in the belt. Informally we say that the belt has a *sufficiently thick monochromatic left subbelt* (above h_1); monochromatic means that each $\mathbb{G}_{\langle p,q\rangle}$ is constant (either black or white) on the subbelt. Therefore we could choose belts and h_1, h_2 and i so that the following additional condition is satisfied:

6. for each frontier point $\langle f(n), n\rangle$ with $h_1 \le n \le h_2$, the point $\langle f(n), n+(h_2-h_1)\rangle$ is an interior point of the belt in which the respective frontier lies between levels h_1 and h_2.

We now say that a colouring $\mathbb{C}$ has a ***repeatable pattern***, based on h_1, h_2 and i, iff there are belts such that the above conditions 1.–6. are satisfied (where the terms *frontier* and *frontier function* are understood as those related to $\mathbb{C}$). We have thus demonstrated that $\mathbb{G}$ has a repeatable pattern. Our algorithm which constructs $\mathbb{G}^0, \mathbb{G}^1, \mathbb{G}^2, \ldots$ terminates when it finds some $\mathbb{G}^j$ which has a repeatable pattern based on some h_1, h_2 and i with $i<j$; such a condition is clearly decidable; and Lemma 1, together with the fact that $\mathbb{G}$ has a repeatable pattern, guarantees termination of the algorithm. Having discovered a repeatable pattern for $\mathbb{G}^j$ based on h_1, h_2 and i with $i<j$, we define the colouring $\mathbb{G}^*$ by defining its frontier functions $f^*_{\langle p,q\rangle}$ inductively as follows:

$$f^*_{\langle p,q\rangle}(n) = \begin{cases} f^j_{\langle p,q\rangle}(n), & \text{if } n \le h_2 \\ f^*_{\langle p,q\rangle}(n-c) + d, & \text{if } n > h_2 \end{cases}$$

where $c = h_2 - h_1$ and $d = f^j_{\langle p,q\rangle}(h_2) - f^j_{\langle p,q\rangle}(h_1)$. Hence each $f^*_{\langle p,q\rangle}$ is periodic, arising from $f^j_{\langle p,q\rangle}$ by repeating the sequence of steps between h_1 and h_2 forever. Also note that if $f^j_{\langle p,q\rangle}(n)=\omega$ for some $n \le h_2$ then $f^*_{\langle p,q\rangle}=f^j_{\langle p,q\rangle}$. We shall show (Lemma 3) that $\mathbb{G}^*$ is in fact $\mathbb{G}$. To this end, we make some considerations and introduce some auxiliary notions.

First recall that the local correctness of a point $\langle m, n\rangle$ in a colouring $\mathbb{C}$ depends only on the restriction of $\mathbb{C}$ to the neighbourhood of $\langle m, n\rangle$. Also recall that the possible transitions from a state $p(m)$ do not depend on m when $m>0$. Therefore $\mathbb{G}^*$ is surely ω-admissible: each point $\langle m, n\rangle$ in the *verified area*, i.e., with $m<j$ and $n<h_2$, is locally correct since it is (by definition) locally correct in $\mathbb{G}^j$, and $\mathbb{G}^j$ and $\mathbb{G}^*$ coincide on the neighbourhood of $\langle m, n\rangle$. Furthermore, each point outside the verified area obviously has a corresponding point in the verified area whose neighbourhood is coloured identically. By the fact that $\mathbb{G}$ is the maximal ω-admissible colouring, we have $f^*_{\langle p,q\rangle} \le f_{\langle p,q\rangle}$. Since $f_{\langle p,q\rangle} \le f^j_{\langle p,q\rangle}$, we have $f^*_{\langle p,q\rangle}(n) = f_{\langle p,q\rangle}(n)$ for all $n \le h_2$ (where $f^*_{\langle p,q\rangle}$ coincides with $f^j_{\langle p,q\rangle}$). The only possibility that $\mathbb{G}^*$ and $\mathbb{G}$ are not equal is if $f^*_{\langle p,q\rangle}(n) < f_{\langle p,q\rangle}(n)$ for some $n>h_2$. Due to the next result (Lemma 2), this will be lead to a contradiction in the proof of Lemma 3.

Let $\mathbf{v} = \langle v_1, v_2\rangle \in \mathbb{Z}\times\mathbb{Z}$ be a vector with integer entries. A point $\langle m, n\rangle \in \mathbb{N}\times\mathbb{N}$ with $m+v_1, n+v_2 \ge 0$ is ***lit by $\mathbf{v}$ in $\mathbb{G}_{\langle p,q\rangle}$*** iff $\mathbb{G}_{\langle p,q\rangle}(m, n) =$ black and $\mathbb{G}_{\langle p,q\rangle}(m+v_1, n+v_2) =$ white; if $\langle m, n\rangle$ is lit by $\mathbf{v}$ in some $\mathbb{G}_{\langle p,q\rangle}$, then we say that $\langle m, n\rangle$ is ***lit by*** $\mathbf{v}$. For points $\langle m, n\rangle, \langle m', n'\rangle \in \mathbb{N}\times\mathbb{N}$ we write $\langle m, n\rangle \leftrightarrow_{\mathbf{v}} \langle m', n'\rangle$ iff both are lit by $\mathbf{v}$, and $|m-m'| \le 1$ and $|n-n'| \le 1$. The transitive closure of $\leftrightarrow_{\mathbf{v}}$ is denoted by $\leftrightarrow^*_{\mathbf{v}}$. Note that $\langle m, n\rangle \leftrightarrow^*_{\mathbf{v}} \langle m', n'\rangle$ can be demonstrated by giving a ***trajectory***, a sequence of points $\langle m_0, n_0\rangle, \langle m_1, n_1\rangle, \ldots, \langle m_k, n_k\rangle$ such that

$$\langle m, n\rangle = \langle m_0, n_0\rangle \leftrightarrow_{\mathbf{v}} \langle m_1, n_1\rangle \leftrightarrow_{\mathbf{v}} \cdots \leftrightarrow_{\mathbf{v}} \langle m_k, n_k\rangle = \langle m', n'\rangle.$$

Lemma 2. *Let* $h>0$ *and* $\mathbf{v} = \langle v_1, v_2 \rangle$ *with* $v_1 \leq 0$ *and* $v_2 < 0$. *If a point* $\langle m_0, n_0 \rangle$ *with* $n_0 + v_2 > h$ *is lit by* $\mathbf{v}$ *then there is a point* $\langle m'_0, n'_0 \rangle$ *with* $n'_0 + v_2 = h$ *such that* $\langle m_0, n_0 \rangle \leftrightarrow^*_{\mathbf{v}} \langle m'_0, n'_0 \rangle$.

Proof. Suppose $\langle m_0, n_0 \rangle$ satisfies the assumption but there is no required $\langle m'_0, n'_0 \rangle$; then $n' + v_2 > h$ for each $\langle m', n' \rangle$ such that $\langle m_0, n_0 \rangle \leftrightarrow^*_{\mathbf{v}} \langle m', n' \rangle$. Define the colouring $\overline{\mathbb{G}}$ by

$$\overline{\mathbb{G}}_{\langle p,q \rangle}(m,n) = \text{black} \quad \text{iff} \quad \mathbb{G}_{\langle p,q \rangle}(m,n) = \text{black, or}$$
$$\langle m-v_1, n-v_2 \rangle \text{ is lit by } \mathbf{v} \text{ in } \mathbb{G}_{\langle p,q \rangle} \quad \text{and}$$
$$\langle m_0, n_0 \rangle \leftrightarrow^*_{\mathbf{v}} \langle m-v_1, n-v_2 \rangle.$$

$\overline{\mathbb{G}}$ obviously satisfies the monotonicity property of colourings, and we can easily check that each point is locally correct in $\overline{\mathbb{G}}$. Hence $\overline{\mathbb{G}}$ is ω-admissible, which contradicts the fact that $\mathbb{G}$ is the *maximal* ω-admissible colouring. □

Lemma 3. $\mathbb{G}^*$ *is equal to* $\mathbb{G}$.

Proof. We have already shown that each $f^*_{\langle p,q \rangle}$ coincides with $f_{\langle p,q \rangle}$ on the set $\{0, 1, 2, \ldots, h_2\}$, so we only have to exclude the possibility that $f^*_{\langle p,q \rangle}(n) < f_{\langle p,q \rangle}(n)$ for some $n > h_2$.

Recall that our algorithm stops by finding a repeatable pattern, for h_1, h_2, i, in $\mathbb{G}^j$ $(i<j)$. Let us fix a corresponding set of belts required by the definition of a repeatable pattern (note that each frontier of $\mathbb{G}^*$ lies in one of the belts above h_1).

We say that a belt B is *valid* iff $\mathbb{G}^*$ coincides with $\mathbb{G}$ when restricted to B. (In particular, the horizontal belt, if it was chosen, is surely valid.) If all belts are valid, then surely $\mathbb{G}^*$ is equal to $\mathbb{G}$. Otherwise, let B be the *rightmost* belt (i.e., the belt with the least slope) which is not valid. Consider an *invalid point* $\langle m_0, n_0 \rangle$ in B, i.e., $\mathbb{G}^*_{\langle p,q \rangle}(m_0, n_0) = \text{white}$ and $\mathbb{G}_{\langle p,q \rangle}(m_0, n_0) = \text{black}$, for some p, q; moreover we suppose n_0 to be minimal (i.e., B is valid below n_0). Note that $n_0 > h_2$.

Let α be the slope of B, and let $\mathbf{v} = \langle v_1, v_2 \rangle$, where $v_1 = (h_1 - h_2)/\alpha$ and $v_2 = h_1 - h_2$ ($\mathbf{v}$ corresponds to the "period of B" in $\mathbb{G}^*$; see Fig. 3). Due to the choice of $\mathbf{v}$ (as the period of B) we have $\mathbb{G}^*_{\langle p,q \rangle}(m_0+v_1, n_0+v_2) = \text{white}$, and since B is valid below n_0, we have $\mathbb{G}_{\langle p,q \rangle}(m_0+v_1, n_0+v_2) = \text{white}$. This means that the point $\langle m_0, n_0 \rangle$ is lit by $\mathbf{v}$ in $\mathbb{G}_{\langle p,q \rangle}$. Due to Lemma 2 (for h_1 in the place of h) there is a point $\langle m'_0, n'_0 \rangle$ (lit by $\mathbf{v}$) such that $\langle m_0, n_0 \rangle \leftrightarrow^*_{\mathbf{v}} \langle m'_0, n'_0 \rangle$ and $n'_0 + v_2 = h_1$, i.e., $n'_0 = h_2$. Recall that the restrictions of $\mathbb{G}^*$ and $\mathbb{G}$ to $\mathbb{N} \times \{0, 1, 2, \ldots, h_2\}$ coincide. Hence if there is no belt to the right of B then there is clearly no point $\langle m', h_2 \rangle$ which would be lit by $\mathbf{v}$. Otherwise let B$'$ be the first belt to the right of B. Any point $\langle m', h_2 \rangle$ which is lit by $\mathbf{v}$ can lie only in, or to the right of, B$'$. Nevertheless any trajectory demonstrating $\langle m_0, n_0 \rangle \leftrightarrow^*_{\mathbf{v}} \langle m', h_2 \rangle$ would have to cross the (sufficiently thick) monochromatic left subbelt of (the valid) B$'$, which is impossible. (The first point on such a trajectory which is in B$'$, and is thus not an interior point of B$'$, cannot be lit by $\mathbf{v}$.) □

We can summarize the preceding argument in the following.

Theorem 1. *There is an algorithm which, given a one counter net, constructs a description of the respective maximal simulation relation; more concretely, it gives periodicity descriptions for the corresponding frontier functions.*

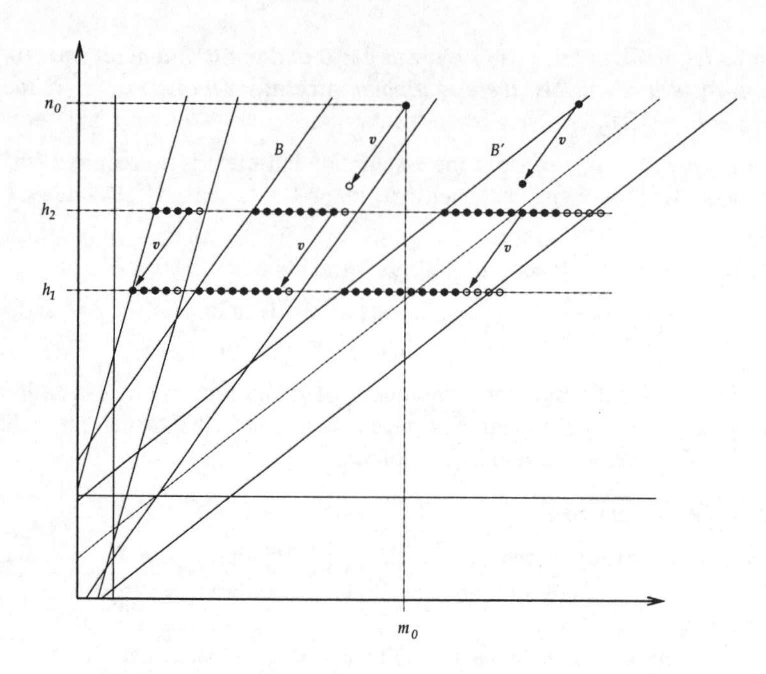

Fig. 3. The assumption $\mathbb{G} \neq \mathbb{G}^*$ leads to a contradiction.

3 Applications

In this section we show how Theorem 1 can be applied to obtain new decidability results for one-counter nets. The following one comes almost for free.

Theorem 2. *The problem of strong* $\preccurlyeq\succcurlyeq$*-regularity of one-counter nets is decidable.*

Proof. Let $p(i)$ be a process of the one-counter net $N = \langle Q, \Sigma, \delta^=, \delta^> \rangle$. Define the set $\mathcal{M} = \{q \in Q \mid p(i) \to^* q(j) \text{ for infinitely many } j \in \mathbb{N}\}$. Observe that $\mathcal{M}$ is effectively constructible using standard techniques for pushdown automata. As Q is finite, we see that $p(i)$ can reach infinitely many pairwise non-equivalent states iff there is $q \in \mathcal{M}$ such that for every $i \in \mathbb{N}$ there is some $j > i$ such that $q(j) \not\preccurlyeq q(i)$. In other words, $p(i)$ is not strongly regular w.r.t. simulation equivalence iff there is $q \in \mathcal{M}$ such that the frontier function $f_{\langle q,q \rangle}$ has no ω-values ($\forall n \in \mathbb{N} : f_{\langle p,q \rangle}(n) < \omega$). □

Next we show that a number of *simulation* problems for processes of one-counter nets can be reduced to the corresponding *bisimulation* problems for processes of one-counter automata. In this way we obtain further (original) decidability results. The basic tool which enables the mentioned reductions is taken from [10] and is described next.

For every image-finite transition system $T = \langle S, Act, \to \rangle$ we define the transition system $\mathcal{B}(T) = \langle S, Act, \mapsto \rangle$ where $\mapsto$ is given by

$$s \overset{a}{\mapsto} t \quad \text{iff} \quad s \overset{a}{\to} t \text{ and } \forall u \in S : (s \overset{a}{\to} u \wedge t \preccurlyeq u) \Longrightarrow u \preccurlyeq t$$

Note that $\mathcal{B}(\mathrm{T})$ is obtained from T by deleting certain transitions (preserving only the "maximal" ones). Also note that T and $\mathcal{B}(\mathrm{T})$ have the same set of states; as we often need to distinguish between processes "s of T" and "s of $\mathcal{B}(\mathrm{T})$", we denote the latter one by $s_{\mathcal{B}}$. A proof of the next (crucial) theorem, relating simulation equivalence and bisimulation equivalence, can be found in [10].

Theorem 3. *Let* s *and* t *be processes of image-finite transition systems* T *and* T′, *respectively. It holds that* $s \precsim\succsim s_{\mathcal{B}}$ *and* $t \precsim\succsim t_{\mathcal{B}}$*; moreover,* $s \precsim\succsim t$ *iff* $s_{\mathcal{B}} \sim t_{\mathcal{B}}$.

The next theorem provides the technical basis for the aforementioned reductions.

Theorem 4. *Let* N *be a one-counter net. Then the transition system* $\mathcal{B}(\mathrm{T_N})$ *is effectively definable within the syntax of one-counter automata, i.e., one can effectively construct a one-counter automaton* M *such that* $\mathrm{T_M}$ *is isomorphic to* $\mathcal{B}(\mathrm{T_N})$*. Moreover, for every state* $s = p(i)$ *of* $\mathrm{T_N}$ *we can effectively construct a state* $p'(i')$ *of* $\mathrm{T_M}$ *which is isomorphic to the state* $s_{\mathcal{B}}$ *of* $\mathcal{B}(\mathrm{T_N})$.

Proof. Let $\mathrm{N} = \langle Q, \Sigma, \delta^{=}, \delta^{>}\rangle$ be a one-counter net, and let $\mapsto$ be the transition relation of $\mathcal{B}(\mathrm{T_N})$. Let us define the function *MaxTran* : $Q \times \Sigma \times \mathbb{N} \to \mathcal{P}(Q \times \{-1, 0, 1\})$ as follows:

$$\langle q, j\rangle \in \mathit{MaxTran}(p, a, i) \;\text{ iff }\; p(i) \overset{a}{\mapsto} q(i+j)$$

where $\mapsto$ is the transition relation of $\mathcal{B}(\mathrm{T_N})$. In fact, *MaxTran*$(p, a, i)$ represents all "maximal" a-transitions of $p(i)$. Our aim is to show that the function *MaxTran* is, in some sense, periodic—we prove that there (effectively) exists $n > 0$ such that for all $p \in Q$, $a \in \Sigma$, and $i \geq n$ we have that *MaxTran*$(p, a, i) =$ *MaxTran*$(p, a, i+n)$. It clearly suffices for our purposes because then we can construct a one-counter automaton $\mathrm{M} = \langle Q \times \{0, \ldots, n-1\}, \Sigma, \gamma^{=}, \gamma^{>}\rangle$ where $\gamma^{=}$ and $\gamma^{>}$ are the least sets satisfying the following conditions:

- if $p(i) \overset{a}{\mapsto} q(j)$ where $0 \leq i, j < n$, then $(\langle q, j\rangle, 0) \in \gamma^{=}(\langle p, i\rangle, a)$
- if $p(n-1) \overset{a}{\mapsto} q(n)$, then $(\langle q, 0\rangle, +1) \in \gamma^{=}(\langle p, n-1\rangle, a)$
- if $p(n+i) \overset{a}{\mapsto} q(n+j)$ where $0 \leq i, j < n$, then $(\langle q, j\rangle, 0) \in \gamma^{>}(\langle p, i\rangle, a)$
- if $p(n) \overset{a}{\mapsto} q(n-1)$, then $(\langle q, n-1\rangle, -1) \in \gamma^{>}(\langle p, 0\rangle, a)$
- if $p(2n-1) \overset{a}{\mapsto} q(2n)$, then $(\langle q, 0\rangle, +1) \in \gamma^{>}(\langle p, n-1\rangle, a)$

Note that the definition of M is effective, because the constant n can be effectively found and for every transition $p(i) \overset{a}{\to} p(j)$ of $\mathrm{T_N}$ we can effectively decide whether $p(i) \overset{a}{\mapsto} p(j)$ (here we need the decidability of simulation for one-counter nets). The fact that $\mathrm{T_M}$ is isomorphic to $\mathcal{B}(\mathrm{T_N})$ is easy to see as soon as we realize that $\mathcal{B}(\mathrm{T_N})$ can be viewed as a sequence of "blocks" of height n, where all "blocks" except for the initial one are the same. The structure of the two (types of) blocks is encoded in the finite control of M, and the number of "current" blocks is stored in its counter (see Fig. 4). Note that M indeed needs the test for zero in order to recognize that the initial block has been entered.

Now we show how to construct the constant n. First, we prove that for all $p \in Q$, $a \in \Sigma$ one can effectively find two constants $k(p, a)$ and $l(p, a)$ such that for every

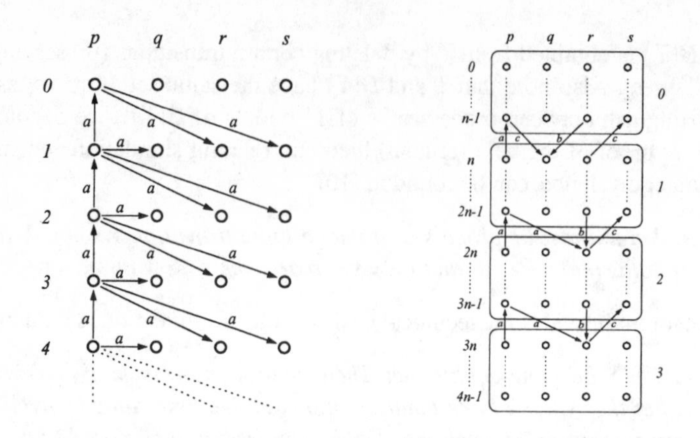

Fig. 4. The structure of T_N (left) and $\mathcal{B}(T_N)$ (right)

$i > k(p, a)$ we have $MaxTran(p, a, i) = MaxTran(p, a, i + l(p, a))$. We start by reminding ourselves that the out-going transitions of $p(i)$ and $p(j)$, where $i, j \geq 1$, are the "same" in the following sense (see Fig. 4):

$$p(i) \xrightarrow{a} q(i+m) \text{ iff } p(j) \xrightarrow{a} q(j+m) \text{ iff } (q, m) \in \delta^{>}(p, a).$$

Hence, the set $MaxTran(p, a, i)$, where $i \geq 1$, is obtained by selecting certain elements from $\delta^{>}(p, a)$. In order to find these elements, we must (by the definition of $\mathcal{B}(T)$) take all pairs $\langle\langle q, m\rangle, \langle r, n\rangle\rangle \in \delta^{>}(p, a) \times \delta^{>}(p, a)$, determine whether $q(i+m) \preccurlyeq r(i+n)$, and select only the "maximals". For each such pair $\langle\langle q, m\rangle, \langle r, n\rangle\rangle$ we define an infinite binary sequence $\mathcal{S}$ as follows: $\mathcal{S}(i) = 1$ if $\mathbb{G}_{\langle q,r\rangle}(i+m, i+n) =$ black, and $\mathcal{S}(i) = 0$ otherwise. As (a description of) $\mathbb{G}_{\langle q,r\rangle}$ can be effectively constructed, and the frontier function $f_{\langle q,r\rangle}$ is periodic (see Theorem 1), we can conclude that $\mathcal{S} = \alpha\beta^{\omega}$ where α, β are finite binary strings. Note that α and β can be "read" from the constructed description of $\mathbb{G}_{\langle q,r\rangle}$ and thus they are effectively constructible. As $\delta^{>}(p, a)$ is finite, there are only finitely many pairs to consider and hence we obtain only finitely many α's and β's. Now we let $k(p, a)$ be the length of the longest α, and let $l(p, a)$ be the product of lengths of all β's. In this way we achieve that the whole information which determines the selection of "maximal" elements of $\delta^{>}(p, a)$ during the construction of $MaxTran(p, a, i)$ is periodic (w.r.t. i) with period $l(p, a)$ after a finite "initial segment" of length $k(p, a)$. Let $K = \max\{k(p, a) \mid p \in Q, a \in \Sigma\}$, and $L = \prod_{p\in Q, a\in\Sigma} l(p, a)$. Finally, let $n = K \cdot L$.

To finish the proof, we need to show that for every state $s = p(i)$ of T_N one can construct a state $p'(i')$ of T_M which is isomorphic to the state $s_{\mathcal{B}}$ of $\mathcal{B}(T_N)$. This is straightforward; we simply take $p' = \langle p, i \bmod n\rangle$ and $i' = i \operatorname{div} n$. □

Two concrete examples of how Theorems 3 and 4 can be applied to obtain (new and nontrivial) positive decidability results on one-counter nets are given next.

Corollary 1. *The problem of $\preccurlyeq\succcurlyeq$-regularity of one-counter nets is decidable.*

Proof. It suffices to realize that a process s of a transition system T is $\preceq\succeq$-regular iff the process $s_{\mathcal{B}}$ of $\mathcal{B}(T)$ is $\sim$-regular. As $\sim$-regularity is decidable for processes of one-counter automata [3], we are done. □

Corollary 2. *Let* $p\alpha$ *be a process of a deterministic pushdown automaton* Γ *and* $q(i)$ *be a process of a one-counter net* N. *The problem whether* $p\alpha \preceq\succeq q(i)$ *is decidable.*

Proof. First, realize that if T is a deterministic transition system then $\mathcal{B}(T) = T$. Hence, $p\alpha \preceq\succeq q(i)$ iff $p\alpha \sim q'(i')$ where $q'(i')$ is the process of Theorem 4. As one-counter automata are (special) pushdown automata, we can apply the result of [13] which says that bisimilarity is decidable for pushdown processes. □

The previous corollary touches, in a sense, the decidability/undecidability border for simulation equivalence, because the problem whether $p\alpha \preceq\succeq q(i)$ where $p\alpha$ is a process of a deterministic PDA Γ and $q(i)$ is a process of a one-counter automaton M is undecidable [7] (in fact, it is undecidable even if we require Γ to be a deterministic one-counter automaton).

References

1. P.A. Abdulla and K. Čerāns. Simulation is decidable for one-counter nets. In *Proceedings of CONCUR'98*, volume 1466 of *LNCS*, pages 253–268. Springer, 1998.
2. O. Burkart, D. Caucal, and B. Steffen. Bisimulation collapse and the process taxonomy. In *Proceedings of CONCUR'96*, volume 1119 of *LNCS*, pages 247–262. Springer, 1996.
3. P. Jančar. Bisimulation equivalence is decidable for one-counter processes. In *Proceedings of ICALP'97*, volume 1256 of *LNCS*, pages 549–559. Springer, 1997.
4. P. Jančar and J. Esparza. Deciding finiteness of Petri nets up to bisimilarity. In *Proceedings of ICALP'96*, volume 1099 of *LNCS*, pages 478–489. Springer, 1996.
5. P. Jančar and F. Moller. Checking regular properties of Petri nets. In *Proceedings of CONCUR'95*, volume 962 of *LNCS*, pages 348–362. Springer, 1995.
6. P. Jančar and F. Moller. Simulation of one-counter nets via colouring. In *Proceedings of Workshop Journées Systèmes Infinis*, LSV-ENS Cachan, pages 1–6, November 1998. (Revised version, Research Report 159, Computing Science Department, Uppsala University, 1999.)
7. P. Jančar, F. Moller, and Z. Sawa. Simulation problems for one-counter machines. In *Proceedings of SOFSEM'99*, volume 1725 of *LNCS*, pages 398–407. Springer, 1999.
8. A. Kučera. On finite representations of infinite-state behaviours. *Information Processing Letters*, 70(1):23–30, 1999.
9. A. Kučera and J. Esparza. A logical viewpoint on process-algebraic quotients. In *Proceedings of CSL'99*, volume 1683 of *LNCS*, pages 499–514. Springer, 1999.
10. A. Kučera and R. Mayr. Simulation preorder on simple process algebras. In *Proceedings of ICALP'99*, volume 1644 of *LNCS*, pages 503–512. Springer, 1999.
11. S. Mauw and H. Mulder. Regularity of BPA-systems is decidable. In *Proceedings of CONCUR'94*, volume 836 of *LNCS*, pages 34–47. Springer, 1994.
12. F. Moller. Infinite results. In *Proceedings of CONCUR'96*, volume 1119 of *LNCS*, pages 195–216. Springer, 1996.
13. G. Sénizergues. Decidability of bisimulation equivalence for equational graphs of finite out-degree. In *Proceedings of 39th Annual Symposium on Foundations of Computer Science*, pages 120–129. IEEE Computer Society Press, 1998.
14. R.J. van Glabbeek. The linear time – branching time spectrum. In *Proceedings of CONCUR'90*, volume 458 of *LNCS*, pages 278–297. Springer, 1990.

Decidability of Reachability Problems for Classes of Two Counters Automata

Alain Finkel and Grégoire Sutre

LSV, ENS Cachan & CNRS UMR 8643, France.
{finkel, sutre}@lsv.ens-cachan.fr

Abstract. We present a global and comprehensive view of the properties of subclasses of two counters automata for which counters are only accessed through the following operations: increment (+1), decrement (−1), reset ($c := 0$), transfer (the whole content of counter c is transfered into counter c'), and testing for zero. We first extend Hopcroft-Pansiot's result (an algorithm for computing a finite description of the semilinear set *post**) to two counters automata with only one test for zero (and one reset and one transfer operations). Then, we prove the semilinearity and the computability of *pre** for the subclass of 2 counters automata with one test for zero on c_1, two reset operations and one transfer from c_1 to c_2. By proving simulations between subclasses, we show that this subclass is the maximal class for which *pre** is semilinear and effectively computable. All the (effective) semilinearity results are obtained with the help of a new symbolic reachability tree algorithm for counter automata using an `Acceleration` function. When `Acceleration` has the so-called stability property, the constructed tree computes exactly the reachability set.

1 Introduction

Context. The highly successful model-checking approach for finite systems is mainly a consequence that finite state automata enjoy a lot of good properties like strong relations with logics (first order and monadic second order, linear temporal logic (LTL)), decidability and complexity results, efficient algorithmics, closure properties, etc. Thus, finite automata are a nice framework for the model-checking of programs which operate on bounded and finite variables. But the knowledge of the domains of variables is sometimes impossible to compute and this is still more true when the domain is infinite.

Before we survey the decidability results about extended automata, we give a precise view of the properties we are interested in. One of the most important class of properties is called safety properties. Model-checking safety properties often reduces to the effective computation of the set of predecessors (pre^*) or/and the set of successors ($post^*$). We thus want to find an infinite-state model for which the pre^* or/and $post^*$ images always belong to a "good" class. In our framework, these two sets are two (infinite) subsets of $Q \times \mathbb{N}^n$ (where Q is

H. Reichel and S. Tison (Eds.): STACS 2000, LNCS 1770, pp. 346–357, 2000.

the finite set of control states and n is the number of counters on which the automaton operates). Semilinear sets of $Q \times \mathbb{N}^n$ form a "good" class since (1) they are closed under $\cap$ and under projection, (2) inclusion is decidable, and (3) they are also the sets expressible in Presburger arithmetics over integers and also the set of regular subsets in the commutative monoid $Q \times \mathbb{N}^n$. Hence, our objective is to find an extension of finite automata for which *pre** and/or *post** are semilinear and, if possible, effectively computable.

Related work. Automata with counters, but without any zero-testing primitive, are known as Vector Addition Systems with States (VASS) and they are equivalent to Petri nets. The reachability problem is thus decidable but the comparison between two reachability sets is undecidable. Moreover, in dimension n, $n > 2$, the reachability set is not always semilinear [HP79].

Stack automata are another well-known extension of finite automata. They have regular reachability sets (in $Q \times \Sigma^*$); this property has been recently rediscovered [BS86, Cau92, FWW97, BEM97] and used for model-checking of stack automata. Of course, stack automata contain 1-counter automata (with the capability of zero-testing). At last, it has been proved that lossy FIFO automata have regular non-computable reachability sets [CFP96] while half-duplex and quasi-stable fifo automata have effectively computable regular reachability sets [CF97]. For flat counters automata (flat means that there is no nested loop in the control transition graph), the reachability relation can be expressed in the Presburger logics [CJ98].

A result of Hopcroft and Pansiot [HP79] said that the reachability set of a 2-dim Vector Addition Systems with States (2-dim VASS) is semilinear and that it is effectively computable. Hopcroft and Pansiot gave an algorithm which computes a description of the semilinear reachability set. The complexity of Hopcroft-Pansiot's algorithm has been studied later by Rosier and Yen [HRHY86] who showed that the deterministic complexity is in 2^{2^n} where n is the size of the automaton with its initial state.

More recently, some papers appeared dealing with verification of reset/transfer Petri nets (or counters automata for which counters are only accessed through the following operations: increment (+1), decrement (-1), reset ($c := 0$), transfer (the whole content of counter c is transfered into counter c'). For these weak counter automata (that do not have a full-fledged test for zero), reachability of a control state is decidable for reset/transfer Petri nets, thus also for counters automata with the following operations {+1, -1, reset, transfer} [DFS98]. Reachability and boundedness are undecidable in dimension 3 (since there are three counters) [DFS98] ; the limit between decidability/undecidability for the boundedness problem is precised as follows: boundedness is decidable for {+1, -1, reset}-counters automata for which only two counters can be reset [DJS99]. Reachability is decidable for counters automata such that only a unique counter can be tested for zero [Rei95]. These extensions except [Rei95] do not contain the complete test for zero.

These results from the litterature do not give a global and comprehensive view of the properties of classes of automata with two {+1, -1, reset, transfer, zero-test}-counters. Can we extend the result of [HP79] to more powerful models of two counters automata ? Are pre^* and/or $post^*$ also semilinear for extended automata ?

Our contribution. We present three types of results:

(1) A *new general symbolic reachability tree (semi-)algorithm* for automata with {+1, -1, reset, transfer, zero-test}-counters.

This algorithm is *generic* and *modular*, and it uses an `Acceleration` function. It exactly computes the reachability set ($post^*$) when the `Acceleration` function satisfies a so-called *stability* property. Our algorithm, given for any n-dim Extended VASS, both generalizes Hopcroft-Pansiot's algorithm (given for 2-dim VASS) and refines Karp-Miller's algorithm (given for n-dim VASS) [KM69]. It gives an exact description of $post^*$ (as Hopcroft-Pansiot's algorithm and not only a coverability set like Karp-Miller's algorithm). We prove termination and correctness of this algorithm for some subclasses of automata with two {+1, -1, reset, transfer, zero-test}-counters.

(2) A *hierarchy result.*

We systematically study all the subclasses of two {+1, -1, reset, transfer, zero-test}-counters automata and we give a complete hierarchy between classes using new simulations. A "maximal decidable model" appears to be 2-counters automata ($\mathbf{T}_1\mathbf{R}_{1,2}\mathbf{Tr}_{12}$) with the following extended operations: test for zero on the first counter, reset on the two counters, and, transfer from the first to the second counter (see Figure 1 on page 356).

(3) *Three main technical results.*

We prove the three following theorems: pre^* is an effective semilinear set for the class $\mathbf{T}_1\mathbf{R}_{1,2}\mathbf{Tr}_{12}$ (Theorem 4) ; $post^*$ is an effective semilinear set for the class $\mathbf{T}_1\mathbf{R}_1\mathbf{Tr}_{12}$ (Theorem 3) ; and, $post^*$ is a semilinear set for the class $\mathbf{T}_1\mathbf{R}_{1,2}\mathbf{Tr}_{12}$ (Theorem 5).

Plan. Section 2 introduces Extended Vector Addition Systems with States (E-VASS), and we present in Section 3 semilinear sets and projections. In Section 4, we give a new symbolic reachability tree algorithm for E-VASS which uses as a parameter a function `Acceleration` and we define the stability property. Section 5 extends Hopcroft-Pansiot's result (effective semilinearity of $post^*$) to 2-dim Extended VASS with one test for zero (and with one reset and one transfer operations). Section 6 proves the effective semilinearity of pre^* for the maximal class $\mathbf{T}_1\mathbf{R}_{1,2}\mathbf{Tr}_{12}$. Section 7 shows that $post^*$ is still semilinear for the maximal class $\mathbf{T}_1\mathbf{R}_{1,2}\mathbf{Tr}_{12}$.

The proofs are technically nontrivial and are omitted here due to space constraints. They can be found in the full version available form the authors and at the URL `http://www.lsv.ens-cachan.fr/Publis/`.

2 Extended Vector Addition Systems with States

Let $\mathbb{Z}$ (resp. $\mathbb{N}$, $\mathbb{N}^*$, $\mathbb{Q}^+$) denote the set of integers (resp. nonnegative integers, positive integers, nonnegative rational numbers). If $i, j \in \mathbb{N}$, we write $[i, j]$ for the set $\{k \in \mathbb{N} \ / \ i \leq k \leq j\}$. Let $\mathbb{N}^n$ (resp. $\mathbb{Z}^n$) denote the set of n-tuples of elements of $\mathbb{N}$ (resp. $\mathbb{Z}$). If x is an n-tuple and $i \in [1, n]$, $x(i)$ is the ith *component* of x. The ith *unit vector* is the n-tuple e_i defined by $e_i(j) = 0$ if $j \neq i$ and $e_i(i) = 1$. Operations on n-tuples are componentwise extensions of the usual operations and when 0 is used as an n-tuple, it denotes the all zero n-tuple. These operations are classically extended on sets of n-tuples (e.g. for $P, P' \subseteq \mathbb{N}^n$, $P + P' = \{p + p' \ / \ p \in P$ and $p' \in P'\}$). Moreover, in an operation involving sets of n-tuples, we shortly write v for the singleton $\{v\}$ (e.g. for $P \subseteq \mathbb{N}^n$ and $x \in \mathbb{N}^n$, we write $x + P$ for $\{x\} + P$).

For every set X, we write $\wp(X)$ for the set of subsets of X and $|X|$ for the cardinal of X. For any infinite sequence $(x_i)_{i \in \mathbb{N}}$ in X, an infinite *subsequence* (of (x_i)) is any infinite sequence $(x_{i_k})_{i_k \in \mathbb{N}}$ with $i_0 < i_1 < i_2 \cdots i_k < i_{k+1} \cdots$.

An *ordering* is any reflexive, transitive and antisymmetric relation $\preceq$ over some set X. A *well ordering* is any ordering $\preceq$ such that, for any infinite sequence $(x_i)_{i \in \mathbb{N}}$ in X, there exists an infinite increasing subsequence $x_{i_0} \preceq x_{i_1} \preceq x_{i_2} \cdots x_{i_k} \preceq x_{i_{k+1}} \cdots$. If $\preceq$ is an ordering on X, then for every subset $Y \subseteq X$, the set $\mathrm{Min}(Y)$ of *minimal elements* of Y is defined by $\mathrm{Min}(Y) = \{y \in Y \ / \ \forall y' \in Y \setminus \{y\}, y' \not\preceq y\}$ and it is finite when $\preceq$ is a well ordering.

The relation $\leq$ between n-tuples in $\mathbb{Z}^n$ is defined by $x \leq y$ if for all $i \in [1, n]$ we have $x(i) \leq y(i)$. This relation is an ordering on $\mathbb{Z}^n$ and it is a well ordering on $\mathbb{N}^n$ (it is not a well ordering on $\mathbb{Z}^n$).

A *labelled transition system* is a structure $\mathcal{TS} = (S, A, \rightarrow)$ where S is a set of *states*, A is a finite set of *actions* and $\rightarrow \subseteq S \times A \times S$ is a set of *transitions*. When S is finite, $\mathcal{TS}$ is a *finite labelled transition system*. We note $\xrightarrow{*}$ the reflexive and transitive closure of $\rightarrow$. For every subset X of S, we write *post*(X) for the set $\{s \in S \ / \ \exists \ r \in X, r \rightarrow s\}$ of *immediate successors* of X, *post*$^*(X)$ for the set $\{s \in S \ / \ \exists \ r \in X, r \xrightarrow{*} s\}$ of *successors* of X, and *pre*$^*(X)$ for the set $\{s \in S \ / \ \exists \ r \in X, s \xrightarrow{*} r\}$ of *predecessors* of X.

We present Extended Vector Addition Systems with States (E-VASS) in two steps : we first describe the finite control structure of an E-VASS and we then define an E-VASS as an operational semantics associated with such a control structure, which leads to an infinite labelled transition system.

An *n-dim E-VASS Control* is any finite labelled transition system $\mathsf{A} = (Q, Op, \rightarrow_{\mathsf{A}})$ such that $Op \subseteq \{add(v) \ / \ v \in \mathbb{Z}^n\} \cup \{test(i), weaktest(i), reset(i), transfer(i \rightarrow j) \ / \ i, j \in [1, n], i \neq j\}$.

Definition 1. An *n-dim E-VASS* $\mathcal{A}$ is a labelled transition system $\mathcal{A} = (S, Op, \rightarrow_{\mathcal{A}})$ based on an *n-dim E-VASS Control* $\mathsf{A} = (Q, Op, \rightarrow_{\mathsf{A}})$, where:

- $S = Q \times \mathbb{N}^n$ is the set of states, and,
- Op is the set of actions, and,
- the set of transitions $\rightarrow_{\mathcal{A}}$ is the smallest subset of $S \times Op \times S$ verifying:

1. if $q \xrightarrow{add(v)}_{\mathcal{A}} q'$ then for all $x \in \mathbb{N}^n$ such that $x+v \geq 0$, $(q,x) \xrightarrow{add(v)}_{\mathcal{A}} (q', x+v)$, and
2. if $q \xrightarrow{test(i)}_{\mathcal{A}} q'$ then for all $x \in \mathbb{N}^n$ such that $x(i) = 0$, $(q,x) \xrightarrow{test(i)}_{\mathcal{A}} (q', x)$, and
3. if $q \xrightarrow{weaktest(i)}_{\mathcal{A}} q'$ then for all $x \in \mathbb{N}^n$ such that $x(i) = 0$ and for all $\alpha \in \mathbb{N}$, $(q,x) \xrightarrow{weaktest(i)}_{\mathcal{A}} (q', x + \alpha e_i)$, and
4. if $q \xrightarrow{reset(i)}_{\mathcal{A}} q'$ then for all $x \in \mathbb{N}^n$, $(q,x) \xrightarrow{reset(i)}_{\mathcal{A}} (q', x')$ where $x'(i) = 0$ and for all $j \neq i, x'(j) = x(j)$, and
5. if $q \xrightarrow{transfer(i \to j)}_{\mathcal{A}} q'$ then for all $x \in \mathbb{N}^n$, $(q,x) \xrightarrow{transfer(i \to j)}_{\mathcal{A}} (q', x')$ where $x'(i) = 0$, $x'(j) = x(i) + x(j)$ and for all $k \notin \{i,j\}, x'(k) = x(k)$.

A $weaktest(i)$ transition is the inverse of a $reset(i)$ in the sense that we have $(q,x) \xrightarrow{weaktest(i)} (q',x')$ iff $(q',x') \xrightarrow{reset(i)} (q,x)$. We will use $weaktest(i)$ transitions in section 6 to compute pre^* for E-VASS with $reset(i)$ transitions. Notice that any ε-transition may be simulated by an $add(0)$ transition. An *n-dim* (non extended) *VASS* is any n-dim E-VASS $\mathcal{A} = (S, Op, \rightarrow_{\mathcal{A}})$ such that $Op \subseteq \{add(v) \ / \ v \in \mathbb{Z}^n\}$.

We define the following classes of 2-dim E-VASS. For every $I \subseteq \{1,2\}$, $J \subseteq \{1,2\}$ and $K \subseteq \{(1,2),(2,1)\}$, we write $\mathbf{T}_I\mathbf{R}_J\mathbf{Tr}_K$ for the class of 2-dim E-VASS $\mathcal{A}$ such the set of actions Op of $\mathcal{A}$ satisfies:

$$\begin{aligned} Op \subseteq \ & \{add(v) \ / \ v \in \mathbb{Z}^n\} \\ & \cup \{test(i), reset(j), transfer(k \to k') \ / \ i \in I, j \in J \text{ and } (k,k') \in K\} \end{aligned}$$

Notice that $\mathbf{T}_\emptyset$ (resp. $\mathbf{R}_\emptyset$, $\mathbf{Tr}_\emptyset$) means that there is no zero-test transition (resp. no reset transition, no transfer transition) and we will omit to write $\mathbf{T}_\emptyset$ (resp. $\mathbf{R}_\emptyset$, $\mathbf{Tr}_\emptyset$). For instance, $\mathbf{T}_1\mathbf{R}_\emptyset\mathbf{Tr}_{12,21}$ (shortly written $\mathbf{T}_1\mathbf{Tr}_{12,21}$) is the class of 2-dim E-VASS $\mathcal{A}$ where the allowed extended transitions are labelled by $test(1)$, $transfer(1 \to 2)$ or $transfer(2 \to 1)$.

Simulations

Let $\mathbf{C}$ and $\mathbf{D}$ be two classes of E-VASS. We say $\mathbf{C}$ is *effectively simulable* by $\mathbf{D}$ if for every E-VASS $\mathcal{A}$ in $\mathbf{C}$ with control states Q_A, there exists an E-VASS $\mathcal{B}$ in $\mathbf{D}$ with control states Q_B computable from $\mathcal{A}$ such that:

1. we have $Q_B \supseteq \overline{Q_A}$, where $\overline{Q_A}$ is a copy of Q_A, and,
2. for any states (q,x) and (q',x') in $\mathcal{A}$, we have $(q,x) \xrightarrow{*}_A (q',x')$ iff $(\overline{q},x) \xrightarrow{*}_B (\overline{q'},x')$.

We will use the notion of effective simulation (which is different from the classical notion of simulation) to have simpler proofs for the effective computation of *post** and *pre**. Assume that $\mathbf{C}$ is effectively simulable by $\mathbf{D}$. If there exists an algorithm which computes for every E-VASS in $\mathbf{D}$ a finite description of *post** (resp. *pre**) then we get that there exists an algorithm which computes for every E-VASS in $\mathbf{C}$ a finite description of *post** (resp. *pre**).

Theorem 1. *We have the following simulations between classes of 2-dim E-VASS:*

1. *the class* $\mathbf{T}_{1,2}$ *is effectively simulable by the class* $\mathbf{T}_1\mathbf{Tr}_{21}$*,*
2. *the class* $\mathbf{T}_1\mathbf{R}_{1,2}\mathbf{Tr}_{12}$ *is effectively simulable by the class* $\mathbf{T}_{1,2}$*,*
3. *the class* $\mathbf{T}_1\mathbf{R}_{1,2}\mathbf{Tr}_{12}$ *is effectively simulable by the class* $\mathbf{T}_1\mathbf{R}_2$*,*
4. *the class* $\mathbf{T}_1\mathbf{R}_1\mathbf{Tr}_{12}$ *is effectively simulable by the class* $\mathbf{T}_1$*.*
5. *the class* $\mathbf{R}_{1,2}\mathbf{Tr}_{12,21}$ *is effectively simulable by the class* $\mathbf{T}_1\mathbf{R}_2$*,*

3 Semilinear Sets and Projections

Let $P = \{p_1, p_2, \cdots, p_k\}$ be a finite subset of $\mathbb{N}^n$. We write P^* for the set $P^* = \{\Sigma_{i=1}^k \alpha_i p_i \;/\; \forall i \in [1,k], \alpha_i \in \mathbb{N}\}$. For every $x \in \mathbb{N}^n$, $(x + P^*)$ is called a *linear* set. A *semilinear* set is a finite union of linear sets. A *description* of a semilinear set L is any finite family $\{(x_i, P_i) \;/\; i \in I\}$ such that $L = \bigcup_{i \in I}(x_i + {P_i}^*)$. We will say that a semilinear set L is *effectively computable* if one can give *explicitly* an algorithm which computes a description of L.

Let us remark that for every $x \in \mathbb{N}^n$ and $v \in \mathbb{Z}^n$, the set $L = ((x+P^*)+v)\cap\mathbb{N}^n$ is semilinear as it can be written as $L = (B + P^*)$ where $B = \mathrm{Min}(((x + P^*) + v) \cap \mathbb{N}^n)$ is finite. Moreover, it can be shown that B is computable [HP79].

For every $i \in [1, n]$, we write H_i for the linear set $H_i = \{x \in \mathbb{N}^n \;/\; x(i) = 0\}$. We define two kinds of projection on H_i, extended on subsets of $\mathbb{N}^n$ in the usual way:

- for every $i \in [1, n]$, $proj_i : \mathbb{N}^n \to \mathbb{N}^n$ is the *orthogonal projection* on H_i defined by $proj_i(x) = x'$ where $x'(i) = 0$, and $\forall k \neq i, x'(k) = x(k)$.
- for every $i, j \in [1, n]$ such that $i \neq j$, $proj_{ij} : \mathbb{N}^n \to \mathbb{N}^n$ is the *diagonal projection* on H_i defined by $proj_{ij}(x) = x'$ where $x'(i) = 0, x'(j) = x(i)+x(j)$ and $\forall k \notin \{i, j\}, x'(k) = x(k)$.

4 Computation of a Symbolic Reachability Tree

Let us consider an n-dim E-VASS $\mathcal{A}$. The algorithm we propose later is based on the construction of a tree labelled by 3-tuples (q, x, P) where q is a control state of $\mathcal{A}$, x is in $\mathbb{N}^n$ and P is a finite subset of $\mathbb{N}^n$. Intuitively, the label $l = (q, x, P)$ of a node t (written $t : l$) represents the set of states $[\![l]\!] = \{q\} \times (x + P^*)$ of $\mathcal{A}$.

Notation. For every control state q of an n-dim E-VASS $\mathcal{A}$, for every $x \in \mathbb{N}^n$ and for every finite subset P of $\mathbb{N}^n$, we write $[\![q, x, P]\!]$ for the set of states $[\![q, x, P]\!] = \{q\} \times (x + P^*)$ of $\mathcal{A}$.

We are now able to define a notion of Symbolic Reachability Tree as follows.

Definition 2. A *Symbolic Reachability Tree* for an n-dim E-VASS $\mathcal{A}$ with a set S_0 of initial states is any rooted directed tree T labelled with 3-tuples (q, x, P) where q is a control state of $\mathcal{A}$, $x \in \mathbb{N}^n$ and P is a finite subset of $\mathbb{N}^n$, such that:

1. we have $\text{post}^*(S_0) = \bigcup_{t:l\ node\ of\ T} [\![l]\!]$, and,
2. for any node $t : l$ in T such that t has a son, then $post([\![l]\!]) \subseteq \bigcup_{i \in I} [\![l_i]\!] \subseteq post^*([\![l]\!])$ where $\{t_i : l_i\}_{i \in I}$ is the set of sons of t.

The `SymbolicTree` algorithm essentially computes a reachability tree. But as the reachability tree of an E-VASS may be infinite in general, an `Acceleration` function is used to accelerate the computation. Similarly to the Karp-Miller strategy [KM69], the `Acceleration` function basically computes the iteration of a repetitive sequence. This `Acceleration` function may depend on the input E-VASS, for instance we will define specific `Acceleration` functions suited to specific classes of E-VASS.

Algorithm 1 `SymbolicTree`($\mathcal{A}, S_0,$ `Acceleration`)

Input: an n-dim E-VASS $\mathcal{A}$, a linear set $S_0 = \{q_0\} \times (x_0 + P_0{}^*)$ of initial states of $\mathcal{A}$ and an `Acceleration` function

1: create root labelled (q_0, x_0, P_0)
2: **while** there are unmarked nodes **do**
3: pick an unmarked node $t : l$, where $l = (q, x, P)$
4: mark t
5: **if** there is an ancestor $t' : l'$ of t such that $[\![l]\!] \subseteq [\![l']\!]$ **then**
6: skip
7: **else**
8: $W(t) \leftarrow$ `Acceleration`$(t, \mathcal{A})$
9: {We now compute the set $post([\![q, x, W]\!])$ of immediate successors of $[\![q, x, W]\!]$}
10: **for each** transition $q \xrightarrow{add(v)} q'$ in A **do**
11: **for each** $m \in \text{Min}(((x + W^*) + v) \cap \mathbb{N}^n)$ **do**
12: construct a son $t' : (q', m, W)$ of t and label the arc $t \xrightarrow{add(v)} t'$
13: **for each** transition $q \xrightarrow{test(i)} q'$ in A **do**
14: **if** $x_i = 0$ **then**
15: construct a son $t' : (q', x, W \cap H_i)$ of t and label the arc $t \xrightarrow{test(i)} t'$
16: **for each** transition $q \xrightarrow{weaktest(i)} q'$ in A **do**
17: **if** $x_i = 0$ **then**
18: construct a son $t' : (q', x, (W \cap H_i) \cup \{e_i\})$ of t and label the arc $t \xrightarrow{test(i)} t'$
19: **for each** transition $q \xrightarrow{reset(i)} q'$ in A **do**
20: construct a son $t' : (q', proj_i(x), proj_i(W) \setminus \{0\})$ of t and label the arc $t \xrightarrow{reset(i)} t'$
21: **for each** transition $q \xrightarrow{transfer(i \to j)} q'$ in A **do**
22: construct a son $t' : (q', proj_{ij}(x), proj_{ij}(W))$ of t and label the arc $t \xrightarrow{transfer(i \to j)} t'$

We will often use the following stability property for the different considered `Acceleration` functions.

Definition 3. An `Acceleration` function is *n-stable* if for every n-dim E-VASS $\mathcal{A}$ with a linear set of initial states S_0 and for every node $t : (q, x, P)$ in the tree constructed by $\texttt{SymbolicTree}(\mathcal{A}, S_0, \texttt{Acceleration})$, we have the following property:

$$[\![q, x, P]\!] \subseteq [\![q, x, W(t)]\!] \subseteq post^*([\![q, x, P]\!])$$

where $W(t) = \texttt{Acceleration}(t, \mathcal{A})$.

Theorem 2. *Let $\mathcal{A}$ be an n-dim E-VASS and $S_0 = \{q_0\} \times (x_0 + {P_0}^*)$ be a linear set of initial states of $\mathcal{A}$. If an* `Acceleration` *function is n-stable then the* `SymbolicTree` *algorithm, applied to $(\mathcal{A}, S_0, \texttt{Acceleration})$, constructs a Symbolic Reachability Tree for $\mathcal{A}$ with initial states S_0.*

Notice that if the `SymbolicTree` algorithm terminates on some n-dim E-VASS $\mathcal{A}$ with a linear set of initial states S_0 and an n-stable `Acceleration` function, then post$^*(S_0)$ is semilinear and effectively computable. Because 3-dim VASS can have non-semilinear reachability set [HP79], we now restrict our study to 2-dim E-VASS. For simplicity, we will shortly write *stable* for 2-stable. In the following of the paper, our aim is to find, for each analyzed class of 2-dim E-VASS, a dedicated stable `Acceleration` function ensuring the termination of the `SymbolicTree` algorithm for this class.

5 Effective Semilinearity of *post** for the Class $\mathbf{T}_1\mathbf{R}_1\mathbf{Tr}_{12}$

We now may generalize Hopcroft-Pansiot's result in showing that *post** is still semilinear and effectively computable for any E-VASS $\mathcal{A}$ in $\mathbf{T}_1\mathbf{R}_1\mathbf{Tr}_{12}$. We extract from Hopcroft-Pansiot's algorithm an $\texttt{Acceleration}_{HP}$ function [1] which is stable and such that for every 2-dim VASS $\mathcal{A}$ with a linear set of initial states S_0, the `SymbolicTree` algorithm applied to $(\mathcal{A}, S_0, \texttt{Acceleration}_{HP})$ terminates. This $\texttt{Acceleration}_{HP}$ function is then used to define a $\texttt{Acceleration}_{\mathbf{T}_1}$ function which allows us to show that *post** is still semilinear and effectively computable for any E-VASS $\mathcal{A}$ in $\mathbf{T}_1\mathbf{R}_1\mathbf{Tr}_{12}$.

Proposition 1. *The function* $\texttt{Acceleration}_{\mathbf{T}_1}$ *is stable. Moreover, for every 2-dim E-VASS $\mathcal{A}$ in the class $\mathbf{T}_1$ with a linear set of initial states S_0, the* `SymbolicTree` *algorithm applied to $(\mathcal{A}, S_0, \texttt{Acceleration}_{\mathbf{T}_1})$ terminates.*

By using the effective simulation of $\mathbf{T}_1\mathbf{R}_1\mathbf{Tr}_{12}$ by $\mathbf{T}_1$ (Theorem 1), and Proposition 1, we obtain the following theorem.

Theorem 3. *For any 2-dim E-VASS $\mathcal{A}$ in the class $\mathbf{T}_1\mathbf{R}_1\mathbf{Tr}_{12}$ with a semilinear set S_0 of initial states, post$^*(S_0)$ is semilinear and it is effectively computable.*

[1] The definition of this function is omitted here, but it can be found in the full version.

Algorithm 2 $\texttt{Acceleration}_{\mathbf{T}_1}(t : (q, x, P), \mathcal{A})$

Input: a node $t : (q, x, P)$ and a 2-dim E-VASS $\mathcal{A}$

1: **if** the branch from the root r to t may be written as $r \xrightarrow{*} u \xrightarrow{add(v)} t$ for some v **then**
2:    **return** $\texttt{Acceleration}_{HP}(t, \mathcal{A})$
3: **else if** the branch from the root r to t may be written as $r \xrightarrow{*} u \xrightarrow{test(1)} t$ **then**
4:    **if** there is an ancestor $t' : (q, x', P')$ of t such that $(x - x') \in \{0\} \times \mathbb{N}^*$ and there is no extended arc except $test(1)$ arcs between t' and t **then**
5:       **return** $P \cup \{x - x'\}$
6: **else**
7:    **return** P

6 Effective Semilinearity of *pre** for the Class $\mathbf{T}_1\mathbf{R}_{1,2}\mathbf{Tr}_{12}$

We present in this section an algorithm for computing *pre** for the largest class (non equivalent to $\mathbf{T}_{1,2}$) of 2-dim E-VASS. We first prove that *post** of a 2-dim E-VASS $\mathcal{A}$ in $\mathbf{T}_1\mathbf{W}_2$ is semilinear and effectively computable (accordingly to previous notations, we note $\mathbf{T}_1\mathbf{W}_2$ the class of 2-dim E-VASS where the allowed extended transitions are labelled by $test(1)$ or $weaktest(2)$.

Algorithm 3 $\texttt{Acceleration}_{\mathbf{T}_1\mathbf{W}_2}(t : (q, x, P), \mathcal{A})$

Input: a node $t : (q, x, P)$ and a 2-dim E-VASS $\mathcal{A}$

1: **if** the branch from the root r to t may be written as $r \xrightarrow{*} u \xrightarrow{weaktest(2)} t$ **then**
2:    **if** there is an ancestor $t' : (q, x', P')$ of t such that $(x - x') \in \mathbb{N}^* \times \{0\}$ and there is no extended arc except $weaktest(2)$ arcs between t' and t **then**
3:       **return** $P \cup \{x - x'\}$
4: **else**
5:    **return** $\texttt{Acceleration}_{\mathbf{T}_1}(t, \mathcal{A})$

Proposition 2. *The function* $\texttt{Acceleration}_{\mathbf{T}_1\mathbf{W}_2}$ *is stable. Moreover, for every 2-dim E-VASS* $\mathcal{A}$ *in the class* $\mathbf{T}_1\mathbf{W}_2$ *with a linear set of initial states* S_0*, the* `SymbolicTree` *algorithm applied to* $(\mathcal{A}, S_0, \texttt{Acceleration}_{\mathbf{T}_1\mathbf{W}_2})$ *terminates.*

Theorem 4. *For any 2-dim E-VASS* $\mathcal{A}$ *in the class* $\mathbf{T}_1\mathbf{R}_{1,2}\mathbf{Tr}_{12}$ *with a semilinear set* S_0 *of initial states, pre**(S_0) *is semilinear and it is effectively computable.*

Corollary 1. *The reachability problem is decidable for the class* $\mathbf{T}_1\mathbf{R}_{1,2}\mathbf{Tr}_{12}$.

7 Semilinearity of *post** for the Class $\mathbf{T}_1\mathbf{R}_{1,2}\mathbf{Tr}_{12}$

Let us recall that in section 5, we proved that *post** is semilinear and effectively computable for any $\mathcal{A}$ in $\mathbf{T}_1\mathbf{R}_1\mathbf{Tr}_{12}$. We now extend this semilinearity result to the class $\mathbf{T}_1\mathbf{R}_{1,2}\mathbf{Tr}_{12}$.

Let $\Omega(\mathcal{A}, S_0)$ be the set defined by $\Omega(\mathcal{A}, S_0) = \bigcup_{s \in S_0} \Omega(\mathcal{A}, s)$ where for every state s of $\mathcal{A}$, $\Omega(\mathcal{A}, s)$ is given by:

$$\Omega(\mathcal{A}, s) = \{q \in Q \;/\; |\{x \in \mathbb{N} \;/\; \exists\, \sigma,\; s \xrightarrow{\sigma test(1)} (q, (0, x))\}| = \infty\}$$

where $\mathcal{A}$ is a 2-dim E-VASS with a set Q of control states and S_0 is a set of initial states of $\mathcal{A}$.

The main problem to ensure termination is that we may have on a branch of the constructed tree infinitely many interleavings between $reset(2)$ and $test(1)$ transitions. Hence, the $\texttt{Acceleration}_{\mathbf{T}_1\mathbf{R}_2}$ function uses the set $\Omega(\mathcal{A}, S_0)$ in order to iterate some special loops in this case. Intuitively, if we have just fired a $reset(2)$ transition, then we do not care about the actual value of the second component before the $reset(2)$. Hence if we know that this value may be arbitrarily large, then we can iterate a simple control loop that increases the first component before the firing of the $reset(2)$ transition. Formally, a *simple control loop* l in an E-VASS $\mathcal{A}$ is a path $l = q \xrightarrow{add(v_1)} q_1 \xrightarrow{add(v_2)} q_2 \cdots q_{n-1} \xrightarrow{add(v_n)} q$ in the E-VASS Control A of $\mathcal{A}$ with no repeating state except that the first state and the last state are the same ; moreover we say that l is a control loop *on* q with *displacement* $d(l) = v_1 + v_2 + \cdots + v_n$.

Notice that the set $\Omega(\mathcal{A}, S_0)$ is finite, as it is a subset of Q. Hence, even if we actually do not know whether we can effectively compute this set, we can use this set as an "oracle" in our $\texttt{Acceleration}_{\mathbf{T}_1\mathbf{R}_2}$ function.

Algorithm 4 $\texttt{Acceleration}_{\mathbf{T}_1\mathbf{R}_2}(t : (q, x, P), \mathcal{A})$

Input: a node $t : (q, x, P)$ and a 2-dim E-VASS $\mathcal{A}$

1: **if** the branch from the root r to t may be written as $r \xrightarrow{*} u \xrightarrow{reset(2)} t$ **then**
2: **if** there is an ancestor $t' : (q, x', P')$ of t such that $(x - x') \in \mathbb{N}^* \times \{0\}$ and there is no extended arc except $reset(2)$ arcs between t' and t **then**
3: **return** $P \cup \{x - x'\}$
4: **else if** the branch from the root r to t may be written as $r \xrightarrow{*} s \xrightarrow{test(1)} t' \xrightarrow{*} t'' \xrightarrow{reset(2)} t$ with no extended arc between t' and t'' **then**
5: **if** the label (q', x', P') of t' is such that $q' \in \Omega(\mathcal{A}, S_0)$ **then**
6: **if** there exists a node $u : (r, y, Q)$ between t' and t'' and a simple control loop on r with displacement $(a, b) \in \mathbb{N}^* \times \mathbb{Z}$ **then**
7: **return** $P \cup \{(a, 0)\}$
8: **else**
9: **return** $\texttt{Acceleration}_{\mathbf{T}_1}(t, \mathcal{A})$

Proposition 3. *The function* $\texttt{Acceleration}_{\mathbf{T}_1\mathbf{R}_2}$ *is stable. Moreover, for every 2-dim E-VASS* $\mathcal{A}$ *in the class* $\mathbf{T}_1\mathbf{R}_2$ *with a linear set of initial states* S_0*, the* `SymbolicTree` *algorithm applied to* $(\mathcal{A}, S_0, \texttt{Acceleration}_{\mathbf{T}_1\mathbf{R}_2})$ *terminates.*

From Theorem 1, the class $\mathbf{T}_1\mathbf{R}_{1,2}\mathbf{Tr}_{12}$ is effectively simulated by $\mathbf{T}_1\mathbf{R}_2$. From Proposition 3, we conclude that $post^*$ is semilinear.

Theorem 5. *For any 2-dim E-VASS $\mathcal{A}$ in the class $\mathbf{T}_1\mathbf{R}_{1,2}\mathbf{Tr}_{12}$ with a semilinear set S_0 of initial states, post*(S_0) is semilinear. Moreover, if $\Omega(\mathcal{A}, S_0)$ is effectively computable, then post*(S_0) is effectively computable.*

It remains an open problem to know whether it is possible to compute a finite description of the semilinear reachability set for the class $\mathbf{T}_1\mathbf{R}_{1,2}\mathbf{Tr}_{12}$.

8 Conclusion

We have systematically studied all the subclasses of two {+1, -1, reset, transfer, zero-test}-counters automata and given a complete hierarchy between classes using new simulations. It turns out that a "maximal decidable model" consists of the class of 2-counters automata ($\mathbf{T}_1\mathbf{R}_{1,2}\mathbf{Tr}_{12}$) with the following extended operations: test for zero on the first counter, reset on the two counters, and transfer from the first to the second counter. For this model, the *pre** image is effectively semilinear.

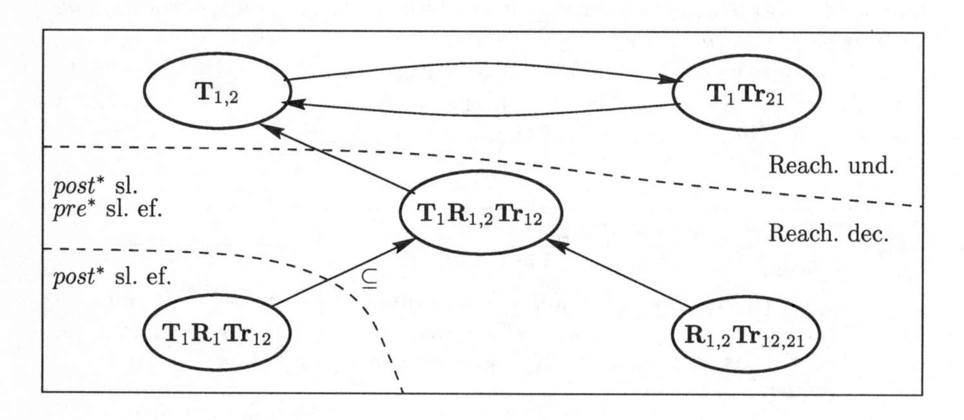

Fig. 1. Graph of inclusions and simulations

Our results cannot be generalized to three {+1, -1}-counters automata because they may have non-semilinear reachability sets; moreover, reachability is undecidable for three {+1, -1, reset}-counters automata such that one of the counters never uses the reset operation [DFS98, Duf98].

Our symbolic algorithm has been defined for any {+1, -1, reset, transfer, zero-test}-counters automata. It can also be used as a semi-algorithm when termination cannot be guaranteed.

A lot of other open problems still arise: for example, we have no answer whether *post** is *effectively* semilinear for the class $\mathbf{T}_1\mathbf{R}_{1,2}\mathbf{Tr}_{12}$. The complexity of the symbolic algorithm is certainly at least in 2^{2^n} [HRHY86] but we don't know if it is primitive recursive or not. Another point which has to be stated is the comparison between classes of 2-counters automata and stack automata.

References

[BEM97] A. Bouajjani, J. Esparza, and O. Maler. Reachability analysis of pushdown automata: Application to model-checking. In *Proc. 8th Int. Conf. Concurrency Theory (CONCUR'97), Warsaw, Poland, Jul. 1997*, volume 1243 of *Lecture Notes in Computer Science*, pages 135–150. Springer, 1997.

[BS86] M. Benois and J. Sakarovitch. On the complexity of some extended word problems defined by cancellation rules. *Information Processing Letters*, 23(6):281–287, 1986.

[Cau92] D. Caucal. On the regular structure of prefix rewriting. *Theoretical Computer Science*, 106(1):61–86, 1992.

[CF97] G. Cécé and A. Finkel. Programs with quasi-stable channels are effectively recognizable. In *Proc. of the* 9th *Conference on Computer-Aided Verification (CAV)*, volume 1254, pages 304–315. LNCS, June 1997.

[CFP96] G. Cécé, A. Finkel, and I. S. Purushothaman. Unreliable channels are easier to verify than perfect channels. *Information and Computation*, 124(1):20–31, 1996.

[CJ98] H. Comon and Y. Jurski. Multiple counters automata, safety analysis and Presburger arithmetic. In *Proc. 10th Int. Conf. Computer Aided Verification (CAV'98), Vancouver, BC, Canada, June-July 1998*, volume 1427 of *Lecture Notes in Computer Science*, pages 268–279. Springer, 1998.

[DFS98] C. Dufourd, A. Finkel, and Ph. Schnoebelen. Reset nets between decidability and undecidability. In *Proc. 25th Int. Coll. Automata, Languages, and Programming (ICALP'98), Aalborg, Denmark, July 1998*, volume 1443 of *Lecture Notes in Computer Science*, pages 103–115. Springer, 1998.

[DJS99] C. Dufourd, P. Jančar, and Ph. Schnoebelen. Boundedness of Reset P/T nets. In *Proc. 26th Int. Coll. Automata, Languages, and Programming (ICALP'99), Prague, Czech Republic, July 1999*, volume 1644 of *Lecture Notes in Computer Science*, pages 301–310. Springer, 1999.

[Duf98] C. Dufourd. Réseaux de Petri avec reset/transfert: Décidabilité et indécidabilité. Thèse de Docteur en Sciences de l'École Normale Supérieure de Cachan, October 1998.

[FWW97] A. Finkel, B. Willems, and P. Wolper. A direct symbolic approach to model checking pushdown systems. *Electronic Notes in Theoretical Computer Science*, 9, 1997.

[HP79] J. Hopcroft and J. J. Pansiot. On the reachability problem for 5-dimensional vector addition systems. *Theoretical Computer Science*, 8:135–159, 1979.

[HRHY86] R. R. Howell, L. E. Rosier, D. T. Huynh, and H-C. Yen. Some complexity bounds for problems concerning finite and 2-dimensional vector addition systems with states. *Theoretical Computer Science*, 46:107–140, 1986.

[KM69] R. M. Karp and R. E. Miller. Parallel program schemata. *Journal of Computer and System Sciences*, 3(2):147–195, 1969.

[Rei95] K. Reinhardt. Reachability in Petri nets with inhibitor arcs, November 1995. Unpublished manuscript. See `www-fs.informatik.uni-tuebingen.de/~reinhard`.

Hereditary History Preserving Bisimilarity Is Undecidable

Marcin Jurdziński and Mogens Nielsen

BRICS*,
Department of Computer Science
University of Aarhus,
Ny Munkegade, Building 540, 8000 Aarhus C, Denmark.
{mju,mn}@brics.dk

Abstract History preserving bisimilarity (hp-bisimilarity) and hereditary history preserving bisimilarity (hhp-bisimilarity) are behavioural equivalences taking into account causal relationships between events of concurrent systems. Their prominent feature is being preserved under action refinement, an operation important for the top-down design of concurrent systems. We show that—unlike hp-bisimilarity—checking hhp-bisimilarity for finite labelled asynchronous transition systems is not decidable, by a reduction from the halting problem of 2-counter machines. To make the proof more transparent we introduce an intermediate problem of checking domino bisimilarity for origin constrained tiling systems, whose undecidability is interesting in its own right. We also argue that the undecidability of hhp-bisimilarity holds for finite labelled 1-safe Petri nets.

1 Introduction

The notion of behavioural equivalence that has attracted most attention in concurrency theory is bisimilarity, originally introduced by Park [20] and Milner [15]; concurrent programs are considered to have the same meaning if they are bisimilar. The prominent role of bisimilarity is due to many pleasant properties it enjoys; we mention a few of them here.

A process of checking whether two transition systems are bisimilar can be seen as a two player game which is in fact an Ehrenfeucht-Fraïssé type of game for modal logic. More precisely, there is a winning strategy for a player who wants to show that the systems are bisimilar if and only if the systems cannot be distinguished by the formulas of the logic; the result due to Hennessy and Milner [9].

Another notable property of bisimilarity is its computational feasibility; see for example the overview note [16]. Let us illustrate this on the examples of finite transition systems and a class of infinite-state transition systems generated

* Basic Research in Computer Science,
Centre of the Danish National Research Foundation.

H. Reichel and S. Tison (Eds.): STACS 2000, LNCS 1770, pp. 358–369, 2000.

by context free grammars. For finite transition systems there are very efficient polynomial time algorithms for checking bisimilarity [13, 19], in sharp contrast to **PSPACE**-completeness of the classical language equivalence. For transition systems generated by context free grammars, while language equivalence is undecidable, bisimilarity is decidable [3], and if the grammar has no redundant nonterminals, even in polynomial time [10]. Furthermore, as the results of [8] indicate, bisimilarity has a very rare status of being a decidable equivalence for context free grammars: all the other equivalences in the linear/branching time hierarchy [6] are indeed undecidable. The algorithmic tractability makes bisimilarity especially attractive for automatic verification of concurrent systems.

The essence of bisimilarity, quoting [9], "is that the behaviour of a program is determined by how it communicates with an observer." Therefore, the notion of what can be observed of a behaviour of a system affects the notion of bisimilarity. An abstract definition of bisimilarity for arbitrary categories of models due to Joyal *et al.* [12] formalizes this idea. Given a category of models where objects are behaviours and morphisms correspond to extension of behaviours, and given a subcategory of observable behaviours, the abstract definition yields a notion of bisimilarity for all behaviours with respect to observable behaviours. For example, for rooted labelled transition systems, taking synchronization trees [15] into which they unfold as their behaviours, and sequences of actions as the observable behaviours, we recover the standard strong bisimilarity of Park and Milner [12].

In order to model concurrency more faithfully several models have been introduced (see [23] for a survey) that make explicit the distinction between events that can occur concurrently, and those that are causally related. Then a natural choice is to replace sequences, *i.e.*, linear orders as the observable behaviours, by partial orders of occurrences of events with causality as the ordering relation. For example, taking unfoldings of labelled asynchronous transition systems into event structures as the behaviours, and labelled partial orders as the observations, Joyal *et al.* [12] obtained from their abstract definition the hereditary history preserving bisimilarity (hhp-bisimilarity), independently introduced and studied by Bednarczyk [1].

A similar notion of bisimilarity has been studied before, namely history preserving bisimilarity (hp-bisimilarity), introduced by Rabinovich and Trakhtenbrot [21] and van Glabbeek and Goltz [7]. For the relationship between hp- and hhp-bisimilarity see for example [1, 12, 5].

One of the important motivations to study partial order based equivalences was the discovery that hp-bisimilarity has a rare status of being preserved under action refinement [7], an operation important for the top-down design of concurrent systems. Bednarczyk [1] has extended this result to hhp-bisimilarity.

There is a natural logical characterization of hhp-bisimilarity checking games as shown by Nielsen and Clausen [17]: they are characteristic games for an extension of modal logic with backwards modalities, interpreted over event structures.

Hp-bisimilarity has been shown to be decidable for 1-safe Petri nets by Vogler [22], and to be **DEXP**-complete by Jategaonkar, and Meyer [11]; let us just mention here that 1-safe Petri nets can be regarded as a proper subclass of finite asynchronous transition systems (see [23] for details), and that decidability of hp-bisimilarity can be easily extended to all finite asynchronous transition systems using the methods of [11].

Hhp-bisimilarity seems to be only a slight strengthening of hp-bisimilarity [12], and hence many attempts have been made to extend the above mentioned algorithms to the case of hhp-bisimilarity. However, decidability of hhp-bisimilarity has remained open, despite several attempts over the years [17, 18, 2, 5]. Fröschle and Hildebrandt [5] have discovered an infinite hierarchy of bisimilarity notions refining hp-bisimilarity, and coarser than hhp-bisimilarity, such that hhp-bisimilarity is the intersection of all the bisimilarities in the hierarchy. They have shown all these bisimilarities to be decidable for 1-safe Petri nets. Fröschle [4] has shown hhp-bisimilarity to be decidable for BPP-processes, a class of infinite state systems.

In this paper, we finally settle the question of decidability of hhp-bisimilarity by showing it to be undecidable for finite 1-safe Petri nets. In order to make the proof more transparent we first introduce an intermediate problem of domino bisimilarity and show its undecidability by a direct reduction from the halting problem of 2-counter machines.

2 Hereditary History Preserving Bisimilarity

Definition 1 (Labelled asynchronous transition system)
A *labelled asynchronous transition system* is a tuple $A = (S, s^{\text{ini}}, E, \rightarrow, L, \lambda, I)$, where S is its set of *states*, $s^{\text{ini}} \in S$ is the *initial state*, E is the set of *events*, $\rightarrow \subseteq S \times E \times S$ is the set of *transitions*, L is the set of labels, and $\lambda : E \rightarrow L$ is the *labelling function*, and $I \subseteq E^2$ is the *independence relation* which is irreflexive and symmetric. We often write $s \xrightarrow{e} s'$, instead of $(s, e, s') \in \rightarrow$. Moreover, the following conditions have to be satisfied:

1. if $s \xrightarrow{e} s'$, and $s \xrightarrow{e} s''$, then $s' = s''$,
2. if $(e, e') \in I$, $s \xrightarrow{e} s'$, and $s' \xrightarrow{e'} t$, then $s \xrightarrow{e'} s''$, and $s'' \xrightarrow{e} t$ for some $s'' \in S$.

An asynchronous transition system is *coherent* if it satisfies one further condition:

3. if $(e, e') \in I$, $s \xrightarrow{e} s'$, and $s \xrightarrow{e'} s''$, then $s' \xrightarrow{e'} t$, and $s'' \xrightarrow{e} t$ for some $t \in S$.

[Definition 1] □

Winskel and Nielsen [23, 18] give a thorough survey and establish formal relationships between asynchronous transition systems and other models for concurrency, such as Petri nets, and event structures. The independence relation is meant to model concurrency: independent events can occur concurrently, while those that are not independent are causally related or in conflict.

Let $A = (S, s^{\text{ini}}, E, \rightarrow, L, \lambda, I)$ be a labelled asynchronous transition system. A sequence of events $\overline{e} = \langle e_1, e_2, \ldots, e_n \rangle \in E^n$ is a *run* of A if there are states $s_1, s_2, \ldots, s_{n+1} \in S$, such that $s_1 = s^{\text{ini}}$, and for all $i \in [n]$ we have $s_i \xrightarrow{e_i} s_{i+1}$. We denote the set of runs of A by $\text{Runs}(A)$. We extend the labelling function λ

to runs in the standard way. We say that $k \in [n]$ is *most recent* in $\overline{e}$, and we denote it by $k \in \mathrm{MR}(\overline{e})$, if and only if $(e_k, e_\ell) \in I$ for all ℓ such that $k < \ell \leq n$. Note that if $k \in \mathrm{MR}(\overline{e})$ then $\overline{e} \oslash k = \langle e_1, \ldots, e_{k-1}, e_{k+1}, \ldots, e_n \rangle \in \mathrm{Runs}(A)$.

Definition 2 (Hereditary history preserving bisimulation)
Let $A_i = (S_i, s_i^{\mathrm{ini}}, E_i, \rightarrow_i, L, \lambda_i, I_i)$ for $i \in \{1, 2\}$ be labelled asynchronous transition systems. A relation $B \subseteq \mathrm{Runs}(A_1) \times \mathrm{Runs}(A_2)$ is a *hereditary history preserving* (hhp-) bisimulation relating A_1 and A_2 if the following conditions are satisfied:

1. $(\varepsilon, \varepsilon) \in B$,

and if $(\overline{e_1}, \overline{e_2}) \in B$ then $\lambda_1(\overline{e_1}) = \lambda_2(\overline{e_2})$, and:

2. for all $e_1 \in E_1$, if $\overline{e_1} \cdot e_1 \in \mathrm{Runs}(A_1)$, then there exists $e_2 \in E_2$, such that $\overline{e_2} \cdot e_2 \in \mathrm{Runs}(A_2)$, and $\lambda_1(e_1) = \lambda_2(e_2)$, and $(\overline{e_1} \cdot e_1, \overline{e_2} \cdot e_2) \in B$,
3. for all $e_2 \in E_2$, if $\overline{e_2} \cdot e_2 \in \mathrm{Runs}(A_2)$, then there exists $e_1 \in E_1$, such that $\overline{e_1} \cdot e_1 \in \mathrm{Runs}(A_1)$, and $\lambda_1(e_1) = \lambda_2(e_2)$, and $(\overline{e_1} \cdot e_1, \overline{e_2} \cdot e_2) \in B$,
4. $k \in \mathrm{MR}(\overline{e_1})$, if and only if $k \in \mathrm{MR}(\overline{e_2})$,
5. if $k \in \mathrm{MR}(\overline{e_1}) = \mathrm{MR}(\overline{e_2})$, then $(\overline{e_1} \oslash k, \overline{e_2} \oslash k) \in B$. [Definition 2] □

Two asynchronous transition systems A_1, and A_2 are hereditary history preserving (hhp-) *bisimilar*, if there is an hhp-bisimulation relating them.

Remark 1 The term hereditary history preserving bisimulation originates from the fact that this notion of bisimulation has an alternative definition, which is formally a small strengthening of the standard definition of history preserving bisimulation [21, 7], based explicitly on partial order behaviours [1, 12]. Note that Definition 2 does not mention partial order behaviours explicitly, but they are implicit in the notion of most recent occurrences of events. For the proof of equivalence of our definition and the other ones see [17].

The main result of this paper is the following theorem proved in section 4.

Theorem 3 (Undecidability of hhp-bisimilarity)
Hhp-bisimilarity is undecidable for finite labelled asynchronous transition systems.

3 Domino Bisimilarity Is Undecidable

3.1 Domino Bisimilarity

Definition 4 (Origin constrained tiling system)
An *origin constrained tiling system* $T = \big(D, D^{\mathrm{ori}}, (H, H^0), (V, V^0), L, \lambda\big)$ consists of a set D of *dominoes*, its subset $D^{\mathrm{ori}} \subseteq D$ called the *origin constraint*, two *horizontal compatibility* relations $H, H^0 \subseteq D^2$, two *vertical compatibility* relations $V, V^0 \subseteq D^2$, a set L of *labels*, and a *labelling function* $\lambda : D \rightarrow L$.

[Definition 4] □

A *configuration* of T is a triple $(d, x, y) \in D \times \mathbb{N} \times \mathbb{N}$, such that if $x = y = 0$ then $d \in D^{\mathrm{ori}}$. In other words, in the "origin" position $(x, y) = (0, 0)$ of the non-negative integer grid only dominoes from the origin constraint D^{ori} are allowed.

Let (d, x, y), and (d', x', y') be configurations of T such that $|x' - x| + |y' - y| = 1$, *i.e.*, the positions (x, y), and (x', y') are neighbours in the non-negative integer grid. Without loss of generality we may assume that $x + y < x' + y'$. We say that configurations (d, x, y), and (d', x', y') are *compatible* if either of the two conditions below holds:

- $x' = x$, and $y' = y + 1$, and
  if $y = 0$, then $(d, d') \in V^0$, and if $y > 0$, then $(d, d') \in V$, or
- $x' = x + 1$, and $y' = y$, and
  if $x = 0$, then $(d, d') \in H^0$, and if $x > 0$, then $(d, d') \in H$.

Definition 5 (Domino bisimulation)
Let $T_i = \left(D_i, D_i^{\mathrm{ori}}, (H_i, H_i^0), (V_i, V_i^0), L_i, \lambda_i\right)$ for $i \in \{1, 2\}$ be origin constrained tiling systems. A relation $B \subseteq D_1 \times D_2 \times \mathbb{N} \times \mathbb{N}$ is a *domino bisimulation* relating T_1 and T_2, if $(d_1, d_2, x, y) \in B$ implies that $\lambda_1(d_1) = \lambda_2(d_2)$, and the following conditions are satisfied for all $i \in \{1, 2\}$:

1. for all $d_i \in D_i^{\mathrm{ori}}$, there is $d_{3-i} \in D_{3-i}^{\mathrm{ori}}$, so that $\lambda_1(d_1) = \lambda_2(d_2)$, and $(d_1, d_2, 0, 0) \in B$,
2. for all $x, y \in \mathbb{N}$, such that $(x, y) \neq (0, 0)$, and $d_i \in D_i$, there is $d_{3-i} \in D_{3-i}$, such that $\lambda_1(d_1) = \lambda_2(d_2)$, and $(d_1, d_2, x, y) \in B$,
3. if $(d_1, d_2, x, y) \in B$, then for all neighbours $(x', y') \in \mathbb{N} \times \mathbb{N}$ of (x, y), and $d'_i \in D_i$, if configurations (d_i, x, y), and (d'_i, x', y') of T_i are compatible, then there exists $d'_{3-i} \in D_{3-i}$, such that $\lambda_1(d'_1) = \lambda_2(d'_2)$, and configurations (d_{3-i}, x, y), and (d'_{3-i}, x', y') of T_{3-i} are compatible, and $(d'_1, d'_2, x', y') \in B$.

[Definition 5] □

We say that two tiling systems are *domino bisimilar* if and only if there is a domino bisimulation relating them.

Theorem 6 (Undecidability of domino bisimilarity)
Domino bisimilarity is undecidable for origin constrained tiling systems.

The proof is a reduction from the halting problem for deterministic 2-counter machines. For a deterministic 2-counter machine M we define in section 3.3 two origin constrained tiling systems T_1, and T_2, enjoying the following property.

Proposition 7 Machine M does not halt, if and only if there is a domino bisimulation relating T_1 and T_2.

3.2 Counter Machines

A 2-counter machine M consists of a finite program with the set L of instruction labels, and instructions of the form:

- ℓ: $\mathtt{c}_i$:= $\mathtt{c}_i$ + 1; goto m
- ℓ: if $\mathtt{c}_i$ = 0 then $\mathtt{c}_i$:= $\mathtt{c}_i$ + 1; goto m
  else $\mathtt{c}_i$:= $\mathtt{c}_i$ - 1; goto n
- halt:

where $i = 1, 2$; $\ell, m, n \in L$, and $\{\mathtt{start}, \mathtt{halt}\} \subseteq L$. A configuration of M is a triple $(\ell, x, y) \in L \times \mathbb{N} \times \mathbb{N}$, where ℓ is the label of the current instruction, and x, and y are the values stored in counters $\mathtt{c}_1$, and $\mathtt{c}_2$, respectively; we denote the set of configurations of M by Confs(M). The semantics of 2-counter machines is standard: let $\vdash_M \subseteq \text{Confs}(M) \times \text{Confs}(M)$ be the usual one-step derivation relation on configurations of M; by $\vdash_M^+$ we denote the reachability (in at least one step) relation for configurations, *i.e.*, the transitive closure of $\vdash_M$.

Before we give a reduction from the halting problem of 2-counter machines to origin constrained domino bisimilarity let us take a look at the directed graph $(\text{Confs}(M), \vdash_M)$, with configurations of M as vertices, and edges denoting derivation in one step. Since machine M is deterministic, for each configuration there is at most one outgoing edge; moreover only halting configurations have no outgoing edges. It follows that connected components of the graph $(\text{Confs}(M), \vdash_M)$ are either trees with edges going to the root which is the unique halting configuration in the component, or have no halting configuration at all. This observation implies the following proposition.

Proposition 8 Let M be a 2-counter machine. The following conditions are equivalent:

1. machine M halts on input $(0, 0)$, *i.e.*, $(\mathtt{start}, 0, 0) \vdash_M^+ (\mathtt{halt}, x, y)$ for some $x, y \in \mathbb{N}$,
2. $(\mathtt{start}, 0, 0) \sim_M (\mathtt{halt}, x, y)$ for some $x, y \in \mathbb{N}$, where the relation $\sim_M \subseteq \text{Confs}(M) \times \text{Confs}(M)$ is the symmetric and transitive closure of $\vdash_M$.

3.3 The Reduction

Now we go for a proof of Proposition 7. The idea is to design a tiling system which "simulates" behaviour of a 2-counter machine.

Let M be a 2-counter machine. We construct a tiling system T_M with the set L of instruction labels of M as the set of dominoes, and the identity function on L as the labelling function. Note that this implies that all tuples belonging to a domino bisimulation relating copies of T_M are of the form (ℓ, ℓ, x, y), so we can identify them with configurations of M, *i.e.*, sometimes we will make no distinction between (ℓ, ℓ, x, y) and $(\ell, x, y) \in \text{Confs}(M)$ for $\ell \in L$.

We define the horizontal compatibility relations $H_M, H_M^0 \subseteq L \times L$ of the tiling system T_M as follows:

- $(\ell, m) \in H_M$ if and only if either of the instructions below is an instruction of machine M:
  - ℓ: `c1 := c1 + 1; goto` m
  - m: `if c1 = 0 then c1 := c1 + 1; goto` n `else c1 := c1 - 1; goto` ℓ
- $(\ell, m) \in H^0_M$ if and only if $(\ell, m) \in H_M$, or the instruction below is an instruction of machine M:
  - ℓ: `if c1 = 0 then c1 := c1 + 1; goto` m `else c1 := c1 - 1; goto` n

Vertical compatibility relations V_M, and V^0_M are defined in the same way, with c_1 instructions replaced with c_2 instructions. We also take $D^{\text{ori}}_M = L$, *i.e.*, all dominoes are allowed in position $(0,0)$. Note that the identity function is a 1-1 correspondence between configurations of M, and configurations of the tiling system T_M; from now on we will hence identify configurations of M and T_M. It follows immediately from the construction of T_M, that two configurations $c, c' \in \text{Confs}(M)$ are compatible as configurations of T_M, if and only if $c \vdash_M c'$, or $c' \vdash_M c$, *i.e.*, compatibility relation of T_M coincides with the symmetric closure of $\vdash_M$. By $\approx_M$ we denote the symmetric and transitive closure of the compatibility relation of configurations of T_M. The following proposition is then straightforward.

Proposition 9 The two relations $\sim_M$, and $\approx_M$ coincide.

Now we are ready to define the two origin constrained tiling systems T_1, and T_2, postulated in Proposition 7. The idea is to have two independent and slightly pruned copies of T_M in T_2: one without the initial configuration $(\texttt{start}, 0, 0)$, and the other without any halting configurations $(\texttt{halt}, x, y)$. The other tiling system T_1 is going to have three independent copies of T_M: the two of T_2, and moreover, another full copy of T_M.

More formally we define $D_2 = (L \times \{1,2\}) \setminus \{(\texttt{halt}, 2)\}$, and $D^{\text{ori}}_2 = D_2 \setminus \{(\texttt{start}, 1)\}$, and $V_2 = ((V_M \otimes 1) \cup (V_M \otimes 2)) \cap (D_2 \times D_2)$, where for a binary relation R we define $R \otimes i$ to be the relation $\{ ((a,i),(b,i)) : (a,b) \in R \}$. The other compatibility relations V^0_2, H_2, and H^0_2 are defined analogously from the respective compatibility relations of T_M.

The tiling system T_1 is obtained from T_2 by adding yet another independent copy of T_M, this time a complete one: $D_1 = D_2 \cup (L \times \{3\})$, and $D^{\text{ori}}_1 = D^{\text{ori}}_2 \cup (L \times \{3\})$, and $V_1 = V_2 \cup (V_M \otimes 3)$, *etc.* The labelling functions of T_1, and T_2 are defined as $\lambda_i((\ell, i)) = \ell$.

In order to show Proposition 7, and hence conclude the proof of Theorem 6, it suffices to establish the following two claims.

Claim 10 If machine M halts on input $(0,0)$, then there is no domino bisimulation relating T_1 and T_2.

Claim 11 If machine M does not halt on input $(0,0)$, then there is a domino bisimulation relating T_1 and T_2.

4 Hhp-Bisimilarity Is Undecidable

The proof of Theorem 3 is a reduction from the problem of deciding domino bisimilarity for origin constrained tiling systems. A method of encoding a tiling system on an infinite grid in the unfolding of a finite asynchronous transition system is due to Madhusudan and Thiagarajan [14]; we use a modified version of a gadget invented by them. For each origin constrained tiling system T we define an asynchronous transition system $A(T)$, such that the following proposition holds.

Proposition 12 There is a domino bisimulation relating origin constrained tiling systems T_1 and T_2, if and only if there is a hhp-bisimulation relating the asynchronous transition systems $A(T_1)$ and $A(T_2)$.

Let $T = \big(D, D^{\mathrm{ori}}, (H, H^0), (V, V^0), L, \lambda\big)$ be an origin constrained tiling system. We define the asynchronous transition system $A(T)$. The schematic structure of $A(T)$ can be seen in Figure 1. The set of events is defined as:

$$\begin{aligned} E_{A(T)} = &\left\{ x_i, y_i \ : \ i \in \{0,1,2,3\} \right\} \\ &\cup \left\{ d_{ij}, \overline{d}_{ij} \ : \ i,j \in \{0,1,2\}, d \in D, \text{ and } d \in D^{\mathrm{ori}} \text{ if } (i,j) = (0,0) \right\}. \end{aligned}$$

The rough idea behind the construction of $A(T)$ is best explained in terms of its event structure unfolding [23], in which the configurations of x- and y-transitions simply represent the grid structure of a tiling system, following [14]. Configurations in general consist of such a grid point plus at most two "d"- and "$\overline{d}$"-events, where the vertical (horizontal) compatibility of the tiling system is represented by the independence between a d_{ij}- and a $\overline{d}_{i(j+1)}$- ($\overline{d}_{(i+1)j}$-) event.

Notation: By abuse of notation we sometimes write d_{xy} or $\overline{d}_{xy}$ for $x, y \in \mathbb{N}$; we always mean by that the events $d_{\hat{x}\hat{y}}$ or $\overline{d}_{\hat{x}\hat{y}}$, respectively, where for $z \in \mathbb{N}$ we define $\hat{z}$ to be z if $z \leq 2$, and 2 for even z, and 1 for odd z if $z > 2$. [Notation] ◇

The labelling function replaces dominoes in "d"-, and "$\overline{d}$"-events, with their labels in the tiling system:

$$\lambda_{A(T)}(e) = \begin{cases} e & \text{if } e \in \left\{ x_i, y_i \ : \ i \in \{0,\ldots,3\} \right\}, \\ \lambda(d)_{ij} & \text{if } e = d_{ij}, \text{ for some } d \in D, \\ \overline{\lambda(d)}_{ij} & \text{if } e = \overline{d}_{ij}, \text{ for some } d \in D. \end{cases}$$

The states, events, and transitions of $A(T)$ can be read from Figure 1; we briefly explain below how to do it.

There are sixteen states in the bottom layer of the structure in Figure 1(a). Let us identify these sixteen states with pairs of numbers shown on the vertical macro-arrows originating in these states shown in Figure 1(a). Each of these macro-arrows denotes a bundle of d_{ij}-, and $\overline{d}_{ij}$-event transitions sticking out of the state below, arranged in the fashion shown in Figure 1(b). For each state (i,j), and domino $d \in D$, there are d_{ij}-, and $\overline{d}_{ij}$-event transitions sticking out, and moreover for each state (i',j') from which there is an arrow in Figure 1(a)

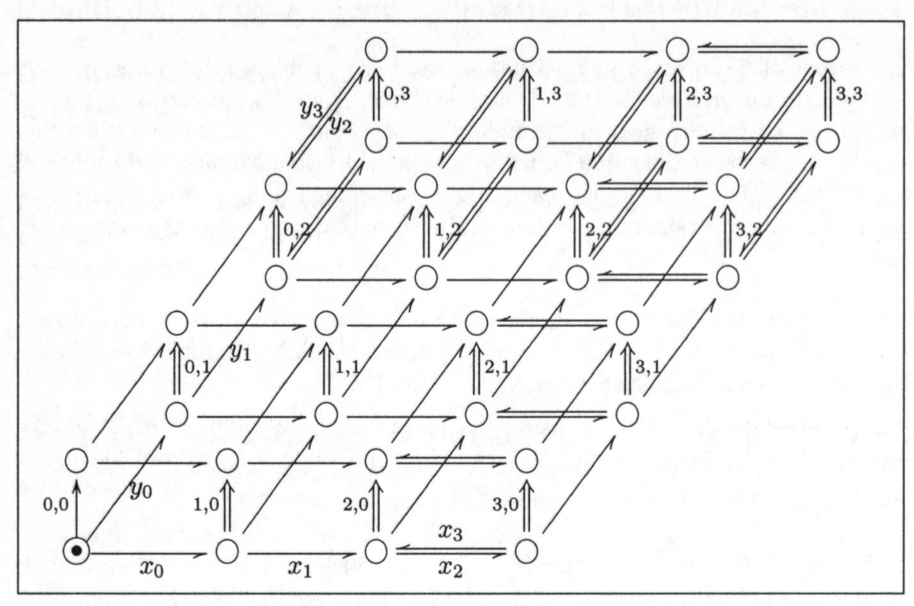

(a) The structure of $A(T)$ in the large.

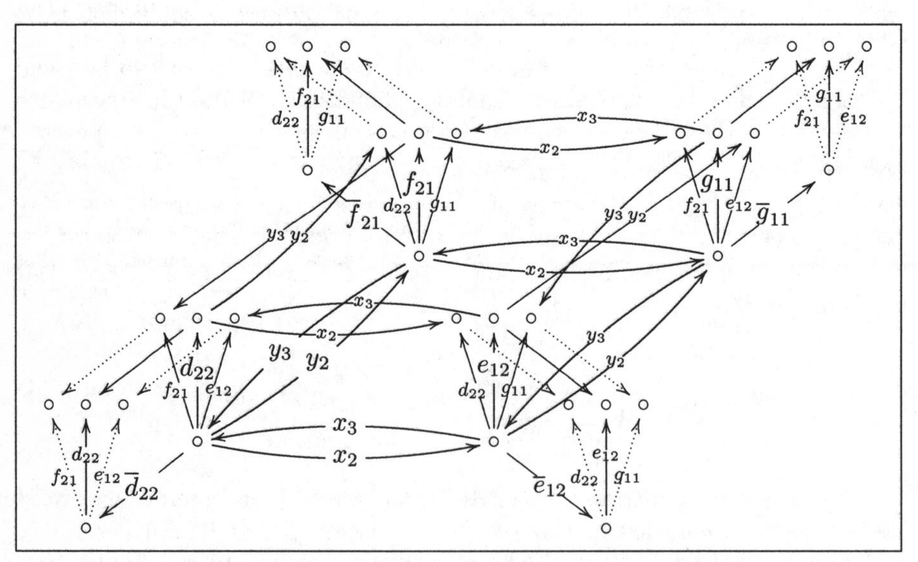

(b) The fine structure of the upper-right cube of $A(T)$.

Fig. 1. The structure of the asynchronous transition system $A(T)$.

to state (i, j), there is a $d_{i'j'}$-event transition sticking out of (i, j). The state $(0, 0)$ is exceptional: only dominoes from the origin constraint D^{ori} are allowed as events of transitions sticking out of it. It is also the initial state of $A(T)$.

As can be seen in Figure 1(b), from both ends of the d_{ij}-event transition rooted in state (i, j), there is an x_i-event transition to the corresponding (bottom, or top) $(i \oplus 1, j)$ state, and an y_i-event transition to the corresponding $(i, j \oplus 1)$ state, where $i \oplus 1 = i + 1$ if $i < 3$, and $i \oplus 1 = 2$ if $i = 3$.

For each $d_{i'j'}$-event transition t sticking out of state (i, j), and each $e \in D$, there can be a pair of transitions which together with t and the $\overline{e}_{ij}$-event transition form a "diamond" of transitions; the events of the transitions lying on the opposite sides of the diamond coincide then. This type of transitions is shown in Figure 1(b) as dotted arrows. The condition for the two transitions closing the diamond to exist is that configurations (d, i', j') and $\big(e, i' + |i' - i|, j' + |j' - j|\big)$ of T are compatible, or $(i', j') = (i, j)$ and $e = d$. We define the independence relation $I_{A(T)} \subseteq E_{A(T)} \times E_{A(T)}$, to be the symmetric closure of the set:

$$
\begin{aligned}
&\big\{\, (x_i, y_j), (x_i, d_{ij}), (y_j, d_{ij}) \;:\; i, j \in \{0, \ldots, 3\}, \text{ and } d \in D \,\big\} \;\cup \\
&\big\{\, (d_{ij}, \overline{d}_{ij}) \;:\; i, j \in \{0, 1, 2\}, \text{ and } d \in D \,\big\} \;\cup \\
&\big\{\, (d_{0j}, \overline{e}_{1j}) \;:\; j \in \{0, 1, 2\}, \text{ and } (d, e) \in H^0 \,\big\} \;\cup \\
&\big\{\, (d_{ij}, \overline{e}_{(i+1)j}) \;:\; i \in \{1, 2\}, j \in \{0, 1, 2\}, \text{ and } (d, e) \in H \,\big\} \;\cup \\
&\big\{\, (d_{i0}, \overline{e}_{i1}) \;:\; i \in \{0, 1, 2\}, \text{ and } (d, e) \in V^0 \,\big\} \;\cup \\
&\big\{\, (d_{ij}, \overline{e}_{i(j+1)}) \;:\; i \in \{0, 1, 2\}, j \in \{1, 2\}, \text{ and } (d, e) \in V \,\big\}.
\end{aligned}
$$

Note that it follows from the above that all diamonds of transitions in $A(T)$ are in fact independence diamonds.

Proof sketch (of Proposition 12)**:** The idea is to show that every domino bisimulation for T_1 and T_2 gives rise to an hhp-bisimulation for $A(T_1)$ and $A(T_2)$, and *vice versa*. First observe, that a run of $A(T_i)$ for $i \in \{1, 2\}$ consists of a number of occurrences of x_j- and y_k-events, x and y of them respectively, and a set of "d"- and "$\overline{d}$"-events, which is of size at most two. In other words, we can map runs of $A(T_i)$ into triples (F_i, x, y), where $F_i \subseteq E_{A(T_i)}$ contains at most two "d"- and "$\overline{d}$"-events, and $x, y \in \mathbb{N}$. Define $\mathrm{Confs}\big(A(T_1), A(T_2)\big)$ to be the set of quadruples (F_1, F_2, x, y) where F_i's are as above and $x, y \in \mathbb{N}$. Then it is a matter of routine verification to see that there exists an hhp-bisimulation between $A(T_1)$ and $A(T_2)$, if and only if there exists a relation $B \subseteq \mathrm{Confs}\big(A(T_1), A(T_2)\big)$, such that $\big\{\, (\overline{e_1}, \overline{e_2}) \;:\; (F_1, F_2, x, y) \in B, \text{where } \overline{e_i} \text{ is mapped to } (F_i, x, y) \,\big\}$ is an hhp-bisimulation relating $A(T_1)$ and $A(T_2)$. Hence, in the following we identify an hhp-bisimulation with such a relation B. The following claim immediately implies Proposition 12.

Claim 13 1. Let $B \subseteq \mathrm{Confs}\big(A(T_1), A(T_2)\big)$ be an hhp-bisimulation relating $A(T_1)$ and $A(T_2)$. Then the set $\big\{\, (d, e, x, y) \;:\; \big(\{\overline{d}_{xy}\}, \{\overline{e}_{xy}\}, x, y\big) \in B \,\big\}$ is a domino bisimulation for T_1 and T_2.

2. Let $B \subseteq \text{Confs}(T_1, T_2)$ be a domino bisimulation relating T_1 and T_2. Then the set $\{ (\{\overline{d}_{xy}\}, \{\overline{e}_{xy}\}, x, y) : (d, e, x, y) \in B \}$ can be extended to an hhp-bisimulation for $A(T_1)$ and $A(T_2)$.

This concludes the proof of Theorem 3. [Proposition 12] ∎

As a corollary of the above proof we get the following strengthening of our main theorem.

Corollary 14 Hhp-bisimilarity is undecidable for finite labelled 1-safe Petri nets.

Proof sketch (of Corollary 14): An attentive reader might have noticed, that the asynchronous transition system $A(T)$ as described in section 4, and sketched in Figure 1, is not coherent, while all asynchronous transition systems derived from (1-safe) Petri nets are [23, 18]. It turns out, however, that $A(T)$ is not far from being coherent: it suffices to close all the diamonds with events d_{ij}, and x_i in positions $(i, j \oplus 1)$, and with events d_{ij}, and y_j in positions $(i \oplus 1, j)$, for $i, j \in \{0, \ldots, 3\}$; note that runs ending at the top of these diamonds are maximal runs. This completion of the transition structure of $A(T)$ does not affect the arguments used to establish Claim 13, and hence Theorem 3, but since it would obscure the picture in Figure 1(b), we have decided not to draw it there. It is laborious but routine to construct a 1-safe Petri net whose derived asynchronous transition system is isomorphic to the completion of $A(T)$ mentioned above.

[Corollary 14] ∎

References

[1] Marek A. Bednarczyk. Hereditary history preserving bisimulations or what is the power of the future perfect in program logics. Technical report, Polish Academy of Sciences, Gdańsk, April 1991. Available at `http://www.ipipan.gda.pl/~marek`.

[2] Gian Luca Cattani and Vladimiro Sassone. Higher dimensional transition systems. In *Proceedings, 11th Annual IEEE Symposium on Logic in Computer Science*, pages 55–62, New Brunswick, New Jersey, 27–30 July 1996. IEEE Computer Society Press.

[3] Søren Christensen, Hans Hüttel, and Colin Stirling. Bisimulation equivalence is decidable for all context-free processes. *Information and Computation*, 121(2):143–148, 1995.

[4] Sibylle Fröschle. Decidability of plain and hereditary history-preserving bisimilarity for BPP. Presented at 6th International Workshop on Expressiveness in Concurrency, EXPRESS'99, Eindhoven, The Netherlands, August 1999.

[5] Sibylle Fröschle and Thomas Hildebrandt. On plain and hereditary history-preserving bisimulation. In Mirosław Kutyłowski, Leszek Pacholski, and Tomasz Wierzbicki, editors, *Mathematical Foundations of Computer Science 1999, 24th International Symposium, MFCS'99*, volume 1672 of *LNCS*, pages 354–365, Szklarska Poręba, Poland, September 6-10 1999. Springer.

[6] R. J. van Glabbeek. The linear time-branching time spectrum (Extended abstract). In J. C. M. Baeten and J. W. Klop, editors, *CONCUR '90, Theories of*

Concurrency: Unification and Extension, volume 458 of *LNCS*, pages 278–297, Amsterdam, The Netherlands, 27–30 August 1990. Springer-Verlag.

[7] Rob van Glabbeek and Ursula Goltz. Equivalence notions for concurrent systems and refinement of actions (Extended abstract). In A. Kreczmar and G. Mirkowska, editors, *Mathematical Foundations of Computer Science 1989*, volume 379 of *LNCS*, pages 237–248, Porąbka-Kozubnik, Poland, August/September 1989. Springer-Verlag.

[8] Jan Friso Groote and Hans Hüttel. Undecidable equivalences for basic process algebra. *Information and Computation*, 115(2):354–371, 1994.

[9] Matthew Hennessy and Robin Milner. Algebraic laws for nondeterminism and concurrency. *Journal of the ACM*, 32(1):137–161, 1985.

[10] Yoram Hirshfeld, Mark Jerrum, and Faron Moller. A polynomial algorithm for deciding bisimilarity of normed context-free processes. *Theoretical Computer Science*, 158(1–2):143–159, 1996.

[11] Lalita Jategaonkar and Albert R. Meyer. Deciding true concurrency equivalences on safe, finite nets. *Theoretical Computer Science*, 154:107–143, 1996.

[12] André Joyal, Mogens Nielsen, and Glynn Winskel. Bisimulation from open maps. *Information and Computation*, 127(2):164–185, 1996.

[13] Paris C. Kanellakis and Scott A. Smolka. CCS expressions, finite state processes, and three problems of equivalence. *Information and Computation*, 86(1):43–68, 1990.

[14] P. Madhusudan and P. S. Thiagarajan. Controllers for discrete event systems via morphisms. In Davide Sangiorgi and Robert de Simone, editors, *CONCUR'98, Concurrency Theory, 9th International Conference, Proceedings*, volume 1466 of *LNCS*, pages 18–33, Nice, France, September 1998. Springer.

[15] R. Milner. *A Calculus of Communicating Systems*, volume 92 of *LNCS*. Springer, 1980.

[16] Faron Moller and Scott A. Smolka. On the computational complexity of bisimulation. *ACM Computing Surveys*, 27(2):287–289, 1995.

[17] Mogens Nielsen and Christian Clausen. Games and logics for a noninterleaving bisimulation. *Nordic Journal of Computing*, 2(2):221–249, 1995.

[18] Mogens Nielsen and Glynn Winskel. Petri nets and bisimulation. *Theoretical Computer Science*, 153(1–2):211–244, 1996.

[19] Robert Paige and Robert E. Tarjan. Three partition refinement algorithms. *SIAM Journal on Computing*, 16(6):973–989, 1987.

[20] D. M. R. Park. Concurrency and automata on infinite sequences. In P. Deussen, editor, *Theoretical Computer Science: 5th GI-Conference*, volume 104 of *LNCS*, pages 167–183. Springer-Verlag, 1981.

[21] A. Rabinovich and B. Trakhtenbrot. Behaviour structures and nets of processes. *Fundamenta Informaticae*, 11:357–404, 1988.

[22] Walter Vogler. Deciding history preserving bisimilarity. In Javier Leach Albert, Burkhard Monien, and Mario Rodríguez-Artalejo, editors, *Automata, Languages and Programming, 18th International Colloquium, ICALP'91*, volume 510 of *LNCS*, pages 493–505, Madrid, Spain, 8–12 July 1991. Springer-Verlag.

[23] Glynn Winskel and Mogens Nielsen. Models for concurrency. In S. Abramsky, Dov M. Gabbay, and T. S. E. Maibaum, editors, *Handbook of Logic in Computer Science*, volume 4, Semantic Modelling, pages 1–148. Oxford University Press, 1995.

The Hardness of Approximating Spanner Problems

Michael Elkin* and David Peleg**

Department of Computer Science and Applied Mathematics,
The Weizmann Institute of Science, Rehovot, 76100 Israel.
{elkin,peleg}@wisdom.weizmann.ac.il.

Abstract. This paper examines a number of variants of the sparse k-spanner problem, and presents hardness results concerning their approximability. Previously, it was known that most k-spanner problems are *weakly inapproximable*, namely, are NP-hard to approximate with ratio $O(\log n)$, for every $k \geq 2$, and that the unit-length k-spanner problem for *constant* stretch requirement $k \geq 5$ is *strongly inapproximable*, namely, is NP-hard to approximate with ratio $O(2^{\log^{\epsilon} n})$ [19].
The results of this paper significantly expand the ranges of hardness for k-spanner problems. In general, strong hardness is shown for a number of k-spanner problems, for certain ranges of the stretch requirement k depending on the particular variant at hand. The problems studied differ by the types of edge weights and lengths used, and include also directed, augmentation and client-server variants of the problem.
The paper also considers k-spanner problems in which the stretch requirement k is relaxed (e.g., $k = \Omega(\log n)$). For these cases, no inapproximability results were known at all (even for a constant approximation ratio) for *any* spanner problem. Moreover, some versions of the k-spanner problem are known to enjoy the *ratio degradation* property, namely, their complexity decreases exponentially with the inverse of the stretch requirement. So far, no hardness result existed precluding *any* k-spanner problem from enjoying this property. This paper establishes strong inapproximability results for the case of relaxed stretch requirement (up to $k = o(n^{\delta})$, for any $0 < \delta < 1$), for a large variety of k-spanner problems. It is also shown that these problems do not enjoy the ratio degradation property.

Classification: Approximation algorithms, Hardness of approximation

1 Introduction

1.1 The Sparse Spanner Problem

The concept of *graph spanners* has been studied in several recent papers, in the context of communication networks, distributed computing, robotics and

* Supported in part by a Leo Frances Gallin Scholarship.
** Supported in part by grants from the Israel Ministry of Science and Art.

H. Reichel and S. Tison (Eds.): STACS 2000, LNCS 1770, pp. 370–381, 2000.

computational geometry [1,4,8,11,9,21,24,25]. Consider a connected simple graph $G = (V, E, \omega, l)$, with $|V| = n$ vertices, where $\omega : E \rightarrow R^+$ is a *weight* function on the edge set of the graph, and $l : E \rightarrow R^+$ is a *length* function on the edge set of the graph. For every pair of vertices $u, v \in V$, let $P(u, v, G)$ be the set of all simple paths from u to v in G. We define the distance between u and v in G to be $dist(u, v, G) = \min_{P \in P(u,v,G)} \sum_{e \in P} l(e)$. A subgraph $G' = (V, E')$ of G is a $k - spanner$ if for every $u, v \in V$, $\frac{dist(u,v,G')}{dist(u,v,G)} \leq k$. We refer to k as the *stretch factor* of G'.

Spanners for general graphs were first introduced in [25], and used to construct a new type of synchronizer for an asynchronous network. For most applications, it is desirable that the spanner be as *sparse* or as *light* as possible, namely, has few edges or small total weight. This leads to the following problem. The *cost* of a subgraph G' is its weight, $\omega(G') = \sum_{e \in E'} \omega(e)$. The goal of the *k-spanner* problem is to find a k-spanner $G' = (V, E')$ with the smallest cost $\omega(G')$.

A number of variants of the sparse spanner problem have been considered in the literature. The *general* k-spanner problem allows arbitrary edge weights and lengths. However, the most basic variant of the sparse spanner problem deals with the simple unweighted uniform case, where $\omega(e) = l(e) = 1$ for every edge $e \in E$ [24,25]. We call this variant the *unweighted* (or *basic*) k-spanner problem.

In-between, one may consider a number of intermediate variants. The first is the *unit-length* k-spanner problem studied in [19]. In this case, the weight function ω may be arbitrary, but the length function l assigns $l(e) = 1$ to every edge $e \in E$.

An important special case of the unit-length k-spanner problem is when the weight function ω may assign only 0 and 1 values. This problem is called the *light-edges* (LE) k-spanner problem, and it is equivalent to the k-spanner *augmentation* problem studied in [14]. Intuitively, such a Boolean function $\omega : E \rightarrow \{0, 1\}$ captures the situation where in addition to the target graph G, we are given also an initial partially constructed subnetwork H', whose edges are assumed to be given in advance for free, and it is required to augment the subnetwork H' into a k-spanner H for G, where edges not in H' must be "paid for" in order to be included in the spanner. We denote the set of zero-weight edges by $\mathcal{L}$. The aim is to minimize the number of new edges needed in order to obtain a k-spanner for the given graph.

A second variant, which can be thought of as the dual of the unit-length k-spanner problem, is the *unit-weight* k-spanner problem, studied in [2,7]. In this case, the length function l may be arbitrary, but the weight function ω assigns $\omega(e) = 1$ to every edge $e \in E$.

Finally, a third variant considered in the literature is the *uniform* k-spanner [1,2,7,26]. In this case, the weight and length functions coincide, i.e. $\omega(e) = l(e)$ for every edge $e \in E$, but that function ω may be arbitrary.

Any one of the above versions may be also generalized to the *client-server* (C-S) k-spanner problem, see [14]. This is a generalization of the k-spanner problem which distinguishes between the two different roles of edges in the problem, i.e., the input specifies also a subset $\mathcal{C}$ of *client* edges which have to be spanned, and a subset $\mathcal{S}$ of *server* edges which may be used for spanning the client edges. We also distinguish three subcases of the C-S k-spanner problem. The first is the *disjoint C-S* k-spanner (hereafter, DJ k-spanner) problem. In this variant the client and server sets are disjoint. The second is the *all-client C-S* k-spanner (hereafter, AC k-spanner) problem, in which the server set is a subset of the client set. Finally, the last variant is the *all-server C-S* k-spanner (hereafter, AS k-spanner) problem, in which the client set is a subset of the server set.

Following [19], we define the *MIN-REP* problem as follows. We are given a bipartite graph $G(V_1, V_2, E)$, where V_1 and V_2 are each split into a disjoint union of r sets; $V_1 = \bigcup_{i=1}^r A_i$ and $V_2 = \bigcup_{i=1}^r B_i$. The sets A_i, B_i all have size N. An instance of the problem consists of the 5-tuple $(V_1, V_2, E, \{A_i\}, \{B_i\})$. The bipartite graph and the partition of V_1 and V_2 induce a supergraph $\mathcal{H}$, whose vertices are the sets A_i and B_j, where $i, j \in \{1, .., r\}$. Two sets A_i and B_j are adjacent in $\mathcal{H}$ iff there exists some $a_i \in A_i$ and $b_j \in B_j$ which are adjacent in G. We assume that $\mathcal{H}$ is regular and (but its degree, $d = deg(\mathcal{H})$, need not be $O(1)$). A set of vertices C is a *REP-cover* for $\mathcal{H}$ if for each super-edge (A_i, B_j) there is a pair $a_i \in A_i$ and $b_j \in B_j$, both belonging to C, such that $(a_i, b_j) \in E$. It is required to select a minimal REP-cover C for $\mathcal{H}$. Note that it is easy to test whether a MIN-REP instance admits a REP-cover, just by checking whether the all vertex set $V_1 \cup V_2$ REP-covers all the superedges. Thus we can assume without loss of generality, that the given instance admits a REP-cover.

Consider also the maximization version of this problem, called *MAX-REP*. In this version the REP-cover C may contain at most one vertex from each supernode.

Another close problem is the *Label-Cover* problem [18]. In this problem a superedge (A_i, B_j) is covered if for *every* vertex $a_i \in A_i \cap C$ there is a vertex $b_j \in B_j \cap C$ such that $(a_i, b_j) \in E$. It also has the minimization and maximization versions called $Label\text{--}Cover_{MIN}$ and $Label\text{--}Cover_{MAX}$ respectively. Note that the $Label - Cover_{MAX}$ problem is equivalent to the MAX-REP problem.

Consider also the Symmetric Label Cover problem [12,5]. It is the MIN-REP problem restricted to a complete bipartite supergraph.

1.2 Previous Results

It is shown in [24] that the problem of determining, for a given unweighted graph $G = (V, E)$ and an integer m, whether there exists a 2-spanner with m or fewer edges is NP-complete. This indicates that it is unlikely to find an exact solution for the sparsest $k-$spanner problem even in the case $k = 2$. Consequently, two possible remaining courses of action for investigating the problem are establishing global bounds on the number of edges required for an unweighted k-spanner of various graph classes and devising approximation algorithms for the problem.

In [24] it is shown that every unweighted $n-$vertex graph G has a polynomial time constructible $(4k + 1)-$spanner with at most $O(n^{1+1/k})$ edges. Hence in particular, every graph G has an $O(\log n)-$spanner with $O(n)$ edges. These results are close to the best possible in general, as implied by the lower bound given in [24].

The results of [24] were improved and generalized in [1,2] to the uniform case, in which the edges weights and lengths coincide. Specifically, it is shown in [1] that given an n−vertex graph and an integer $k \geq 1$, there is a polynomially constructible $(2k+1)$−spanner G' such that $|E(G')| < n \cdot \lceil n^{\frac{1}{k}} \rceil$. Again, this result is shown to be asymptotically the best possible. In [7] it was shown that the weight of the uniform k-spanner obtained by the construction of [1] is bounded by $\omega(G') = O(n^{\frac{2+\epsilon}{k-1}} \cdot \omega(MST))$. They also show how the construction can be used to provide uniform $\log^2 n$-spanners with weight bounded by $\omega(G') = O(\omega(MST))$.

The algorithms of [1,24,2] provide us with *global* upper bounds for sparse k−spanners, i.e., general bounds that hold *for every graph.* However, it may be that for specific graphs, considerably sparser spanners exist. Furthermore, the upper bounds on sparsity given by these algorithms are small (i.e., close to n) only for large values of k. It is therefore interesting to look for *approximation algorithms*, that yield near-optimal bounds for the specific graph at hand.

In [20], a $\log \frac{|E|}{|V|}$ approximation algorithm was presented for the unweighted 2-spanner problem. In [19] the result was extended to an $O(\log n)$-approximation algorithm for the unit-length 2-spanner problem. A $\log \frac{|C|}{|V(C)|}$-approximation algorithm for the unit-length C-S 2-spanner problem was presented in [14]. Approximation algorithms with ratio $\log \frac{|C|}{|V(C)|}$ are given also in [14] for a number of other variants of the problem, such as the unit-length 2-spanner augmentation problem and directed unit-length 2-spanner (augmentation) problems.

Also, since any k-spanner for an n-vertex graph requires at least $n-1$ edges, the results of [1,2,24] cited above can be interpreted as providing an $O(n^{1/k})$-ratio approximation algorithm for the (unweighted or weighted) uniform k-spanner problem. This implies that once the required stretch guarantee is relaxed, i.e., k is allowed to be large, the problem becomes easier to approximate. In particular, the unweighted k-spanner problem admits $O(1)$ approximation once the stretch requirement becomes $k = \Omega(\log n)$, and the uniform k-spanner problem admits an $O(1)$ approximation ratio once the stretch requirement becomes $k = \Omega(\log^2 n)$. We call this property *ratio degradation.*

Previously known hardness results for spanner problems were of two types. First, it is shown in [19] that it is NP-hard to approximate the basic unweighted k-spanner problem by an $O(\log n)$ ratio for $k \geq 2$. This type of $\Omega(\log n)$ inapproximability is henceforth referred to as *weak inapproximability.* This result applies to the majority of the problems studied in this paper. Secondly, it is shown in [19] that the unit-length k-spanner problem for *constant* stretch requirement $k \geq 5$ is hard to approximate with $O(2^{\log^\epsilon n})$ ratio, unless $NP \subseteq DTIME(n^{\text{polylog } n})$ (or, in short, that approximating it with this ratio is quasi-NP-hard). This type of $\Omega(2^{\log^\epsilon n})$ inapproximability is henceforth referred to as *strong inapproximability.* This result was recently extended to $k \geq 3$ in [12] and independently by us, in the current paper, using a different reduction. It is shown in [19] that the MIN-REP problem is strongly inapproximable. As recently shown in [10], for every $0 < \epsilon < 1$ it is NP-hard to approximate the $Label - Cover_{MIN}$ problem with $O(2^{\log^\epsilon n})$ approximation ratio. As discussed in [12,5] this result applies to the Symmetric Label Cover problem and thus to the MIN-REP problem.

1.3 Summary of Our Results

Our results significantly expand the ranges of hardness for spanner problems. Our main results can be classified as follows (see also Table 1).

The Type of k-spanner problem	The Range of the Strong Hardness Proven in the Paper	Previously Known Hardness Results	Ratio Degradation Property
Uniform	$1 < k \leq 3$	Weak hardness for $k \geq 2$ [19]	YES
Unit-weight	$1 < k < 3$	Weak hardness for $k \geq 2$ [19]	YES
Directed	$3 \leq k \leq o(n^\delta)$	Weak hardness for $k \geq 2$ [19]	NO
DJ	$3 \leq k \leq o(n^\delta)$	No previous results	NO
AC	$3 \leq k \leq o(n^\delta)$	Weak hardness for $k \geq 2$ [19]	NO
C-S	$3 \leq k \leq o(n^\delta)$	Weak hardness for $k \geq 2$ [19]	NO
Augmentation	$4 \leq k \leq o(n^\delta)$	Weak hardness for $k \geq 2$ [19]	NO
Unit-weight DJ	$0 < k < \infty$	No previous results	NO
Unit-weight AS	$0 < k < 3$	No previous results	YES
Unit-length	$3 \leq k \leq o(n^\delta)$	Strong hardness for $5 \leq k = O(1)$ [19]. Extended to $3 \leq k = O(1)$ by [12] independently from us.	NO

Table 1. The Summary of Results.

To begin with, we obtain strong inapproximability results for a number of variants of the k-spanner problem. In particular, we prove the first strong inapproximability results for *uniform* k-spanner problems, for the range of stretch requirement $1 < k \leq 3$. Specifically, we prove that

Theorem 1. *For any* $0 < \epsilon < 1$ *and* $1 < k \leq 3 + n^{-3/2}$ *it is quasi-NP-hard to approximate the uniform* k*-spanner problem with ratio* $2^{\log^\epsilon n}$.

The uniform k-spanner problem was intensively studied during the last decade [1,2,7,26], but as mentioned above, the only previous hardness results were for weak inapproximability for the range of $k \geq 2$ [19].

Also, we obtain a strong inapproximability result for the *unit-weight* k-spanner problem, for the range of stretch requirement $1 < k < 3$. Specifically,

Theorem 2. *For any* $\epsilon > 0$ *and stretch requirement* $1 < k < 3$, *it is quasi-NP-hard to approximate the unit-weight* k*-spanner problem with ratio* $2^{\log^\epsilon n}$.

This problem was also studied in [1,7], but again the only previous hardness results were for weak inapproximability, for the range of $k \geq 2$. We note that for the range of the stretch requirement $1 < k < 2$, no hardness result at all was known for the above two problems (even for a constant ratio).

Moreover, we obtain a strong inapproximability results for certain versions of the *unweighted* k-spanner problem as well, for the range of stretch requirement $k > 2$. In particular, we obtain strong inapproximability results for the directed unweighted k-spanner problem, for the DJ unweighted k-spanner problem and for the AC unweighted k-spanner problem, and thus for the C-S k-spanner problem in general. Specifically, we have

Theorem 3. *For any $\epsilon > 0$ and constant integer $k \geq 3$, it is quasi-NP-hard to approximate the unweighted directed k-spanner, unweighted DJ k-spanner or unweighted AC k-spanner problems with ratio $2^{\log^{\epsilon} n}$.*

Note that for the DJ unweighted k-spanner problem no hardness result was known *at all*, since no reduction from the unweighted k-spanner problem to the DJ unweighted k-spanner problem is known. For the AC unweighted k-spanner problem, the only hardness result known is weak inapproximability for $k \geq 2$ [19]. For $k = 2$ a $\log \frac{|C|}{|V(C)|}$-approximation algorithm for both problems is provided in [14].

The *directed* unweighted k-spanner problem was presented already in the first paper defining the notion of k-spanner [24], but the only hardness result known for the problem is weak inapproximability for $k \geq 2$ [19]. We significantly improve the threshold showing strong inapproximability for $k > 2$ (which is the best possible, since for $k = 2$ the problem enjoys a $\log \frac{|E|}{|V|}$-approximation algorithm [14]).

We also obtain strong inapproximability results for the k-spanner augmentation problem for the range of stretch requirement $k > 3$.

Theorem 4. *For any $\epsilon > 0$ and constant integer $k \geq 4$, it is quasi-NP-hard to approximate the unit-length k-spanner augmentation problem with ratio $2^{\log^{\epsilon} n}$.*

The only previously known lower bound for the approximability of the problem was $\Omega(\log n)$ for $k \geq 2$ (see [19]).

Our second contribution involves hardness results for relaxed stretch requirements and ratio degradation. All the previous hardness results [19,12] have shown hardness for k-spanner problems only for certain constant values of the stretch requirement k. For relaxed stretch requirement $k = \Omega(\log n)$ no hardness results (even for a constant approximation ratio) were known for any spanner problem. Furthermore, the uniform k-spanner problem and the unweighted k-spanner problem are known [24,1,7,26] to enjoy the ratio degradation property, i.e., their complexity decreases exponentially with the inverse of the stretch requirement (and, in particular, admit $O(\log n)$ and $O(1)$ approximation ratios, respectively, whenever the stretch requirement becomes $\Omega(\log n)$). No hardness result existed precluding some k-spanner problem from enjoying this property.

In this paper we establish the first strong inapproximability results for the relaxed stretch requirement $k = o(n^\delta)$, for any $0 < \delta < 1$ for a number of spanner problems. Specifically we have

Theorem 5. *For any $0 < \epsilon,\ \delta < 1$ and $3 \le k = o(n^\delta)$ it is quasi-NP-hard to approximate the unit-length k-spanner, unweighted DJ k-spanner, unweighted AC k-spanner or unweighted directed k-spanner problems with ratio $2^{\log^\epsilon n}$. Moreover, for any $0 < \epsilon,\ \delta < 1$ and $4 \le k = o(n^\delta)$ it is quasi-NP-hard to approximate the unit-length k-spanner augmentation problem with ratio $2^{\log^\epsilon n}$.*

Also we show that all these problems do not enjoy the ratio degradation property. These problems include the directed unweighted k-spanner problem, the DJ unweighted k-spanner problem (and thus C-S unweighted k-spanner problem), the AC unweighted k-spanner problem, the unit-length k-spanner problem and the k-spanner augmentation problem. Specifically,

Theorem 6. *For any $\alpha, \beta > 0$ and $0 < \epsilon_1, \epsilon_2 < 1$ there is no algorithm $A(G, k)$ that approximates the unit-length k-spanner or unit-length k-spanner augmentation problems on every n-vertex graph G and for every sufficiently large k with ratio $O(k^\alpha \cdot w^{*\frac{1}{k^\beta}})$ or $O(2^{\log^{\epsilon_1} k} \cdot w^{*\frac{1}{\log^{\epsilon_2} k}})$.*

Theorem 7. *For any $\alpha, \beta > 0$ and $0 < \epsilon_1, \epsilon_2 < 1$, and for any problem Π from among the unit-length k-spanner, unit-length k-spanner augmentation, DJ k-spanner, AC k-spanner, C-S k-spanner or directed k-spanner problems, there is no algorithm $A(G, k)$ that approximates Π on every n-vertex graph G and for every sufficiently large k with ratio $O(k^\alpha \cdot n^{\frac{1}{k^\beta}})$ or $O(2^{\log^{\epsilon_1} k} \cdot n^{\frac{1}{\log^{\epsilon_2} k}})$.*

For weighted k-spanner problems, another potential direction one may consider is to look for an approximation algorithm whose ratio guarantee is some function of the optimal weight w^* (and not only the number of vertices n). We show that the unit-length k-spanner problem, k-spanner augmentation problem and uniform k-spanner problem cannot be approximated neither with ratio $2^{\log^\epsilon n}$ nor with ratio $2^{\log^\epsilon w^*}$ (within the corresponding ranges), unless $NP \subseteq DTIME(n^{\mathrm{polylog}\, n})$.

Theorem 8. *For any $0 < \epsilon,\ \delta < 1$ and $3 \le k = o(n^\delta)$ it is quasi-NP-hard to approximate the unit-length k-spanner problem with ratio $2^{\log^\epsilon n}$ or $2^{\log^\epsilon w^*}$.*

Theorem 9. *For any $0 < \epsilon < 1$ and $1 < k \le 3$ it is quasi-NP-hard to approximate the uniform k-spanner problem with ratio $2^{\log^\epsilon n}$ or $2^{\log^\epsilon w^*}$.*

We obtain also several secondary results. First, we show a strong inapproximability result for the Red-Blue problem. This result was obtained independently by [5]. Also we present several reductions from some k-spanner problems to the Red-Blue problem. Specifically, we prove that

Theorem 10. *For any $0 < \epsilon, \delta < 1$, the RB problem is quasi-NP-hard to approximate with ratio $2^{\log^{\epsilon} n}$, even when restricted to exactly one blue element and two red elements in each set, $|S|$ and $|R|$ are both bounded by $n^{1+\delta}$ and $|B|$ is bounded by n.*

Theorem 11. *The general k-spanner problem such that the ratio between the $\max_{e \in E}\{l(e)\}$ and the $\min_{e \in E}\{l(e)\}$ is bounded by a constant is reducible to the RB problem.*

Finally, in addition to our strong inapproximability result concerning the unit-weight k-spanner problem we obtain strong inapproximability results for the widest possible range of the stretch requirement $0 < k < \infty$ for the unit-weight DJ k-spanner problem.

Theorem 12. *For any $0 \leq k < \infty$ and $0 < \epsilon < 1$ it is quasi-NP-hard to approximate the DJ unit-weight k-spanner problem with ratio $2^{\log^{\epsilon} n}$, even when restricted to two possible edge lengths.*

Also we obtain for the unit-weight AS k-spanner problem a strong inapproximability result for a wider range than for the unit-weight k-spanner problem, specifically, for $0 < k < 3$.

Theorem 13. *For any $0 < k < 3$ and $0 < \epsilon < 1$, it is quasi-NP-hard to approximate the unit-weight AS k-spanner problem with ratio $2^{\log^{\epsilon} n}$, even when restricted to two possible edge lengths.*

All of our results are established by reductions from the MIN-REP problem of [19]. Specifically, we discuss the relationship between the MIN-REP problem and the Label-Cover problem (cf. [18]). Moreover, we show that the MIN-REP problem and the Label-Cover problem admit a $\sqrt{n}$-approximation ratio and that the MIN-REP and the Label-Cover problems restricted to the cases where the girth of the induced supergraph is greater than t, admit an $n^{\frac{2}{t}}$ approximation ratio. In particular, it follows that the MIN-REP and the Label-Cover problems with girth greater than $\log^{\epsilon} n$ (for some constant $\epsilon > 0$) are not strongly inapproximable, i.e., admit an $O(2^{\log^{\epsilon'} n})$-approximation ratio, for some $0 < \epsilon' < 1$.

Specifically, we prove

Theorem 14. *1. MIN-REP and $Label - Cover_{MIN}$ with $girth(\mathcal{H}) \geq t$ admit an $O(n^{\frac{2}{t}})$-approximation algorithm.*
2. MIN-REP and $Label - Cover_{MIN}$ with $girth(\mathcal{H}) = O(\log n)$ admit an $O(1)$ approximation ratio.
3, For any $0 < \epsilon < 1$ there exists $0 < \epsilon' < 1$ such that MIN-REP and $Label - Cover_{MIN}$ with $girth(\mathcal{H}) = O(\log^{\epsilon} n)$ admit an $O(2^{\log^{\epsilon'} n})$ approximation algorithm.

Corollary 1. *The MIN-REP and $Label - Cover_{MIN}$ problems admit a $\sqrt{n}$-approximation algorithm.*

Very recently we resolved the main question left open in this paper, concerning the inapproximability of the basic k-spanner problem. Specifically, we have proved that the basic k-spanner problem is strongly inapproximable for any $k \geq 3$ [16]. In particular, this and the results presented here imply that the uniform k-spanner problem and the unit-length k-spanner problem are strongly inapproximable for any constant value of the stretch requirement $k > 1$ and that the k-spanner augmentation problem is strongly inapproximable for $3 \leq k = o(n^{\delta})$. We have also shown there that MIN-REP and the $Label - Cover_{MIN}$ problems with girth greater than $\log^{\mu} n$ (for some constant $0 < \mu < 1$) are inapproximable within a ratio of $\Omega(2^{\log^{\epsilon} n})$, for any $0 < \epsilon < 1 - \mu$.

In the remainder of the paper, we illustrate our proof techniques by establishing one part of Theorem 1. The remaining proofs are deferred to the full paper. One can also find most of them in the technical report version of the paper [13].

2 The Uniform Spanner Problem

Our proof of Theorem 1 is comprised of four parts. In the first part we prove the result for the stretch requirement $1 < k \leq 2$, in the second we extend it to $1 < k \leq 3$, in the third we prove it for $k = 3$ and finally in the fourth part we slightly improve the result, by proving it for $k = 3 + n^{-3/2}$. In this section we sketch the proof of the first part of the theorem, i.e., for $k = 1 + \epsilon$, $0 < \epsilon \leq 1$. This is done by a reduction from MIN-REP problem.

The reduction is as follows. Fix $0 < \epsilon \leq 1$ and let

$$x = \max\{\epsilon^{-1}, |E| + 2rN^2 + 1\} .$$

Given an instance of MIN-REP problem $(V_1, V_2, E, \{A_i\}, \{B_i\})$, we construct an instance $\bar{G} = (\bar{V}, \bar{E})$ of the general $(1+\epsilon)$-spanner problem as follows.

$$\begin{aligned}
\bar{V} &= \bigcup_{i=1}^{r} A_i \cup \bigcup_{i=1}^{r} B_i \cup \{s_i, t_i\}_{i=1}^{r} \\
&\quad \cup \{d_1^{(i,j)}, d_2^{(i,j)}, c_1^{(i,j)}, c_2^{(i,j)} \mid 1 \leq i \leq r, 1 \leq j \leq N\} , \\
\bar{E} &= E \cup D \cup E_{sA} \cup E_{tB} \cup E_{\mathcal{H}} ,
\end{aligned}$$

with

$$D = \bigcup_{i=1}^{r} \{(a_i^l, a_i^{l'}) \mid a_i^l, a_i^{l'} \in A_i\} \cup \bigcup_{j=1}^{r} \{(b_j^m, b_j^{m'}) \mid b_j^m, b_j^{m'} \in B_j\} ,$$

$$E_{sA} = \bigcup_{i=1}^{r} \{(s_i, a_i^l) \mid a_i^l \in A_i\} ,$$

$$E_{tB} = \bigcup_{i=1}^{r} \{(b_i^l, t_i) \mid b_i^l \in B_i\} ,$$

$$E_{\mathcal{H}} = \{(s_i, t_j) \mid (A_i, B_j) \in \mathcal{H}\} .$$

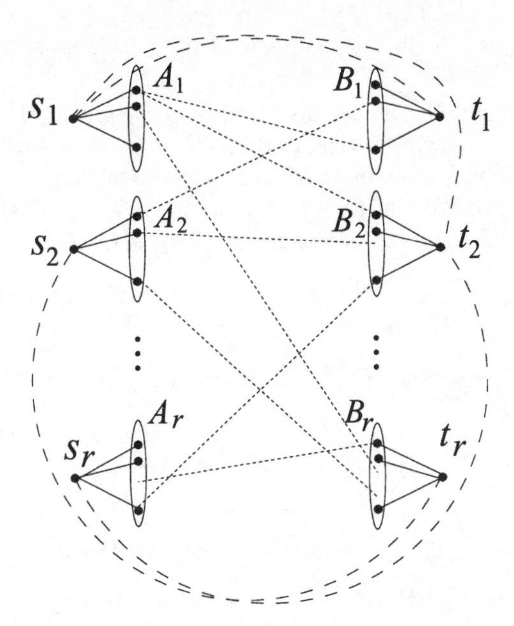

Fig. 1. Dotted lines represent unit weight edges, solid lines represent edges of weight x, dashed lines represent edges of weight $(2x+1)/(1+\epsilon)$.

The weight assignment on the edges is as follows (see Figure 1).

$$\omega(e) = \begin{cases} 1, & e \in E \cup D \\ x, & e \in E_{sA} \cup E_{tB} \\ (2x+1)/(1+\epsilon), & e \in E_{\mathcal{H}} \end{cases}$$

Let us briefly provide some intuition for the way the reduction operates. The vertices s_i and t_j represent the supernodes A_i and B_j, respectively. The edges of $E_{\mathcal{H}}$, between s_i and t_j, correspond exactly to the edges of the supergraph $\mathcal{H}$. We assign these edges the biggest weight in the construction (specifically, $(2x+1)/(1+\epsilon)$) in order to prevent the optimal spanner from using them. On the other hand, the spanner will use intensively the edges of E, from the original graph. In the MIN-REP problem we pay only for the vertices of the original graph which are taken into the REP-cover. For this reason, we assign the edges of E the lowest weight in the construction (the unit weight). Finally, taking any edge of E_{sA}, connecting s_i and some vertex of A_i (or any edge of E_{tB}, connecting t_j and some vertex of B_j) into the spanner represents taking this vertex from A_i (or B_j) into the REP-cover. Since we are interested in minimizing the number of vertices taken into the REP-cover, we assign these edges a weight that is much bigger than the weight of edges of E, for which we are interested not to pay at all, but significantly smaller than the weight of edges of $E_{\mathcal{H}}$, which we try to make too "expensive" for the spanner to use. Hence, the weight of any near-optimal spanner approximately equals the number of $E_{sA} \cup E_{tB}$ edges it uses, multiplied by x.

Given the spanner H for the instance $\bar{G}$ of weight $\omega(H)$, we construct a MIN-REP cover C of size approximately $\omega(H)/x$ in two stages. First, for every spanner H we construct a spanner H' of approximately the same size that does not use (s_i, t_j) edges. We call such a spanner a *proper* spanner. Next, from H' we build a MIN-REP cover of size approximately $\omega(H')/x$.

Lemma 1. *For every $(1+\epsilon)$-spanner H, there is a polynomial time constructible proper $(1+\epsilon)$-spanner H' such that $\omega(H') \leq (1+\epsilon) \cdot \omega(H)$.*

Given a proper spanner H' we construct a REP-cover C of size close to $\omega(H')/x$ by letting

$$C = \{a_i^l, b_j^m \mid (s_i, a_i^l), (b_j^m, t_j) \in H'\} . \quad (1)$$

Lemma 2. *C defined by (1) is a REP-cover for $\bar{G}$, and $|C| \leq (1+\epsilon)\omega(H)/x$.*

Conversely, given a REP-cover C we construct a $(1+\epsilon)$-spanner H for $\bar{G}$ by letting

$$H = E \cup D \cup \{(s_i, a_i^l), (b_j^m, t_j) \in \bar{E} \mid a_i^l, b_j^m \in C\} .$$

Lemma 3. *H is a $(1+\epsilon)$-spanner for $\bar{G}$, and $\omega(H) < 2x \cdot |C|$.*

This establishes Theorem 1 for $k = 1+\epsilon$, where $0 < \epsilon \leq 1$.

References

1. I. Althöfer, G. Das, D. Dobkin and D. Joseph, Generating sparse spanners for weighted graphs, *Proc. 2nd Scandinavian Workshop on Algorithm Theory*, Lect. Notes in Comput. Sci., Vol. 447, pp. 26-37, Springer-Verlag, New York/Berlin, 1990.
2. Baruch Awerbuch, Alan Baratz, and David Peleg. Efficient broadcast and light-weight spanners. Unpublished manuscript, November 1991.
3. B. Bollobas, *Extremal Graph Theory*. Academic Press, 1978.
4. L. Cai, *Tree-2-Spanners*, Technical Report 91-4, Simon Fraser University,1991.
5. R. D. Carr, S. Doddi, G. Konjevod, M. V. Marathe On the Red-Blue Set Cover Problem, Los Alamos National Laboratory.
6. P. Chanas, *Dimensionnement de réseaux ATM*, PhD thesis, CNET Sophia, Sept. 1998.
7. B. Chandra, G. Das, G. Narasimhan and J. Soares, New Sparseness Results on Graph Spanners, *Proc 8th ACM Symposium on Computational Geometry*, pages 192-201, 1992.
8. L.P. Chew, There is a planar graph almost as good as the complete graph, *Proc. ACM Symp. on Computational Geometry*, 1986, pp. 169-177
9. G. Das and D. Joseph, Which triangulation approximates the complete graphs, *Proc. Int. Symp. on Optimal Algorithms*, Lect. Notes in Comput. Sci., Vol. 401, pp. 168-192, Springer-Verlag, New York/Berlin, 1989

10. I. Dinur and S. Safra, On the hardness of Approximating Label Cover, *Electronic Colloquium on Computational Complexity*, Report No. 15 (1999)
11. D.P. Dobkin, S.J. Friedman and K.J. Supowit, Delaunay graphs are almost as good as complete graphs, *Proc. 31st IEEE Symp. on Foundations of Computer Science*, 1987, pp. 20-26.
12. Y. Dodis and S. Khanna, Designing Networks with Bounded Pairwise Distance, *Proc. 30th ACM Ann. Symp. of Theory of Computing*, 1999.
13. M.-L. Elkin and D. Peleg, The Hardness of Approximating Spanner Problems, Technical Report MCS99-14, the Weizmann Institute of Science, 1999.
14. M.-L. Elkin and D. Peleg, The Client-Server 2-Spanner Problem and Applications to Network Design, Technical Report MCS99-24, the Weizmann Institute of Science, 1999.
15. M.-L. Elkin and D. Peleg, Classification of Spanner Problems, in preparation.
16. M.-L. Elkin and D. Peleg, Strong Inapproximability of the Basic k-Spanner Problem, Technical Report MCS99-23, the Weizmann Institute of Science, 1999.
17. U. Feige and L. Lovasz, Two prover one-round proof systems: Their power and their problems, *Proc. 24th ACM Symp. on Theory of Computing*, 733-741, 1992
18. D. Hochbaum *Approximation Algorithms for NP-hard Problems*, PWS Publishing Company, Boston, 1997.
19. G. Kortsarz, On the Hardness of Approximating Spanners, *Proc. APPROX.*, Lect. Notes in Comput. Sci., Vol. 1444, pp. 135-146, Springer-Verlag, New York/Berlin, 1998.
20. G. Kortsarz and D. Peleg, Generating Sparse 2-Spanners. *J. Algorithms*, **17** (1994) 222-236.
21. C. Levcopoulos and A. Lingas, There are planar graphs almost as good as the complete graphs and as short as minimum spanning trees, *Proc. Int. Symp. on Optimal Algorithms*, Lect. Notes in Comput. Sci., Vol. 401, pp. 9-13, Springer-Verlag, New York/Berlin, 1989
22. A.L. Liestman and T.Shermer, Grid Spanners, *Networks* **23** (1993),123-133.
23. D. Peleg, Locality-Sensitive Distributed Computing, unpublished manuscript, 1999.
24. D. Peleg and A. Schäffer, Graph Spanners, *J. Graph Theory* **13** (1989), 99-116.
25. D. Peleg and J.D. Ullman, An optimal synchronizer for the hypercube, *SIAM J. Computing* **18** (1989), pp. 740-747.
26. H. Regev, *The weight of the Greedy Graph Spanner*, Technical Report CS95-22, July 1995.

An Improved Lower Bound on the Approximability of Metric TSP and Approximation Algorithms for the TSP with Sharpened Triangle Inequality
(Extended Abstract)

Hans-Joachim Böckenhauer, Juraj Hromkovič, Ralf Klasing, Sebastian Seibert, and Walter Unger

Lehrstuhl für Informatik I (Algorithmen und Komplexität)
RWTH Aachen, 52056 Aachen, Germany
{hjb,jh,rak,seibert,quax}@i1.informatik.rwth-aachen.de

Abstract. The traveling salesman problem (TSP) is one of the hardest optimization problems in NPO because it does not admit any polynomial time approximation algorithm (unless $P = NP$). On the other hand we have a polynomial time approximation scheme (PTAS) for the Euclidean TSP and the $\frac{3}{2}$-approximation algorithm of Christofides for TSP instances satisfying the triangle inequality. The main contributions of this paper are the following:

(i) We essentially modify the method of Engebretsen [En99] in order to get a lower bound of $\frac{3813}{3812} - \varepsilon$ on the polynomial-time approximability of the metric TSP for any $\varepsilon > 0$. This is an improvement over the lower bound of $\frac{5381}{5380} - \varepsilon$ in [En99]. Using this approach we moreover prove a lower bound δ_β on the approximability of Δ_β-TSP for $\frac{1}{2} < \beta < 1$, where Δ_β-TSP is a subproblem of the TSP whose input instances satisfy the β-sharpened triangle inequality $cost(\{u,v\}) \leqslant \beta \cdot (cost(\{u,x\}) + cost(\{x,v\}))$ for all vertices u, v, x.

(ii) We present three different methods for the design of polynomial-time approximation algorithms for Δ_β-TSP with $\frac{1}{2} < \beta < 1$, where the approximation ratio lies between 1 and $\frac{3}{2}$, depending on β.

Keywords: Approximation algorithms, Traveling Salesman Problem

1 Introduction

The traveling salesman problem (TSP) is one of the hardest optimization problems in NPO. It is considered to be intractable[1] because it does not admit any polynomial time $p(n)$-approximation algorithm for any polynomial p in the input size n. On the other hand there are large subclasses of input instances of the

[1] This holds even if one considers the current view of tractability exchanging exact solutions for their approximations and determinism for randomization.

H. Reichel and S. Tison (Eds.): STACS 2000, LNCS 1770, pp. 382–394, 2000.

TSP that admit polynomial-time approximation algorithms with a reasonable approximation ratio. The Euclidean TSP (also called geometric TSP) admits even a polynomial time approximation scheme and the Δ-TSP (TSP with triangle inequality, also called metric TSP) can be approximated by Christofides algorithm with an approximation ratio of $\frac{3}{2}$.[2] Generally, the recent research has shown that the "relation" of an input instance of the TSP to the triangle inequality may be essential for estimating the hardness of this particular input instance. We say, for every $\beta \geqslant \frac{1}{2}$, that an input instance of the general TSP satisfies the β-triangle inequality if

$$cost(\{u,v\}) \leqslant \beta \cdot (cost(\{u,x\}) + cost(\{x,v\}))$$

for all vertices u, v, x. By Δ_β-TSP we denote the TSP whose input instances satisfy the β-triangle inequality. If $\beta > 1$ then we speak about relaxed triangle inequality and if $\beta < 1$ we speak about sharpened triangle inequality.

Considering the relaxed triangle inequality in [AB95, BC99, BHKSU99] it has been proved that

(i) Δ_β-TSP can be approximated in polynomial time with approximation ratio $\min\{4\beta, \frac{3}{2}\beta^2\}$, and
(ii) unless $P = NP$, Δ_β-TSP cannot be approximated with approximation ratio $1 + \varepsilon \cdot \beta$ for some $\varepsilon > 0$.

Thus, these results enable us to partition all input instances of TSP into infinitely many classes according to their approximability, and the membership to a class can be efficiently decided.

In this paper we consider for the first time the sharpened triangle inequality, i.e. with $\beta < 1$. This does not seem to be as natural as to consider the Δ_β-TSP for $\beta \geqslant 1$, but there are some good reasons to do so. In what follows we list the two main reasons.

1. We have PTASs for the geometrical TSP [Ar96, Ar97, Mi96], but none of them is practical because to achieve an approximation ratio $1 + \varepsilon$ one needs $O(n^{30 \cdot \varepsilon^{-1}})$ time (in the randomized case $O(n \cdot (\log_2 n)^{30 \cdot \varepsilon^{-1}})$). So, from the user's point of view the best algorithm for the geometrical TSP is the Christofides algorithm for Δ-TSP with its approximation ratio of $\frac{3}{2}$. Thus, it is of interest to search for nontrivial subclasses of the input instances of Δ-TSP (or even of the geometrical TSP) for which a better approximation ratio than $\frac{3}{2}$ is achievable. The sharpened triangle inequality has a nice geometrical interpretation: The direct connection *must* be shorter than any connection via a third vertex. So, the problem instances of Δ_β-TSP for $\beta < 1$ have the property that no vertex (point of the plane) lies on or almost on the direct connection between another two vertices (points).

[2] In what follows an α-approximation algorithm for a minimization problem is any algorithm that produces feasible solutions whose costs divided by the cost of optimal solutions is at most α.

2. The problem $\Delta_{\frac{1}{2}}$-TSP is simple because all edges must have the same cost. How hard is then the problem for values of β that are very close to (but different from) $\frac{1}{2}$? Are these problems NP-hard or even APX-hard? If the answers were positive, one could consider to partition the set of input instances of Δ-TSP into an infinite spectrum of Δ_β-TSP instances for $\beta \in (\frac{1}{2}, 1]$ according to the polynomial-time achievable approximation ratio. This could provide a similar picture as partitioning the input instances of the general TSP into classes according to the relaxed triangle inequality in [AB95, BC99, BHKSU99].

Our first and main result is the improvement of the explicit lower bound on the polynomial-time approximability from $\frac{5381}{5380} - \varepsilon$ [En99] to $\frac{3813}{3812} - \varepsilon$ for any $\varepsilon > 0$. Our proof is based on the idea of Engebretsen, who reduced the LinEq2-2(3) problem to the TSP subproblem with edge costs 1 and 2 only. We modify this proof technique by considering the reduction to input instances of Δ-TSP whose edge costs are from $\{1, 2, 3\}$. This modification requires some crucial changes in the construction of Engebretsen as well as some essentially new technical considerations. We apply the obtained lower bound on Δ-TSP to get explicit lower bounds on the approximability of Δ_β-TSP for every β, $\frac{1}{2} < \beta < 1$. So, the answer to our motivation (2) for the study of Δ_β-TSP problems for $\beta < 1$ is that these problems are APX-hard for every $\beta > \frac{1}{2}$.

In the second part of our paper, we present three different approaches for investigating Δ_β-TSP with $\beta < 1$. First, we analyze the behavior of the Christofides algorithm and prove that it is a $\left(1 + \frac{2\beta-1}{3\beta^2-2\beta+1}\right)$-approximation algorithm for Δ_β-TSP with $\frac{1}{2} \leqslant \beta < 1$. Secondly we present a general idea how to modify every α-approximation algorithm for Δ-TSP to achieve an $\left(\frac{\alpha \cdot \beta^2}{\beta^2 + (\alpha-1)\cdot(1-\beta)^2}\right)$-approximation algorithm for Δ_β-TSP. The last approach designs a new special $\left(\frac{2}{3} + \frac{1}{3} \cdot \frac{\beta}{1-\beta}\right)$-approximation algorithm for Δ_β-TSP. This algorithm is the best one for $\frac{1}{2} \leqslant \beta \leqslant \frac{2}{3}$, the first two approaches dominate for $\beta > \frac{2}{3}$. In this way together with the lower bounds we achieve our aim to partition the set of input instances of Δ-TSP into an infinite spectrum according to the polynomial-time approximability. Moreover, our algorithms are efficient (the first runs in time $O(n^{2.5}(\log n)^{1.5})$ [GT91]), and so we have practical solutions for large subclasses of Δ-TSP with an approximation ratio better than $\frac{3}{2}$.

This paper is organized as follows. In Section 2 we present our improved lower bound on the polynomial-time approximability of Δ-TSP. Section 3 contains some elementary fundamental observations about Δ_β-TSP that are useful for all three approaches and the claim that Δ_β-TSP is APX-complete for every $\beta > \frac{1}{2}$. The subsequent sections present the three approaches in the above mentioned order.

2 An Improved Lower Bound for the TSP with Triangle Inequality

In this section we will present an improved lower bound on the approximation ratio of Δ-TSP. We will show that it is NP-hard to approximate Δ-TSP within $\frac{3813}{3812} - \varepsilon$ for any $\varepsilon > 0$. The best previously known lower bound is that of Engebretsen [En99] which states that it is NP-hard to approximate Δ-TSP within $\frac{5381}{5380} - \varepsilon$ for any $\varepsilon > 0$.

We will prove our lower bound even for the following special case of Δ-TSP: Let $\Delta_{\{1,2,3\}}$-TSP denote the special case of Δ-TSP such that all edge costs are from $\{1, 2, 3\}$.

To obtain the lower bound we use a reduction from the LinEq2-2(3) problem that is defined as follows:

Definition 1. *LinEq2-2(3) is the following maximization problem. Given a system of linear equations mod 2 with exactly two variables in each equation and exactly three occurrences of each variable, maximize the number of satisfied equations.*

The idea of our proof is an extension of the proof idea in [En99] where a reduction from LinEq2-2(3) to $\Delta_{\{1,2\}}$-TSP (i.e. TSP with edge costs from $\{1, 2\}$) was used.

To this end, we introduce the notion of gap problems which means to cut out of the original problems a subset of input instances such that there is a certain gap in the allowed quality of the optimal solutions. Then the existence of an NP-hard gap problem with gap (α, β) implies that no polynomial time $\frac{\alpha}{\beta}$-approximation algorithms exists for the underlying optimization problem, unless P=NP (see e.g. [Ho96], chapter 10, or [MPS98], chapter 8).

By (α, β)-LinEq2-2(3), we describe the decision problem which has as input only instances of LinEq2-2(3) where either at least a fraction α or at most a fraction β of the given equations can be satisfied at the same time, for some $0 \leqslant \beta < \alpha \leqslant 1$.

Similarly, we define (α, β)-$\Delta_{\{1,2,3\}}$-TSP. Here, only instances are admitted where the number of vertices divided by the cost of a cheapest tour (between $\frac{1}{3}$ and 1 in general) is either above α or below β.

Please note that we have normalized the considered solution qualities to lie between 0 and 1, a larger number always signaling a better solution for maximization as well as for minimization problems. This normalization could be omitted in principle, but we find it more convenient for comparing different problems and instances of different size.

Below, we will give a reduction from $\left(\frac{(332-\varepsilon_2')}{336}, \frac{(331+\varepsilon_1')}{336}\right)$-LinEq2-2(3) to $\left(\frac{7616}{7626-\varepsilon_1}, \frac{7616}{7624+\varepsilon_2}\right)$-$\Delta_{\{1,2,3\}}$-TSP (for any small $\varepsilon_1, \varepsilon_2 > 0$, and suitable $\varepsilon_1', \varepsilon_2'$).

This reduction together with an inapproximability result by Berman and Karpinski [BK98] (the NP-hardness of $\left(\frac{(332-\varepsilon_2')}{336}, \frac{(331+\varepsilon_1')}{336}\right)$-LinEq2-2(3) in our terms) implies the following theorem.

Theorem 1. *For any $\varepsilon > 0$, approximating $\Delta_{\{1,2,3\}}$-TSP within $\frac{3813}{3812} - \varepsilon$ is NP-hard.*

Corollary 1. *For any $\varepsilon > 0$, approximating Δ-TSP within $\frac{3813}{3812} - \varepsilon$ is NP-hard.*

Due to space limitations we are unable to present the full proof of Theorem 1 in this extended abstract, but we will give a sketch of the proof here.

Sketch of the proof of Theorem 1. For a given LinEq2-2(3) instance, we first construct an undirected graph G_0 which consists of $68n + 1$ vertices, if the given LinEq2-2(3) instance has $3n$ equations and $2n$ variables. Then we construct a $\Delta_{\{1,2,3\}}$-TSP instance G from G_0 by setting the edge costs for all edges in G_0 to one and setting all other edge costs to the maximal possible value from $\{2, 3\}$ such that the triangle inequality is still satisfied. This means that the cost of an edge (u, v) is $\min\{3, dist_0(u, v)\}$ where $dist_0(u, v)$ is the distance of u and v in G_0.

Then we will show that an optimal Hamiltonian tour in G uses $2e$ edges with cost $\neq 1$, iff an optimal assignment in the LinEq2-2(3) instance satisfies all but e equations.

The main technical difficulty of the proof lies in additionally showing that all these expensive edges must have a cost of 3, that is they connect vertices of a distance at least 3 in G_0. More precisely, we show that any tour can be modified without increasing the cost in such a way that the resulting tour has the desired property.

After all, a LinEq2-2(3) instance with $336n$ equations is converted into a Δ-TSP instance with $7616n+1$ vertices such that the following holds. If $(332-\varepsilon_2')n$ equations can be satisfied at the same time, there exists a Hamiltonian tour of cost at most $(7624 + \varepsilon_2)n + 1$, and if at most $(331 + \varepsilon_1')n$ equations can be satisfied at the same time, there exists no Hamiltonian tour of cost less than $(7626 - \varepsilon_1)n + 1$.

The general picture of the construction, that is, the structure of G_0, is shown in Figure 1.

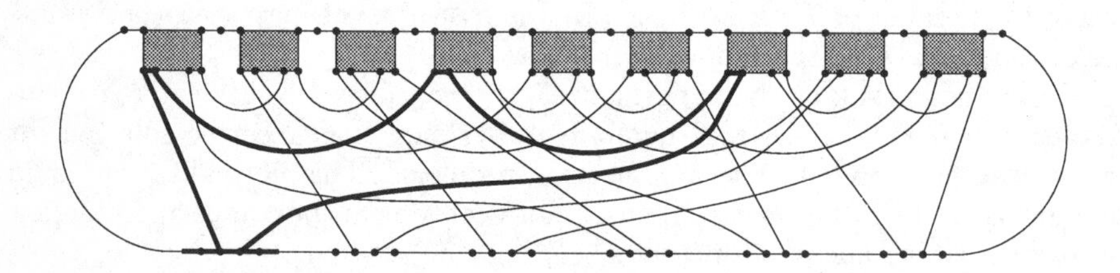

Fig. 1. The complete construction for a sample instance with 9 equations and 6 variables. The first variable cluster is drawn with bold lines.

For each equation, we construct an *equation gadget* which is one of the grey boxes in the upper part of the picture. A *variable cluster* consists of some vertices which are also part of equation gadgets together with some additional vertices. The bold edges in Figure 1 mark a single variable cluster. Finally, all these parts are connected in a large circle as shown in Figure 1.

There are two types of equation gadgets corresponding to the two types of equations, being $x + y = 0$, and $x + y = 1$ respectively. For the first type, an *equation gadget of type 0* is shown in Figure 2 (a), and Figure 2 (b) depicts an *equation gadget of type 1*.

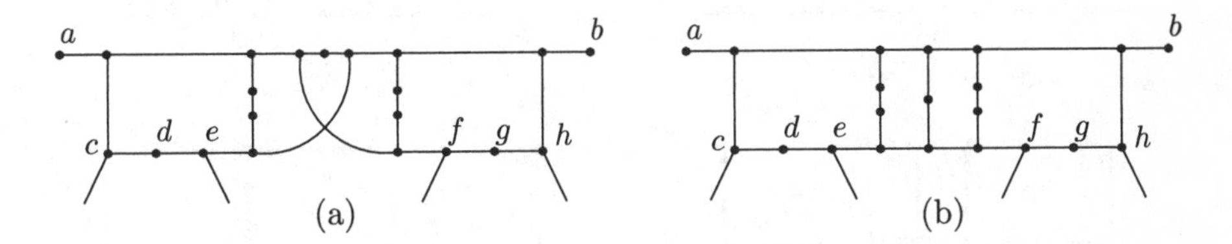

Fig. 2. An equation gadget of type 0 is shown in (a), a gadget of type 1 in (b).

Please note, that vertices a and b are used in common by different gadgets, that is, vertex b of one gadget is vertex a of the next one.

By definition, a LinEq2-2(3) instance has $2m$ variables and $3m$ equations, for some m. Counting the number of vertices used in the equation gadgets and variable clusters, we obtain a graph having $68m + 1$ vertices.

The LinEq2-2(3) instances of [BK98] use only numbers $m = 112n$, for some natural number n, such that out of the resulting $336n$ equations either at least $(332 - \varepsilon_2')n$ or at most $(331 + \varepsilon_1')n$ equations can be satisfied at the same time.

We will show that in our Δ-TSP instances, e non-satisfied equations translate into cost $|V| + 2e$ for an optimal tour. Thus, starting from the LinEq2-2(3) instances of [BK98], we obtain Δ-TSP instances having $68 \cdot (112n) + 1 = 7616n + 1$ vertices. Hence, the cost of an optimal tour will be either at most $7616n + 1 + 2(4 + \varepsilon_2')n = 7624 + \varepsilon_2$ or at least $7616n + 1 + 2(5 - \varepsilon_1')n = 7625 - \varepsilon_1$.

The main technical part of the proof is to show the following claim.

Claim. In G, the cost of an optimal Hamiltonian tour is $|V| + 2e$ iff in the underlying LinEq2-2(3) instance at most all but e equations can be satisfied at the same time.

One direction of this claim is straightforward. Starting from an assignment to the variables of the LinEq2-2(3) instance, we obtain a tour of the claimed maximal cost as follows. It traverses the graph of Figure 1 essentially along the outer cycle, taking some detours through variable clusters and equation gadgets. If a variable is set to 1 in the given assignment, the tour uses the edges of the variable cluster which visit all three gadgets of those equations where the variable occurs. Otherwise it uses the the shortcut below in Figure 1.

Thus, for a satisfied equation of type $x + y = 1$, in the corresponding gadget exactly one of the vertex sequences (c, d, e) resp. (f, g, h) is visited as part of a

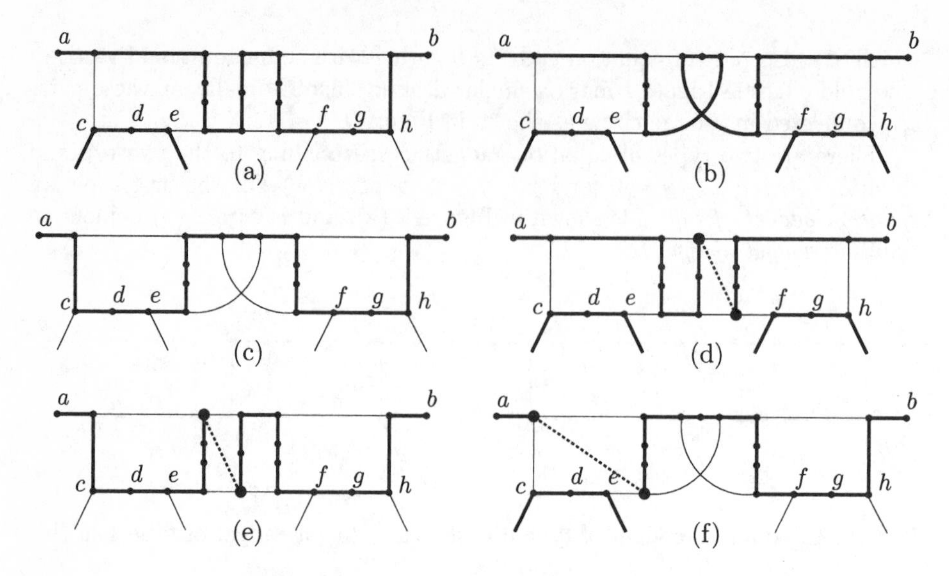

Fig. 3. The traversal of equation gadgets.

variable cluster. The rest of that gadget can be traversed as shown in Figure 3 (a). Similarly, in a gadget for an equation of type $x + y = 0$, both or none of the mentioned sequences are traversed as part of a variable cluster. Thus, it admits a traversal as depicted in Figure 3 (b) and (c).

It remains to include the vertices of equation gadgets for unsatisfied variables in the tour. Here, the traversal of the variable cluster implies that, for a gadget of type 1, both or none of the sequences (c, d, e) resp. (f, g, h) are left open. And for a gadget of type 0, exactly one of those sequences remains open. In this case, we add pieces to the tour as depicted in Figure 3 (d)-(f).

Only in the last step, we have used edges which are not part of G_0 (the dashed edges in Figure 3). These connect vertices having distance at least 3 in G_0, thus they have cost 3. All other edges are part of G_0, i.e., they have cost 1 in G.

Overall, an assignment leaving e equations unsatisfied results in a tour of cost $|V| + 2e$.

For the opposite direction, we have to show that an arbitrary Hamiltonian tour through G can be modified, without increasing the cost, into a tour which has the structure of a tour constructed from an assignment as above. Then, an assignment can be inferred from that tour having the claimed quality in a direct reversal of the above procedure.

The mentioned transformation of an arbitrary tour consists of a lengthy procedure with many case distinctions which makes up the most part of the full proof. In the following, we give an overview of the transformations which are needed, and we demonstrate the method in one exemplary case.

In the sequel, we use the notion of an *endpoint* in the considered tour. An endpoint is a vertex with the property that at least one of its two incident edges used in the tour does not belong to G_0. A *connector* of a gadget denotes the pairs of edges leaving the gadget from vertices c and e, or from vertices f and h respectively. It is *semi-traversed* by a tour, if exactly one of its edges is used by the tour, and it is *traversed* if both belong to the tour. Distances will always refer to G_0.

Now, the task at hand is to modify a given tour, without increasing the cost, in a way such that

1. the distance between endpoints will always be at least 3,
2. there will be no semitraversed connectors, and
3. in the equation gadgets there are always 0 or 2 endpoints.

In the resulting tour, an equation gadget will have no endpoints iff the number of traversed connectors equals the gadget type (modulo 2).

We will illustrate the method by sketching the proof of the following claim.

Claim. Assume a tour having no semi-traversed connectors and a gadget of type 0 where the tour traverses exactly one connector. Then the tour can be modified without increasing the cost, and without changing it on the connector edges, in such a way that there are exactly two endpoints at distance $\geqslant 3$ in that gadget, and it is impossible to modify it such that there are only two endpoints at distance 2.

Roughly speaking, this states that an optimal tour has to look like Figure 3 (f) in a gadget of an unsatisfied equation of type 0.

To show this claim, first one can easily verify that the fixed use of connector edges implies the use of at least two endpoints. Then, the essential insight is that there is no way of managing it with two endpoints of distance 2.

Assuming the contrary, we have to look for a pair of vertices at distance 2 which may be used as endpoints. Distance 2 implies that there is a vertex adjacent to both of these in G_0. That vertex cannot be of degree 2 in G_0 since then the tour could be simply improved to have no endpoints in the gadget at all. It remains to check all pairs of vertices where the common neighbor is of degree 3. That means to check whether we can get an Hamiltonian path from vertex a to b in the gadget of type 0 without using the left connector (i.e. vertices c, d, e) by adding exactly one of the edges shown in Figure 4. It turns out that

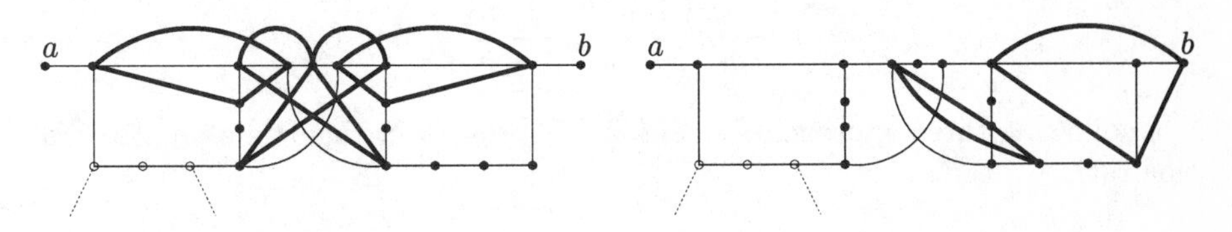

Fig. 4. The traversal of equation gadgets.

this is impossible. □

3 Fundamental Observations about the TSP with Sharpened Triangle Inequality

First, we observe that also the Δ_β-TSP with $\frac{1}{2} < \beta < 1$ is a hard optimization problem.

Lemma 1. *For any $\varepsilon > 0$ and for any $\frac{1}{2} < \beta < 1$ it is NP-hard to approximate the Δ_β-TSP within $\frac{7611+10\beta^2+5\beta}{7612+8\beta^2+4\beta} - \varepsilon$.*

Proof. The claim can be shown analogously to the proof of Theorem 1, using the edge costs $\{1, 2\beta, 2\beta^2 + \beta\}$ instead of $\{1, 2, 3\}$. □

In the sequel, we will use the following notations. Let $G = (V, E)$ be a complete graph with a cost function $cost : E \to \mathbb{R}^{>0}$. Then $c_{\min} = \min_{e \in E} cost(e)$ and $c_{\max} = \max_{e \in E} cost(e)$. Furthermore, H_{opt} will denote an optimal Hamiltonian tour in G.

Lemma 2. *Let $\frac{1}{2} \leqslant \beta < 1$, let $(G, cost)$ be a problem instance of Δ_β-TSP.*

(a) For any two edges e_1, e_2 with a common endpoint $cost(e_1) \leqslant \frac{\beta}{1-\beta} \cdot cost(e_2)$.
(b) $c_{\max} \leqslant \frac{2\beta^2}{1-\beta} \cdot c_{\min}$.

Proof. **(a):** For any triangle consisting of the edges e_1, e_2, e_3 the β-triangle inequality implies $cost(e_1) \leqslant \beta(cost(e_2) + cost(e_3))$ and $cost(e_3) \leqslant \beta(cost(e_1) + cost(e_2))$ and thus $cost(e_1) \leqslant \beta(cost(e_2) + \beta(cost(e_1) + cost(e_2)))$. This implies $cost(e_1) \leqslant \frac{\beta+\beta^2}{1-\beta^2} \cdot cost(e_2) = \frac{\beta}{1-\beta} \cdot cost(e_2)$.

(b): Let $\{a, b\}$ be an edge with cost $c_{\min}$ and let $\{c, d\}$ be an edge with cost $c_{\max}$. If these edges have a common endpoint, the claim follows immediately from (a) since $\frac{2\beta^2}{1-\beta} \geqslant \frac{\beta}{1-\beta}$. If these edges do not have a common endpoint, we know from (a) that $cost(\{a, c\}), cost(\{a, d\}) \leqslant \frac{\beta}{1-\beta} \cdot c_{\min}$. With the triangle inequality we have $c_{\max} \leqslant \beta(cost(\{a, c\}) + cost(\{a, d\})) \leqslant \beta \cdot 2 \cdot \frac{\beta}{1-\beta} \cdot c_{\min} = \frac{2\beta^2}{1-\beta} \cdot c_{\min}$. □

Part (b) of Lemma 2 directly implies that, for every Hamiltonian tour H of an input instance for Δ_β-TSP,

$$\frac{cost(H)}{cost(H_{\text{opt}})} \leqslant \frac{2\beta^2}{1-\beta}.$$

The following three approaches essentially improve over this trivial approximation ratio.

4 Using Δ-TSP Approximation Algorithms for Δ_β-TSP

In the first part of this section, we analyze the approximation ratio of the Christofides algorithm for Δ_β-TSP. The main idea is that we can improve the

$\frac{3}{2}$ approximation ratio due to shortening[3] of paths in two different parts of the Christofides algorithm, namely by building the matching and by constructing the Hamiltonian tour from the Eulerian tour. In fact we statistically prove that we save some non-negligible part of the costs of a "lot" of edges of the constructed Eulerian graph.

Theorem 2. *For $\frac{1}{2} \leqslant \beta < 1$ the Christofides algorithm is a $(1 + \frac{2\beta-1}{3\beta^2-2\beta+1})$-approximation algorithm for Δ_β-TSP.*

We will omit the proof of Theorem 2 in this extended abstract.

Observe, that $\delta_\beta = \frac{2\beta-1}{3\beta^2-2\beta+1}$ is equal to 0 if $\beta = \frac{1}{2}$ (i. e. if all edges have the same cost), and that $\delta_\beta = \frac{1}{2}$ for $\beta = 1$. Thus, δ_β continuously converges to 0 with the degree of the sharpening of the triangle inequality.

Now, we show how to modify any α-approximation algorithm for Δ-TSP to obtain a $\delta_{\alpha,\beta}$-approximation algorithm for Δ_β-TSP with $\delta_{\alpha,\beta} < \alpha$ for every $\beta < 1$. The advantage of this approach is that any improvement on the approximation of Δ-TSP automatically results in an improvement of the approximation ratio for Δ_β-TSP. The idea of this approach is to reduce an input instance of Δ_β-TSP to an input instance of Δ-TSP by subtracting a suitable cost from all edges.

Theorem 3. *Let A be an approximation algorithm for Δ-TSP with approximation ratio α, and let $\frac{1}{2} < \beta < 1$. Then A is an approximation algorithm for Δ_β-TSP with approximation ratio[4] $\frac{\alpha \cdot \beta^2}{\beta^2 + (\alpha-1)\cdot(1-\beta)^2}$.*

Proof. Let $I = (G, cost)$ be a problem instance of Δ_β-TSP, $\frac{1}{2} < \beta < 1$. Let $c = (1-\beta) \cdot 2 \cdot c_{\min}$. For all $e \in E(G)$, let $cost'(e) = cost(e) - c$. Then the TSP instance $I' = (G, cost')$ still satisfies the triangle inequality: Let x, y, z be the costs of the edges of an arbitrary triangle of G. Then $z \leqslant \beta \cdot (x+y)$ holds. Since $c = (1-\beta) \cdot 2 \cdot c_{\min} \leqslant (1-\beta) \cdot (x+y)$ it follows that $z \leqslant \beta \cdot (x+y) \leqslant x + y - c$ and thus $z - c \leqslant x - c + y - c$.

Furthermore we know that a Hamiltonian tour is optimal for I' if and only if it is optimal for I. Let H_{opt} be an optimal Hamiltonian tour for I. Let H be the Hamiltonian tour that is produced by the algorithm A on the input I'. Then $cost'(H) \leqslant \alpha \cdot cost'(H_{\text{opt}})$ holds and thus

$$cost(H) - n \cdot c \leqslant \alpha \cdot (cost(H_{\text{opt}}) - n \cdot c).$$

[3] Note that exchanging a path for an edge saves a positive amount of costs because of the β-triangle inequality.

[4] Observe that the approximation ratio tends to 1 with β approaching $\frac{1}{2}$ and it tends to α with β approaching 1.

This leads to

$$\begin{aligned}
cost(H) &\leqslant \alpha \cdot cost(H_{\text{opt}}) - (\alpha - 1) \cdot n \cdot c \\
&= \alpha \cdot cost(H_{\text{opt}}) - (\alpha - 1) \cdot n \cdot (1 - \beta) \cdot 2 \cdot c_{\min} \\
&\leqslant \alpha \cdot cost(H_{\text{opt}}) - (\alpha - 1) \cdot n \cdot (1 - \beta) \cdot 2 \cdot \frac{1-\beta}{2\beta^2} \cdot c_{\max} \\
&= \alpha \cdot cost(H_{\text{opt}}) - (\alpha - 1) \cdot \frac{(1-\beta)^2}{\beta^2} \cdot n \cdot c_{\max}
\end{aligned}$$

Let $\Gamma = \{\gamma \geqslant 1 \mid cost(H) \leqslant \gamma \cdot cost(H_{\text{opt}}) \leqslant n \cdot c_{\max}\}$. Then, for any $\gamma \in \Gamma$,

$$\begin{aligned}
cost(H) &\leqslant \alpha \cdot cost(H_{\text{opt}}) - (\alpha - 1) \cdot \frac{(1-\beta)^2}{\beta^2} \cdot \gamma \cdot cost(H_{\text{opt}}) \\
&= \left(\alpha - \gamma \cdot (\alpha - 1) \cdot \frac{(1-\beta)^2}{\beta^2}\right) \cdot cost(H_{\text{opt}}).
\end{aligned} \tag{1}$$

As a consequence

$$cost(H) \leqslant \min_{\gamma \in \Gamma} \min \left\{\gamma, \alpha - \gamma \cdot (\alpha - 1) \cdot \frac{(1-\beta)^2}{\beta^2}\right\} \cdot cost(H_{\text{opt}}). \tag{2}$$

This minimum is achieved for $\gamma = \alpha - \gamma \cdot (\alpha - 1) \cdot \frac{(1-\beta)^2}{\beta^2}$. This leads to

$$\gamma = \frac{\alpha \cdot \beta^2}{\beta^2 + (\alpha - 1) \cdot (1 - \beta)^2} \tag{3}$$

which completes the proof. □

Corollary 2. *For $\frac{1}{2} \leqslant \beta < 1$ the Christofides algorithm is a $(1 + \frac{2\beta - 1}{3\beta^2 - 2\beta + 1})$-approximation algorithm for Δ_β-TSP.*

Note that Corollary 2 provides the same approximation ratio as Theorem 2.

5 The Cycle Cover Algorithm

In this section we design a new, special algorithm for Δ_β-TSP with $\beta < 1$. This algorithm provides a better approximation ratio than the previous approaches for $\frac{1}{2} \leqslant \beta < \frac{2}{3}$.

Cycle Cover Algorithm

Input: A complete graph $G = (V, E)$ with a cost function $cost : E \to \mathbb{R}^{\geqslant 0}$ satisfying the β-triangle inequality.

Step 1: Construct a minimum cost cycle cover $C = \{C_1, \ldots, C_k\}$ of G, i. e. a covering of all nodes in G by cycles of length $\geqslant 3$.

Step 2: For $1 \leqslant i \leqslant k$, find the cheapest edge $\{a_i, b_i\}$ in every cycle C_i of C.

Step 3: Obtain a Hamiltonian cycle H of G from C by replacing the edges $\{\{a_i, b_i\} \mid 1 \leqslant i \leqslant k\}$ by the edges $\{\{b_i, a_{i+1}\} \mid 1 \leqslant i \leqslant k-1\} \cup \{\{b_k, a_1\}\}$.
Output: H.

Theorem 4. *For $\frac{1}{2} \leqslant \beta < 1$ the cycle cover algorithm is a $(\frac{2}{3} + \frac{1}{3} \cdot \frac{\beta}{1-\beta})$-approximation algorithm for Δ_β-TSP.*

Proof. Obviously the cost of the minimal cycle cover is a lower bound on the cost of an optimal Hamiltonian cycle. Such a cycle cover can be found in polynomial time [EJ70].

For every cycle C_i, $1 \leqslant i \leqslant k$, the cheapest edge is removed from C_i. Thus, the cost of the remaining edges is at least $\frac{2}{3} \cdot cost(C_i)$ since every cycle of the cover has a length of at least 3. The removed edges are replaced by adjacent edges. According to Lemma 2 the costs of these new edges can exceed the costs of the removed edges in C by a factor of at most $\frac{\beta}{1-\beta}$. Thus, we have

$$cost(H) \leqslant \left(\frac{2}{3} + \frac{1}{3} \cdot \frac{\beta}{1-\beta}\right) \cdot cost(C)$$

which completes the proof. □

References

[Ar96] S. Arora: Polynomial Time Approximation Schemes for Euclidean Traveling Salesman and Other Geometric Problems. *J. ACM* 45(5), 1998, pp. 753–782.

[Ar97] S. Arora: Nearly linear time approximation schemes for Euclidean TSP and other geometric problems. In: *Proc. 38th IEEE FOCS*, 1997, pp. 554–563.

[AB95] T. Andreae, H.-J. Bandelt: Performance guarantees for approximation algorithms depending on parametrized triangle inequalities. *SIAM J. Discr. Math.* 8 (1995), pp. 1–16.

[BC99] M. A. Bender, C. Chekuri: Performance guarantees for the TSP with a parameterized triangle inequality. In: *Proc. WADS'99, LNCS 1663*, Springer 1999, pp. 80–85.

[BHKSU99] H.-J. Böckenhauer, J. Hromkovič, R. Klasing, S. Seibert, W. Unger: Towards the Notion of Stability of Approximation Algorithms and the Traveling Salesman Problem. Extended abstract in: *Proc. CIAC 2000, LNCS*, to appear. Full version in: *Electronic Colloquium on Computational Complexity*, Report No. 31 (1999).

[BK98] P. Berman, M. Karpinski: On some tighter inapproximability results. Technical Report TR98-029, *Electronic Colloquium on Computational Complexity*, 1998.

[Chr76] N. Christofides: Worst-case analysis of a new heuristic for the traveling salesman problem. Technical Report 388, Graduate School of Industrial Administration, Carnegie-Mellon University, Pittsburgh, 1976.

[En99] L. Engebretsen: An explicit lower bound for TSP with distances one and two. Extended abstract in: *Proc. STACS'99, LNCS 1563*, Springer 1999, pp. 373–382. Full version in: *Electronic Colloquium on Computational Complexity*, Revision 1 of Report No. 46 (1999).

[EJ70] J. Edmonds, E. L. Johnson: Matching: A Well-Solved Class of Integer Linear Programs. In: *Proc. Calgary International Conference on Combinatorial Structures and Their Applications*, Gordon and Breach 1970, pp. 89–92.

[GT91] H. N. Gabov, R. E. Tarjan: Faster Scaling Algorithms for General Graph-Matching Problems. In: *J. ACM* 38 (4), 1991, pp. 815–853.

[Ho96] D. S. Hochbaum (Ed.): *Approximation Algorithms for NP-hard Problems.* PWS Publishing Company 1996.

[LLRS85] E. L. Lawler, J. K. Lenstra, A. H. G. Rinnooy Kan, D. B. Shmoys (Eds.): *The Traveling Salesman Problem.* John Wiley & Sons, 1985.

[Mi96] I. S. B. Mitchell: Guillotine subdivisions approximate polygonal subdivisions: Part II — a simple polynomial-time approximation scheme for geometric k-MST, TSP and related problems. Technical Report, Dept. of Applied Mathematics and Statistics, Stony Brook 1996.

[MPS98] E. W. Mayr, H. J. Prömel, A. Steger (Eds.): *Lectures on Proof Verification and Approximation Algorithms. LNCS* 1967, Springer 1998.

[PY93] Ch. Papadimitriou, M. Yannakakis: The traveling salesman problem with distances one and two. *Mathematics of Operations Research* 18 (1993), 1–11.

λ-Coloring of Graphs

Hans L. Bodlaender[1], Ton Kloks[2], Richard B. Tan[1,3], and Jan van Leeuwen[1]

[1] Department of Computer Science. Utrecht University
Padualaan 14, 3584 CH Utrecht, The Netherlands
[2] Department of Mathematics and Computer Science
Vrije Universiteit, 1081 HV Amsterdam, The Netherlands
[3] Department of Computer Science
University of Sciences & Arts of Oklahoma
Chickasha, OK 73018, U.S.A.

Abstract. A λ-coloring of a graph G is an assignment of colors from the set $\{0, \ldots, \lambda\}$ to the vertices of a graph G such that vertices at distance at most two get different colors and adjacent vertices get colors which are at least two apart. The problem of finding λ-colorings with small or optimal λ arises in the context of radio frequency assignment. We show that the problems of finding the minimum λ for planar graphs, bipartite graphs, chordal graphs and split graphs are NP-Complete. We then give approximation algorithms for λ-coloring and compute upperbounds of the best possible λ for outerplanar graphs, planar graphs, graphs of treewidth k, permutation and split graphs. With the exception of the split graphs, all the above bounds for λ are linear in Δ, the maximum degree of the graph. For split graphs, we give a bound of $\lambda \leq \Delta^{1.5}+2\Delta+2$ and show that there are split graphs with $\lambda = \Omega(\Delta^{1.5})$. Similar results are also given for variations of the λ-coloring problem.

1 Introduction

Radio frequency assignment is a widely studied area of research. The task is to assign radio frequencies to transmitters at different locations without causing interference. The problem is closely related to graph coloring where the vertices of a graph represent the transmitters and adjacencies indicate possible interferences.

In [16], Griggs and Yeh introduced a problem proposed by Roberts which they call the $L(2,1)$-labeling problem. It is the problem of assigning radio frequencies (integers) to transmitters such that transmitters that are close (distance 2 apart) to each other receive different frequencies and transmitters that are *very* close together (distance 1 apart) receive frequencies that are at least two apart. To keep the frequency bandwidth small, they are interested in computing the difference between the highest and lowest frequencies that have been assigned to the radio network. They call the *minimum* difference of the range of frequencies, λ. The problem is then equivalent to assigning an integer from $\{0, \ldots, \lambda\}$ to the nodes of the networks satisfying the $L(2,1)$-labeling constraint.

H. Reichel and S. Tison (Eds.): STACS 2000, LNCS 1770, pp. 395–406, 2000.

Subsequently, different bounds of λ were obtained for various graphs. A common parameter used is Δ, the maximum *degree* of a graph. The obvious lower bound for λ is $\Delta+1$, achieved for the tree $K_{1,\Delta}$. In [16] it was shown that for every graph G, $\lambda \leq \Delta^2+2\Delta$. This upperbound was later improved to $\lambda \leq \Delta^2+\Delta$ in [8].

For some special classes of graphs, tight bounds are known and can be computed efficiently. These include paths, cycles, wheels and complete k-partite graphs [16], trees [8,16], cographs [8], k-almost trees [12], cacti, unicycles and bicycles [18], and grids, hexagons and hexagon-meshes [4]. Other types of graphs have also been studied, but only approximate bounds are known for them. These are chordal graphs and unit interval graphs [23], interval graphs [8], hypercubes [13,14,18], bipartite graphs [25], and outerplanar and planar graphs [18].

In this paper, we extend the upperbounds of λ to other graphs and also improve some existing bounds for some classes of graphs. Precisely, new bounds are provided for graphs of treewidth k, permutation graphs and split graphs. We also improve the bounds in [18] for planar graphs and outerplanar graphs. Efficient algorithms for labeling the graphs achieving these bounds are also given. With the exception of split graphs, all the above bounds are linear in Δ. For split graphs, we give a bound of $\lambda \leq \Delta^{1.5}+2\Delta+2$ and show that there are split graphs with $\lambda = \Omega(\Delta^{1.5})$. This is the first bound for λ that we know of that is neither linear in Δ nor Δ^2.

In [16], it was shown that determining λ of a graph is an NP-Complete problem, even for graphs with diameter two. And in [12], it was further shown that it is also NP-Complete to determine if $\lambda \leq k$ for every fixed integer $k \geq 4$ (the case when $\lambda \leq 3$ occurs only when G is a disjoint union of paths of length at most 3). In this paper, we show that the problem remains NP-Complete when restricted to planar graphs, bipartite graphs, chordal graphs and split graphs.

The $L(2,1)$-labeling problem proposed by Roberts is basically a problem of avoiding *adjacent-band interferences* – adjacent bands must have frequencies sufficiently far apart. There are several variations of the λ-coloring problems in the context of frequency assignment in multihop radio networks. Two other common type of collisions (frequencies interference) that have been studied are: direct and hidden collisions. In *direct collisions*, a radio station and its neighbors must have different frequencies, so their signals will not collide (overlap). This is just the normal vertex-coloring problem with its associated chromatic number $\mathcal{X}(G)$. In *hidden collisions*, a radio station must not receive signals of the same frequency from any of its adjacent neighbors. Thus, the only requirement here is that for each station, all its neighbors must have distinct frequencies (colors), but there is no requirement on what the color of the station itself.

In [3,22], the special case of avoiding *hidden collisions* in multihop radio networks were studied. We call this the $L(0,1)$-labeling problem (this notation was not used in [3,22]). In [2,9,10], the problem is to avoid both *direct* and *hidden collisions* in the radio network. Thus, a station and all of its neighbors must all have distinct colors. This is called $L(1,1)$-labeling in [25]. It is also known as distance-2 coloring problem and is equivalent to the normal coloring of the square of a graph, G^2, and has also been well-studied. These variations of λ-coloring are NP-Complete even for planar graphs [3].

Perhaps a more applicative term for all these λ-variations is the one given by Harary [17]: *Radio-Coloring.* We apply our algorithms to these variations as well and obtain similar bounds.

The paper is organized as follows. We give some definitions of graphs and generalizations of the λ-coloring problem in the next section. Then different upperbounds and algorithms for outerplanar graphs, planar graphs, graphs of treewidth k, permutation graphs and split graphs are presented in Section 3. The complexity results for planar graphs, bipartite graphs, chordal graphs and split graphs are given in Section 4. Finally, in the last section we mention some open problems.

2 Preliminaries

2.1 λ-Coloring

Let $G = (V, E)$ be a graph with vertex set V and edge set E. The number of vertices in G is denoted by n and the maximum degree of G by Δ.

Definition 1. *Let G be a graph and d_1, d_2 be two non-negative integers. A λ-*coloring *is an assignment of colors from a set $\{0, \ldots, \lambda\}$ to the vertices of the graph. The λ-coloring satisfies the $L(d_1, d_2)$-constraint if each pair of vertices at distance $i, 1 \leq i \leq 2$, in the graph gets colors that differ by at least d_i. If a λ-coloring of G satisfies the $L(d_1, d_2)$-constraint, then we say that G has an $L(d_1, d_2)$-labeling. The minimum value λ for which G admits a λ-coloring satisfying the $L(d_1, d_2)$-constraint is denoted by $\lambda_{d_1,d_2}(G)$, or, when G is clear from the context, by λ_{d_1,d_2}.*

In this paper, we shall focus mainly on particular $L(d_1, d_2)$-labelings which have been studied in the literature: $L(2, 1)$-labeling ([16]), $L(1, 1)$-labeling ([2,25]) and $L(0, 1)$-labeling ([3,22]).

Fact 1. *For any graph G, the following lower bounds hold:*

1. $\lambda_{0,1} \geq \Delta - 1$ [3],
2. $\lambda_{1,1} \geq \Delta$ [25],
3. $\lambda_{2,1} \geq \Delta + 1$ [16].

All these bounds are easily obtained by considering the tree $K_{1,\Delta}$, which is contained in any graph of maximum degree Δ.

2.2 Special Graphs

Definition 2. *A k-*tree *is a graph of $n \geq k + 1$ vertices defined recursively as follows. A clique of $k + 1$ vertices is a k-tree. A k-tree with $n + 1$ vertices can be formed from a k-tree with n vertices by making a new vertex adjacent to exactly all vertices of a k-clique in the k-tree with n vertices.*

Definition 3. *A graph is a* partial k-tree *if it is a subgraph of a k-tree.*

Definition 4. *The* treewidth *of a graph is the minimum value k for which the graph is a subgraph of a k-tree.*

A useful way of dealing with the treewidth of a graph is via its tree-decomposition.

Definition 5. *A* tree decomposition *of a graph $G = (V, E)$ is a pair $(\{X_i \mid i \in I\}, T = (I, F))$ with $\{X_i \mid i \in I\}$ a collection of subsets of V, and $T = (I, F)$ a tree, such that*

- $\bigcup_{i \in I} X_i = V$
- *for all edges $(v, w) \in E$ there is an $i \in I$ with $v, w \in X_i$*
- *for all $i, j, k \in I$: if j is on the path from i to k in T, then $X_i \cap X_k \subseteq X_j$.*

The width *of a tree decomposition $(\{X_i \mid i \in I\}, T = (I, F))$ is $\max_{i \in I} |X_i| - 1$. The* treewidth *of a graph $G = (V, E)$ is the minimum width over all tree decompositions of G.*

It can be shown that the above definitions of treewidth are equivalent and that every graph with treewidth $\leq k$ is a partial k-tree and conversely, that every partial k-tree has treewidth $\leq k$. For more details on treewidths, k-trees and other equivalent definitions, consult, for example, [20].

We now define a few more special graphs. [15] and [7] are good references for other definitions and results concerning these special graphs.

Definition 6. *A graph is* Chordal *or* Triangulated *iff every cycle of length ≥ 4 has a chord (i.e., there is no induced cycle of length ≥ 4).*

Definition 7. *A vertex of a graph G is* simplicial *if its neighbors induce a clique.*

Fact 2. [11] *Let G be a chordal graph and $|V| = n$. There is an* elimination scheme*(an ordering of vertices) for G, $[v_1, \ldots, v_n]$, such that for each i, $1 \leq i \leq n$, the vertex v_i is a simplicial vertex in the subgraph induced by $[v_{i+1}, \ldots, v_n]$. Such an elimination scheme is called a* perfect elimination scheme.

Definition 8. *A* split graph *is a graph G of which the vertex set can be split into two sets K and S, such that K induces a clique and S induces an independent set in G.*

A *permutation graph* can be obtained from a permutation $\pi = [\pi_1 \ldots \pi_n]$ of integers from 1 to n in the following visual manner. Line up the numbers 1 to n horizontally on a line. On the line below it, line up the corresponding permutation so that π_i is right below i. Now connect each i and π_j such that $\pi_j = i$ with an edge. Such edges are called *matching edges* and the resulting diagram is referred to as a *matching diagram*. The *inversion graph* is the graph $G_\pi = (V, E)$ with $V = \{1, \ldots, n\}$ and $(i, j) \in E$ iff the matching lines of i and j in the matching diagram intersect. Formally, one can define a permutation graph as follows.

Definition 9. *Let $\pi = [\pi_1 \ldots \pi_n]$ be a permutation of integers from 1 to n. Then the permutation graph determined by π is the graph $G_\pi = (V, E)$ with $V = \{1, \ldots, n\}$ and $(i, j) \in E$ iff $(i - j)(\pi_i^{-1} - \pi_j^{-1}) < 0$, where π_i^{-1} is the* inverse *of π_i (i.e., the position of the number i in the sequence π). A graph G is a* permutation graph *if there exists a permutation π such that G is isomorphic to the inversion graph G_π.*

3 Bounds and Algorithms

We use the following heuristic, or small modifications of it, to λ-color graphs G [16,18,23,25]. First we find an *elimination sequence*, an ordering of the vertices, $[v_1, \ldots, v_n]$, satisfying certain conditions for G. In order to do this, we rely on the fact that all the graphs considered have the *hereditary* property, i.e., when a special vertex is eliminated from a graph considered, the induced subgraph remains the same type of graph. Then we simply apply the *greedy* algorithm to color each vertex in the sequence by using the smallest available color in $[0, \ldots, \lambda]$, satisfying the $L(d_1, d_2)$-constraint. For each graph G, we estimate the total number of vertices at distance two that a vertex can have among the vertices that have been colored so far. Finally we compute the upperbound for λ.

3.1 Outerplanar and Planar Graphs

We first give an algorithm for labeling an outerplanar graph and then use the result to obtain a bound for planar graphs. We need the following lemma.

Lemma 1. *In an outerplanar graph, there exists a vertex of degree at most one or a vertex of degree two which has one neighbor that has at most* 4 *neighbors.*

Proof. First, we show that a biconnected outerplanar graph G has a vertex of degree two with a neighbor of degree at most four. (We ignore trivial cases like a graph with one vertex.) The *inner dual* G^* of biconnected outerplanar graph G is formed by taking the dual of G and then removing the vertex that represents the outer region. It is easy to see that G^* is a tree [18]. Note that all leaves in G^* correspond to a face with at least one vertex of degree two in G. Consider two leaves, u and v, of the inner dual with maximum distance in G^*. Suppose to the contrary that all neighbors of a vertex of degree ≤ 2 have more than 4 neighbors in the G. The face represented by u has a vertex x of degree two, and both neighbors x and y of u have degree more than v. Thus, there are at least four faces adjacent to x, and these form a path with length at least three starting from u in G^*. Similarly, there is a different path of length three starting from u in G^* for the faces that contain y. One now can observe that there is a path in G^* of greater length than the path from u to v, contradiction. Hence G has a vertex of degree two that has a neighbor of degree at most four.

Finally, every outerplanar graph $G = (V, E)$ is a subgraph of a biconnected outerplanar graph $H = (V, F)$ [19]. A vertex v that has degree at most two and has a neighbor of degree at most 4 in H has degree at most one in G or has degree two and a neighbor of degree at most 4 in G. □

Algorithm 1. *Algorithm for Outerplanar Graphs*

For $i := 1$ **to** n
  Find *a vertex* v_i *of degree* ≤ 2 *with a neighbor of degree at most 4 or of degree* ≤ 1.
  If *neighbors of* v_i *are not adjacent*
    Then add *a* virtual *edge between them*
  Temporarily *remove* v_i *from* G.

For $i := n$ **to** 1
  Label v_i *with the* smallest *available color in* $\{0, \ldots, \lambda\}$
    satisfying the $L(2,1)$-constraint
  If *neighbors of* v_i *have a virtual edge*
    Then remove *the edge*

In Algorithm 1, we first find an elimination sequence, $[v_1, \ldots, v_n]$, using the condition in Lemma 1. Note that this can always be done due to the hereditary property of outerplanarity. Then, we use this order to color the vertices in a greedy manner, i.e., we color a vertex v_i with the first available color such that the color of vertex v_i differs by at least two from the colors of its already colored neighbors, and differs by at least one from the colors of already colored vertices at distance two. In this way, we make sure that the $L(2,1)$-constraint is fulfilled. The virtual edge (if there is one) is there to guarantee that the colors of v_i's two neighbors are different when they are colored. Note that an operation that removes v_i and adds a virtual edge between its two neighbors does not increase the maximum degree Δ.

We can now compute the value of the maximum color used in Algorithm 1.

Theorem 3. *There is an algorithm for finding an* $L(2,1)$*-labeling of an outerplanar graph with* $\lambda_{2,1} \leq \Delta + 8$.

Proof. We use induction. The first vertex v_n can simply be colored with color 0. Suppose we have colored the vertices in the elimination sequence $[v_{i+1}, \ldots, v_n]$, $i < n$. When we want to color the vertex v_i, v_i can have at most two colored neighbors. First, suppose v_i has two colored neighbors. Each of these two neighbors can account for at most 3 more colors: if we color one of these nodes by color c, then colors $c-1$ and $c+1$ are forbidden for v_i. This means possibly six colors that are unavailable for coloring v_i. Now v_i has at most $\Delta - 1 + 3$ vertices at distance two, which means another $\Delta + 2$ colors that possibly cannot be used for v_i. If there are at least $\Delta + 9$ colors, then there is always at least one color available for v, i.e., $\lambda_{2,1} \leq \Delta + 8$. A similar analysis can be used if v_i has one colored neighbor. □

Corollary 1. *There is an algorithm for finding an* $L(2,1)$*-labeling of a triangulated outerplanar graph with* $\lambda_{2,1} \leq \Delta + 6$.

Proof. In a triangulated graph, there are at most $\Delta - 2 + 2 = \Delta$ distance-2 neighbors of v_i. The total number of colors needed is then $\Delta+7$ or $\lambda_{2,1} \leq \Delta+6$. □

This improves the bound of $\lambda_{2,1} \leq 2\Delta + 2$ in [18] for outerplanar graphs, for sufficiently large Δ.

We now use the above bound for outerplanar graphs to obtain a bound for planar graphs.

Theorem 4. *There is an algorithm for finding an $L(2,1)$-labeling of a planar graph with $\lambda_{2,1} \leq 3\Delta + 28$.*

Proof. Any planar graph can be partitioned into *layers* in the following way: all vertices at the exterior face are in layer 0, and all vertices not in layers $0, \ldots, i$ that are adjacent to a vertex in layer i are in layer $i+1$. Each layer induces an outerplanar graph. (See [1].) Now, we can color the vertices on layers 0, 3, 6, etc. with colors $0, \ldots, \Delta+8$; vertices on layers 1, 4, 7, etc. with colors $\Delta+10, \ldots, 2\Delta+18$; and vertices on layers 2, 5, 8, etc. with colors $2\Delta+20, \ldots, 3\Delta+28$; each layer is colored with the algorithm for outerplanar graphs. Because of the separations between layers, this is an $L(2,1)$-coloring. □

Corollary 2. *For any triangulated planar graph, $\lambda_{2,1} \leq 3\Delta + 22$.*

This improves the bound of $\lambda_{2,1} \leq 8\Delta - 13$ for planar graphs from [18], for sufficiently large Δ.

Theorem 5. *For outerplanar graphs, there are polynomial time algorithms for labeling the graphs such that $\lambda_{0,1} \leq \Delta + 2$ and $\lambda_{1,1} \leq \Delta + 4$.*

Proof. We apply the same algorithm as in Algorithm 1 to find an eliminating sequence $[v_n, \ldots, v_1]$ and then greedily color each v_i in the sequence with the smallest available color satisfying the $L(0,1)$-constraint. As in the proof of Theorem 3, each v_i in the eliminating sequence has at most $\Delta + 2$ neighbors at distance-2, which must have different colors from v_i. Now v_i can have at most two neighbors, but they can have the same color as v_i. So we can color v_i and its neighbors with an extra color. The bound for $\lambda_{0,1}$ now follows.

Similar argument applies for $\lambda_{1,1}$. □

Theorem 6. *For planar graphs, there are polynomial time algorithms for labeling the graphs such that $\lambda_{0,1} \leq 2\Delta + 5$ and $\lambda_{1,1} \leq 3\Delta + 14$.*

It is not hard to see that the algorithms mentioned above can be implemented to run in time $O(n\Delta)$, n the number of vertices.

3.2 Graphs of Treewidth k

For a graph $G = (V,E)$ of treewidth k, we first take a tree-decomposition $(\{X_i \mid i \in I\}, T = (I,F))$ with $\{X_i \mid i \in I\}$ a collection of subsets of V, and $T = (I,F)$ a tree. Let $d(v,w)$ be the distance between vertices v and w.

Algorithm 2. *Algorithm for Graphs of Treewidth k*

1: Add *a set of virtual edges* E'*:*
   Here $(v,w) \in E'$ *iff* $(v,w) \notin E$ *and* $\exists i : v,w \in X_i$.

2: Find *a Perfect Elimination Sequence:* $[v_1,\dots,v_n]$ in $G' = (V, E \cup E')$.

3: **For** $i := n$ **to** 1
   Label v_i *with the smallest available color, such that for all already colored vertices* w*:*
   If $(v,w) \in E$, then *the color of* v *and* w *differ by at least 2,*
   If $(v,w) \in E'$, then *the color of* v *and* w *differ by at least 1,*
   If $d(v,w) = 2$, then *the color of* v *and* w *differ by at least 1.*

Theorem 7. *There is a polynomial time algorithm for labeling a graph G of treewidth k with* $\lambda_{2,1} \leq k\Delta + 2k$.

Proof. After all the virtual edges are added, the new graph $G' = (V, E \cup E')$ is chordal, hence has a perfect elimination sequence (from Fact 2). Also, G' has the same tree decomposition as G so has treewidth at most k. When we are ready to color a vertex v, v has at most k colored neighbors because v with its neighbors forms a clique in G'. Now by the perfect elimination sequence property, an already colored vertex at distance two of v must be adjacent to an already colored neighbor of v in G'. Hence, at most $3k$ colors are unavailable due to the neighbors of v, and at most $(\Delta-1)k$ colors are unavailable due to the vertices at distance two. If we have $k\Delta+2k+1$ colors, then a color for v is always available. The bound now follows. □

Corollary 3. *There is a polynomial time algorithm for labeling a k-tree with* $\lambda_{2,1} \leq k\Delta - k^2 + 3k$.

Proof. As a k-tree is always triangulated, there are at most $k(\Delta - 1 - (k-1)) = k\Delta - k^2$ distance-2 neighbors of v_i. The total number of colors needed is then $k\Delta - k^2 + 3k + 1$ or $\lambda_{2,1} \leq k\Delta - k^2 + 3k$. □

We can apply our algorithm of treewidth k to the other λ-variants.

Theorem 8. *For graphs of treewidth k,* $\lambda_{0,1} \leq k\Delta - k$ *and* $\lambda_{1,1} \leq k\Delta$.

The labeling algorithms given in this section can be implemented in time $O(kn\Delta)$, assuming that we are given the tree-decomposition, which can be found in linear time for treewidth of constant size k [5].

Recently, in [26] it has been shown that $\lambda_{1,1}$ can be computed in polynomial time for graphs with constant treewidth k. A similar argument would yield the result for $\lambda_{0,1}$ as well.

3.3 Permutation Graphs

Theorem 9. *There is a polynomial time algorithm for labeling a permutation graph with* $\lambda_{2,1} \leq 5\Delta - 2$.

Proof. Suppose the vertices are numbered $1, 2, \ldots, n$, and we have a permutation π, with $(i, j) \in E$ iff $(i - j)(\pi_i^{-1} - \pi_j^{-1}) < 0$ (i.e., the lines cross). We color the vertices from 1 to n in order, using the smallest color available satisfying the usual $L(2, 1)$-constraint. To show that the stated bound is sufficient, we make use of the following two claims.

Suppose we are in the midst of this algorithm, ready to color a vertex v. Let w be a vertex at distance 2 from v. Note that v and w can have distance two via a path across a colored vertex or via a path across an uncolored vertex.

Claim. Suppose there is a path $[v, y, w]$ with y a vertex that is not yet colored. Let x be the neighbor of v such that π_x^{-1} is minimal. Then either w is a neighbor of v or x, or w is not colored.

Proof. Suppose w is not a neighbor of v and w is already colored. Then we have $w < v$ and $\pi_w^{-1} < \pi_v^{-1}$. As y is not yet colored, $y > v$, hence $\pi_y^{-1} < \pi_v^{-1}$. Also $y > v > w$, so (y, w) is an edge and $\pi_y^{-1} < \pi_w^{-1}$. By assumption, $\pi_x^{-1} \leq \pi_y^{-1}$, $\pi_x^{-1} < \pi_v^{-1}$. Now $x > v > w$, so $\pi_w^{-1} > \pi_y^{-1} \geq \pi_x^{-1}$. Therefore (w, x) is an edge. □

The proof of the next claim is similar to the proof above.

Claim. Suppose there is a path $[v, y, w]$ with y a vertex that is already colored. Let x be the neighbor of v with x having a minimal π_x^{-1} among the neighbors. Then either w is a neighbor of v or x, or w is not colored.

We now count the total number of colors needed. The vertices x in the above two claims are adjacent to v, so they both have at most $\Delta - 1$ neighbors that are at distance two to v. We thus get a total of $3\Delta + 2(\Delta - 1) + 1 = 5\Delta - 1$ colors. □

Theorem 10. *For permutation graphs, there are polynomial time algorithms for labeling the graphs such that* $\lambda_{0,1} \leq 2\Delta - 2$ *and* $\lambda_{1,1} \leq 3\Delta - 2$.

All the above algorithms can also be implemented in $O(n\Delta)$ time.

3.4 Split Graphs

So far all the bounds for λ that we have obtained are linear in Δ. For split graphs we give a non-linear bound for λ and show that there are split graphs that require this bound.

Theorem 11. *There is a polynomial time algorithm for labeling a split graph with* $\lambda_{2,1} \leq \Delta^{1.5} + 2\Delta + 2$.

Proof. Let S be the independent set and K the clique that split G. $|K| \leq \Delta+1$. We use colors $0, 2, \ldots, 2\Delta$ to color the vertices in K. For S, we will use colors from the set $\{2\Delta+2, \ldots, \Delta^{1.5}+2\Delta+2\}$. If $|S| \leq \Delta^{1.5}$, we just give every vertex in S a distinct color and we are done. Suppose $|S| > \Delta^{1.5}$. We claim that there is always a vertex v in S with degree $\leq \Delta^{0.5}$. For if this is not the case, then all vertices in S have degree $> \Delta^{0.5}$. This implies that the total number of edges emanating from S is greater than Δ^2. By the Pigeonhole Principle, there would be some vertex in the clique K that has degree $> \Delta$; a contradiction. Now, let v be a vertex in S with degree $\leq \Delta^{0.5}$. Recursively color the graph obtained by removing v from G. v has at most $\Delta^{1.5}$ vertices in S at distance two, but as we have $\Delta^{1.5} + 1$ colors to color S, we always have a color left for v. Finally, note that adding v back cannot decrease the distances between other vertices in G as the neighborhood of v is a clique. □

Theorem 12. *For split graphs, there are polynomial time algorithms for labeling the graphs such that* $\lambda_{0,1} \leq \Delta^{1.5}$ *and* $\lambda_{1,1} \leq \Delta^{1.5} + \Delta + 1$.

We now show that the above bounds for λ is actually tight (within constant factor).

Theorem 13. *For every* Δ, *there is a split graph with* $\lambda_{2,1} \geq \lambda_{1,1} \geq \lambda_{0,1} \geq \frac{1}{3}\sqrt{\frac{2}{3}}\Delta^{1.5}$.

Proof. Consider the following split graph. We take an independent set of $\frac{1}{3}\sqrt{\frac{2}{3}}\Delta^{1.5}$ vertices. This set is partitioned into $\sqrt{\frac{2}{3}\Delta}$ groups, each consisting of $\Delta/3$ vertices. The clique consists of $\Delta/3+1$ vertices. Note that we have less than $(\sqrt{\frac{2}{3}\Delta})^2/2 = \Delta/3$ distinct pairs of groups. For each such pair of groups, we take one unique vertex in the clique, and make that vertex adjacent to each vertex in these two groups. In this way, the maximum degree is exactly Δ: each vertex in the clique is adjacent to $\Delta/3$ vertices in the clique and at most $2\Delta/3$ vertices in the independent set.

Now, the resulting graph has diameter two, and any pair of vertices in the independent set have distance exactly two. So, in any $L(0,1)$-labeling of the graph (or $L(1,1)$- or $L(2,1)$-labeling), all vertices in the independent set must receive different colors. □

As split graphs are also chordal graphs [15], the above theorem provides a non-linear lower bound for the upperbound of $\frac{3}{4}\Delta^2$ in [23].

4 Complexity

Without proofs we mention that it is NP-complete to decide whether $\lambda_{2,1} \leq 10$ for a given bipartite planar graph $G = (V, E)$, and that it is NP-complete to decide whether $\lambda_{2,1} \leq |V|$ for a given split graph (and hence this problem also is NP-complete for chordal graphs.)

The proofs can be found in [6]. The result for planar bipartite graphs uses a reduction from 3-coloring 4-regular planar graphs. The result for split graphs uses a reduction from Hamiltonian Path with a modification of a technique from [16].

The problems whether $\lambda_{0,1} \leq |V|$ and $\lambda_{1,1} \leq |V|$ are also NP-complete for the split graphs and chordal graphs.

5 Concluding Remarks

We have given upperbounds of λ for some of the well-known graphs. However, we lack examples of graphs where these bounds are matched. It should be possible to tighten the constant factors in the bounds somewhat. For example, in outerplanar graphs, the conjecture is that $\lambda_{2,1} \leq \Delta+2$ and we have $\lambda_{2,1} \leq \Delta+8$. Also, it is not clear that for planar graphs we need 3Δ in the bounds. Surely the constant 28 in $\lambda_{2,1} \leq 3\Delta + 28$ is much too large. Similar comments apply to the other graphs studied in this paper as well.

For graphs of treewidth k, the $L(0,1)$-labeling and $L(1,1)$-labeling problems are polynomial for constant k. The corresponding problem for $L(2,1)$-labelings appears to be an interesting (but apparently not easy) open problem. The corresponding problem for interval graphs and outerplanar graphs also remain open.

It is conjectured in [16] that $\lambda_{2,1} \leq \Delta^2$ for any graph. This is true for all the graphs that have been studied, but the problem remains open. For chordal graphs, [23] has shown that $\lambda_{2,1} \leq O(\Delta^2)$. We have shown for split graphs (a special case of chordal graphs) that $\lambda_{2,1} = \Theta(\Delta^{1.5})$. What is the best bound for chordal graphs?

Acknowledgement

We thank Hans Zantema for useful discussions and the referees for valuable suggestions.

References

1. B. S. Baker, Approximation algorithms for NP-complete problems on planar graphs, *J. ACM* **41** (1994) pp. 153–180.
2. Battiti, R., A.A. Bertossi and M.A. Bonuccelli, Assigning codes in wireless networks: Bounds and scaling properties, *Wireless Networks* **5** (1999) pp. 195–209.
3. Bertossi A.A. and M.A. Bonuccelli, Code assignment for hidden terminal interference avoidance in multihop packet radio networks, *IEEE/ACM Trans. on Networking* **3** (1995) pp. 441–449.

4. Bertossi A.A., C. Pinotti and R.B. Tan, $L(2,1,1)$-Labeling problem on graphs, In preparation (1999).
5. Bodlaender, H., A linear time algorithm for finding tree-decompositions of small treewidth, *SIAM Journal on Computing* **25** (1996) pp. 1305–1317.
6. Bodlaender, H., T. Kloks, R.B. Tan and J. van Leeuwen, λ-coloring of graphs, *Tech. Report*, Dept. of Computer Science, Utrecht University (1999), in preparation.
7. Brandstädt, A., V.B. Le and J.P. Spinrad, *Graph Classes: a Survey*, SIAM Monographs on Discrete Mathematics and Applications. Society for Industrial and Applied Mathematics, Philadelphia (1999).
8. Chang, G. J. and D. Kuo, The $L(2,1)$-labeling problem on graphs, *SIAM J. Disc. Math.* **9** (1996) pp. 309–316.
9. Chlamtac, I. and S. Kutten, On broadcasting in radio networks - problem analysis and protocol design, *IEEE Trans. on Comm.* **33** No. 12 (1985) pp. 1240–1246.
10. Chlamtac, I. and S. Pinter, Distributed nodes organization algorithm for channel access in a multihop dynamic radio network, *IEEE Trans. on Computers* **36** No. 6 (1987) pp. 728–737.
11. Dirac, G.A., On rigid circuit graphs, *Abh. Math. Sem. Univ. Hamburg* **25** (1961) pp. 71–76.
12. Fiala J., T. Kloks and J. Kratochvíl, Fixed-parameter complexity of λ-labelings, In: *Graph-Theoretic Concept of Computer Science, Proceedings* 25*th WG '99*, Lecture Notes in Computer Science vol. 1665, Springer Verlag (1999), pp. 350–363.
13. Georges, J. P. and D. W. Mauro, On the size of graphs labeled with a condition at distance two, *Journal of Graph Theory* **22** (1996) pp. 47–57.
14. Georges, J. P., D. W. Mauro and M. A. Whittlesey, Relating path coverings to vertex labelings with a condition at distance two, *Discrete Mathematics* **135** (1994) pp. 103–111.
15. M. C. Golumbic, *Algorithmic graph theory and perfect graphs*, Academic Press, New York (1980).
16. Griggs, J. R. and R. K. Yeh, Labeling graphs with a condition at distance 2, *SIAM J. Disc. Math.* **5** (1992) pp. 586–595.
17. Harary, F., Private communication to Paul Spirakis.
18. Jonas, K., Graph colorings analogues with a condition at distance two: $L(2,1)$-labelings and list λ-labelings, Ph.D. thesis, University of South Carolina (1993).
19. Kant, G., Augmenting Outerplanar Graphs, *J. Algorithms* **21** (1996) pp. 1–25.
20. Kloks, T., *Treewidth-computations and approximations*, Springer-Verlag, LNCS **842** (1994).
21. Liu, D. D.-F. and R. K. Yeh, On distance two labelings of graphs, *ARS Combinatorica* **47** (1997) pp. 13–22.
22. Makansi, T., Transmitter-oriented code assignment for multihop packet radio, *IEEE Trans. on Comm.* **35** No. 12 (1987) pp. 1379–1382.
23. Sakai, D., Labeling chordal graphs: distance two condition, *SIAM J. Disc. Math.* **7** (1994) pp. 133–140.
24. Whittlesey, M. A., J. P. Georgess and D. W. Mauro, On the lambda-coloring of Q_n and related graphs, *SIAM J. Disc. Math.* **8** (1995) pp. 499–506.
25. Yeh, R.K., Labeling graphs with a condition at distance two, Ph.D. Thesis, University of South Carolina (1990).
26. Zhou, X., Y. Kanari and T. Nishizeki, Generalized Vertex-Colorings of Partial k-Trees, *IEICE Trans. Fundamentals* (2000), to appear.

Optimal Proof Systems and Sparse Sets*

Harry Buhrman[1], Steve Fenner**[2], Lance Fortnow***[3], and Dieter van Melkebeek†[4]

[1] CWI, INS4, P.O. Box 94079, 1090 GB Amsterdam, The Netherlands.
buhrman@cwi.nl

[2] Department of Computer Science, The University of South Carolina, Columbia, SC 29208.
fenner@cs.sc.edu

[3] University of Chicago,
Current Address: NEC Research, 4 Independence Way, Princeton, NJ 08540.
fortnow@research.nj.nec.com

[4] University of Chicago and DIMACS,
Current Address: DIMACS Center, Rutgers University,
96 Frelinghuysen Road, Piscataway, NJ 08854.
dieter@dimacs.rutgers.edu

Abstract. We exhibit a relativized world where $\mathbf{NP} \cap \mathbf{SPARSE}$ has no complete sets. This gives the first relativized world where no optimal proof systems exist.
We also examine under what reductions $\mathbf{NP} \cap \mathbf{SPARSE}$ can have complete sets. We show a close connection between these issues and reductions from sparse to tally sets. We also consider the question as to whether the $\mathbf{NP} \cap \mathbf{SPARSE}$ languages have a computable enumeration.

1 Introduction

Computer scientists study lower bounds in proof complexity with the ultimate hope of actual complexity class separation. Cook and Reckhow [CR79] formalize this approach. They create a general notion of a proof system and show that polynomial-size proof systems exist if and only if $\mathbf{NP} = \mathbf{coNP}$.

Cook and Reckhow also ask about the possibility of whether optimal proof systems exist. Informally an optimal proof system would have proofs which are no more than polynomially longer than any other proof system.

An optimal proof system would play a role similar to **NP**-complete sets. There exists a polynomial-time algorithm for Satisfiability if and only if $\mathbf{P} = \mathbf{NP}$. Likewise, if we have an optimal proof system, then this system would have polynomial-size proofs if and only if $\mathbf{NP} = \mathbf{coNP}$.

* Several proofs have been omitted to conserve space. A full version can be found at http://www.neci.nj.nec.com/homepages/fortnow/papers.

** Supported in part by NSF grants CCR-9501794 and CCR-9996310.

*** Supported in part by NSF grant CCR-9732922.

† Supported in part by NSF grant CCR-9732922.

H. Reichel and S. Tison (Eds.): STACS 2000, LNCS 1770, pp. 407–418, 2000.

The existence of optimal proof systems remained an interesting open question. No one could exhibit such a system except under various unrealistic assumptions [KP89, MT98]. Nor has anyone exhibited a relativized world where optimal proof systems do not exist.

We construct such a world by building the first oracle relative to which **NP** ∩ **SPARSE** does not have complete sets. Messner and Torán [MT98] give a relativizable proof that if an optimal proof system exists than **NP** ∩ **SPARSE** does have complete sets.

We also consider whether **NP** ∩ **SPARSE**-complete sets exist under other more general reductions than the standard many-one reductions. We show several results such as:

- There exists a relativized world where **NP** ∩ **SPARSE** has no disjunctive-truth-table complete sets.
- There exists a relativized world where **NP** ∩ **SPARSE** has no complete sets under truth-table reductions using $o(n/\log n)$ queries.
- For any positive constant c, there exists an oracle relative to which the class **NP** ∩ **SPARSE** has no complete sets under truth-table reductions using $o(n/\log n)$ queries and $c \cdot \log n$ bits of advice.
- Under a reasonable assumption for all values of $k > 0$, **NP** ∩ **SPARSE** has a complete set under conjunctive truth-table reductions that ask $\frac{n}{k \log n}$ queries and use $O(\log n)$ bits of advice.

The techniques used for relativized results on **NP** ∩ **SPARSE**-complete sets also apply to the question of reducing sparse sets to tally sets. We show several results along these lines as well.

- Every sparse set S is reducible to some tally set T under a 2-round truth-table reduction asking $O(n)$ queries.
- Let c be any positive constant. There exists a sparse set S that does not reduce to any tally set T under truth-table reductions using $o(n/\log n)$ queries even with $c \cdot \log n$ bits of advice.
- Under a reasonable assumption for every sparse set S and every positive constant k, there exists a tally set T and a ctt-reduction from S to T that asks $\frac{n}{k \log n}$ queries and $O(\log n)$ bits of advice. We can also have a 2-round truth-table reduction using $\frac{n}{k \log n}$ queries and no advice.

We use the "reasonable assumptions" to derandomize some of our constructions building on techniques of Klivans and Van Melkebeek [KvM99]. The assumption we need is that there exists a set in **DTIME**$[2^{O(n)}]$ that requires circuits of size $2^{\Omega(n)}$ even when the circuits have access to an oracle for **SAT**. Under this assumption we get tight bounds as described above.

We also examine how **NP** ∩ **SPARSE** compares with other promise classes such as **UP** and **BPP** in particular looking at whether **NP** ∩ **SPARSE** has a uniform enumeration.

The proofs in our paper heavily use techniques from Kolmogorov complexity. We recommend the book of Li and Vitányi [LV97] for an excellent treatment of this subject.

1.1 Reductions and Relativizations

We measure the relative power of sets using reductions. In this paper all reductions will be computed by polynomial-time machines.

We say a set A reduces to a set B if there exists a polynomial-time computable function f such that for all strings x, x is in A if and only if $f(x)$ is in B. We also call this an m-reduction, "m" for many-one.

For more general reductions we need to use oracle machines. The set A Turing-reduces to B if there is a polynomial-time oracle Turing machine M such that $M^B(x)$ accepts exactly when x is in A. A tt-reduction (truth-table) requires that all queries be made before any answers are received.

A 2-round tt-reduction allows a second set of queries to be made after the answers from the first set of queries is known. This can be generalized to k-round tt-reductions but we will not need $k > 2$ in this paper.

We can think of a (one-round) tt-reduction R as consisting of two polynomial-time computable functions: One that creates a list of queries to make and an evaluator that takes the input and the value of B on those queries and either accepts or rejects. We use the notation $Q_R(x)$ to denote the set of queries made by reduction R on input x. For a set of inputs X, we let $Q_R(X) = \cup_{x \in X} Q_R(x)$.

A dtt-reduction (disjunctive-truth-table) means that $M^B(x)$ accepts if any of the queries it makes are in B. A ctt-reduction (conjunctive-truth-table) means that $M^B(x)$ accepts if all of the queries it makes are in B. A $q(n)$-tt reduction is a tt-reduction that makes at most $q(n)$ queries. A btt-reduction (bounded-truth-table) is a k-tt reduction for some fixed k.

We say a language L is r-hard for a class $\mathcal{C}$ if every language in $\mathcal{C}$ r-reduces to L. If L also sits in $\mathcal{C}$ then we say L is r-complete for $\mathcal{C}$.

All the results mentioned and cited in this paper relativize, that is they hold if all machines involved can access the same oracle. If we show that a statement holds in a relativized world that means that proving the negation would require radically different techniques. Please see the survey by Fortnow [For94] for a further discussion on relativization.

1.2 Optimal Proof Systems

A proof system is simply a polynomial-time function whose range is the set of tautological formulae, i.e., formulae that remain true for all assignments. Cook and Reckhow [CR79] developed this concept to give a general proof system that generalizes proof systems such as resolution and Frege proofs. They also give an alternate characterization of the **NP** versus **coNP** question:

Theorem 1 (Cook-Reckhow). **NP** $=$ **coNP** *if and only if there exists a proof system f and a polynomial p such that for all tautologies ϕ, there is a y, $|y| \leq p(|\phi|)$ and $f(y) = \phi$.*

Cook and Reckhow [CR79] also defined optimal and p-optimal proof systems.

Definition 1. *A proof system g is* optimal *if for all proof systems f, there is a polynomial p such that for all x, there is a y such that* $|y| \leq p(|x|)$ *and* $g(y) = f(x)$. *A proof system g is* p-optimal *if y can be computed in polynomial time from x.*

Messner and Torán [MT98] building on work of Krajíček and Pudlák [KP89] show that if **NEE** = **coNEE** then optimal proof systems exist and if **NEE** = **EE** then p-optimal proof systems exist. Here **EE**, double exponential time, is equal to $\mathbf{DTIME}[2^{O(2^n)}]$. The class **NEE** is the nondeterministic version of **EE**.

Messner and Torán [MT98] show consequences of the existence of optimal proof systems.

Theorem 2 (Messner-Torán).

- *If p-optimal proof systems exist then* **UP** *has complete sets.*
- *If optimal proof systems exist then* **NP** ∩ **SPARSE** *has complete sets.*

Hartmanis and Hemachandra [HH84] give a relativized world where **UP** does not have complete sets. Since all of the results mentioned here relativize, Messner and Torán get the following corollary.

Corollary 1 (Messner-Torán). *There exists an oracle relative to which p-optimal proof systems do not exist.*

However Messner and Torán leave open the question as to whether a relativized world exists where there are no optimal proof systems. Combining our relativized world where **NP** ∩ **SPARSE** has no complete sets with Theorem 2 answers this question in the positive.

1.3 Reducing SPARSE to TALLY

A tally set is any subset of 1^*. Given a set S, the census function $c_S(n)$ is the number of strings of length n in S. A set S is sparse if the census function is bounded by a polynomial.

In some sense both sparse sets and tally sets contain the same amount of information but in sparse sets the information may be harder to find. Determining for which kind of reductions **SPARSE** can reduce to **TALLY** is an exciting research area.

Book and Ko [BK88] show that every sparse set tt-reduces to some tally set but there is some sparse set that does not btt-reduce to any tally set.

Ko [Ko89] shows that there is a sparse set that does not dtt-reduce to any tally set. He left open the conjunctive case.

Buhrman, Hemaspaandra and Longpré [BHL95] give the surprising result that every sparse set ctt-reduces to some tally set. Later Saluja [Sal93] proves the same result using slightly different techniques.

Schöning [Sch93] uses these ideas to show that **SPARSE** many-one reduces to **TALLY** with randomized reductions. In particular he shows that for every sparse set S and polynomial p there is a tally set T and a probabilistic polynomial-time computable f such that

– If x is in S then $f(x)$ is always in T.
– If x is not in S then $\Pr[f(x) \in T] \leq 1/p(|x|)$.

We say that S *co-rp-reduces* to T. Schöning notes that his reduction only requires $O(\log n)$ random bits.

1.4 Complete Sets for NP ∩ SPARSE

Hartmanis and Yesha [HY84] first considered the question as to whether the class **NP** ∩ **SPARSE** has complete sets. They show that there exists a tally set T that is Turing-complete for **NP** ∩ **SPARSE**. They also give a relativized world where there is no tally set that is m-complete for **NP** ∩ **SPARSE**.

We should note that **NP** ∩ **TALLY** has m-complete sets. Let M_i be an enumeration of polynomial-time nondeterministic machines and consider

$$\{1^{\langle i,n,k\rangle} \mid M_i(1^n) \text{ accepts in } k \text{ steps}\}. \tag{1}$$

Also there exists a set in $\mathbf{D_p} \cap$ **SPARSE** that is m-hard for **NP** ∩ **SPARSE**. The class $\mathbf{D_p}$ contains the sets that can be written as the difference of two **NP** sets. For the **NP** ∩ **SPARSE**-hard language we need to consider the difference $A - B$ where:

$$A = \{\langle x, 1^i, 1^k\rangle \mid M_i(x) \text{ accepts in } k \text{ steps}\}$$
$$B = \{\langle x, 1^i, 1^k\rangle \mid M_i \text{ accepts more than } k \text{ strings of length } |x| \text{ in } k \text{ steps}\}$$

As a simple corollary we get that if **NP** = **coNP** then **NP** ∩ **SPARSE** has complete sets. However the results mentioned in Section 1.2 imply that one only needs the assumption of **NEE** = **coNEE**.

Schöning [Sch93] notes that from his work mentioned in Section 1.3 if the sparse set S is in **NP** then the corresponding tally set T is also in **NP**. Since **NP** ∩ **TALLY** has complete sets we get that **NP** ∩ **SPARSE** has a complete set under co-rp-reductions. The same argument applied to Buhrman-Hemaspaandra-Longpré shows that **NP** ∩ **SPARSE** has complete sets under ctt-reductions.

2 NP ∩ SPARSE-Complete Sets

In this section, we establish our main result.

Theorem 3. *There exists a relativized world where* **NP** ∩ **SPARSE** *has no complete sets under many-one reductions.*

Proof. Let M_i be a standard enumeration of nondeterministic polynomial-time Turing machines and f_i be an enumeration of polynomial-time reductions where M_i and f_i use at most time n^i.

Let $t(m)$ be the tower function, i.e., $t(0) = 1$ and $t(m+1) = 2^{t(m)}$.

We will build an oracle A. For each i we will let

$$L_i(A) = \{x \mid \text{There is some } y, |y| = 2|x| \text{ and } \langle i, x, y\rangle \in A\}. \tag{2}$$

The idea of the proof is that for each i and j, we will guarantee that either $L(M_i^A)$ has more than n^j elements at some input length n or $L_i(A)$ is sparse and f_j^A does not reduce $L_i(A)$ to $L(M_i^A)$.

We start with the oracle A empty and build it up in stages. At each stage $m = \langle i, j\rangle$ we will add strings of the form $\langle i, x, y\rangle$ to A where $|x| = n = t(m)$ and $|y| = 2n$. For each stage m we will do one of the following:

1. Put more than r^j strings into $L(M_i^A)$ for some length r, or
2. Make $L_i(A) \cap \Sigma^n$ have exactly one string and for some x in Σ^n, have

$$x \in L_i(A) \Leftrightarrow f_j^A(x) \notin L(M_i^A). \tag{3}$$

By the usual tower arguments we can focus only on the strings in A of length n: Smaller strings can all be queried in polynomial-time; larger strings are too long to be queried.

Pick a string z of length $2n2^n$ that is Kolmogorov random conditioned on the construction of A so far. Read off 2^n strings y_x of length $2n$ for each x in Σ^n. Consider $B = \{\langle i, x, y_x\rangle \mid x \in \Sigma^n\}$.

If $L(M_i^B)$ has more than r^j strings of any length r then we can fulfill the requirement for this stage by letting $A = B$. So let us assume this is not the case.

Note that $f_j^B(x)$ for x of length n cannot query any string y_w in B or we would have a shorter description of z by describing y_w by x and the index of the query made by $f_j^B(x)$. Our final oracle will be a subset of B so we can just use $f_j^\emptyset$ as the reduction.

Suppose $f_j^\emptyset(x) = f_j^\emptyset(w)$ for some x and w of length n. We just let A contain the single string $\langle i, x, y_x\rangle$ and $f_j^\emptyset$ cannot be a reduction. Let us now assume that there is no such x and w.

So by counting there must be some $x \in \Sigma^n$ such that $f_j^\emptyset(x) \notin L(M_i^B)$. Let $v = f_j^\emptyset(x)$. We are not done yet since $L_i(B)$ has too many strings.

Now let A again consist of the single string $\langle i, x, y_x\rangle$. If we still have $v \notin L(M_i^A)$ then we have now fulfilled the requirement.

Otherwise it must be the case that $M_i^A(v)$ accepts but $M_i^B(v)$ rejects. Thus every accepting path (and in particular the lexicographically least) of $M_i^A(v)$ must query some string in $B - A$. Since we can describe v by x this allows us a short description of some y_w given y_x for $w \neq x$ which gives us a shorter description of z, so this case cannot happen. □

Corollary 2. *There exists a relativized world where optimal proof systems do not exist.*

Proof. Messner and Torán [MT98] give a relativizable proof that if optimal proof systems exist then **NP** ∩ **SPARSE** has complete sets. □

3 More Powerful Reductions

In the previous section, we constructed a relativized world where the class **NP** $\cap$ **SPARSE** has no complete sets under m-reductions. We now strengthen that construction to more powerful reductions. Using the same techniques as well as other ones, we will also obtain new results on the reducibility of **SPARSE** to **TALLY**.

3.1 Relativized Worlds

We start by extending Theorem 3 to dtt-reductions. We remind the reader that the proofs for these and all theorems in our paper can be found in the full version as noted in the footnote on the first page.

Theorem 4. *There exists a relativized world where* **NP** $\cap$ **SPARSE** *has no dtt-complete sets.*

The proof of Theorem 4 works for any subexponential density bound. In particular, it yields a relativized world where the class of **NP** sets with no more than $2^{n^{o(1)}}$ strings of any length n has no dtt-complete sets.

We can handle polynomial-time tt-reductions with *arbitrary* evaluators provided the number of queries remains in $o(n/\log n)$.

Theorem 5. *There exists a relativized world where* **NP** $\cap$ **SPARSE** *has no complete sets under* $o(n/\log n)$*-tt-reductions.*

For sets of subexponential density the proof of Theorem 5 yields a relativized world where the class of **NP** sets containing no more than $2^{n^{o(1)}}$ strings of any length n, has no complete sets under tt-reductions of which the number of queries is at most n^{α} for some $\alpha < 1$.

On the positive side, recall from Section 1.4 that **NP** $\cap$ **SPARSE** has complete sets under ctt-reductions as well as under co-rp-reductions.

3.2 SPARSE to TALLY

The techniques used in the proofs of Theorems 3, 4, and 5 also allow us to construct a sparse set S that does not reduce to any tally set under the type of reductions considered. As mentioned in Section 1.3, such sets were already known for m-reductions and for dtt-reductions. For $o(n/\log n)$-tt-reductions we provide the first construction.

Theorem 6. *There exists a sparse set* S *that does not* $o(n/\log n)$*-tt-reduce to any tally set.*

On the other side, $O(n)$ queries suffice to reduce any sparse set to a tally set. Previously, it was known that **SPARSE** ctt- and co-rp-reduces to **TALLY** (see Section 1.3). We give the first deterministic reduction for which the degree of the polynomial bounding the number of queries does not depend on the density of the sparse set.

Theorem 7. *Every sparse set S is reducible to some tally set T under a 2-round tt-reduction asking $O(n)$ queries.*

Proof. Schöning [Sch93] shows that for any constant $k > 0$ there exists a tally set T_1 and a polynomial-time reduction R such that for any string x of any length n

$$\begin{aligned} x \in S &\Rightarrow \Pr[R(x, \rho) \in T_1] = 1 \\ x \notin S &\Rightarrow \Pr[R(x, \rho) \in T_1] < \frac{1}{n^k}, \end{aligned} \tag{4}$$

where the probabilities are uniform over strings ρ of length $O(\log n)$.

By picking $\frac{n}{k \log n}$ independent samples ρ_i, we have for any $x \in \Sigma^n$:

$$\begin{aligned} x \in S &\Rightarrow \Pr[(\forall\, i) R(x, \rho_i) \in T_1] = 1 \\ x \notin S &\Rightarrow \Pr[(\forall\, i) R(x, \rho_i) \in T_1] < (\frac{1}{n^k})^{\frac{n}{k \log n}} = \frac{1}{2^n}. \end{aligned}$$

Therefore, there exists a sequence $\tilde{\rho}_i$, $i = 1, \ldots, \frac{n}{k \log n}$, such that

$$\forall x \in \Sigma^n : x \in S \Leftrightarrow (\forall\, i) R(x, \tilde{\rho}_i) \in T_1. \tag{5}$$

Since each $\tilde{\rho}_i$ is of length $O(\log n)$, we can encode them in a tally set T_2 from which we can recover them using $O(\frac{n}{k \log n} \cdot \log n)$ nonadaptive queries. This way, we obtain a 2-round tt-reduction from S to $T_1 \oplus T_2$ using $O(n)$ queries: The first round determines the $\tilde{\rho}_i$'s, and the second round applies (5). Since $T_1 \oplus T_2$ m-reduces to a tally set T, we are done. □

In Section 4.1, we will show that under a reasonable hypothesis we can reduce the number of queries in Theorem 7 from $O(n)$ to $\frac{n}{k \log n}$ for any constant $k > 0$. See Corollary 3.

We do not know whether the **NP** ∩ **SPARSE** equivalent of Theorem 7 holds: Does **NP** ∩ **SPARSE** have a complete set under reductions asking $O(n)$ queries? See Section 6 for a discussion.

4 Reductions with Advice — Tight Results

Our results in Section 3 pointed out a difference in the power of reductions making $o(n/\log n)$ queries and reductions making $O(n)$ queries. In this section we close the remaining gap between $o(n/\log n)$ and $O(n)$ by considering reductions that take some advice. The approach works for both the **NP** ∩ **SPARSE** setting and the **SPARSE**-to-**TALLY** setting.

4.1 SPARSE to TALLY

We first observe that Theorem 6 also holds when we allow the reduction $O(\log n)$ bits of advice.

Theorem 8. *Let c be any positive constant. There exists a sparse set S that does not reduce to any tally set T under $o(n/\log n)$-tt-reductions that take $c \cdot \log n$ bits of advice.*

Theorem 8 is essentially optimal under a reasonable assumption as the next result shows.

Theorem 9. *Suppose there exists a set in* $\mathbf{DTIME}[2^{O(n)}]$ *that requires circuits of size $2^{\Omega(n)}$ even when the circuits have access to an oracle for* **SAT**. *Then for all relativized worlds, every sparse set S and every positive constant k, there exists a tally set T and a ctt-reduction from S to T that asks $\frac{n}{k \log n}$ queries and $O(\log n)$ bits of advice.*

Proof. Let S be a sparse set. The construction in the proof of Theorem 7 can be seen as a ctt-reduction of S to the tally set T_1 that makes $\frac{n}{k \log n}$ queries and gets $O(n)$ bits as advice, namely the sequence of $\frac{n}{k \log n}$ $\tilde{\rho}_i$'s, each of length $\ell(n) \in O(\log n)$.

We will now show how the hypothesis of Theorem 9 allows us to reduce the required advice from $O(n)$ to $O(\log n)$ bits.

The requirement the $\tilde{\rho}_i$'s have to fulfill is condition (5). By a slight change in the parameters of the proof of Theorem 7 (namely, by replacing k by $2k$ in (4)), we can guarantee that most sequences $\tilde{\rho}_i$ actually satisfy (5). Since the implication from left to right in (5) holds for any choice of $\tilde{\rho}_i$'s, we really only have to check

$$\forall x \in \Sigma^n : x \notin S \Rightarrow (\exists i) R(x, \tilde{\rho}_i) \notin T_1. \qquad (6)$$

Without loss of generality, we can assume that $Q_R(\Sigma^n) \cap T_1 = Q_R(S \cap \Sigma^n) \cap T_1$, where $Q_R(X) = \{R(x, \rho) \mid x \in X \text{ and } |\rho| = \ell(|x|)\}$. Therefore, we can replace (6) by the condition

$$\forall x \in \Sigma^n : x \notin S \Rightarrow (\exists i) R(x, \tilde{\rho}_i) \notin Q_R(S \cap \Sigma^n). \qquad (7)$$

Since S is sparse, this condition on the $\tilde{\rho}_i$'s can be checked by a polynomial-size family of circuits with access to an oracle for **SAT**: The circuit has a enumeration of the elements of $S \cap \Sigma^n$ built in, and once a polynomial-time enumeration of $S \cap \Sigma^n$ is available, (7) becomes a **coNP** predicate.

Under the hypothesis of Theorem 9, Klivans and Van Melkebeek [KvM99, Theorem 4.2] construct a polynomial-time computable function f that maps strings of $O(\log n)$ bits to sequences $\tilde{\rho}_i$ such that most of the inputs map to sequences satisfying (7). An explicit input to f for which this holds, suffices as advice for our reduction from S to $T = T_1$. □

Since we can encode the advice in a tally set and recover it from the tally set using $O(\log n)$ queries, we obtain the following in the terminology of Theorem 7.

Corollary 3. *Under the same hypothesis as in Theorem 9, for any constant $k > 0$ every sparse set S is reducible to some tally set T under a 2-round tt-reduction asking $\frac{n}{k \log n}$ queries.*

4.2 Relativized Worlds

Our tight results about the reducibility of **SPARSE** to **TALLY** carry over to the **NP** ∩ **SPARSE** setting.

Theorem 10. *For any constant $c > 0$, there exists a relativized world where* **NP** ∩ **SPARSE** *has no complete sets under $o(n/\log n)$-tt reductions that take $c \cdot \log n$ bits of advice.*

We also note that Theorem 4 can take up to $n - \omega(\log n)$ bits of advice.

Theorem 11. *There exists a relativized world where* **NP** ∩ **SPARSE** *has no complete sets under dtt-reductions that take $n - \omega(\log n)$ bits of advice.*

On the positive side, we obtain:

Theorem 12. *Suppose there exists a set in* **DTIME**$[2^{O(n)}]$ *that requires circuits of size $2^{\Omega(n)}$ even when the circuits have access to an oracle for* **SAT**. *Then for all relativized worlds and all values of $k > 0$,* **NP** ∩ **SPARSE** *has a complete set under ctt-reductions that ask $\frac{n}{k \log n}$ queries and $O(\log n)$ bits of advice.*

5 NP ∩ SPARSE and Other Promise Classes

Informally, a promise class has a restriction on the set of allowable machines beyond the usual time and space bounds. For example, **UP** consists of languages accepted by **NP**-machines with at most one accepting path. Other common promise classes included **NP** ∩ **coNP**, **BPP** (randomized polynomial time), **BQP** (quantum polynomial time) and **NP** ∩ **SPARSE**.

Nonpromise classes have simple complete sets, for example:

$$\{\langle i, x, 1^j \rangle \mid M_i(x) \text{ accepts in at most } j \text{ steps}\} \tag{8}$$

is complete for **NP** if M_i are nondeterministic machines, but no such analogue works for **UP**.

We say that **UP** has a uniform enumeration if there exists a computable function ϕ such that for each i and input x, $M_{\phi(i)}(x)$ uses time at most $|x|^i$ and has at most one accepting path on every input and $\mathbf{UP} = \cup_i L(M_{\phi(i)})$. Uniform enumerations for the other promise classes are similarly defined.

It turns out that for most promise classes, having a complete set and a uniform enumeration are equivalent. Hartmanis and Hemachandra [HH84] show this for **UP** and their proof easily generalizes to the other classes.

Theorem 13 (Hartmanis-Hemachandra). *The classes* **UP**, **NP** ∩ **coNP**, **BPP** *and* **BQP** *have complete sets under many-one reductions if and only if they have uniform enumerations.*

For **NP** ∩ **SPARSE** neither direction of the proof goes through. In fact despite Theorem 3, **NP** ∩ **SPARSE** has a uniform enumeration (in all relativized worlds).

Theorem 14. *The class* **NP** $\cap$ **SPARSE** *has a uniform enumeration.*

In some sense Theorem 14 is a cheat. In the uniform enumeration, all the sets are sparse but we cannot be sure of the census function at a given input length. To examine this case we extend the definition of uniform enumeration.

Definition 2. *We say* **NP** $\cap$ **SPARSE** *has a* uniform enumeration with size bounds *if there exists a computable function* ϕ *such that* **NP** $\cap$ **SPARSE** $= \cup_i L(M_{\phi(i)})$, *and for all* i *and* n, $M_{\phi(i)}$ *accepts at most* n^i *strings of length* n *using at most* n^i *time.*

Hemaspaandra, Jain and Vereshchagin [HJV93] defined a similar extension for the class **FewP**.

We can use Definition 2 to prove a result similar to Theorem 13 for the class **NP** $\cap$ **SPARSE**.

Theorem 15. **NP** $\cap$ **SPARSE** *has complete sets under invertible reductions if and only if* **NP** $\cap$ **SPARSE** *has a uniform enumeration with size bounds.*

The promise class **NP** $\cap$ **SPARSE** differs from the other classes in another interesting way. Consider the question as to whether there exists a language accepted by a nondeterministic machine using time n^3 which has at most one accepting path on each input that is not accepted by any such machine using time n^2. This remains a murky open question for **UP** and the other usual promise classes.

For **NP** $\cap$ **SPARSE** the situation is quite different as shown by Seiferas, Fischer and Meyer [SFM78] and Žàk [Žàk83].

Theorem 16 (Seiferas-Fischer-Meyer,Žàk). *Let the functions* t_1 *and* t_2 *be time-constructible such that* $t_1(n+1) = o(t_2(n))$. *There exists a tally set accepted by a nondeterministic machine in time* $t_2(n)$ *but not in time* $O(t_1(n))$.

6 Open Problems

Several interesting questions remain including the following.

- Theorem 7 which shows that every sparse set reduces to a tally set using $O(n)$ queries does not seem to give a corresponding result for **NP** $\cap$ **SPARSE**-complete sets. Is there a relativized world where **NP** $\cap$ **SPARSE** does not have complete sets under Turing reductions using $O(n)$ queries? If we can construct the $\tilde{\rho}_i$'s in the proof of Theorem 7 in polynomial time using access to a set in **NP** $\cap$ **coNP**, the answer is yes. However, the best we know is to construct them in polynomial time with oracle access to $\mathbf{NP}^{\mathbf{NP}}$.
- Can we reduce or eliminate the assumption needed for Theorem 9, Corollary 3, and Theorem 12? If we knew how to construct the $\tilde{\rho}_i$'s from the proof of Theorem 9 in polynomial time with $O(\log n)$ bits of advice, we could drop the assumption.
- Does **NP** $\cap$ **SPARSE** having m-complete sets imply **NP** $\cap$ **SPARSE** has a uniform enumeration with size bounds? Can we construct in a relativized world a complete set for **NP** $\cap$ **SPARSE** that is not complete under invertible reductions?

References

[BHL95] H. Buhrman, E. Hemaspaandra, and L. Longpré. SPARSE reduces conjunctively to TALLY. *SIAM Journal on Computing*, 24(3):673–681, 1995.

[BK88] R. Book and K. Ko. On sets truth-table reducible to sparse sets. *SIAM Journal on Computing*, 17(5):903–919, 1988.

[CR79] S. Cook and R. Reckhow. The relative efficiency of propositional proof systems. *Journal of Symbolic Logic*, 44:36–50, 1979.

[For94] L. Fortnow. The role of relativization in complexity theory. *Bulletin of the European Association for Theoretical Computer Science*, 52:229–244, February 1994.

[HH84] J. Hartmanis and L. Hemachandra. Complexity classes without machines: On complete languages for UP. *Theoretical Computer Science*, 34:17–32, 1984.

[HJV93] L. Hemaspaandra, S. Jain, and N. Vereshchagin. Banishing robust turing completeness. *International Journal of Foundations of Computer Science*, 4(3):245–265, 1993.

[HY84] J. Hartmanis and Y. Yesha. Computation times of NP sets of different densities. *Theoretical Computer Science*, 34(1-2):17–32, November 1984.

[Ko89] K. Ko. Distinguishing conjunctive and disjunctive reducibilities by sparse sets. *Information and Computation*, 81(1):62–87, 1989.

[KP89] J. Krajíček and P. Pudlák. Propositional proof systems, the consistency of first order theories and the complexity of computations. *Journal of Symbolic Logic*, 54:1063–1079, 1989.

[KvM99] A. Klivans and D. van Melkebeek. Graph nonisomorhism has subexponential size proofs unless the polynomial-time hierarchy collapses. In *Proceedings of the 31st ACM Symposium on the Theory of Computing*, pages 659–667. ACM, New York, 1999.

[LV97] M. Li and P. Vitányi. *An Introduction to Kolmogorov Complexity and Its Applications*. Graduate Texts in Computer Science. Springer, New York, second edition, 1997.

[MT98] J. Messner and J. Torán. Optimal proof systems for propositional logic and complete sets. In *Proceedings of the 15th Symposium on Theoretical Aspects of Computer Science*, volume 1373 of *Lecture Notes in Computer Science*, pages 477–487. Springer, 1998.

[Sal93] S. Saluja. Relativized limitations of left set technique and closure classes of sparse sets. In *Proceedings of the 8th IEEE Structure in Complexity Theory Conference*, pages 215–223. IEEE, New York, 1993.

[Sch93] U. Schöning. On random reductions from sparse sets to tally sets. *Information Processing Letters*, 46(5):239–241, July 1993.

[SFM78] J. Seiferas, M. Fischer, and A. Meyer. Separating nondeterministic time complexity classes. *Journal of the ACM*, 25(1):146–167, 1978.

[Žàk83] S. Žàk. A Turing machine time hierarchy. *Theoretical Computer Science*, 26(3):327–333, 1983.

Almost Complete Sets

Klaus Ambos-Spies[1], Wolfgang Merkle[1], Jan Reimann[1], and Sebastiaan A. Terwijn[2]*

[1] Universität Heidelberg, Mathematisches Institut,
Im Neuenheimer Feld 294, D-69120 Heidelberg, Germany,
ambos|merkle|reimann@math.uni-heidelberg.de

[2] Ludwig-Maximilians-Universität München, Mathematisches Institut,
Theresienstr. 39, D-80333 München, Germany,
terwijn@rz.mathematik.uni-muenchen.de

Abstract. We show that there is a set which is almost complete but not complete under polynomial-time many-one (p-m) reductions for the class **E** of sets computable in deterministic time 2^{lin}. Here a set A in a complexity class **C** is almost complete for **C** under some reducibility r if the class of the problems in **C** which do not r-reduce to A has measure 0 in **C** in the sense of Lutz's resource-bounded measure theory. We also show that the almost complete sets for **E** under polynomial-time bounded one-one length-increasing reductions and truth-table reductions of norm 1 coincide with the almost p-m-complete sets for **E**. Moreover, we obtain similar results for the class **EXP** of sets computable in deterministic time 2^{poly}.

1 Introduction

Lutz [14] introduced measure concepts for the standard deterministic time and space complexity classes which contain the class **E** of sets computable in deterministic time 2^{lin}. These measure concepts have been used for investigating quantitative aspects of the internal structure of the corresponding complexity classes. Most of this work focussed on the measure for **E**, since the majority of the results obtained there carry over to the larger complexity classes. For recent surveys of the work on resource-bounded measure, see Lutz [16] and Ambos-Spies and Mayordomo [3].

Lutz's measure on **E** does not only allow to measure in **E** the relative size of classes of sets with interesting structural properties – like e.g. the classes of complete sets under various reducibilities or of the **P**-bi-immune sets, i.e., the sets which are intractable almost everywhere – but it also leads to important new concepts. The most investigated concept in this direction is probably that of a weakly complete set introduced by Lutz in [15]. While all sets in **E** can be reduced to a complete set (under some given polynomial time reducibility notion), for a weakly complete set, Lutz only requires that the class of the reducible sets does not have measure 0 in **E**, i.e., is a non-negligible part of **E**.

* Supported by Marie Curie Fellowship ERB–FMBI–CT98-3248.

H. Reichel and S. Tison (Eds.): STACS 2000, LNCS 1770, pp. 419–430, 2000.

Originally, Lutz introduced weak completeness for the polynomial time many-one (p-m) reducibility – the reducibility which is used in most completeness proofs in the literature – and he showed that there actually is a weakly p-m-complete set for **E** which is not p-m-complete for **E** (Lutz [15]). In fact, the class of weakly p-m-complete sets for **E** has measure 1 in **E** (Ambos-Spies, Terwijn, Zheng [6]) whereas the class of p-m-complete sets for **E** has measure 0 in **E** (Mayordomo [17]), whence weak completeness leads to a new large class of provably intractable problems.

The large gap between completeness and weak completeness suggested the search for intermediate concepts. A natural candidate for such a concept in the context of resource-bounded measure is the following. Call a set A in **E** almost p-m-complete for **E** if the class of problems which are p-m-reducible to A has measure 1 in **E**, i.e., if the sets in **E** which are not reducible to A can be neglected with respect to measure. Zheng and others (see e.g. [3], Section 7) raised the question whether there are almost p-m-complete sets for **E** which are not p-m-complete for **E**. Here we answer this question affirmatively by constructing a set with these properties.

Our result is contrasted by a result of Regan, Sivakumar and Cai [18], which implies that for the standard transitive polynomial-time reducibilities allowing more than one oracle query – like bounded truth-table (btt), truth-table (tt), and Turing (T) – completeness and almost completeness coincide. It follows that any almost p-m-complete set for **E** is p-btt-complete for **E**, whence – in contrast to the weakly p-m-complete sets – the class of the almost p-m-complete sets for **E** has measure 0 in **E**.

The above results leave the investigation of almost completeness for the polynomial reducibilities besides many-one which allow only one query. Here we show that the almost completeness notions coincide for the reducibilities ranging from one-to-one, length-increasing reductions to truth-table reductions of norm 1. This parallels previous observations for completeness (see [8] and [10]) and weak completeness (see [4]).

The outline of the paper is as follows. In Section 2 we describe the part of Lutz's measure theory for **E** needed in the paper and we review the limiting result on almost completeness by Regan, Sivakumar, and Cai. Section 3 contains the proof of our main result, while in Section 4 the relations among the various completeness notions are discussed. In Section 5 we consider extensions of our results to other complexity classes, and we pose some open problems.

Our notation is standard. For unexplained notation we refer to [3]. The polynomial time reductions considered here are general reductions of Turing type (p-T), truth-table reductions (p-tt) allowing only non-adaptive queries, bounded truth-table reductions (p-btt) in which in addition the number of queries is bounded by a constant, and the special case hereof where this constant c is fixed (btt(c)). We will represent p-btt(1)-reductions by a pair of polynomial time computable functions g and h where $g(x)$ gives the string queried on input x and the Boolean function $h(x)$ tells how the answer of the oracle is evaluated. If the reduction is positive, i.e., $h(x)(i) = i$ for all strings x and all i in $\{0, 1\}$, we get a p-many-one-reduction (p-m) and in this case we omit h. If in addition g is one-to-one (and length-increasing) we obtain a (length-increasing) one-one reduction (p-1 and p-1-li). For r in $\{\mathrm{T}, \mathrm{tt}, \mathrm{btt}, \mathrm{btt}(1), \mathrm{m}, 1, 1\text{-li}\}$ and any set A, we let the lower p-r-span of A be the class $\{B : B \leq_r^{\mathrm{p}} A\}$.

2 Measure on E and Almost Completeness

In this section we describe the fragment of Lutz's measure theory for the class $\mathbf{E}$ of sets computable in deterministic time 2^{lin} which we will need in the following. For a more comprehensive presentation of this theory we refer the reader to the recent surveys by Lutz [16] and by Ambos-Spies and Mayordomo [3].

The measure on $\mathbf{E}$ is obtained by imposing appropriate resource-bounds on a game theoretical characterization of the classical Lebesgue measure.[1]

Definition 1. *A* betting strategy *s is a function $s : \{0,1\}^* \rightarrow [0,1]$. The* (normed) martingale *$d_s : \{0,1\}^* \rightarrow [0,\infty)$* induced by a betting strategy *s is inductively defined by $d_s(\lambda) = 1$ and*

$$d_s(xi) = 2 \cdot |i - s(x)| \cdot d_s(x)$$

for $x \in \{0,1\}^*$ and $i \in \{0,1\}$. A martingale *is a martingale induced by some strategy. A martingale d* succeeds on a set *A if*

$$\limsup_{n\to\infty} d(A \upharpoonright n) = \infty,$$

and d succeeds on a class $\boldsymbol{C}$ if d succeeds on every member A of $\boldsymbol{C}$.

It can be shown that a class $\boldsymbol{C}$ has Lebesgue measure 0, $\mu(\boldsymbol{C}) = 0$, iff some martingale succeeds on $\boldsymbol{C}$. So, by imposing resource bounds, the martingale concept can be used for defining resource-bounded measure concepts.

Definition 2. *Let $t : \mathbb{N} \rightarrow \mathbb{N}$ be a recursive function. A $t(n)$-*martingale *d is a martingale induced by a rational valued betting strategy s such that $s(x)$ can be computed in $t(|x|)$ steps for all strings x.*

A class $\boldsymbol{C}$ has $t(n)$-measure 0, $\mu_{t(n)}(\boldsymbol{C}) = 0$, if some $t(n)$-martingale succeeds on $\boldsymbol{C}$, and $\boldsymbol{C}$ has $t(n)$-measure 1, $\mu_{t(n)}(\boldsymbol{C}) = 1$, if the complement $\overline{\boldsymbol{C}}$ has $t(n)$-measure 0.

Note that for $i \in \{0,1\}$ and for recursive bounds $t(n), t'(n)$ such that $t(n) \leq t'(n)$ almost everywhere,

$$\mu_{t(n)}(\boldsymbol{C}) = i \quad \Rightarrow \quad \mu_{t'(n)}(\boldsymbol{C}) = i \quad \Rightarrow \quad \mu(\boldsymbol{C}) = i \ .$$

In order to obtain measures for complexity classes, resource-bounded measure concepts are defined not for individual bounds but for families of bounds. In particular, working with polynomial bounds yields a measure on $\mathbf{E}$.

[1] Our presentation follows [3]. The $t(n)$-measure defined there slightly differs from the original definition by Lutz, but both definitions lead to the same notions of p-measure and measure on $\mathbf{E}$.

Definition 3. *A* p-martingale *d is a $q(n)$-martingale for some polynomial q. A class $\boldsymbol{C}$ has* p-measure 0, *$\mu_{\mathrm{p}}(\boldsymbol{C}) = 0$, if $\mu_{q(n)}(\boldsymbol{C}) = 0$ for some polynomial $q(n)$, i.e., if some p-martingale succeeds on $\boldsymbol{C}$, and $\mu_{\mathrm{p}}(\boldsymbol{C}) = 1$ if $\mu_{\mathrm{p}}(\overline{\boldsymbol{C}}) = 0$.*

A class $\boldsymbol{C}$ has measure 0 *in* **E**, *$\mu(\boldsymbol{C}|\mathbf{E}) = 0$, if $\mu_{\mathrm{p}}(\boldsymbol{C} \cap \mathbf{E}) = 0$ and $\boldsymbol{C}$ has measure* 1 *in* **E**, *$\mu(\boldsymbol{C}|\mathbf{E}) = 1$, if $\mu(\overline{\boldsymbol{C}}|J\mathbf{E}) = 0$.*

Lutz [14] has shown that this measure concept for **E** is consistent. In particular, **E** itself does not have measure 0 in **E**, namely

$$\mu_{\mathrm{p}}(\mathbf{E}) \neq 0 \text{ whence } \mu(\mathbf{E}|\mathbf{E}) \neq 0 \ . \tag{1}$$

On the other hand, every slice of **E** has measure 0 in **E**, namely

$$\mu_{\mathrm{p}}(\mathbf{DTIME}(2^{kn})) = 0 \text{ whence } \mu(\mathbf{DTIME}(2^{kn})|\mathbf{E}) = 0. \tag{2}$$

Based on the above measure for **E** we can now introduce the completeness notions for **E** which are central for our paper. Here $r \in \{\text{1-li}, 1, \text{m}, \text{btt}(1), \text{btt}, \text{tt}, \text{T}\}$ denotes any of the reducibilities introduced at the end of Section 1.

Definition 4. *a) (Lutz [15]) A set A is* weakly p-r-hard for **E** *if the lower* p-r *span of A does not have measure* 0 *in* **E**. *If, in addition, A is in* **E** *then A is* weakly p-r-complete.

b) (Zheng) A set A is almost p-r-hard for **E** *if the lower* p-r *span of A has measure* 1 *in* **E**. *If, in addition, A is in* **E** *then A is* almost p-r-complete.

Intuitively, a set A in **E** is weakly p-r-complete for **E** if its lower span contains a non-negligible part of **E** and it is almost p-r-complete for **E** if the part of **E** which is not contained in the lower span of A can be neglected. In particular, every p-r-complete set for **E** is almost p-r-complete for **E** and every almost p-r-complete set for **E** is weakly p-r-complete for **E**. Moreover, since **P** has measure 0 in **E** by (2), every weakly p-r-complete set is provably intractable.

After Lutz [15] demonstrated the existence of weakly p-m-complete sets for **E** which are not p-m-complete for **E**, weak completeness was extensively studied and most relations among the different weak completeness and completeness notions have been clarified (see Section 4 below).

A severe limitation on the existence of nontrivial almost complete sets is imposed by the following observation on classes which have measure 1 in **E**.

Theorem 5 (Regan, Sivakumar, and Cai [18]). *Let $\boldsymbol{C}$ be a class such that $\mu(\boldsymbol{C}|\mathbf{E}) = 1$ and $\boldsymbol{C}$ is either closed under symmetric difference or closed under union and intersection. Then $\boldsymbol{C}$ contains all of* **E**.

Since for r in {btt,tt,T} the lower p-r-span of any set A is closed under symmetric difference, this shows that the concept of almost completeness is trivial for these reducibilities.

Corollary 6. *For r in* {btt,tt,T}, *every almost* p-r-*complete (hard) set for* **E** *is* p-r-*complete (hard) for* **E**.

In general, however, the lower p-m-span of a set is neither closed under symmetric difference nor under union or intersection, whence the above argument does not work for almost p-m-completeness. As an immediate consequence of Corollary 6, however, almost p-m-complete sets for **E** must be p-btt-complete for **E**. Since the class of p-btt-complete sets has p-measure 0 (see [5]), this also shows that almost p-m-complete sets are scarce.

Corollary 7. *Every almost* p-m-*complete (hard) set for* **E** *is* p-btt-*complete (hard) for* **E**. *In particular, the class of the almost* p-m-*complete sets for* **E** *has p-measure 0, hence measure 0 in* **E**.

In fact, Corollary 7 can be strengthened to the following result, which was observed first in [3]. Here the proof follows easily from a result of Wang (see [3, Lemma 6.19]), which implies that every set in **E** can be presented as the symmetric difference of two n^k-random sets (for any $k \geq 1$).

Corollary 8. *Every almost* p-m-*hard set for* **E** *is* p-btt(2)-*hard for* **E**.

Despite these limitations, in the next section we will show that there are almost p-m-complete sets for **E** which are not p-m-complete for **E**. Moreover in Section 4 we will obtain the same results for some other p-reducibilities allowing only one oracle query by showing that all these reducibilities yield the same class of almost complete sets.

Our results and proofs will use the characterization of p-measure and measure in **E** in terms of resource-bounded random sets. In the remainder of this section we shortly describe this approach (from [6]) and state some results on the measure in **E** in terms of random sets, which we will need in the following.

Definition 9. *A set R is $t(n)$*-random *if no $t(n)$-martingale succeeds on R.*

For later use, we observe the following trivial relations among random sets for increasing time bounds.

Proposition 10. *Let $t(n), t'(n)$ be recursive functions such that $t(n) \leq t'(n)$ almost everywhere. Then every $t'(n)$-random set is $t(n)$-random.*

The characterization of the p-measure and the measure in **E** in terms of random sets is as follows.

Lemma 11 (Ambos-Spies, Terwijn, and Zheng [6]). *For any class $\boldsymbol{C}$,*

(i) $\mu_{\mathrm{p}}(\boldsymbol{C}) = 0$ *iff there is a number k such that $\boldsymbol{C}$ does not contain any n^k-random set, and*
(ii) $\mu(\boldsymbol{C}|\mathbf{E}) = 0$ *iff there is a number k such that $\boldsymbol{C} \cap \mathbf{E}$ does not contain any n^k-random set.*

To illustrate how results on the p-measure and measure in **E** can be rephrased in terms of randomness and for later use, we consider Mayordomo's result that the class of $\mathbf{DTIME}(2^{kn})$-bi-immune sets has p-measure 1 and the extension of this result to the class of the $\mathbf{DTIME}(2^{kn})$-incompressible sets due to Juedes and Lutz (see [16] or [3] for details).

Theorem 12. *(a) (Mayordomo [17]) Every n^{k+1}-random set is* $\mathbf{DTIME}(2^{kn})$*-bi-immune.*
(b) (Juedes and Lutz [12]) Every n^{k+1}-random set is $\mathbf{DTIME}(2^{kn})$*-incompressible.*

Finally, for the proof of our main theorem in the next section, we will need the following instance of the Borel-Cantelli-Lemma for p-measure (see Regan and Sivakumar [19] for a more general discussion of this lemma). We omit the easy proof which works by constructing a martingale which for all k, reserves a fraction of $1/2^k$ of the initial capital for betting on the event that the intersection with D_k is empty.

Lemma 13. *Let $\{D_1, D_2, \dots\}$ be a sequence of pairwise disjoint finite sets where D_k has cardinality k. Assume further that given x, in time $\mathcal{O}(2^{2|x|})$, firstly, one can decide whether x is in D_k for some k and, if so, secondly, one can compute k and a list of all strings $y < x$ in D_k. Then every n^3-random set intersects almost all of the sets D_k.*

3 An Almost Complete Set Which Is Not Complete

We now turn to the main result of this paper.

Theorem 14. *There is an almost* p-m-*complete set for* $\mathbf{E}$ *which is not* p-m-*complete for* $\mathbf{E}$.

For a proof of Theorem 14 it suffices to show the following lemma.

Lemma 15. *There are sets A and B in* $\mathbf{E}$ *such that* $B \not\leq^{\mathrm{p}}_{\mathrm{m}} A$ *and*

$$\text{for all } n^3\text{-random sets } R \text{ in } \mathbf{EXP},\ R \leq^{\mathrm{p}}_{\mathrm{m}} A. \tag{3}$$

Then, for such sets A and B, the set A is almost p-m-complete for $\mathbf{E}$ by (3) and Lemma 11, whereas B is not p-m-reducible to A and thus A is not p-m-complete for $\mathbf{E}$. In fact, for this argument it suffices to consider $\mathbf{E}$ in place of $\mathbf{EXP}$ in (3). We will use in Section 5, however, that the extension proved here will lead simultaneously to a corresponding result for the class $\mathbf{EXP}$, i.e., the class of sets computable in time 2^{poly}.

Proof. We construct sets A and B as required in stages. To be more precise we choose a strictly increasing function $h : \mathbb{N} \to \mathbb{N}$ with $h(0) = 0$ and we determine the values of A and B for all strings in the interval $I_k = \{x : h(k) \leq |x| < h(k+1)\}$ at stage k. Here the function h is chosen to be p-constructible and, for technical reasons to be explained below, to satisfy

$$\text{(i) } k^2 < h(k) \quad \text{(ii) } k^2 \cdot p_k(h(k)) < 2^{\sqrt[k]{h(k)}} \quad \text{(iii) } p_k(h(k)) < h(k+1) \tag{4}$$

for all $k > 0$ and $p_k(n) = n^k + k$. Note that given x, by p-constructibility of h, we can compute the index k such that $x \in I_k$, as well as $h(k)$, in poly($|x|$) steps.

Before we define stage k of the construction formally, we first discuss the strategies to ensure the required properties of A and B and simultaneously introduce some notation required in the construction.

In order to ensure (3) we let A sufficiently resemble a p-m-complete set for **EXP**. Let $\{C_e : e \geq 0\}$ be an effective enumeration of **EXP** such that $C_e(x)$ can be computed uniformly in $2^{|x|^e} + e$ steps, and let

$$E = \{1^e 0 1^{|x|^e} 0x : x \in C_e \ \& \ e \in \mathbb{N}\}$$

be the padded disjoint union of these sets. Then E can be computed in time 2^n and, for all e, C_e is p-m-reducible to E via

$$g_e(x) = 1^e 0 1^{|x|^e} 0x,$$

whence E is p-m-complete for **EXP**. So, if we let

$$\mathrm{CODE}_e = \mathrm{range}(g_e)$$

denote the set of strings used for coding C_e into E, then in order to satisfy (3) it suffices to meet for all numbers $e \geq 0$, the requirement

$$R_e^1 : \text{ If } C_e \text{ is } n^3\text{-random then } A \cap \mathrm{CODE}_e \text{ is a finite variant of } E \cap \mathrm{CODE}_e \ .$$

In order to meet the requirements R_e^1 we will let A look like E unless the task of making B not p-m-reducible to A will force a disagreement. Since E is in $\mathbf{DTIME}(2^n)$ this procedure is compatible with ensuring that A is in $\mathbf{E}$ as long as the strings on which A and E differ can be recognized in exponential time. In this context note that the sets CODE_e are pairwise disjoint and that, for given x, poly($|x|$) steps suffice to decide whether x is a member of one of these sets and if so to compute the unique e with $x \in \mathrm{CODE}_e$.

The condition $B \not\leq_{\mathrm{m}}^{\mathrm{p}} A$ is satisfied by diagonalization. Let $\{f_k : k \geq 1\}$ be an effective enumeration of the p-m-reductions such that $f_k(x)$ can be computed uniformly in $p_k(|x|) = |x|^k + k$ steps. Then it suffices to meet the requirements

$$R_k^2 : \ \exists x \in \{0,1\}^* \ (B(x) \neq A(f_k(x)))$$

for all numbers $k \geq 1$. We will meet requirement R_k^2 at stage k of the construction.

For this purpose we will ensure that there is a string x from a set of k^2 designated strings of length $h(k)$ such that $B(x)$ and $A(f_k(x))$ differ, while we will let B be empty and let A equal E on I_k otherwise. We will say that this action *injures* an almost completeness requirement R_e^1 if, for the chosen string x, $f_k(x)$ is in $I_k \cap \mathrm{CODE}_e$ and A and E differ on $f_k(x)$. Since A and E agree on $I_k \cap \mathrm{CODE}_e$ otherwise, the conclusion of R_e^1 will fail if and only if the requirement is injured at infinitely many stages.

To avoid injuries we will attempt to diagonalize in such a manner that injuries to the first k requirements R_e^1, $e < k$ are avoided. If the function f_k is not one-to-one on the designated strings or if $f_k(x)$ is shorter than x for some designated string x then the diagonalization will not affect A on I_k at all, whence no injuries occur. The critical case occurs if, for every designated string x, $f_k(x)$ is longer than x and element of some of the sets CODE_e with $e < k$. By the former, the diagonalization has to make $A(f_k(x))$ differ from the canonical value 0 for $B(x)$ (not vice versa, since otherwise we might fail to make B computable in exponential time) whence by the latter some injuries may occur.

By Lemma 13, however, we will be able to argue that if C_e is n^3-random and if there are infinitely many stages at which we are forced to make $A(f_k(x))$ differ from $B(x) = 0$ for some $f_k(x)$ in CODE_e, then at almost all of these stages letting A look like E on the f_k-images of the designated strings will yield the desired diagonalization. So for n^3-random C_e the requirement R_e^1 will be injured only finitely often.

We now give the formal construction. We let $B \cap I_0 = \emptyset$ and $A \cap I_0 = E \cap I_0$. Given $k > 0$, stage k of the construction is as follows. We assume that A and B have already been defined on the intervals $I_0, \ldots, I_{k-1}$, and we will specify both sets on the interval I_k. For the scope of the description of stage k we call the first k^2 strings in I_k the *designated strings.* The designated strings are the potential diagonalization witnesses for requirement R_k^2, i.e., we will guarantee $B(x) \neq A(f_k(x))$ for some designated string x. Observe that every designated string has length $h(k)$ and is mapped by f_k into the union of the intervals I_0 through I_k, as follows by items (i) and (iii) in (4), respectively.

For the definition of A and B on I_k we distinguish the following four cases with respect to the images of the designated strings under the mapping f_k. Here it is to be understood that on I_k the sets A and B will always look like the set E and the empty set, respectively, unless this specification is explicitly overwritten according to one of the cases below. Moreover, as the cases are not exclusive, always the first applicable case is used.

Case 1: Some designated string is not mapped to I_k.
Let x be the least such string. By the preceding discussion, $f_k(x)$ is contained in some interval I_j with $j < k$ and $A(f_k(x))$ has been defined at some previous stage. We let $B(x) = 1 - A(f_k(x))$ (thereby satisfying R_k^2).

Case 2: Two designated strings are mapped to the same string.
Let x be the least designated string such that $f_k(x) = f_k(x')$ for some designated string $x' \neq x$ and let $B(x) = 1$. (Then $B(x) = 1$ differs from $B(x') = 0$, whereas f_k maps x and x' to the same string, whence R_k^2 is met.)

Case 3: Some designated string is not mapped to the set $\bigcup_{e<k} \mathrm{CODE}_e$.
Let x be the least such designated string and let $A(f_k(x)) = 1$. (Note that, by failure of Case 1, $f_k(x)$ is in I_k, and R_k^2 is met since $B(x) = 0$ by convention.)

Case 4: Otherwise.
In this case the k^2 designated strings are mapped by f_k to k^2 different strings in $\bigcup_{e<k} \mathrm{CODE}_e$, whence we can let e_k be the least $e < k$ such that f_k maps at least k designated strings to CODE_{e_k}. Let J_k be the set of the least k designated strings which are mapped to CODE_{e_k} and let $F_k = \{f_k(x) : x \in J_k\}$ be the f_k-image of J_k. Observe that by case assumption all strings in

F_k are in I_k. In case E does not intersect F_k, we let $A(y) = 1$ where y is the maximal element in F_k and we say that $R^1_{e_k}$ is *injured at stage* k. (Then R^2_k is met, because either there has already been a string x in J_k such that $B(x) = 0$ differs from $A(f_k(x))$ or we enforce such a disagreement for some x where $f_k(x) = y$.)

This completes the construction. It remains to show that the constructed sets have the required properties. We first observe that the constructed sets are in $\mathbf{DTIME}(2^{2n})$. We sketch the proof for the set A and leave the similar proof for the set B to the reader. Given a string y, we can compute in time poly($|y|$) the index k where y is in I_k, as well as $h(k)$. Further it takes time $\mathcal{O}(k^2 p_k(h(k)))$ to compute the list of all pairs $(x, f_k(x))$ such that x is a designated string of stage k and it takes time polynomial in the length of this list to check which of the four cases applies and to determine whether according to this case, $A(y)$ might differ from $E(y)$ at all. If not, we simply have to compute $E(y)$. Otherwise, we know that either Case 3 applies and y is in A or Case 4 applies, y is the maximal string in F_k, and y is in A iff none of the $k - 1$ smaller strings in F_k is also in E. Using item (ii) in (4) it is then a routine task to show that A in fact can be computed in time 2^{2n}.

It remains to show that the requirements R^1_e, $e \geq 0$, and R^2_k, $k \geq 0$, are met. For the latter this is immediate by the comments made in the individual cases of the construction. For a proof that the almost completeness requirements R^1_e are met, fix $e \geq 0$ and for a contradiction assume that R^1_e fails. Then C_e is n^3-random and $A \cap \mathrm{CODE}_e$ and $E \cap \mathrm{CODE}_e$ differ on infinitely many intervals I_k. By construction, the latter implies that there are infinitely many stages k where R^1_e is injured. Now, for stages k where Case 4 applies, let $D_k = \{g_e^{-1}(y) : y \in F_k\}$ and let D_k be the inverse image of the first k strings in $\mathrm{CODE}_e \cap I_k$, otherwise. Then, using the second item in (4), one can easily check that the sequence $\{D_k : k \geq 1\}$ satisfies the hypothesis of Lemma 13. So we may fix k such that R^1_e is injured at stage k – whence in particular Case 4 applies during stage k – and C_e intersects D_k, say $z \in C_e \cap D_k$. Then, by the latter, $g_e(z) \in E \cap F_k$, whence R^1_e is not injured at k contrary to the choice of k. □

4 Comparing Completeness Notions

The polynomial-time reducibilities allowing only one oracle query in the range from one-to-one, length-increasing reductions to truth-table reductions of norm 1 lead to the same class of complete sets for **E**. Namely, Berman [8] has shown that every p-m-complete set for **E** is in fact p-1-li-complete while Homer et al.[10] have proven that every p-btt(1)-complete set for **E** is in fact p-m-complete for **E**. Corresponding results for weak completeness have been shown by Ambos-Spies et al. [4]. By the two following theorems, the same phenomenon occurs for almost completeness. Due to space considerations we state these results without giving a proof.

Theorem 16. *A set is almost* p-m-*complete for* **E** *if and only if it is almost* p-1-li-*complete for* **E**.

Theorem 17. *A set is almost* p-btt(1)*-complete for* **E** *if and only if it is almost* p-m*-complete for* **E**.

Previous results in the literature together with the results of this paper clarify most of the relations among the different completeness notions for **E**. If we let $\mathcal{C}(\mathbf{E}, r)$ denote the class of p-r-complete sets for **E**, and if $\mathcal{AC}(\mathbf{E}, r)$ and $\mathcal{WC}(\mathbf{E}, r)$ denote the corresponding classes of almost and weakly complete sets, respectively, the known relations among the classes are summarized in Figure 1.

$$
\begin{array}{ccccc}
\mathcal{C}(\mathbf{E},\text{1-li}) & \subset & \mathcal{AC}(\mathbf{E},\text{1-li}) & \subset & \mathcal{WC}(\mathbf{E},\text{1-li}) \\
\| & & \| & & \| \\
\mathcal{C}(\mathbf{E},\text{m}) & \subset & \mathcal{AC}(\mathbf{E},\text{m}) & \subset & \mathcal{WC}(\mathbf{E},\text{m}) \\
\| & & \| & & \| \\
\mathcal{C}(\mathbf{E},\text{btt}(1)) & \subset & \mathcal{AC}(\mathbf{E},\text{btt}(1)) & \subset & \mathcal{WC}(\mathbf{E},\text{btt}(1)) \\
\cap & & \cap & & \cap \\
\mathcal{C}(\mathbf{E},\text{btt}) & = & \mathcal{AC}(\mathbf{E},\text{btt}) & \subset & \mathcal{WC}(\mathbf{E},\text{btt}) \\
\cap & & \cap & & \cap \\
\mathcal{C}(\mathbf{E},\text{tt}) & = & \mathcal{AC}(\mathbf{E},\text{tt}) & \subseteq & \mathcal{WC}(\mathbf{E},\text{tt}) \\
\cap & & \cap & & \cap \\
\mathcal{C}(\mathbf{E},\text{T}) & = & \mathcal{AC}(\mathbf{E},\text{T}) & \subseteq & \mathcal{WC}(\mathbf{E},\text{T})
\end{array}
$$

Fig. 1. The figure shows the known relations among the completeness notions discussed in this paper. Here '$\subset$' means that a class is a proper subclass. '$\subseteq$' indicates that it is not known if the inclusion is strict. All classes contained in $\mathcal{AC}(\mathbf{E}, \text{btt})$ have measure 0 in **E**, whereas all the weakly complete classes (i.e. the third column) are known to have measure one in **E**. The measure in **E** of the remaining four classes (the complete and almost complete sets for p-tt- and p-T-reducibility) is hitherto unknown.

Note that in Figure 1 the inclusions from top to bottom and from left to right are immediate by definition. The two equalities in the first column are due to Berman [8] and Homer et al. [10] (see above), while the strictness of the remaining three inclusions in this column has been established by Watanabe [20] who separated the standard completeness notions for reducibilities which allow more than one query. The two equalities in the second column are justified by Theorems 16 and 17 above. It follows with Theorem 14 that the first three inclusions from column 1 to column 2 are proper, while the coincidence of completeness and almost completeness for the other three reducibilities follows from Corollary 6 above due to Regan et al. [18]. This corollary also yields that the last two inclusions in column 2 are proper. That the class $\mathcal{AC}(\mathbf{E}, \text{btt}(1))$ is a proper subclass of the class $\mathcal{AC}(\mathbf{E}, \text{btt})$ follows from Corollary 8, since Watanabe [20] has shown that there is a p-btt-complete set for **E**, which is not p-btt(2)-complete.

The relations stated in the third column have been established by Ambos-Spies et al. in [4] where weak completeness notions are compared. The strictness of the first four inclusions between the second and the third column follows from the observation that $\mathcal{AC}(\mathbf{E}, \mathrm{btt})$ has measure 0 in $\mathbf{E}$ (Corollary 7) whereas $\mathcal{WC}(\mathbf{E}, \mathrm{m})$ has nonzero measure in $\mathbf{E}$ ([11]), in fact measure 1 in $\mathbf{E}$ ([6]).

Finally, the question whether the last two inclusions between the second and the third column are proper is still open. It has been shown, however, that these questions cannot be resolved by relativizable techniques: namely, Allender and Strauss [1] have shown that, relative to some oracle, all n^2-random sets are p-tt-complete whereas Ambos-Spies, Lempp, and Mainhardt [2] and, independently, Buhrman et al. [9] have given oracles relative to which no n^2-random set is p-T-complete for **E**. This also shows that the measure in **E** of the classes of complete and almost complete sets for p-tt- and p-T-reducibility is oracle dependent.

5 Further Results and Open Problems

In this paper we looked at the concept of almost completeness only for the class **E** of sets computable in linear exponential time. Similar results, however, can be obtained for other complexity classes. In particular all of our results can be also shown for Lutz's measure on the class **EXP** of sets computable in time 2^{poly}. The analog of our main theorem (Theorem 14) in this setting follows directly from Lemma 15, while analogs of the other results require only minor changes in the proofs. The relations among the different completeness notions in Figure 1 will remain the same if we replace **E** by **EXP**.

The relation between almost p-m-completeness for **E** and **EXP** is known only in part. While it is well known that, for sets in **E**, p-m-completeness for **E** and **EXP** coincide, Juedes and Lutz [13] have shown that every weakly p-m-complete set for **E** is also weakly p-m-complete for **EXP** but that there are weakly p-m-complete sets for **EXP** in **E** which are not weakly p-m-complete for **E**. By refining the technique used in the proof of our main theorem, Ambos-Spies has shown that there is an almost p-m-complete set for **EXP** in **E** which is not almost – in fact not even weakly – p-m-complete for **E**. We do not know, however, whether every almost p-m-complete set for **E** is also almost p-m-complete for **EXP**.

6 Acknowledgements

We would like to thank two anonymous referees of the STACS 2000 conference for helpful corrections and suggestions.

References

1. E. Allender and M. Strauss. Measure on small complexity classes with applications for BPP. In: *Proceedings of the 35th Annual IEEE Symposium an Foundations of Computer Science*, p.867–818, IEEE Computer Society Press, 1994.

2. K. Ambos-Spies, S. Lempp, and G. Mainhardt. Randomness vs. completeness: on the diagonalization strength of resource bounded random sets. In: *Proceedings of the 23rd International Symposium on Mathematical Foundations of Computer Science*, p. 465–473, Lecture Notes in Computer Science, Vol. 1450, Springer, 1998.
3. K. Ambos-Spies and E. Mayordomo. Resource-bounded measure and randomness. In: A. Sorbi (ed.), *Complexity, logic, and recursion theory*, p.1-47. Dekker, New York, 1997.
4. K. Ambos-Spies, E. Mayordomo and X. Zheng. A comparison of weak completeness notions. In: *Proceedings of the 11th Annual IEEE Conference on Computational Complexity*, p.171-178, IEEE Computer Society Press, 1996.
5. K. Ambos-Spies, H.-C. Neis and S. A. Terwijn. Genericity and measure for exponential time. *Theoretical Computer Science*, 168:3–19, 1996.
6. K. Ambos-Spies, S. A. Terwijn and X. Zheng. Resource bounded randomness and weakly complete problems. *Theoretical Computer Science*, 172:195–207, 1997.
7. J. L. Balcázar, J. Díaz, and J. Gabarró. *Structural Complexity*, volume I. Springer-Verlag, 1995.
8. L. Berman. *Polynomial reducibilities and complete sets.* Ph.D. thesis, Cornell University, 1977.
9. H. Buhrman, D. van Melkebeek, K. Regan, D. Sivakumar, M. Strauss. A generalization of resource bounded measure, with application to the BPP vs. EXP problem. In: *Proceedings of the 15th Annual Symposium on Theoretical Aspects of Computer Science*, p.161–171, Lecture Notes in Computer Science, Vol. 1373, Springer, 1998.
10. S. Homer, S. Kurtz and J. Royer. On 1-truth-table-hard languages. *Theoretical Computer Science*, 115:383–389, 1993.
11. D. W. Juedes. Weakly complete problems are not rare. *Computational Complexity*, 5:267–283, 1995.
12. D. W. Juedes and J. H. Lutz. The complexity and distribution of hard problems. *SIAM Journal on Computing*, 24:279–295, 1995.
13. D. W. Juedes and J. H. Lutz. Weak completeness in E and E_2. *Theoretical Computer Science*, 143:149–158, 1995.
14. J. H. Lutz. Almost everywhere high nonuniform complexity. *Journal of Computer and System Sciences*, 44:220–258, 1992.
15. J. H. Lutz. Weakly hard problems. *SIAM Journal on Computing* 24:1170–1189, 1995.
16. J. H. Lutz. The quantitative structure of exponential time. In: Hemaspaandra, Lane A. (ed.) et al., *Complexity theory retrospective* II, p.225–260. Springer, New York, 1997.
17. E. Mayordomo. Almost every set in exponential time is P-bi-immune. *Theoretical Computer Science*, 136:487–506, 1994.
18. K. Regan, D. Sivakumar and J.-Y. Cai. Pseudorandom generators, measure theory and natural proofs. In: *Proceedings of the 36th Annual IEEE Symposium an Foundations of Computer Science*, p.171–178, IEEE Computer Society Press, 1995.
19. K. Regan, D. Sivakumar. Improved Resource-Bounded Borel-Cantelli and Stochasticity Theorems. Technical Report UBCS–TR 95-08, Department of Computer Science, State University of New York at Buffalo, 1995.
20. O. Watanabe. A comparison of polynomial time completeness notions. *Theoretical Computer Science*, 54:249–265, 1987.

Graph Isomorphism Is Low for ZPP(NP) and Other Lowness Results

Vikraman Arvind[1] and Johannes Köbler[2]

[1] Institute of Mathematical Sciences, C.I.T Campus, Chennai 600113, India
[2] Humboldt-Universität zu Berlin, Institut für Informatik, D-10099 Berlin, Germany

Abstract. We show the following new lowness results for the probabilistic class $\mathrm{ZPP}^{\mathrm{NP}}$.

- The class AM ∩ coAM is low for $\mathrm{ZPP}^{\mathrm{NP}}$. As a consequence it follows that Graph Isomorphism and several group-theoretic problems known to be in AM ∩ coAM are low for $\mathrm{ZPP}^{\mathrm{NP}}$.
- The class IP[P/poly], consisting of sets that have interactive proof systems with honest provers in P/poly, is also low for $\mathrm{ZPP}^{\mathrm{NP}}$.

We consider lowness properties of nonuniform function classes, namely, NPMV/poly, NPSV/poly, NPMV_t/poly, and NPSV_t/poly. Specifically, we show that

- Sets whose characteristic functions are in NPSV/poly and that have program checkers (in the sense of Blum and Kannan [8]) are low for AM and $\mathrm{ZPP}^{\mathrm{NP}}$.
- Sets whose characteristic functions are in NPMV_t/poly are low for Σ_2^p.

1 Introduction

In the recent past the probabilistic class $\mathrm{ZPP}^{\mathrm{NP}}$ has appeared in different results and contexts in complexity theory research. E.g. consider the result $\mathrm{MA} \subseteq \mathrm{ZPP}^{\mathrm{NP}}$ [1, 14] which sharpens and improves Sipser's theorem $\mathrm{BPP} \subseteq \Sigma_2^p$. The proof in [1] uses derandomization techniques based on hardness assumptions [21]. Another example is the result that if SAT ∈ P/poly then $\mathrm{PH} = \mathrm{ZPP}^{\mathrm{NP}}$ [20, 10], which improves the classic Karp-Lipton theorem. [1] Actually, Köbler and Watanabe in [20] prove that every self-reducible set[2] A in (NP ∩ co-NP)/poly is *low* for $\mathrm{ZPP}^{\mathrm{NP}}$, i.e. $\mathrm{ZPP}^{\mathrm{NP}^A} = \mathrm{ZPP}^{\mathrm{NP}}$. This stronger result is in a sense natural, since there is usually an underlying lowness result that implies a collapse consequence result like the Karp-Lipton theorem. We may recall here that the lowness result underlying the Karp-Lipton theorem is that self-reducible sets in P/poly are low for Σ_2^p [24].

The notion of lowness was first introduced in complexity theory by Schöning in [24]. It has since then been an important conceptual tool in complexity theory, see e.g. the survey paper [16].

[1] The Karp-Lipton theorem states that if SAT ∈ P/poly then PH collapses to Σ_2^p.
[2] By self-reducibility we mean word-decreasing self-reducibility which is adequate because standard complexity classes contained in EXP have such self-reducible complete problems.

H. Reichel and S. Tison (Eds.): STACS 2000, LNCS 1770, pp. 431–442, 2000.

1.1 Lowness for $\mathrm{ZPP}^{\mathrm{NP}}$

We recall the formal definition of lowness [24]. For a relativizable complexity class $\mathcal{C}$ such that for all sets A, $A \in \mathcal{C}^A$, let $Low(\mathcal{C})$ denote $\{A \mid \mathcal{C}^A = \mathcal{C}\}$. Clearly, $Low(\mathcal{C})$ is contained in $\mathcal{C}$ and consists of languages that are powerless as oracle for $\mathcal{C}$.

Few complexity classes have their low sets exactly characterized. These are well-known examples: $Low(\mathrm{NP}) = \mathrm{NP} \cap \text{co-NP}$, $Low(\mathrm{AM}) = \mathrm{AM} \cap \mathrm{coAM}$ [25]. For most complexity classes however, a complete characterization of the low sets appears to be a challenging open question. Regarding $Low(\Sigma_2^p)$, Schöning proved [25] that $\mathrm{AM} \cap \mathrm{coAM}$ is contained in $Low(\Sigma_2^p)$, implying that $Low(\mathrm{AM}) \subseteq Low(\Sigma_2^p)$. This containment is anomalous because $\mathrm{AM} \not\subseteq \Sigma_2^p$ in some relativized worlds [23]. Indeed, lowness appears to have other anomalous properties: it is not known to preserve containment of complexity classes, for example $\mathrm{NP} \subseteq \mathrm{PP}$ but $\mathrm{NP} \cap \text{co-NP}$ is not known to be in $Low(\mathrm{PP})$. Similarly, $\mathrm{NP} \subseteq \mathrm{MA}$ but $\mathrm{NP} \cap \text{co-NP}$ is not known to be in $Low(\mathrm{MA})$. Little is known about $Low(\mathrm{MA})$ except that it contains BPP and is contained in $\mathrm{MA} \cap \text{co-MA}$ [18].

Regarding $\mathrm{ZPP}^{\mathrm{NP}}$, it is shown in [20] that $Low(\mathrm{ZPP}^{\mathrm{NP}}) \subseteq Low(\Sigma_2^p)$. No characterization of $Low(\mathrm{ZPP}^{\mathrm{NP}})$ is known. Our aim is to show some inclusions in $Low(\mathrm{ZPP}^{\mathrm{NP}})$ as a first step.

We first show in this paper that $\mathrm{AM} \cap \mathrm{coAM}$ is low for $\mathrm{ZPP}^{\mathrm{NP}}$, i.e. $\mathrm{AM} \cap \mathrm{coAM} \subseteq Low(\mathrm{ZPP}^{\mathrm{NP}})$. Hence we have the inclusion chain

$$Low(\mathrm{MA}) \subseteq Low(\mathrm{AM}) \subseteq Low(\mathrm{ZPP}^{\mathrm{NP}}) \subseteq Low(\Sigma_2^p).$$

It follows that Graph Isomorphism and other group-theoretic problems known to be in $\mathrm{AM} \cap \mathrm{coAM}$ [4] are low for $\mathrm{ZPP}^{\mathrm{NP}}$.

We prove another lowness result for $\mathrm{ZPP}^{\mathrm{NP}}$: Let IP[P/poly] denote the class of languages that have interactive proof systems with honest prover in P/poly. We show that $\mathrm{IP[P/poly]} \subseteq Low(\mathrm{ZPP}^{\mathrm{NP}})$, improving the containment $\mathrm{IP[P/poly]} \subseteq Low(\Sigma_2^p)$ shown in [3]. Our proof has a derandomization component in which the Nisan-Wigderson pseudorandom generator [21] is used to derandomize the verifier in the IP[P/poly] protocol. The rest of the proof is based on the random sampling technique as applied in [10, 17].

1.2 NP/poly $\cap$ co-NP/poly and Subclasses

As shown in [20], self-reducible sets in $(\mathrm{NP} \cap \text{co-NP})/\mathrm{poly}$ are *low* for $\mathrm{ZPP}^{\mathrm{NP}}$. However, there are technical difficulties due to which this result doesn't seem to carry over to NP/poly $\cap$ co-NP/poly. The best known collapse consequence of $\mathrm{NP} \subseteq \mathrm{NP/poly} \cap \text{co-NP/poly}$ (equivalently, $\mathrm{NP} \subseteq \text{co-NP/poly}$) is $\mathrm{PH} \subseteq \mathrm{ZPP}(\Sigma_2^p)$ [20].

In order to better understand this aspect of NP/poly $\cap$ co-NP/poly the authors of [11] introduce two interesting subclasses of NP/poly $\cap$ co-NP/poly which we discuss in Section 5. We notice firstly that NP/poly $\cap$ co-NP/poly and the above-mentioned subclasses are closely connected to the function classes

NPMV/poly, NPSV/poly, NPMV_t/poly, and NPSV_t/poly, which are nonuniform analogues of the function classes NPMV, NPSV, NPMV_t, and NPSV_t introduced and studied by Selman and other researchers [26, 12]. More precisely, we note that $A \in (\text{NP} \cap \text{co-NP})/\text{poly}$ if and only if $\chi_A \in \text{NPSV}_t/\text{poly}$, where χ_A denotes the characteristic function of a language A. Similarly, $A \in \text{NP}/\text{poly} \cap \text{co-NP}/\text{poly}$ if and only if $\chi_A \in \text{NPMV}/\text{poly}$. Likewise, NPSV/poly and NPMV_t/poly capture the two new subclasses of NP/poly $\cap$ co-NP/poly defined in [11].

We prove the following new lowness results for these classes:

- We show that self-reducible sets whose characteristic functions are in the function class NPMV_t/poly are low for Σ_2^p (this result is essentially the lowness result underlying the collapse consequence derived in [11, Theorem 5.2]).
- We show that all self-checkable sets — in the program checking sense of Blum and Kannan [8] — whose characteristic functions are in NPSV/poly are low for AM.

Several proofs are omitted from this extended abstract. A full version of the paper is available as a technical report [2].

2 Preliminaries

Let $\Sigma = \{0,1\}$. We denote the cardinality of a set X by $\|X\|$ and the length of a string $x \in \Sigma^*$ by $|x|$. The characteristic function of a language $L \subseteq \Sigma^*$ is denoted by χ_L. The definitions of standard complexity classes like P, NP, E, EXP etc. can be found in standard books [7, 22]. A relativized complexity class $\mathcal{C}$ with oracle A is denoted by either $\mathcal{C}^A$ or $\mathcal{C}(A)$. Likewise, we denote an oracle Turing machine M with oracle A by M^A or $M(A)$.

For a class $\mathcal{C}$ of sets and a class $\mathcal{F}$ of functions from 1^* to Σ^*, let $\mathcal{C}/\mathcal{F}$ [15] be the class of sets A such that there is a set $B \in \mathcal{C}$ and a function $h \in \mathcal{F}$ such that for all $x \in \Sigma^*$,

$$x \in A \Leftrightarrow \langle x, h(1^{|x|}) \rangle \in B.$$

The function h is called an *advice function* for A.

We recall definitions of AM and MA. A language L is in AM if there exist a polynomial p and a set $B \in \text{P}$ such that for all x, $|x| = n$,

$$x \in A \Rightarrow \text{Prob}_{r \in_R \{0,1\}^{p(n)}}[\exists y, |y| = p(n) : \langle x, y, r \rangle \in B] \; = \; 1,$$
$$x \notin A \Rightarrow \text{Prob}_{r \in_R \{0,1\}^{p(n)}}[\exists y, |y| = p(n) : \langle x, y, r \rangle \in B] \leq 1/4.$$

A language L is in MA if there exist a polynomial p and a set $B \in \text{P}$ such that for all x, $|x| = n$,

$$x \in A \Rightarrow \exists y, |y| = p(n) : \text{Prob}_{r \in_R \{0,1\}^{p(n)}}[\langle x, y, r \rangle \in B] \geq 3/4,$$
$$x \notin A \Rightarrow \forall y, |y| = p(n) : \text{Prob}_{r \in_R \{0,1\}^{p(n)}}[\langle x, y, r \rangle \in B] \leq 1/4.$$

Notice that we have taken the definition of AM with 1-sided error, known to be equivalent to AM with 2-sided error. Definitions for single and multiprover interactive proof systems can be found in standard texts, e.g. [22]. Let MIP denote the class of languages with multiprover interactive protocols and IP denote the class of languages with single-prover interactive protocols. We denote by MIP[$\mathcal{C}$] and IP[$\mathcal{C}$] the respective language classes where the prover complexity is bounded by FP($\mathcal{C}$), which is the class of functions that can be computed by a polynomial-time oracle transducer with oracle in $\mathcal{C}$.

3 AM ∩ coAM Is Low for ZPP^{NP}

In this section we show that AM∩coAM is low for ZPP^{NP}. It follows that Graph Isomorphism and a host of group-theoretic problems known to be in AM∩coAM [4] are all low for ZPP^{NP}. We recall here that it is already known that AM∩coAM is low for Σ_2^p [25] and also for AM [18].

We notice first that although AM ∩ coAM ⊆ ZPP^{NP} (because AM ⊆ coR^{NP} and the equality ZPP = R ∩ coR relativizes) and AM ∩ coAM is low for itself, it doesn't follow that AM ∩ coAM is low for ZPP^{NP}. As mentioned before, NP ∩ co-NP is low for NP but is not known to be low for PP or MA.

Theorem 1. AM ∩ coAM *is low for* ZPP^{NP}.

Proof. Let L be any set in AM ∩ coAM. We need to show that a given ZPP$^{\mathrm{NP}^L}$ machine M can be simulated in ZPP^{NP}. Consider an input x of length bounded by n to the machine M. Suppose the lengths of all the queries made to L during the computation are bounded by m. Since $L \in$ AM ∩ coAM, it follows from standard probability amplification techniques (cf. [25]) that there are NP sets A and B, a polynomial p, and subsets $S_m \subseteq \{0,1\}^{p(m)}$ of size $\|S_m\| \geq 2^{p(m)-1}$ such that for all m and all strings y of length $|y| \leq m$,

$y \in L$ implies

$$\forall w : \langle y, w\rangle \in A \text{ and } \forall w \in S_m : \langle y, w\rangle \notin B$$

and $y \notin L$ implies

$$\forall w : \langle y, w\rangle \in B \text{ and } \forall w \in S_m : \langle y, w\rangle \notin A.$$

In other words, any string $w \in S_m$ can be used as advice to decide membership in L for strings of length $|y| \leq m$ with an NP ∩ co-NP computation. Notice, however, that it would be incorrect for us to claim from here that $L \in$ (NP ∩ co-NP)/poly, because if we use a string from $\{0,1\}^{p(m)} - S_m$ as advice, the resulting combination of machines for A and B may not yield an NP ∩ co-NP computation for some input y of length $|y| \leq m$. In fact, a string $w \in \Sigma^{p(m)}$ is a good advice provided that it does *not* satisfy the NP predicate

$$\exists y, |y| \leq m : \langle y, w\rangle \in A \cap B.$$

We now describe the $\mathrm{ZPP}^{\mathrm{NP}}$ machine N that simulates the given $\mathrm{ZPP}^{\mathrm{NP}^L}$ machine M on some input x. Machine N first randomly guesses a string $w \in \Sigma^{p(m)}$. By assumption, w is a good advice with probability at least 1/2, and this can be checked with a single query to the NP predicate above. In case w is good, N can use w to replace the oracle L with an NP $\cap$ co-NP computation when it simulates M on input x. □

Corollary 1. *Graph Isomorphism is low for* $\mathrm{ZPP}^{\mathrm{NP}}$.

The above corollary follows since Graph Isomorphism is in AM $\cap$ coAM [13]. The lowness result also holds for various group-theoretic problems known to be in AM $\cap$ coAM [4].

Notice that the previous proof essentially shows that we can simulate AM $\cap$ coAM with an NP $\cap$ co-NP computation using a random string in a coNP set as advice for the computation. This observation combined with the result of [20] (that self-reducible sets in (NP $\cap$ co-NP)/poly are low for $\mathrm{ZPP}^{\mathrm{NP}}$) immediately yields the following corollary.

Corollary 2. *Self-reducible sets in* (AM $\cap$ coAM)/poly *are low for* $\mathrm{ZPP}^{\mathrm{NP}}$.

Additionally, we also have the following corollary in the average-case complexity setting. We first recall the definition of $\mathcal{AP}$ (see, e.g. [19] for a detailed treatment): $\mathcal{AP}$ is the class of decision problems A such that for every polynomial-time computable distribution there is an algorithm that decides A and is polynomial-time on the average for that distribution.

Corollary 3. *If* NP $\subseteq \mathcal{AP}$ *then* AM $\cap$ coAM = NP $\cap$ co-NP.

The proof uses the assumption NP $\subseteq \mathcal{AP}$ combined with the fact that for any set in AM $\cap$ coAM a large fraction of strings satisfying a coNP predicate are good advice strings, as we have already seen in the proof of Theorem 1. Thus, we can guess such an advice string and use an $\mathcal{AP}$ algorithm for the *uniform* distribution to verify the coNP predicate. Since the $\mathcal{AP}$ algorithm, with its running time truncated to a suitable polynomial bound, will still accept many good advice strings, we get an NP $\cap$ co-NP simulation of AM $\cap$ coAM. This is an application of ideas from [19].

4 IP[P/poly] Is Low for $\mathrm{ZPP}^{\mathrm{NP}}$

The class IP[P/poly] already figures, though implicitly, in the proof of the result in [5] that if EXP $\subseteq$ P/poly then EXP = MA. We quickly recall the proof: Suppose EXP $\subseteq$ P/poly. Note that each language in EXP has a multiprover interactive protocol in which the provers are in EXP. By assumption, therefore, the honest provers can be simulated by polynomial size circuits. Thus the (MIP) protocol can be simulated by an MA protocol where Merlin simply sends the circuits for the provers to Arthur in the first round. In other words, the proof shows the inclusion chain EXP $\subseteq$ MIP[P/poly] $\subseteq$ MA. Since the MA protocol is a single prover interactive protocol, we also have MIP[P/poly] = IP[P/poly] $\subseteq$ MA.

The above collapse consequence result of [5] motivates the study of lowness properties of IP[P/poly]. Our next result states that IP[P/poly] $\subseteq$ $Low(\mathrm{ZPP}^{\mathrm{NP}})$, improving the containment IP[P/poly] $\subseteq$ $Low(\Sigma_2^p)$ shown in [3]. Our result strengthens the result of [17] that NP sets in P/poly with self-computable witnesses are low for $\mathrm{ZPP}^{\mathrm{NP}}$. IP[P/poly] contains such NP sets, but IP[P/poly] may not even be contained in NP. Although IP[P/poly] $\subseteq$ MA $\subseteq$ AM, IP[P/poly] is not known to be closed under complement, and it is not known if IP[P/poly] is contained in coAM. Thus, IP[P/poly] $\subseteq$ $Low(\mathrm{ZPP}^{\mathrm{NP}})$ appears incomparable to AM $\cap$ coAM $\subseteq$ $Low(\mathrm{ZPP}^{\mathrm{NP}})$ shown in Theorem 1 in the previous section. Our result is also incomparable to the result in [20] that self-reducible sets in P/poly are low for $\mathrm{ZPP}^{\mathrm{NP}}$. An interesting aspect of our proof is that it combines derandomization and almost uniform random sampling.

Theorem 2. IP[P/poly] *is low for* $\mathrm{ZPP}^{\mathrm{NP}}$.

The above lowness result easily extends to IP[(NP $\cap$ co-NP)/poly] by observing that the proof relativizes in the following sense: for any oracle set A, $\mathrm{NP}^{\mathrm{IP}[\mathrm{P}^A/\mathrm{poly}]} \subseteq \mathrm{ZPP}^{\mathrm{NP}^A}$.

We conclude this section with another connection to the average-case complexity setting.

Theorem 3. *If* NP $\subseteq \mathcal{AP}$ *and* NP $\subseteq$ P/poly *then PH collapses to* Δ_2^p.

5 Nonuniform Function Classes and Lowness

We now study lowness properties of NPMV/poly, NPSV/poly, NPMV_t/poly, and NPSV_t/poly. These are nonuniform analogs of the function classes NPMV, NPSV, NPMV_t, and NPSV_t studied by Selman [26] and other researchers, e.g. [12]. These nonuniform classes are interesting because when restricted to characteristic functions of sets, NPSV_t/poly coincides with (NP $\cap$ co-NP)/poly and NPMV/poly coincides with NP/poly $\cap$ co-NP/poly. Likewise, we note that the two subclasses of NP/poly $\cap$ co-NP/poly studied in [11], namely all sets underproductively reducible to sparse sets and all sets overproductively reducible to sparse sets, also coincide with NPSV/poly and NPMV_t/poly, respectively.

Following Selman's notation in [26], a transducer is a nondeterministic Turing machine (NDTM for short) T with a write-only output tape. On input x, machine T outputs $y \in \Sigma^*$ if there is an accepting path on input x along which y is output. Hence, the function defined by T on Σ^* could be multivalued and partial. Given a multivalued function f on Σ^* and $x \in \Sigma^*$ we use the notation

$$set\text{-}f(x) = \{y \mid f : x \mapsto y\}$$

to denote the (possibly empty) set of function values for input x. We recall the basic definitions.

Definition 1. [9]

1. NPMV *is the class of multivalued, partial functions f for which there is a polynomial-time NDTM N such that*
   (a) *$f(x)$ is defined (i.e., set-$f(x) \neq \emptyset$) if and only if $N(x)$ has an accepting path.*
   (b) *$y \in$ set-$f(x)$ if and only if there is an accepting path of $N(x)$ where y is output.*
2. NPSV *is the class of single-valued partial functions in* NPMV.
3. NPMV_t *is the class of total functions in* NPMV.
4. NPSV_t *is the class of total single-valued functions in* NPMV.

The classes NPMV/poly, NPSV/poly, NPMV_t/poly, and NPSV_t/poly are the standard nonuniform analogs of the above classes defined as usual [15]: for $\mathcal{F} \in \{\mathrm{NPMV}, \mathrm{NPSV}, \mathrm{NPMV}_t \mathrm{NPSV}_t\}$, a multivalued partial function f is in $\mathcal{F}$/poly if there is a function $g \in \mathcal{F}$, a polynomial p, and an *advice function* $h : 1^* \mapsto \Sigma^*$ with $|h(1^n)| = p(n)$ for all n, such that for all $x \in \Sigma^*$,

$$set\text{-}f(x) = set\text{-}g(\langle x, h(1^{|x|})\rangle).$$

Before we connect these classes to NP/poly ∩ co-NP/poly and its subclasses defined in [11], we recall definitions from [11]: Consider polynomial-time nondeterministic oracle machines N whose computation paths can have three possible outcomes: accept, reject, or ?. The machine N can also be viewed as a transducer which computes, for given oracle D and input x, a multivalued function. More precisely, if we identify accept with value 1 and reject with 0, and consider the ? computation paths as rejecting paths then N^D defines a partial multivalued function: *set*-$N^D(x) \subseteq \{0,1\}$. Machine N^D is said to be *underproductive* if for each x we have $\{0,1\} \not\subseteq$ *set*-$N^D(x)$, and N is said to be *robustly underproductive* if for each oracle D and input x we have $\{0,1\} \not\subseteq$ *set*-$N^D(x)$. Likewise, N^D is *overproductive* if for each x we have *set*-$N^D(x) \neq \emptyset$, and N is said to be *robustly overproductive* if for each oracle D and input x we have *set*-$N^D(x) \neq \emptyset$.

With standard arguments we can convert a sparse set into a polynomial-size advice string and vice-versa (see, e.g. [7]). It follows that $A \in$ NP/poly ∩ co-NP/poly if and only if there is a sparse set S and a nondeterministic machine N such that N^S is both overproductive and underproductive and $A = L(N^S)$. Similarly, $A \in$ (NP ∩ co-NP)/poly if and only if there is a sparse set S and a nondeterministic machine N such that $A = L(N^S)$ and N is both robustly overproductive and robustly underproductive and $A = L(N^S)$.

Proposition 1. *Let χ_A denote the characteristic function for a set $A \subseteq \Sigma^*$:*

1. *χ_A is in* NPMV/poly *if and only if A is in* NP/poly ∩ co-NP/poly.
2. *χ_A is in* NPSV_t/poly *if and only if A is in* (NP ∩ co-NP)/poly.
3. *χ_A is in* NPSV/poly *if and only if there are a sparse set S and a robustly underproductive machine N such that $A = L(N^S)$.*
4. *χ_A is in* NPMV_t/poly *if and only if there are a sparse set S and a robustly overproductive machine N such that $A = L(N^S)$.*

By abuse of notation, we identify χ_A with A in this section. E.g. we write $A \in$ NPSV/poly when we mean $\chi_A \in$ NPSV/poly. We now turn to lowness questions for the nonuniform function classes. The classes NP/poly$\cap$co-NP/poly and (NP $\cap$ co-NP)/poly are of interest in the context of deriving strong collapse consequences from the assumption that NP (or some other hard complexity class) is contained in one of these classes. We recall the known collapse consequence result shown in [20] for NP/poly $\cap$ co-NP/poly under the assumption that NP is contained therein: If NP $\subseteq$ NP/poly $\cap$ co-NP/poly then PH collapses to $\mathrm{ZPP}^{\Sigma_2^p}$. The open question here is whether the collapse consequence can possibly be improved to $\mathrm{ZPP}^{\mathrm{NP}}$. This is one reason to consider classes that lie between NP/poly $\cap$ co-NP/poly and (NP $\cap$ co-NP)/poly.

5.1 A Lowness Result for NPMV$_t$/poly

It is shown in [11] that if an NP-complete problem is in NPMV$_t$/poly then PH collapses to Σ_2^p. In [11] the authors actually state this result in terms of overproductive reductions to sparse sets. We use ideas in their proof to show the underlying lowness result for functions: all word-decreasing self-reducible functions in NPMV$_t$/poly are low for Σ_2^p. We first recall the definition of word-decreasing self-reducible sets (and define its obvious extension to total single-valued functions).

Definition 2. [6] *For strings $x, y \in \Sigma^*$, $x \prec y$ if $|x| < |y|$ or $|x| = |y|$ and x is lexicographically smaller than y. A set A is* word-decreasing self-reducible *if there is a polynomial-time oracle machine M such that $A = L(M^A)$, where on any input x the machine M queries the oracle only about strings y such that $y \prec x$. Similarly, a total single-valued function f on Σ^* is* word-decreasing self-reducible *if there is a polynomial-time oracle transducer T such that T^f computes f, where on any input x, transducer T can query the oracle only about strings y such that $y \prec x$.*

The definition of lowness extends naturally to total, single-valued functions: A functional oracle f returns $f(x)$ on query x. For any relativizable complexity class $\mathcal{C}$ we say that $f \in \mathit{Low}(\mathcal{C})$ if $\mathcal{C}^f = \mathcal{C}$. We show next that self-reducible sets and self-reducible functions in NPMV/poly have identical lowness properties. Hence it suffices to prove lowness of self-reducible sets in NPMV/poly.

Theorem 4. *Let $\mathcal{F}$ contain all self-reducible functions in any of the four function classes* {NPMV/poly, NPSV/poly, NPMV$_t$/poly, NPSV$_t$/poly}*. Let $\mathcal{C}$ be the subclass of $\mathcal{F}$ consisting of characteristic functions (making $\mathcal{C}$ a language class, essentially). For every self-reducible function $f \in \mathcal{F}$ there is a self-reducible set $A \in \mathcal{C}$ such that f and A are polynomial-time Turing equivalent.*

Proof. Given $f \in \mathcal{F}$, we can define the corresponding set $A \in \mathcal{C}_{\mathcal{F}}$ by suitably encoding, for each x, the bits of $f(x)$ in A. We can easily ensure that the self-reducibility of f carries over to A and f and A are polynomial-time Turing equivalent. □

Theorem 5. *Word-decreasing self-reducible sets in* NPMV_t/poly *are low for* Σ_2^p.

Since Σ_k^p, Π_k^p, PP, $\mathrm{C_=P}$, $\mathrm{Mod}_m\mathrm{P}$, PSPACE, and EXP have many-one complete word-decreasing self-reducible sets [6], the following corollary is immediate.

Corollary 4. *If* $\mathcal{C} \in \{\Sigma_k^p, \Pi_k^p, \mathrm{PP}, \mathrm{C_=P}, \mathrm{Mod}_m\mathrm{P}, \mathrm{PSPACE}, \mathrm{EXP}\}$, *for* $k \geq 1$, *has a complete set in* NPMV_t/poly *then* $\mathcal{C} \subseteq \Sigma_2^p$ *and* $\mathrm{PH} = \Sigma_2^p$.

The proof follows since for each $\mathcal{C} \in \{\Sigma_k^p, \Pi_k^p, \mathrm{PP}, \mathrm{C_=P}, \mathrm{Mod}_m\mathrm{P}, \mathrm{PSPACE}, \mathrm{EXP}\}$ and any set A complete for $\mathcal{C}$ w.r.t. polynomial-time Turing reductions we have $\Sigma_3^p \subseteq \Sigma_2^A$.

We end this section with the observation that AM $\cap$ coAM is contained in NPMV_t/poly. It is interesting to now compare the lowness results (Theorems 1 and 5) for these classes.

Proposition 2. *If* $L \in \mathrm{AM} \cap \mathrm{coAM}$ *then* L *is in* NPMV_t/poly.

Proof. Given $L \in \mathrm{AM} \cap \mathrm{coAM}$, as already observed in the proof of Theorem 1, there are NP sets A and B, a polynomial p, and subsets $S_m \subseteq \{0,1\}^{p(m)}$ of size $\|S_m\| \geq 2^{p(m)-1}$ such that for all m and all strings x of length $|x| \leq m$,
$x \in L$ implies

$$\forall w : \langle x, w\rangle \in A \text{ and } \forall w \in S_m : \langle x, w\rangle \notin B$$

and $x \notin L$ implies

$$\forall w : \langle x, w\rangle \in B \text{ and } \forall w \in S_m : \langle x, w\rangle \notin A.$$

We can combine the NP machines M_A and M_B for A and B and build a transducer I that on input $\langle x, w\rangle$ outputs 1 (0) on any accepting simulation of M_A (resp. M_B) on input $\langle x, w\rangle$. Observe that in case $w \in S_m$ transducer I will always yield a single-valued, total computation for all inputs x of length m, outputing either 1 or 0 depending on the membership of x. On the other hand, no matter which $w \in \{0,1\}^{p(m)}$ is used as advice, $\langle x, w\rangle$ is either in A or in B and so the transducer I always outputs at least one of 0 or 1 for any advice string $w \in \{0,1\}^{p(m)}$ and any input x of length m. Hence it follows that L is in NPMV_t/poly. □

5.2 A Lowness Result for NPSV/poly

In [11] it is left as an open problem to discover new lowness (or collapse consequence) results for NPSV/poly. As noted in [11], nothing better is known for NPSV/poly than the collapse consequence result: if SAT is in NPSV/poly then PH collapses to $\mathrm{ZPP}^{\Sigma_2^p}$, which holds even for the larger class NP/poly $\cap$ co-NP/poly [20].

We show that sets in NPSV/poly that are checkable, in the sense of program checking as defined by Blum and Kannan [8], are low for AM and for $\mathrm{ZPP}^{\mathrm{NP}}$.

Since $\oplus$P, PP, PSPACE, and EXP have checkable complete problems, it follows that for any of these classes inclusion in NPSV/poly implies its containment in AM$\cap$coAM. This result is proved on the same lines as the Babai et al result [5]: If EXP is contained in P/poly then EXP $\subseteq$ MA.

Recall the definitions of MIP[$\mathcal{C}$] and IP[$\mathcal{C}$] for a class $\mathcal{C}$ of languages. We prove a technical lemma that immediately yields the lowness result.

Lemma 1. *If* $A \in$ NPSV/poly *then* MIP[A] $\subseteq$ AM.

Proof. Let $L \in$ MIP[A] for some set $A \in$ NPSV/poly. Let T be a nondeterministic transducer and q be a polynomial witnessing that $A \in$ NPSV/poly. We describe an MAM protocol for L:

1. Let x be an input of length n to the protocol. Let $m = p(n)$, where p is a polynomial bounding the size of the queries to A made by the provers during the MIP[A] protocol for inputs of length n.
2. **Merlin** sends advice w of length $q(m)$ to Arthur.
3. **Arthur** sends a polynomial random string r (used for simulating the original MIP protocol) to Merlin.
4. **Merlin** sends back the list of successive queries to set A (generated by simulating the original MIP protocol with random string r), the list of answers to those queries along with the computation paths of transducer T with advice w that certify the answers to the queries.
5. **Arthur** can verify in polynomial time that Merlin's message is all correct and accept if and only if the original MIP protocol accepts.

By the fact that T computes a single-valued partial function for any advice w, although the verifier is simulating the nondeterministic transducer T, it is guaranteed that each accepting computation path has identical output and hence does identical computation. Thus, what makes the above MAM protocol work is the fact that for any advice w and query q all accepting computation paths of $T(q, w)$ output the same value. So, regardless of which computation paths are sent to Arthur by Merlin in Step 4 of the above protocol, Arthur's decision will be the same. In other words, Arthur's acceptance depends only on the random string r, hence exactly preserving the acceptance probability of the original MIP protocol.

Standard techniques (cf. [4]) can be used to convert the MAM protocol to an AM protocol. This completes the proof. □

We have as immediate consequence the following lowness result.

Theorem 6. *If L is a checkable set in* NPSV/poly *then* $L \in$ AM$\cap$coAM *and hence low for AM and* ZPP^{NP}.

Proof. The assumption in the theorem's statement implies that both L and $\overline{L}$ are in MIP[L] by the checker characterization theorem of [8]. Now, applying Lemma 1 yields that both L and $\overline{L}$ are in AM and the result follows. □

We can derive new collapse consequences as corollary, since the classes $\oplus$P, PP, PSPACE, and EXP all have checkable complete problems. It follows that for any of these classes inclusion in NPSV/poly implies its containment in AM$\cap$ coAM.

Corollary 5. *If any of the classes* $\oplus$P, PP, PSPACE, *and* EXP *is contained in* NPSV/poly *then it is low for* AM *and hence* PH = AM.

Notice that we have the same lowness for checkable functions in NPSV/poly.

Corollary 6. *Checkable functions in* NPSV/poly *are low for* AM *and* $\mathrm{ZPP}^{\mathrm{NP}}$.

Proof. Let f be a checkable function in NPSV/poly. We can suitably encode, for each x, the bits of $f(x)$ in a language A which is polynomial-time Turing equivalent to f and hence A is also checkable. The lowness result now follows by invoking Theorem 6. □

Acknowledgements. The first author was partially supported by an Alexander von Humboldt fellowship in the year 1999, and he is grateful to Prof. Uwe Schöning for hosting his visit to Ulm university where this work was carried out.

References

[1] V. Arvind and J. Köbler. On resource-bounded measure and pseudorandomness. In *Proc. 17th Conference on Foundations of Software Technology and Theoretical Computer Science*, volume 1346 of *Lecture Notes in Computer Science*, pages 235–249. Springer-Verlag, 1997.

[2] V. Arvind and J. Köbler. Graph isomorphism is low for ZPP(NP) and other lowness results. Technical Report TR99-033, Electronic Colloquium on Computational Complexity, 1999.

[3] V. Arvind, J. Köbler, and R. Schuler. On helping and interactive proof systems. *International Journal of Foundations of Computer Science*, 6(2):137–153, 1995.

[4] L. Babai. Bounded round interactive proofs in finite groups. *SIAM Journal of Discrete Mathematics*, 5:88–111, 1992.

[5] L. Babai, L. Fortnow, N. Nisan, and A. Wigderson. BPP has subexponential time simulations unless EXPTIME has publishable proofs. *Computational Complexity*, 3:307–318, 1993.

[6] J. L. Balcázar. Self-reducibility. *Journal of Computer and System Sciences*, 41:367–388, 1990.

[7] J. L. Balcázar, J. Díaz, and J. Gabarró. *Structural Complexity I.* EATCS Monographs on Theoretical Computer Science. Springer-Verlag, second edition, 1995.

[8] M. Blum and S. Kannan. Designing programs to check their work. *Journal of the ACM*, 42(1):269–291, 1995.

[9] R. Book, T. Long, and A. L. Selman. Quanitative relativizations of complexity classes. *SIAM Journal on Computing*, 13:461–487, 1984.

[10] N. Bshouty, R. Cleve, R. Gavaldà, S. Kannan, and C. Tamon. Oracles and queries that are sufficient for exact learning. *Journal of Computer and System Sciences*, 52:421–433, 1996.

[11] J. Cai, L. A. Hemaspaandra, and G. Wechsung. Robust reductions. In *Proc. 4th Annual International Computing and Combinatorics Conference*, volume 1449 of *Lecture Notes in Computer Science*, pages 174–183. Springer-Verlag, 1998.

[12] S. Fenner, L. Fortnow, A. Naik, and J. Rogers. Inverting onto functions. In *Proc. 11th Annual IEEE Conference on Computational Complexity*, pages 213–222. IEEE Computer Society Press, 1996.

[13] O. Goldreich, S. Micali, and A. Wigderson. Proofs that yield nothing but their validity or all languages in np have zero-knowledge proof systems. *Journal of the ACM*, 38:691–729, 1991.

[14] O. Goldreich and D. Zuckerman. Another proof that BPP⊆PH (and more). Technical Report TR97-045, Electronic Colloquium on Computational Complexity, October 1997.

[15] R. M. Karp and R. J. Lipton. Some connections between nonuniform and uniform complexity classes. In *Proc. 12th ACM Symposium on Theory of Computing*, pages 302–309. ACM Press, 1980.

[16] J. Köbler. On the structure of low sets. In *Proc. 10th Structure in Complexity Theory Conference*, pages 246–261. IEEE Computer Society Press, 1995.

[17] J. Köbler and U. Schöning. On high sets for NP. In Ding-Zhu Du and K. Ko, editors, *Advances in Complexity and Algorithms*, pages 139–156. Kluwer Academic Publishers, 1997.

[18] J. Köbler, U. Schöning, and J. Torán. *The Graph Isomorphism Problem: Its Structural Complexity*. Birkhäuser, Boston, 1993.

[19] J. Köbler and R. Schuler. Average-case intractability vs. worst-case intractability. In *Proc. 23rd Symposium on Mathematical Foundations of Computer Science*, volume 1450 of *Lecture Notes in Computer Science*, pages 493–502. Springer-Verlag, 1998.

[20] J. Köbler and O. Watanabe. New collapse consequences of NP having small circuits. *SIAM Journal on Computing*, 28(1):311–324, 1999.

[21] N. Nisan and A. Wigderson. Hardness vs randomness. *Journal of Computer and System Sciences*, 49:149–167, 1994.

[22] C. Papadimitriou. *Computational Complexity*. Addison-Wesley, 1994.

[23] M. Santha. Relativized Arthur-Merlin versus Merlin-Arthur games. *Information and Computation*, 80(1):44–49, 1989.

[24] U. Schöning. A low and a high hierarchy within NP. *Journal of Computer and System Sciences*, 27:14–28, 1983.

[25] U. Schöning. Probabilistic complexity classes and lowness. *Journal of Computer and System Sciences*, 39:84–100, 1989.

[26] A. L. Selman. A taxonomy of complexity classes of functions. *Journal of Computer and System Sciences*, 48(2):357–381, 1994.

An Approximation Algorithm for the Precedence Constrained Scheduling Problem with Hierarchical Communications

Evripidis Bampis[1], Rodolphe Giroudeau[1], and Jean-Claude König[2]

[1] La.M.I., C.N.R.S. EP 738, Université d'Evry Val d'Essonne, Boulevard Mitterrand, 91025 Evry Cedex, France
[2] LIRMM, 161 rue Ada, 34392 Montpellier Cedex 5, France

Abstract. We study the problem of minimizing the makespan for the precedence multiprocessor constrained scheduling problem with hierarchical communications [1]. We propose an $\frac{8}{5}$-approximation algorithm for the UET-UCT (Unit Execution Time Unit Communication Time) hierarchical problem with an unbounded number of biprocessor machines. Moreover, we extend this result in the case where each cluster has m processors (where m is a fixed constant) by presenting an ρ-approximation algorithm where $\rho = (2 - \frac{2}{2m+1})$.

1 Introduction

Task scheduling is one of the the most important problems in parallel computation. In such a context, an application is usually represented as a directed acyclic graph where the vertices represent the tasks to be executed and the arcs the communication delays. The parallel architecture is composed by a set of m identical processors and the problem is to find a feasible scheduling minimizing the makespan. i.e. the time at which the last task of the graph finishes its execution. Formally this problem can be stated as follows:

Let $G = (V, E)$ be a precedence graph with n tasks, and let m be the number of available processors. Every task $i \in V$ has a processing time p_i and every arc (i, j) is associated with a communication delay c_{ij}. Let t_i (resp. t_j) be the starting time of task i (resp. j), then if i and j are executed on the same processor then $t_j \geq t_i + p_i$, otherwise $t_j \geq t_i + p_i + c_{ij}$. In what follows we call this model the scheduling model with *homogeneous communications.*

The objective, is to find a schedule, *i.e.* an allocation of each task to a time interval on one processor, such that the communication delays are taken into account and the completion time (makespan) is minimized (the makespan is denoted by C_{max} and it corresponds to $\max_{i \in V}\{t_i + p_i\}$).

This problem has been extensively studied in the last few years. It is $\mathcal{NP}$-hard even for surprisingly simple cases like the well known UET-UCT case where all the execution times and the communication delays are unitary and an unbounded number of processors is available. Using the notation of [6], this last case is

H. Reichel and S. Tison (Eds.): STACS 2000, LNCS 1770, pp. 443–454, 2000.

denoted as $\bar{P}|prec; c_{ij} = 1; p_i = 1|C_{max}$. It was shown that there is no hope to find a heuristic for $\bar{P}|prec; c_{ij} = 1; p_i = 1|C_{max}$ with relative performance strictly less than 7/6 (unless $\mathcal{P} = \mathcal{NP}$) [7], and the best known approximation algorithm is due to Munier and König with a worst-case relative performance equal to 4/3 [8].

We consider here an extension of this classical scheduling model [3] which takes into account hierarchical communications [1] [2]. This extension is motivated by the advance of hierarchical parallel architectures. Parallel architectures of this type include parallel machines constituted by different multiprocessors; biprocessors connected by myrinet switches, architectures where the processors are connected by hierarchical busses, or point-to-point architectures where each component of the topology is a cluster of processors. Formally, we are given m multiprocessor machines (or clusters) that are used to process n precedence constrained tasks. Each machine (cluster) comprises several identical parallel processors. A couple (c_{ij}, ϵ_{ij}) of communication delays is associated to each arc (i, j) between two tasks of the precedence graph. In what follows, c_{ij} (resp. ϵ_{ij}) is called intercluster (resp. interprocessor) communication, and we consider that $c_{ij} \geq \epsilon_{ij}$. If tasks i and j are executed on different machines, then j must be processed at least c_{ij} time units after the completion of i. Similarly, if i and j are executed on the same machine but on different processors then the processing of j can only start ϵ_{ij} units of time after the completion of i. However, if i and j are executed on the same processor then j can start immediately after the end of i. The communication overhead (intercluster or interprocessor delay) does not interfere with the availability of the processors and all processors may execute other tasks. Our goal is to find a feasible schedule of the tasks minimizing the makespan.

Notice that the hierarchical model that we consider here is a generalization of the scheduling model with homogeneous communication delays. Consider for instance that for every arc (i, j) of the precedence graph we have $c_{ij} = \epsilon_{ij}$. In that case the hierarchical model is exactly the classical scheduling model with homogeneous communications.

We focus on the case where the number of clusters is unrestricted, the number of processors within each cluster is equal to two and the intercluster (resp. interprocessor) communication is equal to $c_{ij} = 1$ (resp. $\epsilon_{ij} = 0$).

Using an extension of the classical notation of Lenstra et al. [6], this problem is denoted as $\bar{P}, 2|prec; (c_{ij}, \epsilon_{ij}) = (1, 0); p_i = 1|C_{max}$. Recently, for the multiprocessor scheduling problem with hierarchical communications, it has been proved in [2] that there is no hope of finding a heuristic with relative performance strictly less than 5/4 (unless $\mathcal{P} = \mathcal{NP}$). This result is an extension of the result of Hoogeveen et al. [7], who proved that there is no polynomial time ρ-approximation algorithm with $\rho < 7/6$ for the well-known UET-UCT scheduling problem with homogeneous communication delays ($\bar{P}|prec; c_{ij} = 1; p_i = 1|C_{max}$). However, in [1] it has been proved that the problem is polynomial if the duplication of tasks is allowed.

In what follows, we propose a new scheduling algorithm based on a LP-relaxation that improves the trivial bound of two.

1.1 Preliminaries

Given a precedence graph $G = (V, E)$ a predecessor (resp. successor) of a task i is a task j such that (j, i) (resp. (i, j)) is an arc of G. For every task $i \in V, \Gamma^+(i)$ (resp. $\Gamma^-(i)$) denotes the set of immediate successors (resp. predecessors) of i. We denote the tasks without predecessor (resp. successor) by Z (resp. U). We call source every task belonging to Z.

A schedule σ is a set of n ordered triples $\{(i, M_i, t_i), i \in V\}$: representing that the task i is performed by one of the processors of the cluster M_i at time t_i. Every feasible schedule must respect the following constraints:

1. at any time, a cluster executes at most two tasks;
2. $\forall (i, j) \in E$, if $M_i = M_j$ then $t_j \geq t_i + 1$, otherwise $t_j \geq t_i + 2$.

The makespan of schedule σ is:

$$C^{\sigma}_{max} = \max_{i \in V} \; (t_i + 1).$$

The problem is to find a feasible schedule with a minimum makespan.

In order to evaluate the worst-case performance of an algorithm, we recall the definition of the *relative performance* of a heuristic h:

$$\rho^h = \max_G \frac{C^h_{max}(G)}{C^{opt}_{max}(G)}$$

where $C^{opt}_{max}(G)$ denotes the optimal makespan of a feasible schedule of the graph G, and $C^h_{max}(G)$ the makespan obtained by the heuristic h.

In the next section, we formulate the problem as an integer linear program (ILP). In the third section, we propose a simple heuristic based on a relaxation of the ILP that we analyze in the fourth section by evaluating its worst-case relative performance. In the last section, we extend this result by providing an $(2 - \frac{2}{2m+1})$-approximation algorithm and we show that this bound is tight.

2 The Integer Linear Program

Let us consider an instance of the problem $\bar{P}, 2|prec; (c_{ij}, \epsilon_{ij}) = (1, 0); p_i = 1|C_{max}$ given by a directed acyclic graph $G = (V, E)$ with n tasks.

The aim of this section is to model the above described scheduling problem by an integer linear program (ILP) denoted, in what follows, by Π.

We model the scheduling problem by a set of equations defined on the starting times vector $(t_1, \ldots, t_n)$: in every feasible schedule, every task $i \in V - U$ has at most two successors, w.l.o.g. call them j_1 and $j_2 \in \Gamma^+(i)$, that can be performed by the same cluster as i at time $t_{j_1} = t_{j_2} = t_i + 1$.

The other successors of i, if any, satisfy: $\forall k \in \Gamma^+(i) - \{j_1, j_2\}, t_k \geq t_i + 2$. So, for every arc $(i,j) \in E$, we introduce a variable $x_{ij} \in \{0,1\}$ and the following constraints:

$$\forall (i,j) \in E, t_i + 1 + x_{ij} \leq t_j$$

and

$$\sum_{j \in \Gamma^+(i)} x_{ij} \geq |\Gamma^+(i)| - 2.$$

Similarly, every task i of $V - Z$ has at most two predecessors, w.l.o.g. call them j_1 and $j_2 \in \Gamma^-(i)$, that can be performed by the same cluster as i at time $t_{j_1} = t_{j_2} = t_i - 1$, so:

$$\sum_{j \in \Gamma^-(i)} x_{ji} \geq |\Gamma^-(i)| - 2.$$

If we denote by C_{max} the makespan of the schedule,

$$\forall i \in V, t_i + 1 \leq C_{max}.$$

The above constraints are necessary but not sufficient conditions in order to get a feasible schedule for our problem. For instance, a solution minimizing $C_{\max}$ for the graph of case (a) in Figure 1 will assign to every arc the value 0. However, since every cluster has two processors, and so at most two tasks can be processed on the same cluster simultaneously, the obtained solution is clearly not feasible. Thus, the relaxation of the integer constraints, by considering $0 \leq x_{ij} \leq 1$, and the resolution of the resulting linear program with objective function the minimization of $C_{\max}$, gives just a lower bound of the value of $C_{\max}$.

In order to improve this lower bound, we consider every subgraph of G that is isomorphic to the graphs given in Figure 1 –cases (a) and (b). It is easy to see that in any feasible schedule of G, at least one of the variables associated to the arcs of each one of these graphs must be set to one. So, we add the following constraints:

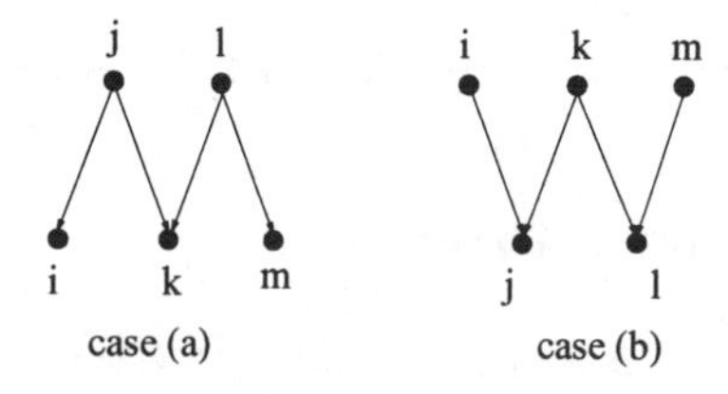

Fig. 1. *Two cases.*

- For the case (a):
  $\forall i, j, k, l, m \in V$, such that $(j,i), (j,k), (l,k), (l,m) \in E, x_{ji} + x_{jk} + x_{lk} + x_{lm} \geq 1$

– For the case (b):
$\forall i, j, k, l, m \in V$, such that $(i, j), (k, j), (k, l), (m, l) \in E, x_{ij} + x_{kj} + x_{kl} + x_{ml} \geq 1$

Thus, in what follows, we consider the following ILP:

$$(\Pi) \begin{cases} & \min C_{max} \\ \forall (i,j) \in E, & x_{ij} \in \{0,1\} \\ \forall i \in V, & t_i \geq 0 \\ \forall (i,j) \in E, & t_i + 1 + x_{ij} \leq t_j \\ \forall (i,j) \in E - U, & \sum_{j \in \Gamma^+(i)} x_{ij} \geq |\Gamma^+(i)| - 2 \\ \forall (i,j) \in E - Z, & \sum_{j \in \Gamma^-(i)} x_{ji} \geq |\Gamma^-(i)| - 2 \\ \forall i,j,k,l,m \in V, \backslash (j,i),(j,k),(l,k),(l,m) \in E, & x_{ji} + x_{jk} + x_{lk} + x_{lm} \geq 1 \\ \forall i,j,k,l,m \in V, \backslash (i,j),(k,j),(k,l),(m,l) \in E, & x_{ij} + x_{kj} + x_{kl} + x_{ml} \geq 1 \\ \forall i \in V, & t_i + 1 \leq C_{max} \end{cases}$$

Once again the integer linear program given above does not always imply a feasible solution for our scheduling problem. For instance, if we consider the precedence graph given in Figure 2, the optimal solution of the integer linear program will set all the arcs to 0. Clearly, this is not a feasible solution for our scheduling problem. However, our goal in this step is to get a good lower bound of the makespan and a solution –eventually not feasible– that we will transform to a feasible one (this transformation is given below).

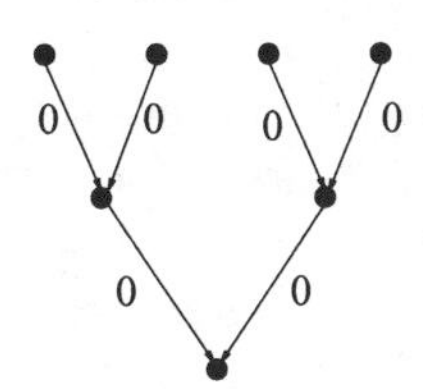

Fig. 2. *Our integer programming formulation does not always imply a feasible solution.*

Let Π^{inf} denotes the linear program corresponding to Π in which we relax the integer constraints $x_{ij} \in \{0, 1\}$ by setting $x_{ij} \in [0, 1]$. Given that the number of variables and the number of constraints are polynomially bounded, this linear program can be solved in polynomial time. The solution of Π^{inf} will assign to every arc $(i, j) \in E$ a value $x_{ij} = e_{ij}$ with $0 \leq e_{ij} \leq 1$ and will determine a lower bound of the value of C_{max} that we denote by Θ^{inf}.

Lemma 1. *Θ^{inf} is a lower bound of an optimal solution for $\bar{P}, 2|prec; (c_{ij}, \epsilon_{ij}) = (1, 0); p_i = 1|C_{max}$.*

Proof. This is true since any optimal feasible solution of the scheduling problem must satisfy all the constraints of the integer linear program Π. □

3 Obtaining a Feasible Solution

The algorithm is divided in two steps:

1. *Step 1* [Rounding]: We transform the solution of the relaxed linear program into an integer one in the following way:
   if $e_{ij} < 0.25$ (resp. $e_{ij} \geq 0.25$) then $x_{ij} = 0$ (resp. $x_{ij} = 1$).
   In the following, we call an arc $(i, j) \in E$ a *0-arc* (resp. *1-arc*) if $x_{ij} = 0$ (resp. $x_{ij} = 1$).
   The solution given by *Step 1* is not necessarily a feasible solution (take for instance the precedence graph of Figure 2), so we must transform it to a feasible one. Notice that the cases given in Figure 1 are eliminated by the linear program.
   In the next step we need the following definition.

   Definition 1. *A critical path with terminal vertex $i \in V$ is the longest path from an arbitrary source of G to task i. The length of a path is defined as the sum of the processing times of the tasks belonging to this path and of the values x_{ij} for every arc in the path.*

2. *Step 2* [Feasible Rounding]: We change the integer solution as follows:
   (a) If i is a source then we keep unchanged the values of x_{ij} obtained in *Step 1*.
   (b) Let i be a task such that all predecessors are already examined. Let A_i be the subset of incoming arcs of i belonging to a critical path with terminal vertex the task i.
      i. If the set A_i contains a *0-arc*, then all the outcoming arcs x_{ij} take the value 1.
      ii. If the set A_i does not contain any *0-arc* (all the critical incoming arcs are valued to 1), then the value of all the outcoming arcs x_{ij} remains the same as in *Step 1*, and all the incoming *0-arcs* are transformed to *1-arcs*.

      Remark: In *Step 2(b)ii* changing the value of an incoming *0-arc* to 1 does not increase the length of any critical path having as terminal vertex i, because it exists at least one critical path with terminal vertex i such that an arc $(j, i) \in E$ is valued by the linear program to at least 0.25 ($e_{ji} \geq 0.25$), and so x_{ji} is already equal to 1.

Feasibility

From the constraints of the linear program one can easily show the following lemma.

Lemma 2. *Every job $i \in V$ has at most two successors (resp. predecessors) such that $e_{ij} < 0.25$ (resp. $e_{ji} < 0.25$).*

From the previous lemma, it is clear that after the rounding procedure of *Step 1*, if we focus on the tasks of two arbitrary consecutive levels of G, we can easily obtain a feasible schedule of the tasks of these levels, by performing the tasks that are connected by 0-arcs on the same cluster and in consecutive times. Unfortunately, this is not the case if we consider the entire graph G (see for instance the example of Figure 2). Since after *Step 2*, there are no consecutive *0-arcs*, we avoid infeasibility, since locally –between two consecutive levels– we can always execute the tasks that are connected by *0-arcs* (in *Step 2* we do not add any new *0-arc*), and globally there are no more consecutive *0-arcs* (hence, we have the time to change cluster and communicate, if necessary).

4 Relative Performance of the Heuristic

First, we prove that 8/5 is an upper bound of ρ^h. Then, we show that this value is reached for a special class of graphs.

4.1 Upper Bound of the Relative Performance

Let us denote by t_i^h the starting time of the task i determined by the heuristic and by t_i^* the starting time of the task i given by the linear program (t_i^* is the longest path from a source to the task i including the processing time of the tasks and the real values of the corresponding arcs).

Lemma 3. *For every task $i \in V, t_i^h \leq \frac{8}{5}t_i^*$*

Proof. We use induction to prove it.

The inequality is true for every task $i \in Z$ (i.e. tasks such that $t_i^h = 0$) and for every task k such that $\Gamma^-(k) \subseteq Z$.

Let us now assume, that the lemma is valid for all the predecessors of the task i.

Let A_i be the set of the critical incoming arcs (i.e. the arcs having i as terminal vertex and belonging to a critical path). We have to consider the following cases:

1. One of the arc(s) of A_i denoted by (j, i) is valued to 0 ($x_{ji} = 0$, which means that $e_{ji} < 0.25$). So $t_i^h = t_j^h + 1$, and $t_i^* \geq t_j^* + 1$. According to the induction hypothesis we have $t_j^h \leq \frac{8}{5}t_j^*$.
   Thus $t_i^h \leq \frac{8}{5}t_j^* + 1$ and consequently $t_i^h \leq \frac{8}{5}(t_i^* - 1) + 1 \leq \frac{8}{5}t_i^*$.
2. One of the arc(s) of A_i denoted by (j, i) is valued to 1 ($x_{ji} = 1$) and $e_{ji} \geq 0.25$. So, $t_i^h = t_j^h + 2$, and $t_i^* \geq t_j^* + 1.25$. According to the induction hypothesis we have $t_j^h \leq \frac{8}{5}t_j^*$.
   Thus $t_i^h \leq \frac{8}{5}t_j^* + 2$ and consequently $t_i^h \leq \frac{8}{5}(t_i^* - 1.25) + 2 \leq \frac{8}{5}t_i^*$.

3. A_i contains an *1-arc* denoted by (j,i) such that $e_{ji} < 0.25$. We have $t_i^h = t_j^h + 2$, and $t_i^* \geq t_j^* + 1$.
   Notice that, the value associated to this *1-arc* has been transformed to 1 after the "study" of task j.
   So in the set A_j it exists a *0-arc*. W.l.o.g. we denote by (k,j) this arc, thus $t_j^h = t_k^h + 1$, and $t_j^* \geq t_k^* + 1$.
   According to the induction hypothesis we have $t_k^h \leq \frac{8}{5}t_k^*$. Thus, $t_i^h = t_k^h + 3$, and $t_i^* \geq t_k^* + 2$.
   Hence, we get $t_i^h \leq \frac{8}{5}t_k^* + 3$, and consequently, $t_i^h \leq \frac{8}{5}(t_i^* - 2) + 3 \leq \frac{8}{5}t_i^*$.

□

Finally, we obtain our main result:

Theorem 1. *The relative performance ρ^h of our heuristic is bounded above by $\frac{8}{5}$.*

Proof. Let us denote by C_{max}^h the makespan of the schedule computed by the heuristic and by C_{max}^{opt} the optimal value of a schedule.

Let us consider a task i of U such that $C_{max}^h = t_i^h + 1$. Then, according to Lemma 3, $C_{max}^h \leq \frac{8}{5}(t_i^* + 1)$. Moreover, $t_i^* + 1 \leq \Theta^{inf}$ and $\Theta^{inf} \leq C_{max}^{opt}$, so we get the theorem.

□

4.2 Tightness of the Bound

We recursively define a sequence of graphs $G_i, i \geq 1$ based on the graphs B_1 and B_2 given in Figures 3 and 4 respectively. The values near each task in Figures 3 and 4 correspond to its starting time.

We compute the value of the makespan obtained by our heuristic, denoted $C_{max}^h(G_i)$, and we propose a schedule σ such that:

$$\lim_{i \to \infty} \frac{C_{max}^h(G_i)}{C_{max}^\sigma(G_i)} = \frac{8}{5}.$$

Notation: In what follows, whenever we write: "$B_1@B_2$", we will consider the graph obtained by the concatenation of the graph B_1 and the graph B_2, in which we identify the tasks of the last level of $B_1$with the tasks of the first level of B_2. More precisely, we have $x_1' = r_1$ and $y_1' = r_2$.

The definition of $G_{i-1}@B_2$ is made in a similar way (the tasks of the last level of G_{i-1} are aggregated with the tasks of the first level of B_2 i.e. $q_1 = r_1$ and $q_2 = r_2$ with $q_1, q_2 \in V(G_{i-1})$ and $r_1, r_2 \in V(B_2)$).

G_i is recursively defined in the following way:

- $G_1 = B_1@B_2$.
- and $G_i = G_{i-1}@B_2$, with $i \geq 2$.

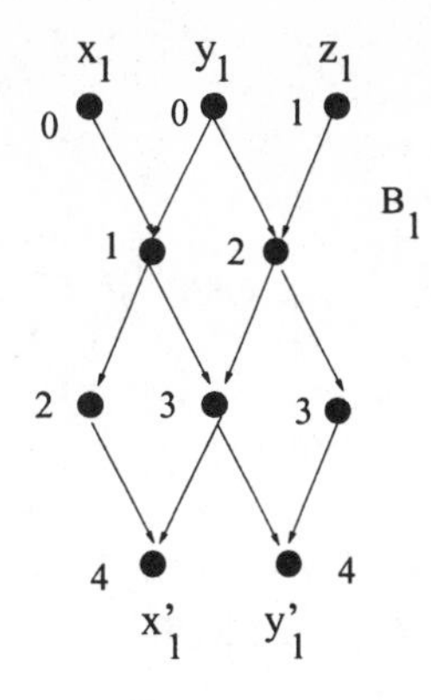

Fig. 3. *The graph B_1 and the associated schedule $\sigma(B_1)$.*

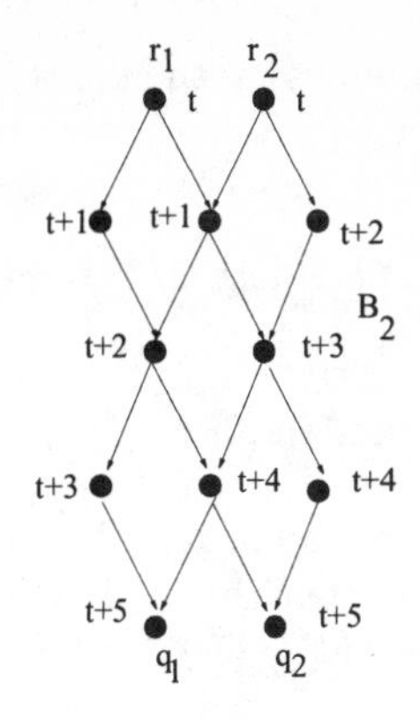

Fig. 4. *The graph B_2 and the associated schedule $\sigma(B_2)$.*

Makespan for Scheduling the Tasks of G_i

Lemma 4. *The makespan for the graph B_1 (resp. B_2) obtained by the heuristic is equal to $C^h_{max}(B_1) = 7$ (resp. $C^h_{max}(B_2) = 9$).*

Proof. The solution of the relaxed linear program will assign the value 0.25 to all the arcs of B_i, $i = 1, 2$. Thus, during the first step, the heuristic will transform all these values to 1, and hence the makespan will be equal to 7 (resp. 9).

□

Lemma 5. *The makespan for G_i given by the heuristic is equal to $C^h_{max}(G_i) = 8i + 7$.*

Proof. By induction on i.

- If $i = 1$ the lemma is valid: $C^h_{max}(G_1) = C^h_{max}(B_1) + C^h_{max}(B_2) - 1 = 15$.
- We assume that the lemma is valid for $i - 1, i \geq 2$ i.e. $C^h_{max}(G_{i-1}) = 8(i - 1) + 7$.
  We have $C^h_{max}(G_i) = C^h_{max}(G_{i-1}) + C^h_{max}(B_2) - 1 = 8(i-1) + 7 + 9 - 1 = 8i + 7$.

□

Let us now construct a better schedule that we call σ. We built σ recursively:

- $\sigma(G_1)$ is obtained by concatenating the schedules of B_1 and B_2 (see Figures 3 and 4) taking of course into account the aggregation of the tasks in $G_1 = B_1 @ B_2$ (given that $x'_1 = r_1$ and $y'_1 = r_2$, we have $t_{x'_1} = t_{r_1}$ and $t_{y'_1} = t_{r_2}$).
- Similary, $\sigma(G_i)$ is obtained by concatenating carefully the schedules of G_{i-1} and of B_2 (taking again into account the aggregation of tasks).

Lemma 6. *The makespan for the graph B_1 (resp. B_2) obtained by σ is equal to $C^\sigma_{max}(B_1) = 5$ (resp. $C^\sigma_{max}(B_2) = 6$).*

Proof. It is obvious by the construction. □

Lemma 7. *The length of the schedule σ for the graph G_i, $C^{\sigma}_{max}(G_i)$, is equal to $5i+5$.*

Proof. By induction on i.

- If $i=1$ the lemma is valid ($C^{\sigma}_{max}(G_1)=10$).
- We assume that the lemma is valid for $i-1, i \geq 2$, i.e. $C^{\sigma}_{max}(G_{i-1}) = 5(i-1)+5$. Since $C^{\sigma}_{max}(B_2)=6$, we obtain $C^{h}_{max}(G_i) = 5(i-1)+5+6-1 = 5i+5$.

□

Theorem 2. *The bound $\rho^h = \frac{8}{5}$ is reached for G_i.*

Proof. By the Lemmas 5 and 7, we have $C^{h}_{max}(G_i) = 8i+7$ and $C^{\sigma}_{max}(G_i) = 5i+5$.

So,

$$\lim_{i\to\infty} \frac{C^{h}_{max}(G_i)}{C^{\sigma}_{max}(G_i)} = \frac{8i}{5i} = \frac{8}{5}.$$

□

5 Extended Model

In this section, we consider an extension of the studied problem where each cluster contains m identical processors, with $m \geq 1$ a fixed constant $(\bar{P}, m|prec; (c_{ij}, \epsilon_{ij}) = (1,0); p_i = 1|C_{max})$.

In order to treat this problem we consider a generalization of the integer linear program presented in Section 2. We have to extend the two cases of Figure 1 by considering the cases given in Figure 5.

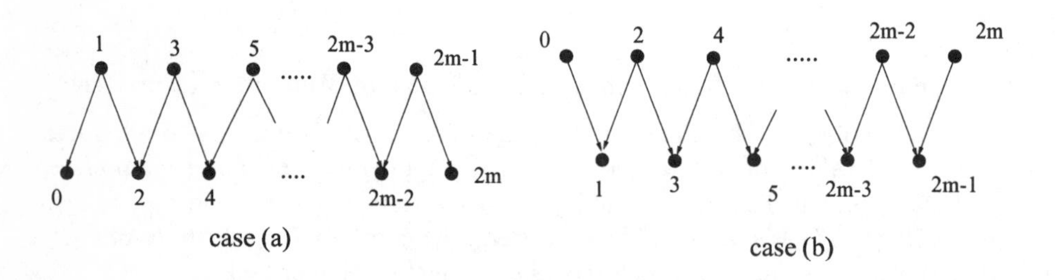

Fig. 5. *Two extended cases.*

- For the case (a):
  For every i even, with $0 \leq i \leq 2m-2$, such that $(i+1, i) \in E$, and i odd with $1 \leq i \leq 2m-1$ such that $(i, i+1) \in E$,

$$\sum_{\substack{0 \leq i \leq 2m-2 \\ i \text{ even}}} x_{(i+1)i} + \sum_{\substack{1 \leq i \leq 2m-1 \\ i \text{ odd}}} x_{i(i+1)} \geq 1.$$

- For the case (b):
  For every i even, with $0 \leq i \leq 2m-2$, such that $(i+1, i) \in E$, and i odd with $1 \leq i \leq 2m-1$ such that $(i, i+1) \in E$,

$$\sum_{\substack{0 \leq i \leq 2m-2 \\ i \text{ even}}} x_{i(i+1)} + \sum_{\substack{1 \leq i \leq 2m-1 \\ i \text{ odd}}} x_{(i+1)i} \geq 1.$$

Thus, in what follows, we consider the following integer linear programming (ILP) problem:

$$(\Pi_1) \begin{cases} & \min C_{max} \\ \forall (i,j) \in E, & x_{ij} \in \{0,1\} \\ \forall i \in V, & t_i \geq 0 \\ \forall (i,j) \in E, & t_i + 1 + x_{ij} \leq t_j \\ \forall (i,j) \in E - U, & \sum_{j \in \Gamma^+(i)} x_{ij} \geq |\Gamma^+(i)| - m \\ \forall (i,j) \in E - Z, & \sum_{j \in \Gamma^-(i)} x_{ji} \geq |\Gamma^-(i)| - m \\ \text{for the cases (a)} & \sum_{\substack{0 \leq i \leq 2m-2 \\ i \text{ even}}} x_{(i+1)i} + \sum_{\substack{1 \leq i \leq 2m-1 \\ i \text{ odd}}} x_{i(i+1)} \geq 1 \\ \text{for the cases (b)} & \sum_{\substack{0 \leq i \leq 2m-2 \\ i \text{ even}}} x_{i(i+1)} + \sum_{\substack{1 \leq i \leq 2m-1 \\ i \text{ odd}}} x_{(i+1)i} \geq 1 \\ \forall i \in V, & t_i + 1 \leq C_{max} \end{cases}$$

We use the same heuristic with two steps except that for the *Step 1*, the rounding is: if $e_{ij} < \frac{1}{2m}$ then $x_{ij} = 0$, otherwise $x_{ij} = 1$.

Notice that, in the case of $m = 1$, the *Step 2* of the algorithm in Section 3 is useless.

Lemma 8. *For every job* $i \in V, t_i^h \leq 2 - \frac{2}{2m+1} t_i^*$

The proof of Lemma 8 is similar to the proof of Lemma 3 by replacing $\frac{8}{5}$ by $2 - \frac{2}{2m+1}$ and 0.25 by $\frac{1}{2m}$ and 1.25 by $1 + \frac{1}{2m}$.
Finally, we obtain the following result:

Theorem 3. *The relative performance* ρ^h *of our heuristic is bounded above by* $2 - \frac{2}{2m+1}$ *and the bound is tight.*

Thus, in the case of $m = 1$ we get the relative performance for the $\bar{P}|prec; c_{ij} = 1; p_i = 1|C_{max}$ problem given by Munier and König [8].

6 Conclusions

In this paper, we gave an approximation algorithm for $\bar{P}, 2|prec; (c_{ij}, \epsilon_{ij}) = (1, 0); p_i = 1|C_{max}$, with relative performance equal to $\frac{8}{5}$. Recall that there is no hope to find a heuristic with relative performance guarantee less than $\frac{5}{4}$ (unless $\mathcal{P} = \mathcal{NP}$) [2]. Our approach is also extended for the more interesting, from a practical point of view, problem $\bar{P}, m|prec; (c_{ij}, \epsilon_{ij}) = (1, 0); p_i = 1|C_{max}$, i.e. for the case where the number of processors in each cluster is any fixed constant. In that case, the performance ratio is a function of the number of processors.

It would be interesting to extend the heuristic presented here in the case where the number of clusters is limited or to develop other heuristics improving the obtained bound. Another extension of this work will concern the well known *small communication case* [4].

References

[1] E. Bampis, R. Giroudeau, and J.-C. König. Using duplication for the precedence constrained multiprocessor scheduling problem with hierarchical communications. In P. Amestoy, P. Berger, M. Daydé, I. Duff, V. Frayssé, L. Giraud, and D. Ruiz, editors, *EuroPar'99 Parallel Processing*, Lecture Notes in Computer Science, No. 1685, pages 369–372. Springer-Verlag, 1999.

[2] E. Bampis, R. Giroudeau, and J.C. König. On the hardness of approximating the precedence constrained multiprocessor scheduling problem with hierarchical communications. Technical Report 34, Université d'Evry Val d'Essonne,submitted, 1999.

[3] B. Chen, C.N. Potts, and G.J. Woeginger. A review of machine scheduling: complexity, algorithms and approximability. Technical Report Woe-29, TU Graz, 1998.

[4] P. Chrétienne, E.J. Coffman Jr, J.K. Lenstra, and Z. Liu. *Scheduling Theory and its Applications.* Wiley, 1995.

[5] M.R. Garey and D.S. Johnson. *Computers and Intractability, a Guide to the Theory of NP-Completeness.* Freeman, 1979.

[6] R.L. Graham, E.L. Lawler, J.K. Lenstra, and A.H.G. Rinnooy Kan. Optimization and approximation in deterministics sequencing and scheduling theory : a survey. *Ann. Discrete Math.*, 5:287–326, 1979.

[7] J.A. Hoogeveen, J.K. Lenstra, and B. Veltman. Three, four, five, six, or the complexity of scheduling with communication delays. *O. R. Lett.*, 16(3), 1994.

[8] A. Munier and J.C. König. A heuristic for a scheduling problem with communication delays. *Operations Research*, 45(1):145–148, 1997.

[9] C. Picouleau. *Etude des problèmes d'optimisation dans les systèmes distribués.* PhD thesis, Université de Paris VI, 1992.

[10] B. Veltman. *Multiprocessor scheduling with communications delays.* PhD thesis, CWI-Amsterdam, Holland, 1993.

Polynomial Time Approximation Schemes for the Multiprocessor Open and Flow Shop Scheduling Problem*

Klaus Jansen[1] and Maxim I. Sviridenko[2]

[1] Institut für Informatik und praktische Mathematik,
Christian-Albrechts-Universität zu Kiel, Olshausenstr. 40, 24098 Kiel, Germany,
kj@informatik.uni-kiel.de

[2] University of Aarhus, BRICS, Ny Munkengade 540, 8000 Aarhus C, Denmark,
sviri@brics.dk

Abstract. We investigate the multiprocessor multi-stage open shop and flow shop scheduling problem. In both problems, there are s stages each consisting of a number m_i of parallel identical machines for $1 \leq i \leq s$. Each job consists of s operations with one operation for each stage. The goal is to find a non-preemptive schedule that minimizes the makespan. We propose polynomial time approximation schemes for the multiprocessor open shop and flow shop scheduling problem when the number of stages s is constant and the numbers of machines m_i are non-constant.

1 Introduction

Problem Definition. A flow shop (or open shop) is a multi-stage production process with the property that all jobs have to pass through the stages. For flow shops the order in which the jobs pass through the stages is the same, whereas for open shops the order is immaterial. There are n jobs J_j, with $j = 1, \ldots, n$, where each job J_j consists of s operations $O_{1j}, \ldots, O_{sj}$. The operation O_{ij}, with $i = 1, \ldots, s$, has to be processed at stage i of the production process, and p_{ij} is the processing time or length of operation O_{ij}.

In the classical open and flow shop problem, there is only one machine available for each stage. In the multiprocessor open and flow shop problem, for every stage i there are m_i identical machines available that can process operations in parallel. Since more than n machines on a stage are not necessary, we may assume that $m_i \leq n$. At any time step, every job is processed by at most one machine and every machine executes at most one job. We assume that preemption is not allowed, i.e. once an operation is started, it must be completed without interruption. The goal is to find a schedule that minimizes the makespan C_{max},

* This work was done while the first author was associated with the research instutute IDSIA Lugano and supported in part by the Swiss National Science Foundation project 21-55778.98, "Resource Allocation and Scheduling in Flexible Manufacturing Systems".

H. Reichel and S. Tison (Eds.): STACS 2000, LNCS 1770, pp. 455–465, 2000.

that is the maximum completion time among all jobs. The minimum makespan among all schedules is denoted by C^*_{max}.

Following the three-field notation scheme [10], the makespan minimization problem in a classical open and flow shop with s stages is denoted by $Os||C_{max}$ and $Fs||C_{max}$ (or $O||C_{max}$ and $F||C_{max}$ depending on whether the number s of stages is constant or not), respectively. The makespan minimization in a multiprocessor s-stage open and flow shop is denoted by $Os(P)||C_{max}$ and $Fs(P)||C_{max}$ (or $O(P)||C_{max}$ and $F(P)||C_{max}$), respectively.

Complexity Results. Gonzales and Sahni [5] proved that $Os||C_{max}$ is NP-hard in the weak sense, and Williamson et al. [15] showed that $O||C_{max}$ is NP-hard in the strong sense. But it is not known whether the problem $Os||C_{max}$ allows a pseudopolynomial time algorithm or whether the problem $Os||C_{max}$ is NP-hard in the strong sense. On the other hand, Garey et al. [4] showed that $F3||C_{max}$ is strongly NP-hard, and Hoogeveen et al. [8] proved that $F2(P2, P1)||C_{max}$ (with two stages, two machines on the first stage and one machine on the second stage) and $F2(P1, P2)||C_{max}$ are already strongly NP-hard. If we have only one stage $s = 1$, then we have the classical strongly NP-hard scheduling problem $P||C_{max}$ of independent jobs on identical machines [3].

A polynomial time approximation scheme (PTAS) for a (minimization) optimization problem P is an algorithm that given any constant value $\varepsilon > 0$ finds in polynomial time a solution of value no larger than $1 + \varepsilon$ times the value of an optimum solution. A fully polynomial time approximation scheme is an approximation scheme that runs in time polynomial in the size of the input and $1/\varepsilon$.

Approximability Results. If the number s of stages is part of the input, Williamson et al. [15] proved that the existence of an approximation algorithm with worst case ratio $< 5/4$ for the problem $O||C_{max}$ or $F||C_{max}$ would imply $P = NP$. On the positive side, Hall [6] and Sevastianov and Woeginger [13] have proposed polynomial time approximation schemes (PTAS) for $Fs||C_{max}$ and $Os||C_{max}$, respectively. Both PTAS's can be generalized to the case where the number of stages and number of machines per stage are all constant. Furthermore, Hochbaum and Shmoys [7] have given a polynomial time approximation scheme for the problem with one stage that is equivalent to $P||C_{max}$.

Chen and Strusevich [2] have developed an approximation algorithm for $O(P)||C_{max}$ (the multiprocessor open shop problem) with worst case ratio $2+\epsilon$. For $O2(P)||C_{max}$, they have derived a worst case ratio of $2 - 2/m^2$ where $m = \max(m_1, m_2) \geq 2$. Schuurman and Woeginger [11] have found an approximation algorithm for $O(P)||C_{max}$ with improved worst case ratio 2. Furthermore, a $(3/2 + \epsilon)$ - approximation algorithm for the problem $O2(P)||C_{max}$ (with two stages) is given in [11]. The existence of an approximations scheme for $Os(P)||C_{max}$ with constant $s \geq 2$ number of stages and arbitrary number of machines per stage was posed as an open problem by Schuurman and Woeginger [11].

Several approximation algorithms have been studied for the two - stage multiprocessor flow shop problem, see e.g. in [1,12,14]. The best result, a polynomial time approximation scheme for $F2(P)||C_{max}$ is given in [12]. Determining the approximability behaviour of $Fs(P)||C_{max}$ with constant $s \geq 3$ number of stages and arbitrary number of machines per stage was posed as an open question in a paper by Hall [6].

New Results. In this paper, we propose polynomial time approximation schemes for both the multiprocessor open shop and flow shop scheduling problem when the number of stages s is constant and the numbers of machines m_i on stage i, $1 \leq i \leq s$ are part of the input. For open shops, this improves even for two stages the best previous known result of $3/2 + \epsilon$ in [11]. Furthermore, we answer the open question by Schuurman and Woeginger [11] for $Os(P)||C_{max}$ and the open question by Hall [6] for $Fs(P)||C_{max}$. Notice that we can not expect a fully polynomial time approximation scheme (since both problems are strongly NP-hard even for a constant number of stages), unless P=NP.

Interestingly, we do not use linear programming. In our approach, we use dynamic programming combined with several ideas from Hall [6], Hochbaum and Shmoys [7], Schuurman and Woeginger [11,12] and Sevastianov and Woeginger [13].

2 Restricted Problem

Let $L_i = \sum_{j=1}^{n} p_{ij}/m_i$ be the average load on stage i, and let $p_i^{(max)} = \max_{1 \leq j \leq n} p_{ij}$ be the maximum processing time of any operation on stage i. For the multiprocessor flow and open shop problem, we can derive the following bounds for the minimum makespan C^*_{max}:

$$\max_{1 \leq i \leq s} \max\left\{L_i, p_i^{(max)}\right\} \leq C^*_{max} \leq \sum_{i=1}^{s}(L_i + p_i^{(max)}).$$

The lower bound should be clear. The ideas for the upper bound are that we can use list scheduling to obtain a schedule with makespan at most $L_i + (1 - \frac{1}{m_i})p_i^{(max)} \leq L_i + p_i^{(max)}$ for operations on stage i and that we can schedule all stages one after another. By dividing all processing times by $2s \max_{1 \leq i \leq s} \max\left\{L_i, p_i^{(max)}\right\}$ we get an instance $\bar{I}$ such that $\frac{1}{2s} \leq \bar{C}^*_{max} \leq 1$. In the rest of the paper we assume without loss of generality that the minimum makespan C^*_{max} satisfies the inequalities $\frac{1}{2s} \leq C^*_{max} \leq 1$.

Following [13], for real numbers $\kappa, \delta > 0$ we define three sets of operations:

$$\begin{aligned} B &= \{O_{ij} | p_{ij} > \kappa\}, \\ M &= \{O_{ij} | \delta\kappa < p_{ij} \leq \kappa\}, \\ S &= \{O_{ij} | p_{ij} \leq \delta\kappa\}. \end{aligned}$$

We will assume that $\delta < 1$ and that $1/\delta^{1/2}$ is an integer; we will choose δ later in Section 4. The operations in B are called *big operations*, the operations in M are called *medium operations* and operations in S are called *small operations*.

In the following, we consider an optimum schedule $\mathcal{S}$ of length $C^*_{max} \leq 1$. We denote with τ_{ij} the *starting time* of operation O_{ij} in $\mathcal{S}$. We partition the schedule $\mathcal{S}$ into intervals of length $\delta\kappa$ and increase each cut (between two intervals) by an interval of length $2\delta^{3/2}\kappa$. Using such enlarged cuts, the processing times of the big and medium operations are increased. We use the following convention: If we cut through the endpoint of an operations or in the middle, then we will enlarge the processing time of the operation by $2\delta^{3/2}\kappa$. If we cut through the startpoint of an operation, then we delay the starting time by $2\delta^{3/2}\kappa$. Let $\bar{\mathcal{S}}$ be the enlarged schedule, and let $\bar{\tau}_{ij}$ be the starting time of O_{ij} in $\bar{\mathcal{S}}$.

Since the number of cuts is at most $\lceil \frac{1}{\delta\kappa} \rceil - 1 \leq \frac{1}{\delta\kappa}$ and each cut is replaced by an interval of length $2\delta^{3/2}\kappa$, the total length of the enlarged schedule $\bar{\mathcal{S}}$ can be bounded by

$$C^*_{max} + \frac{1}{\delta\kappa} \cdot 2\delta^{3/2}\kappa \leq C^*_{max} + 2\delta^{1/2}.$$

If $2\delta^{1/2}$ is small enough (e.g. $2\delta^{1/2} \leq \frac{\epsilon}{8s}$ or equivalent $\delta \leq \frac{\epsilon^2}{(16s)^2}$), then we get an additive factor of at most $\frac{\epsilon}{8s} \leq \frac{\epsilon}{4} C^*_{max}$.

Then, each medium operation O_{ij} is cut at least once (i.e. O_{ij} lies in more than one interval of length $\delta\kappa$). This implies that the processing time of each medium operation O_{ij} is increased by a factor of at least $2\delta^{3/2}\kappa$. In other words, there is now a *time window* of size $\geq p_{ij} + 2\delta^{3/2}\kappa$ where we can shift operation O_{ij} (with length p_{ij}) without generating conflicts between O_{ij} and other operations $O_{i'j}$, $i \neq i'$. The time window of O_{ij} has the form $[\bar{\tau}_{ij}, \bar{\tau}_{ij} + p_{ij} + 2x_{ij}\delta^{3/2}\kappa]$ with $x_{ij} \in \mathbb{N}$ and $x_{ij} \geq 1$. We can shift each medium operation O_{ij} in its corresponding time window such that it starts at a time τ'_{ij} which is a multiple of $\delta^{3/2}\kappa$. Furthermore, we can increase the processing time p_{ij} of O_{ij} to p'_{ij} such that $p'_{ij} \geq p_{ij}$ is a multiple of $\delta^{3/2}\kappa$.

For a big operation $O_{ij} \in B$, the situation is similar. Since $p_{ij} > \kappa$, O_{ij} is cut at least $\frac{1}{\delta}$ times and the corresponding time window is increased by a factor of at least $\frac{1}{\delta} 2\delta^{3/2}\kappa = 2\delta^{1/2}\kappa$. Therefore, we can assume that a big operation starts processing at a time τ'_{ij} which is a multiple of $\delta^{1/2}\kappa$ and has processing length $p'_{ij} \geq p_{ij}$ which is also a multiple of $\delta^{1/2}\kappa$.

We also round the processing times of small operations up to the nearest multiplies of $\frac{2\delta^{1/2}}{ns}$. Since in any critical path there are at most ns operations then the length of the schedule will not increase considerably (we add at most $2\delta^{1/2}$ to the length of the schedule). In the following, we will consider only *restricted schedules* with the properties:

- each medium operation $O_{ij} \in M$ starts at a time which is a multiple of $\delta^{3/2}\kappa$,
- each medium operation $O_{ij} \in M$ has processing time $p'_{ij} = \delta^{3/2}\kappa\ell_{ij}$ with $\ell_{ij} \in \mathbb{N}$,
- each big operation $O_{ij} \in B$ starts at a time which is a multiple of $\delta^{1/2}\kappa$,
- each big operation $O_{ij} \in B$ has processing time $p'_{ij} = \delta^{1/2}\kappa\ell_{ij}$ with $\ell_{ij} \in \mathbb{N}$,
- each small operation $O_{ij} \in S$ has processing time $p'_{ij} = \frac{2\delta^{1/2}}{ns}\ell_{ij}$ with $\ell_{ij} \in \mathbb{N}$.

The processing times satisfy the conditions $p'_{ij} \geq p_{ij}$ for $O_{ij} \in B \cup M \cup S$, $p'_{ij} - p_{ij} \leq \delta^{3/2}\kappa$ for $O_{ij} \in M$ and $p'_{ij} - p_{ij} \leq \delta^{1/2}\kappa$ for $O_{ij} \in B$. Using the fact $C^*_{max} \in [\frac{1}{2s}, 1]$ and the choice of δ above, the optimum restricted schedule has length at most $C^*_{max} + 4\delta^{1/2} \leq (1+\frac{\epsilon}{2})C^*_{max} \leq 2$ (for $\epsilon \leq 2$). In the next Section we present an algorithm to compute restricted schedules based on dynamic programming.

3 Computation of Restricted Schedules

3.1 Starting Intervals of Operations

We have two types of intervals in our restricted schedule (of total length ≤ 2):

- intervals of the *first type* with length $\delta^{1/2}\kappa$,
- intervals of the *second type* with length $\delta^{3/2}\kappa$.

Since $\delta < 1$, the intervals of the second type are smaller. In fact, an interval of the first type consists of $1/\delta$ many intervals of the second type. We assume that $\frac{1}{\delta}$ and $\mu = \frac{2}{\delta^{1/2}\kappa}$ are integer numbers (see also Section 4). The number μ gives an upper bound on the number of intervals of the first type. The number of intervals of the second type is bounded by $\frac{2}{\delta^{3/2}\kappa} \leq \frac{\mu}{\delta}$. The goal is to find a restricted schedule with a minimum number of used intervals of the first type.

For each job J_j, we use vectors $K_j = (\tau_1, \ldots, \tau_s)$ of intervals of the second type for the operations with $\tau_i \in \{0, \ldots, \frac{\mu}{\delta} - 1\}$ for $1 \leq i \leq s$ (to indicate the starting interval). If $O_{ij} \in B \cup S$, then we require that the τ_i is even a multiple of $\frac{1}{\delta}$. These assumptions for big operations arise directly from the restricted schedules. A big operation $O_{ij} \in B$ starts only at the beginning of an interval of the first type. A medium operations $O_{ij} \in M$ starts at the beginning of an interval of the second type. Therefore, we can use integers for the values τ_i. For a small operation $O_{ij} \in S$ the condition 'τ_i is a multiple of $\frac{1}{\delta}$' means that we fix an interval of the first type where O_{ij} has to be started. The number of different vectors K_j is bounded by a constant $O\left((\frac{\mu}{\delta})^s\right)$. A vector $K_j = (\tau_1, \ldots, \tau_s)$ is *feasible* in the flow shop environment, if and only if:

(1) $\tau_1 \leq \tau_2 \leq \ldots \leq \tau_s$
(2) the processing intervals for operations $O_{ij} \in B$ and $O_{i'j} \in B \cup M \cup S$ (or $O_{ij} \in M$ and $O_{i'j} \in M$), $i \neq i'$ do not intersect,
(3) if $O_{ij} \in M$, $O_{i'j} \in S$ and $i < i'$, then $O_{i'j}$ has to start in the interval (of the first type) where O_{ij} ends or afterwards,
(4) if $O_{ij} \in S$ and $O_{i'j} \in M$ (and $i < i'$), then O_{ij} must start in the interval (of the first type) where $O_{i'j}$ begins or before, i.e. $\tau_i < \tau_{i'}$.

In general, these properties do not give directly a feasible schedule: two small operations (or a small and medium operation) may be executed at the same time or in the wrong order. On the other hand, each job in a feasible restricted schedule satisfies these conditions. We show later how we generate a feasible approximate schedule from such a solution. Similar conditions about feasible combinations can be given also for open and job shops.

3.2 Load and Configuration

The *load* $L_{g,i}$ is the total sum of processing times p_{ij} of small operations O_{ij} assigned to start in interval g (of the first type) on stage i. Since in an optimum restricted schedule, a small operation may be executed in two consecutive intervals g and $g+1$, the maximum allowed load $L_{g,i}$ is at most $m_i(\delta^{1/2}\kappa + \delta\kappa)$ for $0 \le g \le \mu - 1$. Using $m_i \le n$ and $\delta < 1$, the values $L_{g,i}$ are bounded above by $2n\delta^{1/2}\kappa$. We have added the value $\delta\kappa$ for the last small operation on each of the m_i machines. Later, we will observe that some of these m_i machines can not be used for a small operation. For instance, a big operation can cover several intervals of the first type. Therefore, we can place a small operation $O_{ij} \in S$ only onto a machine (on stage i) with at least one idle period. Since the processing times of small operations are multiplies of $\frac{2\delta^{1/2}}{ns}$, we have only discrete and at most $O(n^2)$ different load values for an interval (here we use the bounds $L_{g,i} \le 2n\delta^{1/2}\kappa$). We store in our computation a *load vector* $\boldsymbol{L} = (L_{0,1}, \ldots, L_{0,s}, \ldots, L_{\mu-1,s})$. The maximum number of different load vectors is $O(n^{2s\mu})$.

The key observation is that we can have only a constant number of big and medium operations on each machine. The maximum number is at most $\frac{2}{\delta\kappa}$, since $C^*_{max} \le 2$ and the processing times $p'_{ij} \ge p_{ij} \ge \delta\kappa$ for $O_{ij} \in B \cup M$. Furthermore, there is only a constant number $\le \frac{2}{\delta^{3/2}\kappa} = \frac{\mu}{\delta}$ of starting times for big and medium operations (or intervals of the second type) in a restricted schedule. A *schedule type* is described by a set of intervals of the second type (with length $\delta^{3/2}\kappa$) where big and medium operations must be processed. We have only a constant number $T \le 2^{\frac{\mu}{\delta}}$ of different schedule types. Let $S_1, \ldots, S_T$ be the different *schedule types* in a restricted schedule. The first schedule type $S_1 = \emptyset$ corresponds to a free machine (without any assigned big or medium operation). For each stage i, $1 \le i \le s$, in the dynamic program we compute a vector $a^{(i)} = (a^{(i)}_1, \ldots, a^{(i)}_T)$ where $a^{(i)}_k \ge 0$ gives the number of machines with schedule type S_k on stage i. Notice, that $\sum_{k=1}^{T} a^{(i)}_k = m_i$.

A *configuration* given by s vectors $a^{(1)}, \ldots, a^{(s)}$ describes completely the set of intervals on each stage which are occupied by big and medium operations and the set of intervals which may be filled by small operations. Notice also that a configuration does not give an assignment for the big and medium operations. We only fix intervals for processing big and small operations. The total number of different configurations is bounded by $(\max_{1 \le i \le s} m_i)^{T \cdot s} \le n^{T \cdot s}$.

3.3 Dynamic Programming

In this section we present a dynamic programming algorithm for computing restricted schedules. Recall that a restricted schedule may not be feasible for the original shop problem. In the Section 4 we give an algorithm for converting a restricted schedule into feasible one.

Our algorithm consists of two phases. In the first phase we compute a table with elements $(a^{(1)}, \ldots, a^{(s)}, \boldsymbol{L}, k)$. Each element contains an answer on the question: 'Is there an assignment of starting times $K_1, \ldots, K_k$ to jobs $J_1, \ldots, J_k$

and an assignment of machines to the operations of these jobs such that big and medium operations are processed accordingly to the configuration $a^{(1)}, \ldots, a^{(s)}$ and the total length of the small operations assigned to interval g (of the first type) on stage i is $L_{g,i}$?' We will show in the next section that we can obtain a near optimal solution using such an assignment. Notice also, that if there is a restricted schedule of jobs $J_1, \ldots, J_k$ with configuration $a^{(1)}, \ldots, a^{(s)}$ and load vector $\boldsymbol{L}$, then the corresponding element $(a^{(1)}, \ldots, a^{(s)}, \boldsymbol{L}, k)$ must contain answer 'Yes'. In general, the converse statement does not hold. The number of elements in the table is bounded above by $O(n^{2s\mu+sT+1})$. Notice also that (in the first phase) we do not compute the restricted schedules explicitly.

In the second phase we compute an element $(a^{(1)}, \ldots, a^{(s)}, \boldsymbol{L}, n)$ which contains 'Yes' with feasible load vector $\boldsymbol{L}$ (see below) and minimum makespan, i.e. with a minimum number of intervals of the first type used in the configuration and load vector. Let $l_{g,i}$ be the total load in interval g of the first type on stage i reserved for big and medium operations (we can simply compute $l_{g,i}$ from the vector of schedule types $a^{(i)}$). Furthermore, let $m_{g,i} \leq m_i$ be the number of machines with at least one idle time between operations from $B \cup M$ in interval g on stage i. Then a load vector is *feasible* if $L_{g,i} \leq m_i \delta^{1/2}\kappa + m_{g,i}\delta\kappa - l_{g,i}$ for all g and i (we have added $m_{g,i}\delta\kappa$ since some small operations may start at the very end of an interval). Since the number of elements in the table is polynomial, we can find a feasible element with smallest makespan (number of intervals) in polynomial time.

After that we can simply compute an assignment of starting times to jobs and an assignment of machines to operations using an iterative procedure. Given an assignment of starting times $K_1, \ldots, K_n$ and vectors $T_1, \ldots, T_n$ (of indices of schedule types) we can compute a restricted schedule (including an assignment of machines to the big and medium operations) with almost the same makespan. After that we convert the obtained restricted schedule into a feasible schedule of original shop problem (see next Section).

4 Transformation into a Feasible Schedule

4.1 Multiprocessor Flow Shop

Let us consider a machine M on stage i and an interval g (of the first type). The *first step* is to shift all medium operations (which start in this interval) consecutively to the left side of interval g on M (only the last operation will not be shifted if it is executed also in interval $g+1$). Using such a transformation we may generate new infeasibilities between operations of the same job (on different stages). But using this transformation, for every machine M we have at most one time interval in interval g where M is idle.

In the *second step*, we place all small operations into these gaps. Remember that the load $L_{g,i}$ plus the (partial) processing times of medium operations assigned to interval g (on machines of stage i with at least one idle period) is at most $m_{g,i}(\delta^{1/2}\kappa + \delta\kappa)$. If we increase the length of the interval by $2\delta\kappa$, then we can place all small operations completely and without interruptions by a greedy algorithm into the $m_{g,i}$ idle periods of the enlarged interval. The generated schedule is a restricted schedule which is in general not feasible for the original shop problem.

Since the number of intervals of the first type is at most $\mu = 2/(\delta^{1/2}\kappa)$, the total length of the schedule is enlarged by at most $\mu \cdot 2\delta\kappa = 4\delta^{1/2}$. The *last step* is now to delay all machines on stage $1 \leq i \leq s$ by a time of $(\delta^{1/2}\kappa + 2\delta\kappa)(i-1)$. This idea called *sliding* was also used by Hall [6] for classical flow shops. Since there are infeasibilities only between operations of the same job which are processed in the same interval of the first type, (using this sliding) we get a feasible schedule for all jobs and the length of the schedule is increased by at most $(\delta^{1/2}\kappa + 2\delta\kappa)(s-1) \leq 3(s-1)\delta^{1/2}\kappa$. Notice, that the computed restricted schedule has only a minimum number of used interval of the first type and that the load of the small jobs (in the last interval) could also increase the makespan in comparison to the optimum restricted schedule. Therefore, we have to take into account the length $\delta^{1/2}\kappa + 2\delta\kappa \leq 3\delta^{1/2}\kappa$ of the last interval. Finally, we have to add also the value $2\delta^{1/2}$ caused by the consideration of only restricted schedules.

It remains to choose δ and κ and to bound the additive factor $3s\delta^{1/2}\kappa + 4\delta^{1/2}$. Using $\kappa = 1$ (so we do not use big operations for the multiprocessor flow shop problem), we get the additive factor $(3s+4)\delta^{1/2}$. Using $\delta \leq (\frac{\epsilon}{2s(3s+4)})^2$, we get $(3s+4)\delta^{1/2} \leq \frac{\epsilon}{2s} \leq \epsilon C^*_{max}$. Therefore, we define

$$\delta = \left(\frac{1}{\lceil\frac{2s(3s+4)}{\epsilon}\rceil}\right)^2 \leq \left(\frac{\epsilon}{2s(3s+4)}\right)^2$$

and have the property that $\frac{1}{\delta^{1/2}}$ is integral. Since $\kappa = 1$, the number $\mu = \frac{2}{\delta^{1/2}\kappa}$ is also integral.

4.2 Multiprocessor Open Shop

The *first step* for this problem is to find a value for κ such that the average loads of medium operations on each stage are bounded by a constant $\leq \frac{\epsilon}{4s} \leq \frac{\epsilon}{2} C^*_{max}$ or by a smaller constant $\alpha \leq \frac{\epsilon}{4s}$. We define a sequence of blocks as follows:

$$\begin{aligned} B_0 &= \{O_{ij} | p_{ij} \geq \delta\}, \\ B_1 &= \{O_{ij} | \delta^2 \leq p_{ij} < \delta\}, \\ &\dots \\ B_t &= \{O_{ij} | \delta^{t+1} \leq p_{ij} < \delta^t\}, \end{aligned}$$

with $t \geq 1$ where δ is a value that depends on ϵ and s. We will choose δ later. The average load $L_i(B_t)$ of operations in B_t on stage i is given by $\sum_{j|O_{ij} \in B_t} p_{ij}/m_i$. Since

$$\sum_{i=1}^{s} \sum_{t \geq 1} L_i(B_t) = \sum_{i=1}^{s} L_i \leq 1,$$

there exists a constant k with $1 \leq k \leq \lceil \frac{1}{\alpha} \rceil$ such that the total average load $\sum_{i=1}^{s} L_i(B_k) \leq \alpha$. By contradiction, we suppose that $\sum_{i=1}^{s} L_i(B_t) > \alpha$ for every $t = 1, \ldots, \lceil \frac{1}{\alpha} \rceil$. Then, we obtain with

$$1 \geq \sum_{1 \leq t \leq \lceil \frac{1}{\alpha} \rceil} \sum_{i=1}^{s} L_i(B_t) > \lceil \frac{1}{\alpha} \rceil \cdot \alpha \geq 1$$

a contradiction.

Using this idea (that is a generalization of the idea in [13]) and α chosen small enough, we get a block B_k, $k \geq 1$ with average load on each stage bounded by $\alpha \leq \frac{\epsilon}{2} C^*_{max}$. We define $M = B_k$, $S = \bigcup_{\ell \geq k+1} B_\ell$, $B = \bigcup_{\ell < k} B_\ell$ and $\kappa = \delta^k$.

In the multiprocessor open shop problem, all medium operations can be shifted to the end of the schedule. Then, we can apply the algorithm in [11] on the medium operations. It generates a so called *dense schedule*, and the makespan of the generated schedule can be bounded by

$$\leq P_{max} + \max_{1 \leq i \leq s} L_i(B_k)$$

where

$$P_{max} = \max_{1 \leq j \leq n} \sum_{i|1 \leq i \leq s, O_{ij} \in M} p_{ij} \leq s\kappa = s\delta^k.$$

Therefore, the makespan for the medium operations can be bounded by $\leq s\delta^k + \alpha \leq s\delta + \frac{\epsilon}{4s}$ (here we have used $k \geq 1$ and $\delta < 1$).

Using this shifting of the medium operations, we have to place only the small operations into the intervals of the first type. The maximum (rounded) load of the small operations for interval g and stage i is $\leq m_{g,i}(\delta^{1/2}\kappa + \delta\kappa)$ where $m_{g,i} \leq m_i$ is the number of machines with at least one idle period. We increase the length of each interval to $\delta^{1/2}\kappa + (s+1)\delta\kappa$. As consequence, for each interval g we have generated a smaller instance of the multiprocessor open shop problem with

- $m_{g,i}$ machines on stage i,
- maximum processing time p_{max} of an operation $\leq \delta\kappa$,
- total load of small operations (on stage i) $\leq m_{g,i}(\delta^{1/2}\kappa + \delta\kappa)$.

Using the algorithm in [11], we can generate a schedule for such an instance of length at most

$$(s\delta\kappa) + (\delta^{1/2}\kappa + \delta\kappa) \leq \delta^{1/2}\kappa + (s+1)\delta\kappa.$$

Therefore, the enlargment among all intervals of the first type can be bounded by $\frac{2}{\delta^{1/2}\kappa}(s+1)\delta\kappa = 2(s+1)\delta^{1/2}$. The last interval generates in the worst case an additional length (in comparision to the minimum restricted schedule) of $\delta^{1/2}\kappa + (s+1)\delta\kappa$. Furthermore, by the consideration of only restricted schedule we obtain an additional length of $2\delta^{1/2}$. Using $\kappa = \delta^k \leq 1$, we get a total additional length (with length added by medium operations) $\leq C\delta^{1/2} + \frac{\epsilon}{4s}$ where $C \leq 4s + 6$ is a constant independent of δ. We choose $\delta \leq (\frac{\epsilon}{4sC})^2$ and therefore $C\delta^{1/2} + \frac{\epsilon}{4s} \leq \frac{\epsilon}{2s} \leq \epsilon C^*_{max}$. We define

$$\delta = \left(\frac{1}{\lceil \frac{4Cs}{\epsilon} \rceil}\right)^2 \leq \left(\frac{\epsilon}{4Cs}\right)^2$$

and get the property that $\frac{1}{\delta^{1/2}}$ is integral. Using $\kappa = \delta^k$, the number $\mu = \frac{2}{\delta^{1/2}\kappa}$ is also integral.

References

1. B. Chen, Analysis of classes of heuristics for scheduling a two-stage flow shop with parallel machines at one stage, *Journal of the Operational Research Society* 46 (1995), 234-244.
2. B. Chen and V.A. Strusevich, Worst case analysis of heuristics for open shops with parallel machines, *European Journal of Operational Research* 70 (1993), 379-390.
3. M.R. Garey and D.S. Johnson, Strong NP-completeness results: Motivation, examples and implications, *Journal of the ACM* 25 (1978), 499-508.
4. M.R. Garey, D.S. Johnson and R. Sethi, The complexity of flowshop and jobshop scheduling, *Mathematics of Operations Research* 1 (1976), 117-129.
5. T. Gonzales and S. Sahni, Open shop scheduling to minimize finish time, *Journal of the ACM* 23 (1976), 665-679.
6. L.A. Hall, Approximability of flow shop scheduling, *Proceedings of the 36th Annual IEEE Symposium on Foundations of Computer Science* (1995), 82-91 and *Mathematical Programming* 82 (1998), 175-190.
7. D.S. Hochbaum and D.B. Shmoys, Using dual approximation algorithms for scheduling problems: theoretical and practical results, *Journal of the ACM* 34 (1987), 144-162.
8. J.A. Hoogeveen, J.K. Lenstra and B. Veltman, Preemptive scheduling in a two-stage multiprocessor flow shop is NP-hard, *European Journal of Operational Research* 89 (1996), 172-175.
9. K. Jansen, R. Solis-Oba and M.I. Sviridenko, Makespan minimization in job shops: a polynomial time approximation scheme, *Proceedings of the 31th Annual ACM Symposium on Theory of Computing*, to appear, 1999.
10. E.L. Lawler, J.K. Lenstra, A.H.G. Rinnooy Kan and D.B. Shmoys, Sequencing and scheduling: Algorithms and complexity, in: Handbook in Operations Research and Management Science, Vol. 4, North-Holland, 1993, 445-522.
11. P. Schuurman and G.J. Woeginger, Approximation algorithms for the multiprocessor open shop scheduling problem, *Operations Research Letters*, to appear.
12. P. Schuurman and G.J. Woeginger, A polynomial time approximation scheme for the two-stage multiprocessor flow shop problem, *Theoretical Computer Science*, to appear.
13. S.V. Sevastianov and G.J. Woeginger, Makespan minimization in open shops: A polynomial time approximation scheme, *Mathematical Programming* 82 (1998), 191-198.

14. C. Sriskandarajah and S.P. Sethi, Scheduling algorithms for flexible flow shops: worst and average case performance, *European Journal of Operational Research* 43 (1989), 143-160.
15. D.P. Williamson, L.A. Hall, J.A. Hoogeveen, C.A.J. Hurkens, J.K. Lenstra, S.V. Sevastianov and D.B. Shmoys, Short shop schedules, *Operations Research* 45 (1997), 288-294.

Controlled Conspiracy-2 Search*
(Extended Abstract)

Ulf Lorenz

Department of Mathematics and Computer Science
University of Paderborn, Germany

Abstract. When playing board games like chess, checkers, othello etc., computers use game tree search algorithms to evaluate a position. The greatest success of game tree search so far, has been the victory of the chess machine 'Deep Blue' vs. G. Kasparov, the best human chess player in the world.
When a game tree is too large to be examined exhaustively, the standard method for computers to play games is as follows. A partial game tree (envelope) is chosen for examination. This partial game tree may be any subtree of the complete game tree, rooted at the starting position. It is explored by the help of the $\alpha\beta$-algorithm, or any of its variants. All $\alpha\beta$-variants have in common that a single faulty leaf evaluation may cause a wrong decision at the root.
To overcome this insecurity, we propose Cc2s, a new algorithm, which selects an envelope in a way that the decision at the root is stable against a single faulty evaluation. At the same time, it examines this envelope efficiently. We describe the algorithm and analyze its time behavior and correctness. Moreover, we are presenting some experimental results from the domain of chess.
Cc2s is used in the parallel chess program P.ConNerS, which won the 8th International Paderborn Computer Chess Championship 1999.

Keyword: Algorithms and Datastructures

1 Introduction

Some games have been proven to be PSPACE-complete. As a consequence, we cannot do anything better than to examine a complete game tree when we want to find a perfect decision or if we want to know the value of the starting situation. For most of the interesting board games we do not know the correct values of all positions. Therefore, we are forced to base our decisions on heuristic or vague knowledge. An approximation is done by the following method.

First of all, a partial game tree is chosen for examination. This subtree may be a full-width, fixed-depth tree, or any other subtree rooted at the starting position. We call this subtree an *envelope*. Thereafter, a search algorithm assigns

* This work was supported by the DFG research project "Selektive Suchverfahren" under grant Mo 285/12-3.

H. Reichel and S. Tison (Eds.): STACS 2000, LNCS 1770, pp. 466–478, 2000.

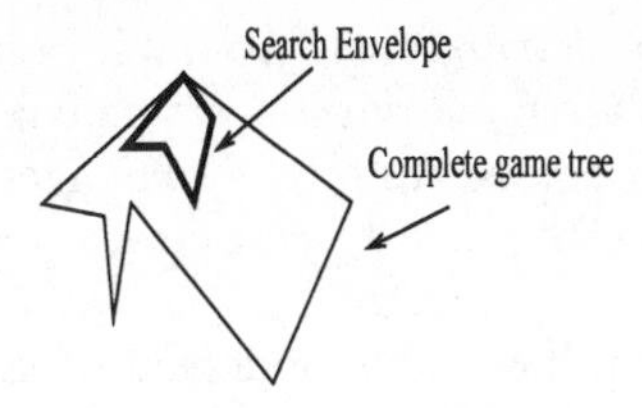

Fig. 1. Only the envelope is examined by a search algorithm.

heuristic evaluations to the leaves and propagates these numbers up the tree by the minimax principle. Usually the envelope is examined by the help of the $\alpha\beta$-algorithm [8], or the MTD(f)-algorithm [1]. As far as the error frequency at the root is concerned, there is no difference, whether or not the envelope is examined by the $\alpha\beta$-algorithm or by a pure minimax algorithm. The result is always the same, only the effort to get the result differs drastically.

The approximation of the real root value by the help of fixed-depth envelopes leads to good results. Nevertheless, there have been found several enhancements that form the envelope more individually. Some of these techniques are domain independent like Singular Extensions [2], Nullmoves [3] [6] or Fail High Reductions [7]. Many others are domain dependent. The form of the envelope strongly determines the quality of the search result.

We distinguish between two classes of game tree search algorithms. On the one hand there are those which are built to determine the minimax value of an envelope. The $\alpha\beta$-algorithm, the SCOUT-algorithm [15] [12] or SSS* [19] have been exhaustively examined in the last 30 years.

A different class is that of the *incremental* searching algorithms [16] which 'grow' the search tree one step a time. At each step a leaf of the current tree is chosen (selection), and the successors of that leaf are added to the tree (expansion). The new leaves are evaluated and the new heuristic values are updated bottom up (update). In contrast to e.g. the $\alpha\beta$-algorithm, these algorithms need linear space in the number of searched nodes. The advantage, however, is that the grown trees need not be of uniform depth and the envelopes need not be determined before the search is finished. Examples of such iterative techniques are the Berliner's B* algorithm [5], Palay's probability-based method [14], and Conspiracy Number Search. Conspiracy Number Search has been introduced by D. McAllester [13]. J. Schaeffer [18] has interpreted the idea and has developed a search algorithm that behaves well on tactical chess positions. Lorenz et al. [11] have presented first ideas of how to build an algorithm which is able to do the same job more efficiently.

The startup point of Conspiracy Number Search (CNS) is the observation that, in a certain sense, the $\alpha\beta$-algorithm computes decisions with low security.

The changing of the value of one single leaf (e.g. because of a fault of the heuristic evaluation function) can change the decision at the root. Thus, the $\alpha\beta$-algorithm takes decisions with security (i.e. conspiracy) one.

The aim of CNS is to distribute the available resources in a way that it is guaranteed that decisions are made with a certain conspiracy $c > 1$. This means that the decision is stable against up to $c - 1$ changes of leaf-values. Schaeffer's algorithm manages this by the help of conspiracy vectors at each node of the game tree.

These vectors inform on how many nodes must change their values in order to change the minimax value of the root to x. As all collected pieces of information must be available at any time, the memory requirement of the method is determined by the number of examined nodes and by the granularity of the evaluation function.

This enormous memory consumption is one of the reasons why the use of CNS has been restricted to tactical positions. With the help of a coarse-grained evaluation function one tries to find decisions which are clearly better than all other alternatives. For tactical positions CNS has been shown to be superior to fixed-depth full-width searches [18].

General inputs do not fulfill the demand that such clearly superior decisions are available. The searching on such instances is called strategic search. There, it is important to come to a decision even when it is only marginally better than the other alternatives. For this purpose, one needs a fine-grained evaluation function, and thus a large amount of memory. It is disappointing to see that the conventional CNS gets severe problems with its termination when the evaluation function is of fine granularity. The CNS sometimes examines large subtrees, only in order to find a decision with low security. In conclusion, these deficiencies have lead to the fact that CNS could not successfully be implemented for general problem instances.

In this paper we discuss a more carefully directed search procedure, which we call Controlled Conspiracy 2 Search (CC2S), and which solves these problems.

1.1 Organization of This Paper

In this paper we are presenting a description of the Controlled Conspiracy 2 Search algorithm. In order to come to a decision at the root, we must determine a lower bound on the value of the best successor and upper bounds on the values of all the other successors of the root. The aim is to base the result on envelopes the leaves of which have a distance of at least t to the root, and which contain at least 2 leaf-disjoint proving-strategies for each of the bounds. (Remark: An error analysis in game trees [10] have lead us to the assumption that 'leaf-disjoint proving strategies' are a key-term in the approximation of game tree values.)

In section 2 we present some basic and general definitions and notations. Section 3 describes the Cc2s-algorithm. We compare it to the $\alpha\beta$-algorithm and to conventional incremental algorithms. The following properties can be observed. a) If the algorithm terminates the result will be based on envelopes with the desired properties, and the outcome is based on minimax values. Thus every minimax-based search algorithm would come to the same result if it examined the same envelope. b) When we examine predefined and finite envelopes (as usually done in the analysis of the $\alpha\beta$-algorithm) our algorithm terminates in finite time. c) At its best case, the new algorithm will examine the minimal number of nodes in order to find a decision, if we slightly change it in a way that it is comparable to the $\alpha\beta$-algorithm.

Section 4 deals with experimental results from the domain of chess.

2 Definitions and Notations

2.1 General Definitions

Definition 1. *In this paper,* $G = (T, h)$ *is a* game tree, *where* $T = (V, K)$ *is a tree (*V *a set of nodes,* $K \subset V \times V$ *the set of edges) and* $h : V \to \mathbb{Z}$ *is a function.* $L(G)$ *is the set of leaves of* T. $\Gamma(v)$ *denotes the set of successors of a node* v.

Remark: We identify the nodes of a game tree G with positions of the underlying game and the edges of T with moves from one position to the next. Moreover, there are two players *MAX and MIN. MAX moves on even and MIN on odd levels. We call the total game tree of a specific game the* universe.

Definition 2. *Let* G *be a game tree. A subtree* E *of* G *is called an* envelope *if, and only if, the root of* E *is the root of* G *and a node* v *of* E *either contains all or none successors of* v *in* G.

Definition 3. *Let* $G = (T, h)$ *be a game tree and* $v \in V$ *a node of* T. *The* Minimax Value, *resp. the function minimax*$: V \to \mathbb{Z}$ *is inductively defined by*

$$minimax(v) := \begin{cases} h(v) & \text{if } v \in L(G) \\ \max\{minimax(v') \mid (v, v') \in K\} & \text{if } v \notin L(G) \text{ and } MAX \text{ to move} \\ \min\{minimax(v') \mid (v, v') \in K\} & \text{if } v \notin L(G) \text{ and } MIN \text{ to move} \end{cases}$$

Definition 4. *Let* G *be a game tree with root* $v \in V$, *and let* $s \in \{MIN, MAX\}$, *Formally,* a strategy for player s, $S_s = (V_s, K_s)$, *is a subtree of* G, *inductively defined by*

- $v \in V_s$
- *If* $u \in V_s$ *is an internal node of* T *where* s *has to move there will exactly be one* $u' \in \Gamma(u)$ *with* $u' \in V_s$ *and* $(u, u') \in K_s$.
- *If* $u \in V_s$ *is an internal node of* T *where the opponent of* s *has to move,* $\Gamma(u) \subset V_s$, *and* $(u, u') \in K_s$ *will hold for all* $u' \in \Gamma(u)$.

Remark: A strategy *is a subtree of G that proves a certain bound of the minimax value of the root of G. A MIN-strategy proves an upper bound of the minimax value of G, and a MAX-strategy a lower one.*

Definition 5. *Two strategies will be called* leaf-disjoint *if they have no leaf in common (cf. Fig. 2).*

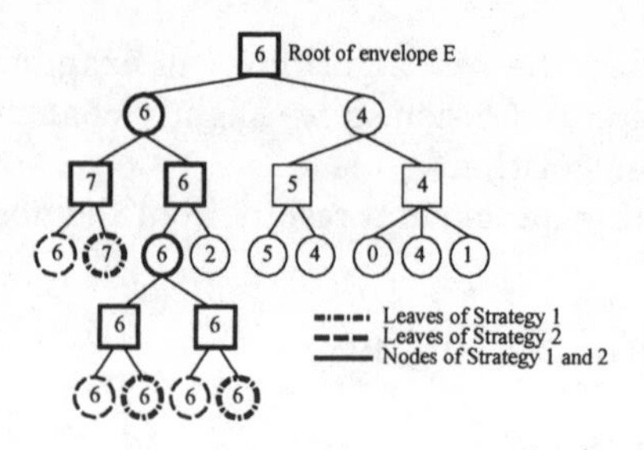

Fig. 2. Two leaf-disjoint Strategies prove the lower bound 6 at the root. This is equivalent to the formulation that the lower bound 6 has the conspiracy number 2 [18].

Definition 6. *Let $G = (T, h) = ((V, K), h)$ be a game tree. A best move is a move from the root to a successor which has the same minimax-value as the root has. Let $m = (v, v')$ be such a move. We say m is* secure *with conspiracy number C and depth d if there exists an $x \in \mathbb{Z}$ so that a) there are at least C leaf disjoint strategies, with leaves at least in depth d, showing the minimax value of v' being greater or equal to x, and b) for all other successors of the root there are at least C leaf disjoint strategies, with leaves at least in depth d, showing the minimax value of them being less or equal to x. C is a lower bound for the number of terminal nodes of G that must change their values in order to change the best move at the root of G.*

Remark: Let $\mathcal{A}$be a game tree search algorithm. We distinguish between the *universe*, an *envelope* and a (current) *search tree.* A search tree is a subtree of the envelope. E.g., the minimax-algorithm and the $\alpha\beta$-algorithm may examine the same envelope, but they usually examine different search trees. Let v be a node. In the following $\Gamma(v)$ is the set of all successors of v, concerning the universe. $\Gamma'(v)$ is the set of those successors of v that are explicitly inspected by the algorithm $\mathcal{A}$

Definition 7. *Let $G = ((V, K), h)$ be a game tree. A* value *is a tuple $w = (a, z) \in \{\text{'}\leq\text{'}, \text{'}\geq\text{'}, \text{'}\#\text{'}\} \times \mathbb{Z}$. a is called the* attribute *of w, z the* number *of w. $W = \{\text{'}\geq\text{'}, \text{'}\leq\text{'}, \text{'}\#\text{'}\} \times \mathbb{Z}$ is the* set of values*. We denote $w_v = (a_v, z_v)$ the value of the node v, with $v \in V$.*

Remark: *Let v be a node. $w_v = (\text{'}\leq\text{'}, x)$ will express that there is a subtree below v the minimax-value of which is $\leq x$. $w_v = (\text{'}\geq\text{'}, x)$ is analogously used. $w_v = (\text{'}\#\text{'}, x)$ implies that there exists a subtree below v the minimax-value of which is $\leq x$, and there is a subtree below v the minimax-value of which is $\geq x$. The two subtrees need not be identical. A value w_1 can be 'in contradiction' to a value w_2 (e.g. $w_1 = (\text{'}\leq\text{'}, 5), w_2 = (\text{'}\geq\text{'}, 6)$), 'supporting' (e.g. $w_1 = (\text{'}\leq\text{'}, 5), w_2 = (\text{'}\leq\text{'}, 6)$), or 'unsettled' (e.g. $w_1 = (\text{'}\geq\text{'}, 5), w_2 = (\text{'}\leq\text{'}, 6)$).*

Definition 8. *A* target *is a tuple* $t = (\omega, \delta, \gamma)$ *with* ω *being a value and* $\delta, \gamma \in \mathbb{N}_0$.

Remark: Let $t_v = (\omega, \delta, \gamma)$ *be a target which is associated with a node* v. δ *expresses the demanded distance from the current node to the leaves of the final envelope.* γ *is the* conspiracy number *of* t_v. *It informs a node on how many leaf-disjoint strategies its result must base. If the demand, expressed by a target, is fulfilled, we say that the target* t_v *is* fulfilled.

3 Description of Cc2s

3.1 The New Search Paradigm

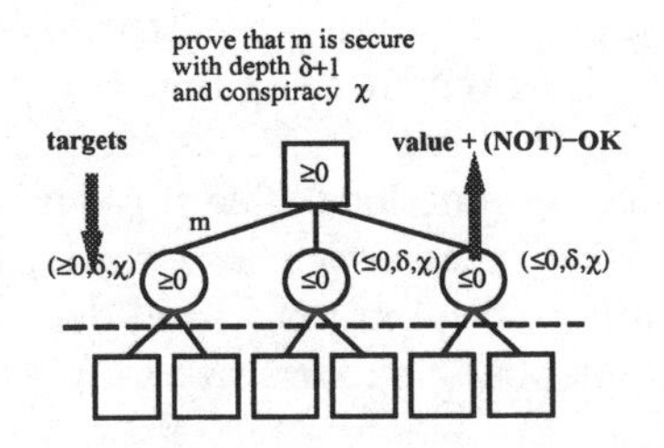

The left figure shows the data flow in our algorithm. In contrast to the minimax-algorithm or the $\alpha\beta$-algorithm we do not look for the minimax value of the root. We try to separate a best move from the others, by proving that there exists a number x so that the minimax value of the successor with the highest payoff is at least x, and the payoffs of the other successors are less or equal to x.

At any point of time the searched tree offers such an x and a best move m. As long as m is not *secure* enough, we take x and m as a hypothesis only, and we commission the successors of the root either to show that new estimations make the hypothesis fail, or to verify it. The terms of 'failing' and 'verifying' are used in a weak sense: they are related to the best possible knowledge at a specific point of time, not to absolute truth. New findings can cancel former 'verifications'. The verification is handled by the help of the targets, which are split and spread over the search tree in a top down fashion. A target t expresses a demand to a node. Each successor of a node, which is supplied by a target, takes its own value as an expected outcome of a search below itself, and commissions its successors to examine some sub-hypotheses, etc. A target t will be fulfilled when t demands a leaf, or when the targets of all of v's successors are fulfilled. When a target is fulfilled at a node v, the result 'OK' is given to the father of v. If the value of v changes in a way that it contradicts the value component of t the result will be 'NOT-OK'.

In the following example, the task is to find the best move, concerning a depth-2 fixed-depth envelope. The figure just below shows the incremental growth of strategies.

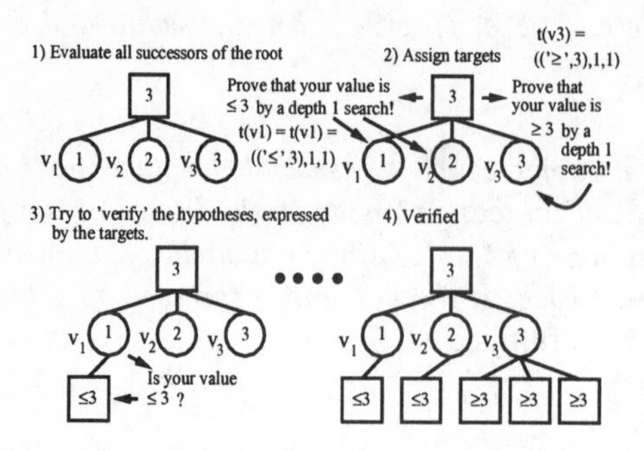

The first step is to evaluate all successors of the root. Concerning a depth one search, v_3 becomes the 'best' successor (i.e. the one with the highest payoff) of the root. Thus, we conjecture that v_3 will also be the best move concerning a depth-2 search. We build the targets $t_1 = (('\leq', 3), 1, 1)$ for v_1, the same for v_2, and $t_3 = (('\geq', 3), 1, 1)$ for v_3.

v_1 starts the verification process as it is the leftmost successor of the root. It generates its first successor and inquires whether or not, its value is ≤ 3. v_2 does the same, and v_3 generates all successors and finds all of them being ≥ 3. Since all partial expansions fit to the targets, v_1, v_2 and v_3 return OK to the root. Thus, we know at the root that the move to v_3 is secure with depth 2.

The targets consume much less memory than the conspiracy vectors of the conventional CNS and, moreover, they allow an efficient cutting mechanism, similar to the $\alpha\beta$-algorithm. We can make use of the fact that it is often expensive and superfluous to compute exact minimax-values, if we only need an upper resp. a lower bound on them. These properties make it possible to use Cc2s even for strategic searches, which need fine-grained evaluation functions. The top-down splitting of the targets offers high flexibility, which allows to realize various security concepts. In the following, we present a hybrid mixture of fixed-depth searches and conspiracy number searching. It is even possible to use tree forming heuristics such as Fail High Reductions [7].

3.2 Algorithmic Details

Values Are Updated Bottom Up A crucial point is how to react when a partial expansion step leads to values which contradicts a target. The task is to gather the new pieces of information, to draw a maximum of benefit from them, and at the same time to guarantee correctness and termination of the whole algorithm. The operator of Figure 3 has been designed for that purpose. We give some examples of contradictions and their solution: Let v be a maxnode with a value $('\leq', 5)$. The value be determined by the fact that all successors of v have values $('\leq', 5)$. Now let us evaluate v by the help of our heuristic evaluation function and the result be that the direct value of v is something > 5, e.g. $('\geq', 7)$. Now, UpdateValue assigns the value $('\#', 5)$ to v and solves the contradiction (ll.9,10). Another example: Let v be a maxnode with value $('\geq', 5)$ and two successors. One successor has got the value $('\leq', 3)$. If the second successor has a value $('\leq', 10)$, or $('\geq', 3)$, or has not been evaluated at all, the value of v will remain $('\geq', 5)$ (l.6). If, however, the second successor gets the value $('\leq', 3)$, the new value will be built by the lines 9, 10, and v will get the value $('\#', 3)$.

```
value UpdateValue(node v)
   ...
 1 /* Let v be a MAX node */
 2 if Γ'(v) ≠ Γ(v) or ∃s ∈ Γ'(v) with a_s = '≥' or
 3    ∃s ∈ Γ'(v) with (a_s = '≤' and z_s ≥ z_v) then {
 4      if a_v = '≤' and ∃s ∈ Γ'(v) with (a_s ∈ {'≥' ,'#' } and z_s ≥ z_v)
 5          then a_v := '#' ;
 6      z_v := max{z_v, max{z_s | s ∈ Γ'(v), a_s ∈ {'#' ,'≥' }};
 7 } else {
 8    if a_v = '≥' or (a_v = '≤' and ∃s ∈ Γ'(v) with (a_s = '#' and z_s ≥ z_v))
 9      then a_v := '#' ;
10    z_v := max{z_s | s ∈ Γ'(v)};
  } /* If v is a MIN node the result is analogously defined. */
    /* You only exchange ≤ and ≥, and max by min. */
```

Fig. 3. Update the heuristic value of a node.

There are some properties of UpdateValue, which make the combined values to more than only heuristic ones.

Theorem 1. *When inner values of nodes are gathered bottom up by the help of UpdateValue we can prove that* **(a)** $w_v = ('\leq', x)$ *implies that there exists a subtree below* v *the minimax-value of which is* $\leq x$, **(b)** $w_v = ('\geq', x)$ *implies that there exists a subtree below* v *the minimax-value of which is* $\geq x$, *and* **(c)** $w_v = ('\#', x)$ *implies that there exists a subtree below* v *whose minimax-value is* $\leq x$ *and there exists a subtree below* v *the minimax-value of which is* $\geq x$. *These two subtrees are not necessarily identical.*

Moreover, it is obvious that there are no longer value contradictions at the nodes. For all proofs of the theorems we refer to the full version of this paper [9].

Forming the Envelope and Handling Security In order to prove a lower bound at a maxnode, or an upper bound at a minnode, it is sufficient that one successor holds the bound. Such a node is called a 'cutnode'. If we want to prove an upper bound at a maxnode, all successors will have to hold the bound. Therefore, we call that type of node an 'allnode'.

We define targets by the help of the following observations: Let v be a maxnode and let x be an upper bound on the minimax value of v. Let b_v be the number of sons of v. Then x is an upper bound on the minimax values of all sons $v_i, i \in \{1 \dots b_v\}$. If there are C-many leaf-disjoint strategies below each son of v, which prove the bound x, there will be C leaf-disjoint strategies which prove the bound x below v, too. To be more general, if C_i is the number of leaf-disjoint strategies below the nodes v_i (for all sons v_i of v), the number of leaf-disjoint strategies that prove the bound x below v will be $\min_{i=1}^{b_v} C_i$.

Now, let v be a maxnode and x a lower bound of the minimax value of v. Let c be the number of sons of v which have a minimax value $\geq x$. (Because of the minimax rule there is at least one such successor.) Let $C_i, i \in \{1 \dots c\}$ be the number of leaf-disjoint strategies below v_i that prove the bound x at the nodes v_i. Then the number of leaf-disjoint strategies that prove the bound x at the node v is $\sum_{i=1}^{c} C_i$. Minnodes are analogously handled.

Last but not least, if there is a strategy below v, the leaves of which have a distance of d to v, there will be strategies below all sons of v, the leaves of which have a distance of $d-1$ to the sons of v.

When our algorithm enters a node v with target t, it decides whether v is a cutnode or an allnode, concerning t. Then it builds sub-targets for all successors of v in a way that t will be fulfilled, when all sub-targets are fulfilled.

The Search Algorithm Dealing with Security In the following, we assume that there is a heuristic evaluation procedure which either can return a heuristic point-value x, or which can answer whether x is smaller or greater than a given number y.

We call the starting routine (no figure) at the root DetermineMove(root r, d, $c=2$). d stands for the remaining depth and c for the conspiracy number which the user wants to achieve. If the successors of r have not been generated yet, DetermineMove will do this, and it will assign heuristic values of the form ('#' ,...) to all successors. It picks up the successor which has got the highest value x, and assigns a lower bound target of the form $(('\geq', x), d, c)$ to the best successor and targets of the form $(('\leq', x), d, c)$ to all other successors. Then it starts the procedure Cc2s on all successors. DetermineMove repeats the previous steps, until Cc2s returns with OK at all of r's successors.

Let $\Gamma'(v)$ be the set of successors of v, as far as the current search tree is concerned. Let t_v be the target for v, w_v the value of v. Let $v_1 \dots v_{|\Gamma'(v)|}$ be the successors of v concerning the current search tree. Let $t_1 \dots t_{|\Gamma'(v)|}$ be the targets of the nodes $v_1 \dots v_{|\Gamma'(v)|}$, and let $w_1 \dots w_{|\Gamma'(v)|}$ be their values. We say that a node is OnTarget(v, t_v) when the value of v is not in contradiction to the value component of t_v. This will express that Cc2s still is on the right way. When Cc2s (figure 4) enters a node v it is guaranteed that v is on target and that the value of v supports t_v (either by DetermineMove, or because of figure 4, ll. 5-6). Firstly, Cc2s checks whether v is a leaf, i.e. whether t_v is trivially fulfilled (l.1). This is the case when the remaining search depth of the target is zero ($\delta_v = 0$) and the demanded conspiracy number i.e. the number of leaf-disjoint bound-proving strategies is 1 ($\gamma_v \leq 1$). If v is not a leaf, the sub-algorithm PartialExpansion (no figure) will try to find successors of v which are well suited for a splitting operation. Therefore, it starts the evaluation of successors which either have not yet been evaluated, or which have an unsettled value in relation to the target $t_v = (\dots, x)$. If a successor s has been evaluated once before and is examined again it will get a point value of the form ('#' $, y$). For an argumentation of progress and termination, it is important that successors which have already been evaluated once before, and now are unsettled, are supplied with a point value. Since '#' -values cannot be unsettled, each node can be evaluated at most twice! For a not yet examined successor s the evaluation function is inquired whether the value of s supports or contradicts the value component of t_v. s gets the value $('\geq', x)$ or $('\leq', x-1)$. PartialExpansion works from left to right, and updates the value of v after each evaluation.

```
bool Cc2s(node v, target t_v = (α_v, β_v, δ_v, γ_v))
1     if (δ_v = 0 and γ_v ≤ 1) or |Γ(v)| = 0 return OK;
2     r := NOT_OK;
3     while r = NOT_OK do {
4       PartialExpansion(v, t_v);
5       if not OnTarget(v, t_v) return NOT_OK;
6       Split(v, t, v_1 ... v_|Γ'(v)|); /* assigns targets to the sons */
7       for i := 1 to |Γ(v)'| do {
8         r := Cc2s(v_i, t_i);
9         w_v := UpdateValue(v);
10        if not OnTarget(v, t_v) return NOT_OK;
11        if r = NOT_OK break ; /* Leave the for-loop, goto l.3 */
12      }
13    } /* while ... */;
14    return OK;
```

Fig. 4. Recursive Search Procedure

If v is an allnode and a partial expansion changes the value of v in a way that it contradicts the target t, Cc2s will immediately stop and leave v by line 11. If v is a cutnode, PartialExpansion will evaluate successors which have not yet been examined or which are unsettled in relation to t, until it will have found γ_v-many successors which support the value of v.

After that, the target of the node v is 'split', i.e. sub-targets are worked out for the successors of v, concerning subsection 3.2. The resulting sub-targets are given to the successors of v, and Cc2s examines the sons of v, until either all sons of v will have fulfilled their targets (some successors may get so called null-targets, i.e. a target that is always fulfilled), or v itself is not 'on target' any longer, which means that the value of v contradicts the current target of v. When a call of Cc2s returns with the result OK at a node $v.i$ (line 8), the node $v.i$ could fulfill its subtarget. When Cc2s returns with NOT-OK, some values below $v.i$ have changed in a way that it seems impossible that the target of $v.i$ can be fulfilled any more. In this case, Cc2s must decide, whether to report a NOT-OK to its father (line 10), or to rearrange new sub-targets to its sons (ll. 11 and 3).

Theorem 2. *If the algorithm terminates the result at the root will be based on an envelope with the desired properties. A minimax-algorithm would come to the same result, concerning that envelope.*

3.3 A Restriction of Cc2s, Compared to the $\alpha\beta$-Algorithm

In this section, we would like to compare our algorithm with the most successful $\alpha\beta$-algorithm. As the $\alpha\beta$-algorithm is not able to deal with conspiracy numbers, we restrict our algorithm to fixed-depth searches. Only do we refrain from the demand of 2 leaf-disjoint proving strategies. Let us call the resulting algorithm

Cc1s. The analysis of the $\alpha\beta$-algorithm is usually restricted to fixed-depth full-width game trees with a uniform branching factor. We follow this restriction.

Let G be a fixed-depth full-width game tree with depth d and breadth b. In order to find the best move, the $\alpha\beta$-algorithm must evaluate the minimax value of the root of G. Therefore it examines at least $b^{\lfloor \frac{d}{2} \rfloor} + b^{\lceil \frac{d}{2} \rceil}$ leaves [8].

Theorem 3. *(Effort) At its best case, the Cc1s-algorithm evaluates* $b^{\lceil (d-1)/2 \rceil} + (b-1) \cdot b^{\lfloor (d-1)/2 \rfloor}$ *leaves of* G*, in order to find the best move at the root. With regard to the number of evaluated leaves in the search tree this is optimal.*

Theorem 4. *(Correctness) The decision-move which Cc1s finds is based on the minimax-value of* G*, i.e. every minimax-based algorithm will come to the same result.*

Theorem 5. *(Termination) If* G *is finite (especially if* G *is a fixed-depth full-width game tree) Cc1s (as well as Cc2s) finishes its work in finite time.*

3.4 Cc2s Compared to Select-Expand-Update Based Algorithms

Our new technique has two advantages over algorithms which expand a leaf node without being informed what the expansion is good for. In practice, it is often faster to decide whether or not the value is below or above a certain bound than to compute a point value. We profit from this. Moreover, we need not generate all sons of a node. We win a constant but decisive factor. E.g. we estimate the factor on about 50 at the game of chess. This is so much that it seems nearly impossible that any algorithm which is based on the Select-Expand-Update paradigm [16] is able to play high level chess.

4 Experimental Results

Firstly, we added some enhancements to the Cc2s algorithm: a) An evaluation is a depth-2 $\alpha\beta$-search plus quiescence search. b) Analogous to the Fail-High-Reduction technique, the remaining depth of a target will be further decreased, if an evaluation, refined by a small threat detection, indicates a cutnode. Moreover, there are some minor important heuristics that speed up the convergence of Cc2s. None of these heuristics is allowed to contain chess specific knowledge.

We are dividing the tests into two parts: The first part is based on the so called BT2630 test-set [4]. It consists of 30 positions. Each one is supplied with a grandmaster-tuned solution move. The machine gets exactly 15 minutes per position. Those positions which have not been solved at the end of the given time count with 900 seconds. Those which are solved correctly, count with the number of seconds needed for computing the solution.

Results on the BT2630 test

Program on Machine	'ELO'
ConNerS on Sparc 144 MHz	2375
ConNerS on Sparc 300 MHz	2408
Cheiron'97 on Sparc 144 MHz	2331
Fritz 4.0 on 200Mhz Pentium	2373
Hiarcs 6.0 on 200Mhz Pentium	2403

The test-set associates a pseudo-ELO-number to the result. (The ELO-system is a statistical measure, which measures the relative playing strength of chess players.) with regard to this test-set, ConNerS stands the comparison with two of the world top level chess programs, Fritz and Hiarcs.

Although the test-set stresses tactical performance, it is important to note that the program finds tactical lines when it uses a fine-grained evaluation function.

The second kind of testing is based on games. Here, the main challenge was to set our Cc2s algorithm into relation with a chess program that uses an $\alpha\beta$ game tree algorithm, supplied with all enhancements: such as transposition tables, sorting heuristics [17], FHR or Nullmove technique etc.
We have selected 25 starting positions and have played two series of 50 games against Cheiron'97. That program uses the negascout algorithm supplied by Fail High Reductions, Transposition tables, killer heuristics, recapture and chess extensions. The program and its predecessors have proven to be successful in several tournaments. The main advantage, however, is that an evaluation function and a depth-2 $\alpha\beta$-search for evaluations of ConNerS are available. Thus, we can exactly compare the remaining search of Cheiron'97 with our new algorithm.

A first series of games started on Sparc 144 MHz machines, each side getting 8 hours for 40 moves. It ended 26.5 to 23.5 for Cheiron'97. In a second series both programs got 4 hours for 40 moves. This fight ended 26.5 to 23.5, too, (although the single results differ). If we take both series together, we have a final result of 53 to 47. We are interpreting this result as Cheiron'97 and ConNerS being equally strong, although we are aware of the fact that these 100 games are not sufficient for a proof in a statistical sense.

The greatest success of P.ConNerS (parallel version of ConNerS) was the winning of the 8th International Paderborn Computer Chess Championship, where we competed with world class programs like Nimzo or Shredder (World Champion 1999). At the World Computer Chess Championship P.ConNerS ended up with 3 to 4 points, but post-tournament analysis revealed that minor opening book variations were the main reasons for the losses.

5 Acknowledgements

I would like to thank Prof. Dr. B. Monien and Dr. R. Feldmann for their long-lasting help and backing. In lots of discussions I got a lot of ideas and hints.

References

[1] A. de Bruin A. Plaat, J. Schaeffer and W. Pijls. A minimax Algorithm better than SSS*. *Artificial Intelligence*, 87:255–293, 1999.

[2] T.S. Anantharaman. Extension heuristics. *ICCA Journal*, 14(2):47–63, 1991.
[3] D.F. Beal. Experiments with the null move. *Advances in Computer Chess 5 (ed. Beal, D.F.)*, pages 65–79, 1989.
[4] H. Bednorz and F. Tönissen. Der neue Bednorz-Tönissen-Test. *Computer Schach und Spiele*, 11(2):24–27, 1994.
[5] H. Berliner. The B* tree search algorithm: A best-first proof procedure. *Artificial Intelligence*, 12(1):23–40, 1979.
[6] C. Donninger. Null move and deep search. *ICCA Journal*, 16(3):137–143, 1993.
[7] R. Feldmann. Fail high reductions. *Advances in Computer Chess 8 (ed. J. van den Herik)*, 1996.
[8] D.E. Knuth and R.W. Moore. An analysis of alpha-beta pruning. *Artificial Intelligence*, 6(4):293–326, 1975.
[9] U. Lorenz. Controlled conspiracy-2 search. Technical report, University of Paderborn, available via http://www.upb.de/cs/ag-monien/PERSONAL/FLULO/publications.html, 1999.
[10] U. Lorenz and B. Monien. The secret of selective game tree search, when using random-error evaluations. Technical report, University of Paderborn, available via http://www.upb.de/cs/ag-monien/PERSONAL/FLULO/publications.html, 1998.
[11] U. Lorenz, V. Rottmann, R. Feldmann, and P. Mysliwietz. Controlled conspiracy number search. *ICCA Journal*, 18(3):135–147, 1995.
[12] T.A. Marsland, A. Reinefeld, and J. Schaeffer. Low overhead alternatives to SSS*. *Artificial Intelligence*, 31(1):185–199, 1987.
[13] D.A. McAllester. Conspiracy Numbers for Min-Max searching. *Artificial Intelligence*, 35(1):287–310, 1988.
[14] A.J. Palay. *Searching with Probabilities.* 1985.
[15] J. Pearl. *Heuristics – Intelligent Search Strategies for Computer Problem Solving.* Addison-Wesley Publishing Co., Reading, MA, 1984.
[16] R.L. Rivest. Game tree searching by min/max approximation. *Artificial Intelligence*, 34(1):77–96, 1987.
[17] J. Schaeffer. The history heuristic. *ICCA Journal*, 6(3):16–19, 1983.
[18] J. Schaeffer. Conspiracy numbers. *Artificial Intelligence*, 43(1):67–84, 1990.
[19] G.C. Stockman. A minimax algorithm better than alpha-beta? *Artificial Intelligence*, 12(2):179–196, 1979.

The Stability of Saturated Linear Dynamical Systems Is Undecidable

Vincent D. Blondel[1], Olivier Bournez[2], Pascal Koiran[3], and John N. Tsitsiklis[4]

[1] Department of Mathematical Engineering, Université catholique de Louvain, Avenue Georges Lemaitre 4, B-1348 Louvain-la-Neuve, Belgium, blondel@inma.ucl.ac.be, http://www.inma.ucl.ac.be/~blondel/

[2] LORIA, Campus Scientifique BP 239, 54506 Vandoeuvre-les-Nancy, France, Olivier.Bournez@loria.fr

[3] LIP, ENS Lyon, 46 allée d'Italie, F-69364 Lyon Cedex 07, France, Pascal.Koiran@ens-lyon.fr

[4] LIDS, MIT, Cambridge, MA 02139, USA, jnt@mit.edu

Abstract. We prove that several global properties (global convergence, global asymptotic stability, mortality, and nilpotence) of particular classes of discrete time dynamical systems are undecidable. Such results had been known only for point-to-point properties. We prove these properties undecidable for saturated linear dynamical systems, and for continuous piecewise affine dynamical systems in dimension three. We also describe some consequences of our results on the possible dynamics of such systems.

1 Introduction

This paper studies problems such as the following: given a discrete time dynamical system of the form $x_{t+1} = f(x_t)$, where $f : \mathbf{R}^n \to \mathbf{R}^n$ is a saturated linear function or, more generally, a continuous piecewise affine function, decide whether all trajectories converge to the origin.

We show in our main theorem that this global convergence problem is undecidable. The same is true for three related problems: Stability (is the dynamical system globally asymptotically stable?), Mortality (do all trajectories go through the origin?), and Nilpotence (does there exist an iterate f^k of f such that $f^k \equiv 0$?).

It is well-known that various types of dynamical systems, such as hybrid systems, piecewise affine systems, or saturated linear systems, can simulate Turing machines, see e.g., [15,12,16,18]. In these simulations, a machine configuration is encoded by a point in the state space of the dynamical system. It then follows that *point-to-point* properties of such dynamical systems are undecidable. For example, given a point in the state space, one cannot decide whether the trajectory starting from this point eventually reaches the origin. The results described in this contribution are of a different nature since they deal with *global* properties of dynamical systems.

H. Reichel and S. Tison (Eds.): STACS 2000, LNCS 1770, pp. 479–490, 2000.

Related undecidability results for such global properties have been obtained in our earlier work [3], but for the case of *discontinuous* piecewise affine systems. The additional requirement of continuity imposed in this paper is a severe restriction, and makes undecidability much harder to establish. Surveys of decidability and complexity results for dynamical systems are given in [1], [12] and [8].

Our main result (Theorem 1) is a proof of Sontag's conjecture [7,19] that global asymptotic stability of saturated linear systems is not decidable. Saturated linear systems are systems of the form $x_{t+1} = \sigma(Ax_t)$ where x_t evolves in the state space $\mathbf{R}^n$, A is a square matrix, and σ denotes componentwise application of the saturated linear function $\sigma : \mathbf{R} \to [-1, 1]$ defined as follows: $\sigma(x) = x$ for $|x| \leq 1$, $\sigma(x) = 1$ for $x \geq 1$, $\sigma(x) = -1$ for $x \leq -1$. These dynamical systems occur naturally as models of neural networks [17,18] or as models of simple hybrid systems [20,5,2].

Theorem 1 is proved in three main steps. First, in Section 4, we prove that any Turing machine can be simulated by a saturated linear dynamical system with a strong notion of simulation. (Turing machines are defined in Section 3.) Then, in Section 5, using a result of Hooper, we prove that there is no algorithm that can decide whether a given continuous piecewise affine system has a trajectory contained in a given hyperplane. Finally, we prove Theorem 1 in Section 6.

In light of our undecidability result, any decision algorithm for the stability of saturated linear systems will be able to handle only special classes of systems. In the full version of this paper [4] we consider two such classes: systems of the form $x_{t+1} = \sigma(Ax_t)$ where A is a nilpotent matrix, or a symmetric matrix. We show that stability remains undecidable for the first class, but is decidable for the second.

Saturated linear systems fall within the class of continuous piecewise affine systems and so our undecidability results equally apply to the latter class of systems. More precise statements for continuous piecewise affine systems are given in Section 7. Finally, some suggestions for further work are made in Section 8.

For some of our results we give complete proofs. For others we provide only a sketch, or we refer to the full version of the paper [4].

2 Dynamical Systems

In the sequel, X denotes a metric space and 0 some arbitrary point of X, to be referred to as the *origin*. When $X \subseteq \mathbf{R}^n$, we assume that 0 is the usual origin of $\mathbf{R}^n$. A *neighborhood* of 0 is an open set that contains 0. Let $f : X \to X$ be a function such that $f(0) = 0$. We say that f is:

(a) *globally convergent* if for every initial point $x_0 \in X$, the trajectory $x_{t+1} = f(x_t)$ converges to 0.
(b) *locally asymptotically stable* if for any neighborhood U of 0, there is another neighborhood V of 0 such that for every initial point $x_0 \in V$, the trajectory $x_{t+1} = f(x_t)$ converges to 0 without leaving U (i.e., $x(t) \in U$ for all $t \geq 0$ and $\lim_{t\to\infty} x_t = 0$).

(c) *globally asymptotically stable* if f is globally convergent and locally asymptotically stable.
(d) *mortal* if for every initial point $x_0 \in X$, there exists $t \geq 0$ with $x_t = 0$. The function f is called *immortal* if it is not mortal.
(e) *nilpotent* if there exists $k \geq 1$ such that the k-th iterate of f is identically equal to 0 (i.e., $f^k(x) = 0$ for all $x \in X$).

Nilpotence obviously implies mortality, which implies global convergence; and global asymptotic stability also implies global convergence. In general, this is all that can be said of the relations between these properties. Note, however, the following simple lemma, which will be used repeatedly.

Lemma 1. *Let X be a metric space with origin 0, and let $f : X \to X$ be a continuous function such that $f(0) = 0$. If f is nilpotent, then it is globally asymptotically stable. Moreover, if X is compact and if there exists a neighbourhood O of 0 and an integer $j \geq 1$ such that $f^j(O) = \{0\}$, the four properties of nilpotence, mortality, global asymptotic stability, and global convergence are equivalent.*

Proof. Assume that f is nilpotent and let k be such that $f^k \equiv 0$. Let U and V be two neighborhoods of 0. A trajectory starting in V never leaves $\bigcup_{i=0}^{k-1} f^i(V)$. By continuity, for any U one can choose V so that $f^i(V) \subseteq U$ for all $i = 0, \ldots, k-1$. A trajectory originating in such a V never leaves U. This shows that f is globally asymptotically stable.

Next, assume that X is compact and that $f^j(O) = \{0\}$ for some neighborhood O of 0 and some integer $j \geq 1$. It suffices to show that if f is globally convergent, then it is nilpotent. If f is globally convergent, then $X = \bigcup_{i \geq 0} f^{-i}(O)$. By compactness, there exists $p \geq 0$ such that $X = \bigcup_{i=0}^{p} f^{-i}(O)$. We conclude that $f^{p+j}(X) = \{0\}$. □

A function $f : \mathbf{R}^n \to \mathbf{R}^{n'}$ is *piecewise affine* if $\mathbf{R}^n$ can be represented as the union of a finite number of subsets X_i where each set X_i is defined by the intersection of finitely many open or closed halfspaces of $\mathbf{R}^n$, and the restriction of f to each X_i is affine. Let $\sigma : \mathbf{R} \to \mathbf{R}$ be the continuous piecewise affine function defined by: $\sigma(x) = x$ for $|x| \leq 1$, $\sigma(x) = 1$ for $x \geq 1$, $\sigma(x) = -1$ for $x \leq -1$. Extend σ to a function $\sigma : \mathbf{R}^n \to \mathbf{R}^n$, by letting $\sigma(x_1, \ldots, x_n) = (\sigma(x_1), \ldots, \sigma(x_n))$. A *saturated affine function* (*σ-function* for short) $f : \mathbf{R}^n \to \mathbf{R}^{n'}$ is a function of the form $f(x) = \sigma(Ax + b)$ for some matrix $A \in \mathbf{Q}^{n' \times n}$ and vector $b \in \mathbf{Q}^{n'}$. Note that we are restricting the entries of A and b to be rational numbers so that we can work within the Turing model of digital computation. A *saturated linear function* (*σ_0-function* for short) is defined similarly except that $b = 0$. Note that the function $\sigma : \mathbf{R}^n \to \mathbf{R}^n$ is piecewise affine, with the polyhedra X_i corresponding to the different faces of the unit cube $[-1, 1]^n$, and so is the linear function $f(x) = Ax$. It is easily seen that the composition of piecewise affine functions is also piecewise affine and therefore σ-functions are piecewise affine.

Our main result is the following theorem.

Theorem 1. *The problems of determining whether a given saturated linear function is (i) globally convergent, (ii) globally asymptotically stable, (iii) mortal, or (iv) nilpotent, are all undecidable.*

Notice that deciding the global asymptotic stability of a saturated linear system is *a priori* no harder than deciding its global convergence, because the local asymptotic stability of saturated linear systems is decidable. (Indeed, a system $x_{t+1} = \sigma(Ax_t)$ is locally asymptotically stable if and only if the system $x_{t+1} = Ax_t$ is, since these systems are identical in a neighborhood of the origin. Furthermore, a linear system is locally asymptotically stable if and only if all of its eigenvalues have magnitude less than one [21].) In fact, we conjecture that for saturated linear systems, global convergence is equivalent to global asymptotic stability. This equivalence is proved for symmetric matrices in the full version of the paper. If this conjecture is true, it is not hard to see that the equivalence of mortality and nilpotence also holds.

Theorem 1 has some "purely mathematical" consequences. For instance:

Corollary 1. *For infinitely many integers n, there exists a nilpotent saturated linear function $f : \mathbf{R}^n \to \mathbf{R}^n$ such that $f^{2^n} \not\equiv 0$.*

Of course, in this corollary, 2^n can be replaced by any recursive function of n. In contrast, if $f : \mathbf{R}^n \to \mathbf{R}^n$ is a nilpotent linear function, then $f^n \equiv 0$. As a side remark, we note that it can be shown that this is not only true for linear functions, but also for polynomials and even more generally for real analytic functions.

We conclude this section with two positive results: globally asymptotically stable saturated linear systems are recursively enumerable and so are saturated linear systems that have a nonzero periodic trajectory. The first observation is due to Eduardo Sontag, the second is due to Alexander Megretski.

Theorem 2. *The set of saturated linear systems that are globally asymptotically stable is recursively enumerable.*

Theorem 3. *The set of saturated linear systems that have a nonzero periodic trajectory is recursively enumerable.*

The proofs of these results are based on elementary arguments, they can be found in the full version of the paper. Combining these two observations with Theorem 1, we deduce that there exist saturated linear systems that are not globally asymptotically stable and have no nonzero periodic trajectories.

Corollary 2. *There exist saturated linear systems that are not globally asymptotically stable and have no nonzero periodic trajectory.*

3 Turing Machines

A *Turing machine* M [14,13] is an abstract deterministic computer with a finite set Q of internal states. It operates on a doubly-infinite tape over some finite alphabet Σ. The tape consists of squares indexed by an integer i, $-\infty < i < \infty$. At any time, the Turing machine scans the square indexed by 0. Depending upon its internal state and the scanned symbol, it can perform one or more of the following operations: replace the scanned symbol with a new symbol, focus attention on an adjacent square (by shifting the tape by one unit), and transfer to a new state.

The instructions for the Turing machine are quintuples of the form

$$[q_i, s_j, s_k, D, q_l]$$

where q_i and s_j represent the present state and scanned symbol, respectively, s_k is the symbol to be printed in place of s_j, D is the direction of motion (left-shift, right-shift, or no-shift of the tape), and q_l is the new internal state. For consistency, no two quintuples can have the same first two entries. If the Turing machine enters a state-symbol pair for which there is no corresponding quintuple, it is said to *halt*.

Without loss of generality, we can and will assume that $\Sigma = \{0, 1, \ldots, n-1\}$, $Q = \{0, 1, \ldots, m-1\}$, $n, m \in \mathbf{N}$, and that the Turing machine halts if and only if the internal state q is equal to zero. We refer to $q = 0$ as the *accepting* state.

The tape contents can be described by two infinite words w_1, $w_2 \in \Sigma^\omega$, where Σ^ω stands for the set of infinite words over the alphabet Σ: w_1 consists of the scanned symbol and the symbols to its right; w_2 consists of the symbols to the left of the scanned symbol, excluding the latter. The tape contents (w_1, w_2), together with an internal state $q \in Q$, constitute a *configuration* of the Turing machine. If a quintuple applies to a configuration (that is, if $q \neq 0$), the result is another configuration, a successor of the original. Otherwise, if no quintuple applies (that is, if $q = 0$), we have a *terminal* configuration. We thus obtain a successor function $\vdash$: $C \to C$, where $C = \Sigma^\omega \times \Sigma^\omega \times Q$ is the set of all configurations (the configuration space). Note that $\vdash$ is a partial function, as it is undefined when $q = 0$. A configuration is said to be *mortal* if repeated application of the function $\vdash$ eventually leads to a terminal configuration. Otherwise, the configuration is called *immortal*. We shall say that a Turing machine M is *mortal* if all configurations are mortal, and that it is *nilpotent* if there exists an integer k such that M halts in at most k steps starting from any configuration.

Theorem 4. *A Turing machine is mortal if and only if it is nilpotent.*

Proof. A nilpotent Turing machine is mortal, by definition. The converse will follow from Lemma 1. In order to apply that lemma, we endow the configuration space of a Turing machine with a topology which makes its successor function $\vdash$ continuous, and its configuration space (X, d) compact. This is a fairly standard construction and we refer the reader to the full version of the paper for a complete description. The constructed function $\vdash$ is identically equal to 0 in a neighborhood of 0. We therefore conclude from Lemma 1 that if M is mortal, then it must be nilpotent. □

The next result is due to Hooper and will play a central role in the sequel.

Theorem 5 ([13]). *The problem of determining whether a given Turing machine is mortal is undecidable.*

In other words, one cannot decide whether a given Turing machine halts for every initial configuration. Equivalently, one cannot decide whether there exists an immortal configuration.

4 Turing Machine Simulation

A *σ^*-function* is a function obtained by composing finitely many σ-functions. It is well known that Turing machines can be simulated by piecewise affine dynamical systems [15,16,18]. Moreover, this simulation can be performed with a σ^*-function (see the full version of the paper for the details of the construction of this function).

Lemma 2 ([15,16,18]). *Let M be a Turing machine and let $C = \Sigma^\omega \times \Sigma^\omega \times Q$ be its configuration space. There exists a σ^*-function $g_M : \mathbf{R}^2 \to \mathbf{R}^2$ and an encoding function $\nu : C \to [0,1]^2$ such that the following diagram commutes:*

$$\begin{array}{ccc} C & \xrightarrow{\vdash} & C \\ \nu\downarrow & & \downarrow\nu \\ \mathbf{R}^2 & \xrightarrow{g_M} & \mathbf{R}^2 \end{array}$$

(i.e. $g_M(\nu(c)) = \nu(c')$ for all configurations $c, c' \in C$ with $c \vdash c'$).

We extend this results by proving that any Turing machine can be simulated by a dynamical system in a stronger sense.

Lemma 3. *Let M be a Turing machine and let $C = \Sigma^\omega \times \Sigma^\omega \times Q$ be its configuration space. Then, there exists a σ^*-function $g_M : \mathbf{R}^2 \to \mathbf{R}^2$, a decoding function $\nu' : [0,1]^2 \to C$, and some subsets $\mathcal{N}^\infty \subset \mathcal{N}^1 \subset [0,1]^2$, $\mathcal{N}^1_{\neg acc} \subset \mathcal{N}^1$ such that the following conditions hold:*

1. *$g_M(\mathcal{N}^\infty) \subseteq \mathcal{N}^\infty$ and $\nu'(\mathcal{N}^\infty) = C$.*
2. *$\mathcal{N}^1_{\neg acc}$ (respectively $\mathcal{N}^1$) is the Cartesian product of two finite unions of closed intervals in $\mathbf{R}$. $\mathcal{N}^1_{\neg acc}$ is at a positive distance from the origin $(0,0)$ of $\mathbf{R}^2$.*
3. *For $x \in \mathcal{N}^1$, the configuration $\nu'(x)$ is nonterminal if and only if $x \in \mathcal{N}^1_{\neg acc}$.*
4. *The following diagram commutes:*

$$\begin{array}{ccc} C & \xrightarrow{\vdash} & C \\ \nu'\uparrow & & \uparrow\nu' \\ \mathcal{N}^1_{\neg acc} & \xrightarrow{g_M} & [0,1]^2 \end{array}$$

(i.e. $\nu'(x) \vdash \nu'(g_M(x))$ for all $x \in \mathcal{N}^1_{\neg acc}$).

Intuitively, ν' is an inverse of the encoding function ν of Lemma 2, in the sense that $\nu'(\nu(c)) = c$ holds for all configurations c. The set $\mathcal{N}^\infty$ is the image of the function ν, consisting of those points $x \in [0,1]^2$ that are unambiguously associated with valid configurations of the Turing machine. The set $\mathcal{N}^1$ consists of those points that lie in some set $B_{\alpha,\beta,q}$ and therefore encode an internal state q, a scanned symbol α, and a symbol β to the left of the scanned one. (However, not all points in $\mathcal{N}^1$ are images of valid configurations. Once it encounters a "decoding failure" our decoding function ν' sets the corresponding tape square, and all subsequent ones to the zero symbol.) Finally, $\mathcal{N}^1_{\neg acc}$ is the subset of $\mathcal{N}^1$ associated with the nonterminal internal states $q \neq 0$. See the full paper for complete details.

Using Lemma 3 and Theorem 5, we can now prove:

Theorem 6. *The problems of determining whether a given (possibly discontinuous) piecewise affine function in dimension 2 is (i) globally convergent, (ii) globally asymptotically stable, (iii) mortal, or (iv) nilpotent, are all undecidable.*

The undecidability of the first three properties was first established in [3]. That proof was based on an undecidability result for the mortality of counter machines, instead of Turing machines.

Proof. We use a reduction from the problem of Theorem 5. Suppose that a Turing machine M is given. Denote by g'_M the discontinuous function which is equal to the function g_M of Lemma 3 on $\mathcal{N}^1_{\neg acc}$, and which is equal to 0 outside of $\mathcal{N}^1_{\neg acc}$.

Since 0 is at a positive distance from $\mathcal{N}^1_{\neg acc}$, we have a neighborhood O of 0 such that $g'_M(O) = \{0\}$. By Lemma 1, all four properties in the statement of the theorem are equivalent.

Assume first that M is mortal. By Theorem 4, there exists k such that M halts on any configuration in at most k steps. We claim that ${g'_M}^{k+1}([0,1]^2) = \{0\}$. Indeed, assume, in order to derive a contradiction, that there exists a trajectory $x_{t+1} = g'_M(x_t)$ with $x_{k+1} \neq 0$. Since g'_M is zero outside $\mathcal{N}^1_{\neg acc}$, we have $x_t \in \mathcal{N}^1_{\neg acc}$ for $t = 0, \ldots, k$. By the commutative diagram of Lemma 3, the sequence $c_t = \nu'(x_t)$ $(t = 0, \ldots, k+1)$ is a sequence of successive configurations of M. This contradicts the hypothesis that M reaches a terminal configuration after at most k steps. It follows that g'_M satisfies properties (i) through (iv).

Conversely, suppose that M has an immortal configuration: there exists an infinite sequence c_t of non-terminal configurations with $c_t \vdash c_{t+1}$ for all $t \in \mathbf{N}$. By condition 1 of Lemma 3, there exists $x_0 \in \mathcal{N}^\infty$ with $\nu'(x_0) = c_0$. We claim that the trajectory $x_{t+1} = g'_M(x_t)$ is immortal: using condition 2 of Lemma 3, it suffices to prove that $x_t \in \mathcal{N}^1_{\neg acc}$ for all t. Indeed, we prove by induction on t that $x_t \in \mathcal{N}^1_{\neg acc} \cap \mathcal{N}^\infty$ and $\nu'(x_t) = c_t$ for all t. Using condition 3 of Lemma 3, the induction hypothesis is true for $t = 0$. Assuming the induction hypothesis for t, condition 1 of Lemma 3 shows that $x_{t+1} \in \mathcal{N}^\infty$. Now, the commutative diagram of Lemma 3 shows that $\nu'(x_{t+1}) = c_{t+1}$, and condition 3 of Lemma 3 shows that $x_{t+1} \in \mathcal{N}^1_{\neg acc}$. This completes the induction. Hence, g'_M is not mortal, and therefore does not satisfy any of the properties (i) through (iv).

5 The Hyperplane Problem

We now reach the second step of our proof. Using the undecidability result of Hooper for the mortality of Turing machines, we prove that it cannot be decided whether a given piecewise affine system has a trajectory that stays forever in a given hyperplane.

Theorem 7. *The problem of determining if a given σ^*-function $f : \mathbf{R}^3 \to \mathbf{R}^3$ has a trajectory $x_{t+1} = f(x_t)$ that belongs to $\{0\} \times \mathbf{R}^2$ for all t is undecidable*

Proof. We reduce the problem of Theorem 5 to this problem.

Suppose that a Turing Machine M is given. Consider the σ^*-function $f : \mathbf{R}^3 \to \mathbf{R}^3$ defined by

$$f(x_1, x_2, x_3) = \begin{pmatrix} \sigma(\sigma(Z_{\mathcal{N}^1_{\neg acc}}(x_2, x_3))) \\ g_M(x_2, x_3) \end{pmatrix}$$

where g_M is the function constructed in Lemma 3 and $Z_{\mathcal{N}^1_{\neg acc}}$ is a σ^*-function that is equal to zero for $x \in \mathcal{N}^1_{\neg acc}$ and is otherwize positive (an explicit construction of this function is provided in the full version of the paper). Note that in the definition of the function f we use a nested application of the function σ. This is to ensure that the definition of f involves an equal number of applications of the σ function on all its components.

Write $(x^1, \ldots, x^d)$ for the components of a point x of $\mathbf{R}^d$.

We prove that f has a trajectory $x_{t+1} = f(x_t)$ with $x^1_t = 0$ for all t, if and only if Turing machine M has an immortal configuration.

Suppose that f has such a trajectory. Since $Z_{\mathcal{N}^1_{\neg acc}}$, and hence $\sigma(\sigma(Z_{\mathcal{N}^1_{\neg acc}}))$, is strictly positive outside of $\mathcal{N}^1_{\neg acc}$, we must have $(x^2_t, x^3_t) \in \mathcal{N}^1_{\neg acc}$ for all $t \geq 0$. By the commutative diagram of Lemma 3, the sequence $\nu'(x^2_t, x^3_t)$, $t \in \mathbf{N}$, is a sequence of successive configurations of M. By condition 3 of Lemma 3, none of these configurations is terminal, i.e. $c_0 = \nu'(x^2_0, x^3_0)$ is an immortal configuration of M.

Conversely, assume that M has an immortal configuration, that is, there exists an infinite sequence of nonterminal configurations with $c_t \vdash c_{t+1}$. The argument here is the same as in the proof of Theorem 6. By condition 1 of Lemma 3, there exists a point $(x^2_0, x^3_0) \in \mathcal{N}^\infty$ with $\nu'(x^2_0, x^3_0) = c_0$. Consider the sequence defined by $(x^2_{t+1}, x^3_{t+1}) = g_M(x^2_t, x^3_t)$ for all t. Since $g_M(\mathcal{N}^\infty) \subseteq \mathcal{N}^\infty$, we have $(x^2_t, x^3_t) \in \mathcal{N}^\infty$ for all $t \geq 0$. Using the assumption that configuration c_t is nonterminal and condition 3 of Lemma 3, we deduce that $(x^2_t, x^3_t) \in \mathcal{N}^1_{\neg acc}$ for all $t \geq 0$, which means precisely that the sequence $x_t = (0, x^2_t, x^3_t)$, $t \in \mathbf{N}$, is a trajectory of f. □

6 Proof of the Main Theorem

We now reach the last step in the proof, which consists of reducing the problem of Theorem 7 to the problems of Theorem 1.

Recall that a *σ-function* is a function of the form $f(x) = \sigma(Ax + b)$ and a *σ_0-function* is a function of the form $f(x) = \sigma(Ax)$. A composition of finitely many σ_0-functions is called a *σ_0^*-function.*

Lemma 4. *The problems of determining whether a given σ_0^*-function $\mathbf{R}^4 \to \mathbf{R}^4$ is (i) globally convergent, (ii) globally asymptotically stable, (iii) mortal, or (iv) nilpotent, are all undecidable.*

Proof. The problem of Theorem 7 can be reduced to the mortality problem for σ_0^*-functions. The construction is such that the σ_0^*-function is equal to zero in an neighborhood of the origin (see the full paper for the construction of the function). It therefore follows from Lemma 1 that for this function, the properties (i)-(iv) are equivalent. These four properties are therefore undecidable. □

We can now prove Theorem 1.

Proof. (of Theorem 1) We reduce the problems in Lemma 4 to the problems in Theorem 1.

Let $f : \mathbf{R}^4 \to \mathbf{R}^4$ be a σ_0^*-function of the form $f = f_k \circ f_{k-1} \circ \ldots \circ f_1$ for some σ_0-functions $f_j(x) = \sigma(A_j x)$, where $f_j : \mathbf{R}^{d_{j-1}} \to \mathbf{R}^{d_j}$ with $d_0, d_1, \ldots, d_k \in \mathbf{N}$, and $d_0 = d_k = 4$.

Let $d = d_0 + d_1 + \cdots + d_k$, and consider the saturated linear function $f' : \mathbf{R}^d \to \mathbf{R}^d$ defined by $f'(x) = \sigma(Ax)$ where

$$A = \begin{pmatrix} 0 & 0 & \ldots 0 & A_k \\ A_1 & 0 & \ldots 0 & 0 \\ 0 & A_2 & \ldots 0 & 0 \\ \vdots & \vdots & 0 & 0 \\ 0 & 0 & \ldots A_{k-1} & 0 \end{pmatrix}$$

Clearly, the iterates of this function simulate the iterates of the function f.

Suppose that f' is mortal (respectively nilpotent, globally convergent, globally asymptotically stable). Then, the same is true for f: indeed, when $x_{t+1} = f(x_t)$ is a trajectory of f, the sequence $(x_t, f_1(x_t), \ldots, f_{k-1} \circ \ldots \circ f_1(x_t))$ is a subsequence of a trajectory of f'.

Conversely, let $x'_{t+1} = f'(x'_t)$ be a trajectory of f'. Write $x'_t = (y_t^1, \ldots, y_t^k)$ with each of the y^j in $\mathbf{R}^{d_{j-1}}$. For every $t_0 \in \{0, \ldots, k-1\}$ and $j \in \{1, \ldots, k\}$, the sequence $t \mapsto y^j_{t_0+kt}$ is a trajectory of f. This implies that the sequence y_t^j, $t \in \mathbf{N}$ is eventually null (respectively, converges to 0) if f is mortal (respectively, globally convergent). For the same reason, the global asymptotic stability of f implies that of f'; and if $f^m \equiv 0$ for some integer m, we have $(f')^{km} \equiv 0$. □

7 Continuous Piecewise Affine Systems

We proved in Theorem 6 that it cannot be decided whether a given *discontinuous* piecewise affine system of dimension 2 is globally convergent, globally asymptotically stable, mortal, or nilpotent. We do not know whether these problems remain undecidable when the systems are of dimension 1.

For continuous systems, we can prove the following.

Theorem 8. *For continuous piecewise affine systems in dimension* 3, *the four properties of global convergence, global asymptotic stability, mortality, and nilpotence are undecidable.*

Proof. The system built in the proof of Lemma 4 is of dimension 4. The construction can be adapted to a system of dimension 3. See the full paper. □

The following proposition is proved in [3].

Theorem 9. *For continuous piecewise affine systems in dimension* 1, *the properties of global convergence, global asymptotic stability, and mortality are decidable.*

One can also show that nilpotence is decidable for this class of systems. Thus, all properties are decidable for continuous piecewise affine systems in dimension 1, and are undecidable in dimension 3. The situation in dimension 2 has not been settled.

Global properties of $f : \mathbf{R}^n \to \mathbf{R}^n$	$n = 1$	$n = 2$	$n = 3$
Piecewise affine	?	Undecidable	Undecidable
Continuous piecewise affine	Decidable	?	Undecidable

8 Final Remarks

In addition to the two question marks in the table of the previous section, several questions which have arisen in the course of this work still await an answer:

1. Does there exist some fixed dimension n such that nilpotence (or mortality, global asymptotic stability and global convergence) of saturated linear systems of dimension n is undecidable? A negative answer would be somewhat surprising since there would be in that case a decision algorithm for each n, but no single decision algorithm working for all n.
2. It would be interesting to study the decidability of these four properties for other special classes of saturated linear systems, as we have already done for nilpotent and symmetric matrices. For instance, is global convergence or global asymptotic stability decidable for systems with invertible matrices? (Note that such a system cannot be nilpotent or mortal.) Are some of the global properties decidable for matrices with entries in $\{-1, 0, 1\}$?
3. For saturated linear systems, is mortality equivalent to nilpotence? Is global convergence equivalent to global asymptotic stability? (This last equivalence is conjectured in Section 2.) We show in the full version of the paper that these equivalences hold for systems with symmetric matrices.
4. For a polynomial map $f : \mathbf{R}^n \to \mathbf{R}^n$ mortality is equivalent to nilpotence; these properties are equivalent to the condition $f^n \equiv 0$, and hence decidable. It is however not clear whether the properties of global asymptotic stability and global convergence are equivalent, or decidable.

5. Does there exist a dimension n such that for any integer k there exists a nilpotent saturated linear system $f : \mathbf{R}^n \to \mathbf{R}^n$ such that $f^k \not\equiv 0$? Note that this question (and some of the other questions) still makes sense if we allow matrices with arbitrary real (instead of rational) entries.

Acknowledgments

Thanks are due to Eduardo Sontag for pointing out Theorem 2 and formulating his undecidability conjecture. P. K. also acknowledges a useful discussion with Eduardo Sontag on the nilpotence of analytic functions; J. T. and V. B. would like to thank Alexander Megretski for helpful comments.

This work was supported in part by the NATO under grant CRG-961115, by the National Science Foundation under grant ECS-9873451 and by the European Commission under the TMR Alapedes network and the Esprit working group NeuroCOLT2.

References

1. R. Alur, C. Courcoubetis, N. Halbwachs, T. A. Henzinger, P. H. Ho, X. Nicollin, A. Olivero, J. Sifakis, and S. Yovine. The algorithmic analysis of hybrid systems. *Theoretical Computer Science*, 138(1):3–34, 6 February 1995.
2. E. Asarin, O. Maler, and A. Pnueli. Reachability analysis of dynamical systems having piecewise-constant derivatives. *Theoretical Computer Science*, 138(1):35–65, February 1995.
3. V. D. Blondel, O. Bournez, P. Koiran, C. Papadimitriou, and J. N. Tsitsiklis. Deciding stability and mortality of piecewise affine dynamical systems. Submitted. Available from `http://www.inma.ucl.ac.be/~blondel/publications`.
4. V. D. Blondel, O. Bournez, P. Koiran, and J. N. Tsitsiklis. The stability of saturated linear dynamical systems is undecidable. Submitted. Available from `http://www.inma.ucl.ac.be/~blondel/publications`.
5. V. D. Blondel and J. N. Tsitsiklis. Complexity of stability and controllability of elementary hybrid systems. *Automatica*, 35(3), 479–489, 1999.
6. V. D. Blondel and J. N. Tsitsiklis. Overview of complexity and decidability results for three classes of elementary nonlinear systems. In Y. Yamamoto and S. Hara, editors, *Learning, Control and Hybrid Systems*, pages 46–58. Springer-Verlag, London, 1999.
7. V. D. Blondel and J. N. Tsitsiklis. Three problems on the decidability and complexity of stability. In V. D. Blondel, E. D. Sontag, M. Vidyasagar, and J. C. Willems, editors, *Open Problems in Mathematical Systems and Control Theory*, pages 45–52. Springer-Verlag, London, 1999.
8. V. D. Blondel and J. N. Tsitsiklis. A survey of computational complexity results in systems and control. To appear in *Automatica*, 1999.
9. O. Bournez. Complexité algorithmique des systèmes dynamiques continus et hybrides. PhD Thesis, Ecole Normale Supérieure de Lyon, 1999.
10. L. Blum, F. Cucker, M. Shub, and S. Smale. *Complexity and Real Computation*. Springer-Verlag, 1998.

11. L. Blum, M. Shub, and S. Smale. On a theory of computation and complexity over the real numbers: NP-completeness, recursive functions and universal machines. *Bulletin of the American Mathematical Society*, 21(1):1–46, July 1989.
12. T. A. Henzinger, P. W. Kopke, A. Puri, and P. Varaiya. What's decidable about hybrid automata? *Journal of Computer and System Sciences*, 57(1):94–124, August 1998.
13. P. K. Hooper. The undecidability of the Turing machine immortality problem. *The Journal of Symbolic Logic*, 31(2): 219–234, June 1966.
14. J. E. Hopcroft and J. D. Ullman. *Introduction to Automata Theory Languages and Computation.* Addison-Wesley, October 1979.
15. P. Koiran, M. Cosnard, and M. Garzon. Computability with low-dimensional dynamical systems. *Theoretical Computer Science*, 132(1-2):113–128, September 1994.
16. C. Moore. Generalized shifts: unpredictability and undecidability in dynamical systems. *Nonlinearity*, 4:199–230, 1991.
17. H. T. Siegelmann and E. D. Sontag. Analog computation via neural networks. *Theoretical Computer Science*, 131(2):331–360, September 1994.
18. H. T. Siegelmann and E. D. Sontag. On the computational power of neural nets. *Journal of Computer and System Sciences*, 50(1):132–150, February 1995.
19. E. D. Sontag. From linear to nonlinear: Some complexity comparisons. In *IEEE Conference on Decision and Control*, pages 2916–2920, New Orleans, December 1995.
20. E. D. Sontag. Interconnected automata and linear systems: A theoretical framework in discrete time. In R. Alur, T. A. Henzinger, and E. D. Sontag, editors, *Hybrid Systems III*, volume 1066 of *Lecture Notes in Computer Science*. Springer-Verlag, 1996.
21. E. D. Sontag. *Mathematical Control Theory: Deterministic Finite Dimensional Systems.* Springer-Verlag, New York, second edition, August 1998.

Tilings: Recursivity and Regularity

Julien Cervelle and Bruno Durand

LIM - CMI, 39 rue Joliot-Curie, 13453 Marseille Cedex 13, France.
{cervelle,bdurand}@cmi.univ-mrs.fr

Abstract. We establish a first step towards a "Rice theorem" for tilings: for non-trivial sets, it is undecidable to know whether two different tile sets produce the same tilings of the place. Then, we study quasiperiodicity functions associated with tilings. This function is a way to measure the regularity of tilings. We prove that, not only almost all recursive functions can be obtained as quasiperiodicity functions, but also, a function which overgrows any recursive function.

1 Introduction

Tilings have been studied for a very long time from different points of views. In 1961, Hao Wang introduced the formalism of colored square tiles (now called Wang tiles) in [19]. He was motivated by the problem of decidability of the satisfiability problem of a class of formula defined by the structure of its prenex normal form: the Kahr class $K = [\forall\exists\forall, (0, \omega)]$ (see [3]). A tile set τ can be recursively transformed into a formula F_τ of the Kahr class such that

- the plane can be tiled by τ if and only if the formula F_τ has a model,
- the plane can be tiled by τ periodically if and only if the formula F_τ has a finite model.

Whether the plane can be tiled by τ was then proved undecidable by Berger in 1966 (domino problem [2]) and the periodic case was proved undecidable by Gurevich and Koriakov in 1972 [12]. These proofs are based on a complicated construction due to Berger and clarified later [2, 17, 12, 1]. We also use this construction in the proofs of our theorems. As far as we know, nobody could prove the undecidability of Kahr class without using tilings, and the undecidability of the 8 other classes (necessary to close Hilbert's *Entscheidungsproblem*) is proved by reduction to the Kahr class. This justifies the importance of studying decision problems over tilings and also their periodicity aspects.

Thus, tilings became basic objects and were often used as tools for proving undecidability results for planar problems (see [14, 6]) and were also broadly used in complexity theory [18, 16, 11, 5, 7]. Later, Wang tiles were used as models for physical constraints in $\mathbb{Z}^2$ or $\mathbb{Z}^3$ mainly for studying quasicrystals (see [13]). Quasicrystals are related to a property called quasiperiodicity which is an extension of periodicity (called uniform recurrence in 1-dimensional language theory).

In this paper, our first goal is to prove a Rice theorem for tilings. In recursivity theory, this theorem says that given any property on functions, the set of

H. Reichel and S. Tison (Eds.): STACS 2000, LNCS 1770, pp. 491–502, 2000.

programs that compute this function is either trivial (*i.e.* empty or full) or non recursive. Intuitively, it seems that problems concerning tilings are also either trivial or undecidable. But the formalisation of this intuitive assertion is not at all trivial and we do not know any formula that could be a good candidate to apprehend this idea. Thus, we restrict to a slightly more restrictive version analogous in recursivity theory to the following proposition: given a program P, the set of programs that compute the same function as P is non recursive. In our tiling framework we prove that given a non-trivial tile set τ, the set of tiles that produce the same tilings of the plane as τ is non recursive (Th. 1). The same construction allows us to prove an analogous theorem concerning limit sets of cellular automata (Th. 2), thus improving Kari's result in [15].

In the sequel, we focus on the notion of quasiperiodicity (which extends periodicity). In [8, 9], an analogous of Furstenberg's lemma (see [10]) was proved for tilings: if a tile set can tile the plane, then it can also be used to form a quasiperiodic tiling of the plane. This means that even with a complicated set of tiles, one cannot force tilings to be very "irregular", "chaotic", or "complex" in an intuitive meaning, because some tilings will be quasiperiodic. Using a (strange) terminology used in physics: *local constraints cannot force chaos.* This result is rather surprising because using an adaptation of Berger's construction one can built some tile sets that can tile the plane, but such that none of the obtained tilings are recursive. The regularity of quasiperiodic tilings can be measured by their quasiperiodicity function. We prove that given any "natural" function f, one can construct a tile set such that the quasiperiodicity function of all tilings obtained is f – where "natural" is formalized near time-constructibility – (Th. 4). Our last result (Th. 5) states that there exists a tile set whose quasiperiodicity function grows to infinity faster than any recursive function. The intuitive meaning of this theorem is that although this tile set produces "regular" (*i.e.* quasiperiodic) tilings, the regularity of these tilings cannot be observed – by computer.

2 Definitions

The classical way to consider tilings of the plane is to use *Wang tiles* [19, 20]. A Wang tile is a unit square tile whose edges are colored. A tile set is a finite set of Wang tiles. A configuration consists of tiles which are placed on a two dimensional infinite grid. A tiling is a configuration in which two juxtaposed tiles have the same color on their common border.

A pattern is a restriction of a configuration to a finite domain of $\mathbb{Z}^2$. A square pattern of size a is a pattern whose domain is $[\![1, a]\!] \times [\![1, a]\!]$.

In order to prove a Rice theorem for tilings, we need to be able to compare tilings. If we use Wang's definition introduced above, we can't compare the tilings produced by two different tile sets, because they may use different colors.

Thus, we chose to use a more general modelization of tilings: *local constraints* (see also [14]). Let us focus on binary maps, *i.e.* configuration of 0 and 1. We "constrain" binary maps by imposing that all patterns of a fixed domain $\mathcal{D}$

extracted from the produced binary maps belong to a certain constraint set (which will be represented by a function from the set of patterns of domain $\mathcal{D}$ to $\{0,1\}$). This is what we will call a local constraint.

Definition 1. *We call* local constraint *a pair* $c = (V, f)$*, where* V *is a vector of* n *elements of* $\mathbb{Z}^2$ *and* f *a function from* $\{0,1\}^n$ *to* $\{0,1\}$*.* V *is called the* neighborhood *and* f *is called the* constraint function.

A binary map *is a function from* $\mathbb{Z}^2$ *to* $\{0,1\}$*. A binary map* p verifies *a local constraint* $c = (f, (v_1, \ldots, v_n))$*, or is* produced *by* c*, if and only if*

$$\forall x \in \mathbb{Z}^2,\ f(p(x+v_1), p(x+v_2), \ldots, p(x+v_n)) = 0.$$

A local constraint is trivial *if and only if it is verified by all binary maps, and* non-trivial *on the other case.*

We will denote by $\mathcal{P}(c)$ *the set of all binary maps verifying* c*.*

Definition 2. *A* binary pattern M *is a pair* $(\mathcal{D}, f)$ *where* $\mathcal{D}$ *is a finite sequence of* $\mathbb{Z}^2$*, and* f *a function from* $\mathcal{D}$ *into* $\{0,1\}$*.* $\mathcal{D}$ *is called the domain of* M*. For all* x *of* $\mathcal{D}$*, we note* $M(x)$ *for* $f(x)$*.*

A binary pattern M *of domain* $\mathcal{D}$ *is* extracted from *a binary map* m *if and only if there exists a* position (i,j) *such that :*

$$\forall z \in \mathcal{D},\ M(z) = m(z + (i,j))$$

Now, we define some key properties of binary patterns.

Definition 3. *We say that a binary pattern* M *of domain* $\mathcal{D}$ cancels out *a constraining function* f *of a local constraint whose neighborhood is* $(n_i)_{1 \le i \le k}$*, if and only if, for all* x *in* $\mathcal{D}$ *such that all the* $x + n_i$ *are in* $\mathcal{D}$*,* $f(M(x+n_1, x+n_2, \ldots, x+n_k) = 0$*.*

A pattern verifies *a local constraint* c *if and only if it cancels out the constraint function of* c*. Otherwise, it is called a* forbidden *pattern for* c*.*

Remark 1. A binary map verifies a local constraint if and only if all its patterns verify this local constraint.

In some way, the approach of planar tilings with local constraint is equivalent to the tiling of the plane using Wang tiles: we can associate to a Wang tile set a local constraint such that each Wang tile corresponds to a certain pattern of 0 and 1, and such that the plane is tilable [resp. periodically tilable] by the Wang tile set if and only if there is a binary map which verifies the local constraint. Conversely, we can associate to each local constraint a Wang tile set such that each Wang tile corresponds to an accepted pattern, and verifying the same equivalence of tilability.

3 Towards a Rice Theorem for Tilings

Thanks to the notion of local constraints, we are able to formulate a theorem which is not exactly a Rice theorem for tilings but is a first step in the direction of this goal. Given two local constraints a and b, we are now able to compare the two sets $\mathcal{P}(a)$ and $\mathcal{P}(b)$ of those tilings they produce since they both are sets of binary maps.

Let c be a local constraint. We call SM(c), the following problem:
Problem: SM(c)
Instance: A local constraint ν
Question: Does the local constraint ν produce exactly the same binary maps as c, (*i.e.* $\mathcal{P}(\nu) = \mathcal{P}(c)$)?

Theorem 1. *Let c be a non-trivial local constraint. Then,* SM(c) *is undecidable (more precisely Σ_1-complete).*

Proof. The idea of the proof is that this problem is at least as difficult to solve as to decide whether there exists a binary map which verifies a given local constraint (domino problem). We choose a forbidden pattern F for c. Then, from a local constraint q, we build a local constraint $\hat{q}$ which produces strictly more maps than c if and only if q produces a binary map. In fact, we use F as a delimiter to code binary maps in the "language" of c.

The following decision problem is called "domino problem" and expressed here for local constrains. It was proved undecidable by Breger in [2]:
Problem: DOMINO
Instance: A local constraint ν
Question: Does a binary map that verifies ν (*i.e.* $\mathcal{P}(\nu) \neq \emptyset$) exist?

This problem has been proved in [2] to be undecidable.

Without loss of generality, we can consider only local constraints whose neighborhood is $[\![1,a]\!] \times [\![1,a]\!]$. Let $c = ([\![1,a]\!] \times [\![1,a]\!], \delta)$ be a non-trivial local constraint. As c is non-trivial, there exists a forbidden binary pattern $F = ([\![1,a]\!] \times [\![1,a]\!], f)$. Thus, any binary map which contains the pattern F does not verify c.

Let's prove that the set of all local constraints which are verified by the same binary maps as c is Σ_1-complete for many-one reductions. To prove that it is recursively enumerable (which is not completely straightforward), we show that the set of all pairs of local constraints which produce exactly the same binary maps is recursively enumerable.

Let f be a program which, given a pair of local constraints (μ, ν), executes the following steps. If $[\![1,m]\!] \times [\![1,m]\!]$ is the domain of μ and $[\![1,n]\!] \times [\![1,n]\!]$, the domain of ν:

- `If` μ `is trivial then halt.`
- `For all forbidden binary pattern` F `of domain` $[\![1,m]\!] \times [\![1,m]\!]$ `of` μ `do`
  - $i = 0$
  - `Repeat (loop` α`)`
    - `Check if all patterns of domain` $[\![1,m+2i]\!] \times [\![1,m+2i]\!]$ `whose centered subpattern is` F `are forbidden for` ν`.`

* **If yes, exit this loop, else increase** i
• **End Repeat**
– **End For**

Suppose μ is non-trivial and assume that $f(\mu,\nu)$ halts. Let P be a binary map that does not verify μ. Then, it contains a forbidden pattern F of domain $[\![1,m]\!] \times [\![1,m]\!]$, at position (x,y). As the program exits the loop α when F is treated, there exists an integer i such that any pattern of domain $[\![1,m+2i]\!] \times [\![1,m+2i]\!]$ whose center subpattern is F is forbidden for ν. Thus, the pattern of P of position $(x-i,y-i)$ is forbidden for ν. Hence, P does not verify ν. All binary maps which don't verify μ don't verify ν.

Conversely, suppose $f(\mu,\nu)$ does not halt. Then, there exists a pattern F for which the loop α never ends. This means that for all integers i, there exists a pattern of domain $[\![1,m+2i]\!] \times [\![1,m+2i]\!]$ whose center subpattern is F verifying ν. Then, by standard diagonal extraction (also called König's lemma, or countable Tychonoff theorem, see [9] for an explanation of these notions in the world of tilings) there exists a binary map R which contains F and is produced by ν. R verifies ν but not μ.

We just proved that $f(\mu,\nu)$ halts if and only if all binary maps that verify ν also verify μ. This is also true when μ is trivial. Let M be the Turing machine which, on input (μ,ν), compute $f(\mu,\nu)$ and then $f(\nu,\mu)$. M halts on (μ,ν) if and only if μ and ν produce the same binary maps. The set of pairs of local constraints which produce the same binary maps is recursively enumerable.

Let's now prove that the set of all local constraints which are verified by the same binary maps as c is Σ_1-hard. Let q be a local constraint whose domain is $[\![1,l]\!] \times [\![1,l]\!]$.

For all binary patterns M of domain $[\![1,x]\!] \times [\![1,x]\!]$, we'll note $\widehat{M}$ the pattern of domain $[\![1,a(2x+1)]\!] \times [\![1,a(2x+1)]\!]$ illustrated by Fig. 1, and build as follows. The pattern $\widehat{M}$ is a juxtaposition of $2x+1$ by $2x+1$ square patterns of size a. Those patterns are:

– For all $i \in [\![0,x]\!]$, and all $j \in [\![0,x]\!]$, the pattern located at position $(2i+1,2j+1)$ is equal to F;
– For all $i \in [\![1,x]\!]$, and all $j \in [\![1,x]\!]$, the pattern located at the position $(2i,2j)$ filled with $M(i,j)$;

For any binary map $\mathcal{P}$, let's define the binary map $\widehat{\mathcal{P}}$ as follows: $\forall(m,n) \in \mathbb{Z}^2$, $\forall(i,j) \in [\![1,a]\!] \times [\![1,a]\!]$,

$$\widehat{\mathcal{P}}(2na+i,2ma+j) = P(n,m);$$

$$\widehat{P}(na+i,(2m+1)a+j) = \widehat{P}((2n+1)a+i,ma+j) = F(i,j).$$

This definition is the extension to binary maps of the previous definition.

Finaly, let's define the local constraint $\widehat{q}$ whose domain is $[\![1,2l+1]\!] \times [\![1,2l+1]\!]$ and whose constraint function is f. The value of f is 0 only on the following binary patterns:

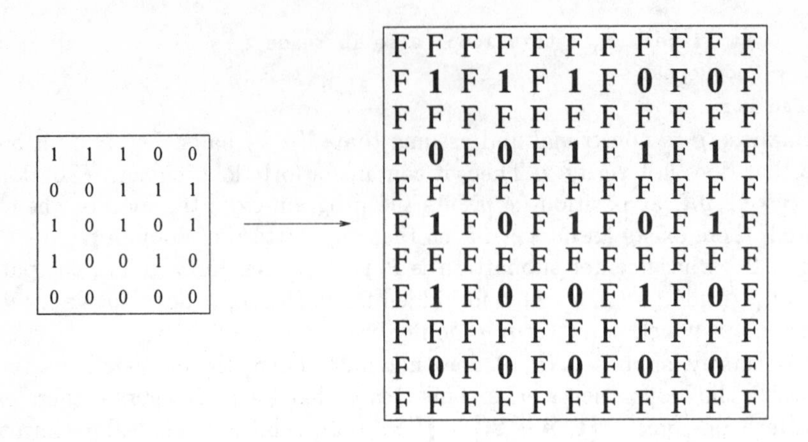

Fig. 1. A sample pattern L, and its associated $\widehat{L}$

- all binary patterns which verify c (rule 1);
- all sub-patterns of domain $[\![1, 2l+1]\!] \times [\![1, 2l+1]\!]$ of all $\widehat{N}$ such that N has $[\![1, l+1]\!] \times [\![1, l+1]\!]$ for domain and verifies q (rule 2).

Any binary map verifying c verifies $\widehat{q}$ because of rule 1. Moreover, if a binary map m verifies q, $\widehat{m}$ verifies $\widehat{q}$ because of rule 2. Hence if there is a binary map verifying q, $\widehat{q}$ verifies strictly more maps than c.

Conversely, if there is a binary map m that verifies $\hat{q}$ but not c, then rule 2 is used at least once, since if not, m would verify c. We deduce that there is a pattern $A = \widehat{M}$ in m, when M is a pattern of domain $[\![1, l]\!] \times [\![1, l]\!]$, and which cancels out the constraint function of q. Since F doesn't cancel out the constraint function of c, the rule 2 is applied to all the patterns that contain F, *i.e.* to all the patterns whose intersection with A is of at least the size of F. Thus only the rule 2 is applied. This implies that $m = \widehat{l}$, for a l which verifies q. Hence, if $\hat{q}$ produces strictly more maps than c, then there is a map which verifies q.

We can conclude that there is a binary map which verifies q if and only if $\hat{q}$ is verified by more maps than c. As $\hat{q}$ can be easily constructed from q by a Turing machine, we have constructed a many-one reduction and the set of all local constraints which verify exactly the same binary maps as c is *many-one-complete*, thus not recursive. □

Thanks to the same kind of proof, we can state an analogous result for cellular automata (see [4] for a modern book on this topic). A part of the proof (enumerability) is more camplicated and uses other arguments; it will be published elswhere.

Definition 4 (limit set). *Let C be a planar (*i.e. *2-dimensionnal) cellular automaton with two states (0 and 1). The limit set $\mathcal{L}(C)$ of C is defined as follows:*

$$A^{(0)} = \{0,1\}^{\mathbb{Z}^2}$$
$$\forall n \geq 1 \ \ A^{(n)} = C(A^{(n-1)})$$
$$\mathcal{L}(C) = \bigcap_{i \in \mathbb{N}} A^{(i)}$$

Let C be a planar cellular automaton, with only two states. We call LS(C), the following problem:
Problem: LS(C)
Instance: A planar cellular automaton A, with exactly two states
Question: Are the limit sets of A and C the same, (*i.e.* $\mathcal{L}(A) = \mathcal{L}(C)$)?

The theorem we prove with almost exactly the same proof as Th. 1 is the following:

Theorem 2. *Let C be a planar cellular automaton, with two states. Then the problem* LS(C) *is undecidable (more precisely Σ_1-complete).*

This theorem improves a result proved in [15], where the number of states of the considered cellular automata is not bounded.

4 Quasiperiodicity

We now present a notion that gives us an idea on the regularity of a tiling. Indeed, a tiling can be more or less regular, that is to say its patterns can be repeated more or less often.

4.1 Definition

We introduce the notion of quasiperiodicity which is a generalisation of the notion of periodicity (called uniform recurrence in language theory). A tiling is quasiperiodic if and only if for each of its subpatterns, there exists a size of a window such that, in any place we put the window on the tiling, we can find a copy of this pattern within the window (see [9]).

Definition 5. *Let P be a configuration made of Wang tiles. P is* quasiperiodic *if and only if for all patterns M extracted from P, there exits an integer n such that M appears in all square patterns of size n extracted from P.*

The following theorem [9] is analogous of Furstenberg lemma in language theory.

Theorem 3. *If τ is a tile set which can tile the plane, then there exists a quasiperiodic tiling of the plane made with tiles of τ.*

We can now define quasiperiodicity functions, which allow us to describe the regularity of quasiperiodic configurations.

Definition 6. *Let c be a quasiperiodic configuration. We call* quasiperiodicity function *of c, the function which maps positive integers n to the smallest integer* $\mathcal{Q}_c(n)$ *such that in all square patterns of size* $\mathcal{Q}_c(n)$ *extracted from c, we can observe a sample of all the square patterns of size n that appear in c.*

4.2 Quasiperiodicity Functions

We now study what kinds of quasiperiodicity functions can be obtained. We first prove that "usual" increasing recursive functions can be obtained.

Definition 7. *A function f is* time-constructible *if and only if there is a Turing machine M over the alphabet* $\{0,1\}$ *which stops after* $f(i)$ *steps of computation on the entry* 1^i.

Theorem 4. Informal version : *any increasing time-constructible functions can be observed as a quasiperiodicity function of a tiling;*
Formal version : *for all time-constructible function f, there exists a tile set* τ *such that, for all tiling p produced by* τ,

$$\mathcal{Q}_p(x) > (\textstyle\int f(x) + 3)^2 + 1 > \frac{\mathcal{Q}_p(x) - 1}{4} + 1$$

where

$$\textstyle\int f(x) = \sum_{i=0}^{x} (i+1) f(i)$$

Proof. Let M be a Turing machine which witnesses that f is time-constructible.

The idea of the proof is to emulate the space-time diagram of the computation of M on the entry i, and then, to include a new pattern of size $i+1$ in the tiling, begining by $i = 0$, then $i = 1$, etc.

In order to do that, we emulate evolution of Turing machines in tilings using Berger's well-known construction described in [2]. This difficult technique was invented for proving the undecidability of the domino problem [2, 17, 1].

Let $\{q_0, q_1, \ldots, q_k\}$ be the states of the machine M and q_f its final state.

In order to be sure that a new small pattern won't appear in the space-time diagram of M, which would increase the value of the quasiperiodicity function, we "stuff" the diagram with #. That is to say we replace this diagram by another n times bigger, in which each character a is replaced by the $n \times n$ following square represented Fig. 2 This is what we'll call an *n-stuffed* diagram.

In each area of computation of the construction, we produce the following diagrams, one above the other, starting with $n = 0$, increasing n by 1 at each step:

$$\begin{array}{cccc} a & \# & \cdots & \# \\ \# & \# & \cdots & \# \\ \vdots & \vdots & \ddots & \vdots \\ \# & \# & \cdots & \# \end{array}$$

Fig. 2. A stuffed letter

$$\begin{array}{l} \left.\begin{array}{l} 1^n\diamond \\ \vdots \\ 1^n\diamond \end{array}\right\} \text{computation of } M \text{ on } 1^n,\ n\text{-stuffed} \\ 1^n \diamond (1\#^n)^n(0\#^n)^\infty \\ \left.\begin{array}{l} 1^n \diamond \#^\infty \\ \quad\vdots \\ 1^n \diamond \#^\infty \end{array}\right\} n \text{ lines} \\ 1^n \diamond 0\#^n 1\#^n q_0\#^n q_1\#^n \cdots q_k\#^n \\ \left.\begin{array}{l} 1^n \diamond \#^\infty \\ \quad\vdots \\ 1^n \diamond \#^\infty \end{array}\right\} n \text{ lines} \end{array}$$

As we list all the letters (0 and 1) and all the states of the Turing machine, we are sure that no new patterns of size at most n can appear.

From step i to $i + 1$, first, the information that a final state has occured is sent to the left margin. Then the number of 1 before the $\diamond$ is increased by 1. This can be done as shown Fig. 3

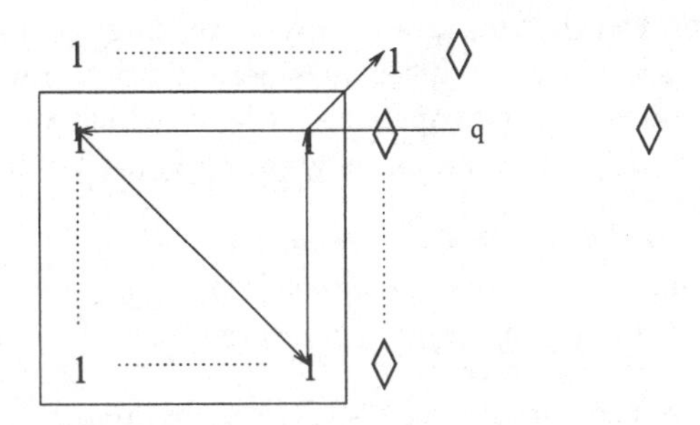

Fig. 3. How to increase the number of 1

Once we have increased the number of 1, we add the step $i + 1$. The $i + 1$-stuffing is illustrated Fig. 4, where the big squares represent the stuffed letters of Fig 2, except for the left column which is the margin of the 1 before the $\diamond$. Nevertheless, the sub-pattern of size i of such squares of size strictly greater than i are the same, even considering the construction arrow. It proves that no new

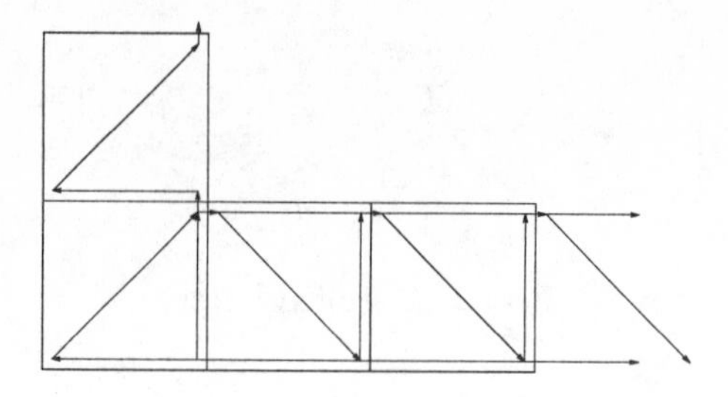

Fig. 4. How to i-stuff the time-space diagram

pattern of size i appears in step $i+1$, and also $i+2$, etc. The minimum size of window which is sure to contain those i first steps is the smallest that is big enough to contain all sub-patterns of size i.

Thanks to Berger's construction, we know that all windows of size $2^{2n+1}+1$ contain the $2^{n+2}-3$ first lines of the construction. Thus, to obtain all patterns of size x, we need x first steps of the construction. The size of these steps is exactly $\int f(x)$. So we need a window size of $2^{2n+1}+1$, where n is the smallest integer such that $2^{n+2}-3 > \int f(x)$. We deduce that if q is the quasiperiodicity function of this tiling,

$$q(x) = 2^{2n+1} + 1$$

and so that

$$q(x) > (\int f(x) + 3)^2 + 1 > \frac{q(x) - 1}{4} + 1$$

□

Since we have proved that for almost any increasing recursive function f, we can construct a tile set which produces quasiperiodic tilings whose all quasiperiodicity functions are nearly f, we construct a tile set which produce quasiperiodic tilings whose quasiperiodicity function is greater than all recursive functions.

Theorem 5. *There exists a tile set τ_0 such that, for all tiling p produced by τ_0, p is quasiperiodic, and no recursive function is an upper bound of $\mathcal{Q}_p$.*

The regularity of any quasiperiodic tiling produced by τ_0 cannot be observed

Proof. Let K be a non-recursive, recursively enumerable subset of $\{0,1\}^\star$ (e.g. the set of all pairs (x, y) such that the Turing machine of number x halts on entry y).

The idea of the proof is to build a tile set which emulates a Turing machine which enumerates K. If we can compute a size of frame in which all patterns of size n appear, then we can obtain an upper bound for the number of steps needed by the machine to output all elements of K of size at most n. This would lead to a decision algorithm for K, which is impossible.

Let M be a Turing machine over the alphabet $\{0, 1, \#\}$, with two tapes, the first one being the working tape, and which, on the empty entry, enumerates the elements of K, and outputs them on the second tape, separated by #, in increasing order.

For instance, if the enumeration of K is $23, 67, 45, 12, 36, 52, 38, \ldots$, the machine will output:

23
23#**67**
23#**45**#67
12#23#45#67
12#23#**36**#45#67
12#23#36#45#**52**#67
12#23#36#**38**#45#52#67
⋮

In the sequel, we use the Berger's construction (see [1]).

Let τ_0 be the tile set that, through Berger's construction emulates this Turing machine M. Let p be a tiling produced by τ_0. Suppose there is a recursive function f such that for all x, we have $f(x) > \mathcal{Q}_p(x)$. Let's prove that K is recursive.

From Berger's construction, we observe that there exists a recursive function ρ, such that n cells of the second tape are represented in any $\rho(n) \times \rho(n)$ patterns of any tiling produced by τ_0. In order to decide whether x belongs to K, we have to know if it is enumerated by M.

Suppose x is enumerated. It is written at most at $a = \rho(1 + \sum_{i=0}^{x} i + 1)$ cells from the beginning of the second tape, since the machine output at most all the numbers between 0 and x to the left of x. As all the square patterns of length a are found in all patterns of length $f(a)$, if x is enumerated by M then $\#x\#$ can be observed in all $f(a) \times f(a)$ patterns found in any tiling produced by τ_0. As in this construction all tilings contain the same (finite) patterns – we say that they are mutually extractible (see [9, 8]) – if $\#x\#$ can be observed in a $f(a) \times f(a)$ pattern of a tiling produced by τ_0, then x is enumerated by M. But thanks to Berger's construction, we can construct square patterns of arbitrary size that can be extended to tilings of the plane produced by τ_0. Hence we can decide whether x is or not enumerated by M which contradicts the non-recursivity of K. □

Acknowledgement

The authors thank Pr. Nikolai K. Vereshchagin and Pr. Alexander Shen for their help and comments.

References

[1] C. Allauzen and B. Durand. Appendix A: "Tiling problems". *The classical decision problem (see [3])*, pages 407–420, 1996.

[2] R. Berger. The undecidability of the domino problem. *Memoirs of the American Mathematical Society*, 66, 1966.

[3] E. Börger, E. Grädel, and Y. Gurevich. *The classical decision problem.* Springer-Verlag, 1996.
[4] M. Delorme and J. Mazoyer, editors. *Cellular automata: a Parallel model.* Kluwer, 1999.
[5] B. Durand. Inversion of 2d cellular automata: some complexity results. *Theoretical Computer Science*, 134:387–401, 1994.
[6] B. Durand. The surjectivity problem for 2D cellular automata. *Journal of Computer and Systems Science*, 49(3):718–725, 1994.
[7] B. Durand. A Random NP-complete problem for inversion of 2D cellular automata. *Theoretical Computer Science*, 148(1):19–32, 1995.
[8] B. Durand. Tilings and quasiperiodicity. In *ICALP'97*, volume 1256 of *Lecture Notes in Computer Science*, pages 65–75. Springer Verlag, July 1997.
[9] B. Durand. Tilings and quasiperiodicity. *Theoretical Computer Science*, 221:61–75, 1999.
[10] H. Furstenberg. *Recurrence in ergodic theory and combinatorial number theory.* Princetown University Press, 1981.
[11] Y. Gurevich. Average case completeness. *Journal of Computer and System Sciences*, 42:346–398, 1991.
[12] Y. Gurevich and I. Koriakov. A remark on Berger's paper on the domino problem. *Siberian Journal of Mathematics*, 13:459–463, 1972. (in Russian).
[13] K. Ingersent. *Matching rules for quasicrystalline tilings*, pages 185–212. World Scientific, 1991.
[14] J. Kari. Reversibility and surjectivity problems of cellular automata. *Journal of Computer and System Sciences*, 48:149–182, 1994.
[15] J. Kari. Rice's theorem for the limit set of cellular automata. *Theoretical Computer Science*, 127(2):229–254, 1994.
[16] L. Levin. Average case complete problems. *SIAM J. Comput*, 15(1):285–286, February 1986.
[17] R.M. Robinson. Undecidability and nonperiodicity for tilings of the plane. *Inventiones Mathematicae*, 12:177–209, 1971.
[18] P. van Embde Boas. Dominoes are forever. Research report 83-04, University of Amsterdam. Department of Mathematics., 1983.
[19] H. Wang. Proving theorems by pattern recognition II. *Bell System Technical Journal*, 40:1–41, 1961.
[20] H. Wang. Dominoes and the $\forall\exists\forall$-case of the decision problem. In *Proc. Symp. on Mathematical Theory of Automata*, pages 23–55. Brooklyn Polytechnic Institute, New York, 1962.

Listing All Potential Maximal Cliques of a Graph

Vincent Bouchitté and Ioan Todinca

LIP-École Normale Supérieure de Lyon
46 Allée d'Italie, 69364 Lyon Cedex 07, France
{Vincent.Bouchitte, Ioan.Todinca}@ens-lyon.fr

Abstract. A potential maximal clique of a graph is a vertex set that induces a maximal clique in some minimal triangulation of that graph. It is known that if these objects can be listed in polynomial time for a class of graphs, the treewidth and the minimum fill-in are polynomially tractable for these graphs. We show here that the potential maximal cliques of a graph can be generated in polynomial time in the number of minimal separators of the graph. Thus, the treewidth and the minimum fill-in are polynomially tractable for all graphs with polynomial number of minimal separators.

1 Introduction

The notion of *treewidth* was introduced at the beginning of the eighties by Robertson and Seymour [25, 26] in the framework of their graph minor theory. A graph H is a minor of a graph G if we can obtain H from G by using the following operations: discard a vertex, discard an edge, merge the endpoints of an edge in a single vertex. Among the deep results obtained by Robertson and Seymour, we can cite the fact that every class of graphs closed by minoration which does not contain all the planar graphs has bounded treewidth.

A graph is *chordal* or *triangulated* if every cycle of length greater or equal to four has a chord, i.e. edge between two non-consecutive vertices of the cycle. A triangulation of a graph is a chordal embedding, that is a supergraph, on the same vertex set, which is triangulated. The treewidth problem consists in finding a triangulation such that the size of the biggest clique is as small as possible. Another closed problem is the minimum fill-in problem. Here we have to find a triangulation of the graph such that the number of the added edges is minimum. In both cases we can restrict to *minimal triangulations*, i.e. triangulations with a set of edges minimal by inclusion.

The treewidth and the minimum fill-in play an important role in various areas of computer science e.g. sparse matrix factorization [27], and algorithmic graph theory [3, 14, 2, 8]. For an extensive survey of these applications see also [5, 7].

Unfortunately the computation of the treewidth and of the minimum fill-in of a graph are NP-hard [1, 30] even for co-bipartite graphs.

There exist several classes of graphs with unbounded treewidth for which we can solve polynomially the problem of the treewidth and the minimum fill-in.

H. Reichel and S. Tison (Eds.): STACS 2000, LNCS 1770, pp. 503–515, 2000.

Among them there are the chordal bipartite graphs [19, 12], circle and circular-arc graphs [28, 23], AT-free graphs with polynomial number of minimal separators [22]. Most of these algorithms use the fact that these classes of graphs have a polynomial number of *minimal separators.* It was conjectured in [17, 18] that the treewidth and the minimum fill-in should be tractable in polynomial time for all the graphs having a polynomial number of minimal separators. We solve here this ESA'93 conjecture.

The crucial interplay between the minimal separators of a graph and the minimal triangulations was pointed out by Kloks, Kratsch and Müller in [21], these results were concluded in Parra and Scheffler [24]. Two minimal separators S and T cross if T intersects two connected components of $G\backslash S$, otherwise they are parallel. The result of [24] states that a minimal triangulation is obtained by considering a maximal set of pairwise parallel separators and by completing them i.e. by adding all the missing edges inside each separator. However this characterization gives no algorithmic information about how we should construct a minimal triangulation in order to minimize the cliquesize or the fill-in.

Trying to solve this later conjecture, we studied in [10, 11] the notion of *potential maximal clique.* A vertex set K is a potential maximal clique if it appears as a maximal clique in some minimal triangulation. In [10], we characterized a potential maximal clique in terms of the *maximal sets of neighbor separators*, which are the minimal separators contained in it. We designed an algorithm which takes as input the graph and the maximal sets of neighbor separators and which computes the treewidth in polynomial time in the size of the input. For all the classes mentioned above we can list the maximal sets of neighbor separators in polynomial time, so we unified all the previous algorithms. Actually, the previous algorithms compute the maximal sets of neighbor separators in an implicit manner. In [11], we gave a new characterization of the potential maximal cliques avoiding the minimal separators. This allowed us to design a new algorithm that, given a graph and its potential maximal cliques, computes the treewidth and the minimum fill-in in polynomial time. Moreover this approach permitted us to solve the two problems for a new class of graphs, namely the weakly triangulated graphs. It was probably the last natural class of graphs with polynomial number of minimal separators for which the two problems remained open.

This paper is devoted to solve the ESA'93 conjecture, that is the treewidth and the minimum fill-in are polynomially tractable for the whole class of graphs having a polynomial number of minimal separators. Recall that if we are able to generate all the potential maximal cliques of any graph in polynomial time in the number of its minimal separators, then the treewidth and the minimum fill-in are also computable in polynomial time in the number of minimal separators. We define the notion of *active separator* for a potential maximal clique which leads to two results. First, the number of potential maximal cliques is polynomially bounded by the number of minimal separators. Secondly, we are able to enumerate the potential maximal cliques in polynomial time in their number.

These results reinforce our conviction that the potential maximal cliques are the pertinent objects to study when dealing with treewidth and minimum fill-in.

2 Preliminaries

Throughout this paper we consider finite, simple, undirected and connected graphs.

Let $G = (V, E)$ be a graph. We will denote by n and m the number of vertices, respectively the number of edges of G. For a vertex set $V' \subseteq V$ of G, we denote by $N_G(V')$ the neighborhood of V' in $G \backslash V'$ - so $N_G(V') \subseteq V \backslash V'$.

A subset $S \subseteq V$ is an *a, b-separator* for two nonadjacent vertices $a, b \in V$ if the removal of S from the graph separates a and b in different connected components. S is a *minimal a, b-separator* if no proper subset of S separates a and b. We say that S is a *minimal separator* of G if there are two vertices a and b such that S is a minimal a, b-separator. Notice that a minimal separator can be strictly included in another one. We denote by Δ_G the set of all minimal separators of G.

Let G be a graph and S a minimal separator of G. We note $\mathcal{C}_G(S)$ the set of connected components of $G \backslash S$. A component $C \in \mathcal{C}_G(S)$ is a *full component associated to* S if every vertex of S is adjacent to some vertex of C, i.e. $N_G(C) = S$. The following lemmas (see [15] for a proof) provide different characterizations of a minimal separator:

Lemma 1. *A set S of vertices of G is a minimal a, b-separator if and only if a and b are in different full components of S.*

Lemma 2. *Let G be a graph and S be an a, b-separator of G. Then S is a minimal a, b-separator if and only if for any vertex x of S there is a path from a to b that intersects S only in x.*

If $C \in \mathcal{C}(S)$, we say that $(S, C) = S \cup C$ is a *block* associated to S. A block (S, C) is called *full* if C is a full component associated to S.

Let now $G = (V, E)$ be a graph and $G' = G[V']$ an induced subgraph of G. We will compare the minimal separators of G and G'.

Lemma 3. *Let G be a graph and $V' \subset V$ a vertex set of G. If S is a minimal a, b-separator of the induced subgraph $G' = G[V']$, then there is a minimal a, b-separator T of G such that $T \cap V' = S$.*

Proof. Let $S' = S \cup (V \backslash V')$. Clearly, S' is an a, b-separator in G. Let T be any minimal a, b-separator contained in S'. We have to prove that $S \subseteq T$. Let x be any vertex of S and suppose that $x \notin T$. Since S is a minimal a, b-separator of G', we have a path μ joining a and b in G' that intersects S only in x (see lemma 2). But μ is also a path of G, that avoids T, contradicting the fact that T is an a, b-separator. It follows that $S \subseteq T$. Clearly, $T \cap V' \subseteq S$ by construction of T, so $T \cap V' = S$. □

The next corollary follows directly from lemma 3.

Corollary 1. *Let $G = (V, E)$ be a graph and a be a vertex of G. Consider the graph $G' = G[V \backslash \{a\}]$. Then for any minimal separator S of G', we have that S or $S \cup \{a\}$ is a minimal separator of G. In particular, $|\Delta_G| \geq |\Delta_{G'}|$.*

3 Potential Maximal Cliques and Maximal Sets of Neighbor Separators

The *potential maximal cliques* are the central object of this paper. We present in this section some known results about the potential maximal cliques of a graph (see also [10, 11, 29]).

Definition 1. *A vertex set Ω of a graph G is called a* potential maximal clique *if there is a minimal triangulation H of G such that Ω is a maximal clique of H.*

We denote by Π_G the set of potential maximal cliques of the graph G.

A potential maximal clique Ω is strongly related to the minimal separators contained in Ω. In particular, any minimal separator of G is contained in some potential maximal clique of G. The number $|\Pi_G|$ of potential maximal cliques of G is at least $|\Delta_G|/n$.

If K is a vertex set of G, we denote by $\Delta_G(K)$ the minimal separators of G included in K.

Definition 2. *A set $\mathcal{S}$ of minimal separators of a graph G is called* maximal set of neighbor separators *if there is a potential maximal clique Ω of G such that $\mathcal{S} = \Delta_G(\Omega)$. We also say that $\mathcal{S}$* borders Ω *in G.*

We proved in [11] that the potential maximal cliques of a graph are sufficient for computing the treewidth and the minimum fill-in of that graph.

Theorem 1. *Given a graph G and its potential maximal cliques Π_G, we can compute the treewidth and the minimum fill-in of G in $\mathcal{O}(n^2|\Delta_G| \times |\Pi_G|)$ time.*

Let now K be a set of vertices of a graph G. We denote by $C_1(K), \ldots, C_p(K)$ the connected components of $G \backslash K$. We denote by $S_i(K)$ the vertices of K adjacent to at least one vertex of $C_i(K)$. When no confusion is possible we will simply speak of C_i and S_i. If $S_i(K) = K$ we say that $C_i(K)$ is a *full component* associated to K. Finally, we denote by $\mathcal{S}_G(K)$ the set of all $S_i(K)$ in the graph G, i.e. $\mathcal{S}_G(K)$ is formed by the neighborhoods, in the graph G, of the connected components of $G \backslash K$.

Consider graph $G = (V, E)$ and a vertex set $X \subseteq V$. We denote by G_X the graph obtained from G by *completing* X, i.e. by adding an edge between every pair of non-adjacent vertices of X. If $\mathcal{X} = \{X_1, \ldots, X_p\}$ is a set of subsets of V, $G_{\mathcal{X}}$ is the graph obtained by completing all the elements of $\mathcal{X}$.

Theorem 2. *Let $K \subseteq V$ be a set of vertices. K is a potential maximal clique if and only if :*

1. *$G \backslash K$ has no full components associated to K.*
2. *$G_{\mathcal{S}_G(K)}[K]$ is a clique.*

Moreover, if K is a potential maximal clique, then $\mathcal{S}_G(K)$ is the maximal set of neighbor separators bordering K, i.e. $\mathcal{S}_G(K) = \Delta_G(K)$.

Remark 1. If K is a potential maximal clique of G, for any pair of vertices x and y of K either x and y are adjacent in G or they are connected by a path entirely contained in some C_i of $G \backslash K$ except for x and y. The second case comes from the fact that if x and y are not adjacent in G they must belong to the same S_i to ensure that K becomes a clique after the completion of $\mathcal{S}_G(K)$. When we will refer to this property we will say that *x and y are connected via the connected component C_i*.

Remark 2. Consider a minimal separator S contained in a potential maximal clique Ω. Let us compare the connected components of $G \backslash S$ and the connected components of $G \backslash \Omega$ (see [11] for the proofs). The set $\Omega \backslash S$ is contained in a full component C_Ω associated to S. All the other connected components of $G \backslash S$ are also connected components of $G \backslash \Omega$. Conversely, a connected component C of $G \backslash \Omega$ is either a connected component of $G \backslash S$ (in which case $N_G(C) \subseteq S$) or it is contained in C_Ω (in which case $N_G(C) \nsubseteq S$).

Remark 3. Unlike the minimal separators, a potential maximal clique Ω' cannot be strictly included in another potential maximal clique Ω. Indeed, for any proper subset Ω' of a potential maximal clique Ω, the difference $\Omega \backslash \Omega'$ is in a full component associated to Ω'.

Theorem 2 leads to a polynomial algorithm that, given a vertex set of a graph G, decides if K is a potential maximal clique of G.

Corollary 2. *Given a vertex set K of a graph G, we can recognize in $\mathcal{O}(nm)$ time if K is a potential maximal clique of G.*

Proof. We can compute in linear time the connected components C_i of $G \backslash K$ and their neighborhoods S_i. We can also verify in linear time that $G \backslash K$ has no full components associated to K.

For each $x \in K$, we compute all the vertices $y \in K$ that are adjacent to x in G or connected to x via a C_i in linear time (we have to search the neighborhood of x and the connected components C_i with $x \in S_i$). So we can verify in $\mathcal{O}(nm)$ time if K satisfies the conditions of theorem 2. □

4 Potential Maximal Cliques and Active Separators

Theorem 2 tells us that if Ω is a potential maximal clique of a graph G, then Ω is a clique in $G_{\Delta_G(\Omega)}$. We will divide the minimal separators of $\Delta_G(\Omega)$ into two classes: those which create edges in $G_{\Delta_G(\Omega)}$, which are called *actives*, and the others, which are called *inactives*. More precisely:

Definition 3. *Let Ω be a potential maximal clique of a graph G and let $S \subset \Omega$ be a minimal separator of G. We say that S is an* active separator *for Ω if Ω is not a clique in the graph $G_{\Delta_G(\Omega)\setminus\{S\}}$, obtained from G by completing all the minimal separators contained in Ω, except S. Otherwise, S is called* inactive *for Ω.*

Proposition 1. *Let Ω be a potential maximal clique of G and $S \subset \Omega$ a minimal separator, active for Ω. Let (S, C_Ω) be the block associated to S containing Ω and let $x, y \in \Omega$ be two non-adjacent vertices of $G_{\Delta_G(\Omega)\setminus\{S\}}$. Then $\Omega\setminus S$ is an minimal x, y-separator in $G[C_\Omega \cup \{x, y\}]$.*

Proof. Remark that the vertices x and y, non-adjacent in $G_{\Delta_G(\Omega)\setminus\{S\}}$, exist by definition of an active separator. Moreover, since $G_{\Delta_G(\Omega)}$ is a clique, we must have $x, y \in S$.

Let us prove first that $\Omega\setminus S$ is a x, y-separator in the graph $G' = G[C_\Omega \cup \{x, y\}]$. Suppose that both x and y are in a same connected component C_{xy} of $G'\setminus(\Omega\setminus S)$. Let $C = C_{xy}\setminus\{x, y\}$. Clearly, $C \subset C_\Omega$ is a connected component of $G\setminus\Omega$. Let T be the neighborhood of C in G. By theorem 2, T is a minimal separator of G, contained in Ω. By construction of T, we have $x, y \in T$. Notice that $T \neq S$, otherwise S would separate C and Ω, contradicting the fact that $C \subset C_\Omega$ (see remark 2). It follows that T is a minimal separator of $\Delta_G(\Omega)$, different from S and containing x and y. This contradicts the fact that x and y are not adjacent in $G_{\Delta_G(\Omega)\setminus\{S\}}$. We can conclude that $\Omega\setminus S$ is an x, y-separator of G'.

We prove now that $\Omega\setminus S$ in a minimal x, y-separator of G'. We will show that, for any vertex $z \in \Omega\setminus S$, there is a path μ joining x and y in G' and such that μ intersects $\Omega\setminus S$ only in z. By theorem 2, x and z are adjacent in $G_{\Delta_G(\Omega)}$, so x and z are adjacent in G or they are connected via a connected component C_i of $G\setminus\Omega$. Notice that $C_i \subset C_\Omega$: indeed, if $C_i \not\subset C_\Omega$, then C_i will be contained in some connected component D of $G\setminus S$, different from C_Ω. According to remark 2, we would have $N_G(C_i) \subseteq N_G(D) \subseteq S$, contradicting $z \in S_i$. In both cases we have a path μ' from x to z in G', that intersects $\Omega\setminus S$ only in z.

For the same reasons, z and y are adjacent in G, or there is a connected component C_j of $G\setminus\Omega$ such that $C_j \subset C_\Omega$ and $z, y \in S_j = N_G(C_j)$. This gives us a path μ'' from z to y in G', such that $\mu''\cap(\Omega\setminus S) = \{z\}$. Remark that $C_i \neq C_j$, otherwise we would have a path from x to y in $C_i \cup \{x, y\}$, contradicting the fact that $\Omega\setminus S$ separates x and y in G'. So the paths μ' and μ'' are disjoint except for z, and their concatenation is a path μ, joining x and y in G' and intersecting $\Omega\setminus S$ only in z. We conclude by lemma 2 that $\Omega\setminus S$ is a minimal separator of G'. □

By proposition 1, the set $T' = \Omega \backslash S$ is a minimal separator of the subgraph of G induced by $C_\Omega \cup \{x, y\}$. By lemma 3, there is a separator T of G such that $T' \subseteq T$ and $T \cap C_\Omega = T'$. We deduce:

Theorem 3. *Let Ω be a potential maximal clique and S be a minimal separator, active for Ω. Let (S, C_Ω) be the block associated to S containing Ω. There is a minimal separator T of G such that $\Omega = S \cup (T \cap C_\Omega)$.*

It follows easily that the number of potential maximal cliques containing at least one active separator is polynomially bounded in the number of minimal separators of G. More exactly number of these potential maximal cliques is bounded by the number of blocks (S, C_Ω) multiplied by the number of minimal separators T, so by $n|\Delta_G|^2$. Clearly, these potential maximal cliques have a simple structure and can be computed directly from the minimal separators of the graph.

Let us make a first observation about the potential maximal cliques containing inactive minimal separators.

Proposition 2. *Let Ω be a potential maximal clique and $S \subset \Omega$ a minimal separator which is inactive for Ω. Let $D_1, \ldots, D_p$ be the full components associated to S that do not intersect Ω. Then Ω is a potential maximal clique of the graph $G \backslash \cup_{i=1}^p D_i$.*

Proof. Let $G' = G \backslash \cup_{i=1}^p D_i$. The connected components of $G' \backslash \Omega$ are exactly the connected components of $G \backslash \Omega$, except for $D_1, \ldots, D_p$, and their neighborhoods in G' are the same as in G. It follows that the set $\mathcal{S}_{G'}(\Omega)$ of the neighborhoods of the connected components of $G' \backslash \Omega$ is exactly $\Delta_G(\Omega) \backslash \{S\}$. Clearly, $G' \backslash \Omega$ has no full components associated to Ω. Since S is not active for Ω, we deduce that Ω is a clique in $G'_{\mathcal{S}_{G'}(\Omega)}$. So, by theorem 2, Ω is a potential maximal clique of G'. □

5 Removing a Vertex

Let $G = (V, E)$ be a graph and a be a vertex of G. We denote by G' the graph obtained from G by removing a, i.e. $G' = G[V \backslash \{a\}]$. We will show here how to obtain the potential maximal cliques of G using the minimal separators of G and G' and the potential maximal cliques of G'. By corollary 1, we know that G has at least as many minimal separators as G': for any minimal separator S of G', either S is a minimal separator of G, or $S \cup \{a\}$ is a minimal separator of G. It will follow that the potential maximal cliques of a graph can be computed in polynomial time in the size of the graph and the number of its minimal separators.

Proposition 3. *Let Ω be a potential maximal clique of G such that $a \in \Omega$. Then $\Omega' = \Omega \backslash \{a\}$ is either a potential maximal clique of G' or a minimal separator of G.*

Proof. Let $C_1, \ldots, C_p$ be the connected components of $G \backslash \Omega$ and S_i be the neighborhood of C_i in G. We denote as usual by $\mathcal{S}_G(\Omega)$ the set of all the S_i's. Remark that the connected components of $G' \backslash (\Omega \backslash \{a\})$ are exactly $C_1, \ldots, C_p$ and their neighborhoods in G' are respectively $S_1 \backslash \{a\}, \ldots, S_p \backslash \{a\}$. Since Ω is a clique in $G_{\mathcal{S}_G(\Omega)}$ (by theorem 2), it follows that $\Omega' = \Omega \backslash \{a\}$ is a clique in $G'_{\mathcal{S}_{G'}(\Omega')}$. If $G' \backslash \Omega'$ has no full components associated to Ω', then Ω' is a potential maximal clique of G', according to theorem 2. Suppose now that C_1 is a full component associated to Ω' in G'. Since C_1 is not a full component associated to Ω in G, it follows that $N_G(C_1) = \Omega'$. Thus, Ω' is a minimal separator of G, by theorem 2. □

Lemma 4. *Let G be a graph and $\tilde{G}$ be any induced subgraph of G. Consider a potential maximal clique Ω of $\tilde{G}$. Suppose that for any connected component C of $G \backslash \tilde{G}$, its neighborhood $N_G(C)$ is strictly contained in Ω. Then Ω is also a potential maximal clique of G.*

Proof. Let C be any connected component of $G \backslash \tilde{G}$. We denote by $\tilde{V}$ the set of vertices of $\tilde{G}$. We want to prove that Ω is a potential maximal clique of the graph $\tilde{G}' = G[\tilde{V} \cup C]$. Indeed, the connected components of $\tilde{G}' \backslash \Omega$ are the connected components of $\tilde{G} \backslash \Omega$ plus C. The set $\mathcal{S}_{\tilde{G}'}(\Omega)$ of their neighborhoods consists in $\{N_G(C)\} \cup \mathcal{S}_{\tilde{G}}(\Omega)$. Since $N_G(C)$ is strictly contained in Ω, $\tilde{G}' \backslash \Omega$ has no full components associated to Ω. Obviously Ω is a clique in $\tilde{G}'_{\mathcal{S}_{\tilde{G}'}(\Omega)}$, so Ω is a potential maximal clique of $\tilde{G}'$.

The result follows by an easy induction on the number of connected components of $G \backslash \tilde{G}$. □

Proposition 4. *Let Ω be a potential maximal clique of G such that $a \notin \Omega$. Let C_a be the connected component of $G \backslash \Omega$ containing a and let S be the minimal separator of Ω such that $S = N(C_a)$.*

If Ω is not a potential maximal clique of $G' = G[V \backslash \{a\}]$, then S is active for Ω. Moreover, S is not a minimal separator of G'.

Proof. Suppose that S is not active for Ω. Let $D_1, \ldots, D_p$ the full components associated to S in G that do not intersect Ω. One of them, say D_1, is C_a. Let G'' be the graph obtained from G by removing the vertices of $D_1 \cup \ldots \cup D_p$. According to proposition 2, Ω is a potential maximal clique of G''. Notice that G'' is also an induced graph of G'. Any connected component C of $G' \backslash G''$ is contained in some D_i, and its neighborhood in G' is included in $S = N_G(D_i)$. Thus, $N_{G'}(C)$ is strictly contained in Ω. It follows from lemma 4 that Ω is a potential maximal clique of G', contradicting our hypothesis. We deduce that, in the graph G, S is an active separator for Ω.

It remains to show that S is not a minimal separator of G'. We prove that if S is a minimal separator of G', then Ω would be a potential maximal clique of G'. Let $C_1, \ldots . C_p, C_a$ be the connected components of $G \backslash \Omega$ and let $S_1, \ldots, S_p, S$ be their neighborhoods in G. Then the connected components of $G' \backslash \Omega$ are

$C_1, \dots, C_p, C'_1, \dots, C'_q$, with $C'_i \subset C_a$. Their neighborhoods in G' are respectively $S_1, \dots, S_p, S'_1, \dots, S'_q$, with $S'_i \subseteq S$. In particular, $G' \backslash \Omega$ has no full component associated to Ω and $\mathcal{S}_{G'}(\Omega)$ contains every element of $\mathcal{S}_G(\Omega)$, except possibly S. Suppose that S is a minimal separator of G' and let D be a full component associated to S in G', different from C_Ω. By remark 2, D is also a connected component of $G' \backslash \Omega$, so $S = N_{G'}(D)$ is an element of $\mathcal{S}_{G'}(\Omega)$. Therefore, $\mathcal{S}_G(\Omega) \subseteq \mathcal{S}_{G'}(\Omega)$, so Ω is a clique in the graph $G'_{\mathcal{S}_{G'}}(\Omega)$. We can conclude by theorem 2 that Ω is a potential maximal clique of G', contradicting our choice of Ω. It follows that S is not a minimal separator of G'. □

The following theorem, that comes directly from propositions 3 and 4 and theorem 3, shows us how to obtain the potential maximal cliques of G from the potential maximal cliques of G' and the minimal separators of G.

Theorem 4. *Let Ω be a potential maximal clique of G and let $G' = G \backslash \{a\}$. Then one of the following cases holds:*

1. *$\Omega = \Omega' \cup \{a\}$, where Ω' is a potential maximal clique of G'.*
2. *$\Omega = \Omega'$, where Ω' is a potential maximal clique of G'.*
3. *$\Omega = S \cup \{a\}$, where S is a minimal separator of G.*
4. *$\Omega = S \cup (C \cap T)$, where S is a minimal separator of G, C is a connected component of $G \backslash S$ and T is a minimal separator of G. Moreover, S does not contain a and S is not a minimal separator of G'.*

Corollary 3. *Let G be a graph, a be a vertex of G and $G' = G \backslash \{a\}$. The number $|\Pi_G|$ of potential maximal cliques of G is polynomially bounded in the number $|\Pi_{G'}|$ of potential maximal cliques of G', the number $|\Delta_G|$ of minimal separators of G and the size n of G.*

More precisely, $|\Pi_G| \leq |\Pi_{G'}| + n(|\Delta_G| - |\Delta_{G'}|)|\Delta_G| + |\Delta_G|$.

Proof. We will count the potential maximal cliques of the graph G corresponding to each case of theorem 4.

Notice that for a potential maximal clique Ω' of G', only one of Ω' and $\Omega' \cup \{a\}$ can be a potential maximal clique of G: indeed, a potential maximal clique of a graph cannot be strictly included in another one (see remark 3). So the number of potential maximal cliques of type 1 and 2 of G is bounded by $|\Pi_{G'}|$.

The number of potential maximal cliques of type 3 is clearly bounded by $|\Delta_G|$.

Let us count now the number of potential maximal cliques of type 4, that can be written as $S \cup (T \cap C)$. By lemma 3, for any minimal separator S' of G', we have that S' or $S' \cup \{a\}$ is a minimal separator of G. Clearly, the number of minimal separators of G of type S' or $S' \cup \{a\}$ with $S' \in \Delta_{G'}$ is at least $|\Delta_{G'}|$. Our minimal separator S does not contain a and is not a minimal separator of G', so S is not of type S' or $S' \cup \{a\}$, with $S' \in \Delta_{G'}$. It follows that the number of minimal separators S that we can choose is at most $|\Delta_G| - |\Delta_{G'}|$. For each minimal separator S, we have at most n connected components C of $G\backslash S$ and at most $|\Delta_G|$ separators T, so the number of potential maximal cliques of type 4 is at most $n(|\Delta_G| - |\Delta_{G'}|)|\Delta_G|$. □

Let now $a_1, a_2, \ldots, a_p$ be an arbitrary ordering of the vertices of G. We denote by G_i the graph $G[\{a_1, \ldots, a_i\}]$, so $G_n = G$ and G_1 has a single vertex. By corollary 3 we have that for any $i, 1 \leq i \leq n$, $|\Pi_{G_{i+1}}| \leq |\Pi_{G_i}| + n(|\Delta_{G_{i+1}}| - |\Delta_{G_i}|)|\Delta_{G_{i+1}}| + |\Delta_{G_{i+1}}|$. Notice that $|\Delta_{G_i}| \leq |\Delta_{G_{i+1}}|$, in particular each graph G_i has at most $|\Delta_G|$ minimal separators. Clearly, the graph G_1 has a unique potential maximal clique. It follows directly that the graph G has at most $n|\Delta_G|^2 + n|\Delta_G| + 1$ potential maximal cliques.

Proposition 5. *The number of the potential maximal cliques of a graph is polynomially bounded in the number of its minimal separators and in the size of the graph.*

More precisely, a graph G has at most $n|\Delta_G|^2 + n|\Delta_G| + 1$ potential maximal cliques.

We give now an algorithm computing the potential maximal cliques of a graph. We suppose that we have a function $IS\_PMC(\Omega, G)$, that returns $TRUE$ if Ω is a potential maximal clique of G, $FALSE$ otherwise.

function ONE_MORE_VERTEX
`Input:` the graphs G, G' and a vertex a such that $G' = G\backslash\{a\}$;
    the potential maximal cliques $\Pi_{G'}$ of G', the minimal separators $\Delta_{G'}$, Δ_G of G' and G.
`Output:` the potential maximal cliques Π_G of G.
`begin`
    $\Pi_G \leftarrow \emptyset$
    `for each` p.m.c. $\Omega' \in \Pi_{G'}$
        `if` $IS\_PMC(\Omega', G)$ `then`
            $\Pi_G \leftarrow \Pi_G \cup \{\Omega'\}$
        `else`
            `if` $IS\_PMC(\Omega' \cup \{a\}, G)$ `then`
                $\Pi_G \leftarrow \Pi_G \cup \{\Omega' \cup \{a\}\}$
    `for each` minimal separator $S \in \Delta_G$
        `if` $IS\_PMC(S \cup \{a\}, G)$ `then`
            $\Pi_G \leftarrow \Pi_G \cup \{S \cup \{a\}\}$
        `if` ($a \notin S$ `and` $S \notin \Delta_{G'}$) `then`
            `for each` $T \in \Delta_G$
                `for each` full component C associated to S in G
                    `if` $IS\_PMC(S \cup (T \cap C), G)$ `then`
                        $\Pi_G \leftarrow \Pi_G \cup \{S \cup (T \cap C)\}$
    `return` Π_G
`end`

Table 1. Computing the p.m.c.'s of G from the p.m.c.'s of $G' = G\backslash\{a\}$

The function ONE_MORE_VERTEX of table 1 computes the potential maximal cliques of a graph G from the potential maximal cliques of a graph $G' = G\backslash\{a\}$. This function is based on theorem 4. The main program, presented in table 2, successively computes the potential maximal cliques of the graphs $G_i = G[\{a_1, \dots a_i\}]$. The algorithm is clearly polynomial in the size of G and $|\Delta_G|$. The complexity proof is omitted due to space restrictions.

main program
`Input`: a graph G
`Output`: the potential maximal cliques Π_G of G
`begin`
let $\{a_1, \dots, a_n\}$ be the vertices of G
$\Pi_{G_1} \leftarrow \{\{a_1\}\}$
$\Delta_{G_1} \leftarrow \emptyset$
`for` $i = 1, n-1$
compute $\Delta_{G_{i+1}}$
$\Pi_{G_{i+1}} = ONE\_MORE\_VERTEX(G_i, G_{i+1}, \Pi_{G_i}, \Delta_{G_i}, \Delta_{G_{i+1}})$
$\Pi_G = \Pi_{G_n}$
`end`

Table 2. Algorithm computing the potential maximal cliques

Theorem 5. *The potential maximal cliques of a graph can be listed in polynomial time in its size and the number of its minimal separators.*

More exactly, the potential maximal cliques of a graph are computable in $\mathcal{O}(n^2m|\Delta_G|^2)$ *time.*

We deduce directly from theorem 1, proposition 5 and theorem 5:

Theorem 6. *The treewidth and the minimum fill-in of a graph can be computed in polynomial time in the size of the graph and the number of its minimal separators. The complexity of the algorithm is* $\mathcal{O}(n^3|\Delta_G|^3 + n^2m|\Delta_G|^2)$.

References

[1] S. Arnborg, D.G. Corneil, and A. Proskurowski. Complexity of finding embeddings in a k-tree. *SIAM J. on Algebraic and Discrete Methods*, 8:277–284, 1987.

[2] S. Arnborg, B. Courcelle, A. Proskurowski, and D. Seese. An algebraic theory of graph reduction. *J. of ACM*, 40:1134–1164, 1993.

[3] S. Arnborg and A. Proskurowski. Linear time algorithms for NP-hard problems restricted to partial k-trees. *Discrete Applied Mathematics*, 23:11–24, 1989.

[4] A. Berry, J.P. Bordat, and O. Cogis. Generating all the minimal separators of a graph. In *Workshop on Graphs WG'99*, Lecture Notes in Computer Science. Springer-Verlag, 1999.

[5] H. Bodlaender. A tourist guide through treewidth. *Acta Cybernetica*, 11:1–23, 1993.

[6] H. Bodlaender. A linear-time algorithm for finding tree-decompositions of small treewidth. *Siam J. Computing*, 25:1305–1317, 1996.

[7] H. Bodlaender. Treewidth: Algorithmic techniques and results. In *Proceedings of MFCS'97*, volume 1295 of *Lecture Notes in Computer Science*, pages 19–36. Springer-Verlag, 1997.

[8] H. Bodlaender and B. de Fluiter. Reduction algorithms for constructing solutions of graphs with small treewidth. In *Proceedings of COCOON'96*, volume 1090 of *Lecture Notes in Computer Science*, pages 199–208. Springer-Verlag, 1996.

[9] H. Bodlaender, J.R. Gilbert, H. Hafsteinsson, and T. Kloks. Approximating treewidth, pathwidth, and minimum elimination tree height. *J. of Algorithms*, 18:238–255, 1995.

[10] V. Bouchitté and I. Todinca. Minimal triangulations for graphs with "few" minimal separators. In *Proceedings 6th Annual European Symposium on Algorithms (ESA'98)*, volume 1461 of *Lecture Notes in Computer Science*, pages 344–355. Springer-Verlag, 1998.

[11] V. Bouchitté and I. Todinca. Treewidth and minimum fill-in of weakly triangulated graphs. In *Proceedings 16th Symposium of Theoretical Aspects in Computer Science (STACS'99)*, volume 1563 of *Lecture Notes in Computer Science*, pages 197–206. Springer-Verlag, 1999.

[12] M. S. Chang. Algorithms for maximum matching and minimum fill-in on chordal bipartite graphs. In *ISAAC'96*, volume 1178 of *Lecture Notes in Computer Science*, pages 146–155. Springer-Verlag, 1996.

[13] B. Courcelle. The monadic second-order logic of graphs III: Treewidth, forbidden minors and complexity issues. *Informatique Théorique*, 26:257–286, 1992.

[14] B. Courcelle and M. Moshbah. Monadic second-order evaluations on tree-decomposable graphs. *Theoretical Computer Science*, 109:49–82, 1993.

[15] M. C. Golumbic. *Algorithmic Graph Theory and Perfect Graphs.* Academic Press, New York, 1980.

[16] T. Hagerup. Dynamic algorithms for graphs of bounded treewidth. In *Proceedings 24th International Colloquium on Automata, Languages, and Programming (ICALP'97)*, Lecture Notes in Computer Science, pages 292–302. Springer-Verlag, 1997.

[17] T. Kloks, H.L. Bodlaender, H. Müller, and D. Kratsch. Computing treewidth and minimum fill-in: all you need are the minimal separators. In *Proceedings First Annual European Symposium on Algorithms (ESA'93)*, volume 726 of *Lecture Notes in Computer Science*, pages 260–271. Springer-Verlag, 1993.

[18] T. Kloks, H.L. Bodlaender, H. Müller, and D. Kratsch. Erratum to the ESA'93 proceedings. In *Proceedings Second Annual European Symposium on Algorithms (ESA'94)*, volume 855 of *Lecture Notes in Computer Science*, page 508. Springer-Verlag, 1994.

[19] T. Kloks and D. Kratsch. Treewidth of chordal bipartite graphs. *J. Algorithms*, 19(2):266–281, 1995.

[20] T. Kloks and D. Kratsch. Listing all minimal separators of a graph. *SIAM J. Comput.*, 27(3):605–613, 1998.

[21] T. Kloks, D. Kratsch, and H. Müller. Approximating the bandwidth for asteroidal triple-free graphs. In *Proceedings Third Annual European Symposium on Algorithms (ESA'95)*, volume 979 of *Lecture Notes in Computer Science*, pages 434–447. Springer-Verlag, 1995.

[22] T. Kloks, D. Kratsch, and J. Spinrad. On treewidth and minimum fill-in of asteroidal triple-free graphs. *Theoretical Computer Science*, 175:309–335, 1997.
[23] T. Kloks, D. Kratsch, and C.K. Wong. Minimum fill-in of circle and circular-arc graphs. *J. Algorithms*, 28(2):272–289, 1998.
[24] A. Parra and P. Scheffler. Characterizations and algorithmic applications of chordal graph embeddings. *Discrete Appl. Math.*, 79(1-3):171–188, 1997.
[25] N. Robertson and P. Seymour. Graphs minors. III. Planar tree-width. *J. of Combinatorial Theory Series B*, 36:49–64, 1984.
[26] N. Robertson and P. Seymour. Graphs minors. II. Algorithmic aspects of tree-width. *J. of Algorithms*, 7:309–322, 1986.
[27] D.J. Rose. Triangulating graphs and the elimination process. *J. Math Anal Appl.*, 32:597–609, 1970.
[28] R. Sundaram, K. Sher Singh, and C. Pandu Rangan. Treewidth of circular-arc graphs. *SIAM J. Discrete Math.*, 7:647–655, 1994.
[29] I. Todinca. *Aspects algorithmiques des triangulations minimales des graphes.* PhD thesis, École Normale Supérieure de Lyon, 1999.
[30] M. Yannakakis. Computing the minimum fill-in is NP-complete. *SIAM Journal on Algebraic and Discrete Methods*, 2:77–79, 1981.

Distance Labeling Schemes for Well-Separated Graph Classes

Michal Katz[1], Nir A. Katz, and David Peleg

[1] Department of Mathematics and Computer Science,
Bar Ilan University, Ramat Gan, 52900, Israel.
(mkatz@isys-oms.com,katz@creditlabs.com)

[2] Department of Applied Mathematics and Computer Science,
The Weizmann Institute, Rehovot 76100, Israel.
(peleg@wisdom.weizmann.ac.il)

Abstract. *Distance labeling schemes* are schemes that label the vertices of a graph with short labels in such a way that the distance between any two vertices can be inferred from inspecting their labels. It is shown in this paper that the classes of interval graphs and permutation graphs enjoy such a distance labeling scheme using $O(\log^2 n)$ bit labels on n-vertex graphs. Towards establishing these results, we present a general property for graphs, called well-(α, g)-separation, and show that graph classes satisfying this property have $O(g(n) \cdot \log n)$ bit labeling schemes. In particular, interval graphs are well-$(2, \log n)$-separated and permutation graphs are well-$(6, \log n)$-separated.

1 Introduction

Traditional graph representations are based on storing the graph topology in a data structure, e.g., an adjacency matrix, enabling one to infer information about the graph by inspecting the data structure. In such a context, the vertices of the graph are usually represented by distinct indices, serving as pointers to the data structure, but otherwise devoid of any meaning or structural significance.

In contrast, one may consider using more "informative" labeling schemes for graphs. The idea is to assign each vertex v a label $L(v)$, selecting the labels in a way that will allow us to infer information about the vertices (e.g., adjacency or distance) *directly* from their labels, without requiring any additional memory.

In particular, a graph family $\mathcal{F}$ is said to have an $l(n)$ *adjacency-labeling scheme* if there is a function L labeling the vertices of each n-vertex graph in $\mathcal{F}$ with distinct labels of up to $l(n)$ bits, and there exists an algorithm that given the labels $L(v), L(w)$ of two vertices v, w in a graph from $\mathcal{F}$, decides the adjacency of v and w in time polynomial in the length of the given labels. (Note that this algorithm is not given any additional information, other than the two labels, regarding the graph from which the vertices were taken.)

Adjacency-labeling schemes were introduced in [KNR88]. Specifically, it is shown in [KNR88] that a number of graph families enjoy $O(\log n)$ adjacency labeling schemes, including trees, bounded arboricity graphs (including, in particular, graphs of bounded degree and graphs of bounded genus, e.g., planar graphs),

H. Reichel and S. Tison (Eds.): STACS 2000, LNCS 1770, pp. 516–528, 2000.

various intersection-based graphs such as interval graphs, and c-decomposable graphs. It is also easy to encode the ancestry (or descendence) relation in a tree using interval-based schemes (cf. [SK85]).

More recently, *distance labeling schemes* were introduced in [Pel99]. These schemes are similar to adjacency labeling schemes, except that the labels of any two vertices u, v in a graph $G \in \mathcal{F}$ should enable us to compute the *distance* between u and v in G. Such schemes can be useful for various applications in the context of communication network protocols, as discussed in [Pel99]. A distance labeling scheme for trees using $O(\log^2 n)$ bit labels has been given in [Pel99]. This result is complemented by a lower bound proven in [GPPR99], showing that $\Omega(\log^2 n)$ bit labels are necessary for the class of all trees.

The distance labeling scheme given in [Pel99] for trees is based on the notion of separators. Moreover, that scheme has recently been expanded to other graph classes with small separators [GPPR99]. In particular, we say that a class of graphs $\mathcal{G}$ has a recursive $f(n)$-separator if every n-node graph $G \in \mathcal{G}$ has a subset of nodes S such that (1) $|S| \leq f(n)$, and (2) every connected component G' of the graph $G \setminus S$, obtained from G by removing all the nodes of S, belongs to $\mathcal{G}$, and has at most $2n/3$ nodes. Then it is shown in [GPPR99] that every graph class $\mathcal{G}$ with an $f(n)$-separator has an $O(f(n) \log n + \log^2 n)$ distance labeling scheme. This implies, for instance, the existence of an $O(\sqrt{n} \log n)$ distance labeling scheme for planar graphs, and an $O(\log^2 n)$ distance labeling scheme for graphs of bounded treewidth.

The current paper expands the study of the problem, by exploring the possibility of designing efficient distance labeling schemes for graph classes that do *not* enjoy small separators. Specifically, we consider graph classes for which there exist recursive separators which are not necessarily small, but are nevertheless "well-behaved" in a certain sense. Intuitively, in graphs enjoying well-behaved separators, distances between vertices can be inferred from relatively little information concerning the distances from every vertex to a few "representative" vertices in the (potentially large) separator. This property can be guaranteed, for instance, in graph classes enjoying small *diameter* separators.

Towards making this intuition more precise, we introduce the notion of *well-(α, g)-separated* graph classes, which are graph classes enjoying separators with some special properties. For any well-(α, g)-separated graph class, we construct a distance labeling scheme with $O(g(n) \cdot \log n)$ bit labels.

We then demonstrate the applicability of our construction technique by establishing the fact that the families of interval graphs and permutation graphs are all well-$(k, \log n)$-separated for some constant k, so they all have $O(\log^2 n)$ bit distance labeling schemes. (Note that in both families, separators might be of size $\Omega(n)$.) Due to space considerations, we present here only the scheme for interval graphs. (The scheme for permutation graphs can be found in [KKP99].)

2 The Well-Separation Property

A well-(α, g)-separated graph family consists of graphs that can be divided by a special separator, into subgraphs of size at most $n/2$, so $O(\log n)$ labels suffice for calculating the distances in the separator, and from every vertex in the subgraphs to the separator.

Let us first define the following terminology. Consider a graph $G = (V, E)$. For every $v, w \in V$, let $dist(v, w, G)$ denote the distance between v and w in G, namely, the length of the shortest path between them. We abbreviate $dist(v, w, G)$ as $dist(v, w)$ whenever G is understood from the context.

For a subset $U \subseteq V$, let $dist(v, U)$ denote the minimum distance between v and any $w \in U$, and let $dist_U(v, w, G)$ be the shortest path between v and w in G, that passes through at least one vertex of U.

Assume each vertex $v \in V$ has a unique identifier $I(v)$. For every $v, w \in V$, define $M(v, w) = \langle I(w), dist(v, w)\rangle$. For every vertex $v \in V$ and subset $D = \{d_1, \ldots, d_t\} \subseteq V$, $M(v, D)$ is defined to be the t-tuple $\langle M(v, d_1), \ldots, M(v, d_t)\rangle$.

Given an n-vertex graph G, a *separator* is a non-empty set of vertices whose removal breaks G into (zero or more) subgraphs with no interconnecting edges between them, each with at most $n/2$ vertices.

Let us now define a separation property which exists in some natural graph families, and whose existence is later used as the basis for the design of a distance labeling scheme.

Well-separation: A graph family $\mathcal{G}$ is *well-(α, g)-separated* for an integer $\alpha > 0$ and a function $g : \mathbb{N} \mapsto \mathbb{N}$, if there exists an identifier function I assigning unique identifiers to the vertices of every graph in $\mathcal{G}$, and for every n-vertex graph G in $\mathcal{G}$, there exists a set of vertices C, called the *α-separator* of G, with the following properties:

1. Deleting C from G disconnects it into (zero or more) subgraphs $G_1, \ldots, G_m$, with no interconnecting edges between them, such that for every $1 \le i \le m$:
   (a) $|V(G_i)| \le n/2$.
   (b) $G_i \in \mathcal{G}$ (hence in particular G_i is well-(α, g)-separated).
2. For every $v \in V(G)$, the identifier $I(v)$ is of size $g(n)$.
3. There exist polynomial time computable functions f^{gg}, f^{gs}, f^{ss} and f^{gsg}, and for every $v \in G$ there exists a *reference set* of vertices $\hat{D}^{v,G} = d_1^v, \ldots, d_\alpha^v \subseteq V(G)$, such that given

$$\xi(v, w, G) = \langle I(v), I(w), M(v, \hat{D}^{v,G}), M(w, \hat{D}^{w,G})\rangle,$$

   (a) the function f^{ss} computes the distance between two vertices in the separator.
   Formally, for $v, w \in C$, $f^{ss}(\xi(v, w, G)) = dist(v, w, G)$.
   (b) the function f^{gs} computes the distance between two vertices one of which is in the separator and the other is not.
   Formally, for $v \in V(G_i), w \in C$, $f^{gs}(\xi(v, w, G)) = dist(v, w, G)$.

(c) the function f^{gg} computes the distance between every two vertices that are not in the same subgraph, and are not in the separator.
Formally, for $v \in V(G_i)$ and $w \in V(G_j)$, $i \neq j$, $f^{gg}(\xi(v,w,G)) = dist(v,w,G)$.

(d) the function f^{gsg} computes for every two vertices in the same subgraph, the length of the shortest path between them, that passes through at least one of the vertices of the separator.
Formally, for $v,w \in V(G_i)$, $f^{gsg}(\xi(v,w,G)) = dist_C(v,w,G)$.

3 A Labeling Scheme for Well-Separated Graphs

This section describes a distance labeling scheme for n-vertex graphs G taken from a well-(α,g)-separated graph family. The construction makes use of the α-separator of the graph G. We assume that G is a connected graph. Otherwise, if G is not connected, we treat each component separately, and add an index to the resulting label of each vertex to indicate its connected component.

3.1 The Labeling System

The vertices of a given well-(α,g)-separated graph $G = (V,E)$ are labeled as follows. As a preprocessing step, calculate for every vertex $v \in V$ the identifier $I(v)$ whose existence is asserted by the well-separation property.

The actual labeling is constructed by a recursive procedure ASSIGN_LABEL, that applied to G, returns the label $L(v)$ of every vertex $v \in V$. The procedure, based on recursively partitioning the graph by finding *α-separators*, is presented in Fig. 1. The procedure generates for every $v \in G$ a label of the form

$$L(v) = \mathcal{J}_1(v) \circ \ldots \circ \mathcal{J}_q(v) \circ I(v) .$$

1. Let C be an α-separator of G.
2. **For** every vertex $v \in C$ set $L(v) \leftarrow \langle M(v,\hat{D}^{v,G}),0\rangle \circ I(v)$
3. **If** $C \neq V(G)$ **then do:**
   (a) Delete C from the graph and partition G into m mutually disconnected subgraphs, $G_1 \ldots G_m$
   (b) **For** every $1 \leq t \leq m$ **do:**
      i. Recursively invoke ASSIGN_LABEL(G_t) to get $L'(v)$ for every $v \in G_t$
      ii. **For** every vertex $v \in G_t$ **do:**
         I. Find $\hat{D}^{v,G} \subseteq C$
         II. $\mathcal{J}(v) = \langle M(v,\hat{D}^{v,G}),t\rangle$
         III. $L(v) = \mathcal{J}(v) \circ L'(v)$
4. **Return** $L(v)$ for every $v \in V(G)$.

Fig. 1. Algorithm ASSIGN_LABEL(G).

3.2 Computing the Distances

Let us next describe a recursive procedure DIST_COMPUT for computing the distance between two vertices v, w in G. Consider two vertices v, w in G, with labels

$$L(v) = \mathcal{J}_1(v) \circ \ldots \circ \mathcal{J}_p(v) \circ I(v) \quad \text{and} \quad L(w) = \mathcal{J}_1(w) \circ \ldots \circ \mathcal{J}_q(w) \circ I(w) ,$$

respectively, where

$$\mathcal{J}_i(v) = \langle M(v, \hat{D}^{v,G^i}), t_i^v \rangle \quad \text{for } 1 \le i \le p ,$$

and

$$\mathcal{J}_i(w) = \langle M(w, \hat{D}^{w,G^i}), t_i^w \rangle \quad \text{for } 1 \le i \le q .$$

Note that during the first few partitions, the vertices v and w may belong to the same subgraph. Let $G^1 = G$. Suppose both v and w belong to the same subgraph G^{i+1} of G^i on level i, for every $1 \le i < k$, but the graph G^k on level k, while still containing both of them, is separated by C^k into subgraphs in such a way that v and w are not in the same subgraph.

Procedure DIST_COMPUT receives the labels $L(v)$ and $L(w)$ of the vertices v and w. It starts by calculating the first level k on which v and w both belong to G^k but are no longer in the same subgraph of G^k. (This is indicated by the fact that in the kth fields of the two labels, $t_k^v \neq t_k^w$ or $t_k^v = 0$ or $t_k^w = 0$.) Let C^i be the separator that separates G^i, for every $1 \le i < k$. The procedure calculates the length of the shortest path $p_i(v, w)$ between v and w that goes through the level-i separator C^i, for every $1 \le i < k$. It then calculates also the length of the shortest path connecting them in the subgraph G^k. Finally, it returns the minimum of these lengths. The procedure is described formally in Figure 2.

3.3 Analysis

We now analyze the correctness and cost of the resulting distance labeling scheme. (Some proofs are omitted; see [KKP99].)

Lemma 1. *In a well-(α, g)-separated graph G, if C is an α-separator of G, and the vertices v, w are in the same subgraph G_i of G induced by C, then*

$$dist(v, w, G) = \min\{dist(v, w, G_i), dist_C(v, w, G)\}.$$

Lemma 2. *For every well-(α, g)-separated graph G, and vertices $v, w \in V$, the output of Algorithm* DIST_COMPUT *is $dist(v, w, G)$.*

Proof. Consider an arbitrary pair of vertices $v, w \in V$, with labels

$$L(v) = \mathcal{J}_1(v) \circ \ldots \circ \mathcal{J}_p(v) \circ I(v) \quad \text{and} \quad L(w) = \mathcal{J}_1(w) \circ \ldots \circ \mathcal{J}_q(w) \circ I(w)$$

For every $1 \le i \le \min\{p, q\}$, let $\mathcal{J}_i(v) = (M(v, \hat{D}^{v,G^i}), t_i^v)$ $\mathcal{J}_i(w) = (M(w, \hat{D}^{w,G^i}), t_i^w)$.

1. **If** $L(v) = L(w)$ **then Return** 0
2. Let k be the minimum index such that $t_k^v \neq t_k^w$ or $t_k^v = 0$ or $t_k^w = 0$.
   /* v and w belong to the same subgraph G^i on each level $1 \leq i \leq k-1$, and are separated on level k. */
3. Initialize $Dist^{\text{sep}}(v, w) \leftarrow \infty$
4. **For** $i = 1$ to $k - 1$ **do:**
   (a) Let $\xi(v, w, G^i) = \langle I(v), I(w), M(v, \hat{D}^{v,G^i}), M(w, \hat{D}^{w,G^i})\rangle$.
   /* G^i is the subgraph containing both v and w on level i */
   (b) $Dist^{\text{sep}}(v, w) \leftarrow \min\{Dist^{\text{sep}}(v, w), f^{gsg}(\xi(v, w, G^i)\}$
   /* Calculate the length of the shortest path between v and w that passes through one of the separators */
5. /* On the kth level */
   Let $\xi(v, w, G^k) = \langle I(v), I(w), M(v, \hat{D}^{v,G^k}), M(w, \hat{D}^{w,G^k})\rangle$.
   (a) **If** $t_k^v = t_k^w = 0$ /* Both v and w are in the separator on level k */
   then Return $DistInG_k \leftarrow f^{ss}(\xi(v, w, G^k))$
   (b) **Else if** $t_k^v = 0$ /* Only v is in the separator */
   then Return $DistInG_k \leftarrow f^{gs}(\xi(w, v, G^k))$
   (c) **Else if** $t_k^w = 0$ /* Only w is in the separator */
   then Return $DistInG_k \leftarrow f^{gs}(\xi(v, w, G^k))$
   (d) **Else if** $t_k^v \neq t_k^w$ /* v and w are in different level k subgraphs of G */
   then Return $DistInG_k \leftarrow f^{gg}(\xi(v, w, G^k))$
6. Return $\min\{DistInG_k, Dist^{\text{sep}}(v, w)\}$.

Fig. 2. Algorithm DIST_COMPUT $(L(v), L(w))$.

Let $1 \leq k \leq \min\{p, q\}$ be the minimum index such that v and w belong to the same subgraph on each level $1 \leq i \leq k-1$, and are separated on level k or both in the separator of this level. For this k, one of the conditions (a)-(d) in the procedure DIST_COMPUT must hold.

Let us examine the cases considered by the procedure one by one. If $t_k^v = t_k^w = 0$, then both v and w are in the separator C^k, and according to the definition of the function f^{ss}, $dist(v, w, G^k) = f^{ss}(\xi(v, w, G^k))$

If $t_k^v = 0$ and $t_k^w \neq 0$, then v is in the separator C^k, and w is in one of the subgraphs of G^k on level k, and by the definition of the function f^{gs}, $dist(v, w, G^k) = f^{gs}(\xi(v, w, G^k))$, The case $t_k^w = 0$ and $t_k^v \neq 0$ is analogous to the previous case.

Finally, if $t_k^v \neq t_k^w \neq 0$, then v and w are in different subgraphs of G^k on level k according to the separator C^k, and according to the definition of the function f^{gg}, $dist(v, w, G^k) = f^{gg}(\xi(v, w, G^k))$.

The last step of the procedure returns the minimum of $dist(v, w, G^k)$ and of $Dist^{\text{sep}}(v, w)$, which holds the length of the shortest path between v and w through one of the higher levels separators C^i, for $1 \leq i \leq k-1$. By Lemma 1, the minimum of $dist(v, w, G^k)$ and of $Dist^{\text{sep}}(v, w)$, is the distance between v and w. ∎

Lemma 3. *The labeling scheme uses $O(g(n)\log n)$ bit labels.*

Proof. On each level of the recursive labeling procedure, the sublabel $\mathcal{J}(v)$ is of the form $\langle M(v,\hat{D}^v), t^v\rangle$. The index t^v clearly requires at most $\log n$ bits.

$$M(v,\hat{D}^v) = \langle M(v,d_1),\ldots,M(v,d_\alpha)\rangle = \langle\langle I(d_1), dist(v,d_1)\rangle,\ldots,\langle I(d_\alpha), dist(v,d_\alpha)\rangle\rangle$$

is of size $O(\alpha\cdot g(n)+\alpha\log n) = O(g(n))$ as $g(n)\geq\log n$. Since the maximum graph size is halved in each application of the recursive labeling procedure, there are at most $\log n$ levels, hence the size of the labels after the recursion is $O(g(n)\cdot\log n)$. Finally, the size of the initial identifiers assigned at the preprocessing step is $|I(v)| = g(n)$, Which is negligible. ∎

Theorem 1. *Any well-(α,g)-separated graph family has a distance labeling scheme of size $O(g(n)\log n)$.*

4 Distance Labeling Scheme for Interval Graphs

4.1 Definitions

We need to introduce some preliminary definitions concerning interval graphs. Given a finite number of intervals on a straight line, a graph associated with this set of intervals can be constructed in the following manner. Each interval corresponds to a vertex of the graph, and two vertices are connected by an edge if and only if the corresponding intervals overlap at least partially [Gol80].

Let $G=(V,E)$, be a connected n-vertex interval graph. For every $v\in V$, we use the interval representation $\mathcal{I}(v)=[l(v),r(v)]$ where $l(v)$ (respectively, $r(v)$) is the left (resp., right) coordinate of v.

Example: Figure 3 describes the intervals corresponding to an 8-vertex interval graph G_{int} and Figure 4 describes the intervals corresponding to an 6-vertex interval graph G_{path}, which will be used throughout what follows to illustrate our basic notions and definitions. □

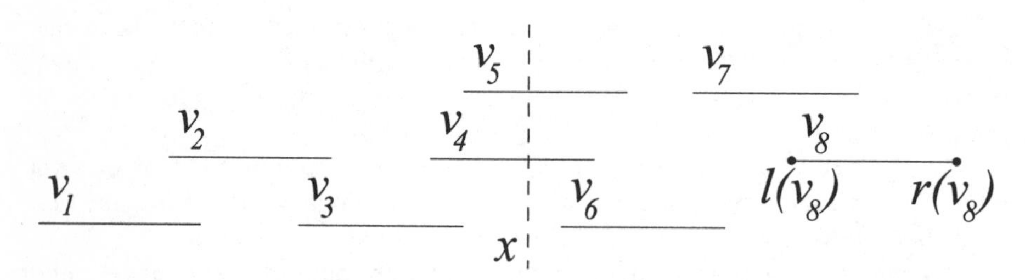

Fig. 3. An 8-vertex interval graph G_{int} in its interval representation.

For every real $x\in\mathbb{R}$, let $V_L(x)$ denote the set of all vertices whose intervals end before x (scanning the intervals from left to right), let $V_R(x)$ denote the set of vertices whose intervals start after x and let $C(x)$ denote the set of vertices whose intervals contains x, i.e.,

$$\begin{aligned} V_L(x) &= \{v \in V \mid r(v) < x\} , \\ V_R(x) &= \{v \in V \mid l(v) > x\} , \\ C(x) &= \{v \in V \mid l(v) \leq x \leq r(v)\} . \end{aligned}$$

Example (cont.): In Figure 3, $V_L(x) = \{v_1, v_2, v_3\}$, $V_R(x) = \{v_6, v_7, v_8\}$ and $C(x) = \{v_4, v_5\}$. □

For every set of vertices U, let $L(U)$ be the leftmost vertex in U according to the left endpoint of its interval, that is, $L(U) = v$ if $v \in U$ and $l(v) \leq l(w)$ for all $w \in U$. The rightmost vertex of U, $R(U)$, is defined in an analogous way, for the right side.

For every vertex $v \in V_L(x)$ to the left of x, we identify a special *contact* vertex for v in $C = C(x)$, denoted $Far(v, C)$, which is the "rightmost" possible vertex in C (w.r.t. its right endpoint) still within distance $dist(v, C)$ from v. Formally, for every $v \in V_L(x)$ and $C \subseteq V$, $V_L(x) \cap C = \emptyset$, $Far(v, C)$ is the vertex satisfying $Far(v, C) \in C$, $dist(v, Far(v, C)) = dist(v, L(C))$, and for every $w \in C$, $dist(v, w) = dist(v, L(C)) \Rightarrow r(Far(v, C)) \geq r(w)$. We abbreviate this as $Far(v)$ whenever C is understood from the context.

Our algorithm also makes use of a vertex slightly closer to v than $Far(v, C)$; let $Far^-(v, C)$ be the rightmost vertex in $V_L(x)$ that can be reached from v in one step less than $Far(v, C)$, i.e., $dist(v, Far^-(v, C))$ is the distance between v and C minus 1. Formally, for every $v \in V_L(x)$, $Far^-(v, C)$ is the vertex satisfying $dist(v, Far^-(v)) = dist(v, L(C)) - 1$, and for every $w \in V_L(x)$, $dist(v, w) = dist(v, L(C)) - 1 \Rightarrow r(Far^-(v, C)) \geq r(w)$. Again, we use $Far^-(v)$ for short whenever no confusion may arise.

For every $v \in V_R(x)$, define $Far(v)$ and $Far^-(v)$ in an analogous way:

1. $Far(v, C)$ is the vertex satisfying $Far(v, C) \in C$, $dist(v, Far(v, C)) = dist(v, R(C))$, and for every $w \in C$, $dist(v, w) = dist(v, R(C)) \Rightarrow l(Far(v, C)) \leq l(w)$.
2. $Far^-(v, C) \in V$ is the vertex satisfying $dist(v, Far^-(v, C)) = dist(v, R(C)) - 1$, and for every $w \in V_R(x)$, $dist(v, w) = dist(v, R(C)) - 1 \Rightarrow l(Far^-(v, C)) \leq l(w)$.

Finally, let $\delta(v) = dist(v, Far(v))$ and $\rho(v, w) = \delta(v) + \delta(w)$.

Example (cont.): In Figure 3, if $C = \{v_4, v_5\}$ then $L(C) = v_4$ and $R(C) = v_5$. Given this separator, $Far(v_1) = v_5$, $Far(v_8) = v_4$, $Far^-(v_1) = v_3$ and $Far^-(v_8) = v_6$. Therefore $\delta(v_1) = 3$, $\delta(v_8) = 3$ and $\rho(v_1, v_8) = 3 + 3 = 6$. □

4.2 Well-$(2, \log n)$-Separation of Interval Graphs

Let us show that interval graphs are well-$(2, \log n)$-separated. Let us start with a high-level overview of our method. The separator C consists of the "middle" intervals in the graph. For enabling distance calculations, the labels of all the other intervals in the graph must encode the distance to the nearest interval in the separator C, and also to the immediate previous interval in the path to C.

The distances between the intervals correspond to the distances between vertices in the graph.

A useful property of this kind of separator is that the vertices in it create a clique in the graph, so the distance between them is 1. Therefore, the distance between two intervals from different sides of the separator C will be either the sum of their distances to C, or, in case there is no overlap between these paths, the sum of the distances plus 1.

If the intervals are in the same side of C,then their distance is calculated recursively, as we continue to divide the subgraph by separators, and calculate the distances to these new separators.

For calculating the distance from an interval $\mathcal{I}(w)$ in C to an interval $\mathcal{I}(v)$ that is not, we check the distance from $\mathcal{I}(v)$ to C; this is the distance between the two intervals if $\mathcal{I}(w)$ overlaps the path of $\mathcal{I}(v)$ to C. If there is no such overlap, the distance increases by 1.

Let us now proceed with a more formal description of the construction.

The separator: Given a graph $G = (V, E)$, we choose the separator C as follows. Choose a real x such that $|V_L(x)| \le n/2$ and $|V_R(x)| \le n/2$. Let $C = \{v \in V \mid l(v) \le x \le r(v)\}$.

Example (cont.): In Figure 3, $C = \{v_4, v_5\}$. □

The identifiers: Define the identifier $I(v)$ as follows. For every $v \in V$, set $I(v) = \langle K(v), l(v), r(v)\rangle$ where $K(v)$ is a distinct number between 1 to n. The size of $I(v)$ is $O(\log n)$.

The reference sets: The set $\hat{D}^{v,G}$ is defined as:

$$\hat{D}^{v,G} = \begin{cases} \langle Far(v), Far^-(v)\rangle, & v \in V(G) \setminus C, \\ \emptyset, & \text{otherwise.} \end{cases}$$

Hence the tuple $\xi(v, w, G)$ for $v \in V(G) \setminus C$ becomes

$$\begin{aligned} \xi(v,w,G) &= \langle I(v), I(w), M(v, \hat{D}^{v,G}), M(w, \hat{D}^{w,G})\rangle \\ &= \langle I(v), I(w), \\ &\quad \langle\langle I(Far(v)), dist(v, Far(v))\rangle,\ \langle I(Far^-(v)), dist(v, Far^-(v))\rangle\rangle, \\ &\quad \langle\langle I(Far(w)), dist(w, Far(w))\rangle,\ \langle I(Far^-(w)), dist(w, Far^-(w))\rangle\rangle\rangle \\ &= \langle I(v), I(w), \\ &\quad \langle\langle I(Far(v)), \delta(v)\rangle,\ \langle I(Far^-(v)), \delta(v) - 1\rangle\rangle, \\ &\quad \langle\langle I(Far(w)), \delta(w)\rangle,\ \langle I(Far^-(w)), \delta(w) - 1\rangle\rangle\rangle\ . \end{aligned}$$

The distance functions: Finally, we define the functions f^{ss}, f^{gs}, f^{gg} and f^{gsg}.

- For $v, w \in C$, $v \neq w$, $f^{ss}(\xi(v,w,G))$ **returns** 1.
- For $v \in V_L(x)$ and $w \in C$, $f^{gs}(\xi(v,w,G))$ is computed as follows.
  Extract $M(v, \hat{D}^{v,G})$, and calculate $\delta(v)$.
  If $l(w) \leq r(Far^-(v))$ **then return** $\delta(v)$,
  Else return $\delta(v) + 1$.
- For $v \in V_L(x)$ and $w \in V_R(x)$, $f^{gg}(\xi(v,w,G))$ is computed as follows.
  Extract $M(v, \hat{D}^{v,G})$ and $M(w, \hat{D}^{w,G})$ from $\xi(v,w,G)$, and calculate $\rho(v,w)$.
  If $l(Far^-(w)) \leq r(Far(v))$ or $l(Far(w)) \leq r(Far^-(v))$
  then return $\rho(v,w)$,
  Else return $\rho(v,w) + 1$.
- For $v, w \in V_L(x)$, $f^{gsg}(\xi(v,w,G))$ is computed as follows.
  Extract $M(v, \hat{D}^{v,G})$ and $M(w, \hat{D}^{w,G})$ from $\xi(v,w,G)$.
  Return $\rho(v,w)$.

4.3 Correctness Proof

We next show that the functions f^{gg}, f^{gs}, f^{ss} and f^{gsg} defined above obey the requirements of the well-separation property. As a consequence, Algorithm Dist_Comput calculates the right distance between every two vertices. To simplify the scheme, we comment that the function f^{gsg} is not necessary for interval graphs, because in these graphs the distances grow monotonely, so for any two vertices in a subgraph G', the shortest path between their internals into G' is never shorter than the shortest path that contains also vertices from outside G'. So the distances according to the scheme are calculated in a similar way, except the use of the function f^{gsg}. The return value of Algorithm Dist_Comput is always the return values of one of the functions f^{gg}, f^{gs} and f^{ss}.

We rely on the following basic property of interval graphs.

Fact 2 [Gol80] *Any induced subgraph of an interval graph is an interval graph.*

Lemma 4. *C is a separator.*

Thus, for every interval graph $G = (V, E)$, $V_L(x)$ and $V_R(x)$ are interval graphs and therefore if G is well-separated, $V_L(x)$ and $V_R(x)$ are well-separated too.

By the definition of $Far(v)$ and $Far^-(v)$ we have:

Lemma 5. *For every $v \in V$,*

1. $\delta(v) = dist(v, Far(v)) = dist(v, Far^-(v)) + 1$,
2. $dist(Far(v), Far^-(v)) = 1$.

Our analysis makes use of the following technical lemma.

Lemma 6. *For every $v \in V_L(x)$ and $w \in V_R(x)$, $dist(v,w) \geq \rho(v,w)$.*

Lemma 7. *For every $v \in V_L(x)$ and $w \in V_R(x)$ such that $l(Far^-(w)) \leq r(Far(v))$, the corresponding intervals of $Far(v)$ and $Far^-(w)$ have a point in common.*

Lemma 8. *For every $v, w \in C$, $dist(v,w) = f^{ss}(\xi(v,w,G))$.*

Lemma 9. *For every $v \in V_L(x)$ and $w \in C$, $dist(v,w) = f^{gs}(\xi(v,w,G))$.*

Proof. There are two cases to consider. If $l(w) \leq r(Far^-(v))$, then the claim holds because the overlap between the intervals $\mathcal{I}(Far^-(v))$ and $\mathcal{I}(w)$ implies that

$$dist(v,w) \leq dist(v, Far^-(v)) + dist(Far^-(v), w) = (\delta(v) - 1) + 1 = \delta(v),$$

and on the other hand $w \in C$, and thus $dist(v,w) \geq \delta(v)$. Hence $dist(v,w) = \delta(v)$ which is the value returned by f^{gs} in this case.

Otherwise, if $l(w) > r(Far^-(v))$, then the claim holds because

$$dist(v,w) \leq dist(v, Far(v)) + dist(Far(v), w) = \delta(v) + 1,$$

and on the other hand there is no overlap between the intervals $\mathcal{I}(w)$ and $\mathcal{I}(Far^-(v))$, so by the properties of $Far(v)$ and $Far^-(v)$, we have $dist(v,w) = \delta(v) + 1$, which is the value returned by f^{gs} in this case. ▮

Example (cont.): In Figure 3, for $v = v_1$ and $w = v_4$, the first case occurs, as $Far(v_1) = v_5$ and $Far^-(v_1) = v_3$. $l(v_4) \leq r(Far^-(v_1)) = r(v_3)$. $\delta(v_1) = 3$, therefore $dist(v_1, v_4) = 3$.

In Figure 4, for $v = v_1$ and $w = v_4$, the second case occurs. $Far(v_1) = v_3$ and $Far^-(v_1) = v_2$. $l(v_4) > r(Far^-(v_1)) = r(v_2)$. $\delta(v_1) = 2$, therefore $dist(v_1, v_4) = \delta(v_1) + 1 = 3$. □

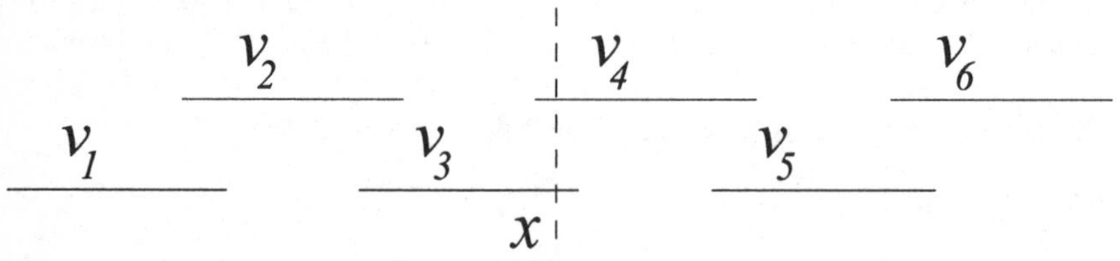

Fig. 4. An interval representation of a 6-vertex path G_{path}.

Lemma 10. *For every $v \in V_L(x)$ and $w \in V_R(x)$, $dist(v,w) = f^{gg}(\xi(v,w,G))$.*

Proof. There are four cases to be examined. When $Far(v) = Far(w)$, by the triangle inequality,

$$dist(v,w) \leq dist(v, Far(v)) + dist(Far(w), w) = \rho(v,w).$$

By Lemma 6 $dist(v,w) \geq \rho(v,w)$, so $dist(v,w) = \rho(v,w)$, and in this case, f^{gg} returned the right distance.

Now, assume $Far(v) \neq Far(w)$. When $l(Far^-(w)) \leq r(Far(v))$, by Lemma 7, the intervals $\mathcal{I}(Far(v))$ and $\mathcal{I}(Far^-(w))$ have a point in common, in particular, the point $l(Far^-(w))$, therefore $dist(Far(v), Far^-(w)) = 1$. By the triangle inequality,

$$\begin{aligned} dist(v,w) &\leq dist(v, Far(v)) + dist(Far(v), Far^-(w)) + dist(Far^-(w), w) \\ &= \delta(v) + 1 + (\delta(w) - 1) \ = \ \rho(v,w) \ . \end{aligned}$$

By Lemma 6, $dist(v,w) = \rho(v,w)$.
The case when $l(Far(w)) \leq r(Far^-(v))$ is handled in the same way.

The remaining case is when $Far(v) \neq Far(w)$, $l(Far^-(w)) > r(Far(v))$ and $l(Far(w)) > r(Far^-(v))$. Because there is no overlap between the intervals $\mathcal{I}(Far(v)), \mathcal{I}(Far^-(w))$ and between the intervals $\mathcal{I}(Far(w)), \mathcal{I}(Far^-(v))$, we have $dist(Far^-(v), Far(w)) = 2$ and $dist(Far(v), Far^-(w)) = 2$, thus $dist(v,w) = dist(v, Far(v)) + dist(Far(v), Far(w)) + dist(Far(w), w) = \rho(v,w) + 1$. ∎

Example (cont.): In Figure 3, if $v = v_1$ and $w = v_8$, then $Far(v_1) = v_5$, $Far(v_8) = v_4$, $Far^-(v_1) = v_3$ and $Far^-(v_8) = v_6$. In this case, $Far(v) \neq Far(w)$ and $l(Far^-(w)) \leq r(Far(v))$. $\delta(v_1) = 3$, $\delta(v_8) = 3$, and $\rho(v_1, v_8) = 3 + 3 = 6$. $l(Far^-(v_8)) \leq r(Far(v_1))$ therefore $dist(v_1, v_8) = 6$.

In Figure 4, for $v = v_1$ and $w = v_6$. Here, $Far(v) \neq Far(w)$,$l(Far^-(w)) > r(Far(v))$ and $l(Far(w)) > r(Far^-(v))$, because $Far(v_1) = v_3$, $Far(v_6) = v_4$, $Far^-(v_1) = v_2$ and $Far^-(v_6) = v_5$. $\delta(v_1) = 2$, $\delta(v_6) = 2$, and $\rho(v_1, v_6) = 2 + 2 = 4$. $Far(v_1) \neq Far(v_6)$, $l(Far^-(v_6)) > r(Far(v_1))$ and $l(Far(v_6)) > r(Far^-(v_1))$. therefore $dist(v_1, v_6) = \rho(v_1, v_6) + 1 = 7$. □

As consequence of Lemmas 4, 8, 9 and 10 we get

Corollary 1. *The class of interval graphs is well-*$(2, \log n)$*-separated.* ∎

Theorem 3. *The class of interval graphs enjoys an* $O(\log^2 n)$ *distance labeling scheme.* ∎

References

[GPPR99] C. Gavoille, D. Peleg, S. Pérennes, and R. Raz. Distance labeling in graphs. In preparation, 1999.

[Gol80] M.C. Golumbic. *Algorithmic Graph Theory and Perfect Graphs.* Academic Press, 1980.

[KNR88] S. Kannan, M. Naor, and S. Rudich. Implicit representation of graphs. In *Proc. 20th ACM Symp. on Theory of Computing*, pages 334–343, May 1988.

[KKP99] M. Katz, N. Katz and D. Peleg. Distance Labeling Schemes for Well-Separated Graph Classes. Technical Report MCS99-26, the Weizmann Institute of Science, 1999.

[Pel99] D. Peleg. Proximity-preserving labeling schemes and their applications. In *Proc. 25th Int. Workshop on Graph-Theoretic Concepts in Computer Science*, June 1999.

[SK85] N. Santoro and R. Khatib. Labelling and implicit routing in networks. *The Computer Journal*, 28:5–8, 1985.

Pruning Graphs with Digital Search Trees. Application to Distance Hereditary Graphs

Jean-Marc Lanlignel, Olivier Raynaud, and Eric Thierry

LIRMM, 161 rue Ada, 34392 Montpellier Cedex 5, France
(lanligne,raynaud,thierry)@lirmm.fr

Abstract. Given a graph, removing pendant vertices (vertices with only one neighbor) and vertices that have a twin (another vertex that has the same neighbors) until it is not possible yields a reduced graph, called the "pruned graph". In this paper, we present an algorithm which computes this "pruned graph" either in linear time or in linear space. In order to achieve these complexity bounds, we introduce a data structure based on digital search trees. Originally designed to store a family of sets and to test efficiently equalities of sets after the removal of some elements, this data structure finds interesting applications in graph algorithmics. For instance, the computation of the "pruned graph" provides a new and simply implementable algorithm for the recognition of distance-hereditary graphs, and we improve the complexity bounds for the complete bipartite cover problem on bipartite distance-hereditary graphs.

Keywords: graph algorithms, distance-hereditary graphs, sets, digital search trees, amortized complexity.

1 Introduction

Distance-hereditary graphs have been introduced by E. Howorka [How77] in 1977. Originally defined as graphs in which every induced path is isometric, many other characterizations have been found for this class: forbidden subgraphs, properties of cycles, metric characterizations ... (see [BM86] for a survey). We focus on one of them: G is a distance-hereditary graph iff every subgraph of G with at least two vertices contains a pendant vertex (a vertex with only one neighbor) or a vertex that has a twin (another vertex that has the same neighbors). Thus repeatedly removing from a graph G a pendant vertex or a vertex that has a twin, whenever possible (a process called "to ***prune*** the graph"), leads to a graph reduced to one single vertex iff G is a distance-hereditary graph.

This pruning process has a special property, which enables us to define a reduction on graphs thanks to the following theorem: given a graph, whatever the order is chosen to remove pendant vertices and vertices with twins, leads to a unique graph, up to an isomorphism. We will call this reduced graph the ***pruned graph*** and the problem of computing the "pruned graph" of any graph appears

H. Reichel and S. Tison (Eds.): STACS 2000, LNCS 1770, pp. 529–541, 2000.

as a generalization of the recognition of distance-hereditary graphs (graphs with a single vertex as "pruned graph").

Figure 1 shows a sequence of deletions which leads to the "pruned graph".

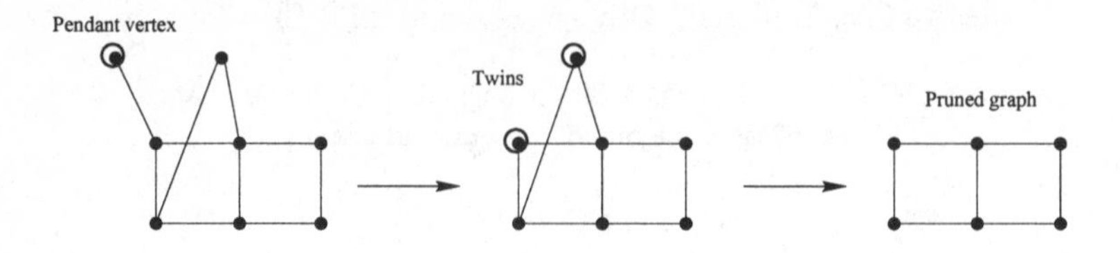

Fig. 1. Pruning process

The aim of this article is to provide algorithms to compute the pruned graph of any graph, with best upper bounds on the complexity. To give a first bound, note that if the graph is given by its adjacency lists, the algorithm which consists of finding pendant vertices and twins by comparing adjacency lists and removing these vertices one by one, would lead to an $O(mn)$ time complexity, where n is the number of vertices and m the number of edges. Our algorithms are based on another representation of the graph: instead of keeping the adjacency lists, we consider the neighbourhoods of the vertices as a family of subsets on the set of all vertices, and we work on the digital search tree associated to this family. Storing the graph in such a data structure, also called ***trie***, enables us to perform, with good complexities, the deletion of one element in all the subsets, the deletion of a subset (corresponding to the deletion of a vertex), as well as the detection of equal subsets (corresponding to the detection of twin vertices). We will present three slightly different variants of this data structure and the associated algorithms, which lead to the following upper bounds for the computation of the pruned graph (n is the initial number of vertices, m the initial number of edges):

Time bound	$m+n$	$m \log n$	n^2
Space bound	n^2	$m \log n$	$m+n$

In section 2, all the operations we need in order to compute the "pruned graph" are restated in terms of sets and operations on sets. We choose this point of view because our algorithms and the complexity analyses we give apply to any family of sets.

In section 3, we give the applications to graph problems. The computation of the "pruned graph" provides a new algorithm for the recognition of distance-hereditary graphs, with the complexities mentioned above, depending on which variant is chosen. They are not linear in both time and space, but the only known linear algorithm is very recent and complex (for instance it uses the modular decomposition of cographs as a subroutine), to the point that a first release was not fully functional ([HM90],[Dam97]).

Willing to take advantage of the graph structure to have faster algorithms, we also prove that some good orderings of the vertices yield a linear algorithm both in time and space in some special pruning cases, for instance the computation of the "pruned graph" of bipartite graphs.

As a final example, we improve the complexity bounds for the complete bipartite cover problem on bipartite distance-hereditary graphs, with a linear algorithm, whereas the only known algorithm proposed in [Mül96] runs with a $O(mn)$ time complexity.

2 Data Structure

2.1 The General Problem

Let $G = (V, E)$ be a non oriented graph. If $x \in V$, $\mathcal{N}(x) = \{y \in V, y \neq x | (x, y) \in E\}$ is the open neighborhood of x in G, and $\mathcal{N}_c(x) = \mathcal{N}(x) \bigcup \{x\}$ its closed neighborhood. An $x \in V$ such that $\#\mathcal{N}(x) = 1$ is called a ***pendant vertex***. Some $x, y \in V$ such that $\mathcal{N}(x) \backslash \{y\} = \mathcal{N}(y) \backslash \{x\}$ are called ***twin vertices***. Thus twins share the same vertices as neighbors, and may or may not be linked together. We will call non-adjacent twins ***true twins***, and adjacent twins ***false twins***.

To compute the "pruned graph" of G, we mainly have to detect false twins, namely vertices x, y such that $\mathcal{N}(x) = \mathcal{N}(y)$, and true twins, namely vertices x, y such that $\mathcal{N}_c(x) = \mathcal{N}_c(y)$. The detection of pendant vertices only needs a data structure to keep track of the degree $\#\mathcal{N}(x)$ of each vertex x. Then we need to delete vertices from the graph, and try to detect new removable vertices, and so on.

This process may be expressed in another way. Let us consider the family of sets $\{\mathcal{N}(x) | x \in V\} \bigcup \{\mathcal{N}_c(x) | x \in V\}$ which completely characterizes the graph G. We mainly have to detect equal sets to detect twins (note that for any x, y, $\mathcal{N}_c(x) = \mathcal{N}(y)$ is never possible). After the deletion of a vertex x, we have to update the family of sets by deleting the sets $\mathcal{N}(x)$ and $\mathcal{N}_c(x)$, and suppress all the occurrences of x in the other sets. In the whole section 2, we deal with these operations on sets, and then we describe the complete algorithm on graphs in section 3. We now reformulate our main problem in terms of sets.

Let X be a set of cardinality n. Suppose we are given k sets on X, $S_1, \ldots, S_k$. The problem is to find a data structure that will enable us, with the best amortized complexity, to perform the following operations:

Operation Equality: Find (i, j) such that $S_i = S_j$, $i \neq j$, or detect that all sets are different.

Operation Set delete: Given i, delete S_i.

Operation Element delete: Given $x \in X$, delete all occurrences of x from all sets.

2.2 General Survey of the Data Structure

The input thus consists of k sets, $S_1, \ldots, S_k$, $S_i \subseteq X$, $\#X = n$. Let l_i be the cardinal of S_i, or its length if we view sets as words. Let d be the size of the input: we suppose that

$$d \in \Omega(k + \sum_{i=1}^{k} l_i).$$

This is obviously true if the sets are given as lists of elements. In particular, in our pruning graph problem, $d \in \Theta(m+n)$ and $k = n$.

We need an order on X, since sorting will be involved. For the general results there is no need to compute a particular order, so the implicit order (from the input) will do. But in some particular cases we are able to guarantee linear time and space algorithms if we compute a well-chosen order.

Note that the order identifies X and $[\![1, n]\!]$: we can identify the sets on X (e.g. the S_i's) with the words of $[\![1, n]\!]$ whose letters are sorted.

As a second step, we store in a *trie* (a tree coding words thanks to labelled edges [Knu73]) the words with letters sorted that represent our sets. We call this trie a ***lexicographic tree***. This structure is best understood with an example: Fig. 2 shows how each edge is labelled with an element of X, and some nodes with some of the S_i's. Note that the edges are sorted from left to right; that each path from the root meets edge labels in increasing order; and that when S_i labels a node, then the path from the root to this node is exactly S_i.

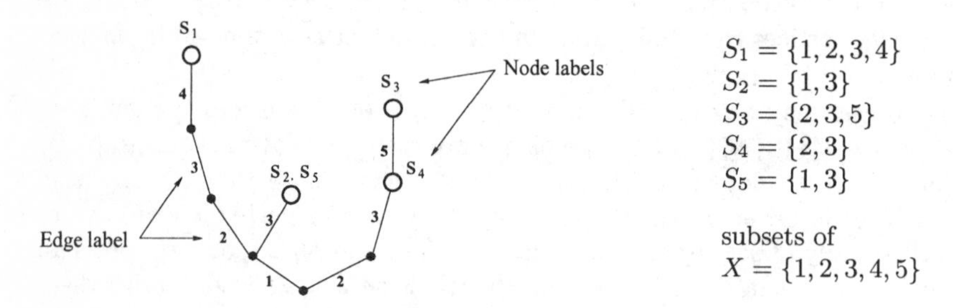

Fig. 2. A lexicographic tree

No two edges with the same parent bear the same label. Thus for a given i exactly one node is labelled with S_i, and this node is also labelled with all S_j's equal to S_i. Note also that all leaves correspond to at least one S_i, and that there are $O(d)$ edges.

The name of the structure comes from the observation that the leaves are lexicographically sorted.

Our data structure for the tree will have some variants, because the coding of the tree structure (parent, children and siblings) is a key point for the complexity analysis, and different choices for kinship will lead to different complexities in time and space.

Concerning the implementation, the basic item is an edge together with the node it leads to. When speaking of the data structure we call this a ***nodge*** (see Fig. 3). It holds the following information:

- A 0 (for instance) marking this structure as being a nodge.
- Two pointers at sets, the "first" and the "last" sets labelling this node. See below how the node labels in our tree are organized.
- The label $x \in X$ of its edge (say 0 for the root nodge).
- Pointers at the next and previous nodges with the same edge label (or NIL if this is the last nodge). The nodges with the same edge label are thus organized in a doubly linked list.
- A pointer at the parent nodge (if this nodge is not the root).
- Some more coding of the tree structure for kinship, to be described later.

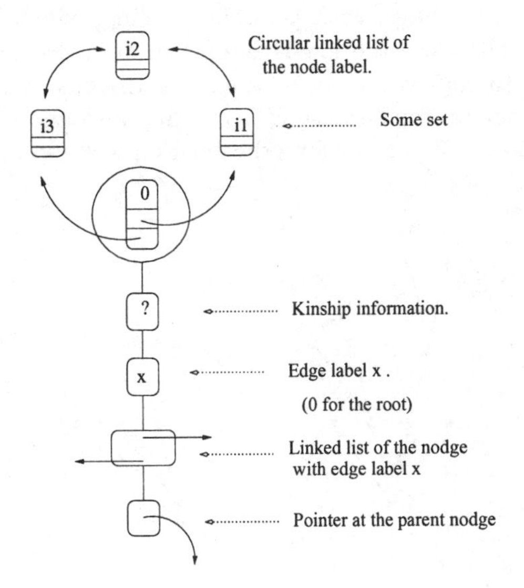

Fig. 3. A nodge

There are two kinds of sets: the ones that are first label of their nodge, and the others, that are duplicates of the firsts. The duplicates are held in a global doubly linked list, the entry of which we call D. If all the nodges of the tree are labelled by at most one S_i then the list D will be empty.

The information concerning the sets is stored in an array with k entries, with at entry i:

- An i marking this entry to be concerning S_i (to tell this entry from a nodge),
- Two pointers at the next and previous equal sets (or to the nodge it labels), as explained,

– Two pointers for the doubly linked list D if S_i is a duplicate (say two NIL's if not).

Finally we need an array of size n for the letters; the entry number x simply consists of a pointer at the first nodge of edge label x.

2.3 Operations

We first have to build the lexicographic tree from sets given by lists of elements: we call that the ***initialization***. This can be done in linear time and space, subject to the choice of the kinship coding : generate the couples (i, x) for $x \in S_i$, bucket sort them according to x and build the tree from the root (see [WWW]). We will check in section 2.4 that the extra time and space needed to initialize the kinship coding remains in suitable bounds.

We will need, as a base step for the operations we eventually want, to know how to delete a nodge. Given a nodge N, we have to delete it, and for every child to put it among N's siblings, with possible merging, which leads to recursive sub-tree merging. Therefore the low-level operation is the merging of a nodge into another one. An edge can be deleted for two reasons: either through direct deletion (`delete-nodge`), or because of a merging into another edge (recursive calls to `merge-nodges`). Figure 4 shows the deletion of a nodge, and the resulting merging of nodges.

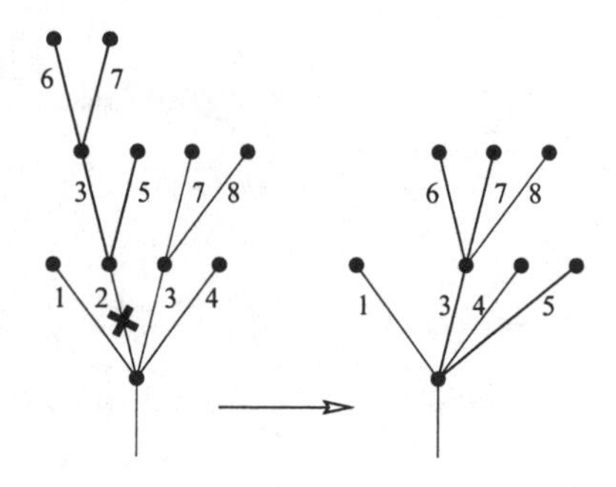

Fig. 4. Deletion of a nodge

These two small procedures (algorithm 1 and 2) are self-describing. Only the "critical loop" of `merge-nodges` needs a more thorough description and a complexity analysis, since it strongly depends on the data structure to be described later. Everything else leads to $O(1)$ operations per deleted nodge, that is, $O(d)$ for all `delete-nodge` calls. The global operations are now easy to describe.

Operation Equality: Use the duplicates list D to know whether there are some equal sets, and if so, to find two equal ones;

Operation Set delete: Remove the set as a node label. Perhaps this creates a leaf without a node label, and a branch is to be deleted. Each nodge deleted this way (with a `delete-nodge` call) costs only $O(1)$ time, since it is childless;
Operation Element delete: While there are some nodges of the given edge label, call `delete-nodge` .

Algorithm 1: `delete-nodge(N)`

Let N' be the parent of N
Remove N from the children of N' (removing a child *given by a pointer* will be an $O(1)$ operation) /*Removes N from the tree*/
`Merge-nodges(N,N')` /*Updates the tree*/

Algorithm 2: `Merge-nodges`(N, N')

/*Merges the subtree above N with the subtree above N' (after recursive calls)*/
Merge the node labels of N into those of N'
if *both N and N' had labels* **then**
  Add the set that is no longer first label to the duplicates list
Remove N from its list of nodges with the same edge label
foreach N_1 *child of* N **do**
  /***Critical loop**: to be fully described later*/
  Let x be the edge label of N_1
  if N' *has a child* N_1' *of edge label* x **then**
    `Merge-nodges`(N_1, N_1')
  else
    Insert N_1 among the children of N'

2.4 Complexity Analysis

Because of a possible long merging of subtrees and because of the repeated deletions of elements and sets, we are rather interested in amortized complexity for the operations. The core of the analysis being the "critical loop" of algorithm 2, we introduce the maximum time T spent on each child during this loop. We now give a time bound for any sequence of `element delete` and `set delete` operations.

Lemma 1. *The time complexity of all operations* `element delete` *and* `set delete` *is in* $O(dT)$.

Idea of the proof: We have to show that the total number of children of the deleted nodges is in $O(d)$. To each child we are able to associate a couple (i, x) with $x \in S_i$, namely: x is the label of the parent, and S_i labels a descendant. □

This lemma is useful to evaluate the complexity of our first two variants of the data structure: we simply compute T and check the initialization complexity.

The three variants described next depend on the data structure chosen for kinship: array, binary search tree or linked list.

The time bounds correspond to the total of all operations `element delete`, `set delete` and `equality` whatever their order (we suppose that less than $O(d)$ operations `equality` are performed, each one being in $O(1)$).

Time $O(d+n)$, Space $O(d+kn)$. It means linear time, but with extra space needed (for the pruning problem this is $O(n^2)$ space). Here we choose arrays to code kinship so that $T \in O(1)$, and the children of a nodge are linked, making a doubly linked list. More precisely, we use a somewhat redundant data structure for the kinship in the tree:

- An array A of size $k \times n$.
- A linked list of the unused lines of A, thus of size less than k.
- Each nodge N holds:
  - 4 more pointers, at the left sibling and right sibling which makes a linked list for siblings' structure, and at the first and last child (we already have the parent; of course NIL pointer are used for inexistent kin). The children of a nodge will not be kept sorted.
  - An index $a \in [\![0, k]\!]$, such that $a = 0$ if N has one child at most. If N has at least two children, then the ath line of A consists of pointers at the children of N, that is, $A[a, x]$ points at the child of N of edge label x if any, or else contains NIL.

The main point is that there are at most k leaves, so at most k nodes have degree more than 1.

It is easy to delete or insert a given child (given by its edge label) in $O(1)$ time thanks to these structures. In particular, changing a nodge with several children into a nodge with a single child as well as the inverse are $O(1)$ operations.

It remains to show that the extra data structures can be initialized in the required time and space; this is obvious if we are given an empty (filled with NIL) kn amount of space, in which to store A.

If not, we cannot afford the time to empty all that space. However it is possible to resort to the indirection trick [AHU71]: this simulates empty space and costs only linear time.

Time $O(d+kn)$, Space $O(d+n)$. (for the pruning problem this means linear space but $O(n^2)$ time). Here we choose to link the children of a nodge with a simple linked list, where the children are kept sorted, the smaller on the left. Each nodge has a pointer at his right sibling and has other pointers at his first and last child. The initialization does not need extra time, as the children are appearing sorted. This is the easiest algorithm, but the amortized cost is somewhat trickier to prove. The analysis relies on the ***holes***.

Definition 1. *An* x-hole *in the lexicographic tree is a couple* (N, x) *such that* N *is a nodge with edge label* x_0, $x_0 \leq x$, N *has no son of edge label* x, *and* N *has a son of edge label* x_1, $x_1 > x$.

Removing a child given by a pointer is also easy in $O(1)$. Thus the global time complexity is $O(d+n)$ plus the time spent on the critical loop. We use the credits method to prove this time to be in $O(d+kn)$: we keep the invariant that each hole is credited with 1 credit, and each nodge with 2 credits plus 1 credit per descendant **leaf**.

Lemma 2. *This makes at most* $3d + kn$ *credits initially.*

The critical loop is a simple sweep of the children of N and N' at the same time, inserting or merging as you advance. All credits correspond to the same amount of time. We check precisely what credit is used for each operation, so that the invariant will be maintained. As a result, the total time is bound by the initial number of credits (see [WWW]).

Time and Space $O(n+d\log n)$. (this is $O(m\log n)$ for the pruning problem). It is easy to find a structure for which $T \in O(\log n)$: for instance a binary search tree, with the edge label as a key. Each nodge simply needs two more pointers, left and right "sub-child".
Each child adds an $O(\log n)$ space, and that makes a global extra space $O(d\log n)$. Removing a child or inserting one with possible merging is easy in time $O(\log n)$.

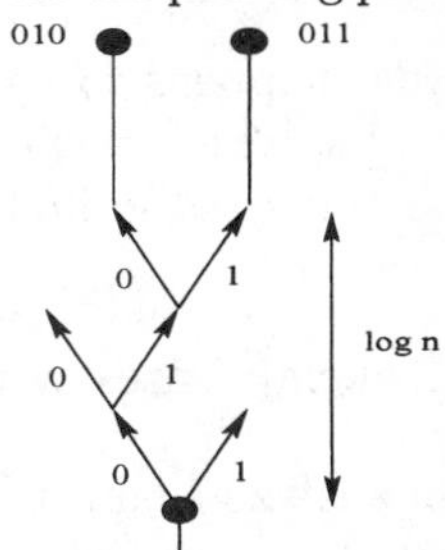

3 Application to Graph Problems

3.1 Pruning a Graph

Let G be a graph with n vertices and m edges. The input is assumed to consist of the adjacency lists, it is of size $\Theta(n+m)$. To each $v \in [\![1,n]\!]$, we associate two sets, $S_{2v-1} = \mathcal{N}(v)$ and $S_{2v} = \mathcal{N}_c(v)$. We apply to our sets $(S_i)_{i\in[\![1,2n]\!]}$ the operations defined earlier.

Deleting a vertex v from the current graph is equivalent to removing S_{2v-1}, S_{2v}, and every edge with label v, from the tree. This is exactly two `set delete` operations, and one `element delete` operation.

An `equality` operation is like scanning the graph for a couple of twins, so we simply lack the data structures necessary to detect pendant vertices. We simply need to maintain the graph structure with the degrees of the vertices. This can be done in linear time.

Algorithm 3 computes the "pruned graph". We suppose that the `equality` operation returns $(0,0)$ if there are no equal sets.

So we can use the results of section 2 with $d = 4m + 2n$ and $k = n$.

Theorem 1. *Given a graph with n vertices and m edges, we can compute its pruned graph in the following bounds:*

Time bound	$n+m$	$m\log n$	n^2
Space bound	n^2	$m\log n$	$n+m$

Algorithm 3: Pruning a graph

```
Initialize the data structures
repeat
    Let (i,j) be the output of an equality operation
    if i ≠ 0 then v ⟵ ⌈i/2⌉
    else v ⟵ a pendant vertex or 0 (No twins)
    if v ≠ 0 then
        Perform a set delete operation on S_{2v-1}
        Perform a set delete operation on S_{2v}
        Perform an element delete operation on v
        Update the degree data structure
until v = 0
```

Remark 1. The order chosen on the vertices may be important: for example, if this order happens to be a reversed pruning order, we might be deleting only leaf nodges, which would make the third algorithm linear in time too. We will thus be interested in finding a good order on the graph.

3.2 Special Cases with Linear Space and Time

We will analyze again the complexity of the algorithm linear in space (the third variant) in the special case when we are not dealing with both true and false twins. We associate to each vertex v only one set S_v: $\mathcal{N}(v)$ if we just remove true twins, $\mathcal{N}_c(v)$ if we just remove false twins. To ensure linear time, we will need two slight modifications:

- Use of a special order on the vertices: a LexBFS order.
- When operation `equality` supplies us with $(i,j) \neq (0,0)$, instead of deleting the vertex corresponding to i, we will delete the vertex corresponding to $\max(i,j)$. Each such vertex is guaranteed to have a smaller twin in the graph.

LexBFS means "lexicographic breadth-first search" [RTL76]. This is a particular breadth-first search, where neighbours to already chosen vertices are favoured. The vertices are numbered, beginning at 1. When a vertex is picked as next vertex, it marks its unnumbered neighbours with its number (see Fig. 5). Lexicographic order is used to compare two marks, and the next vertex is chosen with lowest mark. See also [HMPV97] for a simple algorithm (based on partition refinement, not on marks) to generate a LexBFS order.

Precomputing a LexBFS order on the graph is interesting for two reasons. First, it gives strong properties to the lexicographic tree. The main one concerns the removal of twins: for the `element deletion` operation, merging two nodges can be done in $O(1)$. And all the other operations also have good complexities (see [WWW]). Secondly, after the removal of a pendant vertex or a vertex that has a smaller twin, we still have a LexBFS order on the graph.

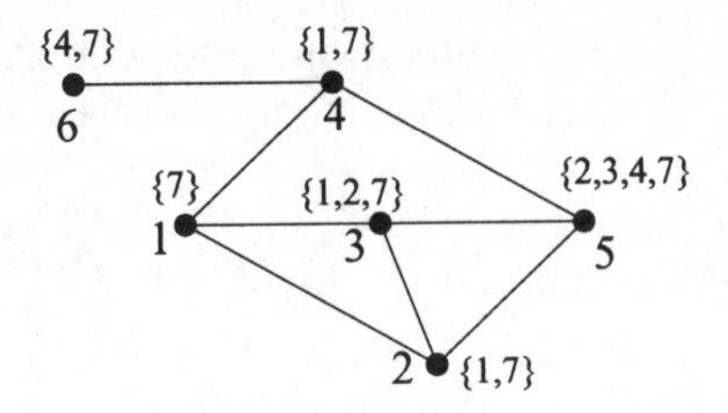

Fig. 5. A LexBFS ordering with the mark of each vertex

Theorem 2.
- *The modified algorithm prunes from a graph its pendant vertices and true twins in linear time and space.*
- *The modified algorithm prunes from a graph its pendant vertices and false twins in linear time and space.*

So we have linear time instead of $O(n^2)$ time.

Example: Bipartite Graphs. *In such graphs, two vertices are false twins iff they are pendant to each other (an isolated edge). Thus we can use our algorithm (deleting pendant and true twin vertices) for an easy linear recognition of distance-hereditary bipartite graphs.*

4 Another Application: Complete Bipartite Cover

The "covering by complete bipartite subgraphs" problem is to find how many complete bipartite subgraphs we need (at the minimum) to cover every edge of a given bipartite graph. This problem is $\mathcal{NP}$-hard in the general case (Problem GT18 in [GJ79]); however [Mül96] showed that the complexity falls to $\mathcal{P}$ for some particular graph classes, like the distance-hereditary graphs. The algorithm solving this particular case is not studied further than being polynomial. The base step is to find a "bisimplicial" edge, which leads to a time complexity in $O(nm)$ on a bipartite graph with n vertices and m edges.

However, it is possible to find an algorithm that runs in linear time and space. We will again prune the input graph.

If we are given an optimal complete bipartite cover on the pruned graph, it is easy to find an optimal one on the original graph. Suppose we add a vertex to a bipartite graph with an optimal cover:

Adding u as a true twin of v: every complete bipartite graph in which v occurs is changed by duplicating v into u. This is still a complete bipartite graph and we still cover all the edges.

Adding u as a pendant vertex: we cannot extend an existing complete bipartite graph so that it includes the new edge ending by u, for that would require u to have another neighbor. Thus we have to add a new graph in the cover, for instance the star centered in v, the unique neighbor of u.

Thus the number sought is simply increased by one for every "pendant operation".

Computing the number of "pendant operations" during the pruning gives a simple algorithm to solve the problem on the class of bipartite distance-hereditary graphs.

5 Conclusion

The lexicographic tree is an interesting representation of a graph, and gives good results for the "pruned graph" problem. Since we can compute it in linear time and space[1] from adjacency lists, this tree is an interesting tool for graph problems.

Besides twin vertices, we are convinced that there must be other properties of graphs that can be read on this data structure, knowing that special orders on vertices and special classes of graphs have a strong influence on the structure of the tree.

As far as we know, it is still an open problem to compute the pruned graph in both linear time and space, in the general case.

6 Acknowledgements

The authors wish to thank Lhouari Nourine for many ideas.

References

[AHU71] A. V. Aho, I. E. Hopcroft, and J. D. Ullman. *The design and analysis of computer algorithms.* Addison-Welsey, 1974, exercise 2.12 on page 71.

[BM86] H. J. Bandelt and H. M. Mulder. Distance-hereditary graphs. *J. Combin. Theory*, Ser. B, 41:182–208, 1986.

[Dam97] G. Damiand. Quelques propriétés des graphes distances héréditaires. Master's thesis, Université de Montpellier II, LIRMM, 1997.

[GJ79] M. R. Garey and D. S. Johnson. *Computers and Intractability: A guide to the theory of NP-completeness.* Freeman, 1979.

[HM90] P. Hammer and F. Maffray. Completely separable graphs. *Discrete Applied Mathematics*, 27:85–99, 1990.

[HMPV97] M. Habib, R. McConnell, C. Paul, and L. Viennot. LexBFS and partition refinement, with applications to transitive orientation, interval graph recognition and consecutive ones testing. *Theor. Comp. Sci.*, 1999. to appear.

[How77] E. Howorka. A characterization of distance-hereditary graphs. *Quart. J. Math. Oxford*, Ser. 2, 28:417–420, 1977.

[Knu73] D. E. Knuth. *The Art of Computer Programming: Sorting and Searching*, volume 3. Addison-Wesley, 1973.

[Mül96] H. Müller. On edge perfectness and classes of bipartite graphs. *Discrete Mathematics*, 149:159–187, 1996.

[1] if we use linked lists for kinship information

[RTL76] D. J. Rose, R. E. Tarjan, and G. S. Leuker. Algorithmic aspects of vertex elimination on graphs. *SIAM J. of Computing*, 5(2):266–283, June 1976.

[WWW] A more detailed version of this article http://www.lirmm.fr/~thierry

Characterizing and Deciding MSO-Definability of Macro Tree Transductions*

Joost Engelfriet and Sebastian Maneth

Leiden University, LIACS, PO Box 9512, 2300 RA Leiden, The Netherlands
{engelfri, maneth}@liacs.nl

Abstract. A macro tree transduction is MSO definable if and only if it is of linear size increase. Furthermore, it is decidable for a macro tree transduction whether or not it is MSO definable.

1 Introduction

Macro tree transducers (MTTs) are a well-known model of syntax-directed semantics that combines top-down tree transducers and context-free tree grammars (see, e.g., [EV85, CF82, FV98]). Their (tree-to-tree) transductions form a large class, containing the translations of attribute grammars. More recently, tree transductions that can be specified in monadic second-order (MSO) logic have been considered (see, e.g., [Cou94, BE98, EM99, KS94]). It is shown in [BE98] that these MSO definable tree transductions can be computed by (a special type of) attribute grammars. Thus, as stated in [EM99], every MSO definable tree transduction can be realized by an MTT. However, not every macro tree transduction is MSO definable. We have considered the question, *which* macro tree transductions are MSO definable?

There are different ways of answering this question. As shown in [EM99; Section 7], every MSO definable tree transduction can be realized in particular by a 'finite copying' MTT, and vice versa, if an MTT M is finite copying, then its transduction is MSO definable. In this paper we prove a characterization in terms of a property which is independent of M: M's transduction is MSO definable if and only if it is of linear size increase, i.e., the size of the output tree is linearly bounded by the size of the input tree. Moreover we prove that it is decidable whether M's transduction is of linear size increase, and hence whether it is MSO definable. Note that the MSO definable tree transductions have several nice features that the macro tree transductions do not possess: by definition they can be specified in MSO logic, they can be computed in linear time [BE98], and they are closed under composition [Cou94].

The idea for our characterization stems from [AU71]; there it is shown that a generalized syntax-directed translation (gsdt) scheme can be realized by a tree-walking transducer if and only if it is of linear size increase. Since gsdt schemes are a variation of top-down tree transducers, and tree-walking transducers are

* This work was supported by the EC TMR Network GETGRATS.

H. Reichel and S. Tison (Eds.): STACS 2000, LNCS 1770, pp. 542–554, 2000.

closely related to finite copying top-down tree transducers [ERS80], our result can be viewed as a generalization of the result of [AU71], from top-down tree transducers to macro tree transducers. Moreover, we obtain a characterization of the MSO definable top-down tree transductions that depends on the transducer: they are exactly the transductions that can be realized by finite copying top-down tree transducers (but to be precise, this only holds if the transducers have regular look-ahead, see, e.g., [GS97; Section 18]).

Note that very often membership in a subclass is undecidable (such as regularity of a context-free language). In cases of decidability there is often a characterization of the subclass that is independent of the device that defines the whole class, analogous to our linear size increase characterization (as an example, in [Cou95] it is shown that a vertex replacement graph language can be generated by a hyperedge replacement graph grammar if and only if the number of edges of its graphs is linearly bounded by the number of nodes).

Structure of the paper: In Section 3 we recall MTTs (with regular look-ahead); they are total deterministic. In Section 4 the two finite copying (fc) properties for MTTs are defined (as in [EM99]) and their decidability is proved; furthermore two pumping lemmas for non-fc MTTs are stated. Based on these pumping lemmas and two normal forms it is shown in Section 5 that non-fc MTTs in normal form are not of linear size increase. From [EM99] we know that fc MTTs realize precisely the MSO definable tree transductions, which are obviously of linear size increase. Together this yields our characterization in Section 6, where the main results are given. The proofs are just sketched; details can be found in the preliminary version of [EM].

2 Trees, Tree Automata, and Tree Transductions

We assume the reader to be familiar with trees, tree automata, and tree transductions (see, e.g., [GS97]). For $m \geq 0$ let $[m] = \{1, \dots, m\}$. A set Σ together with a mapping rank: $\Sigma \to \mathbb{N}$ is called a *ranked* set. For $k \geq 0$, $\Sigma^{(k)}$ is the set $\{\sigma \in \Sigma \mid \text{rank}(\sigma) = k\}$; we also write $\sigma^{(k)}$ to denote that $\text{rank}(\sigma) = k$. For a set A, $\langle \Sigma, A\rangle$ is the ranked set $\{\langle \sigma, a\rangle \mid \sigma \in \Sigma, a \in A\}$ with $\text{rank}(\langle \sigma, a\rangle) = \text{rank}(\sigma)$. The set of all trees over Σ is denoted T_Σ. For a set A, $T_\Sigma(A)$ is the set of all trees over $\Sigma \cup A$, where all elements in A have rank zero. The size of a tree s, denoted size(s), is the number of its nodes. A node of s is denoted by its Dewey notation in $\mathbb{N}^*$ (e.g., the b-labeled node in $\sigma(a, \gamma(b))$ is denoted by 21). The set of all nodes of s is denoted by $V(s)$. For $u \in V(s)$, the subtree of s rooted at u is denoted s/u, and for $s' \in T_\Sigma$ the tree obtained by replacing s/u in s by s' is denoted $s[u \leftarrow s']$. For $\sigma \in \Sigma$, $\#_\sigma(s)$ denotes the number of occurrences of σ in s and for $\Delta \subseteq \Sigma$, $\#_\Delta(s) = \sum\{\#_\sigma(s) \mid \sigma \in \Delta\}$. We fix the set X of *input variables* $x_1, x_2, \dots$ and the set Y of *parameters* $y_1, y_2, \dots$. For $k \geq 0$, $X_k = \{x_1, \dots, x_k\}$ and $Y_k = \{y_1, \dots, y_k\}$.

A *finite tree automaton* is a tuple (P, Σ, h), where P is a finite set of states, Σ is a ranked alphabet, and h is a collection of mappings $h_\sigma : P^k \to P$, for

$\sigma \in \Sigma^{(k)}$. We also use h to denote its (usual) extension to a mapping from T_Σ to P. For $p \in P$ the set $\{s \in T_\Sigma \mid h(s) = p\} = h^{-1}(p)$ is denoted by L_p.

A *tree transduction* is a mapping $\tau\colon T_\Sigma \to T_\Delta$, for ranked alphabets Σ and Δ. The tree transduction τ is *MSO definable* if there exist a finite set C and MSO(Σ)-formulas $\nu_c(x)$, $\psi_{\delta,c}(x)$, and $\chi_{i,c,d}(x,y)$, with $c,d \in C$, $\delta \in \Delta$, and $i \geq 1$, such that for every $s \in T_\Sigma$, $V(\tau(s)) = \{(c,x) \in C \times V(s) \mid s \models \nu_c(x)\}$, node (c,x) has label δ iff $s \models \psi_{\delta,c}(x)$, and (d,y) is the i-th child of (c,x) iff $s \models \chi_{i,c,d}(x,y)$. An MSO($\Sigma$)-formula is a formula of monadic second-order logic that uses atomic formulas $\mathrm{lab}_\sigma(x)$ and $\mathrm{child}_i(x,y)$, with $\sigma \in \Sigma$ and $i \geq 1$, to express that x has label σ and y is the i-th child of x, respectively. The class of all MSO definable tree transductions is denoted *MSOTT*. For examples and more details, see, e.g., [Cou94, BE98].

A tree transduction τ is of *linear size increase* (for short, lsi), if there is a $c \geq 1$ such that, for every $s \in T_\Sigma$, $\mathrm{size}(\tau(s)) \leq c \cdot \mathrm{size}(s)$. The class of all tree transductions of linear size increase is denoted by *LSI*. Note that every MSO definable tree transduction τ is lsi (with constant $c = |C|$), i.e., *MSOTT* $\subseteq$ *LSI*.

3 Macro Tree Transducers

A macro tree transducer is a syntax-directed translation device in which the translation of an input tree may depend on its context. The context information is handled by parameters. We will consider *total deterministic* MTTs only. For technical reasons we add *regular look-ahead* to MTTs (recall from [EV85; Theorem 4.21] that this does not change the class of transductions).

A *macro tree transducer with regular look-ahead* (for short, MTT^{R}) is a tuple $M = (Q, P, \Sigma, \Delta, q_0, R, h)$, where Q is a ranked alphabet of *states*, Σ and Δ are ranked alphabets of *input* and *output symbols*, respectively, $q_0 \in Q^{(0)}$ is the *initial state*, (P, Σ, h) is a finite tree automaton, called the *look-ahead automaton*, and R is a finite set of *rules*. For every $q \in Q^{(m)}$, $\sigma \in \Sigma^{(k)}$, and $p_1, \ldots, p_k \in P$ with $m, k \geq 0$ there is exactly one rule of the form

$$\langle q, \sigma(x_1, \ldots, x_k)\rangle(y_1, \ldots, y_m) \to \zeta \quad \langle p_1, \ldots, p_k\rangle \qquad (*)$$

in R, where $\zeta \in T_{\langle Q, X_k\rangle \cup \Delta}(Y_m)$.

A rule r of the form $(*)$ is called the $(q, \sigma, \langle p_1, \ldots, p_k\rangle)$-rule and its right-hand side ζ is denoted by $\mathrm{rhs}(r)$ or by $\mathrm{rhs}_M(q, \sigma, \langle p_1, \ldots, p_k\rangle)$; it is also called a q-rule. A *top-down tree transducer with regular look-ahead* (for short, T^{R}) is an MTT^{R} all states of which are of rank zero. If the look-ahead automaton is trivial, i.e., $P = \{p\}$ and $h_\sigma(p, \ldots, p) = p$ for all $\sigma \in \Sigma$, then M is called a *macro tree transducer* (for short, MTT) and if M is a T^{R}, then M is called a *top-down tree transducer*. In such cases we omit the look-ahead automaton and simply denote M by $(Q, \Sigma, \Delta, q_0, R)$; we also omit the look-ahead part $\langle p_1, \ldots, p_k\rangle$ in rule $(*)$. If, for every q-rule r with $q \in Q^{(m)}$ and every $j \in [m]$: $\#_{y_j}(\mathrm{rhs}(r)) \leq 1$, then M is *linear in the parameters* (for short, linp). We also use 'linp' as subscript for the corresponding transducers and classes of transductions.

The rules of M are used as term rewriting rules in the usual way with the additional restriction that the rule $(*)$ can be applied at a subtree $\sigma(s_1, \ldots, s_k)$ only if, for every $i \in [k]$, $s_i \in L_{p_i}$. The derivation relation of M (on $T_{\langle Q,T_\Sigma \rangle \cup \Delta}$) is denoted by $\Rightarrow_M$. The *transduction realized by M*, denoted τ_M, is the total function $\{(s,t) \in T_\Sigma \times T_\Delta \mid \langle q_0, s\rangle \Rightarrow_M^* t\}$. The class of all transductions which can be realized by MTT^{R}s (MTTs, T^{R}s) is denoted by MTT^R (MTT, T^R).

Let $s \in T_\Sigma$ and $u \in V(s)$. For $s_1 = s/u$, $\tau_M(s)$ can be obtained by the derivation $\langle q_0, s\rangle \Rightarrow_M^* \ldots \langle q', s_1\rangle \ldots \langle q'', s_1\rangle \cdots \Rightarrow_M^* \tau_M(s)$, in which first the part of s outside of s_1 is translated and then s_1 is translated. The intermediate sentential form $\xi = \ldots \langle q', s_1\rangle \ldots \langle q'', s_1\rangle \ldots$ shows in which states the subtree s_1 of s at u is processed. We will now extend $\Rightarrow_M$ in such a way that ξ can be generated as 'final' tree. Moreover, we allow the translation to start with an arbitrary state instead of q_0. For $q \in Q^{(m)}$ and $s' \in T_\Sigma(P)$, $M_q(s')$ is defined to be the (unique) tree t in $T_{\langle Q,P\rangle \cup \Delta}(Y_m)$ such that $\langle q, s'\rangle(y_1, \ldots, y_m) \Rightarrow_M^* t$, where $\Rightarrow_M$ is extended to $T_{\langle Q,T_\Sigma(P)\rangle \cup \Delta}(Y_m)$ in the obvious way, extending the look-ahead automaton with $h_p = p$ for every $p \in P$. Now we get $\langle q_0, s[u \leftarrow p]\rangle \Rightarrow_M^* \cdots \langle q', p\rangle \ldots \langle q'', p\rangle \cdots = M_{q_0}(s[u \leftarrow p])$, which is ξ with s_1 replaced by $h(s_1) = p$. Note that in particular, $M_{q_0}(s) = \tau_M(s)$.

We say that $\langle q, p\rangle \in \langle Q, P\rangle$ is *reachable*, if there are $s \in T_\Sigma$ and $u \in V(s)$ such that $\langle q, p\rangle$ occurs in $M_{q_0}(s[u \leftarrow p])$.

Assumptions: For an MTT^{R} M we assume from now on that (i) q_0 does not occur in any right-hand side of a rule of M, (ii) if $\langle q, p\rangle$ is not reachable, then there is a $\delta \in \Delta^{(m)}$ such that $\langle q, s\rangle(y_1, \ldots, y_m) \Rightarrow_M \delta(y_1, \ldots, y_m)$ for every $s \in L_p$, (iii) M has no "erasing rules", i.e., rules with right-hand side y_1, and (iv) M is *nondeleting*, i.e., for every q-rule r with $q \in Q^{(m)}$ and every $j \in [m]$: $\#_{y_j}(\mathrm{rhs}(r)) \geq 1$. The proof that (iii) and (iv) may be assumed without loss of generality can be done by the use of regular look-ahead (see [EM99; Lemma 7.11]).

4 Finite Copying

A rule of an MTT^{R} M is copying, if its right-hand side is (i) nonlinear in the input variables or (ii) nonlinear in the parameters. We now want to put a bound on both of these ways of copying. For (ii) this is simple: M is *finite copying in the parameters* (for short, fcp), if for every $q \in Q^{(m)}$, $s \in T_\Sigma$, and $j \in [m]$, $\#_{y_j}(M_q(s)) \leq k$ for a fixed $k \geq 0$. For (i), we generalize the notion of finite copying from top-down tree transducers (cf. [ERS80, AU71]) to MTT^{R}s. The *state sequence of s at node u*, denoted by $\mathrm{sts}_M(s, u)$, contains all states which process the subtree s/u. Formally, $\mathrm{sts}_M(s, u) = \pi(M_{q_0}(s[u \leftarrow p]))$, where $p = h(s/u)$ and π changes every $\langle q, p\rangle$ into q and deletes everything else. If for every $s \in T_\Sigma$ and $u \in V(s)$, $|\mathrm{sts}_M(s, u)| \leq k$ for a fixed $k \geq 0$, then M is called *finite copying in the input* (for short, fci). We also use 'fci' and 'fcp' as subscripts for the corresponding transducers and classes of transductions. The following proposition follows from Lemmas 6.3 and 6.7, and Theorem 7.1 of [EM99] (where 'surp' is used instead of 'linp').

Proposition 1. $MTT^R_{\text{fcp}} = MTT^R_{\text{linp}}$ and $MTT^R_{\text{fci,fcp}} = MTT^R_{\text{fci,linp}} = MSOTT$.

For technical reasons we introduce another way of bounding the copying of input variables: associate with every MTT^{R} M the T^{R} $\text{top}(M) = (Q, P, \Sigma, \{\sigma^{(2)}, e^{(0)}\}, q_0, R', h)$ by changing rule $(*)$ into rule $\langle q, \sigma(x_1, \ldots, x_k)\rangle \to \zeta'\ \langle p_1, \ldots, p_k\rangle$, where $\zeta' = \sigma(\langle q_1, x_{i_1}\rangle, \sigma(\langle q_2, x_{i_2}\rangle, \ldots \sigma(\langle q_n, x_{i_n}\rangle, e) \ldots))$ if $\langle q_1, x_{i_1}\rangle, \ldots, \langle q_n, x_{i_n}\rangle$ are all elements of $\langle Q, X_k\rangle$ that occur in ζ. Then M is *globally fci* (for short, gfci), if $\text{top}(M)$ is fci. We also use 'gfci' as subscript for the corresponding transducers and classes of transductions. Intuitively, the T^{R} $\text{top}(M)$ is carrying out the pure state behavior of M (in particular the copying of input variables) without the parameter behavior of M. Note that $MTT^R_{\text{gfci,fcp}}$ does *not* equal $MTT^R_{\text{fci,fcp}}$ (cf. Example 10), but that $MTT^R_{\text{gfci,linp}}$ does! The following lemma shows the latter, by relating the number of occurrences of a state in $\text{sts}_M(s, u)$ and $\text{sts}_{\text{top}(M)}(s, u)$.

Lemma 2. Let $M = (Q, P, \Sigma, \Delta, q_0, R, h)$ be an MTT^{R}. For every $q, q' \in Q$, $s \in T_\Sigma$, $u \in V(s)$, and $p \in P$:

$$\#_{\langle q',p\rangle}(M_q(s[u \leftarrow p])) \geq \#_{\langle q',p\rangle}(\text{top}(M)_q(s[u \leftarrow p])),$$

and they are equal if M is linp. On the other hand, for every path π in the tree $M_q(s[u \leftarrow p])$, the number of $\langle q', p\rangle$-labeled nodes on π is $\leq$ the number of $\langle q', p\rangle$-labeled nodes of $\text{top}(M)_q(s[u \leftarrow p])$.

Proof. By Assumption (iv), M is nondeleting. If an actual parameter ξ is copied in a derivation step of M, then the states that occur in ξ are copied too, and so the number of occurrences of q' can grow more than in the corresponding (noncopying) step of $\text{top}(M)$. Considering a path π only, there is no copying of states; thus, the application of a rule r of M increases the number of occurrences of q' on π at most by $\#_{\langle q',X\rangle}(\text{rhs}(r))$ which equals $\#_{\langle q',X\rangle}(\text{rhs}(r'))$ for the corresponding rule r' of $\text{top}(M)$. □

Note that Lemma 2 implies that a linp MTT^{R} (and in particular a T^{R}) is gfci iff it is fci. Let us now prove the decidability of the fci and fcp properties.

Lemma 3. For an MTT^{R} M it is decidable (i) whether or not M is fci and (ii) whether or not M is fcp.

Proof. (i) It is straightforward to construct an MTT^{R} M' such that $\tau_{M'}(s[u \leftarrow p]) = M_{q_0}(s[u \leftarrow p])$, and to construct an MTT M'' which translates every tree t in $T_{\langle Q,P\rangle \cup \Delta}$ into a monadic tree $q_1(q_2(\ldots q_n(e)))$, where $\langle q_1, p_1\rangle, \ldots, \langle q_n, p_n\rangle$ are all elements of $\langle Q, P\rangle$ that occur in t. Thus, for $t = \tau_{M'}(s[u \leftarrow p])$, $\tau_{M''}(t)$ equals $\text{sts}_M(s, u)$, seen as a monadic tree. Hence, for $K = \{s \in T_\Sigma(P) \mid \#_P(s) = 1\}$, M is fci iff $\tau_{M''}(\tau_{M'}(K))$ is finite. This is decidable, because finiteness of the range of a composition of MTT^{R}s, restricted to a regular tree language K, is decidable [DE98].
(ii) It is straightforward to construct an MTT^{R} M' with input alphabet $\Sigma \cup \{q^{(1)} \mid q \in Q\}$ and output alphabet $\Delta \cup Y_m$ (for an appropriate m) such that

$\tau_{M'}(q(s)) = M_q(s)$ for every $q \in Q$ and $s \in T_\Sigma$, and to construct an MTT^{R} M'' such that for every $t \in T_\Delta(Y_m)$, $\mathrm{size}(\tau_{M''}(t)) = 1 + \sum\{\#_{y_j}(t) \mid j \in [m]\}$. Hence, for $K = \{q(s) \mid q \in Q, s \in T_\Sigma\}$, M is fcp iff $\tau_{M''}(\tau_{M'}(K))$ is finite. As above this is decidable by [DE98]. □

Pumping Lemmas We now present two pumping lemmas for non-fci T^{R}s and for non-fcp MTT$^{\mathrm{R}}_{\mathrm{gfci}}$s, respectively. They are the core of the proof, in Section 5, that linear size increase implies gfci and fcp. The first lemma is similar to Lemma 4.2 of [AU71]. We use the following notation (to "pump" a tree). For $s \in T_\Sigma$, $u \in V(s)$, $p \in P$, and $s' \in T_\Sigma(P)$, let $s[u \leftarrow p] \bullet s'$ denote $s[u \leftarrow s']$. Thus, $(s[u \leftarrow p])^2 = (s[u \leftarrow p]) \bullet (s[u \leftarrow p]) = s[u \leftarrow s[u \leftarrow p]]$.

Lemma 4. Let $M = (Q, P, \Sigma, \Delta, q_0, R, h)$ be a T^{R}. If M is not fci, then there are $q_1, q_2 \in Q$, $s_0 \in T_\Sigma$, $u_0, u_0u_1 \in V(s_0)$, and $p \in P$ such that the following four conditions hold, with $s_1 = s_0/u_0$.

(1) $\langle q_1, p\rangle$ occurs in $M_{q_0}(s_0[u_0 \leftarrow p])$,
(2) $\langle q_1, p\rangle$ and $\langle q_2, p\rangle$ occur at distinct nodes of $M_{q_1}(s_1[u_1 \leftarrow p])$,
(3) $\langle q_2, p\rangle$ occurs in $M_{q_2}(s_1[u_1 \leftarrow p])$, and
(4) $p = h(s_1) = h(s_1/u_1)$.

Proof. Let $t \in T_\Sigma$ and $v \in V(t)$. Then for every ancestor u of v let $\mathrm{csts}(u, v)$ denote $\mathrm{sts}_M(t, u)$ restricted to states q which *contribute to* $\mathrm{sts}_M(t, v)$, i.e., for which $\xi = M_q(t/u[v' \leftarrow h(t/v)]) \notin T_\Delta(Y)$ with $v = uv'$. For any state r that occurs in ξ we write $q \rightarrow_{u,v} r$. Note that $q \rightarrow_{u,v} q' \rightarrow_{v,w} q''$ implies $q \rightarrow_{u,w} q''$. Since M is not fci, there are arbitrary long state sequences; in particular, for ρ the

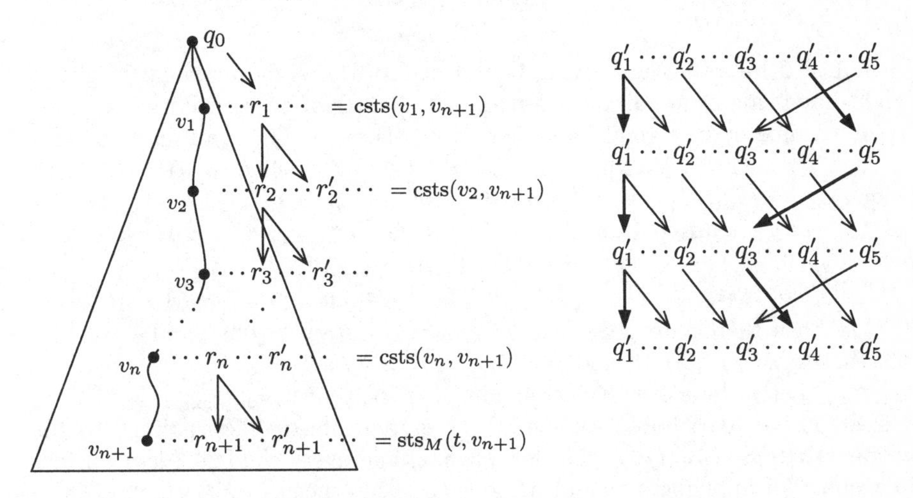

Fig. 1. a. tree t with contributing states **b.** (2) and (3) for $q_1 = q_1'$ and $q_2 = q_4'$

maximal number of occurrences of x_i in a right-hand side of the rules of M, and every $n \geq 1$, there are $t \in T_\Sigma$ and $v_{n+1} \in V(t)$ such that $|\mathrm{sts}_M(t, v_{n+1})| \geq \rho^n$.

This means that (on the path to v_{n+1} in t) at least n times rules that copy the corresponding input variable have been applied. Thus, there are different ancestors $v_1, \ldots, v_n$ of v_{n+1} and states $r_1, \ldots, r_n, r_{n+1}, r'_2, \ldots, r'_n, r'_{n+1}$ such that, for every $i \in [n+1]$, r_i and r'_i occur in $\mathrm{csts}(v_i, v_{n+1})$ (at different positions) and, for every $i \in [n]$, $r_i \rightarrow_{v_i, v_{i+1}} r_{i+1}$ and $r_i \rightarrow_{v_i, v_{i+1}} r'_{i+1}$ (see Fig. 1a, where the nodes v_i in t with their csts's are shown, and arrows mean $\rightarrow_{v_i, v_{i+1}}$).

Take $n = |Q| \cdot |P| \cdot 2^{|Q|}$ and let t, v_i, and r_i, r'_i be as above (for $i \in [n+1]$). Clearly this means that there are distinct indices $i, j \in [n+1]$ such that $r_i = r_j$, $p = h(t/v_i) = h(t/v_j)$, and the same states appear in $\mathrm{csts}(v_i, v_{n+1})$ and in $\mathrm{csts}(v_j, v_{n+1})$. Now let $q'_1 = r_i$ and $q'_2 \in Q$ with $r'_{i+1} \rightarrow_{v_{i+1}, v_j} q'_2$ (q'_2 exists because r'_{i+1} contributes to $\mathrm{sts}_M(t, v_{n+1})$). Then $q'_1 \rightarrow_{v_i, v_j} q'_1$ and $q'_1 \rightarrow_{v_i, v_j} q'_2$, which shows (1), (2), and (4), for $s_0 = t$, $u_0 = v_i$, $u_0 u_1 = v_j$, $q_1 = q'_1$, and $q_2 = q'_2$. To realize (3), we will pump the tree $t_1[v'_j \leftarrow p]$ in t, where $t_1 = t/v_i$ and $v_j = v_i v'_j$. Since the same states appear in $\mathrm{csts}(v_i, v_{n+1})$ and $\mathrm{csts}(v_j, v_{n+1})$, there is a sequence $q'_2 \rightarrow_{v_i, v_j} q'_3 \rightarrow_{v_i, v_j} \cdots \rightarrow_{v_i, v_j} q'_m \rightarrow_{v_i, v_j} q'_{m-\nu}$ with $m \leq |Q|$ and $0 \leq \nu < m$. Let $d \geq m - (\nu + 1)$ be a multiple of $\nu + 1$. For $s_0 = (t[v_i \leftarrow p]) \bullet (t_1[v'_j \leftarrow p])^d \bullet (t_1/v'_j)$, $q_1 = q'_1$, $q_2 = q'_{d+1}$, and $w = v_i (v'_j)^d$: $q_1 \rightarrow_{v_i, w} q_1$, $q_1 \rightarrow_{v_i, w} q_2$, and $q_2 \rightarrow_{v_i, w} q_2$ in s_0, which shows (3) for $u_0 = v_i$ and $u_1 = (v'_j)^d$ and thus the lemma (because the pumping preserves (1),(2), and (4)). Figure 1b outlines the choice of q_2 for $m = 5$ and $\nu = 2$ (thus $d = 3$ and $q_2 = q'_4$). □

Lemma 5. Let $M = (Q, P, \Sigma, \Delta, q_0, R, h)$ be an $\mathrm{MTT}^{\mathrm{R}}_{\mathrm{gfci}}$. If M is not fcp, then there are $m \geq 1$, $q \in Q^{(m)}$, $j \in [m]$, $s \in T_\Sigma$, $u \in V(s)$, and $p \in P$ such that

(1) $\#_{y_j}(M_q(s[u \leftarrow p])) \geq 2$,
(2) $M_q(s[u \leftarrow p])$ has a subtree $\langle q, p\rangle(\xi_1, \ldots, \xi_m)$ such that $\#_{y_j}(\xi_j) \geq 1$, and
(3) $p = h(s) = h(s/u)$.

Proof. Let β be an input copying bound of $\mathrm{top}(M)$, η the maximal height of a right-hand side of M, and κ the maximal number of occurrences of one particular parameter in a right-hand side of M. For $t \in T_\Sigma$, u an ancestor of a $v \in V(t)$, $q \in Q^{(m)}$, $\mu \in [m]$, $r \in Q^{(m')}$, $\nu \in [m']$, define $(q, \mu) \rightarrow_{u,v} (r, \nu)$ if, for $\xi_{quv} = M_q(t/u[v' \leftarrow p])$ with $v = uv'$ and $p = h(t/v)$: $\#_{y_\mu}(\xi_{quv}) \geq 2$ and ξ_{quv} has a subtree $\langle r, p\rangle(\xi_1, \ldots, \xi_{m'})$ such that $\#_{y_\mu}(\xi_\nu) \geq 1$. Note that $(q, \mu) \rightarrow_{u,v} (q', \mu') \rightarrow_{v,w} (q'', \mu'')$ implies $(q, \mu) \rightarrow_{u,w} (q'', \mu'')$.

Claim: For every $N \geq 1$, $u \in V(t)$, $q \in Q^{(m)}$, and $\mu \in [m]$, if $\#_{y_\mu}(M_q(t/u)) > N^{\beta\eta} \cdot \kappa^\beta$ then there exist a descendant v of u, a state $r \in Q^{(m')}$, and a $\nu \in [m']$ such that $(q, \mu) \rightarrow_{u,v} (r, \nu)$ and $\#_{y_\nu}(M_r(t/v)) > N$.

Proof. Let w be a lowest descendant of u such that $\#_{y_\mu}(\xi_{quw}) = 1$. Then at least one of its children satisfies the requirements. In fact, assume to the contrary that $\#_{y_\nu}(M_r(t/v)) \leq N$ for all (r, ν) and every child v of w. Consider the result ζ of applying a rule of M to each occurrence of $\langle r', t/w\rangle$, $r' \in Q$, in the sentential form corresponding to ξ_{quw}. Note that $\zeta \Rightarrow^*_M M_q(t/u)$. It follows from Assumption (ii) and the second part of Lemma 2 that there are at most β occurrences of states on a path of ξ_{quw}. Hence $\#_{y_\mu}(\zeta) \leq \kappa^\beta$, and at most $\beta\eta$ states occur on each path of ζ. Thus, $\#_{y_\mu}(M_q(t/u)) \leq N^{\beta\eta} \cdot \kappa^\beta$, a contradiction.

From the fact that M is not fcp and this claim it follows that, for every $n \geq 1$, there are $t \in T_\Sigma$, different ancestors $v_1, \ldots, v_n$ of a node v_{n+1} of t, and state-parameter pairs $(q_1, \mu_1), \ldots, (q_{n+1}, \mu_{n+1})$ such that $(q_i, \mu_i) \to_{v_i, v_{i+1}} (q_{i+1}, \mu_{i+1})$ for every $i \in [n]$. Take $n = |Q| \cdot \overline{m} \cdot |P|$ where $\overline{m}$ is the maximal rank of a state of M. Then there are distinct indices $i, i' \in [n+1]$ such that $q = q_i = q_{i'}$, $j = \mu_i = \mu_{i'}$, and $p = h(t/v_i) = h(t/v_{i'})$. Then $(q, j) \to_{v_i, v_{i'}} (q, j)$. Let $s = t/v_i$ and $v_i u = v_{i'}$ (and so (3) holds). Then, in s, $(q, j) \to_{v,u} (q, j)$ where v is the root of s, which means that (1) and (2) hold (with $\xi = \xi_{qvu}$). □

5 MTTs of Linear Size Increase

In the remainder of the paper we want to show that if the transduction of an MTT^{R} M is of linear size increase, then M is equivalent to an MTT^{R} which is fci and linp (this suffices to prove our characterization, by Proposition 1 and the inclusion $MSOTT \subseteq LSI$). To do this, we first show how to obtain an equivalent MTT^{R} which is gfci (using Lemma 4). By applying Lemma 5 we can show that it is fcp and thus linp, by (the first part of) Proposition 1. Finally we apply Lemma 4 again to get gfci and linp.

Consider the MTT $M_1 = (Q, \Sigma, \Delta, q_0, R)$ with $Q = \{q_0^{(0)}, q^{(0)}, q'^{(0)}\}$, $\Sigma = \{\gamma^{(1)}, a^{(0)}, b^{(0)}\}$, $\Delta = \{\sigma^{(2)}, a^{(0)}, b^{(0)}\}$, and R consisting of the following rules.

$$
\begin{array}{ll}
\langle q_0, \gamma(x_1)\rangle & \to \sigma(\langle q, x_1\rangle, \langle q', x_1\rangle) \\
\langle q, \gamma(x_1)\rangle & \to \langle q, x_1\rangle \\
\langle q', \gamma(x_1)\rangle & \to \sigma(\langle q, x_1\rangle, \langle q', x_1\rangle) \\
\langle r, \alpha\rangle & \to \alpha \quad \text{(for every } r \in Q \text{ and } \alpha \in \{a, b\}\text{)}
\end{array}
$$

Note that M_1 is a top-down tree transducer; thus M_1 is gfci iff it is fci (cf. Lemma 2). Intuitively, M_1 translates a monadic tree s of height n into a comb t of height n; the leaves of t have the same label as the leaf of s. Thus, $\mathrm{size}(\tau_{M_1}(s)) \leq 2 \cdot \mathrm{size}(s)$ for every $s \in T_\Sigma$ and so τ_{M_1} is lsi. Clearly, M_1 is not fci because $\mathrm{sts}_{M_1}(\gamma^n(a), u) = q^n q'$ for $n \geq 1$ and u the leaf of $\gamma^n(a)$. The reason for this is that M_1 generates many copies of q, but q generates only a finite number of different trees (viz. the trees a and b). How can we change M_1 into an equivalent MTT^{R} which is fci? The idea is to simply delete the state q and to determine by regular look-ahead the appropriate tree in $\{a, b\}$. In this example we just need $L_p = \{\gamma^n(a) \mid n \geq 0\}$ and $L_{p'} = \{\gamma^n(b) \mid n \geq 0\}$ and then the q_0-rule of M_1 is replaced by two q_0-rules with right-hand sides $\sigma(a, \langle q', x_1\rangle)$ and $\sigma(b, \langle q', x_1\rangle)$ for look-ahead p and p', respectively, and similarly for the q'-rule.

In general, we require that every state of an MTT^{R} M, except possibly the initial one, generates infinitely many output trees (in $T_\Delta(Y)$). More precisely, for every $p \in P$, M should generate infinitely many output trees for input trees from L_p. Formally, the MTT^{R} $M = (Q, P, \Sigma, \Delta, q_0, R, h)$ is *input proper* (for short, i-proper), if for every $q \in Q$ and $p \in P$ such that $q \neq q_0$ and $\langle q, p\rangle$ is reachable, the set $\mathrm{Out}(q, p) = \{M_q(s) \mid s \in L_p\}$ is infinite.

This notion was defined in [AU71] for generalized syntax-directed translation schemes (which are a variant of top-down tree transducers) and was there called 'reduced'. The construction in the proof of the following lemma is similar to the one of Lemma 5.5 of [AU71] except that we apply it repeatedly to obtain an i-proper $\mathrm{MTT^R}$ as opposed to their single application which is insufficient (also in their formalism, which means that their proof of the lemma is incorrect).

Lemma 6. For every $\mathrm{MTT^R}$ M there is (effectively) an i-proper $\mathrm{MTT^R}$ M' equivalent to M. If M is linp, then so is M'. Also, if M is a $\mathrm{T^R}$, then so is M'.

Proof. Let $M = (Q, P, \Sigma, \Delta, q_0, R, h)$ be an $\mathrm{MTT^R}$. For each $p \in P$, let $F_p = \{q \in Q \mid \mathrm{Out}(q,p) \text{ is finite}\}$. Note that F_p can be constructed effectively, because it is decidable whether $\mathrm{Out}(q,p)$ is finite. In fact, it is easy to construct an $\mathrm{MTT^R}$ N_q such that for every $s \in T_\Sigma$, $\tau_{N_q}(s) = M_q(s)$ if $s \in L_p$, and e otherwise. Then the range of N_q equals $\mathrm{Out}(q,p) \cup \{e\}$ and its finiteness is decidable by [DE98] (and if it *is* finite, it can be computed).

The $\mathrm{MTT^R}$ M' is constructed in such a way that, if $\langle r, x_i\rangle$ occurs in $\mathrm{rhs}_{M'}(q, \sigma, \langle p_1, \dots, p_k\rangle)$, then $r \notin F_{p_i}$. Clearly, this implies i-properness of M'. We first construct the $\mathrm{MTT^R}$ $\pi(M)$ by simply deleting occurrences of $\langle r, x_i\rangle$ with $r \in F_{p_i}$ and replacing them by the correct tree in $\mathrm{Out}(r, p_i)$, which is determined by regular look-ahead. Due to the change of look-ahead automaton, an occurrence of $\langle r, x_i\rangle$ in the $(q, \sigma, \langle p_1, \dots, p_k\rangle)$-rule of M with $r \notin F_{p_i}$ might produce only finitely many trees for the new look-ahead states (p_i, φ_i), which means that $\pi(M)$ is not i-proper yet. For this reason we have to iterate the application of π until the sets F_p do not change anymore. This results in the desired $\mathrm{MTT^R}$ M'.

For each $p \in P$ let Φ_p be the set of all mappings $\varphi : F_p \to T_\Delta(Y)$ such that there is an $s \in L_p$ with $\varphi(q) = M_q(s)$ for every $q \in F_p$. The (finite) set Φ_p can be constructed effectively: for every possible mapping φ which associates with every $q \in F_p$ a tree in $\mathrm{Out}(q,p)$, φ is in Φ_p iff $K = L_p \cap \bigcap\{\tau_{N_q}^{-1}(\varphi(q)) \mid q \in F_p\}$ is nonempty. This is decidable, because K is regular by [EV85; Theorem 7.4(i)]. Since each mapping in Φ_p fixes an output tree for every state $q \in F_p$, the mappings in Φ_p partition the set of input trees L_p. The sets in this (finite) partition are regular (viz. the sets K) and thus can be determined by regular look-ahead.

The $\mathrm{MTT^R}$ $\pi(M)$ has look-ahead states $\{(p, \varphi) \mid p \in P, \varphi \in \Phi_p\}$ with $h'_\sigma((p_1, \varphi_1), \dots, (p_k, \varphi_k)) = (h_\sigma(p_1, \dots, p_k), \varphi)$ and $\varphi(q) = \zeta_q$ for every $q \in F_p$, where $\zeta_q = \mathrm{rhs}_{\pi(M)}(q, \sigma, \langle(p_1, \varphi_1), \dots, (p_k, \varphi_k)\rangle)$ is obtained from ζ in rule $(*)$ by replacing every $\langle r, x_i\rangle$ by $\varphi_i(r)$, for $r \in F_{p_i}$. □

We can now apply our first pumping lemma (Lemma 4), to non-gfci $\mathrm{MTT^R}$s.

Lemma 7. Let M be an i-proper $\mathrm{MTT^R}$. If $\tau_M \in \mathit{LSI}$, then M is gfci.

Proof. Let $M = (Q, \Sigma, \Delta, q_0, R, P, h)$ and let $c \geq 0$ be such that for every $s \in T_\Sigma$, $\mathrm{size}(\tau_M(s)) \leq c \cdot \mathrm{size}(s)$. Assume now that M is not gfci. This leads to a contradiction as follows.

Since $\mathrm{top}(M)$ is not fci we can apply Lemma 4 to it. Let $q_1, q_2 \in Q$, $s_0, s_1 \in T_\Sigma$, $u_0, u_0u_1 \in V(s_0)$, and $p \in P$ such that (1)–(4) of Lemma 4 hold (with $s_1 = s_0/u_0$). By Lemma 2, (1)–(4) also hold for M. Note that since q_0 does not occur in any right-hand side of M by Assumption (i), (2) implies that $q_2 \neq q_0$.

Thus, since M is i-proper and $\langle q_2, p\rangle$ is reachable (by (1) and (2)), the set $M_{q_2}(L_p)$ is infinite and hence contains arbitrarily large trees. Let $s_2 \in L_p$ such that $\#_\Delta(M_{q_2}(s_2)) > c \cdot (c_0 + c_1 + 1)$, where $c_0 = \text{size}(s_0[u_0 \leftarrow p]) - 1$ and $c_1 = \text{size}(s_1[u_1 \leftarrow p]) - 1$. We now pump the tree $s_1[u_1 \leftarrow p]$ in the tree $(s_0[u_0 \leftarrow p]) \bullet (s_1[u_1 \leftarrow p]) \bullet s_2$: for $i \geq 0$, let $t_i = (s_0[u_0 \leftarrow p]) \bullet (s_1[u_1 \leftarrow p])^i \bullet s_2$. It follows from (2)–(4) that for every $i \geq 0$, $\text{sts}_M(t_i, u_0 u_1^i)$ contains at least i occurrences of q_2. Thus (by nondeletion), $\text{size}(\tau_M(t_i)) \geq i \cdot \#_\Delta(M_{q_2}(s_2))$ which is greater than $i \cdot c \cdot (c_0 + c_1 + 1)$, by the choice of s_2.

Now let $i = c_2 = \text{size}(s_2)$. Since $\text{size}(t_i) = c_0 + ic_1 + c_2$ this means that $\text{size}(\tau_M(t_i)) > c \cdot \text{size}(t_i)$ because $\text{size}(\tau_M(t_i)) > ic(c_0 + c_1 + 1) \geq c(c_0 + ic_1 + i) = c(c_0 + ic_1 + c_2) = c \cdot \text{size}(t_i)$ which contradicts the choice of c. □

Next we show how to get from gfci to fcp (for lsi MTT^Rs). Consider the MTT $M_2 = (Q, \Sigma, \Delta, q_0, R)$ with $Q = \{q_0^{(0)}, q^{(1)}\}$, $\Sigma = \{\sigma^{(2)}, \gamma^{(2)}, \alpha^{(0)}, \beta^{(0)}\}$ and $\Delta = \{\sigma^{(2)}, \gamma^{(2)}, \alpha^{(1)}, \beta^{(1)}, \bar{\sigma}^{(0)}, \bar{\gamma}^{(0)}\}$. For $\delta \in \{\sigma, \gamma\}$ and $a \in \Sigma^{(0)}$, let the following rules be in R.

$$\begin{aligned}
\langle q_0, \delta(x_1, x_2)\rangle &\to \delta(\langle q, x_1\rangle(\bar{\delta}), \langle q, x_2\rangle(\bar{\delta}))\\
\langle q, \delta(x_1, x_2)\rangle(y_1) &\to \delta(\langle q, x_1\rangle(y_1), \langle q, x_2\rangle(y_1))\\
\langle q_0, a\rangle &\to a(\bar{a})\\
\langle q, a\rangle(y_1) &\to a(y_1)
\end{aligned}$$

Intuitively, M_2 moves the root symbol of the input tree to each of its leaves; e.g., for $s = \sigma(\gamma(\alpha, \beta), \alpha)$ we get $\tau_{M_2}(s) = \sigma(\gamma(\alpha(\bar{\sigma}), \beta(\bar{\sigma}), \alpha(\bar{\sigma})))$. Thus, τ_{M_2} is lsi (because $\text{size}(\tau_{M_2}(s)) \leq 2 \cdot \text{size}(s)$). Clearly, M_2 is not fcp, because $\#_{y_1}(M_q(s))$ equals the number of leaves of s. This time, the reason is that M_2 generates a lot of parameter occurrences which contain only finitely many trees (viz., $\bar{\sigma}$ and $\bar{\gamma}$). Again the idea is to eliminate them: An MTT^R M is *parameter proper* (for short, p-proper), if for every $m \geq 1$, $q \in Q^{(m)}$, $j \in [m]$, and $p \in P$, if $\langle q, p\rangle$ is reachable then $\text{Arg}(q, j, p)$ is infinite, where $\text{Arg}(q, j, p)$ is the set $\{\xi_j \mid \exists s \in T_\Sigma, u \in V(s) : M_{q_0}(s[u \leftarrow p])$ has a subtree $\langle q, p\rangle(\xi_1, \ldots, \xi_m)\}$. An MTT^R is *proper*, if it is both i-proper and p-proper.

Lemma 8. For every MTT^R M there is (effectively) a proper MTT^R $\text{prop}(M)$ equivalent to M. If M is linp then so is $\text{prop}(M)$.

The idea of the (quite involved) proof is to start with an i-proper MTT^R and then to make it p-proper by a construction similar to the one in the proof of Lemma 6; but now not the look-ahead but the states of $\pi(M)$ are used to code information (about the content of 'finite-valued' parameters).

We now apply our second pumping lemma (Lemma 5), to non-fcp $\text{MTT}^\text{R}_\text{gfci}$s.

Lemma 9. Let M be a p-proper MTT^R_gfci. If $\tau_M \in \mathit{LSI}$, then M is fcp.

Proof. Let $M = (Q, \Sigma, \Delta, q_0, R, P, h)$ and let $c \geq 0$ be such that for every $s \in T_\Sigma$, $\text{size}(\tau_M(s)) \leq c \cdot \text{size}(s)$. Assume that M is not fcp.

Let $m \geq 1$, $q \in Q^{(m)}$, $j \in [m]$, $s \in T_\Sigma$, $u \in V(s)$, and $p \in P$ such that (1)–(3) of Lemma 5 hold. Clearly, (1) implies that $\#_{y_j}(M_q(s)) \geq 2$, and hence $\langle q, p\rangle$ is reachable (by Assumptions (iv) and (ii), respectively). Thus, by p-properness, $\mathrm{Arg}(q,j,p)$ is infinite. Let $c_1 = \mathrm{size}(s)$ and let t be a tree in $\mathrm{Arg}(q,j,p)$ with $\mathrm{size}(t) > c \cdot (c_1+1)$. Then there exist $s_0 \in T_\Sigma$ and $u_0 \in V(s_0)$ such that $M_{q_0}(s_0[u_0 \leftarrow p])$ has a subtree $\langle q,p\rangle(\zeta_1,\dots,\zeta_m)$ with $\zeta_j = t$. Let $c_0 = \mathrm{size}(s_0)$. We now pump the tree $s[u \leftarrow p]$ in the tree $(s_0[u_0 \leftarrow p]) \bullet (s[u \leftarrow p])$: for $i \geq 0$, let $t_i = (s_0[u_0 \leftarrow p]) \bullet (s[u \leftarrow p])^i$. It is straightforward to show, using (1)–(3) and the fact that M is nondeleting, that $M_q(s[u \leftarrow p]^i)$ contains more than i occurrences of y_j. Thus, since M has no erasing rules by Assumption (iii), $\mathrm{size}(\tau_M(t_i)) > \mathrm{size}(t) \cdot i > c(c_1+1)i$.

Let $i = c_0$. Since $\mathrm{size}(t_i) \leq c_0 + c_1 i = (c_1+1)i$, this means that $\mathrm{size}(\tau_M(t_i)) > c \cdot \mathrm{size}(t_i)$, which contradicts the choice of c. □

6 Main Results

From Lemmas 7 and 9 we conclude that if M is a proper $\mathrm{MTT}^{\mathrm{R}}$ and τ_M is lsi, then M is both gfci and fcp. Unfortunately this is not our desired characterization yet. The class $MTT^R_{\mathrm{gfci,fcp}}$ is too large: it contains transductions that are not lsi.

Example 10. Consider the MTT $M_3 = (Q, \Sigma, \Delta, q_0, R)$ with $Q = \{q_0^{(0)}, q_1^{(1)}, q_2^{(0)}, i^{(0)}\}$, $\Sigma = \{\sigma^{(2)}, \alpha^{(0)}\}$, $\Delta = \{\delta^{(3)}, \sigma^{(2)}, \alpha^{(0)}\}$, and the following rules:

$$\begin{array}{lll}
\langle q_\nu, \sigma(x_1,x_2)\rangle & \to \langle q_1, x_1\rangle(\langle q_2, x_2\rangle) & (\text{for } \nu \in \{0,2\}) \\
\langle q_1, \sigma(x_1,x_2)\rangle(y_1) & \to \delta(y_1, y_1, \langle i, x_1\rangle) & \\
\langle i, \sigma(x_1,x_2) & \to \sigma(\langle i, x_1\rangle, \langle i, x_2\rangle) & \\
\langle q_1, \alpha\rangle(y_1) & \to \delta(y_1, y_1, \alpha) & \\
\langle r, \alpha\rangle & \to \alpha & (\text{for } r \in Q - \{q_1\})
\end{array}$$

As it turns out, M_3 is indeed gfci and fcp: corresponding input and parameter copying bounds are 1 and 2, respectively. It is also proper. Take $s_n = \sigma(\alpha, \sigma(\alpha, \dots \sigma(\alpha,\alpha)\dots))$, where n is the number of σ's. This tree is translated by M_3 into the monadic tree $\langle q_1, \alpha\rangle^n(\langle q_2, \alpha\rangle)$ and then into the full binary tree of height n over $\delta^{(3)}$ and α, in which every δ has an additional right-most leaf α. Thus, τ_{M_3} is not lsi: $\mathrm{size}(\tau_{M_3}(s_n)) = 3 \cdot 2^n - 2$ and $\mathrm{size}(s_n) = 2n+1$.

If M is fcp, we can apply Proposition 1 to get an equivalent $\mathrm{MTT}^{\mathrm{R}}$ $\mathrm{linp}(M)$ that is linp (this is effective [EM99; Lemma 6.3]). To apply Lemma 7 again, we first have to make $\mathrm{linp}(M)$ proper again (by applying 'prop' to it), because its construction does not preserve properness. We are ready to prove our main theorem.

Theorem 11. Let M be an MTT. Then the following statements are equivalent:

(1) τ_M is MSO definable.
(2) τ_M is of linear size increase.
(3) $\mathrm{prop}(M)$ is fcp and $\mathrm{prop}(\mathrm{linp}(\mathrm{prop}(M)))$ is fci.

Proof. Since $MSOTT \subseteq LSI$, (1) $\Rightarrow$ (2). To show (2) $\Rightarrow$ (3), let τ_M be lsi. Then, by Lemmas 7 and 9, prop(M) is fcp and, by Lemma 7 again, prop(linp(prop(M))) is gfci and hence fci (by Lemmas 8 and 2). Finally, (3) $\Rightarrow$ (1): If prop(M) is fcp and prop(linp(prop(M))) $= M'$ is fci then M' is fci and linp, by Lemma 8. Thus τ_M is MSO definable by Proposition 1. □

From Theorem 11, Lemma 3, and the effectivity of prop and linp we obtain the following decidability result.

Theorem 12. It is decidable for an MTT M whether τ_M is MSO definable.

From Theorem 11 and Proposition 1 we get our characterization of *MSOTT* (recall that $MTT^R = MTT$, cf. [EV85; Theorem 4.21]):

Theorem 13. $MSOTT = MTT \cap LSI$.

Finally, we obtain the characterization of MSO definable top-down tree transductions (with regular look-ahead) discussed in the Introduction.

Theorem 14. $T^{\mathrm{R}} \cap MSOTT = T^R_{\mathrm{fci}} = T^R \cap LSI$.

Proof. From Lemmas 6 and 7 we know that $T^R \cap LSI \subseteq T^R_{\mathrm{gfci}}$, which equals T^R_{fci} by Lemma 2. The inclusion $T^R_{\mathrm{fci}} \subseteq T^R \cap MSOTT$ is immediate from Proposition 1, and $T^R \cap MSOTT \subseteq T^R \cap LSI$ follows from the fact that $MSOTT \subseteq LSI$. □

Open Problems It is not clear how MSO definability could be generalized in order to obtain the full class of macro tree transductions or maybe just the class of polynomial size increase (psi) macro tree transductions. Is psi decidable for MTTs? If so, what is the complexity? (cf. [Dre99]). Finally, we do not know whether our result generalizes to nondeterministic MTTs, and for compositions of MTTs we would like to know: is $LSI \cap \bigcup_n MTT^n = MSOTT$?

References

[AU71] A. V. Aho and J. D. Ullman. Translations on a context-free grammar. *Inform. and Control*, 19:439–475, 1971.

[BE98] R. Bloem and J. Engelfriet. A comparison of tree transductions defined by monadic second order logic and by attribute grammars. Technical Report 98-02, Leiden University, 1998. To appear in *J. of Comp. Syst. Sci.*

[CF82] B. Courcelle and P. Franchi-Zannettacci. Attribute grammars and recursive program schemes. *Theoret. Comput. Sci.*, 17:163–191 and 235–257, 1982.

[Cou94] B. Courcelle. Monadic second-order definable graph transductions: a survey. *Theoret. Comput. Sci.*, 126:53–75, 1994.

[Cou95] B. Courcelle. Structural properties of context-free sets of graphs generated by vertex replacement. *Inform. and Comput.*, 116:275–293, 1995.

[DE98] F. Drewes and J. Engelfriet. Decidability of finiteness of ranges of tree transductions. *Inform. and Comput.*, 145:1–50, 1998.

[Dre99] F. Drewes. The complexity of the exponential output size problem for top-down tree transducers. In G. Ciobanu and Gh. Păun, editors, *Proc. FCT 99*, volume 1684 of *LNCS*, pages 234–245. Springer-Verlag, 1999.

[EM] J. Engelfriet and S. Maneth. Macro tree transducers of linear size increase. In preparation, `http://www.wi.leidenuniv.nl/~maneth/LSI.ps.gz`.

[EM99] J. Engelfriet and S. Maneth. Macro tree transducers, attribute grammars, and MSO definable tree translations. *Inform. and Comput.*, 154:34–91, 1999.

[ERS80] J. Engelfriet, G. Rozenberg, and G. Slutzki. Tree transducers, L systems, and two-way machines. *J. of Comp. Syst. Sci.*, 20:150–202, 1980.

[EV85] J. Engelfriet and H. Vogler. Macro tree transducers. *J. of Comp. Syst. Sci.*, 31:71–146, 1985.

[FV98] Z. Fülöp and H. Vogler. *Syntax-Directed Semantics – Formal Models based on Tree Transducers.* EATCS Monographs on Theoretical Computer Science (W. Brauer, G. Rozenberg, A. Salomaa, eds.). Springer-Verlag, 1998.

[GS97] F. Gécseg and M. Steinby. Tree automata. In G. Rozenberg and A. Salomaa, editors, *Handbook of Formal Languages.* Vol. 3. Chap. 1. Springer-Verlag, 1997.

[KS94] N. Klarlund and M. I. Schwartzbach. Graphs and decidable transductions based on edge constraints. In S. Tison, editor, *Proc. CAAP 94*, volume 787 of *LNCS*, pages 187–201. Springer-Verlag, 1994.

Languages of Dot–Depth 3/2

Christian Glaßer* and Heinz Schmitz**

Theoretische Informatik, Universität Würzburg, 97074 Würzburg, Germany
{glasser,schmitz}@informatik.uni-wuerzburg.de

Abstract. We prove an effective characterization of languages having dot–depth 3/2. Let $\mathcal{B}_{3/2}$ denote this class, i.e., languages that can be written as finite unions of languages of the form $u_0L_1u_1L_2u_2 \cdots L_nu_n$, where $u_i \in A^*$ and L_i are languages of dot–depth one. Let F be a deterministic finite automaton accepting some language L. Resulting from a detailed study of the structure of $\mathcal{B}_{3/2}$, we identify a pattern **P** (cf. Fig. 2) such that L belongs to $\mathcal{B}_{3/2}$ if and only if F does not have pattern **P** in its transition graph. This yields an NL–algorithm for the membership problem for $\mathcal{B}_{3/2}$.
Due to known relations between the dot–depth hierarchy and symbolic logic, the decidability of the class of languages definable by Σ_2–formulas of the logic FO[$<$, min, max, S, P] follows. We give an algebraic interpretation of our result.

1 Introduction

We contribute to the theory of finite automata and regular languages, with consequences in logic as well as in algebra. Particularly, we deal with starfree regular languages. These are languages constructed from alphabet letters using only Boolean operations together with concatenation. Alternating these two kinds of operations in order to distinguish combinatorial and sequential aspects, leads to the definition of concatenation hierarchies that exhaust the class of starfree languages. Prominent examples are the dot–depth hierarchy, first studied in [CB71], and the Straubing–Thérien hierarchy [Str81, Thé81, Str85]. Both are known to be strict [BK78] and closely related to each other [Str85]. Most naturally arising questions concerning these hierarchies are of major interest in different research areas since close connections have been exposed, e.g., to finite model theory, theory of finite semigroups, complexity theory and others.

Here we deal with the dot–depth hierarchy. Let A be some finite alphabet with $|A| \geq 2$. For a class $\mathcal{C}$ of languages over A^+ let POL($\mathcal{C}$) be its polynomial closure, i.e., the class of languages $L \subseteq A^+$ that can be written as a finite union of languages $u_0L_1u_1L_2u_2 \cdots L_nu_n$, where $u_i \in A^*$, $L_i \in \mathcal{C}$ and $n \geq 0$. Denote by BC($\mathcal{C}$) its Boolean closure, i.e., the closure of $\mathcal{C}$ under finite union, finite intersection and complementation. Then the dot–depth hierarchy can be defined as the following family of classes $\mathcal{B}_{n/2}$ (notations are adopted from [PW97]).

* Supported by the Studienstiftung des Deutschen Volkes.
** Supported by the Deutsche Forschungsgemeinschaft (DFG), grant Wa 847/4-1.

H. Reichel and S. Tison (Eds.): STACS 2000, LNCS 1770, pp. 555–566, 2000.

1. $\mathcal{B}_0 =_{\text{def}} \{\emptyset, A^+\}$,
2. $\mathcal{B}_{n+1/2} =_{\text{def}} \text{POL}(\mathcal{B}_n)$ for $n \geq 0$, and
3. $\mathcal{B}_{n+1} =_{\text{def}} \text{BC}(\mathcal{B}_{n+1/2})$ for $n \geq 0$.

For a language $L \subseteq A^+$ and a minimal n with $L \in \mathcal{B}_{n/2}$ we say that L has dot–depth $n/2$. One obtains the same hierarchy classes when setting $\mathcal{B}_{n+1/2} =_{\text{def}} \text{POL}(\text{co}\mathcal{B}_{n-1/2})$ with $\text{co}\mathcal{B}_{n-1/2} =_{\text{def}} \left\{ \overline{L} \mid L \in \mathcal{B}_{n-1/2} \right\}$ [Gla98]. By definition, all $\mathcal{B}_{n+1/2}$ are closed under union and it is known, that these classes are also closed under intersection [Arf91]. The question whether there exists an algorithm deciding '$L(F) \in \mathcal{B}_{n/2}$' for $n \geq 0$ and deterministic finite automata F (dfa F, for short) is known as the dot–depth problem. Although many researchers believe the answer should be yes, some suspect the contrary. To our knowledge, only the classes $\mathcal{B}_0$, $\mathcal{B}_{1/2}$ and $\mathcal{B}_1$ were known to be decidable [Kna83, PW97]. Especially, the case of dot–depth 3/2 was mentioned open in [Pin96, PW97]. This can be seen in contrast to the Straubing–Thérien hierarchy, for which beside levels 0, 1/2 and 1 also level 3/2 is known to be decidable [Arf91, PW97]. Some partial results are known for level 2 of this hierarchy, e.g., its decidability in case of a 2-letter alphabet [Str88].

In this paper we prove an effective characterization of languages having dot–depth 3/2. With an automata–theoretic approach we study the class $\mathcal{B}_{3/2}$ in detail. Fix some $k \geq 0$. We look at a word w as a word over A^{k+1} by taking together each $k+1$ consecutive letters and call this the k-decomposition of w. In this way we obtain classes $\mathcal{B}_{3/2,k}$ for which $\mathcal{B}_{3/2} = \bigcup_{k \geq 0} \mathcal{B}_{3/2,k}$. The k-decomposition approach was used before in several contexts, e.g., when relating the dot–depth hierarchy and the Straubing–Thérien hierarchy [Str85], and for a levelwise analysis of dot–depth one languages [Ste85].

We first look at the family of classes $\mathcal{B}_{3/2,k}$. With the help of a series of technical lemmas we prove a useful normalform representation for languages in $\mathcal{B}_{3/2,k}$. Then we provide a combinatorial lemma which keeps control of the lengths of factors x, u, y of some input w to a dfa such that $w = xuy$ and the state reached after input $x'u$ has a loop with label u for all x'. Iterated applications of this lemma are a basic tool in subsequent proofs.

Let F be a dfa accepting some language L. We show that L belongs to $\mathcal{B}_{3/2,k}$ if and only if F does not have pattern $\mathbf{P}_k$ in its transition graph (cf. Fig. 1). This yields an NL–algorithm for the membership problem for $\mathcal{B}_{3/2,k}$ which looks for the non–existence of $\mathbf{P}_k$ in F. Since we encounter for $k = 0$ level 3/2 of the Straubing–Thérien hierarchy, we provide as a by-product a self-contained reproof of the normalform and the decidability result for this class ([Arf91] and [PW97] use deep results from [Has83] and [Sim90], respectively).

Our generalization to arbitrary k enables us to identify a pattern $\mathbf{P}$ such that L belongs to $\mathcal{B}_{3/2}$ if and only if F does not have pattern $\mathbf{P}$ in its transition graph (cf. Fig. 2). So we can affirmatively answer the decidability question for $\mathcal{B}_{3/2}$ even in an efficient way: looking for the non–existence of $\mathbf{P}$ in F yields an NL–algorithm for the membership problem of this class. Moreover, the proof is such that an algorithm can be derived to determine the exact level k of a given language inside $\mathcal{B}_{3/2}$.

We draw some consequences which are due to various relations of the classes of the dot–depth hierarchy to other fields of research. The connection to first–order logic goes back to [MP71]. The dot–depth hierarchy is related to the first–order logic FO[<, min, max, S, P] having unary relations for the alphabet symbols from A, the binary relation <, the successor (predecessor) function S (P, resp.), and constants min and max. Let Σ_n be the subclass of this logic which is defined by at most $n-1$ quantifier alternations, starting with an existential quantifier. It has been proved in [Tho82] (see also [PP86, PW97]) that Σ_n–formulas describe just the $\mathcal{B}_{n-1/2}$ languages and that the Boolean combinations of Σ_n–formulas describe just the $\mathcal{B}_n$ languages. Due to this characterization we can conclude the decidability of the class of languages definable by Σ_2–formulas of the logic FO[<, min, max, S, P]. We also give an algebraic interpretation of our result and characterize the languages of dot–depth 3/2 by a condition on their ordered syntactic semigroups. Spoken in algebraic terms, this yields an effective characterization of the variety of finite ordered semigroups corresponding to the positive variety of languages, as which $\mathcal{B}_{3/2}$ can be understood.

If one looks at our result in combination with the known forbidden pattern characterization of $\mathcal{B}_{1/2}$ from [PW97] it is easy to discover regularities between both patterns. Continuing them leads to the definition of a pattern $\mathbf{R}_n$ for $n \geq 1$, which defines decidable subclasses $\mathcal{C}_n$ of starfree languages in a forbidden pattern manner. We conclude this paper with some informal arguments supporting the possibility that $\mathcal{C}_n = \mathcal{B}_{n-1/2}$ holds also for all $n \geq 3$.

A comprehensive treatment of the issues presented in this extended abstract is given in a self–contained way in [GS99], see *http://www.informatik.uni-wuerzburg.de/reports/tr.html*.

2 The Classes $\mathcal{B}_{3/2,k}$

Throughout the paper we consider languages as subsets of A^+. Let $k \geq 0$. We denote by $A^{\leq k}$ the set of words from A^+ of length less or equal to k (similarly, $A^{<k}, A^{\geq k}, \ldots$). It will be useful for us to look at $w \in A^+$ as a word over A^{k+1} by taking together each $k+1$ consecutive letters. We denote elements from A^{k+1} as $\alpha, \beta, \gamma, \ldots$ and subsets of A^{k+1} as $\Sigma, \Gamma, \ldots$. Let $w = a_1 a_2 \cdots a_{k+l} \in A^+$ for some $l \geq 1$. We call $\widehat{w} =_{\text{def}} (\beta_1, \beta_2, \ldots, \beta_l)$ the k-decomposition of w if $\beta_i = a_i \cdots a_{i+k}$ for $1 \leq i \leq l$. Intuitively, k indicates by how many letters from A consecutive β_i overlap. We set $\alpha(\widehat{w}) =_{\text{def}} \{\beta_1, \beta_2, \ldots, \beta_l\}$. Next we define languages that admit the same k-decomposition with respect to given elements and subsets of A^{k+1}.

Definition 1. *Let $k, n \geq 0$ and $\alpha_1, \ldots, \alpha_n \in A^{k+1}$, $\Sigma_0, \ldots, \Sigma_n \subseteq A^{k+1}$. For every $w \in A^+$ we say $w \in (\Sigma_0, \alpha_1, \Sigma_1, \ldots, \alpha_n, \Sigma_n)_k$ if and only if $|w| \geq k+1$, $\widehat{w} = (\beta_1, \ldots, \beta_l)$ and there exist $0 = j_0 < j_1 < j_2 < \ldots < j_n < j_{n+1} = l+1$ such that*

(a) $\beta_{j_i} = \alpha_i$ for $1 \leq i \leq n$ and
(b) $\beta_j \in \Sigma_i$ for $0 \leq i \leq n$ and $j_i < j < j_{i+1}$.

If we write the expression $(\Sigma_0, \alpha_1, \Sigma_1, \ldots, \alpha_n, \Sigma_n)_k$ we understand this as a syntactical object describing some language. We do not distinguish between this object and the language it stands for. So the language $(\Sigma_0, \alpha_1, \Sigma_1, \ldots, \alpha_n, \Sigma_n)_k$ consists of those words $w \in A^{\geq k+1}$, whose k-decomposition starts with a number (possibly zero) of elements from Σ_0, then α_1, followed by a number (possibly zero) of elements from Σ_1, then α_2 and so on. Note that in case $k = 0$ we deal with the usual concatenation, e.g., $(A_0)_0 = A_0^+$ and $(A_0, a_1, A_1, a_2, A_2)_0 = A_0^* a_1 A_1^* a_2 A_2^*$. For convenient notations we write $(w|\Sigma_0, \alpha_1, \Sigma_1, \ldots, \alpha_n, \Sigma_n|v)_k$ instead of $\big(wA^* \cap A^*v \cap (\Sigma_0, \alpha_1, \Sigma_1, \ldots, \alpha_n, \Sigma_n)_k\big)$.

Definition 2. *Let $k \geq 0$ and $m \geq 1$. Then $\mathcal{B}_{1,(m,k)}$ is the class of languages $L \subseteq A^+$ that are in the Boolean algebra generated by languages L_i such that $L_i = D$ with $D \subseteq A^{<k+m}$ or*

$$L_i = (w|A^{k+1}, \alpha_1, A^{k+1}, \alpha_2, A^{k+1}, \ldots, \alpha_m, A^{k+1}|v)_k$$

where $\alpha_j \in A^{k+1}$ and $w, v \in A^k$.

The definition of these classes is motivated by a characterization of $\mathcal{B}_1$ in terms of the congruence $\sim_{m,k}$ with $k, m \geq 0$ introduced in [Sim72]. With the next theorem we recall in our notations the fact that the classes $\mathcal{B}_{1,(m,k)}$ refine $\mathcal{B}_1$.

Theorem 1 ([Sim72]). *Let $L \subseteq A^+$. Then $L \in \mathcal{B}_1$ if and only if there exist $k \geq 0$ and $m \geq 1$ such that $L \in \mathcal{B}_{1,(m,k)}$.*

For an overview on hierarchies which result from fixing one or the other parameter see [Brz76]. Among others, a hierarchy in $\mathcal{B}_1$ obtained by fixing k has been studied [Sim72, Ste85].

Definition 3 ([Sim72]). *Let $k \geq 0$. Then $\mathcal{B}_{1,k} =_{\text{def}} \bigcup_{m \geq 1} \mathcal{B}_{1,(m,k)}$.*

Proposition 1. $\mathcal{B}_1 = \bigcup_{k \geq 0} \mathcal{B}_{1,k}$ *and* $\mathcal{B}_{3/2} = \bigcup_{k \geq 0} \mathrm{POL}(\mathcal{B}_{1,k})$.

The first equation follows from Theorem 1, while the latter is due to $\mathcal{B}_{3/2} = \mathrm{POL}(\mathcal{B}_1) = \mathrm{POL}(\bigcup_{k \geq 0} \mathcal{B}_{1,k}) = \bigcup_{k \geq 0} \mathrm{POL}(\mathcal{B}_{1,k})$, which gives rise to the starting point of our investigations.

Definition 4. *Let $k \geq 0$. Then $\mathcal{B}_{3/2,k} =_{\text{def}} \mathrm{POL}(\mathcal{B}_{1,k})$.*

By definition, languages in $\mathcal{B}_{3/2,k}$ are finite unions of concatenations of words with languages from $\mathcal{B}_{1,k}$, which are in turn Boolean combinations of the languages L_i from Definition 2, a somewhat unstructured representation. We give the following normalform.

Theorem 2. *Let $k \geq 0$ and $L \subseteq A^+$. Then $L \in \mathcal{B}_{3/2,k}$ if and only if L can be written as a finite union of languages L_i such that $L_i = D$ for $D \subseteq A^{\leq k}$ or $L_i = (\Sigma_0, \alpha_1, \Sigma_1, \ldots, \alpha_n, \Sigma_n)_k$ where $n \geq 1$, $\alpha_j \in A^{k+1}$ and $\Sigma_j \subseteq A^{k+1}$.*

3 Finding Automata Loops in Words

A useful tool in our proofs is the fact that we can find factors in a word which lead to loops in a given dfa. It is important here to analyse the length needed to find such a factor, depending on the size of the dfa in question. For this end, we define a bounding function $\mathcal{K}(n)$ as

$$\mathcal{K}(n) =_{\text{def}} (n+1)^{(n+1)^{(n+1)}}$$

and prove the following rather technical lemma. Let δ be the transition function of the given dfa F. Then every $w \in A^*$ induces a total mapping $\delta^w : S \to S$ on the set of states S with $\delta^w(s) =_{\text{def}} \delta(s, w)$ for all $s \in S$.

Lemma 1. *For every dfa F and for all $v_0, \ldots, v_n \in A^+$ there exist an $m \geq 0$ and indices $0 = i_0 < i_1 < \cdots < i_{2m+1} = n+1$ such that*

1. *$i_{j+1} - i_j \leq \mathcal{K}(|F|)$ for $0 \leq j \leq 2m$ and*
2. *$\delta^{uu} = \delta^u$ for all $u = v_{i_j} v_{i_j+1} \cdots v_{i_{j+1}-1}$ with $1 \leq j < 2m$ and $j \equiv 1(2)$.*

To give some intuition, this means for factors of length one, that for every $w = v_0 v_1 \cdots v_n$ with $v_i \in A$ there exist words $w_0, \ldots, w_m, u_1, \ldots, u_m \in A^{\leq \mathcal{K}(|F|)}$ such that $w = w_0 u_1 w_1 \cdots u_m w_m$ and $\delta^{u_i u_i} = \delta^{u_i}$ for $1 \leq i \leq m$. We use Lemma 1 for arbitrary factors v_i in the proof of Lemma 2 below.

4 Forbidden Pattern Characterization of $\mathcal{B}_{3/2,k}$

Let F be a dfa accepting a language L. We want to show that L is in $\mathcal{B}_{3/2,k}$ if and only if F does not have pattern $\mathbf{P}_k$ (cf. Fig. 1) in its transition graph. As usual

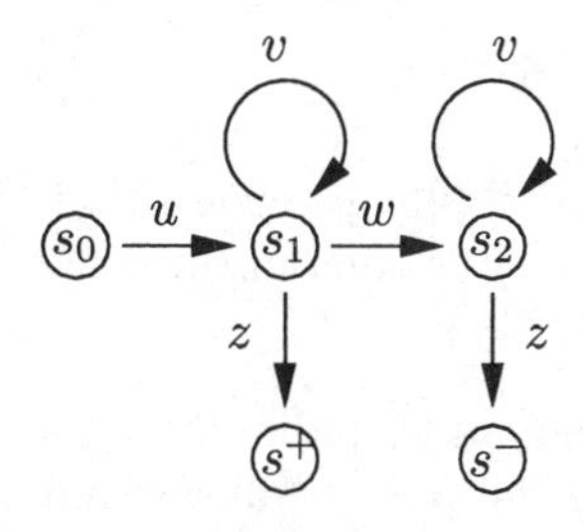

Fig. 1. Pattern $\mathbf{P}_k$ with $u, w, z \in A^*, v \in A^{\geq k+1}$, initial state s_0, accepting state s^+, rejecting state s^- and $\alpha(\widehat{vwv}) \subseteq \alpha(\widehat{vv})$ for k-decompositions of words.

in forbidden pattern proofs, there is on one hand an easier to prove implication (Theorem 3) and on the other hand a more difficult one (Theorem 4).

Theorem 3. *Let $k \geq 0$. If a dfa F has pattern $\mathbf{P}_k$, then $L(F) \notin \mathcal{B}_{3/2,k}$.*

Proof (Sketch). Suppose F has pattern $\mathbf{P}_k$ and $L(F) \in \mathcal{B}_{3/2,k}$. By Theorem 2 we have that $L(F)$ is a finite union of languages L_i such that $L_i = D$ for $D \subseteq A^{\leq k}$ or $L_i = (\Sigma_0, \alpha_1, \Sigma_1, \ldots, \alpha_n, \Sigma_n)_k$. Since we can pump up $uz \in L(F)$ to arbitrary $uv^j z \in L(F)$, we can determine a sufficiently large j, such that $uv^j z \notin A^{\leq k}$ and there is some l such that $\alpha(\widehat{vv}) \in \Sigma_l$. Because $\alpha(\widehat{vwv}) \subseteq \alpha(\widehat{vv})$ by pattern $\mathbf{P}_k$ we can insert w without leaving $L(F)$, a contradiction to pattern $\mathbf{P}_k$. □

The more complicated implication will be a consequence of Lemma 2 given below. There we derive from every $x \in L$ a subset of L which contains x and which can be described by expressions of bounded size. In particular, we consider expressions E of the form

$$w_0 \cdot (v_1|\Sigma_1|v_1')_k \cdot w_1 \cdots (v_n|\Sigma_n|v_n')_k \cdot w_n$$

where $w_i \in A^+$, $v_j, v_j' \in A^k$ and $\Sigma_j \subseteq A^{k+1}$. We define the size of E as $|w_0 w_1 \cdots w_n|$ and identify E with the language described by E. For a fixed $k \geq 0$ let us denote the set of all such expressions by $\mathcal{E}_k$. To analyse the size of expressions in Lemma 2 we make the following definition, where variables a, f, n will be associated with the size of the alphabet A, the size of the dfa F and the cardinality of $\alpha(\widehat{w})$ for a given word w, respectively.

$$\mathcal{L}(k,a,f,n) =_{\mathrm{def}} \begin{cases} k & : \text{ if } n = 0 \\ 2f^f + k + 1 & : \text{ if } n = 1 \\ 3\mathcal{K}(f) \cdot (5f^f a^k + 1) \cdot \mathcal{L}(k,a,f,n-1) & : \text{ otherwise} \end{cases}$$

Lemma 2. *Let $k \geq 0$ and let F be a dfa which does not have pattern $\mathbf{P}_k$. For every $x \in A^+$ there exists an expression $E_x \in \mathcal{E}_k$ of size $\leq \mathcal{L}(k, |A|, |F|, |\alpha(\widehat{x})|)$ with $x \in E_x$ and for all $x', x'' \in A^*$ we have*

$$x'xx'' \in L(F) \Longrightarrow x'E_x x'' \subseteq L(F).$$

In the following, the term 'short' ('long') means that the size of an expression can (can not, resp.) be bounded by a function in $k, |A|, |F|$ and $|\alpha(\widehat{x})|$ (we use 'small' and 'large' for cardinalities).

Proof (Sketch). We prove the lemma by induction on $|\alpha(\widehat{x})|$. If $|\alpha(\widehat{x})| = 0$ then x is short and we are done. If $|\alpha(\widehat{x})| = 1$, it follows that $x = a^{|x|}$ for some letter $a \in A$. Now it is easy to see that either x is short or $a^i \cdot \left(a^k | \left\{a^{k+1}\right\} | a^k\right)_k \cdot a^j$ provides the desired expression for suitable choices of small i and j.

In the induction step we consider $x \in A^+$ with $|\alpha(\widehat{x})| = n + 1 \geq 2$. First of all we decompose x into factors s_i (so-called 'sectors') with $|\alpha(\widehat{s_i})| \leq n$ as follows. Determine the longest prefix s_1 of x such that $|\alpha(\widehat{s_1})| \leq n$. Now we start over with the remaining part of x and determine its longest prefix s_2 such that $|\alpha(\widehat{s_2})| \leq n$ and so on. Observe that we obtain a factorization of x into sectors s_i such that $\alpha(\widehat{s_i s_{i+1}}) = \alpha(\widehat{x})$. Furthermore, the induction hypothesis provides us with short expressions for sectors constructed in this way.

Note that neither the length of sectors, nor their number must be small. The main task of the induction step is to replace the large number of sectors by a small number of terms $(v|\Sigma|v')_k$ in such a way that (i) we do not leave $L(F)$ if we started with $x \in L(F)$ and (ii) we obtain an expression where only a small number of sectors is left.

If the number of sectors is already small, we can replace each sector s with the expression E_s provided by the induction hypothesis and we are done.

If the number of sectors is large, we combine them to pairs $p_i =_{\text{def}} s_{2i-1}s_{2i}$ in order to have $\alpha(\widehat{p_i}) = \alpha(\widehat{x})$. Now we apply Lemma 1 to these pairs and get a partitioning of the sequence of the p_i with $x = w_0u_1w_1 \cdots u_mw_m$ such that (i) each partition u_i (w_i, resp.) consists of a small number of pairs and (ii) every partition u_i leads to an u_i-loop (i.e., $\delta^{u_iu_i} = \delta^{u_i}$) with $\alpha(\widehat{u_i}) = \alpha(\widehat{x})$. Now we assign to each u_i a tag representing the mapping $\delta^{w_0u_1w_1\cdots u_i}$ and the k-suffix of u_i. In a next step we want to find maximal non–intersecting factors between some u_i and u_j having the same tags (so–called 'regions'). Consider the simple greedy algorithm which chooses repeatedly a largest factor between some u_i, u_j having the same tags, such that the region between u_i and u_j does not intersect an existing region (it stops if this is not possible). Since the number of different tags is small, it can be shown that this algorithm returns a small number of regions, such that the number of u_i and w_i in some gap between those regions is also small. It follows that the number of sectors in some gap between regions is bounded. Note that regions may contain a large number of sectors.

We treat all regions of x from right to left. Consider a particular region between u_i and u_j. Then we replace this region with the term $T = u_i \cdot (p|\alpha(\widehat{x})|s)_k \cdot u_j$ where p is the k-prefix and s is the k-suffix of the word $w_iu_{i+1}w_{i+1} \cdots u_{j-1}w_{j-1}$. Now we have reached a situation where we make use of the fact that F does not have pattern $\mathbf{P_k}$. Using Lemma 1 and the tags, it can be observed that u_i and any word from T lead to an u_j-loop. It follows that if we leave $L(F)$ with some word from T, then we would find pattern $\mathbf{P_k}$ in F, a contradiction. We can continue this substitution from right to left, region by region, without leaving $L(F)$, since the tags left to the substitution position remain valid. In this way we obtain an expression $E'_x \in \mathcal{E}_k$ with $x \in E'_x$. Since the whole argument is independent of prefixes x' and suffixes x'', we can show $x'xx'' \in L(F) \Longrightarrow x'E'_xx'' \subseteq L(F)$. Furthermore, the number of terms $(v|\Sigma|v')_k$ and number of remaining sectors in E'_x is small. If we apply the induction hypothesis to these sectors, we obtain the desired expression E_x. This completes the induction. □

Theorem 4. *Let $k \geq 0$. If a dfa F does not have pattern $\mathbf{P_k}$, then $L(F) \in \mathcal{B}_{3/2,k}$.*

Proof (Sketch). Let F be a dfa which does not have pattern $\mathbf{P_k}$. By Lemma 2 we find for every $x \in L(F)$ a short expression $E_x \subseteq L(F)$ with $x \in E_x$. Since there is only a finite number of different expressions of $\mathcal{E}_k$ having the same size, we can write $L(F)$ as a finite union of expressions of $\mathcal{E}_k$. With the help of Theorem 2 it can be shown that languages of $\mathcal{E}_k$ are in $\mathcal{B}_{3/2,k}$. □

Taking together Theorems 3 and 4 we obtain the main result of this section.

Theorem 5. *Let $k \geq 0$ and let F be a dfa. Then $L(F) \in \mathcal{B}_{3/2,k}$ if and only if F does not have pattern $\mathbf{P}_k$.*

To see that this characterization is effective we provide an efficient algorithm to check the non–existence of pattern $\mathbf{P}_k$ in the transition graph of a given dfa. In particular, we show that the occurrence of pattern $\mathbf{P}_k$ can be decided in nondeterministic, logarithmic space (NL) for a fixed $k \geq 0$, which is a class closed under complementation. For this end we guess the states s_1, s_2, s^+, s^- and check the existence of the words u, v, w, z applying the same technique that solves the graph accessibility problem. While guessing v and w we additionally store the last k letters which then enables us to determine all elements of $\alpha(\widehat{vwv})$ and $\alpha(\widehat{vv})$. Since k and the size of the alphabet can be considered as constants, all this can be done in NL.

Furthermore, Theorem 5 allows a concise proof of the strictness of the hierarchy of classes $\mathcal{B}_{3/2,k}$. We obtain $\mathcal{B}_{3/2,k} \subsetneq \mathcal{B}_{3/2,k+1}$ for $k \geq 0$ with help of the witnessing languages $L_k =_{\mathrm{def}} \left(a^{k+1}b, \left\{a^i b a^{k+1-i} : 0 \leq i \leq k+1\right\}, a^{k+1}b\right)_{k+1}$.

5 Forbidden Pattern Characterization of $\mathcal{B}_{3/2}$

We identify the pattern $\mathbf{P}$ given in Fig. 2 which characterizes $\mathcal{B}_{3/2}$.

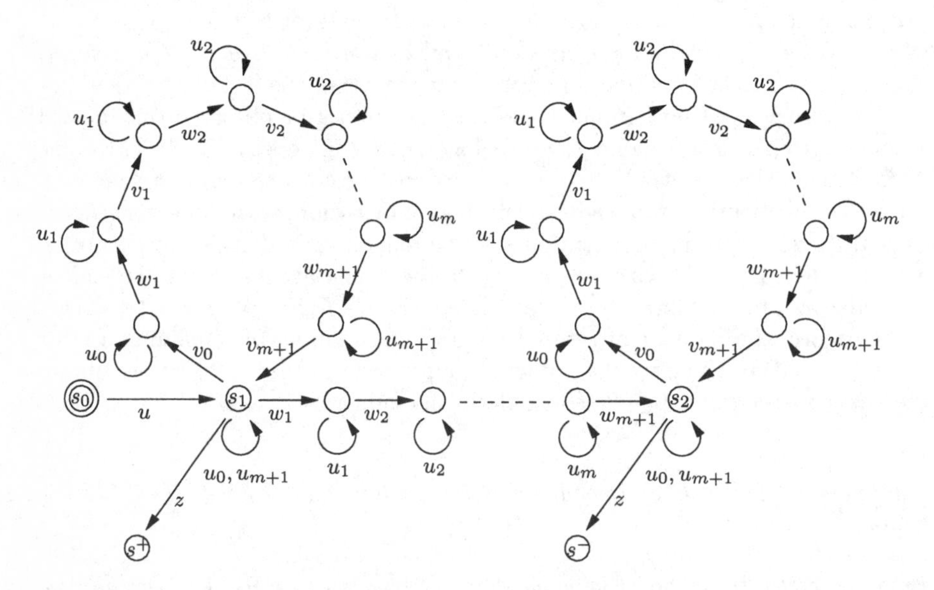

Fig. 2. Pattern **P** with initial state s_0, accepting state s^+, rejecting state s^-, $m \geq 0$, $u_i, w_i \in A^+$ and $u, z, v_i \in A^*$.

Theorem 6. *Let F be a dfa accepting some language L. Then $L \in \mathcal{B}_{3/2}$ if and only if F does not have pattern* $\mathbf{P}$.

Proof (Sketch). We first show that the existence of pattern $\mathbf{P}$ implies the existence of pattern $\mathbf{P_k}$ for every $k \geq 0$. As witnessing words take u, z and $v =_{\text{def}} u_0^k v_0 u_0^k w_1 u_1^k v_1 u_1^k \cdots w_{m+1} u_{m+1}^k v_{m+1} u_{m+1}^k$ and $w =_{\text{def}} u_0^k w_1 u_1^k \cdots w_{m+1} u_{m+1}^k$. This definition ensures that each element of the k-decomposition of vwv overlaps at most two of the u_i^k. It follows that $\alpha(\widehat{vwv}) \subseteq \alpha(\widehat{vv})$.

Now suppose that F has pattern $\mathbf{P_k}$ for every $k \geq 0$. In particular, we find a pattern for $k =_{\text{def}} 3\mathcal{K}(|F|)$ with $|w| \geq k+1$ (take vwv instead of w if necessary). By Lemma 1 we can write w as $w = w_0 u_1 w_1 \cdots u_l w_l$ with words $w_i, u_i \in A^{\leq \mathcal{K}(|F|)}$ such that $\delta^{u_i} = \delta^{u_i u_i}$. Since $\alpha(\widehat{vwv}) \subseteq \alpha(\widehat{vv})$ and $|u_i w_i u_{i+1}| \leq k$ we can find each factor $u_i w_i u_{i+1}$ in vv. This argument leads to pattern $\mathbf{P}$. □

Since this proof establishes a bound on k in the size of the automaton, we can also find an algorithm which determines the minimal k such that $L(F) \in \mathcal{B}_{3/2,k}$ for a given dfa F.

As before in the case of the classes $\mathcal{B}_{3/2,k}$, we exploit now Theorem 6 and construct an efficient algorithm which solves the membership problem for $\mathcal{B}_{3/2}$. Looking for pattern $\mathbf{P}$ can also be done in NL since we may continuously check piecewise the occurrences of the respective subgraphs for $w_i u_i v_i u_i$ and $w_i u_i$. Note that we do not need to bound m since no bound is required for the length of paths in an NL–computation.

Theorem 7. *The membership problem for $\mathcal{B}_{3/2}$ is in* NL.

6 Further Consequences

Due to the various characterizations of the classes of the dot–depth hierarchy, Theorem 7 has immediate consequences in other fields of research. The correspondence of the class of languages definable by Σ_n–formulas of the logic FO$[<, \min, \max, S, P]$ and $\mathcal{B}_{n-1/2}$ from [Tho82] has already been mentioned in the introduction. Due to this characterization we have the following corollary.

Corollary 1. *Given a regular language L, it is decidable whether L is definable by a Σ_2-formula of the logic* FO$[<, \min, \max, S, P]$.

An algebraic interpretation of our Theorem 6 can also be given. For an introduction to the algebraic theory of finite automata we refer to [Pin96]. Let L be a regular language of A^+ and let $F_L = (A, S, \delta, s_0, S')$ be its unique minimal dfa. We define the syntactic semigroup of L via the transition semigroup of F_L, i.e., as $S_L =_{\text{def}} \{\delta^w : w \in A^+\}$ where the composition is defined as $\delta^u \cdot \delta^v =_{\text{def}} \delta^{uv}$. By id_S we denote the identity mapping on S. The syntactic semigroup S_L can be considered as an ordered syntactic semigroup $S_L^{\leq}$ with order relation $\leq$ by setting $\mu \leq \eta$ if and only if $\alpha\eta\beta(s_0) \in S'$ implies $\alpha\mu\beta(s_0) \in S'$ for $\alpha, \beta \in S_L \cup \{\text{id}_S\}$. If μ is an element of a finite semigroup, the minimal idempotent power of μ is denoted as μ^ω.

Theorem 8. *Let $L \subseteq A^+$ be a regular language and $S_L^{\leq}$ be its ordered syntactic semigroup. Then $L \in \mathcal{B}_{3/2}$ if and only if $S_L^{\leq}$ satisfies all inequalities $\{E_m : m \geq 0\}$ for any choice of τ_i, β_i and γ_i from $S_L^{\leq}$ where*

$$E_m =_{\text{def}} \beta^\omega \gamma \beta^\omega \leq \beta^\omega \text{ with}$$
$$\gamma =_{\text{def}} \tau_0^\omega \gamma_1 \tau_1^\omega \gamma_2 \tau_2^\omega \cdots \gamma_{m+1} \tau_{m+1}^\omega \text{ and}$$
$$\beta =_{\text{def}} \tau_0^\omega \beta_0 \tau_0^\omega \gamma_1 \tau_1^\omega \beta_1 \tau_1^\omega \gamma_2 \tau_2^\omega \beta_2 \tau_2^\omega \cdots \gamma_{m+1} \tau_{m+1}^\omega \beta_{m+1} \tau_{m+1}^\omega.$$

Due to [Arf91] the class $\mathcal{B}_{3/2}$ can be understood as a positive +–variety of languages when varying the alphabet A (for the definition of the notion of positive varieties we refer to [PW97]). An Eilenberg–like theorem was given for the case of positive varieties in [Pin95], which states that positive +–varieties of languages and varieties of finite ordered semigroups are in one–one correspondence via the operation of taking ordered syntactic semigroups. So the inequalities $\{E_m : m \geq 0\}$ from Theorem 8 characterize the variety of finite ordered semigroups corresponding to $\mathcal{B}_{3/2}$. It is known that this variety is equal to the Mal'cev product of the variety of finite semigroups corresponding to dot–depth one languages with a certain other variety of finite ordered semigroups, as stated in Theorem 5.8 in [PW97]. The benefit of our characterization is that it is effective as follows from Theorem 7.

7 Conclusions

It was conjectured in [PW97] that the decidability questions for the Straubing–Thérien hierarchy and the dot–depth hierarchy are related not only on levels n for integers n [Str85], but on all levels $n/2$. We confirm the latter with our work now also for $n = 3$, while the general case remains open.

We see the contribution of our paper not only as a stand–alone result providing the decidability of $\mathcal{B}_{3/2}$, but in what we can carry over to the general case. First we note that the nature of our proof is such that it bounds in a computable way (in terms of the automaton size and the alphabet size) the descriptional complexity of a language L, i.e., it bounds the length of an expression that witnesses that L is of dot–depth 3/2. Let us continue on an informal level.

If we compare pattern $\mathbf{P'}$ from [PW97] characterizing $\mathcal{B}_{1/2}$ and our pattern $\mathbf{P}$ characterizing $\mathcal{B}_{3/2}$, we observe that subgraphs of type $\mathbf{P'}$ appear as $u_i v_i u_i$ in pattern $\mathbf{P}$. Now one can repeat inductively this formation procedure using pattern $\mathbf{P}$ in a more complicated pattern in the same way as $\mathbf{P'}$ appears in $\mathbf{P}$. This leads for $n \geq 1$ to the definition of patterns $\mathbf{R_n}$ for which $\mathbf{R_1}$=$\mathbf{P'}$ and $\mathbf{R_2}$=$\mathbf{P}$ holds (This can also be done for the Straubing–Thérien hierarchy with the same formation procedure, but starting with the pattern that characterizes level 1/2 there). Moreover, looking for the existence of $\mathbf{R_n}$ can be effectively carried out with a recursive application of our algorithm for testing pattern $\mathbf{P}$.

Denote by $\mathcal{C}_n$ the class of languages that can be accepted exactly by those dfa's which do not have $\mathbf{R_n}$ in their transition graph. We believe that it is a reasonable conjecture that $\mathcal{B}_{n-1/2} = \mathcal{C}_n$ holds for all $n \geq 1$. As we know now, this is true for $n = 1$ and $n = 2$. We can further support this by some partial results, left without proofs here. First we note that for all $n \geq 1$ it holds that $\mathcal{C}_n$ is a subclass of starfree languages. Moreover, one can show with a generalization of Theorem 3 that $\mathcal{C}_n$ contains $\mathcal{B}_{n-1/2}$. Finally, we see that $\mathcal{C}_n \subsetneq \mathcal{C}_{n+1}$ using the languages L_n from [BK78] witnessing the strictness of the dot–depth hierarchy.

Acknowledgements. For discussions related to our subject we thank Bernd Borchert, Klaus W. Wagner and Thomas Wilke.

References

[Arf91] M. Arfi. Opérations polynomiales et hiérarchies de concaténation. *Theoretical Computer Science*, 91:71–84, 1991.

[BK78] J. A. Brzozowski and R. Knast. The dot-depth hierarchy of star-free languages is infinite. *Journal of Computer and System Sciences*, 16:37–55, 1978.

[Brz76] J. A. Brzozowski. Hierarchies of aperiodic languages. *RAIRO Inform. Theor.*, 10:33–49, 1976.

[CB71] R. S. Cohen and J. A. Brzozowski. Dot-depth of star-free events. *Journal of Computer and System Sciences*, 5:1–16, 1971.

[CK96] C. Choffrut and J. Karhumäki. Combinatorics of words. In G.Rozenberg and A.Salomaa, editors, *Handbook of formal languages*, volume I, pages 329–438. Springer, 1996.

[Gla98] C. Glaßer. A normalform for classes of concatenation hierarchies. Technical Report 216, Inst. für Informatik, Univ. Würzburg, 1998.

[GS99] C. Glaßer and H. Schmitz. Languages of dot-depth 3/2. Technical Report 243, Inst. für Informatik, Univ. Würzburg, 1999.

[Has83] K. Hashiguchi. Representation theorems on regular languages. *Journal of Computer and System Sciences*, 27:101–115, 1983.

[Hig52] G. Higman. Ordering by divisibility in abstract algebras. In *Proc. London Math. Soc.*, volume 3, pages 326–336, 1952.

[Kna83] R. Knast. A semigroup characterization of dot-depth one languages. *RAIRO Inform. Théor.*, 17:321–330, 1983.

[MP71] R. McNaughton and S. Papert. *Counterfree Automata.* MIT Press, Cambridge, 1971.

[Pin95] J. E. Pin. A variety theorem without complementation. *Russian Math.*, 39:74–83, 1995.

[Pin96] J. E. Pin. Syntactic semigroups. In G.Rozenberg and A.Salomaa, editors, *Handbook of formal languages*, volume I, pages 679–746. Springer, 1996.

[PP86] D. Perrin and J. E. Pin. First-order logic and star-free sets. *Journal of Computer and System Sciences*, 32:393–406, 1986.

[PW97] J. E. Pin and P. Weil. Polynomial closure and unambiguous product. *Theory of computing systems*, 30:383–422, 1997.

[Sim72] I. Simon. *Hierarchies of events with dot-depth one.* PhD thesis, University of Waterloo, 1972.

[Sim90] I. Simon. Factorization forests of finite height. *Theoretical Computer Science*, 72:65–94, 1990.

[SS83] J. Sakarovitch and I. Simon. Subwords. In M. Lothaire, editor, *Combinatorics on Words*, Encyclopedia of mathematics and its applications, pages 105–142. Addison-Wesley, 1983.

[Ste85] J. Stern. Characterizations of some classes of regular events. *Theoretical Computer Science*, 35:17–42, 1985.

[Str81] H. Straubing. A generalization of the Schützenberger product of finite monoids. *Theoretical Computer Science*, 13:137–150, 1981.

[Str85] H. Straubing. Finite semigroups varieties of the form V * D. *J.Pure Appl.Algebra*, 36:53–94, 1985.

[Str88] H. Straubing. Semigroups and languages of dot-depth two. *Theoretical Computer Science*, 58:361–378, 1988.

[Thé81] D. Thérien. Classification of finite monoids: the language approach. *Theoretical Computer Science*, 14:195–208, 1981.

[Tho82] W. Thomas. Classifying regular events in symbolic logic. *Journal of Computer and System Sciences*, 25:360–376, 1982.

Random Generation and Approximate Counting of Ambiguously Described Combinatorial Structures*

Alberto Bertoni, Massimiliano Goldwurm, and Massimo Santini

Dipartimento di Scienze dell'Informazione
Università degli Studi di Milano
Via Comelico, 39/41
20135 Milano - Italia
{bertoni,goldwurm,santini}@dsi.unimi.it

Abstract. This paper concerns the *uniform random generation* and the *approximate counting* of combinatorial structures admitting an ambiguous description. We propose a general framework to study the complexity of these problems and present some applications to specific classes of languages. In particular, we give a uniform random generation algorithm for *finitely ambiguous context-free languages* of the same time complexity of the best known algorithm for the unambiguous case. Other applications include a polynomial time uniform random generator and approximation scheme for the census function of (i) languages recognized in polynomial time by *one-way nondeterministic auxiliary pushdown automata* of polynomial ambiguity and (ii) polynomially ambiguous *rational trace languages*.

Keywords: uniform random generation, approximate counting, context-free languages, auxiliary pushdown automata, rational trace languages, inherent ambiguity.

1 Introduction

In this work we propose a general framework to study the complexity of uniform random generation and counting problems for combinatorial structures represented through an ambiguous specification, in the sense that the same object may admit several distinct descriptions. Structures of this kind occur frequently in different contexts. Typical examples are inherently ambiguous context-free languages, usually specified by generating context-free grammars, formal languages accepted by nondeterministic machines, whose words can be represented by the accepting computations, combinatorial objects ambiguously represented by words over a given alphabet as, for instance, traces in free partially commutative monoids. What often happens in these cases is that the counting problem for

* This work has been supported by MURST Research Program "Unconventional computational models: syntactic and combinatorial methods".

H. Reichel and S. Tison (Eds.): STACS 2000, LNCS 1770, pp. 567–580, 2000.

the structure (i.e., determining the number of objects of given size) is a difficult problem, while computing the number of different descriptions of these objects is easy.

As an example, computing the number of words of length n in inherently ambiguous context-free languages is complete for $\sharp P_1$ (the restriction of $\sharp P$ to functions with unary inputs) [6], while determining the number of derivation trees of words of length n in context-free grammars is easily solvable in polynomial time. It is worth noting that the last result holds even assuming the grammar part of the input [16], while the first problem is complete also for inherently ambiguous context-free languages of ambiguity degree 2. As a consequence of these negative results, in the case of ambiguous descriptions, the usual approach to random generation through counting should be avoided whenever we look for efficient algorithms.

We recall that, in the case of unambiguous formal descriptions, the uniform random generation and counting problems have been widely studied in the literature (see, for instance, [14, 8]). In particular, due to their well-known applications, the uniform random generation of unambiguous context-free languages has been considered in several papers [18, 25, 14, 15]. This problem is implicitely treated in [14] as a special case of a more general analysis of algorithms for uniform random generation of combinatorial structures specified by (unambiguous) formal grammars that involve operations of union, product, construction of sets, sequences and cycles. Many classical combinatorial objects can be specified in this way and the same analysis can be carried out for unambiguous context-free languages. The best general routine, assuming a fixed arbitrary grammar, generates an object of size n uniformly at random in $O(n \log n)$ time. This bound is in terms of arithmetic complexity: each step of the algorithm can require an arithmetic operation over $O(n \log n)$-bits integers or it can generate in constant time an integer of $O(n)$ bits uniformly at random.

On the contrary, uniform random generation and counting problems for structures represented by ambiguous formalisms have not received much attention in the literature. An analysis of the complexity of uniform random generation for combinatorial structures defined by polynomial time relations is given in [21]. The authors give some evidence that, under suitable hypotheses, (almost uniform) random generation is easier than counting, but more difficult than recognizing. Recently, following a similar approach, a subexponential time algorithm is presented in [16] for the (almost uniform) random generation of words in a (possibly ambiguous) context-free language.

In this paper we introduce a simple notion of *description* of a combinatorial structure, together with a corresponding notion of *ambiguity*, and study the problem of uniform random generation and approximate counting for structures endowed with such descriptions. We prove a general result (Section 3) stating that if a structure $\boldsymbol{S}$ has a description $\boldsymbol{T}$ with polynomially bounded ambiguity and $\boldsymbol{T}$ admits a polynomial time *uniform random generator (u.r.g.)*, then also $\boldsymbol{S}$ admits a u.r.g. working in polynomial time. If further the counting problem for $\boldsymbol{T}$ is solvable in polynomial time, then $\boldsymbol{S}$ admits a fully polynomial time *randomized approximation scheme (r.a.s.)* for its counting problem. Here, the proofs are based on the Karp-Luby technique for sampling from a union of sets [22] and on Hoeffding's inequality [19], a classical tool for bounding the tail probability of the sum of independent bounded random variables. Moreover, the computation model we use is essentially a RAM, under logarithmic cost criterion, equipped in addition with an unbiased coin tossing device.

These results can be applied to various classes of languages:

(1) We show (Section 4) that, for *finitely ambiguous context-free languages*, a word of length n can be generated uniformly at random in $O(n^2 \log n)$ time and $O(n^2)$ space, using $O(n^2 \log n)$ random bits. We observe that, in our model of computation, the same bounds for time and random bits are obtained for the uniform random generation of unambiguous context-free languages [15]. Analogous bounds are obtained for the corresponding randomized approximation scheme. To prove these results we show in detail a multiplicity version of Earley's algorithm for context-free recognition [12]. We prove that for finitely ambiguous context-free languages, the number of derivation trees of an input word of size n can be computed in $O(n^2 \log n)$ time and $O(n^2)$ space;
(2) We show (Section 5) how to generate, uniformly at random, words from languages accepted by *one-way nondeterministic auxiliary push-down automata* working in polynomial time and using a logarithmic amount of work-space [9, 7]. Also in this case, we obtain polynomial time u.r.g. and r.a.s. whenever the automaton has a polynomial number of accepting computations for each input word;
(3) We consider (Section 6) the uniform random generation and approximate counting of *rational trace languages* [11, 5]. Finitely ambiguous rational trace languages admit u.r.g. and r.a.s. of the same time complexity of the algorithms for their recognition problem. Analogously, we obtain polynomial time u.r.g. and r.a.s. for the rational trace languages that are polynomially ambiguous.

2 Preliminary Definitions

In this work, as model of computation, we assume a Probabilistic Random Access Machine (PrRAM for short) according to which the complexity of a procedure takes into account the number of random bits used by the computation; a similar model is implicitely assumed in [23] where the complexity of random number generation from arbitrary distributions is studied.

Formally, our machine is an augmented version of the standard RAM model [1] equipped in addition with a 1-way read-only *random tape* and an instruction RND. The random tape contains a sequence $r = r_0, r_1 \ldots$ of symbols in $\{0, 1\}$ and, in the initial configuration, its head scans the first symbol. During the computation, the instruction "RND i" transfers the first i unread bits from the random tape into the accumulator and moves the tape head i positions ahead. For a fixed sequence r on the random tape and a given input x, the output $M(x, r)$ of the computation of a PrRAM M, is defined essentially as in the standard RAM model; furthermore, by $M(x)$ we mean the random variable denoting the output $M(x, r)$ assuming r a sequence of independent random variables such that $\Pr\{r_i = 1\} = \Pr\{r_i = 0\} = 1/2$ for every $i \geq 0$. Hence the instruction "RND i" generates an integer in $\{0, \ldots, 2^i - 1\}$ uniformly at random. To evaluate the space and time complexity of a PrRAM computation we adopt the logarithmic cost criterion defined in [1] for the standard RAM model assuming, in addition, a time cost i for every instruction "RND i".

Due to the restriction to unbiased coins, an algorithm in our model may fail to give the correct answer and in this case it outputs a conventional symbol $\perp$. However, good algorithms should reduce as much as possible the probability of such an event.

We think this machine takes the advantages of the two main models considered in connection with the random generation of combinatorial structures, i.e. the "arithmetic" machine assumed in [14] and the probabilistic Turing machine used in [21]. From one side, it is suited for the specification of algorithms at high level allowing an easy analysis of time and space complexity. From the other, it also allows to carry out a somehow realistic analysis of the procedures which does not neglect the size of operands, trying to reasonably satisfy the principle that every elementary step of the machine be implementable in constant time by some fixed hardware.

2.1 Ambiguous Descriptions of Combinatorial Structures

A *combinatorial structure* is a pair $\langle \boldsymbol{S}, |\cdot| \rangle$, where the *domain* $\boldsymbol{S}$ is a finite or denumerable set and the *size* $|\cdot| : \boldsymbol{S} \to \mathbf{N}$ is a function such that $\#\{s \in \boldsymbol{S} : |s| = n\}$ is finite[1] for every $n \in \mathbf{N}$ [13]. Here, we implicitly assume the elements $s \in \boldsymbol{S}$ admit a (recursive) binary representation such that each $|s|$ is polynomially related to the length of its binary representation; this allows our model of computation to manipulate the elements of combinatorial structures. Given such a structure $\langle \boldsymbol{S}, |\cdot| \rangle$, we also denote by $\boldsymbol{S}_n$ the set $\{s \in \boldsymbol{S} : |s| = n\} \subseteq \boldsymbol{S}$ and define the census function $C_{\boldsymbol{S}} : \mathbf{N} \to \mathbf{N}$ by $C_{\boldsymbol{S}}(n) = \#\boldsymbol{S}_n$. Then, we formally introduce the concept of *ambiguous description*:

Definition 1. *$\langle \boldsymbol{T}, |\cdot| \rangle$ is a* description *of $\langle \boldsymbol{S}, |\cdot| \rangle$ via the function $f : \boldsymbol{T} \to \boldsymbol{S}$, if f is a surjective function preserving $|\cdot|$, i.e. $|f(t)| = |t|$ for every $t \in \boldsymbol{T}$. The* ambiguity *of the description is the function $d : \boldsymbol{S} \to \mathbf{N}$ defined by $d(s) = \#\{t \in \boldsymbol{T} : f(t) = s\}$, for every $s \in \boldsymbol{S}$. We say that the description is* ambiguous *if $d(s) > 1$, for some $s \in \boldsymbol{S}$. Moreover, the description is said to be* polynomial *whenever f and d are computable in polynomial time and there exists some $D \in \mathbf{N}$ such that $d(s) = O(|s|^D)$.*

We now introduce the following notions in the spirit of [21].

Definition 2. *An algorithm* A *is a* uniform random generator *(u.r.g.) for a combinatorial structure $\langle \boldsymbol{S}, |\cdot| \rangle$ if, for every $n > 0$ such that $C_{\boldsymbol{S}}(n) > 0$,*

[1] In this paper, to avoid confusion with $|\cdot|$, $\#A$ denotes the cardinality of the set A.

(i) A *on input* n *returns a value* $\mathsf{A}(n) \in \boldsymbol{S} \cup \{\bot\}$,
(ii) $\Pr\{\mathsf{A}(n) = s \mid \mathsf{A}(n) \neq \bot\} = 1/C_{\boldsymbol{S}}(n)$ *for every* $s \in \boldsymbol{S}_n$ *and*
(iii) $\Pr\{\mathsf{A}(n) = \bot\} < 1/4$.

Definition 3. *An algorithm* A *is a* randomized approximation scheme *(r.a.s.) for the census function* $C_{\boldsymbol{S}}$ *of a combinatorial structure* $\langle \boldsymbol{S}, |\cdot|\rangle$ *if, for every* $n > 0$ *such that* $C_{\boldsymbol{S}}(n) > 0$ *and every* $\varepsilon \in (0,1)$,

(i) A *on input* n, ε *returns a value* $\mathsf{A}(n,\varepsilon) \in \mathbf{Q} \cup \{\bot\}$,
(ii) $\Pr\{(1-\varepsilon)C_{\boldsymbol{S}}(n) \leq \mathsf{A}(n,\varepsilon) \leq (1+\varepsilon)C_{\boldsymbol{S}}(n) \mid \mathsf{A}(n,\varepsilon) \neq \bot\} > 3/4$ *and*
(iii) $\Pr\{\mathsf{A}(n,\varepsilon) = \bot\} < 1/4$.

Moreover, a r.a.s. is said to be a fully polynomial time *r.a.s. whenever it works in time polynomial in* n *and* $1/\varepsilon$.

Definition 4. *An algorithm* A *is a* randomized exact counter *(r.e.c.) for a combinatorial structure* $\langle \boldsymbol{S}, |\cdot|\rangle$ *if, for every* $n > 0$ *such that* $C_{\boldsymbol{S}}(n) > 0$,

(i) A *on input* n *returns a value* $\mathsf{A}(n) \in \mathbf{N} \cup \{\bot\}$,
(ii) $\Pr\{\mathsf{A}(n) = C_{\boldsymbol{S}}(n) \mid \mathsf{A}(n) \neq \bot\} > 3/4$ *and*
(iii) $\Pr\{\mathsf{A}(n) = \bot\} < 1/4$.

Observe that the constant $1/4$ in the previous definitions can be replaced by every positive number strictly less than 1, leaving the same notions substantially unchanged. Similarly, in the last two definitions, by taking the median of the outputs of several runs of the algorithm [21], the constant $3/4$ can be replaced by every number strictly between $1/2$ and 1.

3 Uniform Random Generation and Approximate Counting

We start by considering the uniform random generation problem.

Theorem 1. *If a combinatorial structure* $\langle \boldsymbol{S}, |\cdot|\rangle$ *admits a polynomial (ambiguous) description* $\langle \boldsymbol{T}, |\cdot|\rangle$ *and there exists a polynomial time u.r.g. for* $\langle \boldsymbol{T}, |\cdot|\rangle$, *then there exists a polynomial time u.r.g. for* $\langle \boldsymbol{S}, |\cdot|\rangle$.

To prove this theorem, we show a stronger result derived by applying the Karp-Luby technique for sampling from a union of sets [22].

Lemma 1. *Let* $\langle \boldsymbol{S}, |\cdot|\rangle$ *admit a polynomial (ambiguous) description* $\langle \boldsymbol{T}, |\cdot|\rangle$ *via* f *and let* $T_f(n)$ *and* $T_d(n)$ *be respectively the computation time of* f *and its ambiguity* d; *moreover, assume* $d(s) = O(|s|^D)$ *for some* $D \in \mathbf{N}$. *If* B *is a u.r.g. for* $\langle \boldsymbol{T}, |\cdot|\rangle$ *working in* $T_{\mathsf{B}}(n)$ *time and using* $R_{\mathsf{B}}(n)$ *random bits, then there exists a u.r.g. for* $\langle \boldsymbol{S}, |\cdot|\rangle$ *working in time* $O\big(n^{2D} + n^D(T_{\mathsf{B}}(n) + T_f(n) + T_d(n))\big)$ *and using* $O(n^{2D} + n^D R_{\mathsf{B}}(n))$ *random bits.*

Proof. Define, for the sake of brevity, $D(n) = \kappa_1 n^D + \kappa_2$ where κ_1, κ_2 are two integer constants such that $d(s) \leq D(|s|)$; let A be the following algorithm, where $\kappa > 0$ is a suitable integer constant discussed later and $\operatorname{lcm} I$ denotes the least common multiple of a set $I \subseteq \mathbf{N}$.

input n
$m \leftarrow \operatorname{lcm}\{1, \ldots, D(n)\}$, $\ell \leftarrow \lceil \log m \rceil$
$i \leftarrow 0$, $s \leftarrow \perp$
while $i < \kappa D(n)$ and $s = \perp$ **do**
    $i \leftarrow i + 1$
    $t \leftarrow \mathsf{B}(n)$
    if $t \neq \perp$ **then**
        $s \leftarrow f(t)$
        generate $r \in \{1, \ldots, 2^\ell\}$ *uniformly at random*
        if $r > m/d(s)$ **then** $s \leftarrow \perp$
output s

First of all, we focus on the computation time of A on input n. One can show that the time to precompute $D(n)$, m and ℓ is $O(n^{2D})$. The time of each iteration equals $O(T_{\mathsf{B}}(n) + T_f(n) + T_d(n))$ plus the time $O(n^D)$ required for the generation of r. Moreover, each iteration uses at most $R_{\mathsf{B}}(n)$ and $O(n^D)$ random bits to compute t and r respectively, hence the total amount of random bits used by A adds up to $O(n^{2D} + n^D R_{\mathsf{B}}(n))$.

Now, we prove the correctness of A. Assume that $C_{\boldsymbol{S}}(n) > 0$ so that, since f is surjective, $C_{\boldsymbol{T}}(n) > 0$ and let S and T be the random variables representing respectively the value of s and t at the end of a **while** iteration (observe that S and T are well defined since their value is independent of the outcomes of the preceding iterations). Then, S takes value $s \in \boldsymbol{S}_n$ whenever $T \in f^{-1}(s) \subseteq \boldsymbol{T}_n$ and $r \leq m/d(f(T))$. Moreover, by definition of B, there exists some $0 < \delta < 1/4$ such that $\Pr\{T = t\} = (1-\delta) C_{\boldsymbol{T}}(n)^{-1}$ for every $t \in \boldsymbol{T}_n$ and, since r is independent of S and T, $\Pr\{r \leq m/d(f(T)) \mid T = t\} = d(f(t))^{-1} m 2^{-\lceil \log m \rceil}$. Hence, for every $s \in \boldsymbol{S}_n$,

$$
\begin{aligned}
\Pr\{S = s\} &= \sum_{t \in f^{-1}(s)} \Pr\{T = t, r \leq m/d(f(T))\} \\
&= \sum_{t \in f^{-1}(s)} \Pr\{r \leq m/d(f(T)) \mid T = t\} \Pr\{T = t\} \\
&= d(s) \big(d(s)^{-1} m 2^{-\lceil \log m \rceil} (1-\delta) C_{\boldsymbol{T}}(n)^{-1} \big) \\
&= (1-\delta) m 2^{-\lceil \log m \rceil} C_{\boldsymbol{T}}(n)^{-1}
\end{aligned}
$$

which is independent of s. On the other hand, at the end of each **while** iteration, $\Pr\{S = \perp\} = 1 - \Pr\{S \in \boldsymbol{S}_n\} = 1 - C_{\boldsymbol{S}}(n) \left((1-\delta) m 2^{-\lceil \log m \rceil} C_{\boldsymbol{T}}(n)^{-1} \right)$. Then, since $m 2^{-\lceil \log m \rceil} > 1/2$ and $C_{\boldsymbol{T}}(n) \leq C_{\boldsymbol{S}}(n) D(n)$,

$$
\Pr\{S = \perp\} \leq 1 - \frac{3}{8} \frac{1}{D(n)}.
$$

Finally, since $\Pr\{\mathsf{A}(n) = \perp\} = \Pr\{S = \perp\}^{\kappa D(n)}$, an integer $\kappa > 0$ can be chosen such that, for every $n > 0$,

$$\Pr\{\mathsf{A}(n) = \perp\} \leq \left(1 - \frac{3}{8}\frac{1}{D(n)}\right)^{\kappa D(n)} < 1/4.$$

Moreover, for every $s \in \boldsymbol{S}_n$, it holds $\Pr\{\mathsf{A}(n) = s \mid \mathsf{A}(n) \neq \perp\} = \Pr\{S = s \mid S \neq \perp\}$ and, since $\Pr\{S = s\}$ is constant, independent of s, it is immediate to conclude that $\Pr\{\mathsf{A}(n) = s \mid \mathsf{A}(n) \neq \perp\} = 1/C_{\boldsymbol{S}}(n)$. □

Using Hoeffding's inequality [19], a similar result can be proved for the approximate and exact counting problem.

Theorem 2. *Let $\langle \boldsymbol{S}, |\cdot|\rangle$ be a combinatorial structure admitting a polynomial (ambiguous) description $\langle \boldsymbol{T}, |\cdot|\rangle$ and assume there exists a polynomial time u.r.g. for $\langle \boldsymbol{T}, |\cdot|\rangle$. If $C_{\boldsymbol{T}}(n)$ is computable in polynomial time, then there exists a fully polynomial time r.a.s. for $C_{\boldsymbol{S}}$. If further $C_{\boldsymbol{T}}(n)$ is polynomially bounded, then there exists a polynomial time r.e.c. for $\langle \boldsymbol{S}, |\cdot|\rangle$.*

4 Context-Free Languages

Applying our general paradigm to context-free (c.f. for short) grammars, we can design simple u.r.g. and r.a.s. of census functions for inherently ambiguous c.f. languages. To this end, let $G = \langle V, \Sigma, S, P\rangle$ be a c.f. grammar, where V is the set of nonterminal symbols (we also call variables), Σ the alphabet of terminals, $S \in V$ the initial variable and P the family of productions. We assume G in Chomsky normal form [20] without useless variables (clearly, every c.f. language not containing the empty word ϵ can be generated by such a grammar). Moreover, for every $A \in V$, let T_A be the family of derivation trees with root labelled by A deriving a word in Σ^+. It is easy to see that there are finitely many $t \in T_S$ deriving a given $x \in \Sigma^+$; in the following, we denote by $\hat{d}_G(x)$ the number of such trees and call *ambiguity* of G the function $d_G : \mathbf{N} \to \mathbf{N}$ defined by $d_G(n) = \max\{\hat{d}_G(x) : x \in \Sigma^n\}$, for every $n \in \mathbf{N}$. Then, G is said *finitely ambiguous* if there exists a $k \in \mathbf{N}$ such that $d_G(n) \leq k$ for every $n > 0$; in particular, G is said *unambiguous* if $k = 1$. On the other hand, G is said *polynomially ambiguous* if, for some polynomial $p(n)$, we have $d_G(n) \leq p(n)$ for every $n > 0$.

Our idea is to use the structure $\langle T_S, |\cdot|\rangle$ as ambiguous description of the language L generated by G, where, for every $t \in T_S$, $|t|$ is the length of the derived word. To this aim, we need two preliminary procedures: one for generating a tree of given size in T_S uniformly at random, the other for computing the degree of ambiguity $\hat{d}_G(x)$ for any word $x \in \Sigma^+$. The first problem can be solved by adapting the procedure given in [15], based on the general approach to the random generation proposed in [14].

Proposition 1. *Given a context-free grammar $G = \langle V, \Sigma, S, P\rangle$ in Chomsky normal form, there exists a u.r.g. for $\langle T_S, |\cdot|\rangle$ working in $O(n^2 \log n)$ time, $O(n)$ space and using $O(n^2 \log n)$ random bits.*

4.1 Earley's Algorithm for Counting Derivations

The number of derivation trees of a terminal string in a c.f. grammar can be computed by adapting Earley's algorithm [12] for context-free recognition. The main advantage of this procedure, with respect to the well-known CYK algorithm [17], is that in the case of a grammar with bounded ambiguity, the computation only requires quadratic time on a RAM under unit cost criterion [12, 2].

Our algorithm manipulates a *weighted* version of the so-called *dotted productions* of a grammar $G = \langle V, \Sigma, S, P \rangle$ in Chomsky normal form, i.e. expressions of the form $A \rightarrow \alpha \cdot \beta$, where $A \in V$, $\alpha, \beta \in (\Sigma \cup V)^*$ and $A \rightarrow \alpha\beta \in P$. Given an input string $x = a_1 a_2 \ldots a_n$, the algorithm computes a table of entries $S_{i,j}$, for $0 \leq i \leq j \leq n$, each of which is a list of terms of the form $[A \rightarrow \alpha \cdot \beta, t]$, where $A \rightarrow \alpha \cdot \beta$ is a dotted production in G and t is a positive integer. Each pair $[A \rightarrow \alpha \cdot \beta, t]$ is called *state* and t is the *weight* of the state.

The table of lists $S_{i,j}$ computed by the algorithm has the following properties for any pair of indices $0 \leq i \leq j \leq n$:

1) $S_{i,j}$ contains at most one state $[A \rightarrow \alpha \cdot \beta, t]$ for every dotted production $A \rightarrow \alpha \cdot \beta$ in G;
2) a state $[A \rightarrow \alpha \cdot \beta, t]$ belongs to $S_{i,j}$ if and only if there exists $\delta \in V^*$ such that $S \overset{*}{\Rightarrow} a_1 \ldots a_i A \delta$ and $\alpha \overset{*}{\Rightarrow} a_{i+1} \ldots a_j$;
3) if $[A \rightarrow \alpha \cdot \beta, t]$ belongs to $S_{i,j}$, then $t = \#\{\alpha \overset{*}{\Rightarrow} a_{i+1} \ldots a_j\}$, i.e. the number of leftmost derivations $\alpha \overset{*}{\Rightarrow} a_{i+1} \ldots a_j$.

Note that, since there are no ϵ-productions, $[A \rightarrow \alpha \cdot \beta, t] \in S_{i,i}$ implies $\alpha = \epsilon$ for every $0 \leq i \leq n$. Furthermore, once the lists $S_{i,j}$ are completed for any $0 \leq i \leq j \leq n$, the number of derivation trees of the word x can be obtained by the sum $\sum_{[S \rightarrow AB \cdot, t] \in S_{0,n}} t$.

The algorithm first computes the list $S_{0,0}$ of all the states $[A \rightarrow \cdot\, \alpha, 1]$ such that $S \overset{*}{\Rightarrow} A\delta$ for some $\delta \in V^*$. Then, it executes the cycle of *Scanner*, *Predictor* and *Completer* loops given below for $1 \leq j \leq n$, computing at the j-th loop the lists $S_{i,j}$ for $0 \leq i \leq j$. To this end the procedure maintains a family of sets $L_{B,i}$ for $B \in V$ and $1 \leq i \leq j$; each $L_{B,i}$ contains all indices $k \leq i$ such that a state of the form $[A \rightarrow \alpha \cdot B\beta, t]$ belongs to $S_{k,i}$ for some $A \in V$, $\alpha, \beta \in V \cup \{\epsilon\}$, $t \in \mathbf{N}$. Moreover, during the computation every state in $S_{i,j}$ can be unmarked or marked according whether it can still be used to add new states in the table.

The command "ADD D TO $S_{i,j}$" simply appends the state D as unmarked to $S_{i,j}$ and updates $L_{B,i}$ whenever D is of the form $[A \rightarrow \alpha \cdot B\beta, t]$; the command "UPDATE $[A \rightarrow \alpha \cdot \beta, t]$ IN $S_{i,j}$" replaces the state $[A \rightarrow \alpha \cdot \beta, u]$ in $S_{i,j}$ for some $u \in \mathbf{N}$ by $[A \rightarrow \alpha \cdot \beta, t]$, at last, "MARK D IN $S_{i,j}$" transforms the state D in $S_{i,j}$ into a marked state.

1 **for** $j = 1 \ldots n$ **do**

Scanner:
2 **for** $i = j - 1 \ldots 0$ **do**
3 **for** $[A \rightarrow \cdot\, a, t] \in S_{i,j-1}$ **do**

4 MARK $[A \to \cdot a, t]$ IN $S_{i,j-1}$
5 **if** $a = a_j$ **then** ADD $[A \to a \cdot, t]$ TO $S_{i,j}$

Completer:
6 **for** $i = j-1 \ldots 0$ **do**
7 **for** $[B \to \gamma \cdot, t] \in S_{i,j}$ **do**
8 MARK $[B \to \gamma \cdot, t]$ IN $S_{i,j}$
9 **for** $k \in L_{B,i}$ **do**
10 **for** $[A \to \alpha \cdot B\beta, u] \in S_{k,i}$ **do**
11 **if** $[A \to \alpha B \cdot \beta, v] \in S_{k,j}$
12 **then** UPDATE $[A \to \alpha B \cdot \beta, v + tu]$ IN $S_{k,j}$
13 **else** ADD $[A \to \alpha B \cdot \beta, tu]$ TO $S_{k,j}$

Predictor:
14 **for** $i = 0 \ldots j-1$ **do**
15 **for** $[A \to \alpha \cdot B\beta, t] \in S_{i,j}$ **do**
16 MARK $[A \to \alpha \cdot B\beta, t]$ IN $S_{i,j}$
17 **for** $B \to \gamma \in P$ **do**
18 **if** $[B \to \cdot \gamma, 1] \notin S_{j,j}$ **then** ADD $[B \to \cdot \gamma, 1]$ TO $S_{j,j}$
19 **while** $\exists$ UNMARKED $[A \to \cdot B\beta, t] \in S_{j,j}$ **do**
20 MARK $[A \to \cdot B\beta, t]$ IN $S_{j,j}$
21 **for** $B \to \gamma \in P$ **do**
22 **if** $[B \to \cdot \gamma, 1] \notin S_{j,j}$ **then** ADD $[B \to \cdot \gamma, 1]$ TO $S_{j,j}$

It can be shown that, at the end of the computation, properties 1) and 2) above are satisfied (the proof carries over as for the membership problem [2, Section 4.2.2]). Now, let us prove property 3). First note that all states in $S_{i,j}$, for $1 \le i \le j \le n$, are marked during the computation. Hence, we can reason by induction on the order of marking states. The initial condition is satisfied because all states in each $S_{j,j}$, $0 \le j \le n$, are of the form $[A \to \cdot \alpha, t]$ and have weight $t = 1$. Also the states of the form $[A \to a \cdot, t]$ have weight $t = 1$ and again statement 3) is satisfied.

Then, consider a state $D = [A \to \alpha \cdot B\beta, w] \in S_{k,j}$ $(k < j)$. We claim that w is the number of leftmost derivations $\alpha B \stackrel{*}{\Rightarrow} a_{k+1} \ldots a_j$. A state of this form is first added by the *Completer* at line 13. This means that there exists a set of indices I_k such that for every $i \in I_k$ there is $[A \to \alpha \cdot B\beta, u_i] \in S_{k,i}$ with $u_i \in \mathbf{N}$, and a family U_i of states $[B \to \gamma \cdot, t] \in S_{i,k}$ such that

$$w = \sum_{i \in I_k} \sum_{[B \to \gamma \cdot, t] \in U_i} t u_i .$$

Observe that $k \le i < j$ for every $i \in I_k$ and U_i is the subset of all states in $S_{i,j}$ with a dotted production of the form $B \to \gamma \cdot$. Moreover, each state $[A \to \alpha \cdot B\beta, u_i] \in S_{k,i}$ is marked at line 16 or 20 before D is added to $S_{k,j}$. Also all states in U_i, for all $i \in I_k$, are marked during the computation of the weight w. Observe that, due to the form of the grammar, updating such weight w cannot modify the weight of any state in U_i. As a consequence all the sates in U_i are marked before D. Hence, by inductive hypothesis, we have for every $i \in I_k$

$$u_i = \#\{\alpha \overset{*}{\Rightarrow} a_{k+1} \dots a_i\} \tag{1}$$

and, for each $[B \to \gamma \cdot, t] \in U_i$,

$$t = \#\{\gamma \overset{*}{\Rightarrow} a_{i+1} \dots a_j\}. \tag{2}$$

Now the number of leftmost derivations $\alpha B \overset{*}{\Rightarrow} a_{k+1} \dots a_j$ is clearly given by

$$\sum_{k \le i < j} \#\{\alpha \overset{*}{\Rightarrow} a_{k+1} \dots a_i\} \sum_{B \to \gamma \in P} \#\{\gamma \overset{*}{\Rightarrow} a_{i+1} \dots a_j\}$$

and the claim follows from statement 1) and equalities (1) and (2).

Finally, studying the behaviour of the algorithm, one obtains the following

Proposition 2. *Given a context-free grammar G in Chomsky normal form, the number of derivation trees of a string of length n can be computed in polynomial time. If the grammar G is finitely ambiguous, then the algorithm works in $O(n^2 \log n)$ time and $O(n^2)$ space (under logarithmic cost criterion).*

4.2 Inherently Ambiguous Context-Free Languages

Now, let $L \subseteq \Sigma^*$ be a c.f. language. We recall that L is *unambiguous* if it is generated by an unambiguous c.f. grammar, while it is *inherently ambiguous* whenever every c.f. grammar G generating L is ambiguous. We also say that L is *finitely* (respectively, *polynomially*) *ambiguous* if it is generated by a finitely (respectively polynomially) ambiguous c.f. grammar.

Observe that there are natural examples of polynomially ambiguous c.f. languages which are not finitely ambiguous. For instance, if $L = \{ww^R : w \in \{a, b\}^*\}$, then L^2 is inherently ambiguous, but not finitely ambiguous [17, Section 7.3]: however, it is easy to verify that L^2 is generated by a grammar of ambiguity $O(n)$.

Hence, applying Propositions 1 and 2 to Theorems 1 and 2 and Lemma 1, we obtain

Theorem 3. *If L is a polynomially ambiguous context-free language, then there exists a polynomial time u.r.g. for L and a fully polynomial time r.a.s. for its census function C_L. If further C_L is polynomially bounded, then there exists a polynomial time r.e.c. for L. Moreover, if L is a finitely ambiguous, then both the u.r.g. and the r.a.s. work in $O(n^2 \log n)$ time (on a* PrRAM *under logarithmic cost criterion).*

5 One-Way Nondeterministic Auxiliary Pushdown Automata

In this section we apply our scheme to languages accepted by *one-way nondeterministic auxiliary pushdown automata* (1-NAuxPDA, for short). We recall that a 1-NAuxPDA is a nondeterministic Turing machine having a one-way read-only input tape, a pushdown tape and a *log-space bounded* two-way read-write work tape [9, 7]. It is known that the class of languages accepted by polynomial time 1-NAuxPDA corresponds exactly to the class of decision problems that are reducible to context-free recognition via one-way log-space reductions [24].

Given a 1-NAuxPDA M, we define by $\hat{d}_M(x)$ the number of accepting computations of M on input $x \in \Sigma^*$, and call *ambiguity* of M the function $d_M : \mathbf{N} \to \mathbf{N}$ defined by $d_M(n) = \max\{\hat{d}_M(x) : x \in \Sigma^n\}$, for every $n \in \mathbf{N}$. Then, M is said *polynomially ambiguous* if, for some polynomial $p(x)$, we have $d_M(n) \leq p(n)$ for every $n > 0$. Moreover, it is known that, given an integer input $n > 0$, a c.f. grammar G_n can be built, in polynomial time in n, such that $L(G_n) \cap \Sigma^n = L(M) \cap \Sigma^n$, where $L(G_n) \subseteq \Sigma^*$ is the language generated by G_n and $L(M) \subseteq \Sigma^*$ is the language accepted by M. This allows us to apply the results of the previous section to the languages accepted by 1-NAuxPDA.

Here, we describe a modified version of the usual construction of G_n [9] which allows to bound the ambiguity of G_n with respect to the ambiguity of M. First of all, we assume w.l.o.g. that the automaton cannot simultaneously consume input and modify the content of the stack, at most one symbol can be pushed or popped for each single move, there is only a final state and, finally, the input is accepted iff the automaton reaches the final state with the pushdown store and work tape both empty.

A *surface configuration* of a 1-NAuxPDA M on input of length n is a 5-tuple (q, w, i, Γ, j) where q is the state of M, w the content of its work tape, $1 \leq i \leq |w|$ the work tape head position, Γ the symbol on top of the stack and $1 \leq j \leq n+1$ the input tape head position. Observe that there are $n^{O(1)}$ surface configurations on any input of length $n \geq 0$. Two surface configurations C_1, C_2 form a *realizable pair* (C_1, C_2) *(on a word* $y \in \Sigma^+$*)* iff M can move (consuming input y) from C_1 to C_2, ending with its stack at the same height as in C_1, without popping below this height at any step of the computation. If (C_1, D) and and (D, C_2) are realizable pairs on y' and y'' respectively, then (C_1, C_2) is a realizable pair on $y = y'y''$. Let $\mathscr{S}_n$ be the set of surface configurations of M on inputs of length n and define the c.f. grammar $G_n(M) = \langle N, \Sigma, S, P\rangle$ where $N = \{S\} \cup \{(C_1, C_2, \ell) : C_1, C_2 \in \mathscr{S}_n \text{ and } \ell \in \{0, 1\}\}$ and the set P of productions is given by:

(1) $S \to (C_{\mathsf{in}}, C_{\mathsf{fin}}, 0)$ and $S \to (C_{\mathsf{in}}, C_{\mathsf{fin}}, 1)$, where C_{in} and C_{fin} represent respectively the initial and final surface configuration of M;
(2) $(C_1, C_2, 0) \to \sigma \in P$ iff (C_1, C_2) is a realizable pair on $\sigma \in \Sigma \cup \{\varepsilon\}$ via a single move computation;
(3) $(C_1, C_2, 0) \to (C_1, D, 1)(D, C_2, \ell) \in P$, for $\ell \in \{0, 1\}$, iff $C_1, C_2, D \in \mathscr{S}_n$;

(4) $(C_1, C_2, 1) \rightarrow (D_1, D_2, \ell) \in P$, for $\ell \in \{0, 1\}$, iff $C_1, D_1, D_2, C_2 \in \mathscr{S}_n$, D_1 can be reached from C_1 in a single move pushing a symbol a on top of the stack and C_2 can be reached from D_2 in a single move popping the same symbol from the top of the stack.

One can prove that the grammar $G_n(M)$ is computable in polynomial time on input $n > 0$ [3], and the derivation $(C_1, C_2, \ell) \stackrel{*}{\Rightarrow} y$ holds in $G_n(M)$ for some $\ell \in \{0, 1\}$ iff (C_1, C_2) is a realizable pair on y in M; here, the derivations with $\ell = 0$ (respectively, $\ell = 1$) correspond to those computations where the stack height somewhen equals (respectively, never equals) the stack height at the extremes C_1, C_2. This correspondence allows to bound the ambiguity of $G_n(M)$, according to the following proposition the proof of which is omitted.

Proposition 3. *For every polynomial time* 1-NAuxPDA M*, the number of leftmost derivations* $(C_1, C_2, \ell) \stackrel{*}{\Rightarrow} y$ *in* $G_n(M)$ *for* $\ell \in \{0, 1\}$ *is less than or equal to the total number of computations from* C_1 *to* C_2 *consuming* y.

Now, observe that Proposition 1 and 2 can be rephrased assuming the c.f. grammar G as part of the input, still obtaining polynomial time algorithms. Hence, by Theorem 1 and 2, Proposition 3 leads to the following

Theorem 4. *Let* L *be the language accepted by a polynomial time* 1-NAuxPDA *with polynomial ambiguity. Then there exists a polynomial time u.r.g. for* L *and a fully polynomial time r.a.s. for its census function* C_L*. If further* C_L *is polynomially bounded, then there exists a polynomial time r.e.c. for* L.

6 Rational Trace Languages

Another application of our method concerns the uniform random generation and the census function of trace languages. To study this case we refer to [10] for basic definitions. We only recall that, given a trace monoid $\mathbb{M}(\Sigma, I)$ over the independence alphabet (Σ, I), a *trace language*, i.e. a subset of $\mathbb{M}(\Sigma, I)$, is usually specified by considering a string language $L \subseteq \Sigma^*$ and taking the closure $[L] = \{t \in \mathbb{M}(\Sigma, I) : t = [x] \text{ for some } x \in L\}$. In particular, a trace language $T \subseteq \mathbb{M}(\Sigma, I)$ is called *rational* if $T = [L]$ for some regular language $L \subseteq \Sigma^*$. In this case we say that T is *represented* by L and the *ambiguity* of this representation is the function $d_L : \mathbf{N} \to \mathbf{N}$, defined by $d_L(n) = \max_{x \in \Sigma^n} \#(L \cap [x])$. We say that a rational trace language T is *finitely ambiguous* if it is represented by a regular language L such that, for some $k \in \mathbf{N}$, $d_L(n) \leq k$ for every $n > 0$; in this case we say that T has *ambiguity* k.

In the following, assuming a given independence alphabet (Σ, I), we denote by $\mathscr{R}$ the set of all rational trace languages in $\mathbb{M}(\Sigma, I)$ and, by $\mathscr{R}_k$, the subset of trace languages in $\mathscr{R}$ of ambiguity k. Clearly, for every independence alphabet (Σ, I), we have $\mathscr{R}_1 \subseteq \mathscr{R}_2 \subseteq \cdots \subseteq \bigcup_{k=1}^{\infty} \mathscr{R}_k \subseteq \mathscr{R}$.

The properties of these families of languages have been studied in the literature [5]. In particular, it is known that $\mathscr{R}_1 = \mathscr{R}$ if and only if the independence relation I is transitive; on the other hand, if I is not transitive, then we get the following chain of strict inclusions: $\mathscr{R}_1 \subsetneq \mathscr{R}_2 \subsetneq \cdots \subsetneq \bigcup_{k=1}^{\infty} \mathscr{R}_k \subsetneq \mathscr{R}$.

Furthermore, we say that a rational trace language T is *polynomially ambiguous* if it is represented by a regular language L such that, for some polynomial $p(n)$, we have $d_L(n) \leq p(n)$ for every $n > 0$. Observe that there exist examples of polynomially ambiguous rational trace languages which are not finitely ambiguous. For instance, fixing $I = \{(a,b),(b,c)\}$, if $L = (a^*c)^*(ab)^*c(a^*c)^*$, then it turns out that $[L]$ does not belong to $\bigcup_{k=1}^{\infty} \mathscr{R}_k$: however, $[L]$ is polynomially ambiguous with $d_L(n) = O(n)$ since, for every x of the form $x = a^{k_1}ca^{k_2}c \ldots (ab)^{k_s}ca^{k_{s+1}}c \ldots a^{k_t}c$ with $k_1, \ldots, k_s, \ldots k_t \in \mathbf{N}$, it holds

$$L \cap [x] = \{a^{k_1}c \ldots (ab)^{k_i}c \ldots a^{k_t}c : k_i = k_s, 1 \leq i \leq t\}.$$

Now, let us go back to our problem: here, we want to use L as an ambiguous description of $[L]$. Then, also in this case, we first have to design two routines: one for generating a word in L uniformly at random and the other for determining the number of representatives of a trace in a given regular language L. The first routine can be easily obtained in the same vein of [14]. The other algorithm is given by adapting a procedure for solving the membership problem for rational trace language [5, 4]. As a consequence, we get the following

Theorem 5. *Let $T \subseteq \mathbb{M}(\Sigma, I)$ be a finitely ambiguous rational trace language and assume $I \neq \emptyset$. Then, T admits a u.r.g. working in $O(n^\alpha \log n)$ time and using $O(n^2 \log n)$ random bits on a* PrRAM *under logarithmic cost criterion, where α is the size of the maximum clique in (Σ, I). Moreover, there exists a r.a.s. for the census function of T of the same time complexity. On the other hand, if T is polynomially ambiguous, then it admits a polynomial time u.r.g. and a polynomial time r.a.s. for its census function.*

References

[1] A. V. Aho, J. E. Hopcroft, and J. D. Ullman. *The Design and Analysis of Computer Algorithms.* Addison-Wesley, Reading, MA, 1974.

[2] A. V. Aho and J. D. Ullman. *The Theory of Parsing, Translation and Compiling - Vol.I: Parsing.* Prentice Hall, Englewood Cliffs, NJ, 1972.

[3] E. Allender, D. Bruschi, and G. Pighizzini. The complexity of computing maximal word functions. *Computational Complexity*, 3:368–391, 1993.

[4] A. Avellone and M. Goldwurm. Analysis of algorithms for the recognition of rational and context-free trace languages. *RAIRO Informatique théorique et Applications/Theoretical Informatics and Applications*, 32(4-5-6):141–152, 1998.

[5] A. Bertoni, M. Goldwurm, G. Mauri, and N. Sabadini. Counting techniques for inclusion, equivalence and membership problems. In V. Diekert and G. Rozenberg, editors, *The Book of Traces*, chapter 5, pages 131–164. World Scientific, Singapore, 1995.

[6] A. Bertoni, M. Goldwurm, and N. Sabadini. The complexity of computing the number of strings of given length in context-free languages. *Theoretical Computer Science*, 86(2):325–342, 1991.

[7] F.-J. Brandenburg. On one-way auxiliary pushdown automata. In H. Waldschmidt H. Tzschach and H. K.-G. Walter, editors, *Proceedings of the 3rd GI Conference on Theoretical Computer Science*, volume 48 of *Lecture Notes in Computer Science*, pages 132–144, Darmstadt, FRG, March 1977. Springer.
[8] C. Choffrut and M. Goldwurm. Rational transductions and complexity of counting problems. *Mathematical Systems Theory*, 28(5):437–450, 1995.
[9] S. A. Cook. Characterizations of pushdown machines in terms of time-bounded computers. *Journal of the ACM*, 18(1):4–18, January 1971.
[10] V. Diekert and Y. Métivier. Partial commutation and traces. In G. Rozenberg and A. Salomaa, editors, *Handbook on Formal Languages*, volume III, pages 457–527. Springer, Berlin-Heidelberg, 1997.
[11] V. Diekert and G. Rozenberg. *The Book of Traces.* World Scientific, Singapore, 1995.
[12] J. Earley. An efficient context-free parsing algorithm. *Communications of the ACM*, 13(2):94–102, February 1970.
[13] P. Flajolet. Mathematical methods in the analysis of algorithms and data structures. In Egon Börger, editor, *Trends in Theoretical Computer Science*, chapter 6, pages 225–304. Computer Science Press, Rockville, Maryland, 1988.
[14] P. Flajolet, P. Zimmerman, and B. Van Cutsem. A calculus for the random generation of labelled combinatorial structures. *Theoretical Computer Science*, 132(1-2):1–35, 1994.
[15] M. Goldwurm. Random generation of words in an algebraic language in linear binary space. *Information Processing Letters*, 54(4):229–233, 1995.
[16] V. Gore, M. Jerrum, S. Kannan, Z. Sweedyk, and S. Mahaney. A quasi-polynomial-time algorithm for sampling words from a context-free language. *Information and Computation*, 134(1):59–74, 10 April 1997.
[17] M. A. Harrison. *Introduction to Formal Language Theory.* Addison-Wesley, Reading, MA, 1978.
[18] T. Hickey and J. Cohen. Uniform random generation of strings in a context-free language. *SIAM Journal on Computing*, 12(4):645–655, nov 1983.
[19] W. Hoeffding. Probability inequalities for sums of bounded random variables. *Journal of the American Statistical Association*, 58:13–30, 1963.
[20] J. E. Hopcroft and J. D. Ullman. *Introduction to Automata Theory, Language, and Computation.* Addison-Wesley, Reading, MA, 1979.
[21] M. R. Jerrum, L. G. Valiant, and V. V. Vazirani. Random generation of combinatorial structures from a uniform distribution. *Theoretical Computer Science*, 43(2-3):169–188, 1986.
[22] R. M. Karp, M. Luby, and N. Madras. Monte-carlo approximation algorithms for enumeration problems. *Journal of Algorithms*, 10:429–448, 1989.
[23] D. E. Knuth and A. C. Yao. The complexity of nonuniform random number generation. In J. F. Traub, editor, *Algorithms and Complexity: New Directions and Recent Results*, pages 357–428. Academic Press, 1976.
[24] C. Lautemann. On pushdown and small tape. In K. Wagener, editor, *Dirk-Siefkes, zum 50. Geburststag (proceedings of a meeting honoring Dirk Siefkes on his fiftieth birthday)*, pages 42–47. Technische Universität Berlin and Universität Ausgburg, 1988.
[25] H. G. Mairson. Generating words in a context-free language uniformly at random. *Information Processing Letters*, 49(2):95–99, January 1994.
[26] M. Santini. *Random Uniform Generation and Approximate Counting of Combinatorial Structures.* PhD thesis, Dipartimento di Scienze dell'Informazione – Università degli Studi di Milano, 1999.

The CNN Problem and Other k-Server Variants

Elias Koutsoupias* and David Scot Taylor**

University of California at Los Angeles, Los Angeles, CA 90095, USA
{elias, dstaylor}@cs.ucla.edu

Abstract. We study several interesting variants of the k-server problem. In the CNN problem, one server services requests in the Euclidean plane. The difference from the k-server problem is that the server does not have to move to a request, but it has only to move to a point that lies in the same horizontal or vertical line with the request. This, for example, models the problem faced by a crew of a Certain News Network trying to shoot scenes on the streets of Manhattan from a distance; the crew has only to be on a matching street *or* avenue. The CNN problem contains as special cases two important problems: the bridge problem, also known as the cow-path problem, and the weighted 2-server problem in which the 2 servers may have different speeds. We show that any deterministic on-line algorithm has competitive ratio at least $6+\sqrt{17}$. We also show that some successful algorithms for the k-server problem fail to be competitive. In particular, we show that no natural lazy memoryless randomized algorithm can be competitive.
The CNN problem also motivates another variant of the k-server problem, in which servers can move simultaneously, and we wish to minimize the *time* spent waiting for service. This is equivalent to the regular k-server problem under the $\mathcal{L}_\infty$ norm for movement costs. We give a $\frac{1}{2}k(k+1)$ upper bound for the competitive ratio on trees.

1 Introduction

Consider a CNN crew trying to shoot scenes in Manhattan. As long as they are on a matching street or avenue, they can zoom in on a scene. If a scene happens to be at an intersection, the crew has two choices: street or avenue. Of course, the crew must make its choice on-line, without knowing where the subsequent scenes will be.

This is an example of an interesting variant of the k-server problem. We can formulate the CNN problem as follows: there is one server in the plane which services a sequence of requests (points of the plane). To service a request $r = (r_1, r_2)$, the server must align itself with the request either horizontally or vertically, i.e, it must move to a point of the vertical line $x = r_1$ or a point of the horizontal line $y = r_2$. The goal is to minimize the total distance traveled by the server. In the on-line version of the problem the requests are revealed progressively.

* Supported in part by NSF grant CCR-9521606
** Supported in part by the Department of Energy under Contract W-7405-ENG-36

H. Reichel and S. Tison (Eds.): STACS 2000, LNCS 1770, pp. 581–592, 2000.

A more interesting formulation of the CNN problem results by assuming that we have 2 servers, moving in different dimensions, independent from each other. This allows us to generalize the problem as follows: there are two metric spaces M_1 and M_2 with one server in each one. A request is a pair of points (x_1, x_2) with $x_i \in M_i$. To service the request, we have to move only one server to the requested point of its space. We will call this problem the *sum* of two 1-server problems. The CNN problem is the special case where both metric spaces M_1 and M_2 are lines.

More generally, let $\tau_1, \tau_2, \ldots, \tau_n$ be task systems [7] (not necessarily distinct). We can synthesize these task systems to get two new interesting on-line problems: the *sum* and the *product* of $\tau_1, \tau_2, \ldots$. The sum is the problem where we get requests (tasks) for each task system and we have to service only one of them. The product, on the other hand, is the problem where we have to service all of them. When all task systems are identical, $\tau_i \equiv \tau$, the product is related to randomized on-line algorithms for the task system τ. It is a trivial fact that a deterministic algorithm for the product of n copies of τ, with each request the same across all spaces, is not different than a randomized algorithm against an oblivious adversary with exactly n (equiprobable) random choices; these algorithms are called barely random, or mixed strategies, in the literature [6].

The CNN problem belongs to the class of sum problems. In fact, it is one of the simplest conceivable sum problems and a stepping stone towards building a robust (and less ad hoc) theory of on-line computation. Despite its importance, there is no definite positive result for this problem. The lack of results should be attributed to the hardness of the problem rather to the lack of interest. The problem has been known to the research community for quite a few years[1]. The CNN problem and more generally the sum of on-line problems also give flexibility to model problems which the k-server problem cannot. For instance, while the k-server problem has been used to model the behavior of multiple heads on a disk, the CNN problem can be used to model retrieving information which resides on multiple disks. This, for example, happens when we replicate data to achieve higher performance or fault tolerance [1,4,16,20,21]. Each disk may have information in completely different locations, leading to independent costs for information retrieval. We wish to minimize time spent looking for data, but do not actually care which disk the information comes from. In contrast, writing must be performed to all disks; this is closer in spirit to the product on-line problem mentioned above, but in our worst case analysis, there will be little or no writing at all.

In this work, we use *competitive analysis* [5,17,24] to evaluate the quality of on-line algorithms; the competitive ratio is defined to be the worst-case performance of the on-line algorithm, compared to the optimal cost for the same sequence of requests. More precisely, algorithm ALG is c-competitive if there is a constant α such that over any finite input sequence ρ, $\text{ALG}(\rho) \leq c \cdot \text{OPT}(\rho) + \alpha$, where $\text{OPT}(\rho)$ is the optimal cost for ρ. The game-theoretic structure of the

[1] The CNN problem was originally proposed by Mike Saks and William Burley, who obtained some initial results. The name (CNN) was suggested by Gerhard Woeginger.

competitive ratio suggests considering the on-line algorithm as a strategy that competes against an optimal "adversary", who selects the requests and services them too.

In this work, we show some negative results (lower bounds and failed attempts to "import" the k-server theory to this problem). There are no known upper bounds. Compare this with the k-server problem: although the k-conjecture has not been resolved yet, we now know the competitive ratio within a factor of 2 [18]. In particular, the 2-server problem was settled from the beginning [19]. The CNN problem seems very similar to the 2-server problem. Yet, almost all known competitive algorithms for the k-server problem fail for the CNN problem.

We prove that there is no competitive natural lazy memoryless randomized algorithm, (Theorem 1) for the CNN problem. We also observe that if we restrict the requests to a line, the problem is equivalent to the *weighted* 2-server problem [13] in a line. The weighted k-server problem is the variant of the standard k-server problem in which servers have different speeds and the cost is proportional to the time needed to service a request. We show (Theorem 2) that any deterministic on-line algorithm has a competitive ratio at least $6+\sqrt{17}$ for the weighted server problem in a line (and thus for its generalization, the CNN problem). This lower bound holds when one server is arbitrarily faster than the other. We also show that some obvious candidate algorithms are not competitive for the CNN problem (Proposition 1). Some of the results extend directly to the CNN problem in higher dimensions (the sum of more 1-server problems), in which the lower bounds of [13] also apply. It is easy to show that for the sum of any k non-trivial spaces (at least 2 points each) a $2^k - 1$ lower bound on the ratio exists, and that for the simplest spaces (k spaces of 2 points each, 1 unit apart), this bound is tight.

Finally, we study a variant of the k-server problem, motivated by the CNN problem. Instead of trying to minimize the cost of moving servers, we try to minimize the *time* spent waiting for service, but we allow multiple servers to move simultaneously [12]. When a request is made, the on-line algorithm specifies possible movement for each of the k-servers, and tries to minimize the cost of their total movement, under the $\mathcal{L}_\infty$ norm. For k-servers in a tree, we determine the exact ratio $\frac{1}{2}k(k+1)$ of the DC-TREE algorithm of [8,9]. In particular for $k=2$, we show that DC-TREE is optimal with competitive ratio 3.

2 Memoryless Randomized Algorithms

Our first result concerns randomized algorithms for the CNN problem. Within the plane, we define a *natural* memoryless algorithm to be an algorithm which has the following property: *The probability of satisfying any request by moving the shorter distance is only a function of the ratio of the shorter distance to the longer distance.* We break this into several subproperties:

- Translational Invariance: The algorithm does not consider what point its server is at, only relational positions of points within the space (such as the distances to the request).

- Symmetry 1: The algorithm is unbiased towards horizontal and vertical moves. More precisely, let (x, y) be the on-line position and consider two possible next requests at $(x+a, y+b)$ and $(x+b, y+a)$. If the server moves horizontally with probability p to service the first request, it moves vertically with the same probability p to service the second request. Similarly, request $(x-a, y+b)$ also moves horizontally with probability p.
- Scalability 1: Let (x, y) be the on-line position and consider two possible next requests at $(x+a, y+b)$ and $(x+\lambda a, y+\lambda b)$, for some λ. The server moves horizontally with the same probability for both requests.

We call these properties natural because they reflect the fact that for this problem, the plane is symmetric over $y = x$, $y = 0$, and $x = 0$, and is scale and translation invariant. We also require that the algorithm be *lazy*: if the algorithm is at position (x, y), and the request is at (r_x, r_y), it will service the request by moving to either (r_x, y) or (x, r_y). It is standard to require memoryless algorithms to be lazy, otherwise the on-line algorithm can try to encode history information within its non-lazy movement. Although this encoding is impossible for natural algorithms, we nevertheless require the laziness property.

Unlike the k-server problem for which a natural lazy memoryless algorithm, HARMONIC, has finite competitive ratio ($O(2^k \log k)$, [3,15]), no such algorithm for the CNN problem is competitive.

Theorem 1. *There is no competitive natural lazy memoryless randomized algorithm for the* CNN *problem.*

Proof. Fix some memoryless algorithm with optimal competitive ratio. Assume that the adversary is at position $(0, 0)$ (by Translational Invariance). We want to estimate the on-line cost when the adversary produces a worst-case sequence of *lazy requests*, defined as requests for which it does not have to move, i.e., points from the lines $x = 0$ and $y = 0$. Let $\Phi(x, y)$ be the expected cost of the on-line algorithm to converge to $(0, 0)$ starting at position (x, y).

It follows for the above properties of the on-line algorithm that Φ satisfies:

- Symmetry 2: $\Phi(x, y) = \Phi(|y|, |x|)$.
- Scalability 2: $\Phi(\lambda \cdot x, \lambda \cdot y) = \lambda \cdot \Phi(x, y)$.

To prove our theorem, we examine various possible combinations of on-line positions and requests. We start by considering $\Phi(1, \varphi)$, where $\varphi = \frac{1+\sqrt{5}}{2}$, the golden ratio. We try to find recursive bounds on Φ. The following table summarizes combinations on-line positions, requests, and resulting bounds on Φ.

On-Line Position	Next Request	Resulting Inequality	Justification
$(1,\varphi)$	$(\varphi^2,0)$	$\Phi(1,\varphi)$ $\geq \frac{1}{2}(\Phi(\varphi^2,\varphi)+\varphi^2-1)+\frac{1}{2}(\Phi(1,0)+\varphi)$ $=\frac{1}{2}(\varphi\cdot\Phi(\varphi,1)+\varphi)+\frac{1}{2}(\Phi(1,0)+\varphi)$ $=\frac{\varphi}{2}\Phi(1,\varphi)+\frac{1}{2}\Phi(1,0)+\varphi$	Symmetry 1, $\varphi^2-1=\varphi$ Scalability 2, $\varphi^2-1=\varphi$ Symmetry 2
$(1,0)$	$(0,1)$	$\Phi(1,0)$ $\geq \frac{1}{2}(\Phi(1,1)+1)+\frac{1}{2}(\Phi(0,0)+1)$ $=\frac{1}{2}\Phi(1,1)+1$	Symmetry 1
$(1,1)$	$(0,\varphi)$	$\Phi(1,1)$ $\geq p(\Phi(1,\varphi)+\varphi-1)+(1-p)(\Phi(0,1)+1)$ $=p(\Phi(1,\varphi)+\varphi-1)+(1-p)(\Phi(1,0)+1)$	For some p. Symmetry 2
$(1,0)$	$(0,\varphi)$	$\Phi(1,0)$ $\geq p(\Phi(0,0)+1)+(1-p)(\Phi(1,\varphi)+\varphi)$ $=(1-p)\Phi(1,\varphi)+p+(1-p)\varphi$	Symmetry 1, Scalability 1

If we assume that $\Phi(1,\varphi)$ is bounded we get a contradiction: from the first three bounds we get $p<\varphi-1$ and from the first and last $p>\varphi-1$. Therefore, if the initial position is $(0,0)$ and the adversary moves to position $(1,\varphi)$, the on-line cost to converge to $(1,\varphi)$ is unbounded. This shows that the competitive ratio is unbounded. □

3 Server Problems with Different Speeds

We now turn our attention to the restricted CNN problem where all requests are from a line (the server is still allowed to move anywhere in the plane). A lower bound for the restricted problem is naturally a lower bound for the unrestricted one. Without loss of generality, we assume that the line is of the form $y=mx$, for some constant m. For $m=1$, the problem is equivalent to the standard 2-server problem in a line, where moving in the x dimension corresponds to moving one of the servers, and the y dimension the other. Changing m gives a more interesting problem: if all requests are restricted to the form (x, mx), it corresponds to a request in a line at x for a 2-server problem, but this time the servers have different costs for movement. Loosely, this can be interpreted as having servers with different speeds[2] and we wish to minimize the total delay, i.e., the time requests wait for service.

In general, the restriction of the multidimensional CNN problem where all requests are from a line is equivalent to the *weighted k-server problem* in a line. The general weighted server problem was studied in [13]. This work gives a lower bound of $k^{\Omega(k)}$ for any metric space with at least $k+1$ points (and arbitrary speeds). No upper bound is known for arbitrary metric spaces, but [13] gives a doubly exponential upper bound $(2^{2^{O(k)}})$ for uniform metric spaces; this is reduced to exponential $(k^{O(k)})$ when the servers have at most 2 different speeds.

For the restricted version of the CNN problem, the weighted 2-server problem in a line, we show a deterministic lower bound of $6+\sqrt{17}$. The surprising fact exploited in the proof is that the adversary can "simulate" the BRIDGE (or cow-path) problem [2,23]. Thus, the CNN problem contains as a subproblem another fundamental on-line problem. Interestingly, we know of two different ways to view the bridge problem as a special case of the CNN problem. This shows the close connection between these two problems, although the CNN problem is, of course, much more complicated.

The BRIDGE problem is a simple on-line problem, in which an explorer comes to a river. There is a bridge across the river, but it is not known how far away it is, or if it is upstream or downstream. The explorer must try to find the bridge while minimizing movement. The optimal solution involves alternating between the upstream and downstream directions, exploring 1 distance unit downstream, then 2 units (from the original starting position) upstream, then 4 downstream, and continuing in powers of 2 until the bridge is found. This strategy results in total movement no more than 9 times the distance from the original position to the bridge (plus a constant).

Theorem 2. *For the 2-server problem in a line, in which one server is m times faster than the other, the deterministic competitive ratio is at least $6+\sqrt{17}$. More precisely, for any $\epsilon > 0$, there is a sufficiently large m such that the competitive ratio is at least $6+\sqrt{17}-\epsilon$.*

Proof. We first show a weaker lower bound of 9 to exhibit the relation between the CNN problem and the BRIDGE problem. The role of the explorer will be played by the slow server.

[2] To make this interpretation strict, we consider that we are only allowed to move one server at a time, which is natural for our problem motivation (we can only move along streets or avenues). It is also worth considering this question without this restriction, which corresponds to using the Chebychev distance ($\mathcal{L}_\infty$) instead of the Manhattan distance ($\mathcal{L}_1$). The k-server variant of this problem, where servers are allowed to move simultaneously, was introduced in [12] as the min-time server problem. We study this in Section 4.

We shall assume $m >> 1$. Also, without loss of generality, we can assume that the on-line algorithm is lazy (moves a server only to service requests). Let $[l, r]$ be the interval of the line explored (visited so far) by the slow server. Initially, it is safe to assume that the slow server is at $r = 0$ and the fast server is at $l = -1$. When the on-line slow server is at r the adversary's strategy distinguishes two cases: if the fast server is to the right of r, the next request is at l; if the fast server is to the left of r, the next request is at $r + \delta$, where δ is an arbitrarily small positive distance. The adversary's strategy when the slow server is at l is symmetric.

This adversary's strategy forces the slow server to explore larger and larger portion of the line. The slow server at r cannot continually move to the right, because this would result in competitive ratio m. Eventually, to achieve ratio less than m, the slow server will need to go to point l. The adversary can continue to force the slow server to "zig-zag", exploring larger and larger segments of the line; exactly as the explorer does in the BRIDGE problem.

Assume that at the end of the game, the slow server has explored the interval $[-z, y]$, for $z \geq y$, and has just moved from $-z$ to y. Since the competitive ratio of the bridge problem is 9, the total distance moved by the slow server must be at least $9y$ (minus an insignificant term). On the other hand, the adversary can service all requests by moving its slow server to y and its fast server to $-z$. Its total cost is $y + z/m$, which for large m is approximately y. (To be fair, m should be fixed before the on-line algorithm is forced to choose z values, but we can always guarantee that z is no more than some modest constant multiple of y or else we again reach ratio m.)

Thus, the competitive ratio is at least $9 - \epsilon$, where ϵ tends to 0 as m tends to ∞. To simplify the presentation, we henceforth will drop the ϵ term.

It takes a little effort to improve the bound to 10. Observe that we ignored the cost of the on-line fast server. Let x_0 be this cost. The adversary has the alternative strategy to service all requests with its fast server. Its total cost then is x_0. Thus, the competitive ratio is at least

$$\max_{x_0, y}\left(\frac{9y + x_0}{x_0}, \frac{9y + x_0}{y}\right).$$

The minimum value, 10, is clearly obtained when $x_0 = y$.

To get the improved lower bound $6 + \sqrt{17}$, we extend the above sequence of requests. Let ρ be the sequence of requests described above (which completes the "BRIDGE simulation") and consider the request sequence $\rho((0, -z)^{j_i}(y, -z)^{k_i})^n$ where i varies from 1 to n, and j_i (k_i) refers to the number of times the first (second) pair are repeated during the ith repetition of the whole phrase. Let j_i and k_i, be determined as follows: after servicing ρ, the on-line slow server is at y. To service $(0, -z)^{j_1}$, it may use the fast server for a while, but eventually, it must move its slow server to 0. Let j_i be the number of repetitions of $(0, -z)$ needed to make the on-line slow server move to 0 for the ith time, and k_i be the number of repetitions of $(y, -z)$ needed to make the on-line slow server move back to y for the ith time. Let $x_{2i-1} = j_i z/m$, the movement of the on-line fast server before the ith move to 0, and $x_{2i} = k_i z/m$, the movement of the on-line fast server before the ith move back to position y.

The adversary might end the game after any of the slow server moves. The first few terms of the ratio R are:

$$\max_{x_i,y}\left(\frac{9y+x_0}{y}, \frac{10y+x_0+x_1}{x_0}, \frac{11y+x_0+x_1+x_2}{y+x_1}, \frac{12y+x_0+x_1+x_2+x_3}{x_0+x_2}, \ldots\right),$$

and in general:

$$R \geq \max_{x_i,y}\left(\frac{(9+2j)y+\sum_{i=0}^{2j} x_i}{y+\sum_{i=1}^{j} x_{2i-1}}, \frac{(10+2j)y+\sum_{i=0}^{2j+1} x_i}{\sum_{i=0}^{j} x_{2i}}\right).$$

The different denominator types are from the two off-line strategies we have already seen for the ρ requests: moving both servers, or just the fast one. The adversary can choose either of these two[3].

It turns out that to minimize the expression above, we can assume that all values are equal, and the equations simplify to:

$$\frac{x_0+x_1+10y}{x_0} = \frac{x_0+9y}{y},$$

$$\forall i \geq 2, \quad \frac{x_{i-1}+x_i+2y}{x_{i-1}} = \frac{x_0+9y}{y}.$$

Scaling y to be 1, and solving, we get:

$$x_1 = x_0^2 + 8x_0 - 10$$

$$\forall i \geq 2, \; x_i = (x_0+8)^{i-1}x_1 - 2\sum_{j=0}^{i-2}(x_0+8)^j$$

$$\forall i \geq 2, \; x_i = (x_0+8)^{i-1}x_1 - 2\frac{(x_0+8)^{i-1}}{x_0+7} + \frac{2}{x_0+7}$$

and finally

$$\forall i \geq 2, \; x_i = (x_0+8)^{i-1}\left(x_0^2+8x_0-10-\frac{2}{x_0+7}\right) + \frac{2}{x_0+7}.$$

All x_i values must be positive, so the smallest possible value for the equations is when $x_0^2 + 8x_0 - 10 - \frac{2}{x_0+7} = 0$. The only positive root is $x_0 = \sqrt{17} - 3$, which gives the stated bound of $6 + \sqrt{17}$. □

[3] In fact, the off-line servers have a third main strategy to consider, which is moving the slow server to point z and servicing the rest of the queries with the fast server. By eliminating this option, we may be proving a slightly *weaker* lower bound than possible from these three strategies; we can show however that the third strategy does not improve the lower bound by more than 0.22. The equations are complicated by the fact that once the third option is added, it is no longer safe to assume that the optimal BRIDGE solution for ρ will give the best solution to the equations.

To see just how difficult it is to find a competitive ratio for the CNN problem, we notice that some simple algorithms which are competitive for the 2-server problem are not competitive for the case when the servers have different movement costs. The "double coverage" (DC) algorithm in a line is the following simple algorithm: if the request is between the two servers, move both towards it until the request is served. Otherwise, move the closer server (ties broken arbitrarily.) The "balance", or BAL algorithm is also simple: to answer any query, move the server which will have the minimum cumulative cost if it moves to the request. More general balance algorithms base their decision to move a server to a request on two parameters: the cumulative cost of a server and the distance to the request.

Proposition 1. *For 2-servers in a line, one with speed 1 and the other m, no DC or BAL type algorithm has constant competitive ratio (bounded independently of m).*

Obviously these algorithms need to be modified to account for the different speeds of the servers: for instance, in DC it must be possible for the fast server to "pass" the slow server for requests outside of their convex hull, or else it is trivial to achieve competitive ratio at least equal to m. We expect that the statement holds for any variant which maintains the spirit of either algorithm, and use a simple example as an intuitive justification. For DC, consider an on-line configuration at $(0, x)$, where the fast server is at 0, and x large enough that the slow server will reach a request at $x+1$ before the fast one ($(x+1)/m > 1$ will do for the most natural generalization of DC) By repeating the sequence of requests $0, x, 0, x+1$, the on-line servers will pay cost 2 for every 4 requests, while an adversary could satisfy them with just cost $2/m$ (assume $m \geq 1$). By making m large, we can get an arbitrarily large competitive ratio. A similar example can be used for BAL.

Leaving our k-server interpretation behind, we can achieve the same results for the CNN problem by considering requests which lie in the two lines $y = 0$ and $y = 1$. For this special case of the CNN problem, a slightly weaker lower bound was obtained by William Burley and the first author. Now any request can be satisfied at a cost of 1 by moving from one line to the other. We use a strategy similar to before. Suppose the leftmost and rightmost positions of our server so far have been $[l, r]$. If the server is on $y = 0$, place a request at $(l-1, 1)$, and if the server is on $y = 1$, place a request at $(0, r+1)$. We use these requests to again make the server move away from the center just as before, simulating the BRIDGE explorer, and vertical movement corresponds to movement of the fast server in the previous argument. At the end, the adversary can again make the server alternate between the origin and the shorter "arm" of exploration. All equations are the same.

4 The k-Server Problem under the $\mathcal{L}_\infty$ Norm

In this section, we consider the k-server problem, where servers have the same speed, but can be moved simultaneously, and *time* of service is to be minimized.

Once a request is served, the next request is given. By ignoring the fact that our algorithm (and the optimal one) can move more than one server simultaneously, we can achieve a competitive ratio for this problem which is k times larger than for the regular k-server problem. Using the best known bound of [18], this gives a $2k^2 - k$ bound for this ratio, but we expect that this can be improved. For example, within the uniform metric space, moving the servers in order will achieve the (optimal) ratio of k.

We show that in a tree, the competitive ratio is no worse than $\frac{1}{2}k(k+1)$. We employ the DC-TREE algorithm of [9], generalized from DC of [8]. The algorithm is defined as follows: move each of the servers with an unblocked (by other servers) path to the request towards the request at a constant speed. Note that servers may begin moving, and later stop moving as they become blocked by other servers which move onto their path.

Theorem 3. *For k servers in a tree under the $\mathcal{L}_\infty$ norm,* DC-TREE *has competitive ratio $\frac{1}{2}k(k+1)$.*

Proof. Let M_{min} be the distance for the best matching between the on-line and off-line servers, and ΣDC-TREE be the distance between all pairs of on-line servers. To show that DC-TREE is competitive, consider some phase in which a fixed number of servers are moving, and use the following potential:

$$\Phi = \frac{(k+1)M_{min}}{2} + \frac{1}{2}\Sigma\text{DC-TREE}.$$

The analysis is as in [9]. For a move of cost d, the off-line server can increase M_{min} by kd, giving a ratio of $\frac{1}{2}k(k+1)$.

To show this ratio is tight for DC-TREE, consider servers (on-line and off-line) at position $(2, 4, \ldots, 2k)$ of a line. For a cost of 1, the off-line algorithm can move all of its servers to $(1, 3, \ldots, 2k-1)$. The adversary is lazy, and will at each time request its uncovered server which is at the lowest value (i.e., the sequence request will be $1, 3, 1, 5, 3, 1, 7, 5, 3, 1, \ldots$). Each request will cost DC-TREE 1 to serve, and it will take $\frac{1}{2}k(k+1)$ total requests to converge to the off-line position, at which time we are at a position similar to the original one. □

The following lemma shows that the DC-TREE algorithm is optimal for $k = 2$ and Euclidean metric spaces.

Lemma 1. *No on-line algorithm has competitive ratio less than 3 for the Euclidean 2-server problem under the L_∞ norm.*

Proof. It suffices to consider lazy on-line algorithms. Consider requests in a line, with initial configuration (on-line and off-line) $\{0, 2\}$. If there is a request at 1, by symmetry, the on-line algorithm can service it by moving the server at 2, and it can move the other server to any point in the interval $[-1, 1]$ at no extra cost. Now the adversary requests point 3 (and reveals its configuration $\{1, 3\}$). The on-line algorithm must pay at least 2 to service the request. The total on-line cost is 3, but the off-line cost is 1 (move the 2 servers together from $\{0, 2\}$ to $\{1, 3\}$). This configuration is similar to the initial one, and the situation can be repeated indefinitely. Therefore, no on-line algorithm has ratio less than 3. □

In [11], the DC-TREE algorithm for $k = 2$ can be extended to any metric space, not just trees. This is done by considering an extension of the metric space to its "closure" so that any three points are connected by a tree. Consider "virtual" movement through the closure of the space, and only actually move a server when it needs to service a request, which will be a point in the original space. By the triangle inequality, the actual movement will be less than the total virtual movement. Unfortunately, under the $\mathcal{L}_\infty$ norm, this strategy does not work: except for the last move (to the original space), the virtual movement is for free, with the cost for the move being dominated by the movement of the other server. When it comes time to move the server to answer a request, this virtual movement is no longer available for free. Triangle inequality can no longer guarantee that the actual movement is less than the sum of virtual movement.

5 Conclusions and Future Work

We have introduced several interesting variants of the k-server problem, with the power to model new problems. There are numerous open problems left. We mention only few of them here.

The CNN problem is wide open. We conjecture that it has a finite competitive ratio. In fact, we conjecture that the generalized Work Function Algorithm (the one that moves a server which minimizes $\lambda w(A') + d(A, A')$) has constant competitive ratio for the CNN problem for any $\lambda > 1$ ($\lambda = 3$ seems a good candidate); we conjecture the same for the sum of two 1-server problems in general.

The off-line CNN problem seems interesting in its own right and as a stepping stone for the on-line problem. More precisely, we want to find simple and fast approximation algorithms for the off-line CNN problem.

For the Euclidean k-server problem under the $\mathcal{L}_\infty$ norm, the intuition behind DC-TREE suggests that there be an algorithm with ratio better than 4 (which follows from the fact that 2 servers are 2 competitive), though it must be at least 3 (by Lemma 1).

We believe that for all three problems (sum of server problems, weighted k-server, $\mathcal{L}_\infty$ variant of the k-server problem) the Work Function Algorithm has almost optimal competitive ratio. We don't have a candidate randomized algorithm though.

References

1. Y. Azar, A. Z. Broder, A. R. Karlin, and E. Upfal. Balanced allocations. In *Proc. 26th Annual ACM Symposium on the Theory of Computing, (STOC '94)*, pages 593–602, 1994.
2. R. A. Baeza-Yates, J. C. Culberson, and G. J. E. Rawlins. Searching with uncertainty. In *1st Scandinavian Workshop on Algorithm Theory, (SWAT 88)* pages 176–189. Springer-Verlag, Lecture Notes in Computer Science 318, 1988.
3. Y. Bartal and E. F. Grove. The harmonic k-server algorithm is competitive. Unpublished Manuscript, 1994.

4. D. Bitton and J. Gray. Disk Shadowing. In *Proceedings of the 14th International Conference on Very Large Data Bases, (VLDB '88)*, pages 331–338, 1988.
5. J. L. Bentley and C. C. McGeoch. Amortized analyses of self-organizing sequential search heuristics. In *Communications of the ACM, 28(4)* pages 404–411, April 1985.
6. A. Borodin and R. El-Yaniv. *Online Computation and Competitive Analysis*, Cambridge University Press, Cambridge Mass., 1998.
7. A. Borodin, N. Linial, and M. Saks. An optimal online algorithm for metrical task systems. In *Proceedings of the 19th Annual ACM Symposium on Theory of Computing, (STOC '87)* , pages 373–382, 1987.
8. M. Chrobak, H. Karloff, T. Payne, and S. Vishwanathan. New results on server problems. In *SIAM Journal on Discrete Mathematics*, 4(3):323–328, May 1991.
9. M. Chrobak and L. Larmore. An optimal on-line algorithm for k-servers on trees. In *SIAM Journal on Computing*, 20(1):144–148, February 1991.
10. M. Chrobak and L. Larmore. The server problem and on-line games. In *On-Line Algorithms, DIMACS Series in Discrete Mathematics and Theoretical Computer Science*, vol. 7, pages 11–64, 1991.
11. M. Chrobak and L. Larmore. A new approach to the server problem. In *SIAM Journal on Discrete Mathematics*, 4(3):323–328, 1991.
12. A. Fiat, Y. Rabini, and Y. Ravid. Competitive k-server algorithms. In *31st IEEE Annual Symposium on Foundations of Computer Science, (FOCS '90)*, pages 454–463, October 1990.
13. A. Fiat and M. Ricklin. Competitive algorithms for the weighted server problem. In *Theoretical Computer Science, 130(1)* pages 85–99, August 1994.
14. S. Gal. *Search Games*, Academic Press, 1980.
15. E. F. Grove. The harmonic online k-server algorithm is competitive. In *Proc. 23rd Symposium on Theory of Computing, (STOC '91)*, pages 260–266, 1991.
16. F. Kurzweil. Small disk arrays - the emerging approach to high performance. Presentation at COMPCON 88, March 1, 1988.
17. A. R. Karlin, M. S. Manasse, L. Rudolph, and D. D. Sleator. Competitive snoopy caching. In *27th Annual Symposium on Foundations of Computer Science, (FOCS '86)*, pages 244–254, October 1986.
18. E. Koutsoupias and C. Papadimitriou. On the k-server conjecture. In *Proc. 26th Symposium on Theory of Computing, (STOC '94)*, pages 507–511, 1994.
19. M. S. Manasse, L. A. McGeoch, and D. D. Sleator. Competitive algorithms for on-line problems. In *Proceedings of the 20th Annual ACM Symposium on Theory of Computing, (STOC '88)*, pages 322–333, May 1988.
20. R. Muntz, J. R. Santos, and S. Berson. A parallel disk storage system for real-time multimedia applications. In *International Journal of Intelligent Systems, Special Issue on Multimedia Computing System*, v.13, n.12, pages 1137–74, December 1998.
21. D. Patterson, G. Gibson, and R. Katz. A case for redundant arrays of inexpensive disks (RAID). In *ACM SIGMOD Conference Proceedings*, pages 109–116, 1987.
22. C. Papadimitriou and M. Yannakakis. Shortest Paths Without a Map. In *Proc. 16th Internat. Colloq. Automata Lang. Program.*, vol. 372 of *Lecture Notes in Computer Science*, pages 284–296. Springer-Verlag, 1989.
23. C. Papadimitriou and M. Yannakakis. Shortest Paths Without a Map. In *Theoretical Computer Science, 84(1):127–150*, 1991.
24. D. D. Sleator and R. E. Tarjan. Amortized efficiency of list update and paging rules. In *Communications of the ACM, 28(2):202–208*, February 1985.

The Weighted 2-Server Problem

Marek Chrobak[1] and Jiří Sgall[2]

[1] Department of Computer Science, University of California, Riverside, CA 92521.
marek@cs.ucr.edu

[2] Mathematical Inst., AS CR, Žitná 25, CZ-11567 Praha 1, Czech Republic and Dept. of Applied Mathematics, Faculty of Mathematics and Physics, Charles Univ., Praha.
sgall@math.cas.cz, http://www.math.cas.cz/~sgall

Abstract. We consider a generalized 2-server problem in which servers have different costs. We prove that, in uniform spaces, a version of the Work Function Algorithm is 5-competitive, and that no better ratio is possible. We also give a 5-competitive randomized, memoryless algorithm for uniform spaces, and a matching lower bound. For arbitrary metric spaces, we prove that no memoryless randomized algorithm has a constant competitive ratio. We study a subproblem in which a request specifies two points to be covered by the servers, and the algorithm decides which server to move to which point; we give a 9-competitive deterministic algorithm for any metric space (no better ratio is possible).

1 Introduction

In the *weighted k-server problem* we are given a metric space M with k mobile servers. Each server s_i has a given weight $\beta_i > 0$. At each step a request $r \in M$ is issued. In response, one of the servers moves to r, at a cost equal to its weight times the distance from its current location to r. This is an *online problem*, in the sense that it is required that the algorithm decides which server to move to r before the next request is issued. An online algorithm $\mathcal{A}$ is *R-competitive* if, for each request sequence ϱ, the cost incurred by $\mathcal{A}$ is at most R times the optimal service cost for ϱ, plus an additive constant independent of ϱ. The *competitive ratio of* $\mathcal{A}$ is the smallest R for which $\mathcal{A}$ is R-competitive.

The unweighted case, with all $\beta_i = 1$, has been extensively studied during the last decade. The problem was introduced by Manasse, McGeoch and Sleator [21], who gave a 2-competitive algorithm for $k = 2$ and proved that k is a lower bound on the competitive ratio of deterministic algorithms in any metric space. They also conjectured that there exists an k-competitive deterministic algorithm for any metric space; this is now called the *k-Server Conjecture*. For $k \geq 3$, the best known upper bound is $2k-1$, by Koutsoupias and Papadimitriou [18]. An upper bound of k has been proven only in some special cases, including uniform spaces (i.e., spaces with all distances equal to 1), see [9,11,19].

Very little is known about randomized algorithms for k servers. No algorithm with ratio less than k is known for arbitrary spaces. In uniform spaces, the competitive ratio is $H_k \approx \ln k$, the k-th harmonic number [1,22]. For $k = 2$, when

H. Reichel and S. Tison (Eds.): STACS 2000, LNCS 1770, pp. 593–604, 2000.

the metric space is the line, Bartal *et al* [4] give a 1.987-competitive algorithm. The best known lower bound for 2 servers is $1+e^{-1/2} \approx 1.6065$ [14]. Other lower and upper bounds for this problem can be found in [3,7].

The weighted case of the k-server problem turns out to be more difficult. Fiat and Ricklin [17] show that the competitive ratio is at least $k^{\Omega(k)}$ in any space with at least $k+1$ points. They also give a doubly-exponential upper bound for uniform spaces. For $k = 2$, Koutsoupias and Taylor [20], recently proved that no 10.12-competitive algorithm exists if the underlying metric space is the line. For uniform spaces, Feuerstein *et al* [16] gave an 6.275-competitive algorithm.

Our results. We study the case when $k = 2$. In uniform spaces, we improve the upper bound from [16] by proving that a version of the Work Function Algorithm is 5-competitive, and we show that no better ratio is possible. We also give a 5-competitive memoryless randomized algorithm for uniform spaces, as well as a matching lower bound.

For arbitrary spaces, we prove that there is no memoryless randomized algorithm with finite competitive ratio. (A similar result was independently obtained by Koutsoupias and Taylor [20].) This contrasts with the non-weighted case, for which memoryless algorithms exist for any k. For example, the harmonic algorithm is competitive for any k [5,8] and, for $k = 2$, a memoryless 2-competitive algorithm is known [10,15].

Last, we propose a version of the problem in which a request is specified by *two points*, both of which must be covered by the servers, and the algorithm must decide which server to move to which point. For this version, we show a 9-competitive algorithm and we prove that no better ratio is possible. This generalizes the results for the 2-point 1-server request problem [12], as well as for the cow-path problem [2,23].

Adversary arguments and potential functions. We view the computation as a game between two players, the algorithm and the adversary. In each round, the adversary issues a request, serves it using its servers, and then the algorithm serves the request (without knowing the position of the adversary servers). A potential function Φ assigns a real number to the current state. To serve our purpose, Φ must satisfy the following three properties: (i) Φ is bounded from below by some constant, (ii) if the adversary serves the request at cost d, then Φ increases by at most Rd, and (iii) if the algorithm serves the request at cost d, then Φ decreases by at least d. By summing over all requests, it follows that the algorithm is R-competitive. Intuitively, one can think of Φ as the credits that the algorithm has saved in the past and can use to pay for serving future requests.

For randomized algorithms, property (iii) needs to holds on average, over the random choices of the algorithm. There are subtle differences between different adversary models, see [6]. We use two models. The weaker, *oblivious* adversary, has to generate the whole request sequence in advance. The stronger, *adaptive online* adversary generates and serves the requests one by one, with the knowledge of the current positions of the algorithm's servers.

One useful principle for designing potential functions is that of the *lazy potential.* If the adversary continues to request the position of his servers, the potential function has to provide enough credit to pay for all moves before the algorithm converges to the same server positions. Although the potential functions in our paper are not lazy, they are based on a similar idea, namely they are lower bounded by the cost of the algorithm in case the adversary will not move the expensive server. Formulas obtained by these methods are complicated, not easy to analyze, and not even guaranteed to give the best results. The potential functions we actually use are then simplified and improved by trial and error (in some cases also using computer experiments).

Notation. Throughout the paper, without loss of generality, we assume that $\beta_1 = 1$ and $\beta_2 = \beta \leq 1$. Thus s_1 and s_2 denote the expensive and the cheap server of the algorithm, respectively. Similarly, by a_1 and a_2 we denote the expensive and the cheap server of the adversary.

For any points x, y in the given metric space, xy denotes their distance. A metric space is *uniform* if the distance of any two distinct points is equal to 1.

2 Randomized Memoryless Algorithms

In this section we consider randomized memoryless algorithms. Our model of a memoryless randomized algorithm is this: A memoryless algorithm is simply a function that receives on input the distances from each server to the request point r and the distance between the servers, and determines, for each i, the probability that r is served with s_i. The algorithm only moves one server, and only to the request point. This is a natural requirement since, in certain spaces, it may be possible to encode memory states by perturbing the other server position.

First we give a 5-competitive algorithm for uniform spaces and prove that it is optimal. The lower bound holds even for the weaker oblivious adversary, while the upper bound is valid against the stronger adaptive online adversary. For general spaces, we prove that no memoryless algorithm can achieve a finite competitive ratio, even if the underlying metric spaces is the line.

Both lower bounds are based on the following observation. Suppose that initially s_1 is at point a and s_2 at c, and the adversary alternates requests to b and c, that have the same distance from a. Then the probability that s_1 ends at c is at most $1/2$. The reason is that as long as s_1 stays on a, the situation remains identical from a viewpoint of a memoryless algorithm. The first request is on b so s_1 it is more likely to end up at b than at c.

2.1 An Upper Bound for Uniform Spaces

On average, our algorithm moves the expensive server after paying approximately $3/2$ for the moves of the cheap server. More precisely, it is defined as follows:

Algorithm RANDOM: If the request is on s_1 or s_2, do nothing. Otherwise serve the request by s_1 with probability $p = 2\beta/(3 + \beta)$ and by s_2 otherwise.

Theorem 1. *Algorithm RANDOM for the weighted 2-server problem in uniform metric spaces is* $(5 - \beta)$*-competitive against an adaptive online adversary.*

Proof. Consider the algorithm described above. Define a potential function Φ:

$$\Phi(s_1, s_2, a_1, a_2) = \begin{cases} 0 & \text{if } a_1 = s_1 \text{ and } a_2 = s_2, \\ 5\beta - \beta^2 & \text{if } a_1 = s_1 \text{ and } a_2 \neq s_2, \\ 5 - \beta & \text{if } a_1 \neq s_1 \text{ and } a_2 = s_2, \text{ and} \\ 5 + 2\beta/(3 - \beta) & \text{if } a_1 \neq s_1 \text{ and } a_2 \neq s_2. \end{cases}$$

If the adversary moves, it is easy to check that, in each case, the potential increases by at most $(5-\beta)$ times the adversary cost. So it remains to prove that when the adversary requests one of his server positions, the expected change of Φ plus the expected cost of the algorithm is at most 0. This is done by case analysis and straightforward calculation which we omit.

2.2 A Lower Bound for Uniform Spaces

Theorem 2. *For any memoryless randomized algorithm for the weighted 2-server problem in a uniform space of three points, the competitive ratio against an oblivious adversary is at least 5.*

Proof. Let a, b, and c denote the three points. The algorithm has only one parameter, which is the probability p that a request unoccupied by a server is served by s_1. At the beginning, the expensive server is at a and the cheap one at b. We assume that $\beta \to 0$, and all O-notation is relative to this. We prove a lower bound of $5 - o(1)$, the bound of 5 follows by taking β sufficiently small.

First consider what happens if we repeat requests b and c infinitely long. Eventually, the algorithm moves s_1 to b or c and s_2 to the other of these two points. The expected cost of the algorithm is $1+\beta/p$, since it will take on average $1/p$ of moves of the cheap server. We choose k large enough so that after the sequence $(cb)^k$, the probability that s_1 does not move is $o(1)$ and the expected cost of the algorithm is $1 + \beta/p - o(1)$. At the end, the probability that $s_1 = b$ is at most $1/2$, since the first (non-trivial) request was to the point c.

Consider a sequence of requests $((cb)^k(ab)^k)^l$ for $l = \omega(1)$. Let us call a subsequence of requests $(cb)^k$ or $(ab)^k$ a phase. Until the algorithm moves s_1 to b, it pays $1 + \beta/p - o(1)$ for each phase. It takes on average at least $2 - o(1)$ phases to move s_1 to b, so the total cost of the algorithm is $2(1+\beta/p)(1-o(1))$.

The adversary strategy depends on the probability p. If $p \leq 2\beta/3$ then the request sequence is the above sequence $((cb)^k(ab)^k)^l$ for some l satisfying $l = o(1/\beta)$ and $l = \omega(1)$. The adversary serves the sequence by moving a_1 to b and then moving a_2 between a and c $2l$ times, with total cost $1 + 2l\beta = 1 + o(1)$. The cost of the algorithm is $2(1 + \beta/p)(1 - o(1)) \geq 5 - o(1)$.

If $p \geq 2\beta/3$ then the request sequence is $(cb)^m((ca)^k(ba)^k)^l$, where m and l are chosen such that $m = o(1/\beta) = o(1/p)$, $l = o(m)$, $l = \omega(1)$. The adversary does not move a_1 and serves the requests at b and c with a_2, at cost $2(l+m)\beta = 2m\beta(1+o(1))$. The probability that s_1 does not move during the first $2m$ steps is $q = (1-p)^{2m}$. The expected cost of the cheap server during the first $2m$ steps is $\beta/p - q \cdot \beta/p$ (the cost would be β/p for an infinite sequence; with probability

q we stop after m steps and save β/p, which is the expected cost starting after $2m$ steps conditioned on the fact that we have at least $2m$ steps). If s_1 moves, the additional cost is $1+2(1+\beta/p)(1-o(1))$. So the total expected cost of the algorithm is at least $(1-q)\cdot 3(1+\beta/p)(1-o(1))$. Elementary calculations show that $1-q=2mp(1-o(1))$. Thus the competitive ratio is at least

$$\frac{2mp\cdot 3(1+\frac{\beta}{p})(1-o(1))}{2m\beta(1+o(1))} = 3\left(\frac{p}{\beta}+1\right)(1-o(1)) \geq 5-o(1).$$

2.3 A Lower Bound for the Line

Theorem 3. *There exists no competitive randomized memoryless algorithm for the weighted 2-server problem on the real line.*

Proof. Let $R>1$ be an arbitrarily large integer and $\beta=2^{-4R}$. At the beginning, assume that s_2 and a_2 are at 0 and s_1 and a_1 are at 1. We have $4R$ phases. phases. In a phase i, we alternate requests to points 2^i and 0, the total of 2^{8R} requests. The optimal algorithm moves a_1 to 0 and serves all other requests with a_2. The total cost is $1+2^{4R}\beta=2$.

Let C_i be expected cost of the algorithm in phases $i+1, \ldots, 4R$, assuming that after i phases s_2 is at 0 and s_1 is at 2^i. We prove by backwards induction that $C_i \geq (4R-i)2^{i-1}$. By definition, $C_{4R}=0$. For the induction step, we analyze phase i, assuming that s_2 starts at 0 and s_1 at 2^{i-1}. If s_1 stays at 2^{i-1} then the cost is at least $2^{8R}\cdot 2^i\beta \geq 2^{4R} \geq (4R-i)2^{i-1}$ in phase i alone. Assume now that s_1 moves. If s_1 moves to 2^i, then the cost of this and all following phases is at least $2^{i-1}+C_{i+1}$. If s_1 moves to 0 then the cost of this phase is at least 2^{i-1}. Since s_1 is more likely to move to 2^i than to 0, the expected total cost is at least $C_i \geq 2^{i-1}+C_{i+1}/2 \geq (4R-i)2^{i-1}$. Thus the algorithm pays at least $C_0 \geq 2R$, and the competitive ratio is at least R.

3 Deterministic Algorithms

Now we focus on deterministic algorithms. We start by introducing work functions. Then, we give a 5-competitive algorithm for the weighted 2-server problem in uniform spaces and prove that the ratio 5 is optimal. In the last subsection we study a simplified version of the weighted 2-server problem; for this version we give a 9-competitive algorithm for an arbitrary metric space.

Work Functions. A *configuration* is a pair (x,y), where x,y are the locations of the expensive and cheap servers, respectively. By $\omega_\varrho(x,y)$ we denote the *work function* on (x,y), defined as the minimum cost of serving the request sequence ϱ and ending in configuration (x,y). We can compute ω by dynamic programming as follows. Initially, $\omega_\epsilon(x,y)=x_0x+\beta\cdot y_0y$, where (x_0,y_0) is the initial configuration. Let $\varrho=\sigma r$. If $r\in\{x,y\}$ then $\omega_\varrho(x,y)=\omega_\sigma(x,y)$. Otherwise, $\omega_\varrho(x,y)=\min\{\omega_\sigma(r,y)+xr, \omega_\sigma(x,r)+\beta\cdot yr\}$. Function ω_ϱ satisfies the Lipschitz condition: for any points $x,y,u,v\in M$, $\omega_\varrho(u,v)\leq\omega_\varrho(x,y)+xu+\beta\cdot yv$.

We use modified work functions with only one parameter, which is the position of the expensive server, defined by $\omega_\varrho(x) = \min_y \omega_\varrho(x,y)$. Intuitively, as β approaches 0, the position of the cheap server becomes less significant.

For the algorithms based on work function we use the potential functions as follows. The potential is bounded from below by $C - R\min_{(x,y)} \omega(x,y)$, for some constant C. We show that $\Delta cost + \Phi' \leq \Phi$, where $\Delta cost$ is the cost of the algorithm (for one request or one phase), and Φ and Φ' are the old and the new potential, respectively. Summing over all requests, the algorithm's cost is at most the initial constant minus the final potential. This is at most $C' + R\min_{(x,y)} \omega(x,y)$, and R-competitiveness follows.

3.1 The Work-Function Algorithm for Uniform Spaces

For simplicity, we assume that $\beta < 1$ and $1/\beta$ is an integer. This is not a major restriction, since we are interested in the asymptotic behavior for $\beta \to 0$. Further, we assume that the adversary does not make any requests on s_1 and s_2; otherwise the algorithm can simply ignore the request.

The Work Function Algorithm (WFA) minimizes the current cost plus the optimal cost of the new configuration. Adapted to our case, it works as follows.

Algorithm MWFA: Let u be the request to be served, let ω be the work function for the request sequence ending at u. Suppose the current configuration is (x,y). If $\omega(x) = \omega(u) + 1$, move s_1 from x, otherwise move s_2 from y.

Let us now examine $\omega_\varrho(x)$ in more detail. Suppose that $x \neq r$, $\varrho = \sigma r$ and let q be the last request in σ, $q \neq r$. Function ω satisfies the following Lipschitz condition: for all points x, y,

$$\omega_\varrho(x) \leq \omega_\varrho(y) + 1, \tag{1}$$

because if $\omega_\varrho(y) = \omega_\varrho(y,z)$, then $\omega_\varrho(x) \leq \omega_\varrho(x,z) \leq \omega_\varrho(y,z) + 1 = \omega_\varrho(y) + 1$.

For $y \neq r, q$, we have $\omega_\varrho(x,y) = \min\{\omega_\sigma(r,y) + 1, \omega_\sigma(x,r) + \beta\} \geq \omega_\sigma(x,r) = \omega_\varrho(x,r)$. Therefore,

$$\omega_\varrho(x) = \min\{\omega_\varrho(x,r), \omega_\varrho(x,q)\} \;=\; \min\{\omega_\varrho(x,r), \omega_\varrho(r,q) + 1\} \tag{2}$$

We have $\omega_\varrho(x) \leq \omega_\varrho(x,r) = \omega_\sigma(x,r) \leq \omega_\sigma(x) + \beta$. Due to integrality of $1/\beta$, either $\omega_\varrho(x) = \omega_\sigma(x)$ or $\omega_\varrho(x) = \omega_\sigma(x) + \beta$. Note that this also implies that each phase has at least $1/\beta$ requests. The next lemma shows that in typical cases $\omega(x)$ increases.

Lemma 1. *Let $\varrho = \sigma r$ and let q be the last request in σ, $q \neq r$.*
(a) *If $x \notin \{q,r\}$ and $\omega_\varrho(x) < \min\{\omega_\varrho(r), \omega_\varrho(q)\} + 1$ then $\omega_\varrho(x) = \omega_\sigma(x) + \beta$.*
(b) *For $\beta < 1$, if $s \notin \{q,r\}$ and $\omega_\varrho(r) = \min(\omega_\varrho)$, then $\omega_{\varrho s}(r) = \omega_\varrho(r) + \beta$.*

Proof. (a) We have $\omega_\varrho(x) = \omega_\varrho(x,r)$, for otherwise we would get $\omega_\varrho(x) = \omega_\varrho(r,q) + 1 \geq \omega_\varrho(r) + 1$. Thus $\omega_\varrho(x) = \omega_\varrho(x,r) = \min\{\omega_\sigma(x,q) + \beta, \omega_\sigma(q,r) + 1\}$. Since $\omega_\varrho(x) < \omega_\varrho(q) + 1 \leq \omega_\varrho(q,r) + 1 = \omega_\sigma(q,r) + 1$, we get $\omega_\varrho(x) = \omega_\sigma(x,q) + \beta \geq \omega_\sigma(x) + \beta$.

(b) We have $\omega_{\varrho s}(r) = \min\{\omega_\varrho(r,s), \omega_\varrho(s,r)+1\} = \omega_\varrho(r,s) = \min\{\omega_\sigma(r,q)+\beta, \omega_\sigma(q,s)+1\} \geq \min\{\omega_\varrho(r,q)+\beta, \omega_\varrho(q,s)+1-\beta\} \geq \min\{\omega_\varrho(r)+\beta, \omega_\varrho(q)+1-\beta\} \geq \min(\omega_\varrho)+\beta = \omega_\varrho(r)+\beta$.

Theorem 4. *For $\beta < 1$, $1/\beta$ integer, MWFA is a 5-competitive algorithm for the weighted 2-server problem in uniform spaces.*

Proof. We divide the computation into phases, with each phase ending when s_1 moves. We define the potential function for configurations at the beginning of each phase. Let ω be the work function at the beginning of the phase and let r be the last request; thus s_1 is at r. If MWFA moved s_1 to r from z then $\omega(z) = \omega(r) + 1$. Thus, by (1), $\omega(\cdot)$ is minimized at r. Let a and b be the next two minima of ω, that is, $a \neq r$ and $\omega(a) = \min_{x\neq r}\omega(x)$, and $b \notin \{r,a\}$ and $\omega(b) = \min_{x\notin\{r,a\}}\omega(x)$. We define the potential as

$$\Phi = \max\{-\omega(r) - 4\cdot\omega(a),\ 2 - \omega(r) - 4\cdot\omega(b)\}$$

Consider one phase, in which the work function changes from ω to μ, and let s be the last request in this phase. Thus $\mu(r) = \mu(s) + 1$, and s is the minimum of μ. Let c and d be the next two minima of μ.

We need to show that $\Delta cost + \Phi' \leq \Phi$, where $\Delta cost$ is the total cost of the algorithm during the phase, Φ and Φ' are the old and the new potentials.

Without loss of generality, we assume that $\omega(r) = 0$ (since we can uniformly decrease ω and μ by $\omega(r)$). Lemma 1 implies that during the phase the work function on r increases by β on each request except the last (Lemma 1.b applies to the first request in a phase, and Lemma 1.a to all intermediate requests.) Since MWFA pays β for each such request, $\Delta cost \leq \mu(r) - \omega(r) + 1 = 2 + \mu(s)$. Since $s \neq r$, we have $\mu(s) \geq \omega(s) \geq \omega(a)$. We distinguish two cases.

Case 1: $\Phi' = -\mu(s) - 4\mu(c)$. If $c = r$ then $\mu(c) = \mu(s) + 1$ and

$$\Delta cost + \Phi' \leq [2+\mu(s)] + [-4 - 5\mu(s)] \leq -4\mu(s) \leq -4\omega(a) \leq \Phi$$

If $c \neq r$, then $\mu(c) \geq \omega(b)$, since $\mu(c) \geq \omega(c)$ and $\mu(c) \geq \mu(s) \geq \omega(s)$. Thus

$$\Delta cost + \Phi' \leq [2+\mu(s)] + [-\mu(s) - 4\mu(c)] = 2 - 4\mu(c) \leq 2 - 4\omega(b) \leq \Phi$$

Case 2: $\Phi' = 2 - \mu(s) - 4\mu(d)$. If $r \in \{c,d\}$, then $\mu(d) \geq \mu(r) = \mu(s) + 1$, so

$$\Delta cost + \Phi' \leq [2+\mu(s)] + [-2 - 5\mu(s)] = -4\mu(s) \leq -4\omega(a) \leq \Phi$$

Otherwise r, s, c, and d are all distinct. We claim that

$$\mu(d) \geq \min\{\omega(a)+1, \omega(b)+\tfrac{1}{2}\}.$$

This implies that

$$\begin{aligned}\Delta cost + \Phi' &\leq [2+\mu(s)] + [2 - \mu(s) - 4\mu(d)] = 4 - 4\mu(d)\\ &\leq 4 - 4\min\{\omega(a)+1, \omega(b)+\tfrac{1}{2}\} = \max\{-4\omega(a), 2 - 4\omega(b)\} = \Phi\end{aligned}$$

To prove the claim, suppose that $\mu(d) < \omega(a) + 1$. We prove by induction that for any work function ξ during the phase

$$\xi(s) + \xi(c) + \xi(d) \geq \xi(r) + 2\omega(b) \tag{3}$$

Initially, $\xi = \omega$ and $\omega(r) = 0$. Since at least two of $\omega(c)$, $\omega(d)$, $\omega(s)$ are at least $\omega(b)$, the inequality holds.

Consider one step in the phase. Suppose first that some $t \in \{s, c, d\}$ is maxed-out after the request, that is $\xi(t) = 1 + \min_x \xi(x)$. Denote the other two points by y and z, so that $\xi(y) \leq \xi(z)$. Since $\mu(t) \leq \mu(d) < \omega(a)+1$, we have $\xi(t) = \xi(r)+1$. But then $\xi(t) + \xi(y) + \xi(z) = \xi(r) + [\xi(y) + 1] + \xi(z) \geq \xi(r) + 2\omega(b)$.

If the new work function is not maxed out on any of s, c or d, Lemma 1.a implies that the work function must increase on at least one of s, c, d. The right-hand side can increase at most by β, so inequality (3) is preserved.

At the end of the phase we have $\xi = \mu$ and $2\mu(d) \geq \mu(c)+\mu(d) \geq \mu(r)-\mu(s)+2\omega(b) = 1 + 2\omega(b)$. So we showed that $\mu(d) < \omega(a) + 1$ implies $\mu(d) \geq \omega(b) + \frac{1}{2}$.

The proof for a complete phase is now finished. The last phase may be incomplete. Then the cost of the algorithm is at most the increase of the work function, and the theorem follows.

3.2 A Lower Bound for Uniform Spaces

We use a space with three points a, b, c. The expensive server starts at a and the cheap one at b. The adversary always requests the point not occupied by s_1, s_2. We again divide the request sequence into *phases*, where a phase ends when the algorithm moves s_1.

Let t_i be the number of requests served by s_2 in phase i. Thus the algorithm pays $t_i\beta + 1$ in this phase. Let $\omega_i(x)$ be the modified work function (as defined earlier) after i phases.

Now we define $\phi_i = 2[\omega_i(a) + \omega_i(b) + \omega_i(c)] - \min\{\omega_i(a), \omega_i(b), \omega_i(c)\}$. This function is approximately five times the optimal cost. We need to prove that it is a lower bound on the cost of the algorithm.

Lemma 2. *For every phase i, we have $\phi_i - \phi_{i-1} \leq t_i\beta + 1 + 6\beta$.*

Proof. Let u be the position of s_1 at the beginning of the phase. Let $U = \omega_{i-1}(u_i)$, and $V = \omega_{i-1}(v)$, where $v \in \{a, b, c\} - \{u\}$ is the point with the smaller value of ω_{i-1}. During the phase, the adversary alternates the requests to the two other points than u. From the definition of the work function it is easy to see that $\omega_i(u) \leq \min\{V + 1 + \beta, U + (t_i + 1)\beta\}$ and, for $x \neq u$, $\omega_i(x) \leq \omega_{i-1}(x) + \beta$. We distinguish three cases.

Case 1: $U + t_i\beta \leq V$. Then $\omega_i(u) \leq U + (t_i + 1)\beta$ and u achieves the minimum of both ω_{i-1} and ω_i (up to an additive term of β), so $\phi_i - \phi_{i-1} \leq (U + t_i\beta + 6\beta) - U = t_i\beta + 6\beta \leq t_i\beta + 1 + 6\beta$.

Case 2: $V \leq U$. Then V is the minimum of ω_{i-1} and an approximate minimum of ω_i, so $\phi_i - \phi_{i-1} = (2\omega_i(u) + 4\beta) - 2U \leq ((U + t_i\beta) + (V + 1) + 6\beta) - 2U \leq t_i\beta + 1 + 6\beta$.

Case 3: $U \leq V \leq U + t_i\beta$. Now U is the minimum of ω_{i-1} and V is the approximate minimum of ω_i. Thus $\phi_i - \phi_{i-1} \leq (2\omega_i(u) + V + 4\beta) - (2V + U) \leq ((U + t_i\beta) + (V + 1) + V + 6\beta) - (2V + U) = t_i\beta + 1 + 6\beta$.

Theorem 5. *For any deterministic algorithm for the weighted 2-server problem in a uniform space with three points, the competitive ratio is at least* 5.

Proof. Let k be the number of phases, C the cost of the algorithm, and C_{opt} the optimal cost. By summing over all phases, Lemma 2 implies that $C + 6k\beta \geq \phi_k - \phi_0 - 6k\beta \geq 5C_{opt} - 4$. Since the algorithm pays at least 1 in each phase, we have $C \geq k$, and the theorem follows by taking β sufficiently small.

3.3 The Weighted 2-Point Request Problem

In this section we study the modification of the weighted 2-server problem in which each request is specified by two points, say $\{r, s\}$. In response to this request, the algorithm must move one server to r and the other to s. The decision to be made is which server to move to which point.

This can be viewed as a subproblem of the weighted 2-server problem: Replace the 2-point request $\{r, s\}$ by a long sequence $(rs)^*$. Any competitive 2-server algorithm eventually moves his servers to r and s. In this way any R-competitive weighted 2-server algorithm yields a R-competitive algorithm for the weighted 2-point request problem.

On the other hand, in the limit for $\beta \to 0$, we obtain the 2-point request problem with one server studied in [12], which in turn contains the closely related cow-path problem [2,23]. This yields a lower bound of 9. We prove a matching upper bound of 9 for the algorithm WFA_3 which at each step minimizes the cost of the move plus three times the optimal cost of the new configuration.

Algorithm WFA_3**:** Let (x, y) be the current configuration, let $\{r, s\}$ be the new request and ω' the new work function. If $xr + \beta ys + 3\omega'(r, s) \leq xs + \beta yr + 3\omega'(s, r)$, then move to (r, s), otherwise move to (s, r).

Theorem 6. WFA_3 *is R-competitive for the weighted 2-point request problem in any metric space, where* $R = (9 - 3\beta)/(1 + \beta)$.

Proof. Let (x, y) be the current configuration and let ω_x and ω_y the work function values when the expensive server is at x and y, respectively. Let (r, s), ω'_r, and ω'_s be the new configuration and work function values, after serving the request $\{r, s\}$. Denote the distances as follows: $d = xy$, $e = rs$, $a = xr$, $b = xs$, $f = yr$, and $g = ys$. The cost of the algorithm for serving the new request is $\Delta cost = a + \beta g$ and the new work function is $\omega'_r = \min\{\omega_x + a + \beta g, \omega_y + f + \beta b\}$, $\omega'_s = \min\{\omega_x + b + \beta f, \omega_y + g + \beta a\}$.

From the definition, WFA_3 satisfies the invariant $3(\omega_x - \omega_y) \leq (1+\beta)d$. Since in the current step WFA_3 moved to (r, s) and not (s, r), we have $a + \beta g + 3\omega'_r \leq b + \beta f + 3\omega'_s$. We use the following potential function:

$$\Phi = \left[2d + \frac{6}{1+\beta}(\omega_x - \omega_y)\right]^+ - R\omega_x$$

where $[a]^+ = \max(a, 0)$. We need to show that $\Delta cost + \Delta\Phi = a + \beta g + \Phi' - \Phi \le 0$. We distinguish several cases according to the possible values of Φ', ω'_r, and ω'_s. In each case, the last step uses a triangle inequality.

Case 1: $(1+\beta)d + 3(\omega'_r - \omega'_s) \le 0$. Then $\Phi' = -R\omega'_r$.

Case 1.1: $\omega'_r = \omega_x + a + \beta g$. Then

$$\Delta cost + \Delta\Phi \le a + \beta g - R(\omega_x + a + \beta g) + R\omega_x \le 0$$

Case 1.2: $\omega'_r = \omega_y + f + \beta b$. Then

$$\begin{aligned}\Delta cost + \Delta\Phi &\le a + \beta g - R(\omega_y + f + \beta b) - \left[2d + \frac{6}{1+\beta}(\omega_x - \omega_y) - R\omega_x\right]\\ &= a + \beta g - Rf - R\beta b - 2d + \frac{3-3\beta}{1+\beta}(\omega_x - \omega_y)\\ &\le a + \beta g - f - \beta b - 2d + (1-\beta)d\\ &= (a - d - f) + \beta(g - d - b) \le 0\end{aligned}$$

Case 2: $(1+\beta)d + 3(\omega'_r - \omega'_s) \ge 0$.

Case 2.1: $\omega'_r = \omega_x + a + \beta g$ and $\omega'_s = \omega_x + b + \beta f$. Then $a + \beta g \le b + \beta f$ and

$$\begin{aligned}\Delta cost + \Delta\Phi &\le a + \beta g + 2e + \frac{6}{1+\beta}(a + \beta g - b - \beta f) - R(\omega_x + a + \beta g) + R\omega_x\\ &= \frac{2}{1+\beta}[(e - b - a) + \beta(e - f - g) + \beta(a + \beta g) - (b + \beta f)] \le 0\end{aligned}$$

Case 2.2: $\omega'_r = \omega_x + a + \beta g$ and $\omega'_s = \omega_y + g + \beta a$. Then

$$\begin{aligned}\Delta cost + \Delta\Phi &\le a + \beta g - \left[2d + \frac{6}{1+\beta}(\omega_x - \omega_y) - R\omega_x\right]\\ &\quad + \left[2e + \frac{6}{1+\beta}(\omega_x + a + \beta g - \omega_y - g - \beta a) - R(\omega_x + a + \beta g)\right]\\ &= 2(e - d - a - g) - 4(1-\beta)g \le 0\end{aligned}$$

Case 2.3: $\omega'_r = \omega_y + f + \beta b$ and $\omega'_s = \omega_y + g + \beta a$. Then $\Delta cost + \Delta\Phi$ is at most

$$\begin{aligned}& a + \beta g + \left[2e + \frac{6}{1+\beta}(\omega_y + f + \beta b - \omega_y - g - \beta a) - R(\omega_y + f + \beta b)\right]\\ &\quad - \left[2d + \frac{6}{1+\beta}(\omega_x - \omega_y) - R\omega_x\right]\\ &= \frac{1-5\beta}{1+\beta}a - \frac{6-\beta-\beta^2}{1+\beta}g + 2e - 2d + \frac{3-3\beta}{1+\beta}(\omega_x - \omega_y) - \frac{3-3\beta}{1+\beta}f\\ &\le a - \frac{6-\beta-\beta^2}{1+\beta}g + 2e - (1+\beta)d - \frac{3-3\beta}{1+\beta}(f + \beta b)\\ &\le \frac{3-3\beta}{1+\beta}(e - f - g) + (a - e - d - g) \le 0\end{aligned}$$

Case 2.4: $\omega'_r = \omega_y + f + \beta b$ and $\omega'_s = \omega_x + b + \beta f$. Then $\Delta cost + \Delta\Phi$ is at most

$$a + \beta g + \left[2e + \frac{6}{1+\beta}(\omega'_r - \omega'_s) - R\omega'_r\right] - \left[2d + \frac{6}{1+\beta}(\omega_x - \omega_y) - R\omega_x\right]$$
$$= a + \beta g + 2e - 2d + 3(\omega'_r - \omega'_s) - \frac{6}{1+\beta}(\omega'_r - \omega_y) - \frac{3-3\beta}{1+\beta}(\omega'_s - \omega_x)$$
$$\leq a + \beta g + 2e - 2d + (b + \beta f - a - \beta g) - \frac{6}{1+\beta}(f + \beta b) - \frac{3-3\beta}{1+\beta}(b + \beta f)$$
$$= 2(e - d - b - f) - 4(1-\beta)f \leq 0$$

4 Final Comments

We proved mainly results for uniform spaces. Many open problems remain. For example, no competitive algorithm for the weighted 2-server problem in arbitrary spaces is known, and the lower bound of 10.12 from [20] can probably be improved. This lower bound also shows that the optimal competitive ratio for the weighted 2-point request problem is strictly smaller than for the the general weighted 2-server problem. Technically, the 2-point request problem is simpler to analyze, since the work function has only two relevant values.

The weighted 2-server problem is related to the CNN problem from [20]. In this problem, we have one server in the plane. Each request is a point (x, y), and to serve this request we need to move the server to some point with x-coordinate x or with y-coordinate y. The special case when the requests are restricted to some line in the plane is equivalent to a weighted 2-server problem on the line.

Koutsoupias and Taylor [20] also prove a lower bound for memoryless randomized algorithms for the CNN problem. Our lower bound for memoryless algorithms is somewhat stronger, in two respects: the problem is a special case of the CNN problem and, unlike in [20], we do not assume that the algorithm is invariant with respect to scaling distances.

Acknowledgements. Work of the first author was supported by NSF grant CCR-9503498, and an NRC COBASE grant (NSF contract INT-9522667). Work of the second author was partially supported by grant A1019901 of GA AV ČR, postdoctoral grant 201/97/P038 of GA ČR, and cooperative research grant INT-9600919/ME-103 from the NSF (USA) and the MŠMT (Czech Republic).

References

1. D. Achlioptas, M. Chrobak, and J. Noga. Competitive analysis of randomized paging algorithms. In *Proc. 4th European Symp. on Algorithms*, volume 1136 of *Lecture Notes in Computer Science*, pages 419–430. Springer, 1996.
2. R. A. Baeza-Yates, J. C. Culberson, and G. J. E. Rawlins. Searching with uncertainty. In *Proc. 1st Scandinavian Workshop on Algorithm Theory*, Lecture Notes in Computer Science, pages 176–189. Springer, 1988.

3. Y. Bartal, A. Blum, C. Burch, and A. Tomkins. A polylog(n)-competitive algorithm for metrical task systems. In *Proc. 29th Symp. Theory of Computing*, pages 711–719, 1997.
4. Y. Bartal, M. Chrobak, and L. L. Larmore. A randomized algorithm for two servers on the line. In *Proc. 6th European Symp. on Algorithms*, Lecture Notes in Computer Science, pages 247–258. Springer, 1998.
5. Y. Bartal and E. Grove. The harmonic k-server algorithm is competitive. To appear in Journal of the ACM.
6. S. Ben-David, A. Borodin, R. M. Karp, G. Tardos, and A. Widgerson. On the power of randomization in on-line algorithms. In *Proc. 22nd Symp. Theory of Computing*, pages 379–386, 1990.
7. A. Blum, H. Karloff, Y. Rabani, and M. Saks. A decomposition theorem and lower bounds for randomized server problems. In *Proc. 33rd Symp. Foundations of Computer Science*, pages 197–207, 1992.
8. A. Borodin and R. El-Yaniv. *Online Computation and Competitive Analysis*. Cambridge University Press, 1998.
9. M. Chrobak, H. Karloff, T. H. Payne, and S. Vishwanathan. New results on server problems. *SIAM Journal on Discrete Mathematics*, 4:172–181, 1991.
10. M. Chrobak and L. L. Larmore. On fast algorithms for two servers. *Journal of Algorithms*, 12:607–614, 1991.
11. M. Chrobak and L. L. Larmore. An optimal online algorithm for k servers on trees. *SIAM Journal on Computing*, 20:144–148, 1991.
12. M. Chrobak and L. L. Larmore. The server problem and on-line games. In *DIMACS Series in Discrete Mathematics and Theoretical Computer Science*, volume 7, pages 11–64, 1992.
13. M. Chrobak and L. L. Larmore. Metrical task systems, the server problem, and the work function algorithm. In *Online Algorithms: State of the Art*, pages 74–94. Springer-Verlag, 1998.
14. M. Chrobak, L. L. Larmore, C. Lund, and N. Reingold. A better lower bound on the competitive ratio of the randomized 2-server problem. *Information Processing Letters*, 63(2):79–83, 1997.
15. D. Coppersmith, P. G. Doyle, P. Raghavan, and M. Snir. Random walks on weighted graphs and applications to on-line algorithms. *Journal of the ACM*, 40:421–453, 1993.
16. E. Feuerstein, S. Seiden, and A. S. de Loma. The related server problem. Manuscript, 1999.
17. A. Fiat and M. Ricklin. Competitive algorithms for the weighted server problem. *Theoretical Computer Science*, 130:85–99, 1994.
18. E. Koutsoupias and C. Papadimitriou. On the k-server conjecture. *Journal of the ACM*, 42:971–983, 1995.
19. E. Koutsoupias and C. Papadimitriou. The 2-evader problem. *Information Processing Letters*, 57:249–252, 1996.
20. E. Koutsoupias and D. Taylor. Lower bounds for the CNN problem. To appear in STACS 2000 (this volume), 2000.
21. M. Manasse, L. A. McGeoch, and D. Sleator. Competitive algorithms for server problems. *Journal of Algorithms*, 11:208–230, 1990.
22. L. McGeoch and D. Sleator. A strongly competitive randomized paging algorithm. *Algorithmica*, 6(6):816–825, 1991.
23. C. H. Papadimitriou and M. Yannakakis. Shortest paths without a map. *Theoretical Computer Science*, 84:127–150, 1991.

On the Competitive Ratio of the Work Function Algorithm for the k-Server Problem

Yair Bartal[1] and Elias Koutsoupias[2*]

[1] Bell Labs
yair@research.bell-labs.com
[2] Univ of California, Los Angeles
elias@cs.ucla.edu

Abstract. The k-server problem is one of the most fundamental on-line problems. The problem is to schedule k mobile servers to serve a sequence of service points in a metric space to mimize the total mileage. The k-server conjecture [11] that states that there exists an optimal k-competitive on-line algorithm has been open for over 10 years. The top candidate on-line algorithm for settling this conjecture is the Work Function Algorithm (WFA) which was recently shown [7,9] to have competitive ratio at most $2k-1$. In this paper we lend support to the conjecture that WFA is in fact k-competitive by proving that it achieves this ratio in several special metric spaces.

1 Introduction

The k-server problem [11] together with its special case, the paging problem, is probably the most influential on-line problem. The famous k-server conjecture has been open for over 10 years. Yet, the problem itself is very easy to state: There are k servers that can move in a metric space. Their purpose is to service a sequence of requests. A request is simply a point of the metric space and servicing it entails moving a server to the requested point. The objective is to minimize the total distance traveled by all servers. In the on-line version of the problem, the requests are presented one-by-one. The notorious k-server conjecture states that there is an on-line algorithm that has competitive ratio k on any metric space. The top candidate on-line algorithm for settling the k-server conjecture is the Work Function Algorithm (WFA) which was shown [7,9] to have competitive ratio at most $2k-1$.

In this paper, we show three results. The first is that the WFA is k-competitive in the line. Our second result is that the WFA is k-competitive for the "symmetric weighted cache" (represented by weighted star instances). It was known [3,4] that the k-server conjecture holds for these instances, but the algorithm employed was not the WFA, but the Double Coverage algorithm, which has no natural extension for non-tree like metric spaces. Our third result is a new proof of the WFA is k-competitive for metric spaces of $k+2$ points. This was first shown in [7,10] using an involved potential. Our proof here uses a much simpler potential.

* Supported in part by NSF grant CCR-9521606

H. Reichel and S. Tison (Eds.): STACS 2000, LNCS 1770, pp. 605–613, 2000.

There is an interesting connection between the three results of this work. In all cases, the number of minimizers (to be defined later) is at most $k+1$. Although this fact by itself cannot guarantee that the WFA is k-competitive, it is at the heart of our proofs.

2 Preliminaries

We summarize here our notation, conventions and definitions. For a more thorough discussion that includes the history of the problem see [9,2]. Let $\rho = r_1 \ldots r_n$ be a request sequence. The *work function* $w_i(X)$ is defined to be the optimal cost for servicing $r_1 \ldots r_i$ and moving to configuration X. The Work Function Algorithm works as follows: Let A_i be its configuration just before servicing request r_{i+1}. To service r_{i+1}, it moves to configuration A_{i+1} that contains r_{i+1} and minimizes $w_{i+1}(A_{i+1}) + d(A_i, A_{i+1})$.

Chrobak and Larmore [5] introduced the concept of *extended cost* of the WFA (which they call pseudocost): The extended cost for request r_{i+1} is equal to the maximum increase of the work function: $\max_X\{w_{i+1}(X)-w_i(X)\}$. They showed that the extended cost is equal to the on-line plus the off-line cost (see also [9]). Consequently, to prove that the Work Function Algorithm is c-competitive, it suffices to bound the total extended cost by $(c+1)\text{OPT}(\rho)+\text{const}$, where $\text{OPT}(\rho)$ is the optimal (off-line) cost to service ρ.

For general metric spaces, the best known upper bound on the competitive ratio for the k-server problem is $2k-1$ [7,9] (see also [8] for a simpler proof), which improved the previous exponential (in k) bounds [6,1]. Unlike the previous results, the algorithm employed in [7,9] to establish the $2k-1$ bound is the WFA. The proof is based on some fundamental properties (Quasiconvexity and Duality) of work functions. Here we will make use of the Duality property which characterizes the configurations that achieve the maximum $\max_X\{w_{i+1}(X) - w_i(X)\}$.

Lemma 1 (Duality lemma [7,9]). *Let X be a configuration that minimizes*

$$w_i(X) - \sum_{x \in X} d(r_{i+1}, x).$$

Then X minimizes also

$$w_{i+1}(X) - \sum_{x \in X} d(r_{i+1}, x)$$

and maximizes the extended cost

$$\max_X\{w_{i+1}(X) - w_i(X)\}.$$

A configuration X that minimizes $w_i(X) - \sum_{x\in X} d(p,x)$ will be called a *minimizer* of p with respect to w_i.

3 The WFA for the Line

In this section, we will show that the WFA is k-competitive in the line. To simplify the presentation, we assume that all requests are in a fixed interval $[a, b]$. Let us denote the configuration that contains m copies of a and $k - m$ copies of b as $a^m b^{k-m}$. We shall call these configurations *extreme*. Observe that there are exactly $k+1$ extreme configurations ($m = 0, \ldots, k$). The next lemma shows that we can generally assume that minimizers are extreme configurations.

Lemma 2. *Assume that all requests are in the interval $[a, b]$. For any point $p \in [a, b]$ and any work function w_i, there is $m \in \{0, \ldots, k\}$ such that $a^m b^{k-m}$ is a minimizer of p with respect to w_i.*

Proof. Clearly, there is a minimizer X of p with respect to w_i with all points in the interval $[a, b]$. Assume that there is a point $x \in X$ in the interval $[a, p]$. What will happen if we slide x to a? The work function $w_i(X)$ can increase by at most $d(a, x)$ while the distance of x from p will increase by exactly $d(a, x)$. Therefore $X - x + a$ is also a minimizer of p. More precisely,

$$w_i(X - x + a) - \sum_{y \in X - x + a} d(p, y) \leq (w_i(X) + d(a, x)) - (\sum_{y \in X} d(p, y) + d(a, x))$$
$$= w_i(X) - \sum_{y \in X} d(p, y)$$

Similarly, we can slide all points of X to either a or b. If X has m points in $[a, p]$, then $a^m b^{k-m}$ is a minimizer of p. □

Theorem 1. *The WFA is k-competitive in the line.*

Proof. We first show the somewhat simpler result that the WFA is k-competitive in an interval $[a, b]$. The same proof extends to the infinite line.

We define a potential Φ_i to be the sum of w_i on all extreme configurations:

$$\Phi_i = \sum_{j=0}^{k} w_i(a^j b^{k-j}).$$

We will show that Φ_n is an upper bound (within a constant) of the extended cost. By Lemma 2, there is m such that $a^m b^{k-m}$ is a minimizer of r_{i+1} with respect to w_i. The increase of the potential, $\Phi_{i+1} - \Phi_i$, is equal to the increase of the work function on all extreme configurations. Since the work function increases monotonically, i.e., $w_{i+1}(X) \geq w_i(X)$, the increase $\Phi_{i+1} - \Phi_i$ of the potential is at least $w_{i+1}(a^m b^{k-m}) - w_i(a^m b^{k-m})$, which is the extended cost to service r_{i+1}. It follows, by telescoping, that the total extended cost, i.e., the sum of the extended cost for all requests, is bounded from above by $\Phi_n - \Phi_0$.

For a fixed interval $[a, b]$, the values of a work function cannot differ too much: for any work function w and any configurations X and Y: $w(X) - w(Y) \leq$

$d(X,Y) \leq kd(a,b)$. This allows us to conclude that Φ_n is equal (within a constant) to $(k+1)\text{OPT}(\rho_n) = (k+1)\min_X\{w(X)\}$ and that Φ_0 is constant. The total extended cost is therefore bounded above by $(k+1)\text{OPT}(\rho_n) + \text{const}$ which implies the k-competitiveness of WFA.

We now turn to the infinite line. We have to be more careful for this case and to actually compute the constants ignored in the previous paragraph. Let's first observe that we can again assume that all requests are in an interval $[a,b]$ where a is the leftmost request of ρ_n and b is the rightmost one. The difference with the case of a fixed interval $[a,b]$ is that now we cannot assume that $d(a,b)$ is constant. Thus we have to show that the additive term depends only on the initial configuration A_0 and is independent on $d(a,b)$. We can easily compute the initial potential $\Phi_0 = \sum_{j=0}^k w(a^j b^{k-j}) = \sum_{j=0}^k d(a^j b^{k-j}, A_0)$. It is easy to see that the last expression is equal to $\frac{k(k+1)}{2} d(a,b) - |A_0|$, where $|A_0|$ is the sum of the distances between all pairs of points in A_0: $|A_0| = \frac{1}{2}\sum_{a_1,a_2 \in A_0} d(a_1,a_2)$. Similarly, if A_n is the final configuration of the optimal off-line algorithm, then $\Phi_n \leq (k+1)w_n(A_n) + \frac{k(k+1)}{2} d(a,b) - |A_n|$. It follows that the extended cost is bounded above by $\Phi_n - \Phi_0 \leq (k+1)w_n(A_n) - |A_n| + |A_0| \leq (k+1)w_n(A_n) + |A_0|$, which shows that the total extending cost is bounded above by $(k+1)\text{OPT}(\rho) + \text{const}$ and the proof is complete. □

4 The WFA for Weighted Cache

It is well known that the problem of accessing pages in a weighted cache can be modeled by the k-server problem on weighted star instaces (trees of depth 1). The leaves of the star represent pages and the leaves where servers reside correspond to the pages in the cache. The weight on the edge from the leaf to the center is half of the cost for fetching the corresponding page into the cache (since the server has to pay this cost twice per passing thru that leaf). The center of the star is denoted c.

We show that WFA is k-competitive on such instances.

Recall that a minimizer of x is a configuration A that minimizes $m_i(A,x) = w_i(A) - \sum_{a \in A} d(a,x)$. It is easy to see that there is always a minimizer that does not include x. Define $\mu_i(A,x)$ as follows: If $x \notin A$ then $\mu_i(A,x) = w_i(A) - \sum_{a\in A} d(a,c) - d(c,x)$; otherwise, if $x \in A$, let $\mu_i(A,x) = w_i(A) - \sum_{a \in A-x} d(a,c)$. Since $\mu_i(A,x) = m_i(A,x) + (k-1)d(c,x)$, we have that a configuration A is a minimizer if and only if it minimizes $\mu_i(A,x)$.

Let the configuration of an adversary be $U_i = \{u_1, \ldots, u_k\}$. We define:

$$\Phi(U_i, w_i) = \sum_{l=1}^{k} \min_A \mu_i(A, u_l).$$

Let the next request be r_{i+1} and assume that the adversary moves the server from u_j to the request. The new adversary configuration is $U_i - u_j + r_{i+1}$. The next lemma bounds the change in Φ.

Lemma 3. *For any configuration U_i, any $u_j \in U_i$, and any r_{i+1}*

$$\Phi(U_i - u_j + r_{i+1}, w_i) - \Phi(U_i, w_i) \geq -d(u_j, r_{i+1}).$$

Proof. Let A be an arbitrary configuration that does not contain r_{i+1}. We first show that there exists a configuration A' such that $\mu_i(A, r_{i+1}) \geq \mu_i(A', u_j) - d(u_j, r_{i+1})$.

If $u_j \notin A$ then let $A' = A$. We have

$$\begin{aligned} \mu_i(A, r_{i+1}) &= w_i(A) - \sum_{a \in A} d(a, c) - d(c, r_{i+1}) \\ &\geq w_i(A) - \sum_{a \in A} d(a, c) - d(c, u_j) - d(u_j, r_{i+1}) \\ &= \mu_i(A', u_j) - d(u_j, r_{i+1}). \end{aligned}$$

If $u_j \in A$ then let $A' = A - u_j + r_{i+1}$. We have

$$\begin{aligned} \mu_i(A, r_{i+1}) &= w_i(A) - \sum_{a \in A} d(a, c) - d(c, r_{i+1}) \\ &= w_i(A) - \sum_{a \in A - u_j} d(a, c) - d(c, u_j) - d(c, r_{i+1}) \\ &\geq w_i(A - u_j + r_{i+1}) - d(u_j, r_{i+1}) \\ &\quad - \sum_{a \in A - u_j + r_{i+1}} d(a, c) - d(c, u_j) \\ &= \mu_i(A', u_j) - d(u_j, r_{i+1}). \end{aligned}$$

It follows that

$$\Phi(U_i - u_j + r_{i+1}, w_i) - \Phi(U_i, w_i) = \min_A \mu_i(A, r_{i+1}) - \min_A \mu_i(A, u_j) \geq -d(u_j, r_{i+1}).$$

□

Lemma 4. *For any configuration U_{i+1} that contains the last request r_{i+1} of w_{i+1}*

$$\Phi(U_{i+1}, w_{i+1}) - \Phi(U_{i+1}, w_i) \geq \max_X \{w_{i+1}(X) - w_i(X)\}.$$

Proof. Let B be a minimizer of r_{i+1} with respect to w_i that does not contain r_{i+1}. Then by the Duality lemma (Lemma 1), B is also a minimizer of r_{i+1} with respect to w_{i+1}. From the monotinicity property of work functions we have:

$$\mu_{i+1}(A, u_l) \geq \mu_i(A, u_l)$$

for all A and u_l. It follows that

$$\begin{aligned}\Phi(U_{i+1}, w_{i+1}) - \Phi(U_{i+1}, w_i) &\geq \min_A \mu_{i+1}(r_{i+1}) - \min_A \mu_i(r_{i+1}) \\ &= w_{i+1}(B) - \left(\sum_{b\in B} d(b,c)\right) - d(c, r_{i+1}) \\ &\quad - \left[w_i(B) - \left(\sum_{b\in B} d(b,c)\right) - d(c, r_{i+1})\right] \\ &= w_{i+1}(B) - w_i(B)\end{aligned}$$

The proof is complete, since by the Duality lemma:

$$w_{i+1}(B) - w_i(B) = \max_X \{w_{i+1}(X) - w_i(X)\}.$$

□

We can now combine the two above lemmata to get the main result of this section.

Theorem 2. *The work function algorithm is k-competitive for the weighted star.*

Proof. Let w_0, w_n be the initial and final work functions, and U_0, U_n be the initial and final adversary configurations respectively.

Let EXT and OPT denote the total extended cost and the optimal offline cost.

Combining Lemmas 3 and 4 we get that

$$\Phi(U_{i+1}, w_{i+1}) - \Phi(U_i, w_i) \geq \max_X \{w_{i+1}(X) - w_i(X)\} - d(u_j, u'_j),$$

where $u'_j = r_{i+1}$. The distance $d(u_j, u'_j) = d(U_i, U_{i+1})$ is the cost of the adversary to service r_{i+1}.

Summing for all requests and assuming that the adversary moves optimally, we get

$$\Phi(U_n, w_n) - \Phi(U_0, w_0) \geq \text{EXT} - \text{OPT}.$$

Since $\Phi_n = \Phi(U_n, w_n) \leq k \cdot w_n(U_n)$, and $\Phi_0 = \Phi(U_0, w_0) = -|U_0|$ (the sum of the distances between all pairs of points in U_0), we obtain

$$\text{EXT} \leq \Phi_n - \Phi_0 + \text{OPT} \leq (k+1) \cdot \text{OPT} + |U_0|.$$

The total extended cost is bounded above by $k+1$ times the optimal cost plus a constant depending only on the initial configuration. We conclude that the work function algorithm is k-competitive for weighted star metric spaces. □

5 Metric Spaces with $k+2$ Points

In this section, we show that the k-server conjecture holds for metric spaces of $k+2$ points. This result was first shown in [7,10], but we give a simpler proof here. As in [7,10], instead of studying the k-server problem on $k+2$ points, it is simpler to consider the "dual" problem which is called the *2-evader problem.* In the 2-evader problem, 2 evaders occupy *distinct* points of a metric space M of $k+2$ points. The evaders respond to a sequence of ejections (requests) which is simply a sequence of points. If an evader occupies the point of an ejection, it has to move to some other point. The objective is to minimize the total distance traveled by the 2 evaders.

The 2-evader problem is equivalent to the k-server problem: servers occupy the points not occupied by evaders, and an ejection for the evaders is a request for the servers. This equivalence allows the theory of the k-server problem and in particular the notion of the extended cost and the Duality lemma to be transfered to the evader problem. See [10] for a more extensive discussion of the evader problem and its equivalence to the k-server problem. The extended cost is again equal to the maximum increase of the work function. The corresponding Duality lemma is:

Lemma 5 (Duality lemma for the 2-evader problem). *Assume that* $\{x,y\}$ *minimizes the expression* $w_i(x,y)+d(r_{i+1},x)+d(r_{i+1},y)$. *Then* $\{x,y\}$ *minimizes also* $w_{i+1}(x,y)+d(r_{i+1},x)+d(r_{i+1},y)$ *and maximizes the extended cost:*

$$\max_{x,y}\{w_{i+1}(x,y)-w_i(x,y)\}.$$

As in the k-server problem, a configuration $\{x,y\}$ that minimizes $w_i(x,y)+d(p,x)+d(p,y)$ is called a minimizer of p with respect to w_i. It is not hard to show (see [10]) that without loss of generality a minimizer of a point p contains p. In particular, a minimizer of r_{i+1} is a configuration $\{r_{i+1},x\}$ that minimizes $w_i(r_{i+1},x)+d(r_{i+1},x)$.

With the Duality lemma, we are ready to prove the main theorem of this section. We will make use of the following notational convenience: whenever we write $w(x,y)$, we implicitly mean that x and y are distinct.

Theorem 3. *The* WFA *algorithm is k-competitive in every metric space of* $k+2$ *points.*

Proof. The argument again is based on a potential. We want to find a potential Φ_i that "includes" a minimizer of r_{i+1}. It is easy to see that the potential $\hat{\Phi}_i = \sum_a \min_x\{w_i(a,x)+d(a,x)\}$ includes a minimizer of r_{i+1} and can be used to prove that the WFA algorithm is $(k+1)$-competitive. This follows from $\hat{\Phi}_{i+1} - \hat{\Phi}_i \geq \min_x\{w_{i+1}(r_{i+1},x)+d(r_{i+1},x)\} - \min_x\{w_i(r_{i+1},x)+d(r_{i+1},x)\}$; by the Duality lemma, the last expression is equal to the extended cost to service r_{i+1}. Clearly, the total extended cost is $\hat{\Phi}_n - \hat{\Phi}_0$. Since $\hat{\Phi}_n$ is within a constant from $(k+2)\text{OPT}(\rho)$ and $\hat{\Phi}_0$ is constant, it follows that the WFA has competitive ratio at most $k+1$.

How can we alter $\hat{\Phi}_i$ to reduce the competitive ratio to k? Let b_1 and b_2 minimize $w(x,y)+d(x,y)$. The crucial observation is that $\{b_1,b_2\}$ is a minimizer of both b_1 and b_2. Thus, the number of distinct minimizers is at most $k+1$. Equivalently, even if we subtract $\min_{x,y}\{w_i(x,y)+d(x,y)\}$ from $\hat{\Phi}_i$, the resulting expression still contains a minimizer for every point and in particular of r_{i+1}. This suggests the following potential:

$$\Phi_i = \sum_a \min_x\{w_i(a,x)+d(a,x)\} - \min_{x,y}\{w_i(x,y)+d(x,y)\}. \tag{1}$$

Notice that $\Phi_i = \sum_{a\neq b_1} \min_x\{w_i(a,x)+d(a,x)\} = \sum_{a\neq b_2} \min_x\{w_i(a,x)+d(a,x)\}$. Since b_1 and b_2 are distinct, at least one of them is not equal to r_{i+1}; without loss of generality, say $b_1 \neq r_{i+1}$. By expressing

$$\Phi_i = \sum_{a\neq b_1} \min_x\{w_i(a,x)+d(a,x)\},$$

we observe that the sum includes the term corresponding to r_{i+1}. For the potential Φ_{i+1}, we also get

$$\begin{aligned}
\Phi_{i+1} &= \sum_a \min_x\{w_{i+1}(a,x)+d(a,x)\} - \min_{x,y}\{w_{i+1}(x,y)+d(x,y)\} \\
&\geq \sum_a \min_x\{w_{i+1}(a,x)+d(a,x)\} - \min_y\{w_{i+1}(b_1,y)+d(b_1,y)\} \\
&= \sum_{a\neq b_1} \min_x\{w_{i+1}(a,x)+d(a,x)\}.
\end{aligned}$$

Therefore, by subtracting, we get $\Phi_{i+1}-\Phi_i \geq \min_x\{w_{i+1}(r_{i+1},x)+d(r_{i+1},x)\} - \min_x\{w_i(r_{i+1},x)+d(r_{i+1},x)\}$ which is equal to the extended cost to service r_{i+1}. By applying to Φ_i the same argument we used for $\hat{\Phi}_i$, we establish that the WFA algorithm is k-competitive.

Notice an important difference between the potential we use in this proof and the potential of [10]: the potential here involves a max operator (the minus min part of (1)). On the other hand, the potential of [10] has only a min operator (and seems to be the minimal potential). □

6 Conclusions

We showed that the WFA algorithm is k-competitive for the line, the weighted cache and for all metric spaces of $k+2$ points. In all cases, we exploited the fact that the number of different minimizers is $k+1$. This suggests that it may be worth investigating the cardinality of the set of minimizers for other special metric spaces, even for general metric spaces. Even if a metric space is guaranteed to have at most $k+1$ minimizers, we don't know how to use this fact in general to establish that the WFA is k-competitive for this metric space. Is there a simple sufficient condition for this? Finally, as an intermediate step towards establishing the k-server conjecture, can we show that the WFA is k-competitive for trees?

References

1. Yair Bartal and Edward Grove. The Harmonic k-server algorithm is competitive. To appear in Journal of the ACM, 1999.
2. Allan Borodin and Ran El-Yaniv. *Online Computation and Competitive Analysis.* Cambridge University Press, 1998.
3. Marek Chrobak, Howard Karloff, Tom H. Payne, and Sundar Vishwanathan. New results on server problems. *SIAM Journal on Discrete Mathematics*, 4:172–181, 1991.
4. Marek Chrobak and Lawrence L. Larmore. An optimal online algorithm for k servers on trees. *SIAM Journal on Computing*, 20:144–148, 1991.
5. Marek Chrobak and Lawrence L. Larmore. The server problem and on-line games. In *DIMACS Series in Discrete Mathematics and Theoretical Computer Science*, volume 7, pages 11–64, 1992.
6. Amos Fiat, Yuval Rabani, and Yiftach Ravid. Competitive k-server algorithms. *Journal of Computer and System Sciences*, 48:410–428, 1994.
7. Elias Koutsoupias. *On-line algorithms and the k-server conjecture.* PhD thesis, University of California, San Diego, La Jolla, California, June 1994.
8. Elias Koutsoupias. Weak adversaries for the k-server problem. In *Proc 40th Symp. Foundations of Computer Science*, 1999.
9. Elias Koutsoupias and Christos Papadimitriou. On the k-server conjecture. *Journal of the ACM*, 42(5):971–983, September 1995.
10. Elias Koutsoupias and Christos Papadimitriou. The 2-evader problem. *Information Processing Letters*, 57(5):249–252, March 1996.
11. Mark Manasse, Lyle A. McGeoch, and Daniel Sleator. Competitive algorithms for online problems. In *Proc. 20th Symp. Theory of Computing*, pages 322–333, 1988.

Spectral Bounds on General Hard Core Predicates*
(Extended Abstract)

Mikael Goldmann[1] and Alexander Russell[2]**

[1] Numerical Analysis and Computer Science, Royal Institute of Technology, S-100 44 Stockholm, Sweden
[2] Computer Science & Engineering, University of Connecticut, Storrs, CT 06269-3155, USA

Abstract. A Boolean function b is a hard core predicate for a one-way function f if b is polynomial time computable but $b(x)$ is difficult to predict from $f(x)$. A general family of hard core predicates is a family of functions containing a hard core predicate for any one-way function. A seminal result of Goldreich and Levin asserts that the family of parity functions is a general family of hard core predicates. We show that no general family of hard core predicates can consist of functions with $O(n^{1-\epsilon})$ average sensitivity, for any $\epsilon > 0$. As a result, such families cannot consist of monotone functions, functions computed by generalized threshold gates, or symmetric d-threshold functions, for $d = O(n^{1/2-\epsilon})$ and $\epsilon > 0$. This also subsumes a 1997 result of Goldmann and Näslund which asserts that such families cannot consist of functions computable in AC^0. The above bound on sensitivity is obtained by (lower) bounding the high order terms of the Fourier transform.

1 Introduction

A basic assumption on which much of modern (theoretical) cryptography rests is the existence of *one-way functions*. In general, such functions may have quite pathological structure, and the development of useful cryptographic primitives from general one-way functions (often with additional properties) is one of the triumphs of modern cryptography. One of the more troubling ways that a one-way function may be unsatisfactory is that it may "leak" information about x into $f(x)$; in particular, it may be possible to compute nearly all of x from $f(x)$ in polynomial time. The problem of showing that $f(x)$ hides *at least one bit* of information about x is the *hard core predicate* problem. Goldreich and Levin [4], in a seminal 1989 paper, demonstrated that *every one-way function has a hard core predicate.* Specifically, they show that for any one-way function f, there is

* Part of this work was done while visiting McGill University

** Supported by NSF NYI Grant No. CCR-9457799 and a David and Lucile Packard Fellowship for Science and Engineering. This research was done while the author was a postdoc at the University of Texas at Austin.

H. Reichel and S. Tison (Eds.): STACS 2000, LNCS 1770, pp. 614–625, 2000.

a polynomial-time predicate b_f so that $b_f(x)$ is difficult to compute from $f(x)$. A hard core predicate, though a basic primitive, has remarkably potency:

- If f is a permutation, a hard core predicate immediately gives rise to a pseudorandom generator.
- If f is a permutation, a hard core predicate immediately gives rise to a secure bit-commitment scheme.
- If f is a one-way trapdoor permutation, a hard core predicate for f immediately gives rise to a probabilistic encryption scheme (see [5]).
- The Goldreich–Levin construction of a hard core predicate for any one-way function is an important ingredient in the proof that the existence of one-way functions implies the existence of pseudorandom generators [7].

Considering their importance, attention has been given to how *simple* such predicates can be. A 1997 result of Goldmann and Näslund [2] shows that they cannot, in general, be computed in AC^0. We strengthen this result, demonstrating that, in general, hard core predicates must have a non-negligible portion of their Fourier transform concentrated on high-degree coefficients. From this it follows that such predicates

- cannot have small average sensitivity (specifically, they cannot have average sensitivity $O(n^{1-\epsilon})$ for any $\epsilon > 0$),
- cannot be monotone,
- cannot be computed by generalized threshold functions, and
- cannot be computed by symmetric d-threshold functions, with $d = O(n^{1/2-\epsilon})$ for any $\epsilon > 0$.

As mentioned above, this bound on the spectrum also implies that general hard-core predicates cannot be in AC^0, as it is known that the Fourier transform of any AC^0 function is concentrated on coefficients of weight $\log^{O(1)} n$ [10]. It is interesting to note that these results parallel those for universal hash functions obtained by Mansour et. al. in [12].

Section 2 defines the notions of one-way function and hard core predicate. Section 3 briefly erects the framework of Fourier analysis for Boolean functions. Sections 4 and 5 are devoted to proving the main theorem and discussing some applications.

2 One-Way Functions and Hard Core Predicates

A function $f\colon \{0,1\}^* \longrightarrow \{0,1\}^*$ is *length preserving* if $f(\{0,1\}^n) \subset \{0,1\}^n$ for all n. We write $f^{(n)}$ for f restricted to inputs of length n. For convenience, and without loss of generality, we restrict our attention to length preserving one-way functions:

Definition 1. *A (length-preserving) function $f\colon \{0,1\}^* \longrightarrow \{0,1\}^*$ is a* one-way function *if f is computable in polynomial time, and for all functions $A\colon \{0,1\}^* \longrightarrow \{0,1\}^*$ computable by polynomial-size circuits and for all $k > 0$,*

$$\Pr\left[f(A(f^{(n)}(x))) = f^{(n)}(x)\right] = O(n^{-k}) ,$$

where the probability is taken uniformly over all $x \in \{0,1\}^n$.

In this cryptographic setting we consider A to be a polynomially bounded adversary attempting to invert the function f.

As discussed in the introduction, a *hard core predicate* for a one-way function f is a polynomial time predicate b for which the value $b(x)$ is difficult to predict from $f(x)$. For reasons which will become clear later, it will be convenient for us to express Boolean functions as functions taking values in the set $\{\pm 1\}$.

Definition 2. *The Boolean function* $b\colon \{0,1\}^* \longrightarrow \{\pm 1\}$ *is a* hard core predicate *for a length-preserving one-way function* f *if* b *is computable in polynomial time and for all functions* $A\colon \{0,1\}^* \longrightarrow \{\pm 1\}$*, computable by polynomial-size circuits, and for all* $k > 0$,

$$\Pr\left[A(f^{(n)}(x)) = b^{(n)}(x)\right] = \frac{1}{2} + O(n^{-k}) ,$$

this probability taken uniformly over all $x \in \{0,1\}^n$.

For a more detailed discussion of one-way functions, hard core predicates, and their uses in modern cryptography, see [3] and [11].

3 Fourier Analysis of Boolean Functions

Let $L(\mathbb{Z}_2^n) = \{f : \mathbb{Z}_2^n \to \mathbb{R}\}$ denote the set of real valued functions on $\mathbb{Z}_2^n = \{0,1\}^n$. Though our interest shall be in Boolean functions, it will be temporarily convenient to consider this richer space. $L(\mathbb{Z}_2^n)$ is a vector space over $\mathbb{R}$ of dimension 2^n, and has a natural inner product: for $f, g \in L(\mathbb{Z}_2^n)$, we define

$$\langle f, g\rangle = \frac{1}{2^n} \sum_{x\in\{0,1\}^n} f(x)g(x) .$$

For a subset $\alpha \subset \{1,\ldots,n\}$, we define $\chi_\alpha : \{0,1\}^n \to \mathbb{R}$ so that $\chi_\alpha(x) = \prod_{a\in\alpha}(-1)^{x_a}$. These functions χ_α are the *characters* of $\mathbb{Z}_2^n = \{0,1\}^n$. Among their many wonderful properties is the fact that *the characters form an orthonormal basis for* $L(\mathbb{Z}_2^n)$:

Proposition 1.

1. $\forall \alpha \subset [n]$, $\sum_{x\in\{0,1\}^n} \chi_\alpha(x) = \begin{cases} 2^n & \text{if } \alpha = \emptyset \\ 0 & \text{otherwise,} \end{cases}$
2. $\forall \alpha, \beta \subset [n]$, $\chi_\alpha(x)\chi_\beta(x) = \chi_{\alpha \oplus \beta}(x)$, *where* $\alpha \oplus \beta$ *denotes the symmetric difference of* α *and* β, *and*
3. $\forall \alpha, \beta \subset [n]$, $\langle \chi_\alpha, \chi_\beta\rangle = \begin{cases} 1 & \text{if } \alpha = \beta \\ 0 & \text{otherwise.} \end{cases}$

Considering item 3, the characters $\{\chi_\alpha \mid \alpha \subset [n]\}$ are orthogonal and have unit length. Since there are 2^n characters, they span $L(\mathbb{Z}_2^n)$, as promised. Any function $f : \{0,1\}^n \to \mathbb{R}$ may then be written in terms of this basis: $f = \sum_{\alpha \subset [n]} \widehat{f}_\alpha \chi_\alpha$ where $\widehat{f}_\alpha = \langle f, \chi_\alpha \rangle$ is the projection of f onto χ_α. These coefficients $\widehat{f}_\alpha$, $\alpha \subset [n]$, are the *Fourier coefficients* of f, and, as we have above observed, uniquely determine the function f.

Given the above, it is easy to establish the *Plancherel* equality:

Proposition 2. *Let* $f \in L(\mathbb{Z}_2^n)$. *Then* $\|f\|_2^2 = \sum_\alpha \widehat{f}_\alpha^2$, *where* $\|f\|_2^2 = \langle f, f \rangle = \frac{1}{2^n} \sum_{x \in \{0,1\}^n} f(x)^2$.

As always, $\widehat{f}_\emptyset = \mathsf{Exp}[f]$ and, when f is Boolean, $\sum_\alpha \widehat{f}_\alpha^2 = \|f\|_2^2 = 1$.

A prominent theme in the study of (continuous) Fourier analysis is *local-global duality*:

> ... the speed of convergence of a Fourier series improves with the smoothness of f. This reflects the fact that *local* features of f (such as smoothness) are reflected in *global* features of $\widehat{f}$ (such as rapid decay at $n = \pm\infty$). *This local-global duality is one of the major themes of Fourier series and integrals,* ...
>
> –Dym, McKean, [1, p.31]

This very same duality (between smoothness of f and rapid decay of $\widehat{f}$) shall be central for our study. In our framework, a natural measure of *smoothness* for a Boolean function f is its *average sensitivity*:

Definition 3. *The* average sensitivity *of a Boolean function* $f : \{0,1\}^n \to \{\pm 1\}$ *is the quantity*

$$\overline{\mathsf{S}}(f) = \frac{1}{2^n} \sum_{x \in \{0,1\}^n} \sum_{i=1}^{n} \frac{|f(x) - f(x \oplus e_i)|}{2}$$

where $e_i \in \{0,1\}^n$ *denotes the vector containing a single* 1 *at position* i *and* $\oplus$ *denotes coordinatewise sum modulo 2. (The* $\frac{1}{2}$ *factor appearing in the last term here reflects our choice of* $\{\pm 1\}$ *as the range of our Boolean functions.)*

Let us look at some examples. The average sensitivity for the n-input parity function is n, since for any input x, flipping any of the n input bits will change the parity. The n-input OR function has average sensitivity $2n2^{-n}$: if the input is all 0, then flipping any bit changes the value of the function (for this input the inner sum is equal to n), for any of the n inputs with a single input bit being 1 flipping *that* bit will change the value of the function (for each of these n inputs the inner sum is equal to 1), and for all inputs with at least two bits set to 1 the inner sum is equal to 0.

Observe that average sensitivity is proportional to the likelihood that a random pair of neighboring points take on different values: "smooth" functions, where neighboring points are likely to agree, should have small average sensitivity. Functions with small average sensitivity are likely to have the same value on similar looking inputs–it is this property we shall exploit.

The connection between average sensitivity (smoothness) and rapid decay of the Fourier transform is given by the following equality, due to Kahn, Kalai, and Linial [8]:

$$\bar{\mathsf{S}}(f) = \sum_{\alpha} |\alpha| \widehat{f}_{\alpha}^{2} . \tag{1}$$

Considering the above equality, and recalling that $\|f\|_2 = 1$ for a Boolean function f, the average sensitivity of f is exactly determined by the distribution of this unit mass among the terms $\widehat{f}_{\alpha}^{2}$. This is a manifestation of the local-global duality principle mentioned above: functions having their Fourier transform concentrated on small coefficients (those for which $|\alpha|$ is small) have small average sensitivity and, as such, are smooth. In this case, we opt to *define* our notion of smoothness in terms of the Fourier transform as follows:

Definition 4. *We say that a function $f\colon \{0,1\}^n \longrightarrow \{\pm 1\}$ is (t,δ)-smooth iff*

$$\sum_{|\alpha|>t} \widehat{f}_{\alpha}^{2} \leq \delta .$$

A function $g\colon \{0,1\}^* \longrightarrow \{\pm 1\}$ is $(t(n),\delta(n))$-smooth iff there exists $n_0 > 0$ so that for all $n \geq n_0$ $g^{(n)}$ is $(t(n),\delta(n))$-smooth.

A final word on notation: we will frequently study functions $f(x,y)$ that take two strings x,y as input. Using I_x and I_y as the (disjoint) index sets for x and y respectively, it will be convenient to index the Fourier coefficients of f with two sets α,β, where $\alpha \subset I_x$ and $\beta \subset I_y$:

$$f(x,y) = \sum_{\substack{\alpha\subset I_x \\ \beta\subset I_y}} \widehat{f}_{\alpha,\beta}\chi_{\alpha\cup\beta}(x,y) = \sum_{\substack{\alpha\subset I_x \\ \beta\subset I_y}} \widehat{f}_{\alpha,\beta}\chi_{\alpha}(x)\chi_{\beta}(y) ,$$

where $\chi_{\alpha\cup\beta}(x,y) = \chi_{\alpha}(x)\chi_{\beta}(y)$ since $\alpha \cap \beta = \emptyset$.

4 Main Result

We can now begin working toward our main result which asserts that if one-way functions exist, then there are one-way functions for which every hard core predicate is highly non-smooth. In section 5 we explore the consequences of this theorem for general hard core predicates.

The following theorem implies the bound on sensitivity claimed in the introduction.

Theorem 1. *If there exists a one-way function, then for every $\epsilon > 0$ there is a one-way function f_ϵ such that no $(\gamma n^{1-\epsilon},\delta)$-smooth Boolean function $b\colon \{0,1\}^* \longrightarrow \{\pm 1\}$ can be a hard core predicate for f_ϵ if $\gamma + \delta < 1/16$.*

Proof. Let $g : \{0,1\}^* \to \{0,1\}^*$ be a (length preserving) one-way function. Fix an arbitrary constant $\epsilon > 0$. We describe below a one-way function $f_\epsilon = f(x,y)$ of n variables so that if $b^{(n)}(x,y) = \sum_{\alpha,\beta} \widehat{b^{(n)}}_{\alpha,\beta}\chi_\alpha(x)\chi_\beta(y)$ and $\sum_{|\alpha|+|\beta|>\gamma n^{1-\epsilon}}(\widehat{b^{(n)}}_{\alpha,\beta})^2 \leq \delta$ then, given $f(x,y)$, one can guess $b^{(n)}(x,y)$ with high probability. We will henceforth abandon the superscript on b whenever convenient.

Assume that n is even and define $f(x,y)$ where $|x| = |y| = n/2$ as follows. For an element $w \in \{0,1\}^k$ and a subset $S = \{s_1, \ldots, s_l\} \subset \{1, \ldots, k\}$, let $w_S = w_{s_1} \ldots w_{s_l}$, where $s_1 < \cdots < s_l$. The input x is divided into $t_x \stackrel{\text{def}}{=} n^{1-\epsilon}$ blocks, each consisting of $W_x(n) \stackrel{\text{def}}{=} n^\epsilon/2$ bits. Similarly, y is divided into $t_y \stackrel{\text{def}}{=} \log n^{1-\epsilon}$ blocks $B_1, \ldots, B_{t_y}$, each consisting of $W_y(n) \stackrel{\text{def}}{=} n/(2\log n^{1-\epsilon})$ bits. For simplicity, we ignore issues of integrality for these quantities. Then the value $f(x,y)$ is computed as follows. Write $y = y_{B_1} \cdots y_{B_{t_y}}$ and let $J_i = \bigoplus_{k \in B_i} y_k$ be the parity of the bits in y_{B_i}; interpret the result $J_1, \ldots, J_{t_y}$ as a binary coded integer $J(y) \in \{0, \ldots, n^{1-\epsilon} - 1\}$.

We define the set

$$\mathcal{J}(y) = \left\{ J(y)\frac{n^\epsilon}{2}, J(y)\frac{n^\epsilon}{2} + 1, \ldots, [J(y)+1]\frac{n^\epsilon}{2} - 1 \right\} ;$$

these are precisely the indices of the $J(y)$th block of x. Finally, define $f(x,y) = (z,y)$ where $z_i = x_i$ when $i \notin \mathcal{J}(y)$ and $z_{\mathcal{J}(y)} = g(x_{\mathcal{J}(y)})$. Clearly f is a one-way function, since inverting f in polynomial time implies inversion of g on n-bit inputs in time polynomial in $(2n)^{1/\epsilon}$.

Now, let $b : \{0,1\}^* \to \{\pm 1\}$ be a $(\gamma n^{1-\epsilon}, \delta)$-smooth Boolean predicate. Our goal is to show that b cannot be a hard core predicate for f. For the remainder of this section, fix the input length to n, an integer large enough so that $b^{(n)}$ is $(\gamma n^{1-\epsilon}, \delta)$-smooth and $4\gamma < n^\epsilon / \log n^{1-\epsilon}$.

The following lemma then implies the theorem.

Lemma 1. *If $b^{(n)}(x,y)$ is $(\gamma n^{1-\epsilon}, \delta)$-smooth, then there is a probabilistic polynomial time algorithm A_b such that*

$$\Pr[A_b(f(x,y)) = b(x,y)] \geq 1 - 8(\gamma + \delta) ,$$

this probability taken over uniformly random choice of x and y and the coin tosses of A_b.

Remark 1. Hard core predicates are defined with respect to adversaries that are polynomial-size circuits. However, since probabilistic polynomial-time algorithms are less powerful than polynomial-size circuits, the above lemma is sufficient.

Describing A_b is simple. Given $f(x,y)$, y, and hence $\mathcal{J}(y)$ is known. Also, all of x except $x_{\mathcal{J}(y)}$ is known. Form x' by letting $x'_i = x_i$ when $i \notin \mathcal{J}(y)$, and picking

$x'_{\mathcal{J}(y)}$ uniformly at random. Finally, let $A_b(f(x,y)) = b(x',y)$. The guess is correct when $b(x,y) = b(x',y)$. Note that x' and y are independent and uniformly distributed. However, x and x' are of course highly dependent.

In this case, Lemma 1 follows from Lemma 2, below.

Lemma 2. *If $b(x,y)$ is $(\gamma n^{1-\epsilon}, \delta)$-smooth, and x, x', y are generated as described above, then* $\Pr[b(x,y) = b(x',y)] \geq 1 - 8(\gamma + \delta)$.

Let Z be the indicator function

$$Z(A,B) = \begin{cases} 1 & \text{if } A \cap B \neq \emptyset, \text{ and} \\ 0 & \text{otherwise.} \end{cases}$$

and define

$$\begin{aligned} e(x,y) &= \sum_{|\alpha|+|\beta| \leq \gamma n^{1-\epsilon}} \widehat{b}_{\alpha,\beta} \chi_\alpha(x) \chi_\beta(y)(1 - Z(\alpha, \mathcal{J}(y))) , \\ h(x,y) &= \sum_{|\alpha|+|\beta| \leq \gamma n^{1-\epsilon}} \widehat{b}_{\alpha,\beta} \chi_\alpha(x) \chi_\beta(y) Z(\alpha, \mathcal{J}(y)) , \text{ and} \\ r(x,y) &= \sum_{|\alpha|+|\beta| > \gamma n^{1-\epsilon}} \widehat{b}_{\alpha,\beta} \chi_\alpha(x) \chi_\beta(y) . \end{aligned}$$

Now, $b(x,y) = e(x,y) + h(x,y) + r(x,y)$, and note that $e(x,y)$ depends only on inputs *exposed* by $f(x,y)$ (i.e., it does not depend on $x_{\mathcal{J}(y)}$), whereas each term in $h(x,y)$ depends on some *hidden* bits (i.e., bits in $x_{\mathcal{J}(y)}$).

Observe that when x, x', y are generated according to the above procedure, $e(x,y) = e(x',y)$. We will prove that for random (x,y), with high probability both $|r(x,y)|$ and $|h(x,y)|$ are small—this is enough to prove that with high probability $b(x,y) = b(x',y)$. The contributions of $r(x,y)$ and $h(x,y)$ are bounded by the following two lemmas.

Lemma 3. *If $b(x,y)$ is $(\gamma n^{1-\epsilon}, \delta)$-smooth, and x, y are uniformly distributed, then*

$$\Pr\left[|r(x,y)| \geq \lambda\right] \leq \lambda^{-2}\delta .$$

Proof. We bound the probability using a special case of the Chebychev inequality: for a real-valued random variable X such that $\mathrm{Exp}[X] = 0$,

$$\Pr\left[|X - \mathrm{Exp}[X]| \geq \lambda\right] \leq \lambda^{-2}\, \mathrm{Exp}[X^2] . \tag{2}$$

As $\mathrm{Exp}[r(x,y)] = 0$, we have by linearity of expectation

$$\Pr\left[|r(x,y)| \geq \lambda\right] \leq \lambda^{-2} \sum_{|\alpha|+|\beta| > \gamma n^{1-\epsilon}} (\widehat{b}_{\alpha,\beta})^2\, \mathrm{Exp}[(\chi_\alpha(x)\chi_\beta(y))^2] \leq \lambda^{-2}\delta .$$

□

Lemma 4. *If* $4\gamma < n^{\epsilon}/\log n^{1-\epsilon}$ *then* $\Pr\big[|h(x,y)| \geq \lambda\big] \leq \lambda^{-2}\gamma$.

The proof of Lemma 4 is slightly technical, so let us first see that Lemmas 3 and 4 together imply that with high probability $b(x,y) = b(x',y)$.

Proof (of Lemma 2). Since $|b(x,y)| = |b(x',y)| = 1$ it is enough to show that with high probability $|b(x,y) - b(x',y)| < 2$.

Considering that $e(x,y) = e(x',y)$, applying the triangle inequality we have

$$|b(x,y) - b(x',y)| \leq |r(x,y)| + |r(x',y)| + |h(x,y)| + |h(x',y)| \;.$$

Therefore,

$$\begin{aligned}\Pr\left[|b(x,y) - b(x',y)| \geq 2\right] \leq \Pr\big[|r(x,y)| \geq 1/2\big] + \Pr\big[|r(x',y)| \geq 1/2\big] + \\ \Pr\big[|h(x,y)| \geq 1/2\big] + \Pr\big[|h(x',y)| \geq 1/2\big] \\ \leq 8\delta + 8\gamma \;.\end{aligned}$$

The last inequality follows by two applications of Lemma 3 and two applications of Lemma 4 (with $\lambda = 1/2$). □

As mentioned above, from Lemma 2 follows Theorem 1. □

It remains to prove Lemma 4.

Proof (Proof of Lemma 4). By linearity of expectation and independence of x and y,

$$\mathsf{Exp}[h(x,y)] = \sum_{|\alpha|+|\beta| \leq \gamma n^{1-\epsilon}} \widehat{b}_{\alpha,\beta}\, \mathsf{Exp}[\chi_\alpha(x)]\, \mathsf{Exp}[Z(\alpha, \mathcal{J}(y))\chi_\beta(y)] \;.$$

Now, $Z(\emptyset, \mathcal{J}(y)) = 0$ for all y, and when $\alpha \neq \emptyset$ then $\mathsf{Exp}[\chi_\alpha(x)] = 0$; hence each term in $h(x,y)$ has expectation 0 and $\mathsf{Exp}[h(x,y)] = 0$. By (2), it follows that

$$\Pr\big[|h(x,y)| \geq \lambda\big] \leq \lambda^{-2}\, \mathsf{Exp}[(h(x,y))^2] \;.$$

Expanding $\mathsf{Exp}[(h(x,y))^2]$ yields the expression

$$\sum_{\substack{|\alpha|+|\beta| \leq \gamma n^{1-\epsilon} \\ |\alpha'|+|\beta'| \leq \gamma n^{1-\epsilon}}} \widehat{b}_{\alpha,\beta}\widehat{b}_{\alpha',\beta'}\, \mathsf{Exp}[\chi_{\alpha \oplus \alpha'}(x)]\, \mathsf{Exp}[\chi_{\beta \oplus \beta'}(y) Z(\alpha, \mathcal{J}(y)) Z(\alpha', \mathcal{J}(y))] \;.$$

As $\mathsf{Exp}[\chi_\alpha(x)\chi_{\alpha'}(x)] = \delta_{\alpha,\alpha'}$, we have

$$\Pr[|h(x,y)| \geq \lambda] \leq \lambda^{-2} \sum_{\substack{|\alpha|+|\beta| \leq \gamma n^{1-\epsilon} \\ |\alpha|+|\beta'| \leq \gamma n^{1-\epsilon}}} \widehat{b}_{\alpha,\beta}\widehat{b}_{\alpha,\beta'}\, \mathsf{Exp}[\chi_{\beta \oplus \beta'}(y) Z(\alpha, \mathcal{J}(y))] \;. \quad (3)$$

Accept, for the moment, the following claim.

Claim. *If* $\max(|\beta|, |\beta'|) < \gamma n^{1-\epsilon}$, *then* $\mathsf{Exp}[\chi_{\beta \oplus \beta'}(y) Z(\alpha, \mathcal{J}(y))] \leq \gamma\, \delta_{\beta,\beta'}$.

From the claim and (3) we have

$$\Pr[|h(x,y)| \geq \lambda] \leq \lambda^{-2} \sum_{|\alpha|+|\beta| \leq \gamma n^{1-\epsilon}} (\widehat{b}_{\alpha,\beta})^2\, \gamma \leq \lambda^{-2}\gamma\,.$$

The last inequality follows from the Plancherel equality.

It remains to prove the claim. First note that $\chi_{\beta \oplus \beta'}(y)$ and $Z(\alpha, \mathcal{J}(y))$ are independent when $\max(|\beta|, |\beta'|) < \gamma n^{1-\epsilon}$. This follows from the fact that each 'bit' in $J(y)$ is the parity of $n/(2\log n^{1-\epsilon})$ bits, whereas $\chi_{\beta \oplus \beta'}(y)$ only depends on the parity of $|\beta \oplus \beta'| \leq 2\gamma n^{1-\epsilon}$ bits. Thus, even when all the bits in $\beta \oplus \beta'$ are fixed, the bits of $J(y)$ are still uniformly and independently distributed. As $\mathsf{Exp}[\chi_{\beta \oplus \beta'}(y)] = \delta_{\beta,\beta'}$, it remains to show that $\mathsf{Exp}[Z(\alpha, \mathcal{J}(y))] \leq \gamma$, but this follows from the fact that $\mathsf{Exp}[Z(\alpha, \mathcal{J}(y))]$ is the probability that a randomly picked block intersects the fixed set α. This is bounded from above by $|\alpha| / n^{1-\epsilon}$, and as $|\alpha| \leq \gamma n^{1-\epsilon}$, we are done. □

5 General Families of Hard Core Predicates

In this section we consider a slightly different definition for the concept of a hard core predicate. We require only that they are computable in *non-uniform* polynomial time (that is, computable by polynomial-size circuits). The reason for considering this weaker definition is that rather than focusing on the behavior of a single predicate, we wish to explore families of predicates guaranteed to contain a hard core predicate for any one-way function f. Such families are called *general families of hard core predicates.* Typically, as in the Goldreich–Levin construction, a randomly chosen member of the family of predicates is likely to be a hard core predicate for f, and folding this "random choice" into the definition of b (naively) requires non-uniformity. This does not greatly affect the results in the previous section. The only difference is that the algorithm A_b requires a (polynomial-size) circuit for the predicate b so that it can evaluate b on an input x, y.

Definition 5. *A family* $\mathcal{B} \subset \{\pm 1\}^{\{0,1\}^*}$ *is called a* general family of hard core predicates *if for every one-way function* f *there is a (non-uniform) polynomial time computable predicate* $b \in \mathcal{B}$*, such that* b *is a hard core predicate for* f*.*

The theorem of Goldreich and Levin mentioned in the introduction asserts that the collection of functions

$$\mathcal{B}_{\mathrm{GL}} = \left\{ p : \{0,1\}^* \to \{\pm 1\} \,\middle|\, \forall n, p^{(n)} = \chi_{\alpha_n}, \text{for some } \alpha_n \subset \{1, \ldots, n\} \right\}$$

is a general family of hard core predicates.

One consequence of the theorem of the last section is that general families of hard core predicates cannot be smooth:

Corollary 1. *If $\mathcal{B}$ is a general family of hard core predicates, then, for every $\epsilon > 0$, it must contain a function which is not $(n^{1-\epsilon}, 1/17)$-smooth.*

The close connection between smoothness and average sensitivity implies the following.

Corollary 2. *If $\mathcal{B}$ is a general family of hard core predicates then, for every $\epsilon > 0$, $\mathcal{B}$ must contain a function with average sensitivity greater than $n^{1-\epsilon}$ for all sufficiently large n.*

Proof sketch. The lower bound on the sensitivity follows from equation (1) coupled with the lower bound on smoothness. If $S(b) \leq n^{1-\epsilon}$, then for $\epsilon' < \epsilon$, b is $(n^{1-\epsilon'}, n^{\epsilon'-\epsilon})$-smooth. □

A celebrated theorem of Linial, Mansour, and Nisan shows that functions in AC^0 are smooth:

Theorem 2 ([10]). *Let $f : \{0,1\}^* \to \{\pm 1\}$ be a Boolean function with polynomial-size constant depth circuits. Then f is $(\log^{O(1)} n, o(1))$-smooth.*

An immediate corollary is a theorem of Goldmann and Näslund [2] asserting that a general family of hard core predicates must contain predicates outside AC^0:

Corollary 3. *If $\mathcal{B}$ is a general family of hard core predicates, then it must contain a function which is not in AC^0.*

It is interesting to note the folklore theorem [9] which asserts that any monotone function f has small average sensitivity:

Lemma 5. *Let f be a monotone Boolean function, then $\overline{\mathsf{S}}(f) = O(\sqrt{n})$.*

Clearly, the same bound holds for any generalized monotone function (a generalized monotone function is obtained by negating some of the inputs to a monotone function). In light of the above, the following is immediate:

Corollary 4. *If $\mathcal{B}$ is a general family of hard core predicates, then it must contain a non-monotone function.*

A Boolean function $f : \{0,1\}^n \to \{\pm 1\}$ is a *d-threshold function* if there exists a real multivariate polynomial $p \in \mathbb{R}[x_1, \ldots, x_n]$ of total degree d or less so that $\forall (x_1, \ldots, x_n) \in \{0,1\}^n$, $f(x_1, \ldots, x_n) = \operatorname{sign} p(x_1, \ldots, x_n)$. When $d = 1$ such functions are *generalized threshold functions* and are generalized monotone functions; their average sensitivity is addressed in Lemma 5 above.

In general, it has been shown by Gotsman and Linial [6] that d-threshold functions are $(d, 1-\epsilon_d)$-smooth, for a constant $\epsilon_d > 0$ independent of n. Though this is not strong enough for our application, they show that under the added assumption that f is *symmetric*, one has

$$\overline{\mathsf{S}}(f) \leq 2^{-n+1} \sum_{k=0}^{d-1} \binom{n}{[(n-k)/2]} \left(n - \left[\frac{n-k}{2} \right] \right),$$

where $[x]$ is the integer part of x. Observe that when $d = O(n^{\frac{1}{2}-\epsilon})$, this quantity is $O(n^{1-\epsilon})$. Then the following is immediate.

Corollary 5. *If $\mathcal{B}$ is a general family of hard core predicates, then it must contain a function which, for large enough n, cannot be expressed as the sign of a symmetric polynomial of degree $d = O(n^{1/2-\epsilon})$, for any $\epsilon > 0$.*

6 A Remark on the Uniform Version of the Goldreich–Levin Theorem

There is also a uniform perspective on the Goldreich–Levin construction. Given any one-way function f we construct a new one-way function g_f (defined on even length inputs) where

$$g_f^{(2n)} \colon \{0,1\}^n \times \{0,1\}^n \longrightarrow \{0,1\}^n \times \{0,1\}^n .$$

Let an n-bit string y encode a set $\alpha(y) \subset \{1,\dots,n\}$ in the natural way. Define

$$g_f^{(2n)}(x,y) = (f(x),y)$$
$$b_{\mathrm{GL}}^{(2n)}(x,y) = \chi_{\alpha(y)}(x) .$$

Then, for any one-way function f, the predicate b_{GL} is a hard-core predicate for g_f. Note that this is a uniform construction, and that the predicate b_{GL} is independent of f.

A natural way to generalize this would be to consider a construction, where for every one-way function $f(x)$ there is a padded version $g_f(x,y) = (f(x),y)$ where $|y| = p(|x|)$ for some polynomial p and a predicate b such that $b(x,y)$ is a hard core predicate for g_f. What kind of lower bounds can one show for, for instance, the sensitivity of b?

Our results extend to this notion of a general hard core predicate simply by observing that if b is a hard-core predicate for g_f for all f, then the family

$$\left\{c\colon \{0,1\}^* \longrightarrow \{\pm 1\} \,\middle|\, c^{(n)}(x) = b^{(n+p(n))}(x,y_0) \text{ for some } y_0 \in \{0,1\}^{p(n)}\right\}$$

is a general family of hard core predicates. It follows for example that b cannot be monotone and cannot be computable by AC^0-circuits. The results on sensitivity also carry over, but observe that the parameter n refers to the length of the input x, not to the combined length of x and y.

7 Conclusion and Open Questions

The results presented here indicate a certain degree of optimality on behalf of the Goldreich–Levin construction (see also [2]). (Observe that with probability 1 a function selected from B_{GL} will have linear average sensitivity.) Also, it suggests a connection between families of universal hash functions and general hard core predicates. On the one hand, several well-known examples of universal hash functions have been shown to be general hard core predicates [4,13,14], and on the other hand, smooth functions make poor hash functions as well as poor hard core predicates. An interesting (and very open-ended) problem is to determine if there is a nice connection between universal hash functions and hard core predicates.

Acknowledgment

We thank Johan Håstad and Mats Näslund for interesting and helpful discussions.

References

1. H. Dym and H. P. McKean. *Fourier Series and Integrals*, volume 14 of *Probability and Mathematical Statistics*. Academic Press, 1972.
2. M. Goldmann and M. Näslund. The complexity of computing hard core predicates. In B. S. Kaliski Jr., editor, *Advances in Cryptology—CRYPTO '97*, volume 1294 of *Lecture Notes in Computer Science*, pages 1–15. Springer-Verlag, 17–20 Aug. 1997.
3. O. Goldreich. *Modern Cryptography, Probabilistic Proofs and Pseudorandomness.* Springer-Verlag, 1999.
4. O. Goldreich and L. A. Levin. A hard-core predicate for all one-way functions. In *Proceedings of the Twenty First Annual ACM Symposium on Theory of Computing*, pages 25–32, Seattle, Washington, 15–17 May 1989.
5. S. Goldwasser and S. Micali. Probabilistic encryption. *J. Comput. Syst. Sci.*, 28(2):270–299, Apr. 1984.
6. C. Gotsman and N. Linial. Spectral properties of threshold functions. *Combinatorica*, 14(1):35–50, 1994.
7. J. Håstad, R. Impagliazzo, L. A. Levin, and M. Luby. Construction of a pseudorandom generator from any one-way function. *SIAM J. Comp.*, 28(4):1364–1396, 1998.
8. J. Kahn, G. Kalai, and N. Linial. The influence of variables on Boolean functions (extended abstract). In *29th Annual Symposium on Foundations of Computer Science*, pages 68–80, White Plains, New York, 24–26 Oct. 1988. IEEE.
9. N. Linial. Private communication. December, 1998.
10. N. Linial, Y. Mansour, and N. Nisan. Constant depth circuits, Fourier transform, and learnability. *Journal of the ACM*, 40(3): 607–620, 1993.
11. M. Luby. *Pseudorandomness and Cryptographic Applications.* Princeton University Press, 1996.
12. Y. Mansour, N. Nisan, and P. Tiwari. The computational complexity of universal hashing. *Theoretical Computer Science*, 107(1):121–133, 4 Jan. 1993.
13. M. Näslund. Universal hash functions and hard core bits. In *EUROCRYPT: Advances in Cryptology: Proceedings of EUROCRYPT*, 1995.
14. M. Näslund. All bits in $ax + b \bmod p$ are hard (extended abstract). In N. Koblitz, editor, *Advances in Cryptology—CRYPTO '96*, volume 1109 of *Lecture Notes in Computer Science*, pages 114–128. Springer-Verlag, 18–22 Aug. 1996.

Randomness in Visual Cryptography

Annalisa De Bonis and Alfredo De Santis

Dipartimento di Informatica ed Applicazioni Università di Salerno,
84081 Baronissi (SA), Italy,
{debonis,ads}@dia.unisa.it,
http://www.dia.unisa.it/~ads

Abstract. A visual cryptography scheme for a set $\mathcal{P}$ of n participants is a method to encode a secret image into n shadow images called shares each of which is given to a distinct participant. Certain *qualified* subsets of participants can recover the secret image, whereas *forbidden* subsets of participants have no information on the secret image. The shares given to participants in $X \subseteq \mathcal{P}$ are xeroxed onto transparencies. If X is *qualified* then the participants in X can visually recover the secret image by stacking their transparencies without any cryptography knowledge and without performing any cryptographic computation.
This is the first paper which analyzes the amount of randomness needed to visually share a secret image. It provides lower and upper bounds to the randomness of visual cryptography schemes. Our schemes represent a dramatic improvement on the randomness of all previously known schemes.

Keywords: Cryptography, Randomness, Secret Sharing, Visual Cryptography.

1 Introduction

A visual cryptography scheme (VCS) for a set $\mathcal{P}$ of n participants is a method to encode a secret image into n shadow images called shares each of which is given to a distinct participant. Certain *qualified* subsets of participants can recover the secret image, whereas *forbidden* subsets of participants have no information on the secret image. The specification of all qualified and forbidden subsets of participants constitutes an *access structure*. The shares given to participants in $X \subseteq \mathcal{P}$ are xeroxed onto transparencies. If X is *qualified* then the participants in X can visually recover the secret image by stacking their transparencies without any cryptography knowledge and without performing any cryptographic computation. The definition of visual cryptography scheme was given by Naor and Shamir in [14]. They analyzed (k, n)-threshold visual cryptography schemes, that is schemes where any subset of k participants is qualified, whereas groups of less than k participants are forbidden. The model by Naor and Shamir has been extended in [1,2] to general access structures.

All previous papers on visual cryptography mainly focus on two parameters: the *pixel expansion*, which represents the number of subpixels in the encoding of

H. Reichel and S. Tison (Eds.): STACS 2000, LNCS 1770, pp. 626–638, 2000.

the original image, and the *contrast*, which measures the "difference" between a black and a white pixel in the reconstructed image. In particular, several results on the contrast and the pixel expansions of VCSs can be found in [2,5,6,7,11].

This is the first paper which analyzes the amount of randomness needed to visually share a secret image. Random bits are a natural computational resource which must be taken into account when designing cryptographic algorithms. Considerable effort has been devoted to reduce the number of bits used by probabilistic algorithms (see for example [12]) and to analyze the amount of randomness required in order to achieve a given performance. Motivated by the fact that "truly" random bits are hard to generate, it has also been investigated the possibility of using imperfect source of randomness in randomized algorithms [18]. The amount of randomness used in a computation is an important issue in many practical applications. Suppose we want to secretly share an image among four participants in such a way that groups of at most three participants have no information on the secret image. The previously known VCS for such access structure is due to Naor and Shamir [14] and uses $\log(8!) \approx 15.3$ random bits per pixels. In this paper we will present a minimum randomness VCS for that access structure which uses only 3 random bits per pixels. For typical images of tens of thousands or hundreds of thousands of pixels, the difference between the randomness of these two VCSs considerably affects the time and space needed for the encoding. Randomness has played a significant role in a cryptography project recently realized at the University of Salerno. The main goal of such project was the creation of a web server which implements visual secret scheme for arbitrary access structures. To speed up the share generation, the random bits are produced beforehand and stored in a file. Because of space limitations, the file is dynamically managed in such a way that it contains enough random numbers to satisfy a certain number of client requests. Since the file management is very time consuming and adds a considerable overhead to server computation, then it follows that the amount of randomness involved in the share generation greatly affects the efficiency of the server.

This paper provides lower and upper bounds to the randomness of visual cryptography schemes. Our schemes represent a dramatic improvement on the randomness of all previously known schemes.

Outline of the Paper The model we employ to secretly share an image among n participants is described in Section 2. In Section 3 we provide a simple technique to obtain lower bounds on the randomness of any VCS and derive a lower bound on the randomness of (k,n)-threshold VCSs, for any $2 \leq k \leq n$. In Section 4 we give a complete characterization of (k,k)-threshold VCSs with both minimum randomness and minimum pixel expansion. Section 5 deals with visual cryptography schemes for general access structures and provides tools to derive upper bounds on the randomness of VCSs for any access structure. In Section 6 we provide a technique to construct minimum randomness $(2,n)$-threshold VCSs for any $n \geq 2$, and a technique to construct (k,n)-threshold VCSs for any value of k and n, with $n \geq k \geq 2$, which dramatically improves on the previously known constructions with respect to the randomness.

2 The Model

Let $\mathcal{P} = \{1, \ldots, n\}$ be a set of elements called *participants*, and let $2^{\mathcal{P}}$ denote the set of all subsets of $\mathcal{P}$. Let $\Gamma_{\mathsf{Qual}} \subseteq 2^{\mathcal{P}}$ and $\Gamma_{\mathsf{Forb}} \subseteq 2^{\mathcal{P}}$, where $\Gamma_{\mathsf{Qual}} \cap \Gamma_{\mathsf{Forb}} = \emptyset$. We refer to members of Γ_{Qual} as *qualified sets* and we call members of Γ_{Forb} *forbidden sets*. The pair $(\Gamma_{\mathsf{Qual}}, \Gamma_{\mathsf{Forb}})$ is called the *access structure* of the scheme.

Let Γ_0 consist of all the minimal qualified sets: $\Gamma_0 = \{A \in \Gamma_{\mathsf{Qual}} : A' \notin \Gamma_{\mathsf{Qual}}$ for all $A' \subset A\}$. A participant $P \in \mathcal{P}$ is an *essential* participant if there exists a set $X \subseteq \mathcal{P}$ such that $X \cup \{P\} \in \Gamma_{\mathsf{Qual}}$ but $X \notin \Gamma_{\mathsf{Qual}}$. A non-essential participant does not need to participate "actively" in the reconstruction of the image, since the information he has is not needed by any set in $\mathcal{P}$ in order to recover the shared image. In any VCS having non-essential participants, these participants do not require any information in their shares. If a participant P is not essential then we can construct a visual cryptography scheme giving him nothing as his share or, as we will see later, a share completely "white".

In the case where Γ_{Qual} is monotone increasing, Γ_{Forb} is monotone decreasing, and $\Gamma_{\mathsf{Qual}} \cup \Gamma_{\mathsf{Forb}} = 2^{\mathcal{P}}$, the access structure is said to be *strong*, and Γ_0 is termed a *basis*. (This situation is the usual setting for traditional secret sharing.) In a strong access structure, $\Gamma_{\mathsf{Qual}} = \{C \subseteq \mathcal{P} : B \subseteq C$ for some $B \in \Gamma_0\}$, and we say that Γ_{Qual} is the *closure* of Γ_0.

Notice that if a set of participants X is a superset of a qualified set X', then they can recover the shared image by considering only the shares of the set X'. This does not in itself rule out the possibility that stacking all the transparencies of the participants in X does not reveal any information about the shared image.

A (k, n)-threshold structure $(\Gamma_{Qual}, \Gamma_{Forb})$ on a set $\mathcal{P}$ of n participants is any access structure in which $\Gamma_0 = \{B \subseteq \mathcal{P} : |B| = k\}$ and $\Gamma_{Forb} = \{B \subseteq \mathcal{P} : |B| \leq k\}$. A VCS for a (k, n)-threshold structure is called (k, n)-threshold VCS.

We assume that the message consists of a collection of black and white pixels. Each pixel appears in n versions called *shares*, one for each transparency. Each share is a collection of m black and white subpixels. The resulting structure can be described by an $n \times m$ boolean matrix $M = [m_{ij}]$ where $m_{ij} = 1$ iff the j-th subpixel in the i-th transparency is black. Therefore the grey level of the combined share, obtained by stacking the transparencies $i_1, \ldots, i_s$, is proportional to the Hamming weight $w(V)$ of the m-vector $V = OR(R_{i_1}, \ldots, R_{i_s})$ where $R_{i_1}, \ldots, R_{i_s}$ are the rows of M associated with the transparencies we stack. This grey level is interpreted by the visual system of the users as black or as white according with some rule of contrast.

Definition 1. *Let $(\Gamma_{\mathsf{Qual}}, \Gamma_{\mathsf{Forb}})$ be an access structure on a set of n participants. Two collections (multisets) of $n \times m$ boolean matrices $\mathcal{C}_0$ and $\mathcal{C}_1$ constitute a* visual cryptography scheme $(\Gamma_{\mathsf{Qual}}, \Gamma_{\mathsf{Forb}})$-VCS *if there exist a value $\alpha(m)$ and a collection $\{(X, t_X)\}_{X \in \Gamma_{\mathsf{Qual}}}$ satisfying:*

1. Any (qualified) set $X = \{i_1, i_2, \ldots, i_p\} \in \Gamma_{\mathsf{Qual}}$ can recover the shared image by stacking their transparencies.
   Formally, for any $M \in \mathcal{C}_0$, the "or" V of rows $i_1, i_2, \ldots, i_p$ satisfies $w(V) \leq t_X - \alpha(m) \cdot m$; whereas, for any $M \in \mathcal{C}_1$ it results that $w(V) \geq t_X$.

2. Any (forbidden) set $X = \{i_1, i_2, \ldots, i_p\} \in \Gamma_{\mathsf{Forb}}$ has no information on the shared image.
Formally, the two collections of $p \times m$ *matrices* $\mathcal{D}_b$*, with* $b \in \{0, 1\}$*, obtained by restricting each* $n \times m$ *matrix in* $\mathcal{C}_b$ *to rows* $i_1, i_2, \ldots, i_p$ *are indistinguishable in the sense that they contain the same matrices with the same frequencies.*

Each pixel of the original image will be encoded into n pixels, each of which consists of m subpixels. To share a white (black, resp.) pixel, the dealer randomly chooses one of the matrices in $\mathcal{C}_0$ ($\mathcal{C}_1$, resp.) and distributes row i to participant i, for $i = 1, \ldots, n$.

The first property of Definition 1 is related to the contrast of the image. It states that when a qualified set of users stack their transparencies they can correctly recover the shared image. The value $\alpha(m)$ is called *relative difference* and the number $\gamma = \alpha(m) \cdot m$, which is assumed to be an integer, is referred to as the *contrast* of the image. The set $\{(X, t_X)\}_{X \in \Gamma_{\mathsf{Qual}}}$ is called the *set of thresholds* and t_X is the *threshold* associated to $X \in \Gamma_{\mathsf{Qual}}$. We want the contrast to be as large as possible and at least one, that is, $\alpha(m) \geq 1/m$. The second property is called *security*, since it implies that, even by inspecting all their shares, a forbidden set of participants cannot gain any information in deciding whether the shared pixel was white or black.

The model of visual cryptography we consider is the same as that described in [1,2]. This model is a generalization of the one proposed in [14], since with each set $X \in \Gamma_{\mathsf{Qual}}$ we associate a (possibly) different threshold t_X. Further, the access structure is not required to be strong in our model.

Notice that $\mathcal{C}_0$ ($\mathcal{C}_1$) is a multiset of $n \times m$ boolean matrices, therefore we allow a matrix to appear more than once in $\mathcal{C}_0$ ($\mathcal{C}_1$). Moreover, the size of the collections $\mathcal{C}_0$ and $\mathcal{C}_1$ does not need to be the same.

The *randomness* of a visual cryptography scheme represents the number of random bits per pixel used by the dealer to share an image among the participants. The definition of randomness for secret sharing schemes has been introduced in [8]. Since visual cryptography schemes are a special kind of secret sharing schemes, then, following [8], we define the randomness of a VCS realized by $\mathcal{C}_0$ and $\mathcal{C}_1$ as $\mathcal{R}^{(\mathcal{C}_0, \mathcal{C}_1), p} = p \log |\mathcal{C}_0| + (1 - p) \log |\mathcal{C}_1|$, where p denotes the probability (frequency) of the white pixels in the image to be encoded. Let $\Gamma = (\Gamma_{\mathsf{Qual}}, \Gamma_{\mathsf{Forb}})$ be a given access structure. In accordance with [8], the randomness of the access structure Γ is defined as $\mathcal{R}_\Gamma = \inf_{\mathcal{A}, \mathcal{I}} \mathcal{R}^{(\mathcal{C}_0, \mathcal{C}_1), p}$, where $\mathcal{A}$ denotes the set of all pairs of collections $\mathcal{C}_0$ and $\mathcal{C}_1$ realizing a VCS for Γ, and $\mathcal{I} = [0, 1]$ is the range of all values of the probability p. This definition is equivalent to the following $\mathcal{R}_\Gamma = \min_{\mathcal{A}} \log(\min\{|\mathcal{C}_0|, |\mathcal{C}_1|\})$. The above definition implies that, given a pair of matrix collections $\mathcal{C}_0$ and $\mathcal{C}_1$ realizing a VCS for the access structure Γ, we are mainly concerned with the quantity $\log(\min\{|\mathcal{C}_0|, |\mathcal{C}_1|\})$. Hence, we define the randomness $\mathcal{R}(\mathcal{C}_0, \mathcal{C}_1)$ of a VCS realized by $\mathcal{C}_0$ and $\mathcal{C}_1$ as $\mathcal{R}(\mathcal{C}_0, \mathcal{C}_1) = \log(\min\{|\mathcal{C}_0|, |\mathcal{C}_1|\})$. We point out that all VCSs presented in this paper are realized by equal sized matrix collections $\mathcal{C}_0$ and $\mathcal{C}_1$. As a consequence, all our upper bounds hold even if we alternatively define the randomness of the VCS to be the logarithm of the maximum of $|\mathcal{C}_0|$ and $|\mathcal{C}_1|$. Notice that the randomness of any VCS is at least one in that the share assigned to any essential participant has to be chosen in a set of size at least two.

Observe that any subset of a forbidden subset is forbidden, so Γ_{Forb} is necessarily monotone decreasing. Moreover, it is easy to see that no superset of a qualified subset is forbidden. Hence, a strong access structure is simply one in which Γ_{Qual} is monotone increasing and $\Gamma_{\mathsf{Qual}} \cup \Gamma_{\mathsf{Forb}} = 2^{\mathcal{P}}$.

Notice also that, given an (admissible) access structure $(\Gamma_{\mathsf{Qual}}, \Gamma_{\mathsf{Forb}})$, we can "embed" it in a strong access structure $(\Gamma'_{\mathsf{Qual}}, \Gamma'_{\mathsf{Forb}})$ in which $\Gamma_{\mathsf{Qual}} \subseteq \Gamma'_{\mathsf{Qual}}$ and $\Gamma_{\mathsf{Forb}} \subseteq \Gamma'_{\mathsf{Forb}}$. One way to do this is to take $(\Gamma'_{\mathsf{Qual}}, \Gamma'_{\mathsf{Forb}})$ to be the strong access structure whose basis Γ_0 consists of the minimal sets in Γ_{Qual}.

In view of the above observations, upper bounds on the randomness of VCSs for a strong access structure extend also to the access structures which can be embedded in it.

Most of the VCSs presented in literature [1,2,5,14] can be represented by means of two $n \times m$ characteristic matrices, S^0 and S^1, called *basis matrices*. The collections $\mathcal{C}_0$ and $\mathcal{C}_1$ are obtained by permuting the columns of the corresponding basis matrix (S^0 for $\mathcal{C}_0$, and S^1 for $\mathcal{C}_1$) in all possible ways. Hence, the collections $\mathcal{C}_0$ and $\mathcal{C}_1$ have both size equal to $m!$. The algorithm for the VCS based on the previous construction of the collections $\mathcal{C}_0$ and $\mathcal{C}_1$ has small memory requirements (it keeps only the basis matrices S^0 and S^1) and it is efficient (to choose a matrix in $\mathcal{C}_0$ ($\mathcal{C}_1$, resp.) it only generates a permutation of the columns of S^0 (S^1, resp.)).

3 Lower Bounds on the Randomness of VCSs

The following theorem is a useful tool to derive lower bounds on the randomness of VCSs for any access structure.

Theorem 1. *Let $\mathcal{C}_0$ and $\mathcal{C}_1$ realize a VCS for the access structure $(\Gamma_{\mathsf{Qual}}, \Gamma_{\mathsf{Forb}})$ on a set of participants $\mathcal{P}$. Let G be a subset of Γ_{Forb} with the property that for any pair of distinct sets $A, B \in G$, there exists a set $C \in \Gamma_{\mathsf{Qual}}$ such that $C \subseteq A \cup B$. For any $i \in \mathcal{P}$, let $G_i = \{A \in G : i \in A\}$ and let $d_i = |\{R : M[\{i\}] = R$ for some $M \in \mathcal{C}_0\}| \geq 2$. Then, $\mathcal{C}_0$ and $\mathcal{C}_1$ have both size larger than or equal to* $\max\{|G|, \max_{i \in \mathcal{P}}\{d_i \cdot |G_i|\}\}$.

As an application of the above theorem, we derive a lower bound on the size of the matrix collections realizing a VCS for a (k,n)-threshold structure $(\Gamma_{Qual}, \Gamma_{Forb})$. Let G denote the family of all subsets of $\mathcal{P}$ of size $k-1$. It is $G \subseteq \Gamma_{\mathsf{Forb}}$. Moreover, for any $A, B \in G$, with $A \neq B$, one has that $A \cup B$ contains at least a subset of $\mathcal{P}$ of size k. Hence, G satisfies the hypothesis of Theorem 1. Let $G_i = \{A \in G : i \in A\}$, for $i = 1, \ldots, n$. It is $|G_i| = \binom{n-1}{k-2}$. For $i = 1, \ldots, n$, let d_i be defined as in Theorem 1. Then, Theorem 1 implies that $\mathcal{C}_0$ and $\mathcal{C}_1$ have both size larger than or equal to both $\binom{n}{k-1}$ and $(\max_{i \in \{1,\ldots,n\}}\{d_i\})\binom{n-1}{k-2} \geq 2\binom{n-1}{k-2}$.

The following theorem provides a better lower bound on the size of the collections $\mathcal{C}_0$ and $\mathcal{C}_1$ realizing a (k,n)-threshold VCS.

Theorem 2. *Let $\mathcal{C}_0$ and $\mathcal{C}_1$ be two matrix collections realizing a (k,n)-threshold VCS, with $n \geq k \geq 2$. $\mathcal{C}_0$ and $\mathcal{C}_1$ have both size larger than or equal to $(n-k+2)^{k-1}$. Consequently, the randomness of a (k,n)-threshold VCS is at least $(k-1)\log(n-k+2)$.*

4 Minimum Randomness (k,k)-Threshold VCSs

In this section we give a characterization of (k,k)-threshold VCSs, for $k \geq 2$, with both minimum randomness and minimum pixel expansion. Notice that from Theorem 2 it follows that two collections $\mathcal{C}_0$ and $\mathcal{C}_1$ realizing a (k,k)-threshold VCS have both size at least 2^{k-1}. Hence, the randomness of a (k,k)-threshold VCS is at least $k-1$.

In the following we provide a construction for (k,k)-threshold VCSs with minimum randomness $k-1$. In these (k,k)-threshold VCSs each participant is assigned two row vectors as share.

For any $i = 1,\ldots,k$, let R_i and $\tilde{R}_i$ denote two row vectors and let $\mathbf{v}$ be a k-entry binary vector. We denote with $\mathbf{M}(\mathbf{v}, R_1,\ldots,R_k,\tilde{R}_1,\ldots,\tilde{R}_k)$ the k-row matrix whose i-th row, $i = 1,\ldots,k$, is equal to R_i if the i-th entry of $\mathbf{v}$ is 0, and to $\tilde{R}_i$ otherwise.

In the following we will refer to columns having even weight as *even columns* and to those having *odd weight* as odd columns.

We will denote with $\mathcal{M}_{h,k}$, for $k \geq 2$ and $h \geq 1$, a $k \times h2^{k-1}$ matrix which contains all even columns with multiplicity h and no odd column. The following theorem holds.

Theorem 3. *For any $k \geq 2$ and $h \geq 1$, let R_i be the i-th row of $\mathcal{M}_{h,k}$ and let $\overline{R}_i$ denote the bitwise complement of R_i. The matrix collections $\mathcal{C}_0 = \{\mathbf{M}(\mathbf{v}, R_1,\ldots,R_k,\overline{R}_1,\ldots,\overline{R}_k) : w(\mathbf{v})$ is even$\}$ and $\mathcal{C}_1 = \{\mathbf{M}(\mathbf{v}, R_1,\ldots,R_k,\overline{R}_1,\ldots,\overline{R}_k) : w(\mathbf{v})$ is odd$\}$ realize a (k,k)-threshold VCS with pixel expansion $h2^{k-1}$, relative difference $\alpha(m) = 1/2^{k-1}$ and minimum randomness $k-1$.*

In order to assign the shares to each participant, the dealer does not have to construct all matrices of the collections $\mathcal{C}_0$ and $\mathcal{C}_1$ of Theorem 3. Alternatively, every time a white pixel has to be shared among the k participants, the dealer randomly chooses

$$\begin{cases} \text{a } k\text{-entry binary vector } \mathbf{v} \text{ of even weight if the pixel is white} \\ \text{a } k\text{-entry binary vector } \mathbf{v} \text{ of odd weight if the pixel is black,} \end{cases}$$

and for $i = 1,\ldots,k$, selects the i-th row of $\mathbf{M}(\mathbf{v}, R_1,\ldots,R_k,\overline{R}_1,\ldots,\overline{R}_k)$ as share for participant i.

The following lemma provides a first characterization of a minimum randomness (k,k)-threshold VCS.

Lemma 1. *Let $\mathcal{C}_0$ and $\mathcal{C}_1$ be two matrix collections realizing a minimum randomness (k,k)-threshold VCS, $k \geq 2$. Then, for any $i = 1,\ldots,k$, the set $\mathbf{R_i} = \{R$: there is $M \in \mathcal{C}_0 \cup \mathcal{C}_1$ such that the i-th row of M is $R\}$ consists of only two row vectors.*

Let $M_t^0 \in \mathcal{C}_0$ and let $\mathbf{R_i} = \{R_i, \tilde{R}_i\}$, with R_i being the i-th row of M_t^0, for $i = 1, \ldots, k$. Then, it results $\mathcal{C}_0 = \{\mathbf{M}(\mathbf{v}, R_1, \ldots, R_k, \tilde{R}_1, \ldots, \tilde{R}_k) : w(\mathbf{v})$ is even$\}$ and $\mathcal{C}_1 = \{\mathbf{M}(\mathbf{v}, R_1, \ldots, R_k, \tilde{R}_1, \ldots, \tilde{R}_k) : w(\mathbf{v})$ is odd$\}$.

In order to characterize the structure of a minimum randomness (k,k)-threshold VCS, we fix the value of the contrast $\gamma = \alpha(m) \cdot m$. Recall that such a quantity measures the "difference" between a black and a white pixel in the reconstructed image. For any $k \geq 2$ and $\gamma \geq 1$, we will denote with $m^*(k,\gamma)$ the pixel expansion of the (k,k)-threshold VCS with smallest pixel expansion among those having contrast γ. We will show that, for any given value of the contrast γ, the construction of Theorem 3 is the only one providing a (k,k)-threshold VCS with minimum pixel expansion $m^*(k,\gamma) = \gamma 2^{k-1}$ and minimum randomness $k-1$. First we show that in any (k,k)-threshold VCS with contrast γ and with minimum pixel expansion $m^*(k,\gamma)$, each matrix of $\mathcal{C}_0$ contains all even columns with multiplicity γ, whereas each matrix of $\mathcal{C}_1$ contains all odd columns with the same multiplicity.

Theorem 4. *Let $\gamma \geq 1$, and let $\mathcal{C}_0 = \{M_1^0, \ldots, M_{c_0}^0\}$ and $\mathcal{C}_1 = \{M_1^1, \ldots, M_{c_1}^1\}$ be two matrix collections realizing a (k,k)-threshold VCS , $k \geq 2$, with contrast γ and with minimum pixel expansion $m^*(k,\gamma)$. Each matrix in $\mathcal{C}_0$ consists of all even columns each occurring with multiplicity γ, whereas each matrix in $\mathcal{C}_1$ consists of all odd columns each occurring with the same multiplicity γ. Consequently, $m^*(k,\gamma) = \gamma 2^{k-1}$.*

From Theorem 4 it follows that, for any given value of the contrast γ, the (k,k)-threshold VCS of Theorem 3 is optimal with respect to the pixel expansion. Moreover, for any (k,k)-threshold VCS with both minimum randomness and pixel expansion $m^*(k,\gamma)$ one has that the following theorem holds.

Theorem 5. *Let $\mathcal{C}_0$ and $\mathcal{C}_1$ be two matrix collections realizing a minimum randomness (k,k)-threshold VCS, $k \geq 2$, with contrast γ and minimum pixel expansion $m^*(k,\gamma) = \gamma 2^{k-1}$. For any $i = 1, \ldots, k$, the set $\mathbf{R_i} = \{R :$ there is $M \in \mathcal{C}_0 \cup \mathcal{C}_1$ such that the i-th row of M is $R\}$ consists of two row vectors, one being the bitwise complement of the other.*

For any $k \geq 2$ and any value of the contrast $\gamma \geq 1$, Theorems 4 and 5 provide a complete characterization of (k,k)-threshold VCSs with minimum randomness and pixel expansion $m^*(k,\gamma)$, thus proving that, for any given $\gamma \geq 1$, the construction of Theorem 3 is the only one providing a (k,k)-threshold VCS with minimum randomness $k-1$ and contrast $\gamma = h$.

5 Constructions for General Access Structures

5.1 A Construction Using Cumulative Arrays

Let $\Gamma = (\Gamma_{\mathsf{Qual}}, \Gamma_{\mathsf{Forb}})$ be a strong access structure on a set of participants $\mathcal{P}$, and let Γ_{MFS} denote the collection of the maximal forbidden sets of Γ : $\Gamma_{MFS} =$

$\{B \in \Gamma_{\mathsf{Forb}} : B \cup \{i\} \in \Gamma_{\mathsf{Qual}} : \text{ for all } i \in \mathcal{P} \setminus B\}$. In this subsection we will show that the minimum randomness of any visual cryptography scheme is less than or equal to $f-1$ where f is the size of Γ_{MFS}. Indeed, we show how to construct visual cryptography schemes with randomness $f-1$ for any strong access structure. Our technique is a generalization of that given in [1] which allows to obtain the basis matrices of a VCS for a given strong access structure from the basis matrices of a (f,f)-threshold VCS. We show how to obtain a VCS for a given strong access structure from any (f,f)-threshold VCS, not necessarily defined by means of basis matrices. Both our technique and that given in [1] are based on the cumulative array method introduced in [17]. A *cumulative map* (β, T) for Γ_{Qual} is a finite set T along with a mapping $\beta : \mathcal{P} \longrightarrow 2^T$ such that for any $Q \in \mathcal{P}$, it results $\bigcup_{a \in Q} \beta(a) = T \iff Q \in \Gamma_{\mathsf{Qual}}$. Let $\Gamma_{MFS} = \{F_1, \ldots, F_f\}$. Given a set $T = \{T_1, \ldots, T_f\}$, we can construct a cumulative map (β, T) for any Γ_{Qual} by defining, for any $i \in \mathcal{P}$, $\beta(i) = \{T_j | i \notin F_j, 1 \le j \le f\}$.

A cumulative array is a $|\mathcal{P}| \times |T|$ boolean matrix, denoted by CA, such that $CA(i,j) = 1$ if and only if $i \notin F_j$. We can construct a visual cryptography scheme for any strong access structure $\Gamma = (\Gamma_{\mathsf{Qual}}, \Gamma_{\mathsf{Forb}})$ as follows. Let CA be the cumulative array for Γ_{Qual} obtained by using the cumulative map (β, T). Let $\hat{\mathcal{C}}_0 = \{\hat{M}_1^0, \ldots, \hat{M}_{2^{f-1}}^0\}$ and $\hat{\mathcal{C}}_1 = \{\hat{M}_1^1, \ldots, \hat{M}_{2^{f-1}}^1\}$ be two collections of $f \times h2^{f-1}$, for some $h \ge 1$, matrices realizing the (f,f)-threshold VCS of Theorem 3. The collections $\mathcal{C}_0 = \{M_1^0, \ldots, M_{2^{f-1}}^0\}$ and $\mathcal{C}_1 = \{M_1^1, \ldots, M_{2^{f-1}}^1\}$ for a visual cryptography scheme for the strong access structure $(\Gamma_{\mathsf{Qual}}, \Gamma_{\mathsf{Forb}})$ are obtained as follows. For any fixed i let $j_{i,1}, \ldots, j_{i,g_i}$ be the integers j such that $CA(i,j) = 1$. The i-th row of M_i^0 (M_i^1, resp.) consists of the "or" of the rows $j_{i,1}, \ldots, j_{i,g_i}$ of $\hat{M}_i^0$ ($\hat{M}_i^1$, resp.). Hence, the following theorem holds.

Theorem 6. *Let $\Gamma = (\Gamma_{\mathsf{Qual}}, \Gamma_{\mathsf{Forb}})$ be a strong access structure, and let Γ_{MFS} be the family of the maximal forbidden sets in Γ_{Forb}. Then, there exists a $(\Gamma_{\mathsf{Qual}}, \Gamma_{\mathsf{Forb}})$-VCS with randomness $|\Gamma_{MFS}| - 1$, pixel expansion $m = 2^{|\Gamma_{MFS}|-1}$, and $t_X = m$, for any $X \in \Gamma_{\mathsf{Qual}}$.*

5.2 Constructing VCSs from Smaller Schemes

Ateniese *et al.* [1] showed how to construct the VCS for an access structure $(\Gamma_{\mathsf{Qual}}, \Gamma_{\mathsf{Forb}}) = (\Gamma'_{\mathsf{Qual}} \cup \Gamma''_{\mathsf{Qual}}, \Gamma'_{\mathsf{Forb}} \cap \Gamma''_{\mathsf{Forb}})$, on a set of n participants, using the VCSs for the structures $(\Gamma'_{\mathsf{Qual}}, \Gamma'_{\mathsf{Forb}})$ and $(\Gamma''_{\mathsf{Qual}}, \Gamma''_{\mathsf{Forb}})$. In the following we denote with $\circ$ the operator "concatenation" of two matrices. We recall the following theorem from [1].

Theorem 7. *Let $(\Gamma'_{\mathsf{Qual}}, \Gamma'_{\mathsf{Forb}})$ and $(\Gamma''_{\mathsf{Qual}}, \Gamma''_{\mathsf{Forb}})$ be two access structures on the same set of participants. Suppose there exist a $(\Gamma'_{\mathsf{Qual}}, \Gamma'_{\mathsf{Forb}})$-VCS and a $(\Gamma''_{\mathsf{Qual}}, \Gamma''_{\mathsf{Forb}})$-VCS with basis matrices Z^0, Z^1 and T^0, T^1, respectively. Then, the matrices $S^0 = Z^0 \circ T^0$ and $S^1 = Z^1 \circ T^1$ are the the basis matrices of a $(\Gamma'_{\mathsf{Qual}} \cup \Gamma''_{\mathsf{Qual}}, \Gamma'_{\mathsf{Forb}} \cap \Gamma''_{\mathsf{Forb}})$-VCS. The randomness of the resulting VCS is equal to $\log((|Z^0| + |T^0|)!)$. If the original access structures are both strong, then so is the resulting access structure.*

Let $\mathcal{C}_0$ ($\mathcal{C}_1$, resp.) be the collection of all distinct matrices obtained from S^0 (S^1, resp.) by permuting the columns of $\hat{Z}^0$ ($\hat{Z}^1$, resp.) and, independently, those of $\hat{T}^0$ ($\hat{T}^1$, resp.). $\mathcal{C}_0$ and $\mathcal{C}_1$ realize a $(\Gamma'_{\mathsf{Qual}} \cup \Gamma''_{\mathsf{Qual}}, \Gamma'_{\mathsf{Forb}} \cap \Gamma''_{\mathsf{Forb}})$-VCS with randomness $\log(m'! \cdot m''!) = \log(m'!) + \log(m''!)$.

Composition can be applied also to VCSs which are not represented by basis matrices, as the following lemma shows. Let $\mathcal{C}'$ and $\mathcal{C}''$ be two matrix collections. We denote with $\mathcal{C}' \circ \mathcal{C}''$ the matrix collection $\{M_i^{'} \circ M_j^{''} : M_i^{'} \in \mathcal{C}' \text{ and } M_j^{''} \in \mathcal{C}''\}$.

Lemma 2. *Let the two matrix collections* $\mathcal{C}'_0 = \{M_1^{'0}, \ldots, M_{s_0}^{'0}\}$ *and* $\mathcal{C}'_1 = \{M_1^{'1}, \ldots, M_{s_1}^{'1}\}$ *realize a VCS with randomness* $\mathcal{R}(\mathcal{C}'_0, \mathcal{C}'_1)$ *for the access structure* $(\Gamma'_{\mathsf{Qual}}, \Gamma'_{\mathsf{Forb}})$ *on a set of participants* $\mathcal{P}$, *and let the two matrix collections* $\mathcal{C}''_0 = \{M_1^{''0}, \ldots, M_{t_0}^{''0}\}$ *and* $\mathcal{C}''_1 = \{M_1^{''1}, \ldots, M_{t_1}^{''1}\}$ *realize a VCS with randomness* $\mathcal{R}(\mathcal{C}''_0, \mathcal{C}''_1)$ *for the access structure* $(\Gamma''_{\mathsf{Qual}}, \Gamma''_{\mathsf{Forb}})$ *on the same set of participants* $\mathcal{P}$. *Suppose that for any* $X \in \Gamma''_{\mathsf{Qual}} \setminus \Gamma'_{\mathsf{Qual}}$ *and for any* $i \in \{1, \ldots, s_0\}$ *and* $j \in \{1, \ldots, s_1\}$, *it results* $w(M_i^{'0}[X]) = w(M_j^{'1}[X])$, *and that for any* $X \in \Gamma'_{\mathsf{Qual}} \setminus \Gamma''_{\mathsf{Qual}}$ *and for any* $i \in \{1, \ldots, t_0\}$ *and* $j \in \{1, \ldots, t_1\}$, *it results* $w(M_i^{''0}[X]) = w(M_j^{''1}[X])$. *Then, the two matrix collections* $\mathcal{C}_0 = \mathcal{C}'_0 \circ \mathcal{C}''_0$ *and* $\mathcal{C}_1 = \mathcal{C}'_1 \circ \mathcal{C}''_1$ *realize a VCS for the access structure* $(\Gamma'_{\mathsf{Qual}} \cup \Gamma''_{\mathsf{Qual}}, \Gamma'_{\mathsf{Forb}} \cap \Gamma''_{\mathsf{Forb}})$ *with randomness* $\log(\min\{|\mathcal{C}'_0| \cdot |\mathcal{C}''_0|, |\mathcal{C}'_1| \cdot |\mathcal{C}''_1|\}) \geq \mathcal{R}(\mathcal{C}'_0, \mathcal{C}'_1) + \mathcal{R}(\mathcal{C}''_0, \mathcal{C}''_1)$. *If the original access structures are both strong, then so is the resulting access structure.*

Lemma 2 is often used in conjunction with the following theorem to obtain VCSs on a set of participants $\mathcal{P}$ from schemes on sets of participants contained in $\mathcal{P}$.

Theorem 8. *Let* $(\Gamma_{\mathsf{Qual}}, \Gamma_{\mathsf{Forb}})$ *be an access structure on a set of participants* $\mathcal{P}$. *Let* i *be a non-essential participant of* $\mathcal{P}$ *and let* $(\Gamma'_{\mathsf{Qual}}, \Gamma'_{\mathsf{Forb}})$ *be an access structure on* $\mathcal{P} \setminus \{i\}$ *with* $\Gamma'_{\mathsf{Qual}} = \{A \setminus \{i\} : A \in \Gamma_{\mathsf{Qual}}\}$. *Let the two matrix collections* $\mathcal{C}'_0$ *and* $\mathcal{C}'_1$ *realize a VCS* Σ' *for the access structure* $(\Gamma'_{\mathsf{Qual}}, \Gamma'_{\mathsf{Forb}})$. *Then, the two matrix collections* $\mathcal{C}_0$ *and* $\mathcal{C}_1$, *obtained by adding an all-zero row in correspondence of participant* i *to all matrices in* $\mathcal{C}'_0$ *and* $\mathcal{C}'_1$, *realize a VCS for the access structure* $(\Gamma_{\mathsf{Qual}}, \Gamma_{\mathsf{Forb}})$, *with the same randomness as* Σ'.

The following theorem is a consequence of Theorem 3, Lemma 2 and Theorem 8. It provides a technique to derive an upper bound on the randomness of the VCSs for any access structure.

Theorem 9. *Let* $(\Gamma_{\mathsf{Qual}}, \Gamma_{\mathsf{Forb}})$ *be a strong access structure on a set of participants* $\mathcal{P}$ *with basis* Γ_0. *There exists a* $(\Gamma_{\mathsf{Qual}}, \Gamma_{\mathsf{Forb}})$-*VCS with randomness* $\sum_{A \in \Gamma_0}(|A| - 1)$.

6 Upper Bounds on the Randomness of (k, n)-Threshold VCSs

In this section we derive upper bounds on the randomness of (k, n)-threshold VCSs. We start by providing a construction for minimum randomness $(2, n)$-threshold VCSs.

Theorem 10. *Let M be an $n \times m$ matrix whose rows consist of n distinct binary vectors each of weight $w < m$. Let $\sigma_{s,t} = 1 + [(s+t-1) \bmod n]$. For $j = 1, \ldots, n$, let $\hat{M}_j$ denote the $n \times m$ matrix having all n rows equal to $M[\{j\}]$, and let $\tilde{M}_j$ denote the $n \times m$ matrix such that $\tilde{M}_j[\{i\}] = M[\{\sigma_{i,j}\}]$, for $i = 1, \ldots, n$. Then, the two collections of matrices $\mathcal{C}_0 = \{\hat{M}_1, \ldots, \hat{M}_n\}$ and $\mathcal{C}_1 = \{\tilde{M}_1, \ldots, \tilde{M}_n\}$ define a strong $(2,n)$-threshold VCS with minimum randomness* $\log n$.

Both Theorem 6 and Theorem 9 provide a construction for (k,n)-threshold VCSs. The former implies a (k,n)-threshold VCS with randomness $\binom{n}{k-1} - 1$, whereas the latter implies a (k,n)-threshold VCS with randomness $(k-1)\binom{n}{k}$. In the following we prove two upper bounds on the randomness of (k,n)-threshold VCSs which represent an improvement on the above mentioned upper bounds. To this aim, we use a technique based on *starting matrices*, along the same line of [1] where an analogous technique is employed to obtain (k,n)-threshold VCSs realized by means of basis matrices.

Definition 2. *A* starting matrix $SM(n,\ell,k)$ *is a $n \times \ell$ matrix whose entries are elements of a ground set $\{a_1, \ldots, a_k\}$, with the property that, for any subset of k rows, there exists at least one column such that the entries in the k given rows of that column are all distinct.*

Let M^* be a starting matrix $SM(n,\ell,k)$. We can construct a (k,n)-threshold VCS as follows. Let $\mathcal{C}_0 = \{M_1^0, \ldots, M_{2^{k-1}}^0\}$ and $\mathcal{C}_1 = \{M_1^1, \ldots, M_{2^{k-1}}^1\}$ be the two collections of $k \times h2^{k-1}$, for some $h \geq 1$, matrices realizing the (k,k)-threshold VCS of Theorem 3. Let $\mathbf{c}_t$ denote the t-th column of M^*. For $j = 1, \ldots, 2^{k-1}$, let $M_j^0\{\mathbf{c}_t\}$ ($M_j^1\{\mathbf{c}_t\}$, resp.) denote the $n \times m$ matrix obtained by replacing each a_i-entry in $\mathbf{c}_t$, for $i = 1, \ldots, k$, with the i-th row of M_j^0 (M_j^1, resp.). For $j = 1, \ldots, 2^{k-1}$, a row of $M_j^0\{\mathbf{c}_t\}$ ($M_j^1\{\mathbf{c}_t\}$, resp.) corresponding to an a_i-entry of $\mathbf{c}_t$ is called a_i-row.

Theorem 11. *Let $\mathcal{C}_0 = \{M_1^0, \ldots, M_{2^{k-1}}^0\}$ and $\mathcal{C}_1 = \{M_1^1, \ldots, M_{2^{k-1}}^1\}$ be the two collections of $k \times h2^{k-1}$, for some $h \geq 1$, matrices realizing the (k,k)-threshold VCS of Theorem 3, and let $\mathbf{c}_t$ denote the t-th column of a starting matrix $SM(n,\ell,k)$. For $i = 1, \ldots, k$, let $Q_{i,t} = \{v : v \in \{1, \ldots, n\}$ and the v-th entry of $\mathbf{c}_t$ is equal to $a_i\}$. The two matrix collections $\mathcal{D}_{0,t} = \{M_1^0\{\mathbf{c}_t\}, \ldots, M_{2^{k-1}}^0\{\mathbf{c}_t\}\}$ and $\mathcal{D}_{1,t} = \{M_1^1\{\mathbf{c}_t\}, \ldots, M_{2^{k-1}}^1\{\mathbf{c}_t\}\}$ realize a VCS with randomness $k-1$ for the strong access structure $(\Gamma_{\mathsf{Qual}}, \Gamma_{\mathsf{Forb}})$ on participant set $\mathcal{P} = \{1, \ldots, n\}$ with basis $\Gamma_0 = \{X \subseteq \mathcal{P} : |X| = k$ and $|X \cap Q_{i,t}| = 1,$ for $i = 1 \ldots, k\}$.*

Let $\{\mathbf{c}_1, \ldots, \mathbf{c}_\ell\}$ be the set of the columns of the starting matrix M^* and let $\mathcal{D}_{0,t} = \{M_1^0\{\mathbf{c}_t\}, \ldots, M_{2^{k-1}}^0\{\mathbf{c}_t\}\}$ and $\mathcal{D}_{1,t} = \{M_1^1\{\mathbf{c}_t\}, \ldots, M_{2^{k-1}}^1\{\mathbf{c}_t\}\}$, for $t = 1, \ldots, \ell$. Let $\mathcal{D}_0 = \{\mathcal{D}_{0,1} \circ \ldots \circ \mathcal{D}_{0,\ell}\}$ and $\mathcal{D}_1 = \{\mathcal{D}_{1,1} \circ \ldots \circ \mathcal{D}_{1,\ell}\}$. Lemma 2 and Theorem 11 imply that $\mathcal{D}_0$ and $\mathcal{D}_1$ realize a strong (k,n)-threshold VCS. Hence, the following theorem holds.

Theorem 12. *If there exists a starting matrix $SM(n,\ell,k)$ then there exists a strong (k,n)-threshold VCS with randomness $(k-1)\ell$.*

The *SM* matrix is a representation of a *Perfect Hash Family* (or PHF). Fredman and Komlós [10] proved that for any PHF it holds that $\ell = \Omega(k^{k-1}/k!) \log n$. They also proved the weaker but simpler bound $\ell = \Omega(1/\log k) \log n$. Mehlhorn [13] proved that there exist PHFs with $\ell = O(ke^k) \log n$. These bounds are in general non-constructive, but in [16] there can be found a recursive construction which for any constant $k \geq 2$ and for any integer $n \geq k$, yields a PHF with $\ell = O(k^{2 \log^* n} \log n)$. Therefore, the following corollaries of Theorem 12 hold.

Corollary 1. *For any k and n with $2 \leq k \leq n$, there exists a (k,n)-threshold VCS with randomness $O(k^2 e^k) \log n$.*

Corollary 2. *Let $k \geq 2$ be a constant. For any $n \geq k$ there exists a constructible (k,n)-threshold VCS with randomness $O(k^{(2 \log^* n + 1)} \log n)$.*

Naor *et al.* [14]	Ateniese *et al.* [1]	Thm. 6
$n^k \log(2^{k-1}!)$	$\log\left((O(k(2e)^k) \log n)!\right)$	$\binom{n}{k-1} - 1$
Thm. 9	Cor. 1	Cor. 2
$(k-1)\binom{n}{k}$	$O(k^2 e^k) \log n$	$O(k^{(2 \log^* n + 1)} \log n)$, for k constant

The above table reports the values of previously known upper bounds on the randomness of the (k,n)-threshold VCSs, along with those derived in the present paper. The first upper bound reported in the table is relative to the randomness of the very first (k,n)-threshold VCS described in literature [14]. By Stirling approximation formula, the randomness of this scheme is larger than $n^k \log\left(\sqrt{2^k \pi}(2^{k-1}/e)^{2^{k-1}}\right)$. Then, the randomness of this scheme is larger than the randomness of all (k,n)-threshold VCSs described in the present paper.

Since $\binom{n}{k-1} = \frac{k}{n-k+1}\binom{n}{k}$, then the randomness of the scheme implied by Theorem 6 is always smaller than or equal to that of the scheme implied by Theorem 9. Notice that the upper bound of Corollary 1 is asymptotically smaller than that implied by Theorem 6 when k is a sublinear function of n, whereas the bound of Corollary 2 is asymptotically smaller than that implied by Theorem 6 for any fixed k with $2 \leq k \leq n$. In our knowledge, all efficient constructions for (k,n)-threshold VCS, given so far in literature, use basis matrices. The randomness of such a scheme with pixel expansion m is equal to $\log(m!)$. For that reason, it is interesting to compare the (k,n)-threshold VCSs of Corollary 1 and Corollary 2 with the (k,n)-threshold VCS having the smallest pixel expansion among those given so far in literature [1]. This (k,n)-threshold VCS has pixel expansion $O(k(2e)^k) \log n$. By Stirling approximation formula, the asymptotical upper bound on the randomness of this scheme is larger than $O(k(2e)^k)(\log n)(\log(O(k2^k e^{k-1}) \log n))$. As a consequence, this upper bound is larger than those of both corollaries.

7 Conclusions and Open Problems

In this paper the randomness of visual cryptography schemes has been investigated. Lower bounds on the randomness of VCSs for general access structures and (k, n)-threshold VCSs have been derived and general techniques to construct (k, n)-threshold VCSs have been presented and analyzed. We have also provided minimum randomness constructions for $(2, n)$-threshold VCSs and (k, k)-threshold VCSs. All proofs and examples of our results are given in the extended version of the paper. That version also provides minimum randomness VCSs for all strong access structures on at most four participants. Further results on the randomness of visual cryptography schemes can be found in [9].

Randomness is an aspect of visual cryptography which has been considered in this paper for the first time. Nevertheless, it deserves further investigations. Indeed, many problems are left open. The most challenging one is to reduce the gap between our lower bounds and upper bounds to the randomness of the (k, n)-threshold VCSs for $2 < k < n$.

References

1. G. Ateniese, C. Blundo, A. De Santis, and D. R. Stinson, *Visual Cryptography for General Access Structures. Information and Computation* **129-2**, 86–106 (1996).
2. G. Ateniese, C. Blundo, A. De Santis, and D. R. Stinson, *Constructions and Bounds for Visual Cryptography. ICALP 1996*, LNCS **1099**, 416–428.
3. M. Atici, S. S. Magliveras, D. R. Stinson, and W.-D. Wei, *Some Recursive Constructions for Perfect Hash Families. Journal of Comb. Design*, **4**, 352–363 (1996).
4. A. Beimel and B. Chor, *Universally ideal secret sharing schemes. IEEE Trans. Inform. Theory*, **40**(3), 786–794 (1994).
5. C. Blundo, P. D'Arco, A. De Santis, and D. R. Stinson, *Contrast Optimal Threshold Visual Cryptography Schemes.* submitted for publication (1998).
6. C. Blundo and A. De Bonis, *New Constructions for Visual Cryptography. Proc. ICTCS '98*, P. Degano, U. Vaccaro, G. Pirillo, (Eds.), 290–303, World Scientific.
7. C. Blundo, A. De Santis, and D. R. Stinson, *On the Contrast in Visual Cryptography Schemes.* To appear on *Journal of Cryptology*. Available also at *Theory of Cryptography Library* as `ftp://theory.lcs.mit.edu/pub/tcryptol/96-13.ps`.
8. C. Blundo, A. De Santis, and U. Vaccaro, *Randomness in Distribution Protocols. Information and Computation*, **131**, 111–139 (1996).
9. A. De Bonis and A. De Santis, *Visual and Non-Visual Secret Sharing Schemes.* submitted for publication.
10. M. L. Fredman and J. Komlós, *On the Size of Separating System and Families of Perfect Hash Functions. SIAM J. Alg. Disc. Math.*, 5(1), 1984.
11. T. Hofmeister, M. Krause, and H. U. Simon, *Contrast-Optimal k out of n Secret Sharing Schemes in Visual Cryptography. COCOON '97*, LNCS **1276**, 176–185.
12. R. Impagliazzo and D. Zuckerman, *How to Recycle Random Bits. Proc. 21st Annual ACM Symp. on Theory of Computing*, 248–255 (1989).
13. M. Mehlhorn, *On the Program Size of Perfect and Universal Hash Functions. Proc. 23rd IEEE Symp. on Foundations of Comp. Science*, 170–175 (1982).
14. M. Naor and A. Shamir, *Visual Cryptography.* LNCS **950**, 1–12 (1995).

15. D. R. Stinson, *An Introduction to Visual Cryptography.* Presented at *Public Key Solutions '97*, Toronto, Canada, April 28–30 (1997). Available as `http://bibd.unl.edu/ stinson/VKS-PKS.ps`.
16. D. R. Stinson and R. Wei, *New Constructions for Perfect Hash Families and Related Structures using Combinatorial Designs and Codes.* preprint.
17. G. J. Simmons, W.-A. Jackson, and K. Martin, *Decomposition Constructions for Secret Sharing Schemes. Bulletin of the ICA*, 1:72–88 (1991).
18. D. Zuckerman, *Simulating BPP Using a General Weak Random Source. Proc. 32nd IEEE Symp. on Foundations of Comp. Science*, 79–89 (1991).

Online Dial-a-Ride Problems: Minimizing the Completion Time*

Norbert Ascheuer, Sven O. Krumke, and Jörg Rambau

Konrad-Zuse-Zentrum für Informationstechnik Berlin, Department Optimization,
Takustr. 7, D-14195 Berlin-Dahlem, Germany,
{ascheuer,krumke,rambau}@zib.de

Abstract. We consider the following online dial-a-ride problem (OLDARP): Objects are to be transported between points in a metric space. Transportation requests arrive online, specifying the objects to be transported and the corresponding source and destination. These requests are to be handled by a server which starts its work at a designated origin and which picks up and drops objects at their sources and destinations. The server can move at constant unit speed. After the end of its service the server returns to its start in the origin. The goal of OLDARP is to come up with a transportation schedule for the server which finishes as early as possible, i.e., which minimizes the makespan.

We analyze several competitive algorithms for OLDARP and establish tight competitiveness results. The first two algorithms, REPLAN and IGNORE are very simple and natural: REPLAN completely discards its (preliminary) schedule and recomputes a new one when a new request arrives. IGNORE always runs a (locally optimal) schedule for a set of known requests and ignores all new requests until this schedule is completed. We show that both strategies, REPLAN and IGNORE, are 5/2-competitive.

We then present a somewhat less natural strategy SMARTSTART, which in contrast to the other two strategies may leave the server idle from time to time although unserved requests are known. The SMARTSTART-algorithm has an improved competitive ratio of 2, which matches our lower bound.

1 Introduction

Transportation problems where objects are to be transported between given sources and destinations in a metric space are classical problems in combinatorial optimization. In the classical setting, one assumes that the complete input for an instance is available for an algorithm to compute a solution. In many cases this *offline optimization* does not reflect the real-world situation appropriately. For instance, the transportation requests in an elevator system are hardly known in advance. Decisions have to be made *online* without the knowledge of future requests.

* Research supported by the German Science Foundation (grant 883/5-2)

H. Reichel and S. Tison (Eds.): STACS 2000, LNCS 1770, pp. 639–650, 2000.

Online algorithms are tailored to cope with such situations. Whereas offline algorithms work on the complete input sequence, online algorithms only get to see the requests released so far and thus have to account for future requests that may or may not arise at a later time. A common way to evaluate the quality of online algorithms is *competitive analysis* [6]: An algorithm ALG is called c-competitive if its "cost" on any input sequence is at most c-times the optimum offline cost.

In this paper we consider the following online dial-a-ride problem (OLDARP): Objects are to be transported between the points of a metric space. A request consists of the objects to be transported and the corresponding source and destination of the transportation request. The requests arrive online and must be handled by a server which starts and ends its work at a distinguished origin. The server picks up and drops objects at their sources and destinations, respectively. The goal of OLDARP is to come up with a transportation schedule for the server which finishes as early as possible.

Related Work. We do not claim originality for the two online-algorithms IGNORE and REPLAN; instead, we show how to analyze them for OLDARP and how ideas from both strategies can be used to construct a new online strategy SMARTSTART with better competitive ratio.

The first—to the best of our knowledge—occurrence of the strategy IGNORE can be found in the paper by Shmoys, Wein, and Williamson [12]: They show a general result about obtaining competitive algorithms for minimizing the total completion time (also called the makespan) in machine scheduling problems when the jobs arrive over time: If there is a ρ-approximation algorithm for the offline version, then this implies a 2ρ-competitive algorithm for the online-version, which is essentially the IGNORE strategy. The results from [12] show that IGNORE-type strategies are 2-competitive for a number of online-scheduling problems. The strategy REPLAN is probably folklore; it can be found also under different names like reopt or optimal.

The difference of OLDARP studied here to the machine scheduling problems treated in [12] are as follows: For OLDARP, the "execution time" of jobs depends on their execution order. OLDARP can be viewed as a generalized scheduling problem with setup costs and order dependent execution times. It should be stressed that in this paper we do not allow additive constants in the definition of the competitive ratio. If one allowed an additive constant equal to the diameter of the metric space, then the algorithms REPLAN and IGNORE would in fact be also 2-competitive.

In [5,4,3] the authors studied the Online Traveling Salesman Problem (OLTSP) which is obtained as a special case of OLDARP treated in this paper, when for each request its source and destination coincide. It is shown in [5] that there is a metric space (the boundary of the unit square) where any deterministic algorithm for the OLTSP has a competitive ratio of at least 2. For the case that the metric space is the real line, a lower bound of $\frac{9+\sqrt{17}}{8} \approx 1.64$ is given in [4]. A 2-competitive algorithm that works in an arbitrary (symmetric) metric space and a $7/4$-competitive algorithm for the real line are presented in [5].

Recently, in an independent effort Feuerstein and Stougie [9] analyzed the algorithm IGNORE (which they call dlt for "Don't listen while traveling") and established the same competitive ratio as in this paper.

Our Contribution. We provide a number of competitive algorithms for the basic version of the problem OLDARP. The best algorithm, SMARTSTART, has competitive ratio 2 which improves the result of 5/2 established in [9]. This competitiveness result closes the gap to the lower bound.

Our algorithm SMARTSTART induces an alternative 2-competitive algorithm for the OLTSP. Moreover, SMARTSTART can be used to obtain a $\frac{7+\sqrt{13}}{4}$-competitive polynomial time algorithm for the OLTSP. Since $\frac{7+\sqrt{13}}{4} \approx 2.6514$, this improves the result of [5,4], where a 3-competitive polynomial time algorithm for the OLTSP was presented.

IGNORE and SMARTSTART can be applied to the generalization of OLDARP (and OLTSP) where there are k-servers with arbitrary capacities $C_1, \ldots, C_k \in \mathbb{N}$. Their competitive ratios of 5/2 of 2, respectively, hold as well in the generalized case. As a corollary, we obtain 2-competitive algorithms for a number of machine scheduling problems with setup-costs which generalizes a result from [12].

Paper Outline. This paper is organized as follows: In Section 2 we formally define the problem OLDARP and introduce notation. In Section 3 we establish lower bounds for the competitive ratio of deterministic online algorithms for OLDARP. In Section 4 we analyze two simple strategies, REPLAN and IGNORE. Section 5 contains our improved algorithm SMARTSTART. The competitive ratios of our three algorithms are summarized in Table 1.

Table 1. Competitive ratios of algorithms in this paper.

Algorithm	REPLAN	IGNORE	SMARTSTART	Lower bound
Closed Schedules	5/2	5/2	2	2

2 Preliminaries

An instance of the basic online dial-a-ride problem OLDARP consists of a metric space $M = (X, d)$ with a distinguished origin $o \in X$ and a sequence $\sigma = r_1, \ldots, r_m$ of requests. It is assumed that for all pairs (x, y) of points from M, there is a path $p : [0, 1] \to X$ in X with $p(0) = x$ and $p(1) = y$ of length $d(x, y)$ (see [5] for a thorough discussion of the model). Examples of a metric spaces that satisfy the above condition are the Euclidean space $\mathbb{R}^p$ and a metric space induced by an undirected edge-weighted graph.

Each request is a triple $r_i = (t_i, a_i, b_i)$, where t_i is a real number, the time where request r_i is released (becomes known), and $a_i \in V$ and $b_i \in V$ are the source and destination, respectively, between which the new object is to be transported. We assume that the sequence $\sigma = r_1, \ldots, r_m$ of requests is given in order of non-decreasing release times.

A server is located at the origin $o \in X$ at time 0 and can move at constant unit speed. In the basic version the server has *unit-capacity*, i.e., it can carry at most one object at a time. (Extensions to the case of more servers with arbitrary capacities are also considered in this paper.) We do not allow *preemption*: once the server has picked up an object, it is not allowed to drop it at any other place than its destination.

An online algorithm for OLDARP does neither have information about the release time of the last request nor about the total number of requests. The online algorithm must determine the behavior of the server at a certain moment t of time as a function depending on all the requests released up to time t and on the current time t. In contrast, the offline algorithm has information about all requests in the whole sequence σ already at time 0.

Given a sequence σ of requests, a *valid transportation schedule* for σ is a sequence of moves of the server such that the following conditions are satisfied: (a) The server starts its movement in the origin vertex o, (b) each transportation request in σ is served, but starting not earlier than the time it becomes known, and (c) the server returns to the origin vertex after having served the last request.

The objective function of the OLDARP is the *(total) completion time* (also called the "*makespan*" in scheduling) of the server, that is, the time when the server has served all requests and returned to the origin.

Let $\mathsf{ALG}(\sigma)$ denote the completion time of the server moved by algorithm ALG on the sequence σ of requests. We use OPT to denote an optimal offline algorithm. An online algorithm ALG for OLDARP is *c-competitive*, if there exists a constant c such that for any request sequence σ the inequality $\mathsf{ALG}(\sigma) \leq c \cdot \mathsf{OPT}(\sigma)$ holds true. It should be mentioned that in sometimes literature the definition of the competitive ratio allows an additive constant. In that case a c-competitive algorithm with additive constant zero is then referred to as *strictly c-competitive.* In this paper we do not allow additive constants.

3 Lower Bounds

We first address the question how well an online algorithm can perform compared to the optimal offline algorithm. Since OLDARP generalizes the OLTSP we obtain the following result from the lower bound established in [5]:

Theorem 1. *If ALG is a deterministic c-competitive algorithm for* OLDARP, *then* $c \geq 2$. □

For the case that the metric space is the real line, a lower bound of $\frac{9+\sqrt{17}}{8} \approx 1.64$ was given in [4,3]. We provide an improved lower bound for OLDARP below:

Theorem 2. *If ALG is a deterministic c-competitive algorithm for* OLDARP *on the real line, then* $c \geq 1 + \sqrt{2}/2 \approx 1.707$.

Proof. Suppose that ALG is a deterministic online algorithm with competitive ratio $c \leq 1 + \sqrt{2}/2$. At time $t = 0$, the algorithm ALG is faced with two requests $r_1 = (0, o, 2)$ and $r_2 = (0, 2, o)$. The optimum offline cost to serve these two requests is 4.

The server operated by ALG must start serving request r_2 at some time $2 \leq T \leq 4c - 2$, because otherwise ALG could not be c-competitive. At time T the adversary issues another request $r_3 = (T, T, 2)$. Then $\mathsf{OPT}(r_1, r_2, r_3) = 2T$. On the other hand, $\mathsf{ALG}(r_1, r_2, r_3) \geq 3T + 2$. Thus, the competitive ratio c of ALG satisfies

$$c \geq \frac{3T + 2}{2T} \geq \frac{3}{2} + \frac{1}{T} \geq \frac{3}{2} + \frac{1}{4c - 2}.$$

The smallest value $c \geq 1$ such that $c \geq 3/2 + 1/(4c - 2)$ is $c = 1 + \sqrt{2}/2$. □

4 Two Simple Strategies

In this section we present and analyze two very natural online-strategies for for OLDARP and prove both of them to be 5/2-competitive.

Strategy REPLAN As soon as a new request arrives the server completes the current carrying move (if it is performing one), then the server stops and replans: it computes a new shortest schedule which starts at the current position of the server, takes care of all yet unserved requests, and returns to the origin.

Strategy IGNORE The server remains idle until the point in time t when the first requests become known. The algorithm then serves the requests released at time t immediately, following a shortest schedule S. All requests that arrive during the time when the algorithm follows this schedule are temporarily ignored. After S has been completed and the server is back in the origin, the algorithm computes a shortest schedule for all unserved requests and follows this schedule. Again, all new requests that arrive during the time that the server is following the schedule are temporarily ignored. A schedule for the ignored requests is computed as soon as the server has completed its current schedule. The algorithm keeps on following schedules and temporarily ignoring requests this way.

Both algorithms above repeatedly solve "offline instances" of OLDARP. These offline instances have the property that all release times are at least as large as the current time. Thus, the corresponding offline problem is the following: given a number of transportation requests (with release times all zero), find a shortest transportation for them.

For a sequence σ of requests and a point x in the metric space M let $L^*(t, x, \sigma)$ denote the *length of a shortest schedule* (i.e., the time difference between its completion time and the start time t) which starts in x at time t, serves all requests from σ (but not earlier than their release times), and ends in the origin. Clearly, for $t' \geq t$ we have that $L^*(t', x, \sigma) \leq L^*(t, x, \sigma)$. Moreover, $\mathsf{OPT}(\sigma) = L^*(0, o, \sigma)$ and thus $\mathsf{OPT}(\sigma) \geq L^*(t, o, \sigma)$ for any time $t \geq 0$.

Since the optimum offline server OPT cannot serve the last request $r_m = (t_m, a_m, b_m)$ from σ before this request is released we get that

$$\mathsf{OPT}(\sigma) \geq \max\{L^*(t, o, \sigma), t_m + d(a_m, b_m) + d(b_m, o)\} \quad \text{for any } t \geq 0. \tag{1}$$

Lemma 1. *Let $\sigma = r_1, \ldots, r_m$ be a sequence of requests. Then for any $t \geq t_m$ and any request $r_i = (t_i, a_i, b_i)$ from σ*

$$L^*(t, b_i, \sigma \backslash r_i) \leq L^*(t, o, \sigma) - d(a_i, b_i) + d(a_i, o).$$

Here $\sigma \backslash r_i$ denotes the sequence obtained from σ by deleting the request r_i.

Proof. Consider an optimum schedule S^* which starts at the origin o at time t, serves all requests in σ and has length $L^*(t, o, \sigma)$. It suffices to construct another schedule S which starts in b_i no earlier than time t, serves all requests in $\sigma \backslash r_i$ and has length at most $L^*(t, o, \sigma) - d(a_i, b_i) + d(a_i, o)$.

Let S^* serve the requests in the order $r_{j_1}, \ldots, r_{j_m}$ and let $r_i = r_{j_k}$. Notice that if we start in b at time t and serve the requests in the order

$$r_{j_{k+1}}, \ldots, r_{j_m}, r_{j_1}, \ldots, r_{j_{k-1}}$$

and move back to the origin we obtain a schedule S with the desired properties. □

We are now ready to prove the result about the performance of REPLAN:

Theorem 3. *Algorithm REPLAN is 5/2-competitive.*

Proof. Let $\sigma = r_1, \ldots, r_m$ be any sequence of requests. We distinguish between two cases depending on the current load of the REPLAN-server at the time t_m (i.e., the time when the last request is released).

If the server is currently empty it recomputes an optimal schedule which starts at its current position, denoted by $s(t_m)$, serves all unserved requests, and returns to the origin. This schedule has length at most $L^*(t_m, s(t_m), \sigma) \leq d(o, s(t_m)) + L^*(t_m, o, \sigma)$. Thus,

$$\begin{aligned} \mathsf{REPLAN}(\sigma) &\leq t_m + d(o, s(t_m)) + L^*(t_m, o, \sigma) \\ &\overset{(1)}{\leq} t_m + d(o, s(t_m)) + \mathsf{OPT}(\sigma) \end{aligned} \tag{2}$$

New now consider the second case, when the server is currently serving a request $r = (t, a, b)$. The time needed to complete this move is $d(s(t_m), b)$. Then a shortest schedule starting at b serving all unserved requests is computed which has length at most $L^*(t_m, b, \sigma \backslash r)$. Thus in the second case

$$\begin{aligned} \mathsf{REPLAN}(\sigma) &\leq t_m + d(s(t_m), b) + L^*(t_m, b, \sigma \backslash r) \\ &\leq t_m + d(s(t_m), b) + L^*(t_m, o, \sigma) - d(a, b) + d(a, o) \quad \text{by Lemma 1} \\ &\leq t_m + \mathsf{OPT}(\sigma) - d(a, b) + \underbrace{d(s(t_m), b) + d(a, s(t_m))}_{=d(a,b)} + d(s(t_m), o) \\ &= t_m + d(o, s(t_m)) + \mathsf{OPT}(\sigma). \end{aligned}$$

This means that inequality (2) holds in both cases. Since the REPLAN server has traveled to position $s(t_m)$ at time t_m, there must be a request $r_j = (t_j, a_j, b_j)$ in σ where either $d(o, a_i) \geq d(o, s(t_m))$ or $d(o, b_i) \geq d(o, s(t_m))$. By the triangle inequality this implies that the optimal offline server will have to travel at least twice the distance $d(o, s(t_m))$ during its schedule. Thus, $d(o, s(t_m)) \leq \mathsf{OPT}(\sigma)/2$. Plugging this result into inequality (2) we get that the total time the REPLAN server needs is no more than $5/2\,\mathsf{OPT}(\sigma)$. □

We are now going to analyze the competitiveness of the second simple strategy IGNORE.

Theorem 4. *Algorithm IGNORE is 5/2-competitive.*

Proof. Consider again the point in time t_m when the last request r_m becomes known. If the IGNORE server is currently idle at the origin o, then its completes its last schedule no later than $t_m + L^*(t_m, o, \sigma_{=t_m})$, where $\sigma_{=t_m}$ is the set of requests released at time t_m. Since $L^*(t_m, o, \sigma_{=t_m}) \leq \mathsf{OPT}(\sigma)$ and $\mathsf{OPT}(\sigma) \geq t_m$, it follows that in this case IGNORE completes no later than time $2\,\mathsf{OPT}(\sigma)$.

It remains the case that at time t_m the IGNORE-server is currently working on a schedule S for a subset σ_S of the requests. Let t_S denote the starting time of this schedule. Thus, the IGNORE-server will complete S at time $t_S + L^*(t_S, o, \sigma_S)$. Denote by $\sigma_{\geq t_S}$ the set of requests presented after the IGNORE-server started with S at time t_S. Notice that $\sigma_{\geq t_S}$ is exactly the set of requests that are served by IGNORE in its last schedule. The IGNORE-server will complete its total service no later than time $t_S + L^*(t_S, o, \sigma_S) + L^*(t_m, o, \sigma_{\geq t_S})$.

Let $r_f \in \sigma_{\geq t_S}$ be the first request from $\sigma_{\geq t_S}$ served by OPT. Thus

$$\mathsf{OPT}(\sigma) \geq t_f + L^*(t_f, a_f, \sigma_{\geq t_S}) \geq t_S + L^*(t_m, a_f, \sigma_{\geq t_S}). \tag{3}$$

Now, $L^*(t_m, o, \sigma_{\geq t_S}) \leq d(o, a_f) + L^*(t_m, a_f, \sigma_{\geq t_S})$ and $L^*(t_S, o, \sigma_S) \leq \mathsf{OPT}(\sigma)$. Therefore,

$$\begin{aligned}\mathsf{IGNORE}(\sigma) &\leq t_S + \mathsf{OPT}(\sigma) + d(o, a_f) + L^*(t_m, a_f, \sigma_{\geq t_S}) \\ &\overset{(3)}{\leq} 2\,\mathsf{OPT}(\sigma) + d(o, a_f) \\ &\leq \frac{5}{2}\,\mathsf{OPT}(\sigma).\end{aligned}$$

This completes the proof. □

The following instance shows that the competitive ratio of 5/2 proved for IGNORE is asymptotically tight even for the case when the metric space is the real line. At time 0 there is a request $r_1 = (0, 1, 0)$. The next requests are $r_2 = (\varepsilon, 2, 3)$ and $r_3 = (2 + \varepsilon, 2, 1)$. It is easy to see that $\mathsf{IGNORE}(r_1, r_2, r_3) = 10 + 4\varepsilon$, while $\mathsf{OPT}(r_1, r_2, r_3) = 4 + 2\varepsilon$. Thus, the ratio $\mathsf{IGNORE}(r_1, r_2, r_3)/\mathsf{OPT}(r_1, r_2, r_3)$ can be made arbitrarily close to 5/2 by choosing $\varepsilon > 0$ small enough.

4.1 Application to Machine Scheduling Problems

We close this section by showing how the IGNORE strategy can be applied to a generalization of the basic problem OLDARP and how this generalization can be used to model certain machine scheduling problems where there are setup costs for different job types. Consider the generalization k-OLDARP of OLDARP when there are $k \in \mathbb{N}$ servers with arbitrary capacities $C_1, \ldots, C_k \in \mathbb{N}$. Capacity C_j means that server j can carry at most C items at a time. The objective function for k-OLDARP is the time when the last server has returned to the origin (after all requests have been served).

For k-OLDARP the IGNORE strategy always plans schedules for its servers such that the length of the longest schedule is minimized. All schedules are constructed in such a way that they start and end in the origin. New requests are ignored until the last of the servers has returned to the origin. It is not too hard to see that the proof of Theorem 4 remains valid even for k-OLDARP:

Theorem 5. *Algorithm IGNORE is 5/2-competitive even for the extension k-OLDARP of OLDARP where there are $k \in \mathbb{N}$ servers with arbitrary capacities $C_1, \ldots, C_k \in \mathbb{N}$.* □

Suppose there are k uniform machines where jobs can be run and assume that there are ℓ *types* of jobs. For each job type j *setup cost* of s_j is given: if a job of type $k \neq j$ has been processed immediately before on a machine, then an additional cost of s_j is incurred to start processing a job of type j on this machine. This setup cost models for instance the situation where special auxiliary device must be installed at the machine to perform a certain job type. The problem of minimizing the makespan in this machine scheduling problem when jobs arrive over time can be modeled as k-OLDARP (with k unit-capacity servers) on a star shaped metric space with center o: For each of the p job types there is one ray in the metric space emanating from p. On ray R_j there is a special point x_j at distance $s_j/2$ from o. A job of type j with processing time p is modeled by a transportation request from the point x_j to the point $x_j + p/2$.

5 The SMARTSTART Strategy

In this section we present and analyze our algorithm SMARTSTART which achieves a best-possible competitive ratio of 2 (cf. the lower bound given in Theorem 1). The idea of the algorithm is basically to emulate the IGNORE-strategy but to make sure that each sub-schedule is completed "not too late": if a sub-schedule would take "too long" to complete then the algorithm waits for a specified amount of time. Intuitively this construction tries to avoid the worst-case situation for IGNORE where right after the algorithm started a schedule a new request becomes known.

The algorithm SMARTSTART has a fixed "waiting scaling" parameter $\theta > 1$. From time to time the algorithm consults its "work-or-sleep" routine: this subroutine computes an (approximately) shortest schedule S for all unserved requests starting and ending in the origin. If this schedule can be completed no later than time θt (where t is the current time) the subroutine returns (S, work), otherwise it returns (S, sleep).

In the sequel it will be convenient to assume that the "work-or-sleep" subroutine uses a ρ-approximation algorithm for computing a schedule: the approximation algorithm always finds a schedule of length at most ρ times the optimum one. While in online computation one is usually not interested in time complexity (and thus in view of competitive analysis we can assume that $\rho = 1$), employing a polynomial time approximation algorithm will enable us to get a practical algorithm (and in particular to improve the result of [5,4] for the OLTSP).

The server of algorithm SMARTSTART can assume three states:

idle In this case the server has served all known requests, is sitting in the origin and waiting for new requests to occur.

sleeping In this case the server knows of some unserved requests but also knows that they take too long to serve (what "too long" means will be formalized in the algorithm below).

working In this state the algorithm (or rather the server operated by it) is following a computed schedule.

We now formalize the behavior of the algorithm by specifying how it reacts in each of the three states.

Strategy SMARTSTART

- If the algorithm is idle at time T and new requests arrive, calls "work-or-sleep". If the result is (S, work), the algorithm enters the working state where it follows S. Otherwise the algorithm enters the sleeping state with wakeup time t', where $t' \geq T$ is the earliest time such that $t' + l(S) \leq \theta t'$, where $l(S)$ is the length of the just computed schedule S, i.e., $t' = \min\{\, t \geq T : t + l(S) \leq \theta t \,\}$.
- In the sleeping state the algorithm simply does nothing until its wakeup time t'. At this time the algorithm reconsults the "work-or-sleep" subroutine. If the result is (S, work), then the algorithm enters the working state and follows S. Otherwise the algorithm continues to sleep with new wakeup time $\min\{\, t \geq t' : t + l(S) \leq \theta t \,\}$.
- In the working state, i.e, while the server is following a schedule all new requests are (temporarily) ignored. As soon as the current schedule is completed the server either enters the idle-state (if there are no unserved requests) or it reconsults the "work-or-sleep" subroutine which determines the next state (sleeping or working).

Theorem 6. *For all $\theta \geq \rho$, $\theta > 1$, Algorithm SMARTSTART is c-competitive with*

$$c = \max\left\{\theta, \rho\left(1 + \frac{1}{\theta - 1}\right), \frac{\theta}{2} + \rho\right\}.$$

Moreover, the best possible choice of θ is $\frac{1}{2}\left(1 + \sqrt{1 + 8\rho}\right)$ and yields a competitive ratio of $\frac{1}{4}\left(4\rho + 1 + \sqrt{1 + 8\rho}\right)$.

Proof. Let $\sigma_{=t_m}$ be the set of requests released at time t_m, where t_m denotes again the point in time when the last requests becomes known. We distinguish between different cases depending on the state of the SMARTSTART-server at time t_m:

Case 1: The server is idle.

In this case the algorithm consults its "work-or-sleep" routine which computes an approximately shortest schedule S for the requests in $\sigma_{=t_m}$. The SMARTSTART-server will start its work at time $t' = \min\{\, t \geq t_m : t + l(S) \leq \theta t \,\}$, where $l(S) = L^*(t_m, o, \sigma_{=t_m})$ denotes the length of the schedule S.

If $t' = t_m$, then by construction the algorithm completes no later than time $\theta t_m \leq \theta\, \mathsf{OPT}(\sigma)$. Otherwise $t' > t_m$ and it follows that $t' + l(S) = \theta t'$. By the performance guarantee of ρ of the approximation algorithm employed in "work-or-sleep" we have that $\mathsf{OPT}(\sigma) \geq l(S)/\rho = \frac{\theta-1}{\rho} t'$. Thus, it follows that

$$\mathsf{SMARTSTART}(\sigma) = t' + l(S) \leq \theta t' \leq \theta \cdot \frac{\rho\, \mathsf{OPT}(\sigma)}{\theta - 1} = \rho \left(1 + \frac{1}{\theta - 1}\right) \mathsf{OPT}(\sigma).$$

Case 2: The server is sleeping.

Note that the wakeup time of the server is no later than $\min\{\, t \geq t_m : t + l(S) \leq \theta t \,\}$, where S is now a shortest schedule for all the requests in σ not yet served by SMARTSTART at time t_m, and we can proceed as in Case 1.

Case 3: The algorithm is working.

If after completion of the current schedule the server enters the sleeping state then the arguments given above establish that the completion time of the SMARTSTART-server does not exceed $\rho\left(1 + \frac{1}{\theta-1}\right) \mathsf{OPT}(\sigma)$.

The remaining case is that the SMARTSTART-server starts its final schedule S' immediately after having completed S. Let t_S be the time when the server started S and denote by $\sigma_{\geq t_S}$ the set of requests presented after the server started S at time t_S. Notice that $\sigma_{\geq t_S}$ is exactly the set of requests that are served by SMARTSTART in its last schedule S'.

$$\mathsf{SMARTSTART}(\sigma) = t_S + l(S) + l(S'). \tag{4}$$

Here $l(S)$ and $l(S') \leq \rho L^*(t_m, o, \sigma_{\geq t_S})$ denotes the length of the schedule S and S', respectively. We have that

$$t_S + l(S) \leq \theta t_S, \tag{5}$$

since the SMARTSTART only starts a schedule at some time t if it can complete it not later than time θt. Let $r_f \in \sigma_{\geq t_S}$ be the first request from $\sigma_{\geq t_S}$ served by OPT.

Using the arguments given in the proof of Theorem 4 we conclude that

$$\mathsf{OPT}(\sigma) \geq t_S + L^*(t_m, a_f, \sigma_{\geq t_S}). \tag{6}$$

Moreover, since the tour of length $L^*(t_m, a_f, \sigma_{\geq t_S})$ starts in a_f and returns to the origin it follows from the triangle inequality that $L^*(t_m, a_f, \sigma_{\geq t_S}) \geq d(o, a_f)$. Thus, from (6) we get

$$\mathsf{OPT}(\sigma) \geq t_S + d(o, a_f). \tag{7}$$

On the other hand

$$l(S') \leq \rho\,(d(o,a_f) + L^*(t_m, a_f, \sigma_{\geq t_S})) \stackrel{(6)}{\leq} \rho\,(\mathsf{OPT}(\sigma) - t_S + d(o,a_f))\,. \tag{8}$$

Using (5) and (8) in (4) and the assumption that $\theta \geq \rho$, we obtain

$$\begin{aligned}
\mathsf{SMARTSTART}(\sigma) &\leq \theta t_S + l(S') && \text{by (5)}\\
&\leq (\theta-\rho)t_S + \rho\, d(o,a_f) + \rho\,\mathsf{OPT}(\sigma) && \text{by (8)}\\
&\leq \theta\,\mathsf{OPT}(\sigma) + (2\rho-\theta)d(o,a_f) && \text{by (7)}\\
&\leq \begin{cases} \theta\,\mathsf{OPT}(\sigma) + (2\rho-\theta)\frac{\mathsf{OPT}(\sigma)}{2} & \text{, if } \theta \leq 2\rho \\ \theta\,\mathsf{OPT}(\sigma) & \text{, if } \theta > 2\rho \end{cases}\\
&\leq \max\left\{\tfrac{\theta}{2}+\rho, \theta\right\}\,\mathsf{OPT}(\sigma)
\end{aligned}$$

This completes the proof. □

For "pure" competitive analysis we may assume that each schedule S computed by "work-or-sleep" is in fact an optimal schedule, i.e., that $\rho = 1$. The best competitive ratio for SMARTSTART is then achieved for that value of θ where the three terms θ, $1+\frac{1}{\theta-1}$ and $\frac{\theta}{2}+1$ are equal. This is the case for $\theta = 2$ and yields a competitive ratio of 2. We thus obtain the following corollary.

Corollary 1. *For $\rho = 1$, $\theta = 2$ algorithm SMARTSTART is 2-competitive.* □

The offline dial-a-ride problem is also known as the Stacker-Crane-Problem. In [10] the authors present a 9/5-approximation algorithm. On paths the problem can be solved in polynomial time [2,11]. In [7] an approximation algorithm for the single server dial-a-ride problem with performance $\mathcal{O}(\sqrt{C}\log n \log\log n)$ was given, where C denotes the capacity of the server.

For the special case of the Online Traveling Salesman Problem (OLTSP) Christofides' algorithm [8] yields a polynomial time approximation algorithm with $\rho = 3/2$. For this value of ρ, the best competitive ratio of SMARTSTART is attained for $\theta = \frac{1+\sqrt{13}}{2}$ and equals $\frac{7+\sqrt{13}}{4} \approx 2.6514$. Thus, for the OLTSP our algorithm SMARTSTART can be used to obtain a polynomial time competitive algorithm with competitive ratio approximately 2.6514. This improves the result of [5,4] where a 3-competitive polynomial time algorithm for OLTSP was given.

We finally note that the SMARTSTART-strategy inherits some desirable properties from the IGNORE-strategy: The algorithm can also be used for k-OLDARP and provides the same competitive ratio.

Theorem 7. *Algorithm SMARTSTART is 2-competitive for the extension k-OLDARP of OLDARP where there are $k \in \mathbb{N}$ servers with arbitrary capacities $C_1, \ldots, C_k \in \mathbb{N}$.* □

The last result implies a 2-competitive algorithm for the machine scheduling problems with setup costs discussed in Section 4.1.

6 Remarks

Our investigations of the OLDARP were originally motivated by the performance analysis of a large distribution center of Herlitz AG, Berlin [1]. Its automatic pallet transportation system employs several vertical transportation systems (elevators) in order to move pallets between the various floors of the building. The pallets that have to be transported during one day of production are not known in advance. If the objective is chosen as minimizing the completion time (makespan) then this can be modeled by the OLDARP where the metric space is induced by a graph which is a simple path.

References

1. N. Ascheuer, M. Grötschel, S. O. Krumke, and J. Rambau. Combinatorial online optimization. In *Proceedings of the International Conference of Operations Research*, pages 21–37. Springer, 1998.
2. M. J. Atallah and S. R. Kosaraju. Efficient solutions to some transportation problems with applications to minimizing robot arm travel. *SIAM Journal on Computing*, 17(5):849–869, October 1988.
3. G. Ausiello, E. Feuerstein, S. Leonardi, L. Stougie, and M. Talamo. Serving request with on-line routing. In *Proceedings of the 4th Scandinavian Workshop on Algorithm Theory*, volume 824 of *Lecture Notes in Computer Science*, pages 37–48, July 1994.
4. G. Ausiello, E. Feuerstein, S. Leonardi, L. Stougie, and M. Talamo. Competitive algorithms for the traveling salesman. In *Proceedings of the 4th Workshop on Algorithms and Data Structures*, volume 955 of *Lecture Notes in Computer Science*, pages 206–217, August 1995.
5. G. Ausiello, E. Feuerstein, S. Leonardi, L. Stougie, and M. Talamo. Algorithms for the on-line traveling salesman. *Algorithmica*, 1999. To appear.
6. A. Borodin and R. El-Yaniv. *Online Computation and Competitive Analysis*. Cambridge University Press, 1998.
7. M. Charikar and B. Raghavachari. The finite capacity dial-A-ride problem. In *Proceedings of the 39th Annual IEEE Symposium on the Foundations of Computer Science*, 1998.
8. N. Christofides. Worst-case analysis of a new heuristic for the traveling salesman problem. Technical report, Graduate School of Industrial Administration, Carnegie-Mellon University, Pittsburgh, PA, 1976.
9. E. Feuerstein and L. Stougie. On-line single server dial-a-ride problems. *Theoretical Computer Science*, special issue on on-line algorithms, to appear.
10. G. N. Frederickson, M. S. Hecht, and C. E. Kim. Approximation algorithms for some routing problems. *SIAM Journal on Computing*, 7(2):178–193, May 1978.
11. D. Hauptmeier, S. O. Krumke, J. Rambau, and H.-C. Wirth. Euler is standing in line. In *Proceedings of the 25th International Workshop on Graph-Theoretic Concepts in Computer Science, Ascona, Switzerland*, volume 1665 of *Lecture Notes in Computer Science*, pages 42–54, June 1999.
12. D. B. Shmoys, J. Wein, and D. P. Williamson. Scheduling parallel machines on-line. *SIAM Journal on Computing*, 24(6):1313–1331, December 1995.

The Power Range Assignment Problem in Radio Networks on the Plane
(Extended Abstract)

Andrea E.F. Clementi[1], Paolo Penna[1], and Riccardo Silvestri[2]

[1] Dipartimento di Matematica, Università di Roma "Tor Vergata",
{clementi,penna}@mat.uniroma2.it
[2] Dipartimento di Matematica Pura e Applicata, Università de L'Aquila,
silver@dsi.uniroma1.it

Abstract. Given a finite set S of points (i.e. the stations of a radio network) on the plane and a positive integer $1 \leq h \leq |S| - 1$, the 2D MIN h R. ASSIGN. problem consists of assigning transmission ranges to the stations so as to minimize the total power consumption provided that the transmission ranges of the stations ensure the communication between any pair of stations in at most h hops.
We provide a lower bound on the total power consumption $\mathsf{opt}_h(S)$ yielded by an optimal range assignment for *any* instance (S, h) of 2D MIN h R. ASSIGN., for any positive constant $h > 0$. The lower bound is a function of $|S|$, h and the minimum distance over all the pairs of stations in S. Then, we derive a constructive upper bound for the same problem as a function of $|S|$, h and the maximum distance over all the pairs of stations in S (i.e. *the diameter* of S). Finally, by combining the above bounds, we obtain a polynomial-time approximation algorithm for 2D MIN h R. ASSIGN. restricted to *well-spread* instances, for any positive constant h.
Previous results for this problem were known only in special 1-dimensional configurations (i.e. when points are arranged on a line).

Keywords: Approximation Algorithms, Lower Bounds, Multi-Hop Packet Radio Networks, Power Consumption.

1 Introduction

A *Multi-Hop Packet Radio Network* [6] is a finite set of radio stations located on a geographical region that are able to communicate by transmitting and receiving radio signals. A transmission range is assigned to each station s and any other station t within this range can directly (i.e. by one *hop*) receive messages from s. Communication between two stations that are not within their respective ranges can be achieved by *multi-hop* transmissions. In general, Multi-Hop Packet Radio Networks are adopted whenever the construction of more traditional networks is impossible or, simply, too expensive.

H. Reichel and S. Tison (Eds.): STACS 2000, LNCS 1770, pp. 651–660, 2000.

It is reasonably assumed [6] that the power P_t required by a station t to correctly transmit data to another station s must satisfy the inequality

$$\frac{P_t}{d(t,s)^\beta} > \gamma \tag{1}$$

where $d(t,s)$ is the distance between t and s, $\beta \geq 1$ is the *distance-power gradient*, and $\gamma \geq 1$ is the *transmission-quality* parameter. In an ideal environment (see [6]) $\beta = 2$ but it may vary from 1 to more than 6 depending on the environment conditions of the place the network is located. In the rest of the paper, we fix $\beta = 2$ and $\gamma = 1$, however, our results can be easily extended to any $\beta, \gamma \geq 1$.

Given a set $S = \{s_1, \ldots, s_n\}$ of radio stations on an Euclidean space, a *range assignment* for S is a function $r : S \to \mathcal{R}^+$, and the *cost* of r is defined as

$$\mathsf{cost}(r) = \sum_{i=1}^{n} r(s_i)^2.$$

As defined in the abstract, the 2D MIN h R. ASSIGN. problem consists of finding a minimum cost range assignment for a given set S of radio stations on the plane provided that the assignment ensure the communication between any pair of stations in at most h hops, where h is an input integer parameter $(1 \leq h \leq |S| - 1)$.

1.1 Previous Works

Combinatorial optimization problems arising from the design of radio networks have been the subject of several papers over the last years (see [6] for a survey). In particular, NP-completeness results and approximation algorithm for scheduling communication in radio networks have been derived in [1,3,7,8]. Kirousis *et al*, in [4], investigated the complexity of the MIN R. ASSIGN. problem that consists of minimizing the overall transmission power assigned to a set S of stations of a radio network, provided that (multi-hop) communication is guaranteed for any pair of stations (notice that no bounds are required on the maximum number of hops for the communication). It turns out that the complexity of this problem depends on the dimension of the space the stations are located on. In the 1-dimensional case (i.e. when the stations are located along a line) they provide a polynomial-time algorithm that finds a range assignment of minimum cost. As for stations located in the 3-dimensional space they show that MIN R. ASSIGN. is NP-hard. They also provide a polynomial-time 2-approximation algorithm that works for any dimension. Then, Clementi *et al* in [2] proved that the MIN R. ASSIGN. problem in three dimensions is APX-complete thus implying that it does not admit PTAS unless P = NP (see [5] for a formal definition of these concepts). They also prove that the MIN R. ASSIGN. problem is NP-hard in the 2-dimensional case.

All the results mentioned above concern the case in which no restriction on the maximum number h of hops required by the communications among stations is imposed: a range assignment is feasible if it just guarantees a strong connectivity of the network. When, instead, a fixed bound on the number h of hops is imposed, the computational complexity of the corresponding problem is unknown (from [4,2] we know only that the problem is NP-hard for spaces of dimension at least 2 and $h = \Omega(n)$). However, in [4], two tight bounds for the minimum power consumption required by n points arranged on a unit chain are given (notice that in this case, given any n, there is only one instance of the problem).

Theorem 1 (The Unit Chain Case [4]). *Let N be a set of n colinear points at unit distance. Then the order of magnitude of the overall power required by any optimal range assignment of diameter h for N is respectively:*

- $\Theta\left(n^{\frac{2^{h+1}-1}{2^h-1}}\right)$, *for any fixed positive integer h;*
- $\Theta\left(\frac{n^2}{h}\right)$, *for any $h = \Omega(\log n)$.*

Furthermore the two above (implicit) upper bounds are constructive.

1.2 Our Results

We investigate the 2D MIN h R. ASSIGN. problem for constant values of h (i.e. when h is independent from the number of stations). We first provide the following general lower bound on the cost of optimal solutions for this problem.

Theorem 2. *For any set S of stations on the plane, let $\delta(S)$ be the minimum distance between any pair of different stations in S, and let $\mathsf{opt}_h(S)$ be the cost of an optimal range assignment. Then, it holds*

$$\mathsf{opt}_h(S) = \Omega(\delta(S)^2 |S|^{1+1/h}),$$

for any fixed positive integer h.

The second result of this paper is an efficient method to derive a solution for any instance of our problem for fixed values of h. Given a set of stations S, let us define

$$D(S) = \max\{d(s_i, s_j) \mid s_i, s_j \in S\}.$$

Then, our method yields the following result.

Theorem 3. *For any set of stations S on the plane, it is possible to construct in time $O(h|S|)$ a feasible range assignment $r_h(S)$ such that*

$$\mathsf{cost}(r_h(S)) = O(D(S)^2 |S|^{1/h}),$$

for any fixed positive integer h.

The above bounds provide a fast evaluation of the order of magnitude of the power consumption required by (and sufficient to) any radio network on the plane. This may result useful in network design phase in order to efficiently select a good configuration. Indeed, instead of blindly trying and evaluating a huge number of tentative configurations, an easy application of our bounds could allow us to determine whether or not all such configurations are equivalent from the power consumption point of view.

Let us now consider the instance G_n of 2D MIN h R. ASSIGN. in which n stations are placed on a square grid of side $\sqrt{n}$ and the distance between adjacent pairs of stations is 1 (notice that this is the 2-dimensional version of the unit chain case studied in [4] - see Theorem 1).

Since our lower bound holds *for any* station set S, by combining Theorem 2 and 3, we easily obtain that

$$\mathsf{opt}_h(G_n) = \Theta\left(n^{1+1/h}\right). \tag{2}$$

The square grid configuration is the most regular case of *well-spread* instances. In general, we say that a family $\mathcal{S}$ of *well-spread* instances is a family of instances S such that $D(S) = O(\delta(S)\sqrt{|S|})$. Notice that the above property is rather natural: informally speaking, in a well-spread instance, any two stations must be not "too close". This is the typical situation in most of radio network configurations adopted in practice [6]. It turns out that the optimal bound in Eq. 2 holds for any family of well-spread instances. The following two corollaries are thus easy consequences of Theorems 2 and 3.

Corollary 1. *Let $\mathcal{S}$ be a family of well-spread instances. For any $S \in \mathcal{S}$, it holds that*

$$\mathsf{opt}_h(S) = \Theta\left(\delta(S)^2|S|^{1+1/h}\right),$$

for any positive integer constant h.

Corollary 2. *Let $\mathcal{S}$ be any family of well-spread instances. Then, for any positive integer constant h, the* 2D MIN h R. ASSIGN. *problem restricted to $\mathcal{S}$ admits a polynomial-time approximation algorithm with constant performance ratio (i.e. the restriction is in* APX*).*

2 Preliminaries

Let $S = \{s_1, \ldots, s_n\}$ be a set of n points (representing stations) in $\mathcal{R}^2$ with the Euclidean distance $d : \mathcal{R}^2 \times \mathcal{R}^2 \to \mathcal{R}^+$, where $\mathcal{R}^+$ denotes the set of non negative reals. We define

$$\delta(S) = \min\{d(s_i, s_j) \mid s_i, s_j \in S, i \neq j\}$$

and

$$D(S) = \max\{d(s_i, s_j) \mid s_i, s_j \in S\}.$$

A *range assignment* for S is a function $r : S \rightarrow \mathcal{R}^+$. The *cost* $\mathsf{cost}(r)$ of r is defined as

$$\mathsf{cost}(r) = \sum_{i=1}^{n} (r(s_i))^2.$$

Observe that we have set the distance-power gradient β to 2 (see Eq. 1), however our results can be easily extended to any constant $\beta \geq 1$.

The *communication graph* of a range assignment r is the directed graph $G_r(S, E)$ where $(s_i, s_j) \in E$ if and only if $r(s_i) \geq d(s_i, s_j)$. We say that an assignment r for S is of diameter h $(1 \leq h \leq n-1)$ if the corresponding communication graph is strongly connected and has diameter h (in short, an *h-assignment*).

As defined in the Introduction, given a set S of n points in $\mathcal{R}^2$ and a positive integer h, the 2D MIN h R. ASSIGN. problem consists of finding an h-assignment r_{min} for S of minimum cost. The cost of an optimal h-assignment for a given set of stations S is denoted as $\mathsf{opt}_h(S)$.

In the proof of our results we will make use of the well-known Hölder inequality. We thus present it in the following convenient form. Let x_i, $i = 1, \ldots, k$ be a set of k non negative reals and let $p, q \in \mathcal{R}$ such that $p \geq 1$ and $q \leq 1$. Then, it holds that:

$$\sum_{i=1}^{k} x_i^p \geq k \left(\frac{\sum_{i=1}^{k} x_i}{k} \right)^p; \tag{3}$$

$$\sum_{i=1}^{k} x_i^q \leq k \left(\frac{\sum_{i=1}^{k} x_i}{k} \right)^q. \tag{4}$$

3 The Lower Bound

Given a set S of stations and a "base" station $b \in S$, we define $\mathsf{opt}_h(S, b)$ as the minimum cost of any range assignment ensuring that any station $s \in S$ can reach b in at most h hops. By the definition of the 2D MIN h R. ASSIGN. problem, it should be clear that the cost required by any instance S of this problem is at least $\mathsf{opt}_h(S, b)$, for any $b \in S$. So, the main result of this section is an easy consequence of the following lemma.

Lemma 1. *Let S be any set of stations such that $\delta(S) = 1$. For every $b \in S$ and every positive constant integer h, it holds that*

$$\mathsf{opt}_h(S, b) = \Omega(|S|^{1+1/h}).$$

Proof. We first observe that, since $\delta(S) = 1$, for sufficiently large sets S (more precisely, for any S such that $|S| > 16$), the maximum number of stations contained in a disk of radius $R = \sqrt{|S|/3}$ is at most $|S|/2$.

Let $r_h^{all-to-one}$ be a range assignment that ensures that all the stations in S can reach b in at most h hops. We prove that $\mathsf{cost}(r_h^{all-to-one}) = \Omega(|S|^{1+1/h})$ by induction on h.
For $h = 1$, consider the disk of radius R and centered in b. By the above observation, there are at least $|S|/2$ stations at distance greater than R from b. The cost required by such stations to reach b in one hop is at least

$$(|S|/2)R^2 = \Omega(|S|^2).$$

Let $h \geq 2$, we define

$$FAR = \{s \in S \mid d(s,b) > R\}.$$

Clearly, we have that $|FAR| \geq |S|/2$. Every station s in FAR must reach b in $k \leq h$ hops, it thus follows that there exist $k \leq h$ positive reals $x_1, \ldots, x_k$ (where x_i is the distance covered by the i-th hop of the communication from s to b) such that

$$x_1 + x_2 + \cdots + x_k \geq R.$$

So, at least one index j exists for which $x_j \geq R/k \geq R/h$. We can thus define the set of "bridge" stations

$$B = \{s \in S \mid r_h^{all-to-one}(s) \geq R/h)\}.$$

Two cases may arise.

Case $|B| \geq |S|^{\frac{1}{h}}$. In this case, since $|R| = \sqrt{|S|/3}$,

$$\begin{aligned}\sum_{s \in S} (r_h^{all-to-one}(s))^2 &\geq |B|(R/h)^2 \\ &\geq \frac{1}{3h^2}|S|^{1+\frac{1}{h}} \\ &= \Omega\left(|S|^{1+\frac{1}{h}}\right).\end{aligned}$$

Case $|B| < |S|^{\frac{1}{h}}$. By means of the assignment $r_h^{all-to-one}$, every station in FAR reaches in at most $h-1$ hops some bridge station. Let $B = \{b_1, \ldots, b_{|B|}\}$. So, we can partition the set $FAR \cup B$ into $|B|$ subsets $A_1, \ldots, A_{|B|}$ such that all the stations in A_i reach b_i in at most $h-1$ hops[1]. So,

$$\begin{aligned}\sum_{s \in S} (r_h^{all-to-one}(s))^2 &\geq \sum_{i=1}^{|B|} \mathsf{opt}_{h-1}(A_i, b_i) \\ &= \Omega\left(\sum_{i=1}^{|B|} |A_i|^{1+\frac{1}{h-1}}\right)\end{aligned}$$

[1] Notice that if a station reaches two or more bridge stations, we can put the station into any of the corresponding set A_i's. We also assume that $b_i \in A_i$, for $1 \leq i \leq |B|$.

where the last bound is a consequence of the inductive hypothesis. Since

$$\sum_{i=1}^{|B|} |A_i| = |FAR \cup B| \geq |S|/2,$$

the Hölder inequality (see Eq. 3) implies that

$$\begin{aligned}\sum_{i=1}^{|B|} |A_i|^{1+\frac{1}{h-1}} &\geq |B| \left(\frac{|S|/2}{|B|}\right)^{1+\frac{1}{h-1}} \\ &= \Omega\left(\left(\frac{1}{|B|}\right)^{\frac{1}{h-1}} |S|^{1+\frac{1}{h-1}}\right) \\ &= \Omega\left(|S|^{1+\frac{1}{h}}\right)\end{aligned}$$

where the last equivalence is due to the condition $|B| < |S|^{\frac{1}{h}}$.

Proof of Theorem 2.
For $\delta(S) = 1$, the theorem is an immediate consequence of Lemma 1. The general case $\delta(S) > 0$ can be reduced to the previous case by simply rescaling the instance by a factor of $1/\delta(S)$.

□

4 The Upper Bound

Proof of Theorem 3.
The proof consists of a recursive construction of an h-assignment $r_h(S)$ having cost $O(D(S)^2|S|^{1/h})$. For $h = 1$, $r_1(S)$ assigns a range $D(S)$ to each station in S. Thus, $\mathsf{cost}(r_1(S)) = D(S)^2|S|$.

Let us consider the smallest square Q that contains all points in S. Notice that the side l of Q is at most $D(S)$. Let us consider a grid that subdivides Q into k^2 subsquares of the same size l/k (the choice of k will be given later).

Informally speaking, for every non empty subsquare we choose a "base" station and we give power sufficient to let it cover all the stations in S in one hop. Then, in every subsquare we complete the assignment by making any station able to reach the base station in $h-1$ hops. For this task we apply the recursive construction.

The cost of $r_h(S)$ is thus bounded by

$$\mathsf{cost}(r_h(S)) \leq k^2 D(S)^2 + \sum_{i=1}^{k^2} \mathsf{cost}(r_{h-1}(S_i)),$$

where S_i is the set of the stations in the i-th subsquare. Since $D(S_i) = O(D(S)/k)$ we apply the inductive hypothesis and we obtain

$$\begin{aligned}\mathsf{cost}(r_h(S)) &= \mathrm{O}\left(k^2 D(S)^2 + \sum_{i=1}^{k^2} |S_i|^{1/(h-1)} \left(\frac{D(S)}{k}\right)^2\right) \\ &= \mathrm{O}\left(k^2 D(S)^2 + \left(\frac{D(S)}{k}\right)^2 \sum_{i=1}^{k^2} |S_i|^{1/(h-1)}\right) \\ &= \mathrm{O}\left(k^2 D(S)^2 + \left(\frac{D(S)}{k}\right)^2 k^2 \left(\frac{|S|}{k^2}\right)^{1/(h-1)}\right),\end{aligned}$$

where the last equality follows from the Hölder inequality (see Eq. 4) and from the fact that $\sum_{i=1}^{k^2} |S_i| = |S|$. Now we choose

$$k = |S|^{\frac{1}{2h}}$$

in order to equate the additive terms in the last part of the above equation. By replacing this value in the equation we obtain

$$\mathsf{cost}(r_h(S)) = \mathrm{O}\left(D(S)^2 |S|^{1/h}\right).$$

It is easy to verify that the partition of Q into k^2 subsquares and the rest of the computation in each inductive step can be done in time $O(|S|)$. So, the overall time complexity is $O(h|S|)$.

□

5 Tight Bounds and Approximability

Let us consider the simple instance G_n of 2D MIN h R. ASSIGN. in which n stations are placed on a square grid of side $\sqrt{n}$, and the distance between adjacent pairs of stations is 1.

By Combining Theorems 2 and 3, we easily obtain that

$$\mathsf{opt}_h(G_n) = \Theta\left(n^{1+1/h}\right).$$

This also implies that the range assignment constructed in the proof of Theorem 3 yields a constant-factor approximation.

It turns out that the above considerations can be extended to any "well-spread" instance.

Definition 1. *A family $\mathcal{S}$ of* well-spread *instances is a family of instances S such that $D(S) = O(\delta(S)\sqrt{|S|})$.*

The following two corollaries are easy consequences of Theorems 2 and 3.

Corollary 3. *Let $\mathcal{S}$ be a family of well-spread instances. Then, For any $S \in \mathcal{S}$,*

$$\mathsf{opt}_h(S) = \Theta\left(\delta(S)^2|S|^{1+1/h}\right),$$

for any positive integer constant h.

Corollary 4. *Let $\mathcal{S}$ be any family of well-spread instances.*
Then, the 2D MIN h R. ASSIGN. *problem restricted to $\mathcal{S}$ is in* APX, *for any positive integer constant h.*

6 Open Problems

As discussed in the Introduction, finding bounds for the power consumption of general classes of radio networks might result very useful in network design. Thus it is interesting to derive new, tighter bounds for possibly a larger class of radio network configurations.
Another question left open by this paper is whether the 2D MIN h R. ASSIGN. is NP-hard for constant h. We conjecture a positive answer even though we believe that any proof will depart significantly from those adopted for the unbounded cases (i.e. the MIN R. ASSIGN. problem - see [4,2]). More precisely, all the reductions adopted in the unbounded cases start from the minimum vertex cover problem that seems to be very unsuitable for our problem. When the stations are arranged on a line (i.e. the 1-dimensional case), we instead conjecture that the problem is in P for any value of h.

Acknowledgments Part of this work has been done while the authors were visiting the research center of INRIA Sophia Antipolis, partially supported by the INRIA project COMMOBIL.

References

1. E. Arikan, "Some Complexity Results about Packet Radio Networks", IEEE Transactions on Information Theory, 456-461, IT-30, 1984.
2. A.E.F. Clementi, P. Penna, and R. Silvestri, "Hardness Results for the Power Range Assignment Problem in Packet Radio Networks", II International Workshop on Approximation Algorithms for Combinatorial Optimization Problems (APPROX'99), LNCS, 1671, 1999.
3. A. Ephemides and T. Truong, "Scheduling Broadcast in Multihop Radio Networks", IEEE Transactions on Communications, 456-461, 30, 1990.
4. L. M. Kirousis and E. Kranakis and D. Krizanc and A. Pelc, "Power Consumption in Packet Radio Networks", 14th Annual Symposium on Theoretical Aspects of Computer Science (STACS 97), LNCS 1200, 363-374, 1997 (to appear also in TCS).
5. C.H. Papadimitriou, "Computational Complexity", Addison-Wesley Publishing Company, Inc., 1995.

6. K. Pahlavan and A. Levesque, "Wireless Information Networks", Wiley-Interscience, New York, 1995.
7. S. Ramanathan and E. Lloyd, "Scheduling Broadcasts in Multi-hop Radio Networks" IEEE/ACM Transactions on Networking, 166-172, 1, 1993.
8. R. Ramaswami and K. Parhi, "Distributed Scheduling of Broadcasts in Radio Network", INFOCOM'89, 497-504, 1989.

Author Index